STUDENT'S SOLUTIONS MANUAL

PART ONE

ARDIS • BORZELLINO • BUCHANAN • MOGILL • NELSON

to accompany

THOMAS' CALCULUS
ELEVENTH EDITION

BASED ON THE ORIGINAL WORK BY

George B. Thomas, Jr.
Massachusetts Institute of Technology

AS REVISED BY

Maurice D. Weir
Naval Postgraduate School

Joel Hass
University of California, Davis

Frank R. Giordano
Naval Postgraduate School

PEARSON

Addison
Wesley

Boston   San Francisco   New York
London Toronto Sydney Tokyo Singapore Madrid
Mexico City Munich Paris Cape Town Hong Kong Montreal

Reproduced by Pearson Addison-Wesley from electronic files supplied by the authors.

Copyright © 2005 Pearson Education, Inc.
Publishing as Pearson Addison-Wesley, 75 Arlington Street, Boston, MA 02116.

ISBN   0-321-22646-1

5 6 BB 08 07

PREFACE TO THE STUDENT

The Student's Solutions Manual contains the solutions to all of the odd-numbered exercise in the 11th Edition of THOMAS' CALCULUS by Maurice Weir, Joel Hass and Frank Giordano, excluding the Computer Algebra System (CAS) exercises. We have worked each solution to ensure that it

- conforms exactly to the methods, procedures and steps presented in the text

- is mathematically correct

- includes all of the steps necessary so you can follow the logical argument and algebra

- includes a graph or figure whenever called for by the exercise, or if needed to help with the explanation

- is formatted in an appropriate style to aid in its understanding

How to use a solution's manual

- solve the assigned problem yourself

- if you get stuck along the way, refer to the solution in the manual as an aid but continue to solve the problem on your own

- if you cannot continue, reread the textbook section, or work through that section in the Student Study Guide, or consult your instructor

- if your answer is correct by your solution procedure seems to differ from the one in the manual, and you are unsure your method is correct, consult your instructor

- if your answer is incorrect and you cannot find your error, consult your instructor

Acknowledgments

Solutions Writers
    William Ardis, Collin County Community College-Preston Ridge Campus
    Joseph Borzellino, California Polytechnic State University
    Linda Buchanan, Howard College
    Tim Mogill
    Patricia Nelson, University of Wisconsin-La Crosse

Accuracy Checkers
    Karl Kattchee, University of Wisconsin-La Crosse
    Marie Vanisko, California State University, Stanislaus
    Tom Weigleitner, VISTA Information Technologies

Thanks to Rachel Reeve, Christine O'Brien, Sheila Spinney, Elka Block, and Joe Vetere for all their guidance and help at every step.

TABLE OF CONTENTS

10 Conic Sections and Polar Coordinates 303

11 Infinite Sequences and Series 343

CHAPTER 1 PRELIMINARIES

1.1 REAL NUMBERS AND THE REAL LINE

1. Executing long division, $\frac{1}{9} = 0.\overline{1}$, $\frac{2}{9} = 0.\overline{2}$, $\frac{3}{9} = 0.\overline{3}$, $\frac{8}{9} = 0.\overline{8}$, $\frac{9}{9} = 0.\overline{9}$

3. NT = necessarily true, NNT = Not necessarily true. Given: $2 < x < 6$.
   a) NNT. 5 is a counter example.
   b) NT. $2 < x < 6 \Rightarrow 2 - 2 < x - 2 < 6 - 2 \Rightarrow 0 < x - 2 < 2$.
   c) NT. $2 < x < 6 \Rightarrow 2/2 < x/2 < 6/2 \Rightarrow 1 < x < 3$.
   d) NT. $2 < x < 6 \Rightarrow 1/2 > 1/x > 1/6 \Rightarrow 1/6 < 1/x < 1/2$.
   e) NT. $2 < x < 6 \Rightarrow 1/2 > 1/x > 1/6 \Rightarrow 1/6 < 1/x < 1/2 \Rightarrow 6(1/6) < 6(1/x) < 6(1/2) \Rightarrow 1 < 6/x < 3$.
   f) NT. $2 < x < 6 \Rightarrow x < 6 \Rightarrow (x - 4) < 2$ and $2 < x < 6 \Rightarrow x > 2 \Rightarrow -x < -2 \Rightarrow -x + 4 < 2 \Rightarrow -(x - 4) < 2$.
      The pair of inequalities $(x - 4) < 2$ and $-(x - 4) < 2 \Rightarrow |x - 4| < 2$.
   g) NT. $2 < x < 6 \Rightarrow -2 > -x > -6 \Rightarrow -6 < -x < -2$. But $-2 < 2$. So $-6 < -x < -2 < 2$ or $-6 < -x < 2$.
   h) NT. $2 < x < 6 \Rightarrow -1(2) > -1(x) < -1(6) \Rightarrow -6 < -x < -2$

5. $-2x > 4 \Rightarrow x < -2$

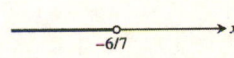

7. $5x - 3 \le 7 - 3x \Rightarrow 8x \le 10 \Rightarrow x \le \frac{5}{4}$

9. $2x - \frac{1}{2} \ge 7x + \frac{7}{6} \Rightarrow -\frac{1}{2} - \frac{7}{6} \ge 5x$
   $\Rightarrow \frac{1}{5}\left(-\frac{10}{6}\right) \ge x$ or $-\frac{1}{3} \ge x$

11. $\frac{4}{5}(x - 2) < \frac{1}{3}(x - 6) \Rightarrow 12(x - 2) < 5(x - 6)$
    $\Rightarrow 12x - 24 < 5x - 30 \Rightarrow 7x < -6$ or $x < -\frac{6}{7}$

13. $y = 3$ or $y = -3$

15. $2t + 5 = 4$ or $2t + 5 = -4 \Rightarrow 2t = -1$ or $2t = -9 \Rightarrow t = -\frac{1}{2}$ or $t = -\frac{9}{2}$

17. $8 - 3s = \frac{9}{2}$ or $8 - 3s = -\frac{9}{2} \Rightarrow -3s = -\frac{7}{2}$ or $-3s = -\frac{25}{2} \Rightarrow s = \frac{7}{6}$ or $s = \frac{25}{6}$

19. $-2 < x < 2$; solution interval $(-2, 2)$

21. $-3 \le t - 1 \le 3 \Rightarrow -2 \le t \le 4$; solution interval $[-2, 4]$

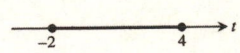

23. $-4 < 3y - 7 < 4 \Rightarrow 3 < 3y < 11 \Rightarrow 1 < y < \frac{11}{3}$;
    solution interval $\left(1, \frac{11}{3}\right)$

25. $-1 \le \frac{z}{5} - 1 \le 1 \Rightarrow 0 \le \frac{z}{5} \le 2 \Rightarrow 0 \le z \le 10$;
    solution interval $[0, 10]$

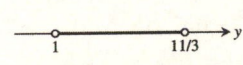

27. $-\frac{1}{2} < 3 - \frac{1}{x} < \frac{1}{2} \Rightarrow -\frac{7}{2} < -\frac{1}{x} < -\frac{5}{2} \Rightarrow \frac{7}{2} > \frac{1}{x} > \frac{5}{2}$
    $\Rightarrow \frac{2}{7} < x < \frac{2}{5}$; solution interval $\left(\frac{2}{7}, \frac{2}{5}\right)$

29. $2s \geq 4$ or $-2s \geq 4 \Rightarrow s \geq 2$ or $s \leq -2$;

solution intervals $(-\infty, -2] \cup [2, \infty)$

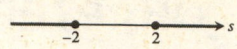

31. $1 - x > 1$ or $-(1 - x) > 1 \Rightarrow -x > 0$ or $x > 2$

$\Rightarrow x < 0$ or $x > 2$; solution intervals $(-\infty, 0) \cup (2, \infty)$

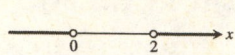

33. $\frac{r+1}{2} \geq 1$ or $-\left(\frac{r+1}{2}\right) \geq 1 \Rightarrow r + 1 \geq 2$ or $r + 1 \leq -2$

$\Rightarrow r \geq 1$ or $r \leq -3$; solution intervals $(-\infty, -3] \cup [1, \infty)$

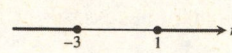

35. $x^2 < 2 \Rightarrow |x| < \sqrt{2} \Rightarrow -\sqrt{2} < x < \sqrt{2}$;

solution interval $\left(-\sqrt{2}, \sqrt{2}\right)$

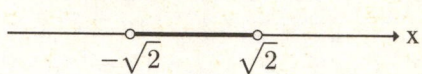

37. $4 < x^2 < 9 \Rightarrow 2 < |x| < 3 \Rightarrow 2 < x < 3$ or $2 < -x < 3$

$\Rightarrow 2 < x < 3$ or $-3 < x < -2$;

solution intervals $(-3, -2) \cup (2, 3)$

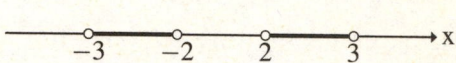

39. $(x - 1)^2 < 4 \Rightarrow |x - 1| < 2 \Rightarrow -2 < x - 1 < 2$

$\Rightarrow -1 < x < 3$; solution interval $(-1, 3)$

41. $x^2 - x < 0 \Rightarrow x^2 - x + \frac{1}{4} < \frac{1}{4} \Rightarrow \left(x - \frac{1}{2}\right)^2 < \frac{1}{4} \Rightarrow \left|x - \frac{1}{2}\right| < \frac{1}{2} \Rightarrow -\frac{1}{2} < x - \frac{1}{2} < \frac{1}{2} \Rightarrow 0 < x < 1$.

So the solution is the interval $(0, 1)$

43. True if $a \geq 0$; False if $a < 0$.

45. (1) $|a + b| = (a + b)$ or $|a + b| = -(a + b)$;

both squared equal $(a + b)^2$

(2) $ab \leq |ab| = |a| \, |b|$

(3) $|a| = a$ or $|a| = -a$, so $|a|^2 = a^2$; likewise, $|b|^2 = b^2$

(4) $x^2 \leq y^2$ implies $\sqrt{x^2} \leq \sqrt{y^2}$ or $x \leq y$ for all nonnegative real numbers x and y. Let $x = |a + b|$ and
$y = |a| + |b|$ so that $|a + b|^2 \leq (|a| + |b|)^2 \Rightarrow |a + b| \leq |a| + |b|$.

47. $-3 \leq x \leq 3$ and $x > -\frac{1}{2} \Rightarrow -\frac{1}{2} < x \leq 3$.

49. Let δ be a real number > 0 and $f(x) = 2x + 1$. Suppose that $|x - 1| < \delta$. Then $|x - 1| < \delta \Rightarrow 2|x - 1| < 2\delta \Rightarrow$
$|2x - 2| < 2\delta \Rightarrow |(2x + 1) - 3| < 2\delta \Rightarrow |f(x) - f(1)| < 2\delta$

51. Consider: i) $a > 0$; ii) $a < 0$; iii) $a = 0$.

i) For $a > 0, |a| = a$ by definition. Now, $a > 0 \Rightarrow -a < 0$. Let $-a = b$. By definition, $|b| = -b$. Since $b = -a$,
$|-a| = -(-a) = a$ and $|a| = |-a| = a$.

ii) For $a < 0, |a| = -a$. Now, $a < 0 \Rightarrow -a > 0$. Let $-a = b$. By definition, $|b| = b$ and thus $|-a| = -a$. So again
$|a| = |-a|$.

iii) By definition $|0| = 0$ and since $-0 = 0, |-0| = 0$. Thus, by i), ii), and iii) $|a| = |-a|$ for any real number.

53. a) $1 = 1 \Rightarrow |1| = 1 \Rightarrow \left|b \cdot \frac{1}{b}\right| = \frac{|b|}{|b|} \Rightarrow |b| \cdot \left|\frac{1}{b}\right| = \frac{|b|}{|b|} \Rightarrow \frac{|b| \cdot \left|\frac{1}{b}\right|}{|b|} = \frac{|b|}{|b| \cdot |b|} \Rightarrow \left|\frac{1}{b}\right| = \frac{1}{|b|}$

b) $\left|\frac{a}{b}\right| = \left|a \cdot \frac{1}{b}\right| = \left|a\right| \cdot \left|\frac{1}{b}\right| = \left|a\right| \cdot \frac{1}{|b|} = \frac{|a|}{|b|}$

1.2 LINES, CIRCLES, AND PARABOLAS

1. $\Delta x = -1 - (-3) = 2, \Delta y = -2 - 2 = -4; d = \sqrt{(\Delta x)^2 + (\Delta y)^2} = \sqrt{4 + 16} = 2\sqrt{5}$

3. $\Delta x = -8.1 - (-3.2) = -4.9, \Delta y = -2 - (-2) = 0; d = \sqrt{(-4.9)^2 + 0^2} = 4.9$

5. Circle with center $(0, 0)$ and radius 1.

7. Disk (i.e., circle together with its interior points) with center $(0, 0)$ and radius $\sqrt{3}$.

9. $m = \frac{\Delta y}{\Delta x} = \frac{-1-2}{-2-(-1)} = 3$

   perpendicular slope $= -\frac{1}{3}$

   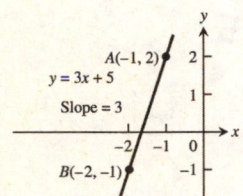

11. $m = \frac{\Delta y}{\Delta x} = \frac{3-3}{-1-2} = 0$

   perpendicular slope does not exist

   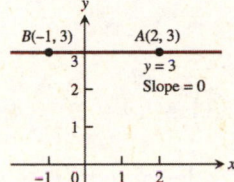

13. (a) $x = -1$
    (b) $y = \frac{4}{3}$

15. (a) $x = 0$
    (b) $y = -\sqrt{2}$

17. $P(-1, 1), m = -1 \Rightarrow y - 1 = -1(x - (-1)) \Rightarrow y = -x$

19. $P(3, 4), Q(-2, 5) \Rightarrow m = \frac{\Delta y}{\Delta x} = \frac{5-4}{-2-3} = -\frac{1}{5} \Rightarrow y - 4 = -\frac{1}{5}(x - 3) \Rightarrow y = -\frac{1}{5}x + \frac{23}{5}$

21. $m = -\frac{5}{4}, b = 6 \Rightarrow y = -\frac{5}{4}x + 6$

23. $m = 0, P(-12, -9) \Rightarrow y = -9$

25. $a = -1, b = 4 \Rightarrow (0, 4)$ and $(-1, 0)$ are on the line $\Rightarrow m = \frac{\Delta y}{\Delta x} = \frac{0-4}{-1-0} = 4 \Rightarrow y = 4x + 4$

27. $P(5, -1), L: 2x + 5y = 15 \Rightarrow m_L = -\frac{2}{5} \Rightarrow$ parallel line is $y - (-1) = -\frac{2}{5}(x - 5) \Rightarrow y = -\frac{2}{5}x + 1$

29. $P(4, 10), L: 6x - 3y = 5 \Rightarrow m_L = 2 \Rightarrow m_\perp = -\frac{1}{2} \Rightarrow$ perpendicular line is $y - 10 = -\frac{1}{2}(x - 4) \Rightarrow y = -\frac{1}{2}x + 12$

31. x-intercept $= 4$, y-intercept $= 3$

   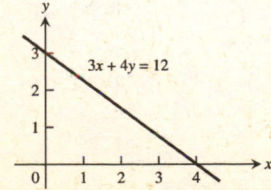

33. x-intercept $= \sqrt{3}$, y-intercept $= -\sqrt{2}$

   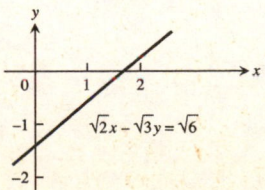

35. $Ax + By = C_1 \Leftrightarrow y = -\frac{A}{B}x + \frac{C_1}{B}$ and $Bx - Ay = C_2 \Leftrightarrow y = \frac{B}{A}x - \frac{C_2}{A}$. Since $\left(-\frac{A}{B}\right)\left(\frac{B}{A}\right) = -1$ is the product of the slopes, the lines are perpendicular.

37. New position $= (x_{old} + \Delta x, y_{old} + \Delta y) = (-2 + 5, 3 + (-6)) = (3, -3)$.

39. $\Delta x = 5, \Delta y = 6, B(3, -3)$. Let $A = (x, y)$. Then $\Delta x = x_2 - x_1 \Rightarrow 5 = 3 - x \Rightarrow x = -2$ and $\Delta y = y_2 - y_1 \Rightarrow 6 = -3 - y \Rightarrow y = -9$. Therefore, $A = (-2, -9)$.

41. $C(0, 2), a = 2 \Rightarrow x^2 + (y - 2)^2 = 4$

43. $C(-1, 5), a = \sqrt{10} \Rightarrow (x + 1)^2 + (y - 5)^2 = 10$

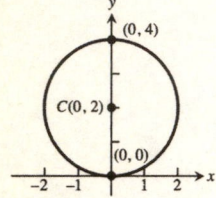

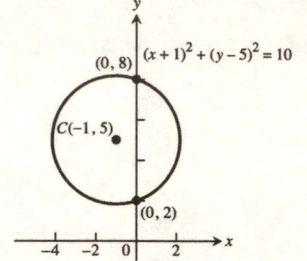

45. $C\left(-\sqrt{3}, -2\right), a = 2 \Rightarrow \left(x + \sqrt{3}\right)^2 + (y + 2)^2 = 4,$

$x = 0 \Rightarrow \left(0 + \sqrt{3}\right)^2 + (y + 2)^2 = 4 \Rightarrow (y + 2)^2 = 1$

$\Rightarrow y + 2 = \pm 1 \Rightarrow y = -1$ or $y = -3$. Also, $y = 0$

$\Rightarrow \left(x + \sqrt{3}\right)^2 + (0 + 2)^2 = 4 \Rightarrow \left(x + \sqrt{3}\right)^2 = 0$

$\Rightarrow x = -\sqrt{3}$

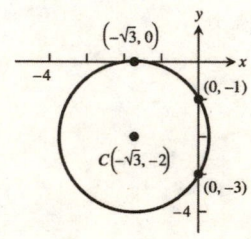

47. $x^2 + y^2 + 4x - 4y + 4 = 0$

$\Rightarrow x^2 + 4x + y^2 - 4y = -4$

$\Rightarrow x^2 + 4x + 4 + y^2 - 4y + 4 = 4$

$\Rightarrow (x + 2)^2 + (y - 2)^2 = 4 \Rightarrow C = (-2, 2), a = 2.$

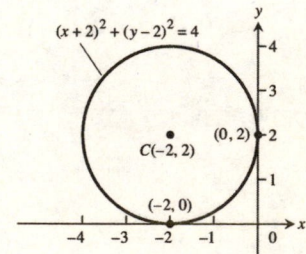

49. $x^2 + y^2 - 3y - 4 = 0 \Rightarrow x^2 + y^2 - 3y = 4$

$\Rightarrow x^2 + y^2 - 3y + \frac{9}{4} = \frac{25}{4}$

$\Rightarrow x^2 + \left(y - \frac{3}{2}\right)^2 = \frac{25}{4} \Rightarrow C = \left(0, \frac{3}{2}\right),$

$a = \frac{5}{2}.$

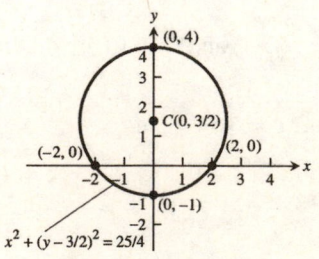

51. $x^2 + y^2 - 4x + 4y = 0$

$\Rightarrow x^2 - 4x + y^2 + 4y = 0$

$\Rightarrow x^2 - 4x + 4 + y^2 + 4y + 4 = 8$

$\Rightarrow (x - 2)^2 + (y + 2)^2 = 8$

$\Rightarrow C(2, -2), a = \sqrt{8}.$

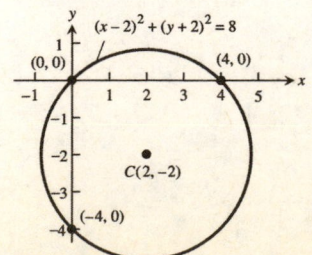

53. $x = -\frac{b}{2a} = -\frac{-2}{2(1)} = 1$

    $\Rightarrow y = (1)^2 - 2(1) - 3 = -4$

    $\Rightarrow V = (1, -4)$. If $x = 0$ then $y = -3$.

    Also, $y = 0 \Rightarrow x^2 - 2x - 3 = 0$

    $\Rightarrow (x - 3)(x + 1) = 0 \Rightarrow x = 3$ or

    $x = -1$. Axis of parabola is $x = 1$.

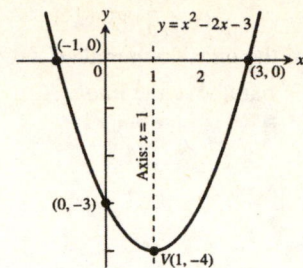

55. $x = -\frac{b}{2a} = -\frac{4}{2(-1)} = 2$

    $\Rightarrow y = -(2)^2 + 4(2) = 4$

    $\Rightarrow V = (2, 4)$. If $x = 0$ then $y = 0$.

    Also, $y = 0 \Rightarrow -x^2 + 4x = 0$

    $\Rightarrow -x(x - 4) = 0 \Rightarrow x = 4$ or $x = 0$.

    Axis of parabola is $x = 2$.

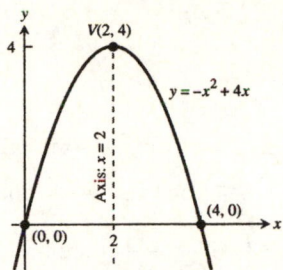

57. $x = -\frac{b}{2a} = -\frac{-6}{2(-1)} = -3$

    $\Rightarrow y = -(-3)^2 - 6(-3) - 5 = 4$

    $\Rightarrow V = (-3, 4)$. If $x = 0$ then $y = -5$.

    Also, $y = 0 \Rightarrow -x^2 - 6x - 5 = 0$

    $\Rightarrow (x + 5)(x + 1) = 0 \Rightarrow x = -5$ or

    $x = -1$. Axis of parabola is $x = -3$.

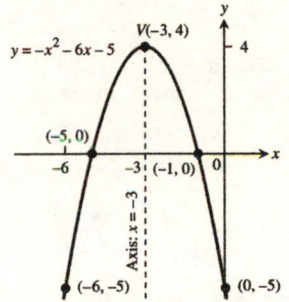

59. $x = -\frac{b}{2a} = -\frac{1}{2(1/2)} = -1$

    $\Rightarrow y = \frac{1}{2}(-1)^2 + (-1) + 4 = \frac{7}{2}$

    $\Rightarrow V = \left(-1, \frac{7}{2}\right)$. If $x = 0$ then $y = 4$.

    Also, $y = 0 \Rightarrow \frac{1}{2}x^2 + x + 4 = 0$

    $\Rightarrow x = \frac{-1 \pm \sqrt{-7}}{1} \Rightarrow$ no x intercepts.

    Axis of parabola is $x = -1$.

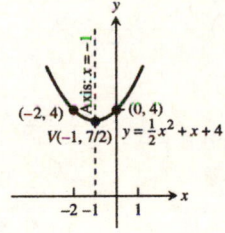

61. The points that lie outside the circle with center $(0, 0)$ and radius $\sqrt{7}$.

63. The points that lie on or inside the circle with center $(1, 0)$ and radius 2.

65. The points lying outside the circle with center $(0, 0)$ and radius 1, but inside the circle with center $(0, 0)$, and radius 2 (i.e., a washer).

67. $x^2 + y^2 + 6y < 0 \Rightarrow x^2 + (y+3)^2 < 9$.
The interior points of the circle centered at
$(0, -3)$ with radius 3, but above the line
$y = -3$.

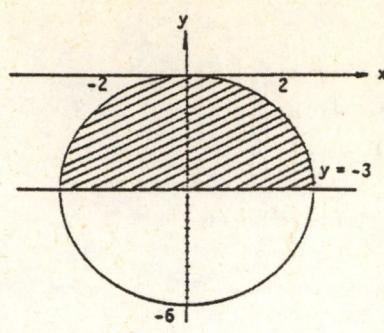

69. $(x+2)^2 + (y-1)^2 < 6$

71. $x^2 + y^2 \le 2, x \ge 1$

73. $x^2 + y^2 = 1$ and $y = 2x \Rightarrow 1 = x^2 + 4x^2 = 5x^2$
$\Rightarrow \left(x = \frac{1}{\sqrt{5}} \text{ and } y = \frac{2}{\sqrt{5}}\right)$ or $\left(x = -\frac{1}{\sqrt{5}} \text{ and } y = -\frac{2}{\sqrt{5}}\right)$.
Thus, A $\left(\frac{1}{\sqrt{5}}, \frac{2}{\sqrt{5}}\right)$, B $\left(-\frac{1}{\sqrt{5}}, -\frac{2}{\sqrt{5}}\right)$ are the
points of intersection.

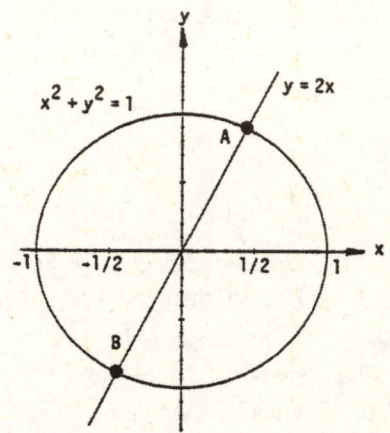

75. $y - x = 1$ and $y = x^2 \Rightarrow x^2 - x = 1$
$\Rightarrow x^2 - x - 1 = 0 \Rightarrow x = \frac{1 \pm \sqrt{5}}{2}$.
If $x = \frac{1+\sqrt{5}}{2}$, then $y = x + 1 = \frac{3+\sqrt{5}}{2}$.
If $x = \frac{1-\sqrt{5}}{2}$, then $y = x + 1 = \frac{3-\sqrt{5}}{2}$.
Thus, A $\left(\frac{1+\sqrt{5}}{2}, \frac{3+\sqrt{5}}{2}\right)$ and B $\left(\frac{1-\sqrt{5}}{2}, \frac{3-\sqrt{5}}{2}\right)$
are the intersection points.

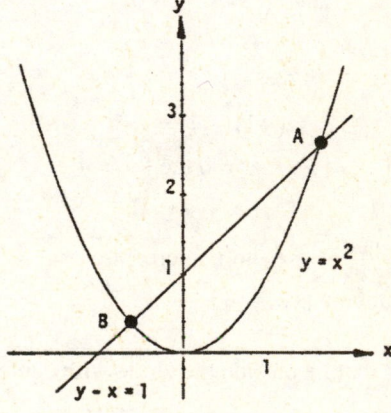

77. $y = 2x^2 - 1 = -x^2 \Rightarrow 3x^2 = 1$
$\Rightarrow x = \frac{1}{\sqrt{3}}$ and $y = -\frac{1}{3}$ or $x = -\frac{1}{\sqrt{3}}$ and $y = -\frac{1}{3}$.
Thus, A $\left(\frac{1}{\sqrt{3}}, -\frac{1}{3}\right)$ and B $\left(-\frac{1}{\sqrt{3}}, -\frac{1}{3}\right)$ are the
intersection points.

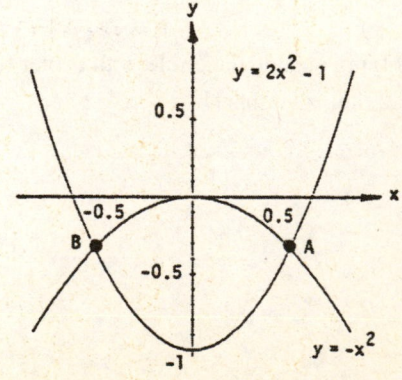

79. $x^2 + y^2 = 1 = (x - 1)^2 + y^2$

$\Rightarrow x^2 = (x - 1)^2 = x^2 - 2x + 1$

$\Rightarrow 0 = -2x + 1 \Rightarrow x = \frac{1}{2}$. Hence

$y^2 = 1 - x^2 = \frac{3}{4}$ or $y = \pm \frac{\sqrt{3}}{2}$. Thus,

$A\left(\frac{1}{2}, \frac{\sqrt{3}}{2}\right)$ and $B\left(\frac{1}{2}, -\frac{\sqrt{3}}{2}\right)$ are the

intersection points.

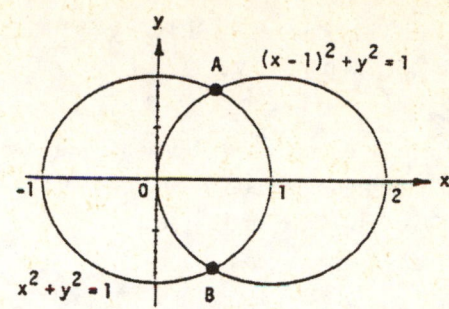

81. (a) $A \approx (69°, 0 \text{ in}), B \approx (68°, .4 \text{ in}) \Rightarrow m = \frac{68° - 69°}{.4 - 0} \approx -2.5°/\text{in.}$

(b) $A \approx (68°, .4 \text{ in}), B \approx (10°, 4 \text{ in}) \Rightarrow m = \frac{10° - 68°}{4 - .4} \approx -16.1°/\text{in.}$

(c) $A \approx (10°, 4 \text{ in}), B \approx (5°, 4.6 \text{ in}) \Rightarrow m = \frac{5° - 10°}{4.6 - 4} \approx -8.3°/\text{in.}$

83. $p = kd + 1$ and $p = 10.94$ at $d = 100 \Rightarrow k = \frac{10.94 - 1}{100} = 0.0994$. Then $p = 0.0994d + 1$ is the diver's

pressure equation so that $d = 50 \Rightarrow p = (0.0994)(50) + 1 = 5.97$ atmospheres.

85. $C = \frac{5}{9}(F - 32)$ and $C = F \Rightarrow F = \frac{5}{9}F - \frac{160}{9} \Rightarrow \frac{4}{9}F = -\frac{160}{9}$ or $F = -40°$ gives the same numerical reading.

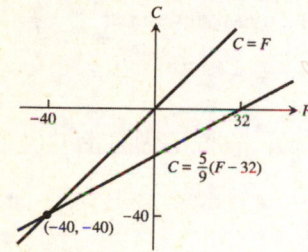

87. length $AB = \sqrt{(5 - 1)^2 + (5 - 2)^2} = \sqrt{16 + 9} = 5$

length $AC = \sqrt{(4 - 1)^2 + (-2 - 2)^2} = \sqrt{9 + 16} = 5$

length $BC = \sqrt{(4 - 5)^2 + (-2 - 5)^2} = \sqrt{1 + 49} = \sqrt{50} = 5\sqrt{2} \neq 5$

89. Length $AB = \sqrt{(\Delta x)^2 + (\Delta y)^2} = \sqrt{1^2 + 4^2} = \sqrt{17}$ and length $BC = \sqrt{(\Delta x)^2 + (\Delta y)^2} = \sqrt{4^2 + 1^2} = \sqrt{17}$.

Also, slope $AB = \frac{4}{-1}$ and slope $BC = \frac{1}{4}$, so $AB \perp BC$. Thus, the points are vertices of a square. The coordinate

increments from the fourth vertex $D(x, y)$ to A must equal the increments from C to $B \Rightarrow 2 - x = \Delta x = 4$ and

$-1 - y = \Delta y = 1 \Rightarrow x = -2$ and $y = -2$. Thus $D(-2, -2)$ is the fourth vertex.

91. Let $A(-1, 1)$, $B(2, 3)$, and $C(2, 0)$ denote the points.

Since BC is vertical and has length $|BC| = 3$, let

$D_1(-1, 4)$ be located vertically upward from A and

$D_2(-1, -2)$ be located vertically downward from A so

that $|BC| = |AD_1| = |AD_2| = 3$. Denote the point

$D_3(x, y)$. Since the slope of AB equals the slope of

CD_3 we have $\frac{y - 3}{x - 2} = -\frac{1}{3} \Rightarrow 3y - 9 = -x + 2$ or

$x + 3y = 11$. Likewise, the slope of AC equals the slope

of BD_3 so that $\frac{y - 0}{x - 2} = \frac{2}{3} \Rightarrow 3y = 2x - 4$ or $2x - 3y = 4$.

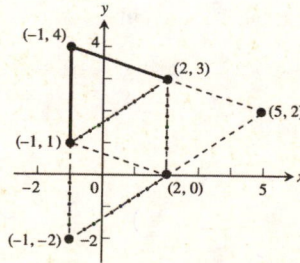

Solving the system of equations $\left.\begin{array}{r} x + 3y = 11 \\ 2x - 3y = 4 \end{array}\right\}$ we find $x = 5$ and $y = 2$ yielding the vertex $D_3(5, 2)$.

93. $2x + ky = 3$ has slope $-\frac{2}{k}$ and $4x + y = 1$ has slope -4. The lines are perpendicular when $-\frac{2}{k}(-4) = -1$ or $k = -8$ and parallel when $-\frac{2}{k} = -4$ or $k = \frac{1}{2}$.

95. Let $M(a, b)$ be the midpoint. Since the two triangles shown in the figure are congruent, the value a must lie midway between x_1 and x_2, so $a = \frac{x_1 + x_2}{2}$.

    Similarly, $b = \frac{y_1 + y_2}{2}$.

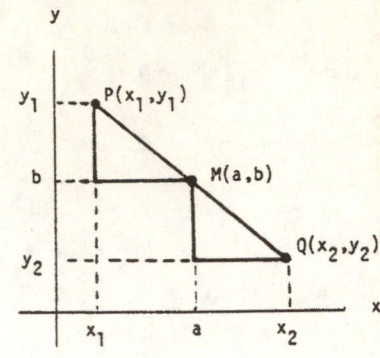

1.3 FUNCTIONS AND THEIR GRAPHS

1. domain $= (-\infty, \infty)$; range $= [1, \infty)$

3. domain $= (0, \infty)$; y in range $\Rightarrow y = \frac{1}{\sqrt{t}}, t > 0 \Rightarrow y^2 = \frac{1}{t}$ and $y > 0 \Rightarrow$ y can be any positive real number
    $\Rightarrow$ range $= (0, \infty)$.

5. $4 - z^2 = (2 - z)(2 + z) \geq 0 \Leftrightarrow z \in [-2, 2] =$ domain. Largest value is $g(0) = \sqrt{4} = 2$ and smallest value is
    $g(-2) = g(2) = \sqrt{0} = 0 \Rightarrow$ range $= [0, 2]$.

7. (a) Not the graph of a function of x since it fails the vertical line test.
    (b) Is the graph of a function of x since any vertical line intersects the graph at most once.

9. $y = \sqrt{\left(\frac{1}{x}\right) - 1} \Rightarrow \frac{1}{x} - 1 \geq 0 \Rightarrow x \leq 1$ and $x > 0$. So,
    (a) No $(x > 0)$;   (b) No; division by 0 undefined;
    (c) No; if $x \geq 1, \frac{1}{x} < 1 \Rightarrow \frac{1}{x} - 1 < 0$;   (d) $(0, 1]$

11. base $= x$; $(\text{height})^2 + \left(\frac{x}{2}\right)^2 = x^2 \Rightarrow$ height $= \frac{\sqrt{3}}{2}x$; area is $a(x) = \frac{1}{2}(\text{base})(\text{height}) = \frac{1}{2}(x)\left(\frac{\sqrt{3}}{2}x\right) = \frac{\sqrt{3}}{4}x^2$;
    perimeter is $p(x) = x + x + x = 3x$.

13. Let $D =$ diagonal of a face of the cube and $\ell =$ the length of an edge. Then $\ell^2 + D^2 = d^2$ and (by Exercise 10)
    $D^2 = 2\ell^2 \Rightarrow 3\ell^2 = d^2 \Rightarrow \ell = \frac{d}{\sqrt{3}}$. The surface area is $6\ell^2 = \frac{6d^2}{3} = 2d^2$ and the volume is $\ell^3 = \left(\frac{d^2}{3}\right)^{3/2} = \frac{d^3}{3\sqrt{3}}$.

15. The domain is $(-\infty, \infty)$.

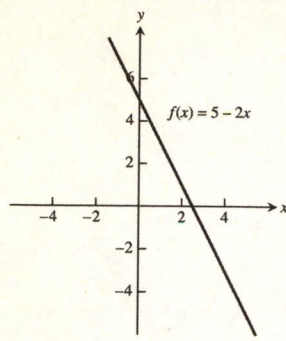

17. The domain is $(-\infty, \infty)$.

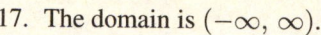
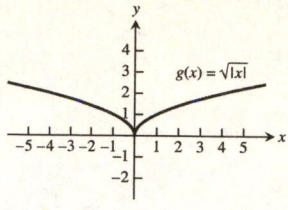

19. The domain is $(-\infty, 0) \cup (0, \infty)$.

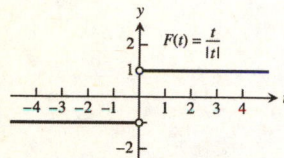

21. Neither graph passes the vertical line test

(a)

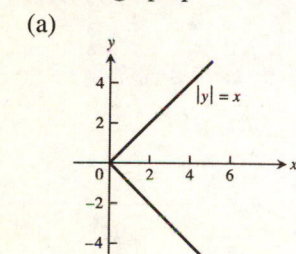

(b)

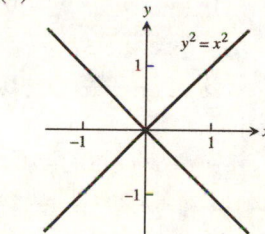

23.

x	0	1	2
y	0	1	0

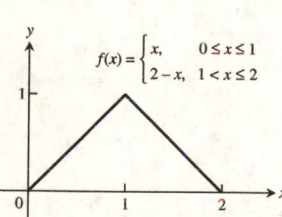

$$f(x) = \begin{cases} x, & 0 \le x \le 1 \\ 2-x, & 1 < x \le 2 \end{cases}$$

25. $y = \begin{cases} 3-x, & x \le 1 \\ 2x, & 1 < x \end{cases}$

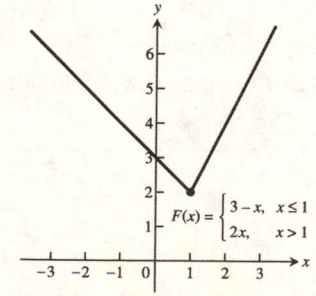

$$F(x) = \begin{cases} 3-x, & x \le 1 \\ 2x, & x > 1 \end{cases}$$

27. (a) Line through $(0, 0)$ and $(1, 1)$: $y = x$

Line through $(1, 1)$ and $(2, 0)$: $y = -x + 2$

$$f(x) = \begin{cases} x, & 0 \le x \le 1 \\ -x + 2, & 1 < x \le 2 \end{cases}$$

(b) $f(x) = \begin{cases} 2, & 0 \le x < 1 \\ 0, & 1 \le x < 2 \\ 2, & 2 \le x < 3 \\ 0, & 3 \le x \le 4 \end{cases}$

29. (a) Line through $(-1, 1)$ and $(0, 0)$: $y = -x$

Line through $(0, 1)$ and $(1, 1)$: $y = 1$

Line through $(1, 1)$ and $(3, 0)$: $m = \frac{0-1}{3-1} = \frac{-1}{2} = -\frac{1}{2}$, so $y = -\frac{1}{2}(x-1) + 1 = -\frac{1}{2}x + \frac{3}{2}$

$f(x) = \begin{cases} -x & -1 \le x < 0 \\ 1 & 0 < x \le 1 \\ -\frac{1}{2}x + \frac{3}{2} & 1 < x < 3 \end{cases}$

(b) Line through $(-2, -1)$ and $(0, 0)$: $y = \frac{1}{2}x$

Line through $(0, 2)$ and $(1, 0)$: $y = -2x + 2$

Line through $(1, -1)$ and $(3, -1)$: $y = -1$

$f(x) = \begin{cases} \frac{1}{2}x & -2 \le x \le 0 \\ -2x + 2 & 0 < x \le 1 \\ -1 & 1 < x \le 3 \end{cases}$

31. (a) From the graph, $\frac{x}{2} > 1 + \frac{4}{x} \Rightarrow x \in (-2, 0) \cup (4, \infty)$

(b) $\frac{x}{2} > 1 + \frac{4}{x} \Rightarrow \frac{x}{2} - 1 - \frac{4}{x} > 0$

$x > 0$: $\frac{x}{2} - 1 - \frac{4}{x} > 0 \Rightarrow \frac{x^2 - 2x - 8}{2x} > 0 \Rightarrow \frac{(x-4)(x+2)}{2x} > 0$

$\Rightarrow x > 4$ since x is positive;

$x < 0$: $\frac{x}{2} - 1 - \frac{4}{x} > 0 \Rightarrow \frac{x^2 - 2x - 8}{2x} < 0 \Rightarrow \frac{(x-4)(x+2)}{2x} < 0$

$\Rightarrow x < -2$ since x is negative;

sign of $(x - 4)(x + 2)$

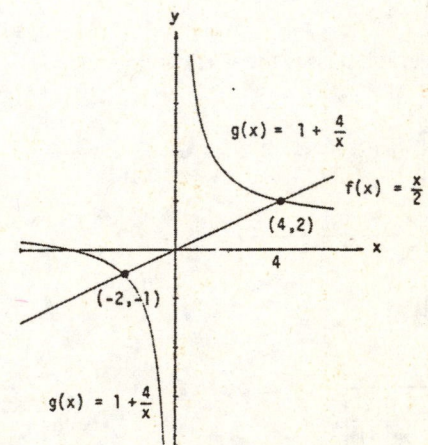

Solution interval: $(-2, 0) \cup (4, \infty)$

33. (a) $\lfloor x \rfloor = 0$ for $x \in [0, 1)$ (b) $\lceil x \rceil = 0$ for $x \in (-1, 0]$

35. For any real number x, $n \le x \le n + 1$, where n is an integer. Now: $n \le x \le n + 1 \Rightarrow -(n + 1) \le -x \le -n$. By definition: $\lceil -x \rceil = -n$ and $\lfloor x \rfloor = n \Rightarrow -\lfloor x \rfloor = -n$. So $\lceil -x \rceil = -\lfloor x \rfloor$ for all $x \in \Re$.

37. $v = f(x) = x(14 - 2x)(22 - 2x) = 4x^3 - 72x^2 + 308x$; $0 < x < 7$.

39. (a) Because the circumference of the original circle was 8π and a piece of length x was removed.

(b) $r = \frac{8\pi - x}{2\pi} = 4 - \frac{x}{2\pi}$

(c) $h = \sqrt{16 - r^2} = \sqrt{16 - \left(4 - \frac{x}{2\pi}\right)^2} = \sqrt{16 - \left(16 - \frac{4x}{\pi} + \frac{x^2}{4\pi^2}\right)} = \sqrt{\frac{4x}{\pi} - \frac{x^2}{4\pi^2}} = \sqrt{\frac{16\pi x}{4\pi^2} - \frac{x^2}{4\pi^2}} = \frac{\sqrt{16\pi x - x^2}}{2\pi}$

(d) $V = \frac{1}{3}\pi r^2 h = \frac{1}{3}\pi \left(\frac{8\pi - x}{2\pi}\right)^2 \cdot \frac{\sqrt{16\pi x - x^2}}{2\pi} = \frac{(8\pi - x)^2 \sqrt{16\pi x - x^2}}{24\pi^2}$

41. A curve symmetric about the x-axis will not pass the vertical line test because the points (x, y) and $(x, -y)$ lie on the same vertical line. The graph of the function $y = f(x) = 0$ is the x-axis, a horizontal line for which there is a single y-value, 0, for any x.

1.4 IDENTIFYING FUNCTIONS; MATHEMATICAL MODELS

1. (a) linear, polynomial of degree 1, algebraic.   (b) power, algebraic.
    (c) rational, algebraic.   (d) exponential.

3. (a) rational, algebraic.   (b) algebraic.
    (c) trigonometric.   (d) logarithmic.

5. (a) Graph h because it is an even function and rises less rapidly than does Graph g.
    (b) Graph f because it is an odd function.
    (c) Graph g because it is an even function and rises more rapidly than does Graph h.

7. Symmetric about the origin
    Dec: $-\infty < x < \infty$
    Inc: nowhere

9. Symmetric about the origin
    Dec: nowhere
    Inc: $-\infty < x < 0$
          $0 < x < \infty$

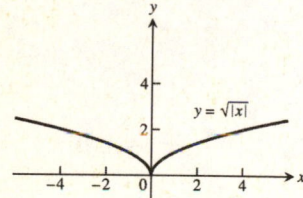

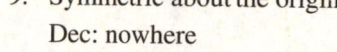
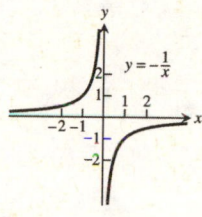

11. Symmetric about the y-axis
    Dec: $-\infty < x \le 0$
    Inc: $0 < x < \infty$

13. Symmetric about the origin
    Dec: nowhere
    Inc: $-\infty < x < \infty$

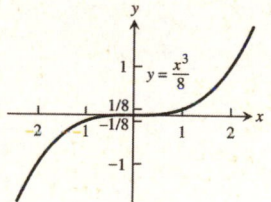

15. No symmetry
    Dec: $0 \le x < \infty$
    Inc: nowhere

17. Symmetric about the y-axis
    Dec: $-\infty < x \le 0$
    Inc: $0 < x < \infty$

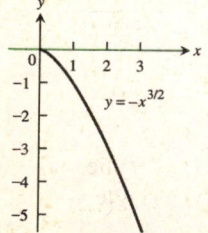

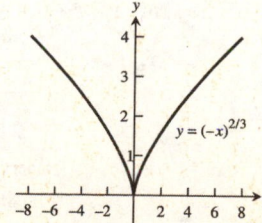

19. Since a horizontal line not through the origin is symmetric with respect to the y-axis, but not with respect to the origin, the function is even.

21. Since $f(x) = x^2 + 1 = (-x)^2 + 1 = -f(x)$. The function is even.

23. Since $g(x) = x^3 + x$, $g(-x) = -x^3 - x = -(x^3 + x) = -g(x)$. So the function is odd.

25. $g(x) = \frac{1}{x^2 - 1} = \frac{1}{(-x)^2 - 1} = g(-x)$. Thus the function is even.

27. $h(t) = \frac{1}{t - 1}$; $h(-t) = \frac{1}{-t - 1}$; $-h(t) = \frac{1}{1 - t}$. Since $h(t) \neq -h(t)$ and $h(t) \neq h(-t)$, the function is neither even nor odd.

29. $h(t) = 2t + 1$, $h(-t) = -2t + 1$. So $h(t) \neq h(-t)$. $-h(t) = -2t - 1$, so $h(t) \neq -h(t)$. The function is neither even nor odd.

31. (a)

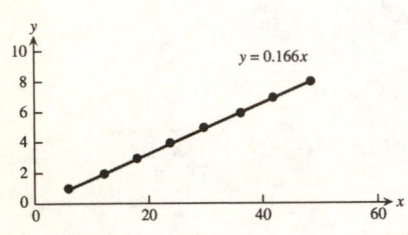

The graph supports the assumption that y is proportional to x. The constant of proportionality is estimated from the slope of the regression line, which is 0.166.

(b)

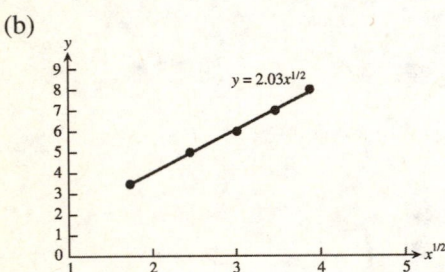

The graph supports the assumption that y is proportional to $x^{1/2}$. The constant of proportionality is estimated from the slope of the regression line, which is 2.03.

33. (a)   The scatterplot of y = reaction distance versus x = speed is

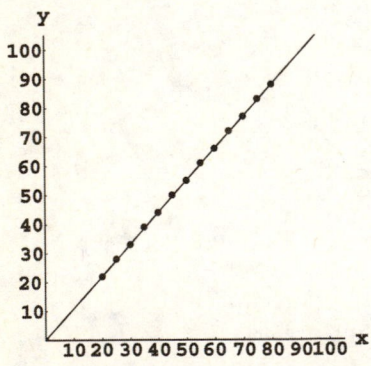

Answers for the constant of proportionality may vary. The constant of proportionality is the slope of the line, which is approximately 1.1.

(b)   Calculate x′ = speed squared. The scatterplot of x′ versus y = braking distance is:

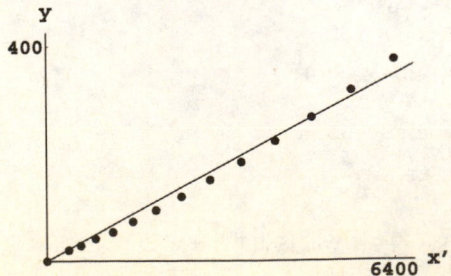

Answers for the constant of proportionality may vary. The constant of proportionality is the slope of the line, which is approximately 0.059.

35. (a)

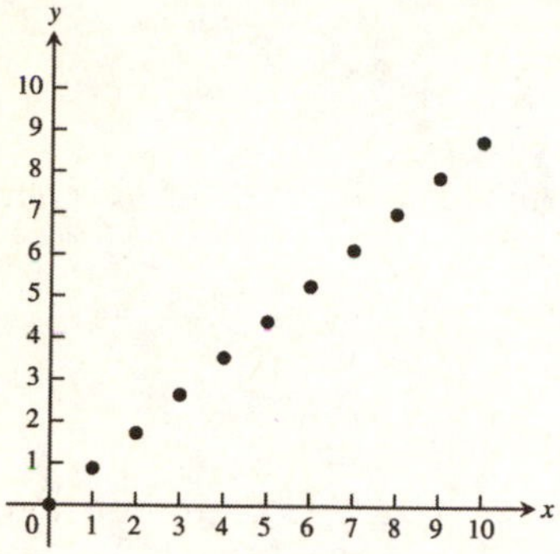

The hypothesis is reasonable.

(b) The constant of proportionality is the slope of the line $\approx \frac{8.741 - 0}{10 - 0}$ in./unit mass $= 0.874$ in./unit mass.

(c) $y(\text{in.}) = (0.87 \text{ in./unit mass})(13 \text{ unit mass}) = 11.31$ in.

1.5 COMBINING FUNCTIONS; SHIFTING AND SCALING GRAPHS

1. $D_f: -\infty < x < \infty, D_g: x \geq 1 \Rightarrow D_{f+g} = D_{fg}: x \geq 1.$ $R_f: -\infty < y < \infty, R_g: y \geq 0, R_{f+g}: y \geq 1, R_{fg}: y \geq 0$

3. $D_f: -\infty < x < \infty, D_g: -\infty < x < \infty \Rightarrow D_{f/g}: -\infty < x < \infty$ since $g(x) \neq 0$ for any x; $D_{g/f}: -\infty < x < \infty$ since $f(x) \neq 0$ for any x. $R_f: y = 2, R_g: y \geq 1, R_{f/g}: 0 < y \leq 2, R_{g/f}: y \geq \frac{1}{2}$

5. (a) $f(g(0)) = f(-3) = 2$

(b) $g(f(0)) = g(5) = 22$

(c) $f(g(x)) = f(x^2 - 3) = x^2 - 3 + 5 = x^2 + 2$

(d) $g(f(x)) = g(x + 5) = (x + 5)^2 - 3 = x^2 + 10x + 22$

(e) $f(f(-5)) = f(0) = 5$

(f) $g(g(2)) = g(1) = -2$

(g) $f(f(x)) = f(x + 5) = (x + 5) + 5 = x + 10$

(h) $g(g(x)) = g(x^2 - 3) = (x^2 - 3)^2 - 3 = x^4 - 6x^2 + 6$

7. (a) $u(v(f(x))) = u\left(v\left(\frac{1}{x}\right)\right) = u\left(\frac{1}{x^2}\right) = 4\left(\frac{1}{x}\right)^2 - 5 = \frac{4}{x^2} - 5$

(b) $u(f(v(x))) = u\left(f\left(x^2\right)\right) = u\left(\frac{1}{x^2}\right) = 4\left(\frac{1}{x^2}\right) - 5 = \frac{4}{x^2} - 5$

(c) $v(u(f(x))) = v\left(u\left(\frac{1}{x}\right)\right) = v\left(4\left(\frac{1}{x}\right) - 5\right) = \left(\frac{4}{x} - 5\right)^2$

(d) $v(f(u(x))) = v(f(4x - 5)) = v\left(\frac{1}{4x - 5}\right) = \left(\frac{1}{4x - 5}\right)^2$

(e) $f(u(v(x))) = f\left(u\left(x^2\right)\right) = f\left(4\left(x^2\right) - 5\right) = \frac{1}{4x^2 - 5}$

(f) $f(v(u(x))) = f(v(4x - 5)) = f\left((4x - 5)^2\right) = \frac{1}{(4x - 5)^2}$

9. (a) $y = f(g(x))$

(b) $y = j(g(x))$

(c) $y = g(g(x))$

(d) $y = j(j(x))$

(e) $y = g(h(f(x)))$

(f) $y = h(j(f(x)))$

11.

	g(x)	f(x)	(f ∘ g)(x)
(a)	$x - 7$	$\sqrt{x}$	$\sqrt{x - 7}$
(b)	$x + 2$	$3x$	$3(x + 2) = 3x + 6$

	g(x)	f(x)	(f ∘ g)(x)
(c)	x^2	$\sqrt{x-5}$	$\sqrt{x^2-5}$
(d)	$\frac{x}{x-1}$	$\frac{x}{x-1}$	$\frac{\frac{x}{x-1}}{\frac{x}{x-1}-1} = \frac{x}{x-(x-1)} = x$
(e)	$\frac{1}{x-1}$	$1+\frac{1}{x}$	x
(f)	$\frac{1}{x}$	$\frac{1}{x}$	x

13. (a) $f(g(x)) = \sqrt{\frac{1}{x}+1} = \sqrt{\frac{1+x}{x}}$

   $g(f(x)) = \frac{1}{\sqrt{x+1}}$

   (b) Domain (f∘g): $(0, \infty)$, domain (g∘f): $(-1, \infty)$

   (c) Range (f∘g): $(1, \infty)$, range (g∘f): $(0, \infty)$

15. (a) $y = -(x+7)^2$ (b) $y = -(x-4)^2$

17. (a) Position 4   (b) Position 1   (c) Position 2   (d) Position 3

19.

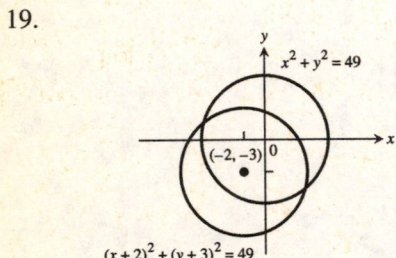

21.

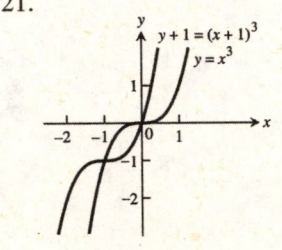

23.

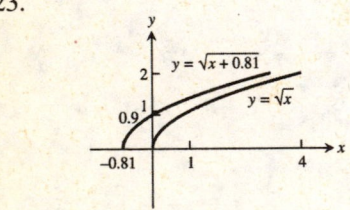

25.

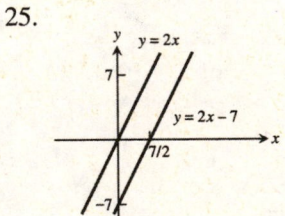

27.

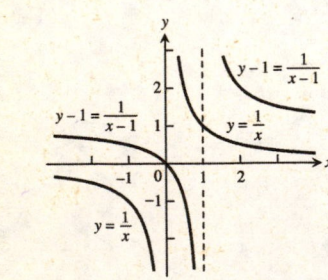

29.

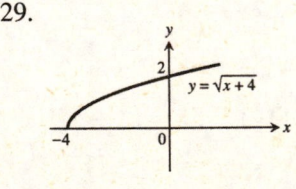

31.

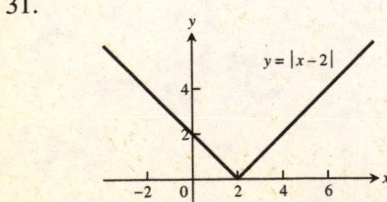

33.

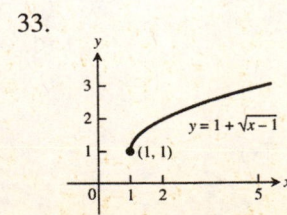

35.

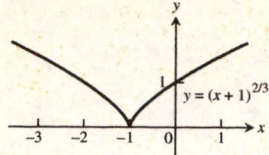

37.

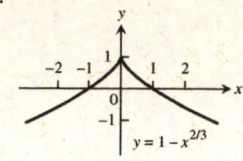

39.

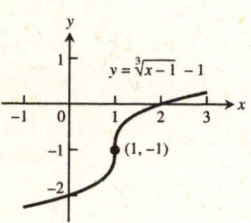

41.

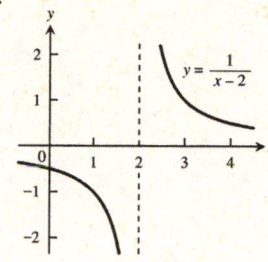

43.

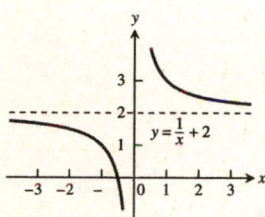

45.

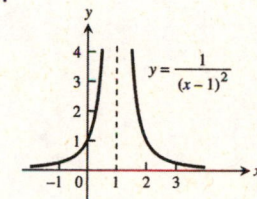

47.

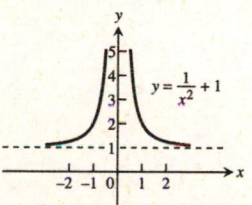

49. (a) domain: $[0, 2]$; range: $[2, 3]$

(b) domain: $[0, 2]$; range: $[-1, 0]$

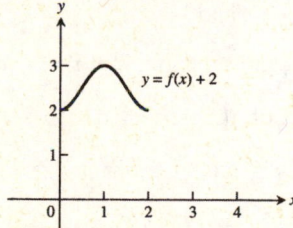

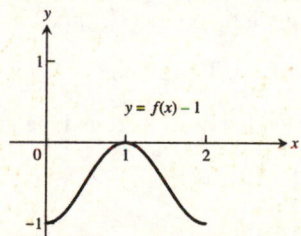

(c) domain: $[0, 2]$; range: $[0, 2]$

(d) domain: $[0, 2]$; range: $[-1, 0]$

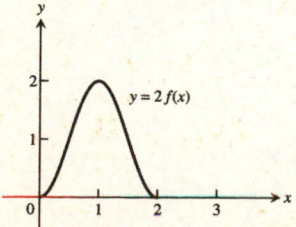

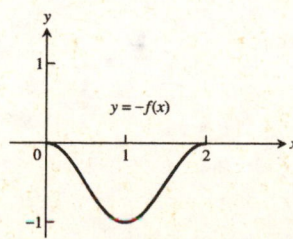

(e) domain: $[-2, 0]$; range: $[0, 1]$

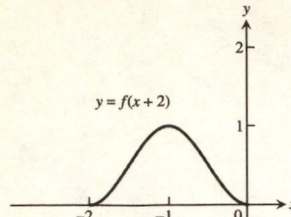

(f) domain: $[1, 3]$; range: $[0, 1]$

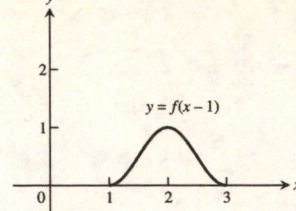

(g) domain: $[-2, 0]$; range: $[0, 1]$

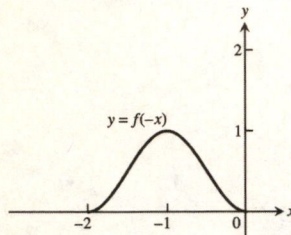

(h) domain: $[-1, 1]$; range: $[0, 1]$

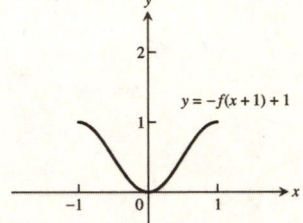

51. $y = 3x^2 - 3$

53. $y = \frac{1}{2}\left(1 + \frac{1}{x^2}\right) = \frac{1}{2} + \frac{1}{2x^2}$

55. $y = \sqrt{4x + 1}$

57. $y = \sqrt{4 - \left(\frac{x}{2}\right)^2} = \frac{1}{2}\sqrt{16 - x^2}$

59. $y = 1 - (3x)^3 = 1 - 27x^3$

61. Let $y = -\sqrt{2x + 1} = f(x)$ and let $g(x) = x^{1/2}$, $h(x) = \left(x + \frac{1}{2}\right)^{1/2}$, $i(x) = \sqrt{2}\left(x + \frac{1}{2}\right)^{1/2}$, and

$j(x) = -\left[\sqrt{2}\left(x + \frac{1}{2}\right)^{1/2}\right] = f(x)$. The graph of $h(x)$ is the graph of $g(x)$ shifted left $\frac{1}{2}$ unit; the graph of $i(x)$ is the graph

of $h(x)$ stretched vertically by a factor of $\sqrt{2}$; and the graph of $j(x) = f(x)$ is the graph of $i(x)$ reflected across the x-axis.

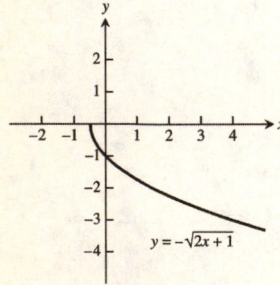

63. $y = f(x) = x^3$. Shift $f(x)$ one unit right followed by a shift two units up to get $g(x) = (x-1)^3 + 2$.

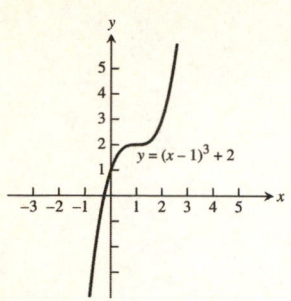

65. Compress the graph of $f(x) = \frac{1}{x}$ horizontally by a factor of 2 to get $g(x) = \frac{1}{2x}$. Then shift $g(x)$ vertically down 1 unit to get $h(x) = \frac{1}{2x} - 1$.

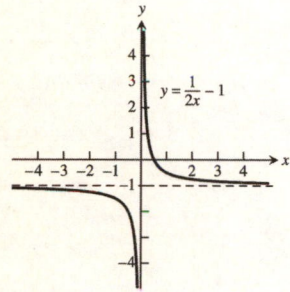

67. Reflect the graph of $y = f(x) = \sqrt[3]{x}$ across the x-axis to get $g(x) = -\sqrt[3]{x}$.

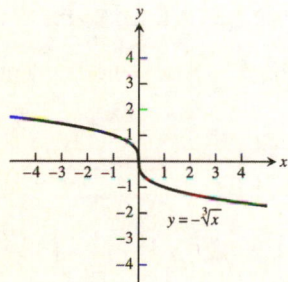

69.

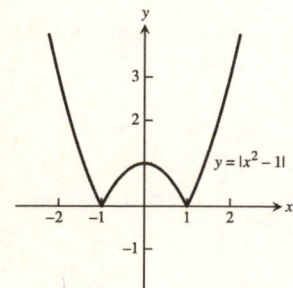

71. $9x^2 + 25y^2 = 225 \Rightarrow \frac{x^2}{5^2} + \frac{y^2}{3^2} = 1$

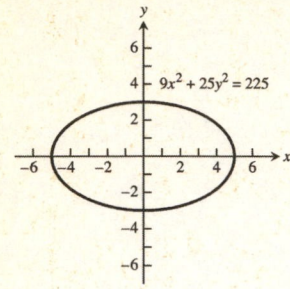

73. $3x^2 + (y-2)^2 = 3 \Rightarrow \frac{x^2}{1^2} + \frac{(y-2)^2}{(\sqrt{3})^2} = 1$

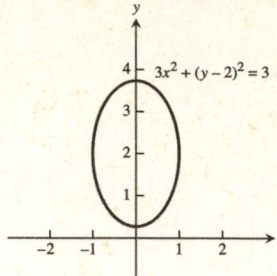

75. $3(x-1)^2 + 2(y+2)^2 = 6$

$\Rightarrow \frac{(x-1)^2}{(\sqrt{2})^2} + \frac{[y-(-2)]^2}{(\sqrt{3})^2} = 1$

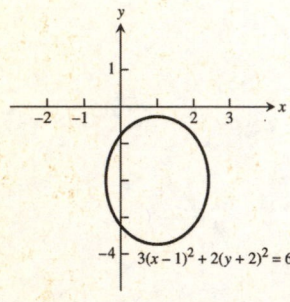

77. $\frac{x^2}{16} + \frac{y^2}{9} = 1$ has its center at $(0, 0)$. Shiftinig 4 units left and 3 units up gives the center at $(h, k) = (-4, 3)$. So the equation is $\frac{[x-(-4)]^2}{4^2} + \frac{(y-3)^2}{3^2} = 1 \Rightarrow \frac{(x+4)^2}{4^2} + \frac{(y-3)^2}{3^2} = 1$. Center, C, is $(-4, 3)$, and major axis, $\overline{AB}$, is the segment from $(-8, 3)$ to $(0, 3)$.

79. (a) $(fg)(-x) = f(-x)g(-x) = f(x)(-g(x)) = -(fg)(x)$, odd

(b) $\left(\frac{f}{g}\right)(-x) = \frac{f(-x)}{g(-x)} = \frac{f(x)}{-g(x)} = -\left(\frac{f}{g}\right)(x)$, odd

(c) $\left(\frac{g}{f}\right)(-x) = \frac{g(-x)}{f(-x)} = \frac{-g(x)}{f(x)} = -\left(\frac{g}{f}\right)(x)$, odd

(d) $f^2(-x) = f(-x)f(-x) = f(x)f(x) = f^2(x)$, even

(e) $g^2(-x) = (g(-x))^2 = (-g(x))^2 = g^2(x)$, even

(f) $(f \circ g)(-x) = f(g(-x)) = f(-g(x)) = f(g(x)) = (f \circ g)(x)$, even

(g) $(g \circ f)(-x) = g(f(-x)) = g(f(x)) = (g \circ f)(x)$, even

(h) $(f \circ f)(-x) = f(f(-x)) = f(f(x)) = (f \circ f)(x)$, even

(i) $(g \circ g)(-x) = g(g(-x)) = g(-g(x)) = -g(g(x)) = -(g \circ g)(x)$, odd

81. (a)

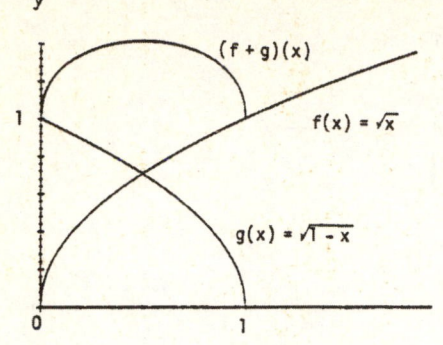

(b)

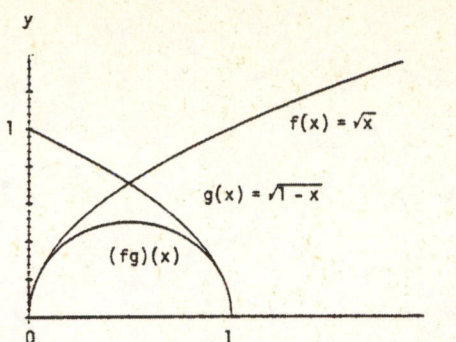

(c)

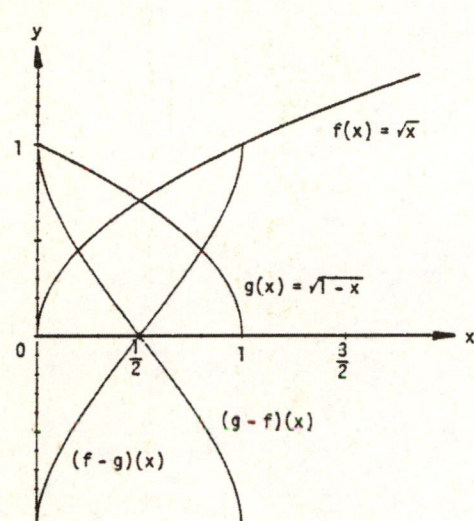

(d)

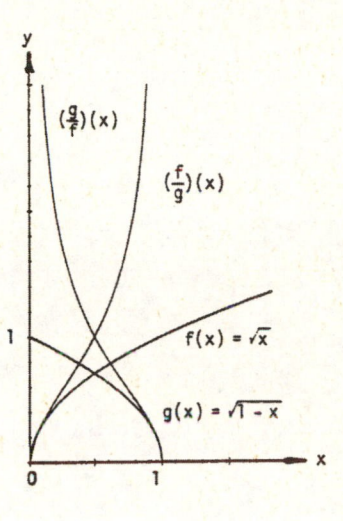

1.6 TRIGONOMETRIC FUNCTIONS

1. (a) $s = r\theta = (10)\left(\frac{4\pi}{5}\right) = 8\pi$ m

    (b) $s = r\theta = (10)(110°)\left(\frac{\pi}{180°}\right) = \frac{110\pi}{18} = \frac{55\pi}{9}$ m

3. $\theta = 80° \Rightarrow \theta = 80°\left(\frac{\pi}{180°}\right) = \frac{4\pi}{9} \Rightarrow s = (6)\left(\frac{4\pi}{9}\right) = 8.4$ in. (since the diameter $= 12$ in. $\Rightarrow$ radius $= 6$ in.)

5.

θ	$-\pi$	$-\frac{2\pi}{3}$	0	$\frac{\pi}{2}$	$\frac{3\pi}{4}$
$\sin\theta$	0	$-\frac{\sqrt{3}}{2}$	0	1	$\frac{1}{\sqrt{2}}$
$\cos\theta$	-1	$-\frac{1}{2}$	1	0	$-\frac{1}{\sqrt{2}}$
$\tan\theta$	0	$\sqrt{3}$	0	und.	-1
$\cot\theta$	und.	$\frac{1}{\sqrt{3}}$	und.	0	-1
$\sec\theta$	-1	-2	1	und.	$-\sqrt{2}$
$\csc\theta$	und.	$-\frac{2}{\sqrt{3}}$	und.	1	$\sqrt{2}$

7. $\cos x = -\frac{4}{5}$, $\tan x = -\frac{3}{4}$

9. $\sin x = -\frac{\sqrt{8}}{3}$, $\tan x = -\sqrt{8}$

11. $\sin x = -\frac{1}{\sqrt{5}}$, $\cos x = -\frac{2}{\sqrt{5}}$

13.

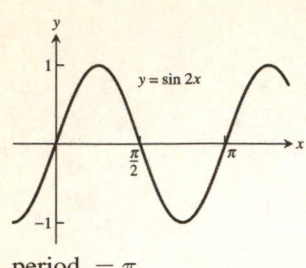

period $= \pi$

15.

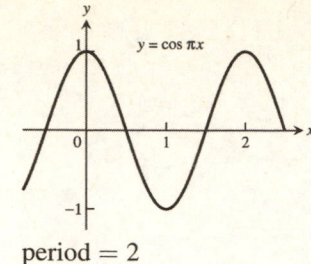

period $= 2$

17.

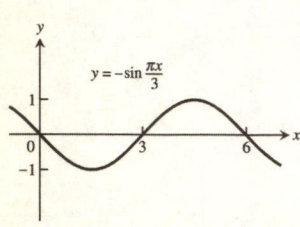

period $= 6$

19.

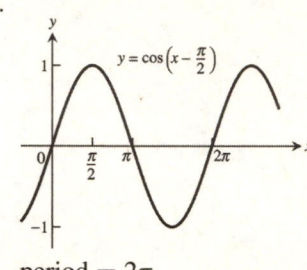

period $= 2\pi$

21.

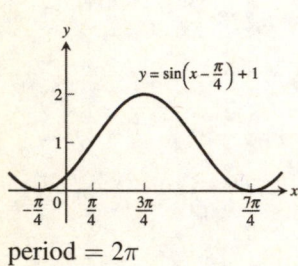

period $= 2\pi$

23. period $= \frac{\pi}{2}$, symmetric about the origin

25. period $= 4$, symmetric about the y-axis

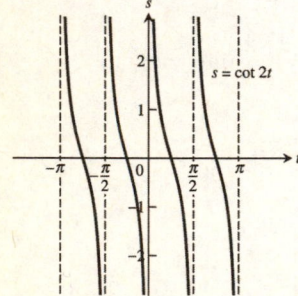

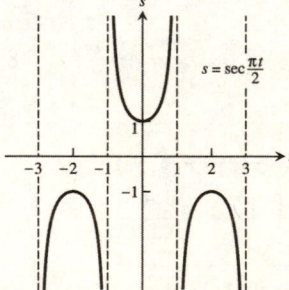

27. (a) Cos x and sec x are positive in QI and QIV and
        negative in QII and QIII. Sec x is undefined when
        cos x is 0. The range of sec x is $(-\infty, -1] \cup [1, \infty)$;
        the range of cos x is $[-1, 1]$.

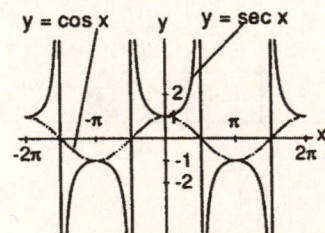

(b) Sin x and csc x are positive in QI and QII and
negative in QIII and QIV. Csc x is undefined when
sin x is 0. The range of csc x is $(-\infty, -1] \cup [1, \infty)$;
the range of sin x is $[-1, 1]$.

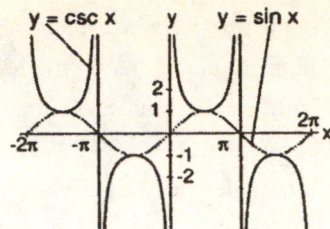

29. D: $-\infty < x < \infty$; R: $y = -1, 0, 1$

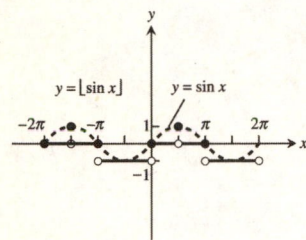

31. $\cos\left(x - \frac{\pi}{2}\right) = \cos x \cos\left(-\frac{\pi}{2}\right) - \sin x \sin\left(-\frac{\pi}{2}\right) = (\cos x)(0) - (\sin x)(-1) = \sin x$

33. $\sin\left(x + \frac{\pi}{2}\right) = \sin x \cos\left(\frac{\pi}{2}\right) + \cos x \sin\left(\frac{\pi}{2}\right) = (\sin x)(0) + (\cos x)(1) = \cos x$

35. $\cos(A - B) = \cos(A + (-B)) = \cos A \cos(-B) - \sin A \sin(-B) = \cos A \cos B - \sin A (-\sin B)$
$= \cos A \cos B + \sin A \sin B$

37. If $B = A$, $A - B = 0 \Rightarrow \cos(A - B) = \cos 0 = 1$. Also $\cos(A - B) = \cos(A - A) = \cos A \cos A + \sin A \sin A$
$= \cos^2 A + \sin^2 A$. Therefore, $\cos^2 A + \sin^2 A = 1$.

39. $\cos(\pi + x) = \cos \pi \cos x - \sin \pi \sin x = (-1)(\cos x) - (0)(\sin x) = -\cos x$

41. $\sin\left(\frac{3\pi}{2} - x\right) = \sin\left(\frac{3\pi}{2}\right)\cos(-x) + \cos\left(\frac{3\pi}{2}\right)\sin(-x) = (-1)(\cos x) + (0)(\sin(-x)) = -\cos x$

43. $\sin\frac{7\pi}{12} = \sin\left(\frac{\pi}{4} + \frac{\pi}{3}\right) = \sin\frac{\pi}{4}\cos\frac{\pi}{3} + \cos\frac{\pi}{4}\sin\frac{\pi}{3} = \left(\frac{\sqrt{2}}{2}\right)\left(\frac{1}{2}\right) + \left(\frac{\sqrt{2}}{2}\right)\left(\frac{\sqrt{3}}{2}\right) = \frac{\sqrt{6}+\sqrt{2}}{4}$

45. $\cos\frac{\pi}{12} = \cos\left(\frac{\pi}{3} - \frac{\pi}{4}\right) = \cos\frac{\pi}{3}\cos\left(-\frac{\pi}{4}\right) - \sin\frac{\pi}{3}\sin\left(-\frac{\pi}{4}\right) = \left(\frac{1}{2}\right)\left(\frac{\sqrt{2}}{2}\right) - \left(\frac{\sqrt{3}}{2}\right)\left(-\frac{\sqrt{2}}{2}\right) = \frac{1+\sqrt{3}}{2\sqrt{2}}$

47. $\cos^2\frac{\pi}{8} = \frac{1+\cos\left(\frac{2\pi}{8}\right)}{2} = \frac{1+\frac{\sqrt{2}}{2}}{2} = \frac{2+\sqrt{2}}{4}$

49. $\sin^2\frac{\pi}{12} = \frac{1-\cos\left(\frac{2\pi}{12}\right)}{2} = \frac{1-\frac{\sqrt{3}}{2}}{2} = \frac{2-\sqrt{3}}{4}$

51. $\tan(A + B) = \frac{\sin(A+B)}{\cos(A+B)} = \frac{\sin A \cos B + \cos A \cos B}{\cos A \cos B - \sin A \sin B} = \frac{\frac{\sin A \cos B}{\cos A \cos B} + \frac{\cos A \sin B}{\cos A \cos B}}{\frac{\cos A \cos B}{\cos A \cos B} - \frac{\sin A \sin B}{\cos A \cos B}} = \frac{\tan A + \tan B}{1 - \tan A \tan B}$

53. According to the figure in the text, we have the following: By the law of cosines, $c^2 = a^2 + b^2 - 2ab \cos \theta$
$= 1^2 + 1^2 - 2\cos(A - B) = 2 - 2\cos(A - B)$. By distance formula, $c^2 = (\cos A - \cos B)^2 + (\sin A - \sin B)^2$
$= \cos^2 A - 2\cos A \cos B + \cos^2 B + \sin^2 A - 2\sin A \sin B + \sin^2 B = 2 - 2(\cos A \cos B + \sin A \sin B)$. Thus
$c^2 = 2 - 2\cos(A - B) = 2 - 2(\cos A \cos B + \sin A \sin B) \Rightarrow \cos(A - B) = \cos A \cos B + \sin A \sin B$.

55. $c^2 = a^2 + b^2 - 2ab \cos C = 2^2 + 3^2 - 2(2)(3)\cos(60°) = 4 + 9 - 12\cos(60°) = 13 - 12\left(\frac{1}{2}\right) = 7$.
Thus, $c = \sqrt{7} \approx 2.65$.

57. From the figures in the text, we see that $\sin B = \frac{h}{c}$. If C is an acute angle, then $\sin C = \frac{h}{b}$. On the other hand, if C is obtuse (as in the figure on the right), then $\sin C = \sin(\pi - C) = \frac{h}{b}$. Thus, in either case, $h = b \sin C = c \sin B \Rightarrow ah = ab \sin C = ac \sin B$.

By the law of cosines, $\cos C = \frac{a^2+b^2-c^2}{2ab}$ and $\cos B = \frac{a^2+c^2-b^2}{2ac}$. Moreover, since the sum of the interior angles of a triangle is π, we have $\sin A = \sin(\pi - (B+C)) = \sin(B+C) = \sin B \cos C + \cos B \sin C$

$= \left(\frac{h}{c}\right)\left[\frac{a^2+b^2-c^2}{2ab}\right] + \left[\frac{a^2+c^2-b^2}{2ac}\right]\left(\frac{h}{b}\right) = \left(\frac{h}{2abc}\right)(2a^2 + b^2 - c^2 + c^2 - b^2) = \frac{ah}{bc} \Rightarrow ah = bc \sin A.$

Combining our results we have $ah = ab \sin C$, $ah = ac \sin B$, and $ah = bc \sin A$. Dividing by abc gives

$\underbrace{\frac{h}{bc} = \frac{\sin A}{a} = \frac{\sin C}{c} = \frac{\sin B}{b}}_{\text{law of sines}}.$

59. From the figure at the right and the law of cosines,
$b^2 = a^2 + 2^2 - 2(2a) \cos B$
$= a^2 + 4 - 4a\left(\frac{1}{2}\right) = a^2 - 2a + 4.$
Applying the law of sines to the figure, $\frac{\sin A}{a} = \frac{\sin B}{b}$
$\Rightarrow \frac{\sqrt{2}/2}{a} = \frac{\sqrt{3}/2}{b} \Rightarrow b = \sqrt{\frac{3}{2}}\,a$. Thus, combining results,
$a^2 - 2a + 4 = b^2 = \frac{3}{2}a^2 \Rightarrow 0 = \frac{1}{2}a^2 + 2a - 4$
$\Rightarrow 0 = a^2 + 4a - 8.$ From the quadratic formula and the fact that $a > 0$, we have
$a = \frac{-4+\sqrt{4^2-4(1)(-8)}}{2} = \frac{4\sqrt{3}-4}{2} \simeq 1.464.$

61. $A = 2, B = 2\pi, C = -\pi, D = -1$

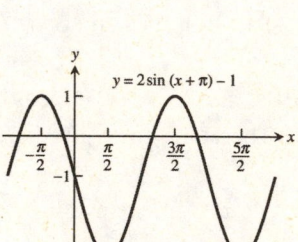

63. $A = -\frac{2}{\pi}, B = 4, C = 0, D = \frac{1}{\pi}$

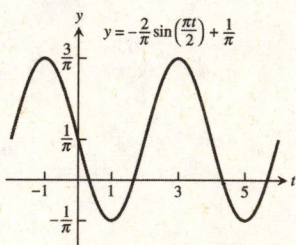

65. (a) amplitude $= |A| = 37$ (b) period $= |B| = 365$
    (c) right horizontal shift $= C = 101$ (d) upward vertical shift $= D = 25$

1.7 GRAPHING WITH CALCULATORS AND COMPUTERS

1-3.   The most appropriate viewing window displays the maxima, minima, intercepts, and end behavior of the graphs and has little unused space.

1. d.

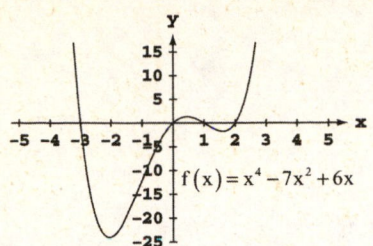

3. d.

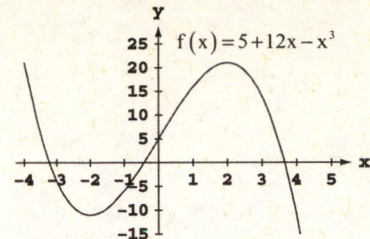

5-29.   For any display there are many appropriate display widows. The graphs given as answers in Exercises 5−30 are not unique in appearance.

5. [−2, 5] by [−15, 40]

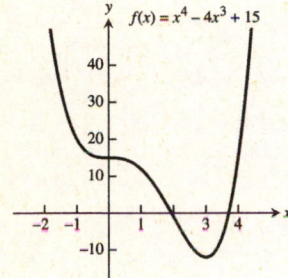

7. [−2, 6] by [−250, 50]

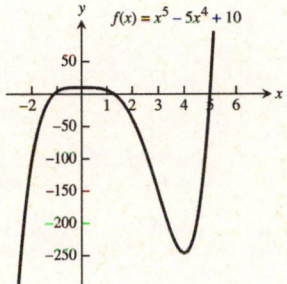

9. [−4, 4] by [−5, 5]

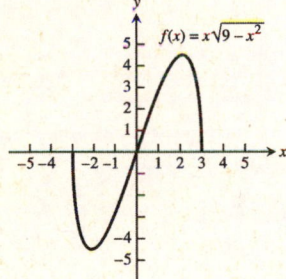

11. [−2, 6] by [−5, 4]

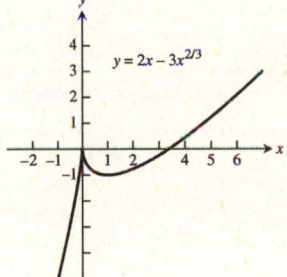

13. [−1, 6] by [−1, 4]

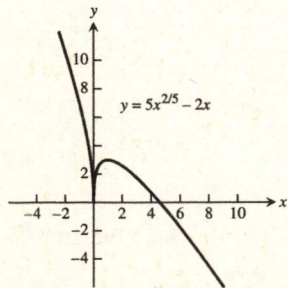

15. [−3, 3] by [0, 10]

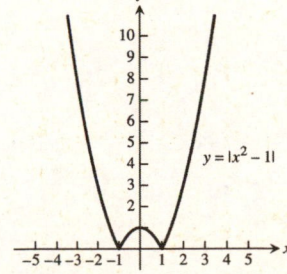

17. $[-5, 1]$ by $[-5, 5]$

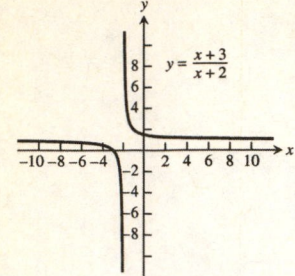

$$y = \frac{x+3}{x+2}$$

19. $[-4, 4]$ by $[0, 3]$

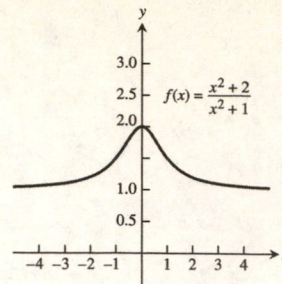

$$f(x) = \frac{x^2+2}{x^2+1}$$

21. $[-10, 10]$ by $[-6, 6]$

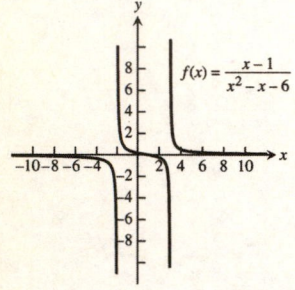

$$f(x) = \frac{x-1}{x^2-x-6}$$

23. $[-6, 10]$ by $[-6, 6]$

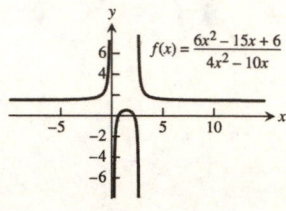

$$f(x) = \frac{6x^2-15x+6}{4x^2-10x}$$

25. $[-0.03, 0.03]$ by $[-1.25, 1.25]$

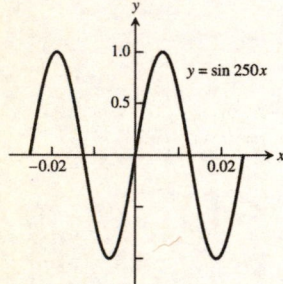

$$y = \sin 250x$$

27. $[-300, 300]$ by $[-1.25, 1.25]$

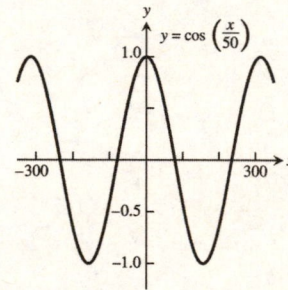

$$y = \cos\left(\frac{x}{50}\right)$$

29. $[-0.25, 0.25]$ by $[-0.3, 0.3]$

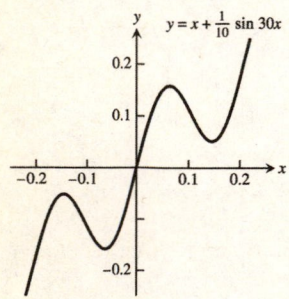

$$y = x + \frac{1}{10}\sin 30x$$

31. $x^2 + 2x = 4 + 4y - y^2 \Rightarrow y = 2 \pm \sqrt{-x^2 - 2x + 8}$.

The lower half is produced by graphing

$y = 2 - \sqrt{-x^2 - 2x + 8}$.

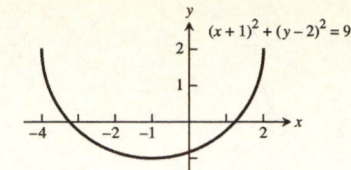

33.

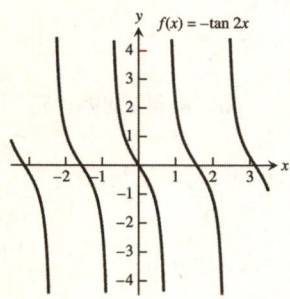

35.

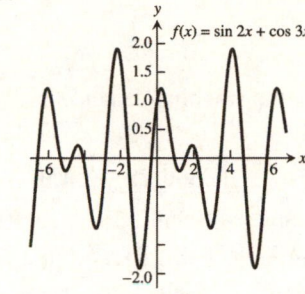

37.

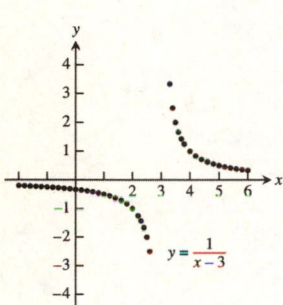

39.

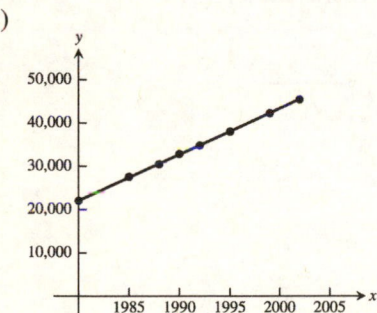

41. (a) $y = 1059.14x - 2074972$

(b) $m = 1059.14$ dollars/year, which is the yearly increase in compensation.

(c)

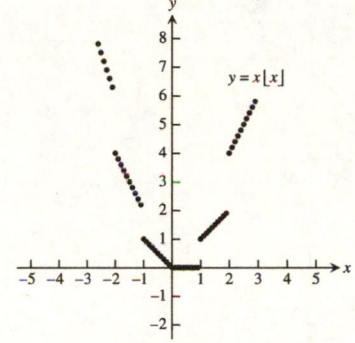

(d) Answers may vary slightly. $y = (1059.14)(2010) - 2074972 = \$53,899$

43. (a) Let x represent the speed in miles per hour and d the stopping distance in feet. The quadratic regression function is

$d = 0.0866x^2 - 1.97x + 50.1$.

(b)

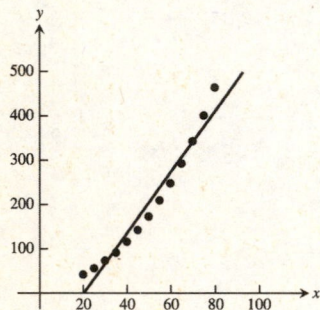

(c) From the graph in part (b), the stopping distance is about 370 feet when the vehicle is 72 mph and it is about 525 feet when the speed is 85 mph.

Algebraically: $d_{quadratic}(72) = 0.0866(72)^2 - 1.97(72) + 50.1 = 367.6$ ft.

$$d_{quadratic}(85) = 0.0866(85)^2 - 1.97(85) + 50.1 = 522.8 \text{ ft.}$$

(d) The linear regression function is $d = 6.89x - 140.4 \Rightarrow d_{linear}(72) = 6.89(72) - 140.4 = 355.7$ ft and $d_{linear}(85) = 6.89(85) - 140.4 = 445.2$ ft. The linear regression line is shown on the graph in part (b). The quadratic regression curve clearly gives the better fit.

CHAPTER 1 PRACTICE EXERCISES

1. $7 + 2x \geq 3 \Rightarrow 2x \geq -4 \Rightarrow x \geq -2$

3. $\frac{1}{5}(x - 1) < \frac{1}{4}(x - 2) \Rightarrow 4(x - 1) < 5(x - 2)$
    $\Rightarrow 4x - 4 < 5x - 10 \Rightarrow 6 < x$

5. $|x + 1| = 7 \Rightarrow x + 1 = 7 \text{ or } -(x + 1) = 7 \Rightarrow x = 6 \text{ or } x = -8$

7. $\left|1 - \frac{x}{2}\right| > \frac{3}{2} \Rightarrow 1 - \frac{x}{2} < -\frac{3}{2} \text{ or } 1 - \frac{x}{2} > \frac{3}{2} \Rightarrow -\frac{x}{2} < -\frac{5}{2} \text{ or } -\frac{x}{2} > \frac{1}{2} \Rightarrow -x < -5 \text{ or } -x > 1$
    $\Rightarrow x > 5 \text{ or } x < -1$

9. Since the particle moved to the y-axis, $-2 + \Delta x = 0 \Rightarrow \Delta x = 2$. Since $\Delta y = 3\Delta x = 6$, the new coordinates are $(x + \Delta x, y + \Delta y) = (-2 + 2, 5 + 6) = (0, 11)$.

11. The triangle ABC is neither an isosceles triangle nor is it a right triangle. The lengths of AB, BC and AC are $\sqrt{53}$, $\sqrt{72}$ and $\sqrt{65}$, respectively. The slopes of AB, BC and AC are $\frac{7}{2}$, -1 and $\frac{1}{8}$, respectively.

13. $y = 3(x - 1) + (-6) \Rightarrow y = 3x - 9$

15. $x = 0$

17. $y = 2$

19. $y = -3x + 3$

21. Since $4x + 3y = 12$ is equivalent to $y = -\frac{4}{3}x + 4$, the slope of the given line (and hence the slope of the desired line) is $-\frac{4}{3}$. $y = -\frac{4}{3}(x - 4) - 12 \Rightarrow y = -\frac{4}{3}x - \frac{20}{3}$

23. Since $\frac{1}{2}x + \frac{1}{3}y = 1$ is equivalent to $y = -\frac{3}{2}x + 3$, the slope of the given line is $-\frac{3}{2}$ and the slope of the perpendicular line is $\frac{2}{3}$. $y = \frac{2}{3}(x + 1) + 2 \Rightarrow y = \frac{2}{3}x + \frac{8}{3}$

25. The area is $A = \pi r^2$ and the circumference is $C = 2\pi r$. Thus, $r = \frac{C}{2\pi} \Rightarrow A = \pi \left(\frac{C}{2\pi}\right)^2 = \frac{C^2}{4\pi}$.

27. The coordinates of a point on the parabola are (x, x^2). The angle of inclination θ joining this point to the origin satisfies the equation $\tan\theta = \frac{x^2}{x} = x$. Thus the point has coordinates $(x, x^2) = (\tan\theta, \tan^2\theta)$.

29.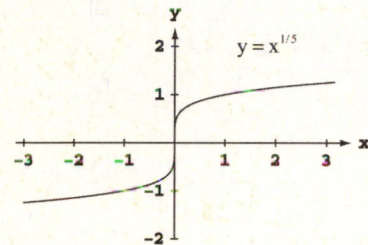

Symmetric about the origin.

31.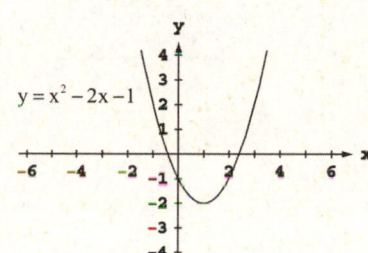

Neither

33. $y(-x) = (-x)^2 + 1 = x^2 + 1 = y(x)$. Even.

35. $y(-x) = 1 - \cos(-x) = 1 - \cos x = y(x)$. Even.

37. $y(-x) = \frac{(-x)^4 + 1}{(-x)^3 - 2(-x)} = \frac{x^4 + 1}{-x^3 + 2x} = -\frac{x^4 + 1}{x^3 - 2x} = -y(x)$. Odd.

39. $y(-x) = -x + \cos(-x) = -x + \cos x$. Neither even nor odd.

41. (a) The function is defined for all values of x, so the domain is $(-\infty, \infty)$.
    (b) Since $|x|$ attains all nonnegative values, the range is $[-2, \infty)$.

43. (a) Since the square root requires $16 - x^2 \geq 0$, the domain is $[-4, 4]$.
    (b) For values of x in the domain, $0 \leq 16 - x^2 \leq 16$, so $0 \leq \sqrt{16 - x^2} \leq 4$. The range is $[0, 4]$.

45. (a) The function is defined for all values of x, so the domain is $(-\infty, \infty)$.
    (b) Since $2e^{-x}$ attains all positive values, the range is $(-3, \infty)$.

47. (a) The function is defined for all values of x, so the domain is $(-\infty, \infty)$.
    (b) The sine function attains values from -1 to 1, so $-2 \leq 2\sin(3x + \pi) \leq 2$ and hence $-3 \leq 2\sin(3x + \pi) - 1 \leq 1$. The range is $[-3, 1]$.

49. (a) The logarithm requires $x - 3 > 0$, so the domain is $(3, \infty)$.

    (b) The logarithm attains all real values, so the range is $(-\infty, \infty)$.

51. (a) The function is defined for $-4 \le x \le 4$, so the domain is $[-4, 4]$.

    (b) The function is equivalent to $y = \sqrt{|x|}$, $-4 \le x \le 4$, which attains values from 0 to 2 for x in the domain. The range is $[0, 2]$.

53. First piece: Line through $(0, 1)$ and $(1, 0)$. $m = \frac{0-1}{1-0} = \frac{-1}{1} = -1 \Rightarrow y = -x + 1 = 1 - x$

    Second piece: Line through $(1, 1)$ and $(2, 0)$. $m = \frac{0-1}{2-1} = \frac{-1}{1} = -1 \Rightarrow y = -(x - 1) + 1 = -x + 2 = 2 - x$

    $f(x) = \begin{cases} 1 - x, & 0 \le x < 1 \\ 2 - x, & 1 \le x \le 2 \end{cases}$

55. (a) $(f \circ g)(-1) = f(g(-1)) = f\left(\frac{1}{\sqrt{-1+2}}\right) = f(1) = \frac{1}{1} = 1$

    (b) $(g \circ f)(2) = g(f(2)) = g\left(\frac{1}{2}\right) = \frac{1}{\sqrt{\frac{1}{2}+2}} = \frac{1}{\sqrt{2.5}}$ or $\sqrt{\frac{2}{5}}$

    (c) $(f \circ f)(x) = f(f(x)) = f\left(\frac{1}{x}\right) = \frac{1}{1/x} = x, x \ne 0$

    (d) $(g \circ g)(x) = g(g(x)) = g\left(\frac{1}{\sqrt{x+2}}\right) = \frac{1}{\sqrt{\frac{1}{\sqrt{x+2}}+2}} = \frac{\sqrt[4]{x+2}}{\sqrt{1+2\sqrt{x+2}}}$

57. (a) $(f \circ g)(x) = f(g(x)) = f(\sqrt{x + 2}) = 2 - (\sqrt{x+2})^2 = -x, x \ge -2$.

    $(g \circ f)(x) = f(g(x)) = g(2 - x^2) = \sqrt{(2 - x^2) + 2} = \sqrt{4 - x^2}$

    (b) Domain of $f \circ g$: $[-2, \infty)$.

    Domain of $g \circ f$: $[-2, 2]$.

    (c) Range of $f \circ g$: $(-\infty, 2]$.

    Range of $g \circ f$: $[0, 2]$.

59.

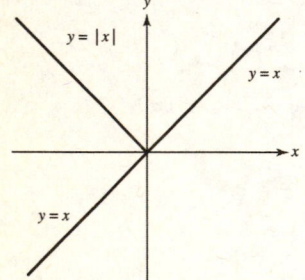

The graph of $f_2(x) = f_1(|x|)$ is the same as the graph of $f_1(x)$ to the right of the y-axis. The graph of $f_2(x)$ to the left of the y-axis is the reflection of $y = f_1(x)$, $x \ge 0$ across the y-axis.

61.

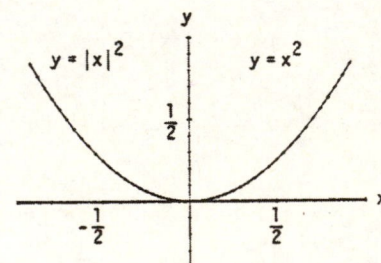

It does not change the graph.

63.

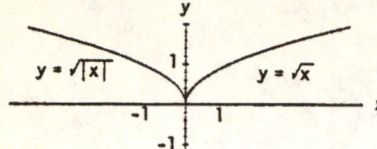

The graph of $f_2(x) = f_1(|x|)$ is the same as the graph of $f_1(x)$ to the right of the y-axis. The graph of $f_2(x)$ to the left of the y-axis is the reflection of $y = f_1(x)$, $x \geq 0$ across the y-axis.

65.

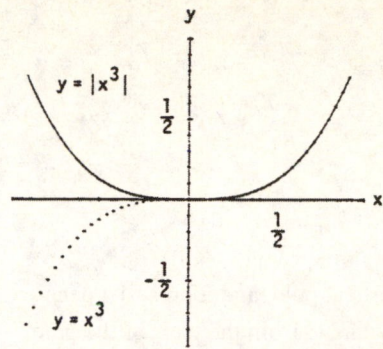

Whenever $g_1(x)$ is positive, the graph of $y = g_2(x) = |g_1(x)|$ is the same as the graph of $y = g_1(x)$. When $g_1(x)$ is negative, the graph of $y = g_2(x)$ is the reflection of the graph of $y = g_1(x)$ across the x-axis.

Whenever $g_1(x)$ is positive, the graph of $y = g_2(x) = |g_1(x)|$ is the same as the graph of $y = g_1(x)$. When $g_1(x)$ is negative, the graph of $y = g_2(x)$ is the reflection of the graph of $y = g_1(x)$ across the x-axis.

67.

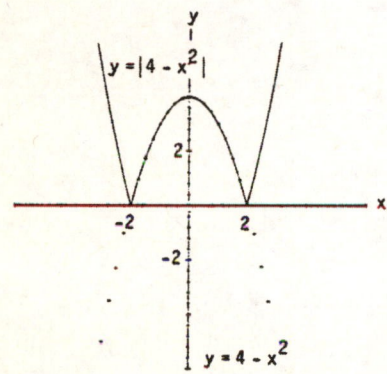

69.

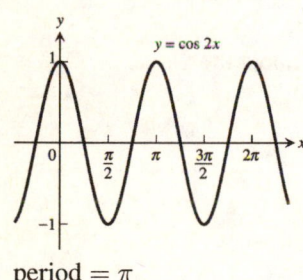

period $= \pi$

71.

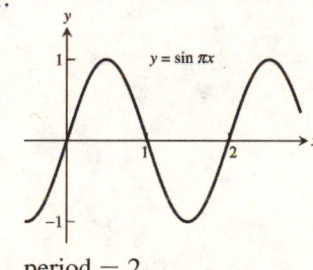

period $= 2$

73.

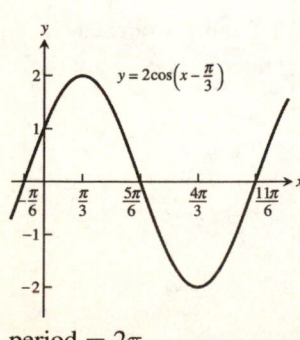

period $= 2\pi$

75. (a) $\sin B = \sin \frac{\pi}{3} = \frac{b}{c} = \frac{b}{2} \Rightarrow b = 2 \sin \frac{\pi}{3} = 2\left(\frac{\sqrt{3}}{2}\right) = \sqrt{3}$. By the theorem of Pythagoras,

$a^2 + b^2 = c^2 \Rightarrow a = \sqrt{c^2 - b^2} = \sqrt{4 - 3} = 1$.

(b) $\sin B = \sin \frac{\pi}{3} = \frac{b}{c} = \frac{2}{c} \Rightarrow c = \frac{2}{\sin \frac{\pi}{3}} = \frac{2}{\left(\frac{\sqrt{3}}{2}\right)} = \frac{4}{\sqrt{3}}$. Thus, $a = \sqrt{c^2 - b^2} = \sqrt{\left(\frac{4}{\sqrt{3}}\right)^2 - (2)^2} = \sqrt{\frac{4}{3}} = \frac{2}{\sqrt{3}}$.

77. (a) $\tan B = \frac{b}{a} \Rightarrow a = \frac{b}{\tan B}$ (b) $\sin A = \frac{a}{c} \Rightarrow c = \frac{a}{\sin A}$

79. Let h = height of vertical pole, and let b and c denote the distances of points B and C from the base of the pole, measured along the flatground, respectively. Then, $\tan 50° = \frac{h}{c}$, $\tan 35° = \frac{h}{b}$, and $b - c = 10$. Thus, $h = c \tan 50°$ and $h = b \tan 35° = (c + 10) \tan 35°$

$\Rightarrow c \tan 50° = (c + 10) \tan 35°$

$\Rightarrow c (\tan 50° - \tan 35°) = 10 \tan 35°$

$\Rightarrow c = \frac{10 \tan 35°}{\tan 50° - \tan 35°} \Rightarrow h = c \tan 50°$

$= \frac{10 \tan 35° \tan 50°}{\tan 50° - \tan 35°} \approx 16.98$ m.

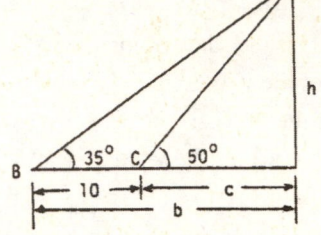

81. (a)

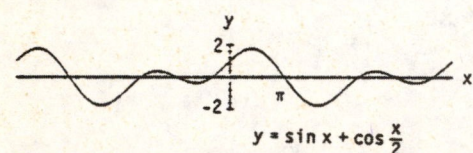

(b) The period appears to be 4π.

(c) $f(x + 4\pi) = \sin(x + 4\pi) + \cos\left(\frac{x + 4\pi}{2}\right) = \sin(x + 2\pi) + \cos\left(\frac{x}{2} + 2\pi\right) = \sin x + \cos \frac{x}{2}$ since the period of sine and cosine is 2π. Thus, $f(x)$ has period 4π.

CHAPTER 1 ADDITIONAL AND ADVANCED EXERCISES

1. (a) The given graph is reflected about the y-axis.   (b) The given graph is reflected about the x-axis.

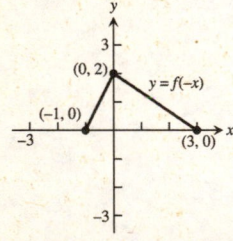

(c) The given graph is shifted left 1 unit, stretched vertically by a factor of 2, reflected about the x-axis, and then shifted upward 1 unit.

(d) The given graph is shifted right 2 units, stretched vertically by a factor of 3, and then shifted downward 2 units.

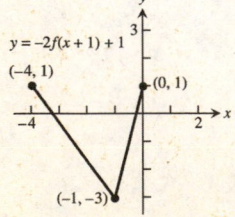

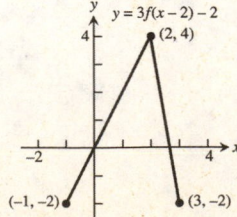

3. There are (infinitely) many such function pairs. For example, $f(x) = 3x$ and $g(x) = 4x$ satisfy
    $f(g(x)) = f(4x) = 3(4x) = 12x = 4(3x) = g(3x) = g(f(x))$.

5. If f is odd and defined at x, then $f(-x) = -f(x)$. Thus $g(-x) = f(-x) - 2 = -f(x) - 2$ whereas
    $-g(x) = -(f(x) - 2) = -f(x) + 2$. Then g cannot be odd because $g(-x) = -g(x) \Rightarrow -f(x) - 2 = -f(x) + 2$
    $\Rightarrow 4 = 0$, which is a contradiction. Also, $g(x)$ is not even unless $f(x) = 0$ for all x. On the other hand, if f is
    even, then $g(x) = f(x) - 2$ is also even: $g(-x) = f(-x) - 2 = f(x) - 2 = g(x)$.

7. For (x, y) in the 1st quadrant, $|x| + |y| = 1 + x$
    $\Leftrightarrow x + y = 1 + x \Leftrightarrow y = 1$. For (x, y) in the 2nd
    quadrant, $|x| + |y| = x + 1 \Leftrightarrow -x + y = x + 1$
    $\Leftrightarrow y = 2x + 1$. In the 3rd quadrant, $|x| + |y| = x + 1$
    $\Leftrightarrow -x - y = x + 1 \Leftrightarrow y = -2x - 1$. In the 4th
    quadrant, $|x| + |y| = x + 1 \Leftrightarrow x + (-y) = x + 1$
    $\Leftrightarrow y = -1$. The graph is given at the right.

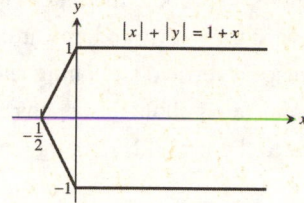

9. By the law of sines, $\frac{\sin \frac{\pi}{3}}{\sqrt{3}} = \frac{\sin A}{a} = \frac{\sin B}{b} = \frac{\sin \frac{\pi}{4}}{b} \Rightarrow b = \frac{\sqrt{3} \sin (\pi/4)}{\sin (\pi/3)} = \frac{\sqrt{3} \left(\frac{\sqrt{2}}{2} \right)}{\frac{\sqrt{3}}{2}} = \sqrt{2}$.

11. By the law of cosines, $a^2 = b^2 + c^2 - 2bc \cos A \Rightarrow \cos A = \frac{b^2 + c^2 - a^2}{2bc} = \frac{2^2 + 3^2 - 2^2}{2(2)(3)} = \frac{3}{4}$.

13. By the law of cosines, $b^2 = a^2 + c^2 - 2ac \cos B \Rightarrow \cos B = \frac{a^2 + c^2 - b^2}{2ac} = \frac{2^2 + 4^2 - 3^2}{(2)(2)(4)} = \frac{4 + 16 - 9}{16}$
    $= \frac{11}{16}$. Since $0 < B < \pi$, $\sin B = \sqrt{1 - \cos^2 B} = \sqrt{1 - \frac{121}{256}} = \frac{\sqrt{135}}{16} = \frac{3\sqrt{15}}{16}$.

15. (a) $\sin^2 x + \cos^2 x = 1 \Rightarrow \sin^2 x = 1 - \cos^2 x = (1 - \cos x)(1 + \cos x) \Rightarrow (1 - \cos x) = \frac{\sin^2 x}{1 + \cos x}$
    $\Rightarrow \frac{1 - \cos x}{\sin x} = \frac{\sin x}{1 + \cos x}$

    (b) Using the definition of the tangent function and the double angle formulas, we have
    $$\tan^2 \left(\frac{x}{2} \right) = \frac{\sin^2 \left(\frac{x}{2} \right)}{\cos^2 \left(\frac{x}{2} \right)} = \frac{\frac{1 - \cos \left(2 \left(\frac{x}{2} \right) \right)}{2}}{\frac{1 + \cos \left(2 \left(\frac{x}{2} \right) \right)}{2}} = \frac{1 - \cos x}{1 + \cos x}.$$

17. As in the proof of the law of sines of Section P.5, Exercise 57, $ah = bc \sin A = ab \sin C = ac \sin B$
    $\Rightarrow$ the area of ABC $= \frac{1}{2}$ (base)(height) $= \frac{1}{2} ah = \frac{1}{2} bc \sin A = \frac{1}{2} ab \sin C = \frac{1}{2} ac \sin B$.

19. 1. $b + c - (a + c) = b - a$, which is positive since $a < b$. Thus, $a + c < b + c$.
    2. $b - c - (a - c) = b - a$, which is positive since $a < b$. Thus, $a - c < b - c$.
    3. $c > 0$ and $a < b \Rightarrow c - 0 = c$ and $b - a$ are positive $\Rightarrow (b - a)c = bc - ac$ is positive $\Rightarrow ac < bc$.
    4. $a < b$ and $c < 0 \Rightarrow b - a$ and $-c$ are positive $\Rightarrow (b - a)(-c) = ac - bc$ is positive $\Rightarrow bc < ac$.
    5. Since $a > 0$, a and $\frac{1}{a}$ are positive $\Rightarrow \frac{1}{a} > 0$.
    6. Since $0 < a < b$, both $\frac{1}{a}$ and $\frac{1}{b}$ are positive. By (3), $a < b$ and $\frac{1}{a} > 0 \Rightarrow a \left(\frac{1}{a} \right) < b \left(\frac{1}{a} \right)$ or $1 < \frac{b}{a}$
        $\Rightarrow 1 \left(\frac{1}{b} \right) < \frac{b}{a} \left(\frac{1}{b} \right)$ by (3) since $\frac{1}{b} > 0 \Rightarrow \frac{1}{b} < \frac{1}{a}$.
    7. $a < b < 0 \Rightarrow \frac{1}{a}$ and $\frac{1}{b}$ are both negative, i.e., $\frac{1}{a} < 0$ and $\frac{1}{b} < 0$. By (4), $a < b$ and $\frac{1}{a} < 0 \Rightarrow b \left(\frac{1}{a} \right) < a \left(\frac{1}{a} \right)$
        $\Rightarrow \frac{b}{a} < 1 \Rightarrow 1 \left(\frac{1}{b} \right) < \frac{b}{a} \left(\frac{1}{b} \right)$ by (4) since $\frac{1}{b} < 0 \Rightarrow \frac{1}{b} < \frac{1}{a}$.

21. The fact that $|a_1 + a_2 + \ldots + a_n| \leq |a_1| + |a_2| + \ldots + |a_n|$ holds for $n = 1$ is obvious. It also holds for
    $n = 2$ by the triangle inequality. We now show it holds for all positive integers n, by induction.
        Suppose it holds for $n = k \geq 1$: $|a_1 + a_2 + \ldots + a_k| \leq |a_1| + |a_2| + \ldots + |a_k|$ (this is the induction
    hypothesis). Then $|a_1 + a_2 + \ldots + a_k + a_{k+1}| = |(a_1 + a_2 + \ldots + a_k) + a_{k+1}| \leq |a_1 + a_2 + \ldots + a_k| + |a_{k+1}|$

(by the triangle inequality) $\leq |a_1| + |a_2| + \ldots + |a_k| + |a_{k+1}|$ (by the induction hypothesis) and the inequality holds for $n = k + 1$. Hence it holds for all n by induction.

23. If f is even and odd, then $f(-x) = -f(x)$ and $f(-x) = f(x) \Rightarrow f(x) = -f(x)$ for all x in the domain of f. Thus $2f(x) = 0 \Rightarrow f(x) = 0$.

25. $y = ax^2 + bx + c = a\left(x^2 + \frac{b}{a}x + \frac{b^2}{4a^2}\right) - \frac{b^2}{4a} + c = a\left(x + \frac{b}{2a}\right)^2 - \frac{b^2}{4a} + c$

   (a) If $a > 0$ the graph is a parabola that opens upward. Increasing a causes a vertical stretching and a shift of the vertex toward the y-axis and upward. If $a < 0$ the graph is a parabola that opens downward. Decreasing a causes a vertical stretching and a shift of the vertex toward the y-axis and downward.

   (b) If $a > 0$ the graph is a parabola that opens upward. If also $b > 0$, then increasing b causes a shift of the graph downward to the left; if $b < 0$, then decreasing b causes a shift of the graph downward and to the right.

       If $a < 0$ the graph is a parabola that opens downward. If $b > 0$, increasing b shifts the graph upward to the right. If $b < 0$, decreasing b shifts the graph upward to the left.

   (c) Changing c (for fixed a and b) by Δc shifts the graph upward Δc units if $\Delta c > 0$, and downward $-\Delta c$ units if $\Delta c < 0$.

27. If $m > 0$, the x-intercept of $y = mx + 2$ must be negative. If $m < 0$, then the x-intercept exceeds $\frac{1}{2}$
   $\Rightarrow 0 = mx + 2$ and $x > \frac{1}{2} \Rightarrow x = -\frac{2}{m} > \frac{1}{2} \Rightarrow 0 > m > -4$.

29. (a) By Exercise #95 of Section 1.2, the coordinates of P are $\left(\frac{a+0}{2}, \frac{b+0}{2}\right) = \left(\frac{a}{2}, \frac{b}{2}\right)$. Thus the slope of $OP = \frac{\Delta y}{\Delta x} = \frac{b/2}{a/2} = \frac{b}{a}$.

   (b) The slope of $AB = \frac{b-0}{0-a} = -\frac{b}{a}$. The line segments AB and OP are perpendicular when the product of their slopes is $-1 = \left(\frac{b}{a}\right)\left(-\frac{b}{a}\right) = -\frac{b^2}{a^2}$. Thus, $b^2 = a^2 \Rightarrow a = b$ (since both are positive). Therefore, AB is perpendicular to OP when $a = b$.

CHAPTER 2 LIMITS AND CONTINUITY

2.1 RATES OF CHANGE AND LIMITS

1. (a) Does not exist. As x approaches 1 from the right, g(x) approaches 0. As x approaches 1 from the left, g(x)
    approaches 1. There is no single number L that all the values g(x) get arbitrarily close to as x $\rightarrow$ 1.

    (b) 1

    (c) 0

3. (a) True $\qquad\qquad$ (b) True $\qquad\qquad$ (c) False

    (d) False $\qquad\qquad$ (e) False $\qquad\qquad$ (f) True

5. $\lim\limits_{x \to 0} \frac{x}{|x|}$ does not exist because $\frac{x}{|x|} = \frac{x}{x} = 1$ if x > 0 and $\frac{x}{|x|} = \frac{x}{-x} = -1$ if x < 0. As x approaches 0 from the left,
    $\frac{x}{|x|}$ approaches -1. As x approaches 0 from the right, $\frac{x}{|x|}$ approaches 1. There is no single number L that all
    the function values get arbitrarily close to as x $\rightarrow$ 0.

7. Nothing can be said about f(x) because the existence of a limit as x $\rightarrow$ x_0 does not depend on how the function
    is defined at x_0. In order for a limit to exist, f(x) must be arbitrarily close to a single real number L when
    x is close enough to x_0. That is, the existence of a limit depends on the values of f(x) for x <u>near</u> x_0, not on the
    definition of f(x) at x_0 itself.

9. No, the definition does not require that f be defined at x = 1 in order for a limiting value to exist there. If f(1)
    is defined, it can be any real number, so we can conclude nothing about f(1) from $\lim\limits_{x \to 1} f(x) = 5$.

11. (a) $f(x) = (x^2 - 9)/(x + 3)$

x	−3.1	−3.01	−3.001	−3.0001	−3.00001	−3.000001
f(x)	−6.1	−6.01	−6.001	−6.0001	−6.00001	−6.000001

x	−2.9	−2.99	−2.999	−2.9999	−2.99999	−2.999999
f(x)	−5.9	−5.99	−5.999	−5.9999	−5.99999	−5.999999

    The estimate is $\lim\limits_{x \to -3} f(x) = -6$.

    (b)

    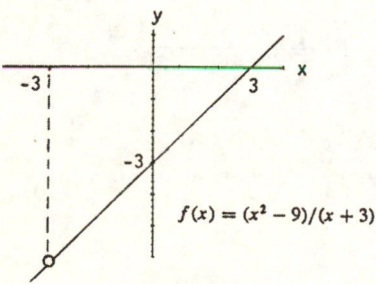

    $f(x) = (x^2 - 9)/(x + 3)$

    (c) $f(x) = \frac{x^2 - 9}{x + 3} = \frac{(x+3)(x-3)}{x+3} = x - 3$ if $x \neq -3$, and $\lim\limits_{x \to -3} (x - 3) = -3 - 3 = -6$.

13. (a) $G(x) = (x + 6)/(x^2 + 4x - 12)$

x	−5.9	−5.99	−5.999	−5.9999	−5.99999	−5.999999
G(x)	−.126582	−.1251564	−.1250156	−.1250015	−.1250001	−.1250000

x	−6.1	−6.01	−6.001	−6.0001	−6.00001	−6.000001
G(x)	−.123456	−.124843	−.124984	−.124998	−.124999	−.124999

(b)

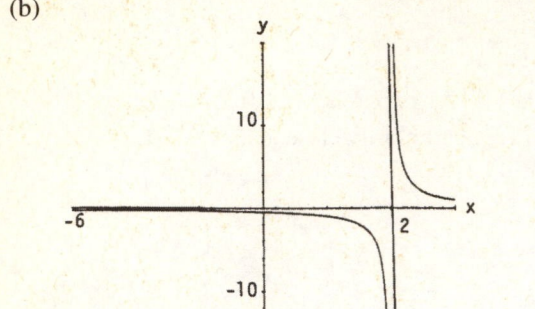

$G(x) = (x + 6)/(x^2 + 4x - 12)$

(c) $G(x) = \frac{x+6}{(x^2 + 4x - 12)} = \frac{x+6}{(x+6)(x-2)} = \frac{1}{x-2}$ if $x \neq -6$, and $\lim\limits_{x \to -6} \frac{1}{x-2} = \frac{1}{-6-2} = -\frac{1}{8} = -0.125$.

15. (a) $f(x) = (x^2 - 1)/(|x| - 1)$

x	−1.1	−1.01	−1.001	−1.0001	−1.00001	−1.000001
f(x)	2.1	2.01	2.001	2.0001	2.00001	2.000001

x	−.9	−.99	−.999	−.9999	−.99999	−.999999
f(x)	1.9	1.99	1.999	1.9999	1.99999	1.999999

(b)

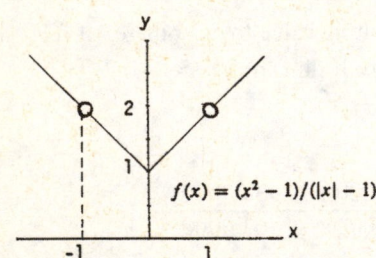

$f(x) = (x^2 - 1)/(|x| - 1)$

(c) $f(x) = \frac{x^2 - 1}{|x| - 1} = \begin{cases} \frac{(x+1)(x-1)}{x-1} = x + 1, & x \geq 0 \text{ and } x \neq 1 \\ \frac{(x+1)(x-1)}{-(x+1)} = 1 - x, & x < 0 \text{ and } x \neq -1 \end{cases}$, and $\lim\limits_{x \to -1} (1 - x) = 1 - (-1) = 2$.

17. (a) $g(\theta) = (\sin \theta)/\theta$

θ	.1	.01	.001	.0001	.00001	.000001
g(θ)	.998334	.999983	.999999	.999999	.999999	.999999

θ	−.1	−.01	−.001	−.0001	−.00001	−.000001
g(θ)	.998334	.999983	.999999	.999999	.999999	.999999

$\lim\limits_{\theta \to 0} g(\theta) = 1$

(b)

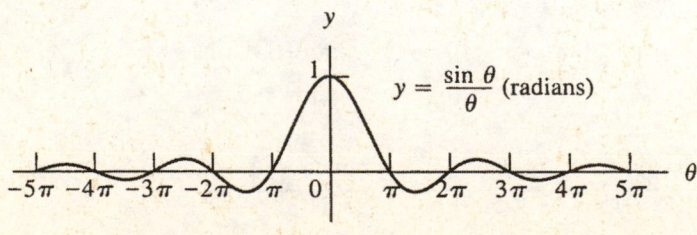

$y = \frac{\sin \theta}{\theta}$ (radians)

NOT TO SCALE

19. (a) $f(x) = x^{1/(1-x)}$

x	.9	.99	.999	.9999	.99999	.999999
f(x)	.348678	.366032	.367695	.367861	.367877	.367879

x	1.1	1.01	1.001	1.0001	1.00001	1.000001
f(x)	.385543	.369711	.368063	.367897	.367881	.367878

$\lim\limits_{x \to 1} f(x) \approx 0.36788$

(b)

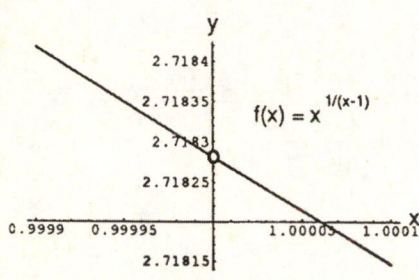

Graph is NOT TO SCALE. Also the intersection of the axes is not the origin: the axes intersect at the point $(1, 2.71820)$.

21. $\lim\limits_{x \to 2} 2x = 2(2) = 4$

23. $\lim\limits_{x \to \frac{1}{3}} (3x - 1) = 3\left(\frac{1}{3}\right) - 1 = 0$

25. $\lim\limits_{x \to -1} 3x(2x - 1) = 3(-1)(2(-1) - 1) = 9$

27. $\lim\limits_{x \to \frac{\pi}{2}} x \sin x = \frac{\pi}{2} \sin \frac{\pi}{2} = \frac{\pi}{2}$

29. (a) $\frac{\Delta f}{\Delta x} = \frac{f(3) - f(2)}{3 - 2} = \frac{28 - 9}{1} = 19$

(b) $\frac{\Delta f}{\Delta x} = \frac{f(1) - f(-1)}{1 - (-1)} = \frac{2 - 0}{2} = 1$

31. (a) $\frac{\Delta h}{\Delta t} = \frac{h\left(\frac{3\pi}{4}\right) - h\left(\frac{\pi}{4}\right)}{\frac{3\pi}{4} - \frac{\pi}{4}} = \frac{-1 - 1}{\frac{\pi}{2}} = -\frac{4}{\pi}$

(b) $\frac{\Delta h}{\Delta t} = \frac{h\left(\frac{\pi}{2}\right) - h\left(\frac{\pi}{6}\right)}{\frac{\pi}{2} - \frac{\pi}{6}} = \frac{0 - \sqrt{3}}{\frac{\pi}{3}} = \frac{-3\sqrt{3}}{\pi}$

33. $\frac{\Delta R}{\Delta \theta} = \frac{R(2) - R(0)}{2 - 0} = \frac{\sqrt{8+1} - \sqrt{1}}{2} = \frac{3 - 1}{2} = 1$

35. (a)

Q	Slope of PQ $= \frac{\Delta p}{\Delta t}$
$Q_1(10, 225)$	$\frac{650 - 225}{20 - 10} = 42.5$ m/sec
$Q_2(14, 375)$	$\frac{650 - 375}{20 - 14} = 45.83$ m/sec
$Q_3(16.5, 475)$	$\frac{650 - 475}{20 - 16.5} = 50.00$ m/sec
$Q_4(18, 550)$	$\frac{650 - 550}{20 - 18} = 50.00$ m/sec

(b) At $t = 20$, the Cobra was traveling approximately 50 m/sec or 180 km/h.

37. (a)

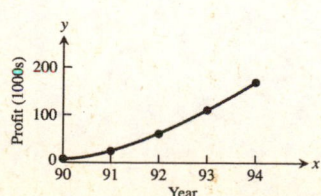

(b) $\frac{\Delta p}{\Delta t} = \frac{174 - 62}{1994 - 1992} = \frac{112}{2} = 56$ thousand dollars per year

(c) The average rate of change from 1991 to 1992 is $\frac{\Delta p}{\Delta t} = \frac{62 - 27}{1992 - 1991} = 35$ thousand dollars per year.

The average rate of change from 1992 to 1993 is $\frac{\Delta p}{\Delta t} = \frac{111 - 62}{1993 - 1992} = 49$ thousand dollars per year.

So, the rate at which profits were changing in 1992 is approximatley $\frac{1}{2}(35 + 49) = 42$ thousand dollars per year.

39. (a) $\frac{\Delta g}{\Delta x} = \frac{g(2) - g(1)}{2 - 1} = \frac{\sqrt{2} - 1}{2 - 1} \approx 0.414213$ $\frac{\Delta g}{\Delta x} = \frac{g(1.5) - g(1)}{1.5 - 1} = \frac{\sqrt{1.5} - 1}{0.5} \approx 0.449489$

$\frac{\Delta g}{\Delta x} = \frac{g(1 + h) - g(1)}{(1 + h) - 1} = \frac{\sqrt{1 + h} - 1}{h}$

(b) $g(x) = \sqrt{x}$

$1 + h$	1.1	1.01	1.001	1.0001	1.00001	1.000001
$\sqrt{1 + h}$	1.04880	1.004987	1.0004998	1.0000499	1.000005	1.0000005
$\left(\sqrt{1 + h} - 1\right)/h$	0.4880	0.4987	0.4998	0.499	0.5	0.5

(c) The rate of change of $g(x)$ at $x = 1$ is 0.5.

(d) The calculator gives $\lim\limits_{h \to 0} \frac{\sqrt{1 + h} - 1}{h} = \frac{1}{2}$.

2.2 CALCULATING LIMITS USING THE LIMIT LAWS

1. $\lim\limits_{x \to -7} (2x + 5) = 2(-7) + 5 = -14 + 5 = -9$

3. $\lim\limits_{x \to 2} (-x^2 + 5x - 2) = -(2)^2 + 5(2) - 2 = -4 + 10 - 2 = 4$

5. $\lim\limits_{t \to 6} 8(t - 5)(t - 7) = 8(6 - 5)(6 - 7) = -8$

7. $\lim\limits_{x \to 2} \frac{x + 3}{x + 6} = \frac{2 + 3}{2 + 6} = \frac{5}{8}$

9. $\lim\limits_{y \to -5} \frac{y^2}{5 - y} = \frac{(-5)^2}{5 - (-5)} = \frac{25}{10} = \frac{5}{2}$

11. $\lim\limits_{x \to -1} 3(2x - 1)^2 = 3(2(-1) - 1)^2 = 3(-3)^2 = 27$

13. $\lim\limits_{y \to -3} (5 - y)^{4/3} = [5 - (-3)]^{4/3} = (8)^{4/3} = \left((8)^{1/3}\right)^4 = 2^4 = 16$

15. $\lim\limits_{h \to 0} \frac{3}{\sqrt{3h + 1} + 1} = \frac{3}{\sqrt{3(0) + 1} + 1} = \frac{3}{\sqrt{1} + 1} = \frac{3}{2}$

17. $\lim\limits_{h \to 0} \frac{\sqrt{3h + 1} - 1}{h} = \lim\limits_{h \to 0} \frac{\sqrt{3h + 1} - 1}{h} \cdot \frac{\sqrt{3h + 1} + 1}{\sqrt{3h + 1} + 1} = \lim\limits_{h \to 0} \frac{(3h + 1) - 1}{h\left(\sqrt{3h + 1} + 1\right)} = \lim\limits_{h \to 0} \frac{3h}{h\left(\sqrt{3h + 1} + 1\right)} = \lim\limits_{h \to 0} \frac{3}{\sqrt{3h + 1} + 1}$

$= \frac{3}{\sqrt{1} + 1} = \frac{3}{2}$

19. $\lim\limits_{x \to 5} \frac{x - 5}{x^2 - 25} = \lim\limits_{x \to 5} \frac{x - 5}{(x + 5)(x - 5)} = \lim\limits_{x \to 5} \frac{1}{x + 5} = \frac{1}{5 + 5} = \frac{1}{10}$

21. $\lim\limits_{x \to -5} \frac{x^2 + 3x - 10}{x + 5} = \lim\limits_{x \to -5} \frac{(x + 5)(x - 2)}{x + 5} = \lim\limits_{x \to -5} (x - 2) = -5 - 2 = -7$

23. $\lim\limits_{t \to 1} \frac{t^2 + t - 2}{t^2 - 1} = \lim\limits_{t \to 1} \frac{(t + 2)(t - 1)}{(t - 1)(t + 1)} = \lim\limits_{t \to 1} \frac{t + 2}{t + 1} = \frac{1 + 2}{1 + 1} = \frac{3}{2}$

25. $\lim\limits_{x \to -2} \frac{-2x-4}{x^3+2x^2} = \lim\limits_{x \to -2} \frac{-2(x+2)}{x^2(x+2)} = \lim\limits_{x \to -2} \frac{-2}{x^2} = \frac{-2}{4} = -\frac{1}{2}$

27. $\lim\limits_{u \to 1} \frac{u^4-1}{u^3-1} = \lim\limits_{u \to 1} \frac{(u^2+1)(u+1)(u-1)}{(u^2+u+1)(u-1)} = \lim\limits_{u \to 1} \frac{(u^2+1)(u+1)}{u^2+u+1} = \frac{(1+1)(1+1)}{1+1+1} = \frac{4}{3}$

29. $\lim\limits_{x \to 9} \frac{\sqrt{x}-3}{x-9} = \lim\limits_{x \to 9} \frac{\sqrt{x}-3}{(\sqrt{x}-3)(\sqrt{x}+3)} = \lim\limits_{x \to 9} \frac{1}{\sqrt{x}+3} = \frac{1}{\sqrt{9}+3} = \frac{1}{6}$

31. $\lim\limits_{x \to 1} \frac{x-1}{\sqrt{x+3}-2} = \lim\limits_{x \to 1} \frac{(x-1)(\sqrt{x+3}+2)}{(\sqrt{x+3}-2)(\sqrt{x+3}+2)} = \lim\limits_{x \to 1} \frac{(x-1)(\sqrt{x+3}+2)}{(x+3)-4} = \lim\limits_{x \to 1} \left(\sqrt{x+3}+2\right)$
$= \sqrt{4}+2 = 4$

33. $\lim\limits_{x \to 2} \frac{\sqrt{x^2+12}-4}{x-2} = \lim\limits_{x \to 2} \frac{\left(\sqrt{x^2+12}-4\right)\left(\sqrt{x^2+12}+4\right)}{(x-2)\left(\sqrt{x^2+12}+4\right)} = \lim\limits_{x \to 2} \frac{(x^2+12)-16}{(x-2)\left(\sqrt{x^2+12}+4\right)}$

$= \lim\limits_{x \to 2} \frac{(x-2)(x+2)}{(x-2)\left(\sqrt{x^2+12}+4\right)} = \lim\limits_{x \to 2} \frac{x+2}{\sqrt{x^2+12}+4} = \frac{4}{\sqrt{16}+4} = \frac{1}{2}$

35. $\lim\limits_{x \to -3} \frac{2-\sqrt{x^2-5}}{x+3} = \lim\limits_{x \to -3} \frac{\left(2-\sqrt{x^2-5}\right)\left(2+\sqrt{x^2-5}\right)}{(x+3)\left(2+\sqrt{x^2-5}\right)} = \lim\limits_{x \to -3} \frac{4-(x^2-5)}{(x+3)\left(2+\sqrt{x^2-5}\right)}$

$= \lim\limits_{x \to -3} \frac{9-x^2}{(x+3)\left(2+\sqrt{x^2-5}\right)} = \lim\limits_{x \to -3} \frac{(3-x)(3+x)}{(x+3)\left(2+\sqrt{x^2-5}\right)} = \lim\limits_{x \to -3} \frac{3-x}{2+\sqrt{x^2-5}} = \frac{6}{2+\sqrt{4}} = \frac{3}{2}$

37. (a) quotient rule
    (b) difference and power rules
    (c) sum and constant multiple rules

39. (a) $\lim\limits_{x \to c} f(x)\,g(x) = \left[\lim\limits_{x \to c} f(x)\right]\left[\lim\limits_{x \to c} g(x)\right] = (5)(-2) = -10$
    (b) $\lim\limits_{x \to c} 2f(x)\,g(x) = 2\left[\lim\limits_{x \to c} f(x)\right]\left[\lim\limits_{x \to c} g(x)\right] = 2(5)(-2) = -20$
    (c) $\lim\limits_{x \to c}\,[f(x)+3g(x)] = \lim\limits_{x \to c} f(x) + 3\lim\limits_{x \to c} g(x) = 5+3(-2) = -1$
    (d) $\lim\limits_{x \to c} \frac{f(x)}{f(x)-g(x)} = \frac{\lim\limits_{x \to c} f(x)}{\lim\limits_{x \to c} f(x) - \lim\limits_{x \to c} g(x)} = \frac{5}{5-(-2)} = \frac{5}{7}$

41. (a) $\lim\limits_{x \to b}\,[f(x)+g(x)] = \lim\limits_{x \to b} f(x) + \lim\limits_{x \to b} g(x) = 7+(-3) = 4$
    (b) $\lim\limits_{x \to b} f(x) \cdot g(x) = \left[\lim\limits_{x \to b} f(x)\right]\left[\lim\limits_{x \to b} g(x)\right] = (7)(-3) = -21$
    (c) $\lim\limits_{x \to b} 4g(x) = \left[\lim\limits_{x \to b} 4\right]\left[\lim\limits_{x \to b} g(x)\right] = (4)(-3) = -12$
    (d) $\lim\limits_{x \to b} f(x)/g(x) = \lim\limits_{x \to b} f(x)/\lim\limits_{x \to b} g(x) = \frac{7}{-3} = -\frac{7}{3}$

43. $\lim\limits_{h \to 0} \frac{(1+h)^2-1^2}{h} = \lim\limits_{h \to 0} \frac{1+2h+h^2-1}{h} = \lim\limits_{h \to 0} \frac{h(2+h)}{h} = \lim\limits_{h \to 0} (2+h) = 2$

45. $\lim\limits_{h \to 0} \frac{[3(2+h)-4]-[3(2)-4]}{h} = \lim\limits_{h \to 0} \frac{3h}{h} = 3$

47. $\lim\limits_{h \to 0} \frac{\sqrt{7+h}-\sqrt{7}}{h} = \lim\limits_{h \to 0} \frac{\left(\sqrt{7+h}-\sqrt{7}\right)\left(\sqrt{7+h}+\sqrt{7}\right)}{h\left(\sqrt{7+h}+\sqrt{7}\right)} = \lim\limits_{h \to 0} \frac{(7+h)-7}{h\left(\sqrt{7+h}+\sqrt{7}\right)}$

$= \lim\limits_{h \to 0} \frac{h}{h\left(\sqrt{7+h}+\sqrt{7}\right)} = \lim\limits_{h \to 0} \frac{1}{\sqrt{7+h}+\sqrt{7}} = \frac{1}{2\sqrt{7}}$

49. $\lim\limits_{x \to 0} \sqrt{5 - 2x^2} = \sqrt{5 - 2(0)^2} = \sqrt{5}$ and $\lim\limits_{x \to 0} \sqrt{5 - x^2} = \sqrt{5 - (0)^2} = \sqrt{5}$; by the sandwich theorem,

$\lim\limits_{x \to 0} f(x) = \sqrt{5}$

51. (a) $\lim\limits_{x \to 0} \left(1 - \frac{x^2}{6}\right) = 1 - \frac{0}{6} = 1$ and $\lim\limits_{x \to 0} 1 = 1$; by the sandwich theorem, $\lim\limits_{x \to 0} \frac{x \sin x}{2 - 2 \cos x} = 1$

(b) For $x \neq 0$, $y = (x \sin x)/(2 - 2 \cos x)$
lies between the other two graphs in the
figure, and the graphs converge as $x \to 0$.

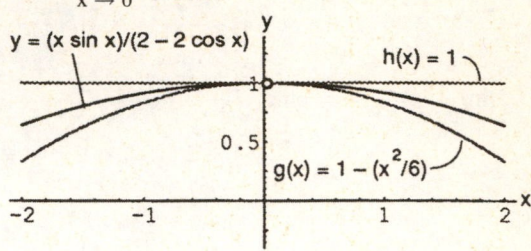

53. $\lim\limits_{x \to c} f(x)$ exists at those points c where $\lim\limits_{x \to c} x^4 = \lim\limits_{x \to c} x^2$. Thus, $c^4 = c^2 \Rightarrow c^2(1 - c^2) = 0$

$\Rightarrow$ c = 0, 1, or −1. Moreover, $\lim\limits_{x \to 0} f(x) = \lim\limits_{x \to 0} x^2 = 0$ and $\lim\limits_{x \to -1} f(x) = \lim\limits_{x \to 1} f(x) = 1$.

55. $1 = \lim\limits_{x \to 4} \frac{f(x) - 5}{x - 2} = \dfrac{\lim\limits_{x \to 4} f(x) - \lim\limits_{x \to 4} 5}{\lim\limits_{x \to 4} x - \lim\limits_{x \to 4} 2} = \dfrac{\lim\limits_{x \to 4} f(x) - 5}{4 - 2} \Rightarrow \lim\limits_{x \to 4} f(x) - 5 = 2(1) \Rightarrow \lim\limits_{x \to 4} f(x) = 2 + 5 = 7.$

57. (a) $0 = 3 \cdot 0 = \left[\lim\limits_{x \to 2} \frac{f(x) - 5}{x - 2}\right]\left[\lim\limits_{x \to 2} (x - 2)\right] = \lim\limits_{x \to 2} \left[\left(\frac{f(x) - 5}{x - 2}\right)(x - 2)\right] = \lim\limits_{x \to 2} [f(x) - 5] = \lim\limits_{x \to 2} f(x) - 5$

$\Rightarrow \lim\limits_{x \to 2} f(x) = 5.$

(b) $0 = 4 \cdot 0 = \left[\lim\limits_{x \to 2} \frac{f(x) - 5}{x - 2}\right]\left[\lim\limits_{x \to 2} (x - 2)\right] \Rightarrow \lim\limits_{x \to 2} f(x) = 5$ as in part (a).

59. (a) $\lim\limits_{x \to 0} x \sin \frac{1}{x} = 0$

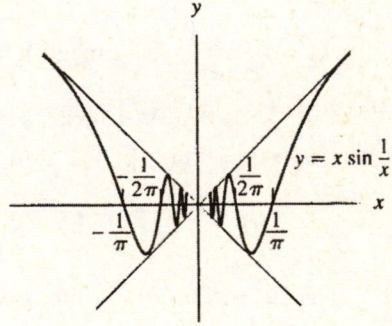

(b) $-1 \leq \sin \frac{1}{x} \leq 1$ for $x \neq 0$:

$x > 0 \Rightarrow -x \leq x \sin \frac{1}{x} \leq x \Rightarrow \lim\limits_{x \to 0} x \sin \frac{1}{x} = 0$ by the sandwich theorem;

$x < 0 \Rightarrow -x \geq x \sin \frac{1}{x} \geq x \Rightarrow \lim\limits_{x \to 0} x \sin \frac{1}{x} = 0$ by the sandwich theorem.

2.3 PRECISE DEFINITION OF A LIMIT

1.

Step 1: $|x - 5| < \delta \Rightarrow -\delta < x - 5 < \delta \Rightarrow -\delta + 5 < x < \delta + 5$

Step 2: $\delta + 5 = 7 \Rightarrow \delta = 2$, or $-\delta + 5 = 1 \Rightarrow \delta = 4$.

The value of δ which assures $|x - 5| < \delta \Rightarrow 1 < x < 7$ is the smaller value, $\delta = 2$.

3.

Step 1: $|x - (-3)| < \delta \Rightarrow -\delta < x + 3 < \delta \Rightarrow -\delta - 3 < x < \delta - 3$

Step 2: $-\delta - 3 = -\frac{7}{2} \Rightarrow \delta = \frac{1}{2}$, or $\delta - 3 = -\frac{1}{2} \Rightarrow \delta = \frac{5}{2}$.

The value of δ which assures $|x - (-3)| < \delta \Rightarrow -\frac{7}{2} < x < -\frac{1}{2}$ is the smaller value, $\delta = \frac{1}{2}$.

5.

Step 1: $\left|x - \frac{1}{2}\right| < \delta \Rightarrow -\delta < x - \frac{1}{2} < \delta \Rightarrow -\delta + \frac{1}{2} < x < \delta + \frac{1}{2}$

Step 2: $-\delta + \frac{1}{2} = \frac{4}{9} \Rightarrow \delta = \frac{1}{18}$, or $\delta + \frac{1}{2} = \frac{4}{7} \Rightarrow \delta = \frac{1}{14}$.

The value of δ which assures $\left|x - \frac{1}{2}\right| < \delta \Rightarrow \frac{4}{9} < x < \frac{4}{7}$ is the smaller value, $\delta = \frac{1}{18}$.

7. Step 1: $|x - 5| < \delta \Rightarrow -\delta < x - 5 < \delta \Rightarrow -\delta + 5 < x < \delta + 5$

Step 2: From the graph, $-\delta + 5 = 4.9 \Rightarrow \delta = 0.1$, or $\delta + 5 = 5.1 \Rightarrow \delta = 0.1$; thus $\delta = 0.1$ in either case.

9. Step 1: $|x - 1| < \delta \Rightarrow -\delta < x - 1 < \delta \Rightarrow -\delta + 1 < x < \delta + 1$

Step 2: From the graph, $-\delta + 1 = \frac{9}{16} \Rightarrow \delta = \frac{7}{16}$, or $\delta + 1 = \frac{25}{16} \Rightarrow \delta = \frac{9}{16}$; thus $\delta = \frac{7}{16}$.

11. Step 1: $|x - 2| < \delta \Rightarrow -\delta < x - 2 < \delta \Rightarrow -\delta + 2 < x < \delta + 2$

Step 2: From the graph, $-\delta + 2 = \sqrt{3} \Rightarrow \delta = 2 - \sqrt{3} \approx 0.2679$, or $\delta + 2 = \sqrt{5} \Rightarrow \delta = \sqrt{5} - 2 \approx 0.2361$; thus $\delta = \sqrt{5} - 2$.

13. Step 1: $|x - (-1)| < \delta \Rightarrow -\delta < x + 1 < \delta \Rightarrow -\delta - 1 < x < \delta - 1$

Step 2: From the graph, $-\delta - 1 = -\frac{16}{9} \Rightarrow \delta = \frac{7}{9} \approx 0.77$, or $\delta - 1 = -\frac{16}{25} \Rightarrow \frac{9}{25} = 0.36$; thus $\delta = \frac{9}{25} = 0.36$.

15. Step 1: $|(x + 1) - 5| < 0.01 \Rightarrow |x - 4| < 0.01 \Rightarrow -0.01 < x - 4 < 0.01 \Rightarrow 3.99 < x < 4.01$

Step 2: $|x - 4| < \delta \Rightarrow -\delta < x - 4 < \delta \Rightarrow -\delta + 4 < x < \delta + 4 \Rightarrow \delta = 0.01$.

17. Step 1: $\left|\sqrt{x + 1} - 1\right| < 0.1 \Rightarrow -0.1 < \sqrt{x + 1} - 1 < 0.1 \Rightarrow 0.9 < \sqrt{x + 1} < 1.1 \Rightarrow 0.81 < x + 1 < 1.21$
$\Rightarrow -0.19 < x < 0.21$

Step 2: $|x - 0| < \delta \Rightarrow -\delta < x < \delta$. Then, $-\delta = -0.19 \Rightarrow \delta = 0.19$ or $\delta = 0.21$; thus, $\delta = 0.19$.

19. Step 1: $\left|\sqrt{19 - x} - 3\right| < 1 \Rightarrow -1 < \sqrt{19 - x} - 3 < 1 \Rightarrow 2 < \sqrt{19 - x} < 4 \Rightarrow 4 < 19 - x < 16$
$\Rightarrow -4 > x - 19 > -16 \Rightarrow 15 > x > 3$ or $3 < x < 15$

Step 2: $|x - 10| < \delta \Rightarrow -\delta < x - 10 < \delta \Rightarrow -\delta + 10 < x < \delta + 10$.
Then $-\delta + 10 = 3 \Rightarrow \delta = 7$, or $\delta + 10 = 15 \Rightarrow \delta = 5$; thus $\delta = 5$.

21. Step 1: $\left|\frac{1}{x} - \frac{1}{4}\right| < 0.05 \Rightarrow -0.05 < \frac{1}{x} - \frac{1}{4} < 0.05 \Rightarrow 0.2 < \frac{1}{x} < 0.3 \Rightarrow \frac{10}{2} > x > \frac{10}{3}$ or $\frac{10}{3} < x < 5$.

Step 2: $|x - 4| < \delta \Rightarrow -\delta < x - 4 < \delta \Rightarrow -\delta + 4 < x < \delta + 4$.
Then $-\delta + 4 = \frac{10}{3}$ or $\delta = \frac{2}{3}$, or $\delta + 4 = 5$ or $\delta = 1$; thus $\delta = \frac{2}{3}$.

23. Step 1: $|x^2 - 4| < 0.5 \Rightarrow -0.5 < x^2 - 4 < 0.5 \Rightarrow 3.5 < x^2 < 4.5 \Rightarrow \sqrt{3.5} < |x| < \sqrt{4.5} \Rightarrow -\sqrt{4.5} < x < -\sqrt{3.5}$,
for x near -2.

Step 2: $|x - (-2)| < \delta \Rightarrow -\delta < x + 2 < \delta \Rightarrow -\delta - 2 < x < \delta - 2$.
Then $-\delta - 2 = -\sqrt{4.5} \Rightarrow \delta = \sqrt{4.5} - 2 \approx 0.1213$, or $\delta - 2 = -\sqrt{3.5} \Rightarrow \delta = 2 - \sqrt{3.5} \approx 0.1292$;
thus $\delta = \sqrt{4.5} - 2 \approx 0.12$.

25. Step 1: $|(x^2 - 5) - 11| < 1 \Rightarrow |x^2 - 16| < 1 \Rightarrow -1 < x^2 - 16 < 1 \Rightarrow 15 < x^2 < 17 \Rightarrow \sqrt{15} < x < \sqrt{17}.$

    Step 2: $|x - 4| < \delta \Rightarrow -\delta < x - 4 < \delta \Rightarrow -\delta + 4 < x < \delta + 4.$

    Then $-\delta + 4 = \sqrt{15} \Rightarrow \delta = 4 - \sqrt{15} \approx 0.1270,$ or $\delta + 4 = \sqrt{17} \Rightarrow \delta = \sqrt{17} - 4 \approx 0.1231;$

    thus $\delta = \sqrt{17} - 4 \approx 0.12.$

27. Step 1: $|mx - 2m| < 0.03 \Rightarrow -0.03 < mx - 2m < 0.03 \Rightarrow -0.03 + 2m < mx < 0.03 + 2m \Rightarrow$

    $2 - \frac{0.03}{m} < x < 2 + \frac{0.03}{m}.$

    Step 2: $|x - 2| < \delta \Rightarrow -\delta < x - 2 < \delta \Rightarrow -\delta + 2 < x < \delta + 2.$

    Then $-\delta + 2 = 2 - \frac{0.03}{m} \Rightarrow \delta = \frac{0.03}{m},$ or $\delta + 2 = 2 + \frac{0.03}{m} \Rightarrow \delta = \frac{0.03}{m}.$ In either case, $\delta = \frac{0.03}{m}.$

29. Step 1: $\left|(mx + b) - \left(\frac{m}{2} + b\right)\right| < c \Rightarrow -c < mx - \frac{m}{2} < c \Rightarrow -c + \frac{m}{2} < mx < c + \frac{m}{2} \Rightarrow \frac{1}{2} - \frac{c}{m} < x < \frac{1}{2} + \frac{c}{m}.$

    Step 2: $\left|x - \frac{1}{2}\right| < \delta \Rightarrow -\delta < x - \frac{1}{2} < \delta \Rightarrow -\delta + \frac{1}{2} < x < \delta + \frac{1}{2}.$

    Then $-\delta + \frac{1}{2} = \frac{1}{2} - \frac{c}{m} \Rightarrow \delta = \frac{c}{m},$ or $\delta + \frac{1}{2} = \frac{1}{2} + \frac{c}{m} \Rightarrow \delta = \frac{c}{m}.$ In either case, $\delta = \frac{c}{m}.$

31. $\lim_{x \to 3} (3 - 2x) = 3 - 2(3) = -3$

    Step 1: $|(3 - 2x) - (-3)| < 0.02 \Rightarrow -0.02 < 6 - 2x < 0.02 \Rightarrow -6.02 < -2x < -5.98 \Rightarrow 3.01 > x > 2.99$ or

    $2.99 < x < 3.01.$

    Step 2: $0 < |x - 3| < \delta \Rightarrow -\delta < x - 3 < \delta \Rightarrow -\delta + 3 < x < \delta + 3.$

    Then $-\delta + 3 = 2.99 \Rightarrow \delta = 0.01,$ or $\delta + 3 = 3.01 \Rightarrow \delta = 0.01;$ thus $\delta = 0.01.$

33. $\lim_{x \to 2} \frac{x^2 - 4}{x - 2} = \lim_{x \to 2} \frac{(x + 2)(x - 2)}{(x - 2)} = \lim_{x \to 2} (x + 2) = 2 + 2 = 4, x \neq 2$

    Step 1: $\left|\left(\frac{x^2 - 4}{x - 2}\right) - 4\right| < 0.05 \Rightarrow -0.05 < \frac{(x + 2)(x - 2)}{(x - 2)} - 4 < 0.05 \Rightarrow 3.95 < x + 2 < 4.05, x \neq 2$

    $\Rightarrow 1.95 < x < 2.05, x \neq 2.$

    Step 2: $|x - 2| < \delta \Rightarrow -\delta < x - 2 < \delta \Rightarrow -\delta + 2 < x < \delta + 2.$

    Then $-\delta + 2 = 1.95 \Rightarrow \delta = 0.05,$ or $\delta + 2 = 2.05 \Rightarrow \delta = 0.05;$ thus $\delta = 0.05.$

35. $\lim_{x \to -3} \sqrt{1 - 5x} = \sqrt{1 - 5(-3)} = \sqrt{16} = 4$

    Step 1: $\left|\sqrt{1 - 5x} - 4\right| < 0.5 \Rightarrow -0.5 < \sqrt{1 - 5x} - 4 < 0.5 \Rightarrow 3.5 < \sqrt{1 - 5x} < 4.5 \Rightarrow 12.25 < 1 - 5x < 20.25$

    $\Rightarrow 11.25 < -5x < 19.25 \Rightarrow -3.85 < x < -2.25.$

    Step 2: $|x - (-3)| < \delta \Rightarrow -\delta < x + 3 < \delta \Rightarrow -\delta - 3 < x < \delta - 3.$

    Then $-\delta - 3 = -3.85 \Rightarrow \delta = 0.85,$ or $\delta - 3 = -2.25 \Rightarrow 0.75;$ thus $\delta = 0.75.$

37. Step 1: $|(9 - x) - 5| < \epsilon \Rightarrow -\epsilon < 4 - x < \epsilon \Rightarrow -\epsilon - 4 < -x < \epsilon - 4 \Rightarrow \epsilon + 4 > x > 4 - \epsilon \Rightarrow 4 - \epsilon < x < 4 + \epsilon.$

    Step 2: $|x - 4| < \delta \Rightarrow -\delta < x - 4 < \delta \Rightarrow -\delta + 4 < x < \delta + 4.$

    Then $-\delta + 4 = -\epsilon + 4 \Rightarrow \delta = \epsilon,$ or $\delta + 4 = \epsilon + 4 \Rightarrow \delta = \epsilon.$ Thus choose $\delta = \epsilon.$

39. Step 1: $\left|\sqrt{x - 5} - 2\right| < \epsilon \Rightarrow -\epsilon < \sqrt{x - 5} - 2 < \epsilon \Rightarrow 2 - \epsilon < \sqrt{x - 5} < 2 + \epsilon \Rightarrow (2 - \epsilon)^2 < x - 5 < (2 + \epsilon)^2$

    $\Rightarrow (2 - \epsilon)^2 + 5 < x < (2 + \epsilon)^2 + 5.$

    Step 2: $|x - 9| < \delta \Rightarrow -\delta < x - 9 < \delta \Rightarrow -\delta + 9 < x < \delta + 9.$

    Then $-\delta + 9 = \epsilon^2 - 4\epsilon + 9 \Rightarrow \delta = 4\epsilon - \epsilon^2,$ or $\delta + 9 = \epsilon^2 + 4\epsilon + 9 \Rightarrow \delta = 4\epsilon + \epsilon^2.$ Thus choose

    the smaller distance, $\delta = 4\epsilon - \epsilon^2.$

41. Step 1:   For $x \neq 1, |x^2 - 1| < \epsilon \Rightarrow -\epsilon < x^2 - 1 < \epsilon \Rightarrow 1 - \epsilon < x^2 < 1 + \epsilon \Rightarrow \sqrt{1 - \epsilon} < |x| < \sqrt{1 + \epsilon}$

    $\Rightarrow \sqrt{1 - \epsilon} < x < \sqrt{1 + \epsilon}$ near $x = 1.$

Step 2: $|x - 1| < \delta \Rightarrow -\delta < x - 1 < \delta \Rightarrow -\delta + 1 < x < \delta + 1.$

Then $-\delta + 1 = \sqrt{1 - \epsilon} \Rightarrow \delta = 1 - \sqrt{1 - \epsilon}$, or $\delta + 1 = \sqrt{1 + \epsilon} \Rightarrow \delta = \sqrt{1 + \epsilon} - 1$. Choose $\delta = \min\left\{1 - \sqrt{1 - \epsilon}, \sqrt{1 + \epsilon} - 1\right\}$, that is, the smaller of the two distances.

43. Step 1: $\left|\frac{1}{x} - 1\right| < \epsilon \Rightarrow -\epsilon < \frac{1}{x} - 1 < \epsilon \Rightarrow 1 - \epsilon < \frac{1}{x} < 1 + \epsilon \Rightarrow \frac{1}{1+\epsilon} < x < \frac{1}{1-\epsilon}.$

Step 2: $|x - 1| < \delta \Rightarrow -\delta < x - 1 < \delta \Rightarrow 1 - \delta < x < 1 + \delta.$

Then $1 - \delta = \frac{1}{1+\epsilon} \Rightarrow \delta = 1 - \frac{1}{1+\epsilon} = \frac{\epsilon}{1+\epsilon}$, or $1 + \delta = \frac{1}{1-\epsilon} \Rightarrow \delta = \frac{1}{1-\epsilon} - 1 = \frac{\epsilon}{1-\epsilon}.$

Choose $\delta = \frac{\epsilon}{1+\epsilon}$, the smaller of the two distances.

45. Step 1: $\left|\left(\frac{x^2-9}{x+3}\right) - (-6)\right| < \epsilon \Rightarrow -\epsilon < (x - 3) + 6 < \epsilon, x \neq -3 \Rightarrow -\epsilon < x + 3 < \epsilon \Rightarrow -\epsilon - 3 < x < \epsilon - 3.$

Step 2: $|x - (-3)| < \delta \Rightarrow -\delta < x + 3 < \delta \Rightarrow -\delta - 3 < x < \delta - 3.$

Then $-\delta - 3 = -\epsilon - 3 \Rightarrow \delta = \epsilon$, or $\delta - 3 = \epsilon - 3 \Rightarrow \delta = \epsilon$. Choose $\delta = \epsilon$.

47. Step 1: $x < 1$: $|(4 - 2x) - 2| < \epsilon \Rightarrow 0 < 2 - 2x < \epsilon$ since $x < 1$. Thus, $1 - \frac{\epsilon}{2} < x < 0$;

$x \geq 1$: $|(6x - 4) - 2| < \epsilon \Rightarrow 0 \leq 6x - 6 < \epsilon$ since $x \geq 1$. Thus, $1 \leq x < 1 + \frac{\epsilon}{6}.$

Step 2: $|x - 1| < \delta \Rightarrow -\delta < x - 1 < \delta \Rightarrow 1 - \delta < x < 1 + \delta.$

Then $1 - \delta = 1 - \frac{\epsilon}{2} \Rightarrow \delta = \frac{\epsilon}{2}$, or $1 + \delta = 1 + \frac{\epsilon}{6} \Rightarrow \delta = \frac{\epsilon}{6}$. Choose $\delta = \frac{\epsilon}{6}.$

49. By the figure, $-x \leq x \sin\frac{1}{x} \leq x$ for all $x > 0$ and $-x \geq x \sin\frac{1}{x} \geq x$ for $x < 0$. Since $\lim_{x \to 0}(-x) = \lim_{x \to 0} x = 0$, then by the sandwich theorem, in either case, $\lim_{x \to 0} x \sin\frac{1}{x} = 0.$

51. As x approaches the value 0, the values of g(x) approach k. Thus for every number $\epsilon > 0$, there exists a $\delta > 0$ such that $0 < |x - 0| < \delta \Rightarrow |g(x) - k| < \epsilon.$

53. Let $f(x) = x^2$. The function values do get closer to -1 as x approaches 0, but $\lim_{x \to 0} f(x) = 0$, not -1. The function $f(x) = x^2$ never gets <u>arbitrarily</u> <u>close</u> to -1 for x near 0.

55. $|A - 9| \leq 0.01 \Rightarrow -0.01 \leq \pi\left(\frac{x}{2}\right)^2 - 9 \leq 0.01 \Rightarrow 8.99 \leq \frac{\pi x^2}{4} \leq 9.01 \Rightarrow \frac{4}{\pi}(8.99) \leq x^2 \leq \frac{4}{\pi}(9.01)$

$\Rightarrow 2\sqrt{\frac{8.99}{\pi}} \leq x \leq 2\sqrt{\frac{9.01}{\pi}}$ or $3.384 \leq x \leq 3.387$. To be safe, the left endpoint was rounded up and the right endpoint was rounded down.

57. (a) $-\delta < x - 1 < 0 \Rightarrow 1 - \delta < x < 1 \Rightarrow f(x) = x$. Then $|f(x) - 2| = |x - 2| = 2 - x > 2 - 1 = 1$. That is, $|f(x) - 2| \geq 1 \geq \frac{1}{2}$ no matter how small δ is taken when $1 - \delta < x < 1 \Rightarrow \lim_{x \to 1} f(x) \neq 2.$

(b) $0 < x - 1 < \delta \Rightarrow 1 < x < 1 + \delta \Rightarrow f(x) = x + 1$. Then $|f(x) - 1| = |(x + 1) - 1| = |x| = x > 1$. That is, $|f(x) - 1| \geq 1$ no matter how small δ is taken when $1 < x < 1 + \delta \Rightarrow \lim_{x \to 1} f(x) \neq 1.$

(c) $-\delta < x - 1 < 0 \Rightarrow 1 - \delta < x < 1 \Rightarrow f(x) = x$. Then $|f(x) - 1.5| = |x - 1.5| = 1.5 - x > 1.5 - 1 = 0.5.$ Also, $0 < x - 1 < \delta \Rightarrow 1 < x < 1 + \delta \Rightarrow f(x) = x + 1$. Then $|f(x) - 1.5| = |(x + 1) - 1.5| = |x - 0.5|$ $= x - 0.5 > 1 - 0.5 = 0.5$. Thus, no matter how small δ is taken, there exists a value of x such that $-\delta < x - 1 < \delta$ but $|f(x) - 1.5| \geq \frac{1}{2} \Rightarrow \lim_{x \to 1} f(x) \neq 1.5.$

59. (a) For $3 - \delta < x < 3 \Rightarrow f(x) > 4.8 \Rightarrow |f(x) - 4| \geq 0.8$. Thus for $\epsilon < 0.8$, $|f(x) - 4| \geq \epsilon$ whenever $3 - \delta < x < 3$ no matter how small we choose $\delta > 0 \Rightarrow \lim_{x \to 3} f(x) \neq 4.$

(b) For $3 < x < 3 + \delta \Rightarrow f(x) < 3 \Rightarrow |f(x) - 4.8| \geq 1.8$. Thus for $\epsilon < 1.8$, $|f(x) - 4.8| \geq \epsilon$ whenever $3 < x < 3 + \delta$

no matter how small we choose $\delta > 0 \;\Rightarrow\; \lim\limits_{x \to 3} f(x) \neq 4.8$.

(c) For $3 - \delta < x < 3 \;\Rightarrow\; f(x) > 4.8 \;\Rightarrow\; |f(x) - 3| \geq 1.8$. Again, for $\epsilon < 1.8$, $|f(x) - 3| \geq \epsilon$ whenever $3 - \delta < x < 3$

no matter how small we choose $\delta > 0 \;\Rightarrow\; \lim\limits_{x \to 3} f(x) \neq 3$.

2.4 ONE-SIDED LIMITS AND LIMITS AT INFINITY

1. (a) True   (b) True   (c) False   (d) True

   (e) True   (f) True   (g) False   (h) False

   (i) False   (j) False   (k) True   (l) False

3. (a) $\lim\limits_{x \to 2^+} f(x) = \frac{2}{2} + 1 = 2$, $\lim\limits_{x \to 2^-} f(x) = 3 - 2 = 1$

   (b) No, $\lim\limits_{x \to 2} f(x)$ does not exist because $\lim\limits_{x \to 2^+} f(x) \neq \lim\limits_{x \to 2^-} f(x)$

   (c) $\lim\limits_{x \to 4^-} f(x) = \frac{4}{2} + 1 = 3$, $\lim\limits_{x \to 4^+} f(x) = \frac{4}{2} + 1 = 3$

   (d) Yes, $\lim\limits_{x \to 4} f(x) = 3$ because $3 = \lim\limits_{x \to 4^-} f(x) = \lim\limits_{x \to 4^+} f(x)$

5. (a) No, $\lim\limits_{x \to 0^+} f(x)$ does not exist since $\sin\left(\frac{1}{x}\right)$ does not approach any single value as x approaches 0

   (b) $\lim\limits_{x \to 0^-} f(x) = \lim\limits_{x \to 0^-} 0 = 0$

   (c) $\lim\limits_{x \to 0} f(x)$ does not exist because $\lim\limits_{x \to 0^+} f(x)$ does not exist

7. (a)

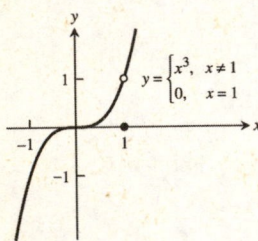

   (b) $\lim\limits_{x \to 1^-} f(x) = 1 = \lim\limits_{x \to 1^+} f(x)$

   (c) Yes, $\lim\limits_{x \to 1} f(x) = 1$ since the right-hand and left-hand limits exist and equal 1

9. (a) domain: $0 \leq x \leq 2$

     range: $0 < y \leq 1$ and $y = 2$

   (b) $\lim\limits_{x \to c} f(x)$ exists for c belonging to

     $(0, 1) \cup (1, 2)$

   (c) $x = 2$

   (d) $x = 0$

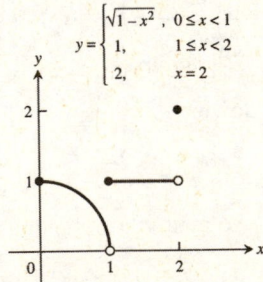

11. $\lim\limits_{x \to -0.5^-} \sqrt{\frac{x+2}{x-1}} = \sqrt{\frac{-0.5+2}{-0.5+1}} = \sqrt{\frac{3/2}{1/2}} = \sqrt{3}$

13. $\lim\limits_{x \to -2^+} \left(\frac{x}{x+1}\right)\left(\frac{2x+5}{x^2+x}\right) = \left(\frac{-2}{-2+1}\right)\left(\frac{2(-2)+5}{(-2)^2+(-2)}\right) = (2)\left(\frac{1}{2}\right) = 1$

15. $\lim\limits_{h \to 0^+} \frac{\sqrt{h^2+4h+5} - \sqrt{5}}{h} = \lim\limits_{h \to 0^+} \left(\frac{\sqrt{h^2+4h+5} - \sqrt{5}}{h}\right)\left(\frac{\sqrt{h^2+4h+5} + \sqrt{5}}{\sqrt{h^2+4h+5} + \sqrt{5}}\right)$

$\quad = \lim\limits_{h \to 0^+} \frac{(h^2+4h+5) - 5}{h\left(\sqrt{h^2+4h+5} + \sqrt{5}\right)} = \lim\limits_{h \to 0^+} \frac{h(h+4)}{h\left(\sqrt{h^2+4h+5} + \sqrt{5}\right)} = \frac{0+4}{\sqrt{5}+\sqrt{5}} = \frac{2}{\sqrt{5}}$

17. (a) $\displaystyle\lim_{x \to -2^+} (x+3) \frac{|x+2|}{x+2} = \lim_{x \to -2^+} (x+3) \frac{(x+2)}{(x+2)}$ $\qquad$ ($|x+2| = x+2$ for $x > -2$)

$\qquad\qquad = \displaystyle\lim_{x \to -2^+} (x+3) = (-2)+3 = 1$

(b) $\displaystyle\lim_{x \to -2^-} (x+3) \frac{|x+2|}{x+2} = \lim_{x \to -2^-} (x+3) \left[\frac{-(x+2)}{(x+2)}\right]$ $\qquad$ ($|x+2| = -(x+2)$ for $x < -2$)

$\qquad\qquad = \displaystyle\lim_{x \to -2^-} (x+3)(-1) = -(-2+3) = -1$

19. (a) $\displaystyle\lim_{\theta \to 3^+} \frac{|\theta|}{\theta} = \frac{3}{3} = 1$ $\qquad\qquad\qquad\qquad$ (b) $\displaystyle\lim_{\theta \to 3^-} \frac{|\theta|}{\theta} = \frac{2}{3}$

21. $\displaystyle\lim_{\theta \to 0} \frac{\sin\sqrt{2\theta}}{\sqrt{2\theta}} = \lim_{x \to 0} \frac{\sin x}{x} = 1$ $\qquad$ (where $x = \sqrt{2\theta}$)

23. $\displaystyle\lim_{y \to 0} \frac{\sin 3y}{4y} = \frac{1}{4}\lim_{y \to 0} \frac{3\sin 3y}{3y} = \frac{3}{4}\lim_{y \to 0} \frac{\sin 3y}{3y} = \frac{3}{4}\lim_{\theta \to 0} \frac{\sin\theta}{\theta} = \frac{3}{4}$ $\qquad$ (where $\theta = 3y$)

25. $\displaystyle\lim_{x \to 0} \frac{\tan 2x}{x} = \lim_{x \to 0} \frac{\left(\frac{\sin 2x}{\cos 2x}\right)}{x} = \lim_{x \to 0} \frac{\sin 2x}{x\cos 2x} = \left(\lim_{x \to 0}\frac{1}{\cos 2x}\right)\left(\lim_{x \to 0}\frac{2\sin 2x}{2x}\right) = 1 \cdot 2 = 2$

27. $\displaystyle\lim_{x \to 0} \frac{x\csc 2x}{\cos 5x} = \lim_{x \to 0}\left(\frac{x}{\sin 2x}\cdot\frac{1}{\cos 5x}\right) = \left(\frac{1}{2}\lim_{x \to 0}\frac{2x}{\sin 2x}\right)\left(\lim_{x \to 0}\frac{1}{\cos 5x}\right) = \left(\frac{1}{2}\cdot 1\right)(1) = \frac{1}{2}$

29. $\displaystyle\lim_{x \to 0} \frac{x + x\cos x}{\sin x\cos x} = \lim_{x \to 0}\left(\frac{x}{\sin x\cos x} + \frac{x\cos x}{\sin x\cos x}\right) = \lim_{x \to 0}\left(\frac{x}{\sin x}\cdot\frac{1}{\cos x}\right) + \lim_{x \to 0}\frac{x}{\sin x}$

$\qquad = \displaystyle\lim_{x \to 0}\left(\frac{1}{\frac{\sin x}{x}}\right)\cdot\lim_{x \to 0}\left(\frac{1}{\cos x}\right) + \lim_{x \to 0}\left(\frac{1}{\frac{\sin x}{x}}\right) = (1)(1) + 1 = 2$

31. $\displaystyle\lim_{t \to 0} \frac{\sin(1 - \cos t)}{1 - \cos t} = \lim_{\theta \to 0} \frac{\sin\theta}{\theta} = 1$ since $\theta = 1 - \cos t \to 0$ as $t \to 0$

33. $\displaystyle\lim_{\theta \to 0} \frac{\sin\theta}{\sin 2\theta} = \lim_{\theta \to 0}\left(\frac{\sin\theta}{\sin 2\theta}\cdot\frac{2\theta}{2\theta}\right) = \frac{1}{2}\lim_{\theta \to 0}\left(\frac{\sin\theta}{\theta}\cdot\frac{2\theta}{\sin 2\theta}\right) = \frac{1}{2}\cdot 1\cdot 1 = \frac{1}{2}$

35. $\displaystyle\lim_{x \to 0} \frac{\tan 3x}{\sin 8x} = \lim_{x \to 0}\left(\frac{\sin 3x}{\cos 3x}\cdot\frac{1}{\sin 8x}\right) = \lim_{x \to 0}\left(\frac{\sin 3x}{\cos 3x}\cdot\frac{1}{\sin 8x}\cdot\frac{8x}{3x}\cdot\frac{3}{8}\right)$

$\qquad = \displaystyle\frac{3}{8}\lim_{x \to 0}\left(\frac{1}{\cos 3x}\right)\left(\frac{\sin 3x}{3x}\right)\left(\frac{8x}{\sin 8x}\right) = \frac{3}{8}\cdot 1\cdot 1\cdot 1 = \frac{3}{8}$

Note: In these exercises we use the result $\displaystyle\lim_{x \to \pm\infty} \frac{1}{x^{m/n}} = 0$ whenever $\frac{m}{n} > 0$. This result follows immediately from

Example 6 and the power rule in Theorem 8: $\displaystyle\lim_{x \to \pm\infty}\left(\frac{1}{x^{m/n}}\right) = \lim_{x \to \pm\infty}\left(\frac{1}{x}\right)^{m/n} = \left(\lim_{x \to \pm\infty}\frac{1}{x}\right)^{m/n} = 0^{m/n} = 0.$

37. (a) -3 $\qquad\qquad\qquad\qquad\qquad\qquad\qquad\qquad$ (b) -3

39. (a) $\frac{1}{2}$ $\qquad\qquad\qquad\qquad\qquad\qquad\qquad\qquad\quad$ (b) $\frac{1}{2}$

41. (a) $-\frac{5}{3}$ $\qquad\qquad\qquad\qquad\qquad\qquad\qquad\qquad$ (b) $-\frac{5}{3}$

43. $-\frac{1}{x} \le \frac{\sin 2x}{x} \le \frac{1}{x} \Rightarrow \displaystyle\lim_{x \to \infty}\frac{\sin 2x}{x} = 0$ by the Sandwich Theorem

45. $\displaystyle\lim_{t \to \infty} \frac{2 - t + \sin t}{t + \cos t} = \lim_{t \to \infty} \frac{\frac{2}{t} - 1 + \left(\frac{\sin t}{t}\right)}{1 + \left(\frac{\cos t}{t}\right)} = \frac{0 - 1 + 0}{1 + 0} = -1$

47. (a) $\displaystyle\lim_{x \to \infty} \frac{2x+3}{5x+7} = \lim_{x \to \infty} \frac{2 + \frac{3}{x}}{5 + \frac{7}{x}} = \frac{2}{5}$ $\qquad\qquad$ (b) $\frac{2}{5}$ (same process as part (a))

49. (a) $\lim\limits_{x \to \infty} \frac{x+1}{x^2+3} = \lim\limits_{x \to \infty} \frac{\frac{1}{x}+\frac{1}{x^2}}{1+\frac{3}{x^2}} = 0$ (b) 0 (same process as part (a))

51. (a) $\lim\limits_{x \to \infty} \frac{7x^3}{x^3-3x^2+6x} = \lim\limits_{x \to \infty} \frac{7}{1-\frac{3}{x}+\frac{6}{x^2}} = 7$ (b) 7 (same process as part (a))

53. (a) $\lim\limits_{x \to \infty} \frac{10x^5+x^4+31}{x^6} = \lim\limits_{x \to \infty} \frac{\frac{10}{x}+\frac{1}{x^2}+\frac{31}{x^6}}{1} = 0$

   (b) 0 (same process as part (a))

55. (a) $\lim\limits_{x \to \infty} \frac{-2x^3-2x+3}{3x^3+3x^2-5x} = \lim\limits_{x \to \infty} \frac{-2-\frac{2}{x^2}+\frac{3}{x^3}}{3+\frac{3}{x}-\frac{5}{x^2}} = -\frac{2}{3}$

   (b) $-\frac{2}{3}$ (same process as part (a))

57. $\lim\limits_{x \to \infty} \frac{2\sqrt{x}+x^{-1}}{3x-7} = \lim\limits_{x \to \infty} \frac{\left(\frac{2}{x^{1/2}}\right)+\left(\frac{1}{x^2}\right)}{3-\frac{7}{x}} = 0$

59. $\lim\limits_{x \to -\infty} \frac{\sqrt[3]{x}-\sqrt[5]{x}}{\sqrt[3]{x}+\sqrt[5]{x}} = \lim\limits_{x \to -\infty} \frac{1-x^{(1/5)-(1/3)}}{1+x^{(1/5)-(1/3)}} = \lim\limits_{x \to -\infty} \frac{1-\left(\frac{1}{x^{2/15}}\right)}{1+\left(\frac{1}{x^{2/15}}\right)} = 1$

61. $\lim\limits_{x \to \infty} \frac{2x^{5/3}-x^{1/3}+7}{x^{8/5}+3x+\sqrt{x}} = \lim\limits_{x \to \infty} \frac{2x^{1/15}-\frac{1}{x^{19/15}}+\frac{7}{x^{8/5}}}{1+\frac{3}{x^{3/5}}+\frac{1}{x^{11/10}}} = \infty$

63. Yes. If $\lim\limits_{x \to a^+} f(x) = L = \lim\limits_{x \to a^-} f(x)$, then $\lim\limits_{x \to a} f(x) = L$. If $\lim\limits_{x \to a^+} f(x) \neq \lim\limits_{x \to a^-} f(x)$, then $\lim\limits_{x \to a} f(x)$ does not exist.

65. If f is an odd function of x, then $f(-x) = -f(x)$. Given $\lim\limits_{x \to 0^+} f(x) = 3$, then $\lim\limits_{x \to 0^-} f(x) = -3$.

67. Yes. If $\lim\limits_{x \to \infty} \frac{f(x)}{g(x)} = 2$ then the ratio of the polynomials' leading coefficients is 2, so $\lim\limits_{x \to -\infty} \frac{f(x)}{g(x)} = 2$ as well.

69. At most 1 horizontal asymptote: If $\lim\limits_{x \to \infty} \frac{f(x)}{g(x)} = L$, then the ratio of the polynomials' leading coefficients is L, so $\lim\limits_{x \to -\infty} \frac{f(x)}{g(x)} = L$ as well.

71. For any $\epsilon > 0$, take $N = 1$. Then for all $x > N$ we have that $|f(x) - k| = |k - k| = 0 < \epsilon$.

73. $I = (5, 5+\delta) \Rightarrow 5 < x < 5 + \delta$. Also, $\sqrt{x-5} < \epsilon \Rightarrow x - 5 < \epsilon^2 \Rightarrow x < 5 + \epsilon^2$. Choose $\delta = \epsilon^2$
   $\Rightarrow \lim\limits_{x \to 5^+} \sqrt{x-5} = 0$.

75. As $x \to 0^-$ the number x is always negative. Thus, $\left|\frac{x}{|x|} - (-1)\right| < \epsilon \Rightarrow \left|\frac{x}{-x} + 1\right| < \epsilon \Rightarrow 0 < \epsilon$ which is always true independent of the value of x. Hence we can choose any $\delta > 0$ with $-\delta < x < 0 \Rightarrow \lim\limits_{x \to 0^-} \frac{x}{|x|} = -1$.

77. (a) $\lim\limits_{x \to 400^+} \lfloor x \rfloor = 400$. Just observe that if $400 < x < 401$, then $\lfloor x \rfloor = 400$. Thus if we choose $\delta = 1$, we have for any number $\epsilon > 0$ that $400 < x < 400 + \delta \Rightarrow |\lfloor x \rfloor - 400| = |400 - 400| = 0 < \epsilon$.

   (b) $\lim\limits_{x \to 400^-} \lfloor x \rfloor = 399$. Just observe that if $399 < x < 400$ then $\lfloor x \rfloor = 399$. Thus if we choose $\delta = 1$, we have for any number $\epsilon > 0$ that $400 - \delta < x < 400 \Rightarrow |\lfloor x \rfloor - 399| = |399 - 399| = 0 < \epsilon$.

   (c) Since $\lim\limits_{x \to 400^+} \lfloor x \rfloor \neq \lim\limits_{x \to 400^-} \lfloor x \rfloor$ we conclude that $\lim\limits_{x \to 400} \lfloor x \rfloor$ does not exist.

79. $\displaystyle\lim_{x \to \pm\infty} x \sin \frac{1}{x} = \lim_{\theta \to 0} \frac{1}{\theta} \sin \theta = 1, \ \left(\theta = \frac{1}{x}\right)$

81. $\displaystyle\lim_{x \to \pm\infty} \frac{3x+4}{2x-5} = \lim_{x \to \pm\infty} \frac{3+\frac{4}{x}}{2-\frac{5}{x}} = \lim_{t \to 0} \frac{3+4t}{2-5t} = \frac{3}{2}, \ \left(t = \frac{1}{x}\right)$

83. $\displaystyle\lim_{x \to \pm\infty} \left(3+\frac{2}{x}\right)\left(\cos \frac{1}{x}\right) = \lim_{\theta \to 0} (3+2\theta)(\cos \theta) = (3)(1) = 3, \ \left(\theta = \frac{1}{x}\right)$

2.5 INFINITE LIMITS AND VERTICAL ASYMPTOTES

1. $\displaystyle\lim_{x \to 0^+} \frac{1}{3x} = \infty$ $\qquad\qquad \left(\frac{\text{positive}}{\text{positive}}\right)$

3. $\displaystyle\lim_{x \to 2^-} \frac{3}{x-2} = -\infty$ $\qquad\qquad \left(\frac{\text{positive}}{\text{negative}}\right)$

5. $\displaystyle\lim_{x \to -8^+} \frac{2x}{x+8} = -\infty$ $\qquad\qquad \left(\frac{\text{negative}}{\text{positive}}\right)$

7. $\displaystyle\lim_{x \to 7} \frac{4}{(x-7)^2} = \infty$ $\qquad\qquad \left(\frac{\text{positive}}{\text{positive}}\right)$

9. (a) $\displaystyle\lim_{x \to 0^+} \frac{2}{3x^{1/3}} = \infty$ $\qquad\qquad$ (b) $\displaystyle\lim_{x \to 0^-} \frac{2}{3x^{1/3}} = -\infty$

11. $\displaystyle\lim_{x \to 0} \frac{4}{x^{2/5}} = \lim_{x \to 0} \frac{4}{\left(x^{1/5}\right)^2} = \infty$

13. $\displaystyle\lim_{x \to \left(\frac{\pi}{2}\right)^-} \tan x = \infty$

15. $\displaystyle\lim_{\theta \to 0^-} (1 + \csc \theta) = -\infty$

17. (a) $\displaystyle\lim_{x \to 2^+} \frac{1}{x^2-4} = \lim_{x \to 2^+} \frac{1}{(x+2)(x-2)} = \infty$ $\qquad \left(\frac{1}{\text{positive·positive}}\right)$

   (b) $\displaystyle\lim_{x \to 2^-} \frac{1}{x^2-4} = \lim_{x \to 2^-} \frac{1}{(x+2)(x-2)} = -\infty$ $\qquad \left(\frac{1}{\text{positive·negative}}\right)$

   (c) $\displaystyle\lim_{x \to -2^+} \frac{1}{x^2-4} = \lim_{x \to -2^+} \frac{1}{(x+2)(x-2)} = -\infty$ $\qquad \left(\frac{1}{\text{positive·negative}}\right)$

   (d) $\displaystyle\lim_{x \to -2^-} \frac{1}{x^2-4} = \lim_{x \to -2^-} \frac{1}{(x+2)(x-2)} = \infty$ $\qquad \left(\frac{1}{\text{negative·negative}}\right)$

19. (a) $\displaystyle\lim_{x \to 0^+} \frac{x^2}{2} - \frac{1}{x} = 0 + \lim_{x \to 0^+} \frac{1}{-x} = -\infty$ $\qquad \left(\frac{1}{\text{negative}}\right)$

   (b) $\displaystyle\lim_{x \to 0^-} \frac{x^2}{2} - \frac{1}{x} = 0 + \lim_{x \to 0^-} \frac{1}{-x} = \infty$ $\qquad \left(\frac{1}{\text{positive}}\right)$

   (c) $\displaystyle\lim_{x \to \sqrt[3]{2}} \frac{x^2}{2} - \frac{1}{x} = \frac{2^{2/3}}{2} - \frac{1}{2^{1/3}} = 2^{-1/3} - 2^{-1/3} = 0$

   (d) $\displaystyle\lim_{x \to -1} \frac{x^2}{2} - \frac{1}{x} = \frac{1}{2} - \left(\frac{1}{-1}\right) = \frac{3}{2}$

21. (a) $\displaystyle\lim_{x \to 0^+} \frac{x^2-3x+2}{x^3-2x^2} = \lim_{x \to 0^+} \frac{(x-2)(x-1)}{x^2(x-2)} = -\infty$ $\qquad \left(\frac{\text{negative·negative}}{\text{positive·negative}}\right)$

   (b) $\displaystyle\lim_{x \to 2^+} \frac{x^2-3x+2}{x^3-2x^2} = \lim_{x \to 2^+} \frac{(x-2)(x-1)}{x^2(x-2)} = \lim_{x \to 2^+} \frac{x-1}{x^2} = \frac{1}{4}, x \neq 2$

   (c) $\displaystyle\lim_{x \to 2^-} \frac{x^2-3x+2}{x^3-2x^2} = \lim_{x \to 2^-} \frac{(x-2)(x-1)}{x^2(x-2)} = \lim_{x \to 2^-} \frac{x-1}{x^2} = \frac{1}{4}, x \neq 2$

(d) $\lim\limits_{x \to 2} \dfrac{x^2 - 3x + 2}{x^3 - 2x^2} = \lim\limits_{x \to 2} \dfrac{(x-2)(x-1)}{x^2(x-2)} = \lim\limits_{x \to 2} \dfrac{x-1}{x^2} = \dfrac{1}{4}, x \neq 2$

(e) $\lim\limits_{x \to 0} \dfrac{x^2 - 3x + 2}{x^3 - 2x^2} = \lim\limits_{x \to 0} \dfrac{(x-2)(x-1)}{x^2(x-2)} = -\infty$ $\left(\dfrac{\text{negative·negative}}{\text{positive·negative}} \right)$

23. (a) $\lim\limits_{t \to 0^+} \left[2 - \dfrac{3}{t^{1/3}} \right] = -\infty$ (b) $\lim\limits_{t \to 0^-} \left[2 - \dfrac{3}{t^{1/3}} \right] = \infty$

25. (a) $\lim\limits_{x \to 0^+} \left[\dfrac{1}{x^{2/3}} + \dfrac{2}{(x-1)^{2/3}} \right] = \infty$ (b) $\lim\limits_{x \to 0^-} \left[\dfrac{1}{x^{2/3}} + \dfrac{2}{(x-1)^{2/3}} \right] = \infty$

(c) $\lim\limits_{x \to 1^+} \left[\dfrac{1}{x^{2/3}} + \dfrac{2}{(x-1)^{2/3}} \right] = \infty$ (d) $\lim\limits_{x \to 1^-} \left[\dfrac{1}{x^{2/3}} + \dfrac{2}{(x-1)^{2/3}} \right] = \infty$

27. $y = \dfrac{1}{x-1}$ 29. $y = \dfrac{1}{2x+4}$

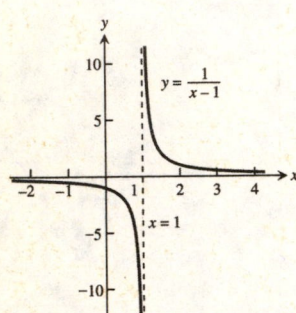   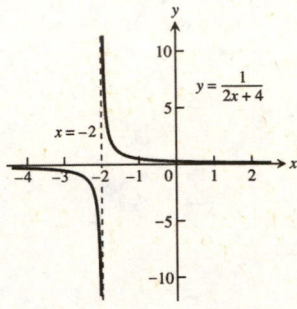

31. $y = \dfrac{x+3}{x+2} = 1 + \dfrac{1}{x+2}$ 33. $y = \dfrac{x^2}{x-1} = x + 1 + \dfrac{1}{x-1}$

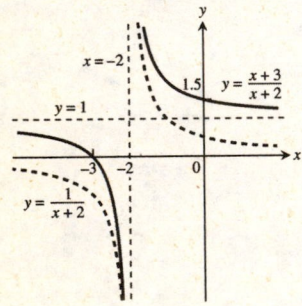   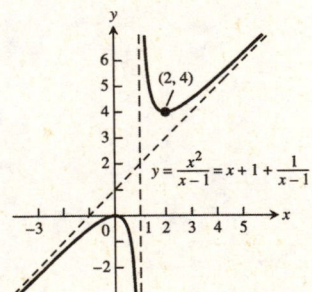

35. $y = \dfrac{x^2 - 4}{x-1} = x + 1 - \dfrac{3}{x-1}$ 37. $y = \dfrac{x^2 - 1}{x} = x - \dfrac{1}{x}$

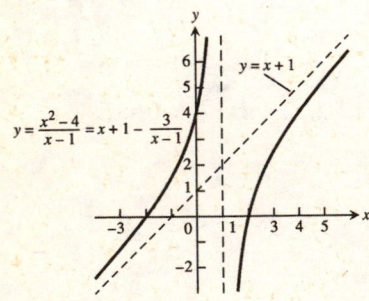   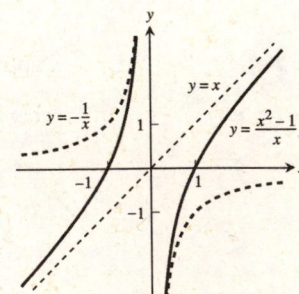

39. Here is one possibility.

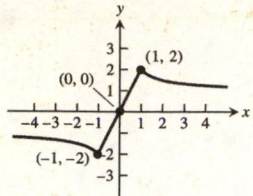

41. Here is one possibility.

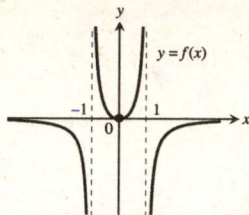

43. Here is one possibility.

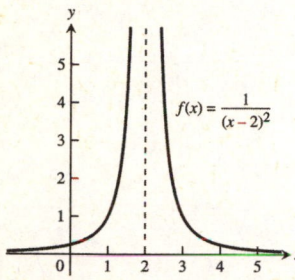

45. Here is one possibility.

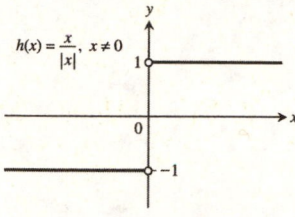

47. For every real number $-B < 0$, we must find a $\delta > 0$ such that for all x, $0 < |x - 0| < \delta \Rightarrow \frac{-1}{x^2} < -B$. Now,

$-\frac{1}{x^2} < -B < 0 \Leftrightarrow \frac{1}{x^2} > B > 0 \Leftrightarrow x^2 < \frac{1}{B} \Leftrightarrow |x| < \frac{1}{\sqrt{B}}$. Choose $\delta = \frac{1}{\sqrt{B}}$, then $0 < |x| < \delta \Rightarrow |x| < \frac{1}{\sqrt{B}}$

$\Rightarrow \frac{-1}{x^2} < -B$ so that $\lim_{x \to 0} -\frac{1}{x^2} = -\infty$.

49. For every real number $-B < 0$, we must find a $\delta > 0$ such that for all x, $0 < |x - 3| < \delta \Rightarrow \frac{-2}{(x-3)^2} < -B$.

Now, $\frac{-2}{(x-3)^2} < -B < 0 \Leftrightarrow \frac{2}{(x-3)^2} > B > 0 \Leftrightarrow \frac{(x-3)^2}{2} < \frac{1}{B} \Leftrightarrow (x-3)^2 < \frac{2}{B} \Leftrightarrow 0 < |x - 3| < \sqrt{\frac{2}{B}}$. Choose

$\delta = \sqrt{\frac{2}{B}}$, then $0 < |x - 3| < \delta \Rightarrow \frac{-2}{(x-3)^2} < -B < 0$ so that $\lim_{x \to 3} \frac{-2}{(x-3)^2} = -\infty$.

51. (a) We say that f(x) approaches infinity as x approaches x_0 from the left, and write $\lim_{x \to x_0^-} f(x) = \infty$, if

    for every positive number B, there exists a corresponding number $\delta > 0$ such that for all x,
    $x_0 - \delta < x < x_0 \Rightarrow f(x) > B$.

    (b) We say that f(x) approaches minus infinity as x approaches x_0 from the right, and write $\lim_{x \to x_0^+} f(x) = -\infty$,

    if for every positive number B (or negative number $-B$) there exists a corresponding number $\delta > 0$ such
    that for all x, $x_0 < x < x_0 + \delta \Rightarrow f(x) < -B$.

    (c) We say that f(x) approaches minus infinity as x approaches x_0 from the left, and write $\lim_{x \to x_0^-} f(x) = -\infty$,

    if for every positive number B (or negative number $-B$) there exists a corresponding number $\delta > 0$ such
    that for all x, $x_0 - \delta < x < x_0 \Rightarrow f(x) < -B$.

53. For $B > 0$, $\frac{1}{x} < -B < 0 \Leftrightarrow -\frac{1}{x} > B > 0 \Leftrightarrow -x < \frac{1}{B} \Leftrightarrow -\frac{1}{B} < x$. Choose $\delta = \frac{1}{B}$. Then $-\delta < x < 0$

$\Rightarrow -\frac{1}{B} < x \Rightarrow \frac{1}{x} < -B$ so that $\lim_{x \to 0^-} \frac{1}{x} = -\infty$.

55. For $B > 0$, $\frac{1}{x-2} > B \Leftrightarrow 0 < x - 2 < \frac{1}{B}$. Choose $\delta = \frac{1}{B}$. Then $2 < x < 2 + \delta \Rightarrow 0 < x - 2 < \delta \Rightarrow 0 < x - 2 < \frac{1}{B}$

$\Rightarrow \frac{1}{x-2} > B > 0$ so that $\lim_{x \to 2^+} \frac{1}{x-2} = \infty$.

57. $y = \sec x + \frac{1}{x}$

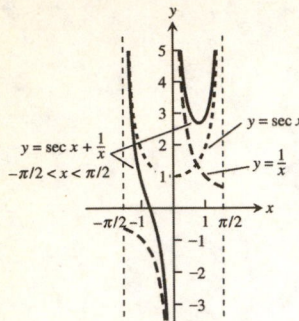

59. $y = \tan x + \frac{1}{x^2}$

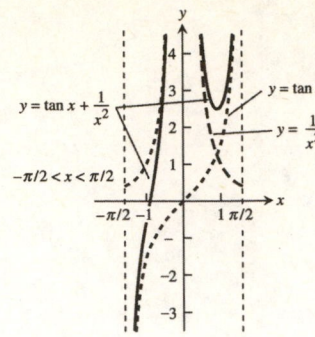

61. $y = \frac{x}{\sqrt{4 - x^2}}$

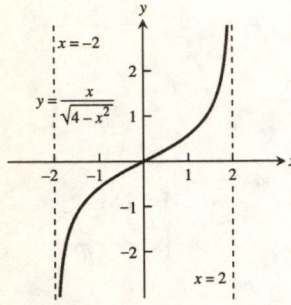

63. $y = x^{2/3} + \frac{1}{x^{1/3}}$

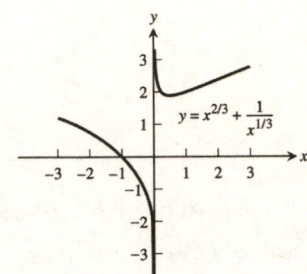

2.6 CONTINUITY

1. No, discontinuous at $x = 2$, not defined at $x = 2$

3. Continuous on $[-1, 3]$

5. (a) Yes   (b) Yes, $\lim\limits_{x \to -1^+} f(x) = 0$

   (c) Yes   (d) Yes

7. (a) No   (b) No

9. $f(2) = 0$, since $\lim\limits_{x \to 2^-} f(x) = -2(2) + 4 = 0 = \lim\limits_{x \to 2^+} f(x)$

11. Nonremovable discontinuity at $x = 1$ because $\lim\limits_{x \to 1} f(x)$ fails to exist ($\lim\limits_{x \to 1^-} f(x) = 1$ and $\lim\limits_{x \to 1^+} f(x) = 0$).
    Removable discontinuity at $x = 0$ by assigning the number $\lim\limits_{x \to 0} f(x) = 0$ to be the value of $f(0)$ rather than
    $f(0) = 1$.

13. Discontinuous only when $x - 2 = 0 \Rightarrow x = 2$

15. Discontinuous only when $x^2 - 4x + 3 = 0 \Rightarrow (x - 3)(x - 1) = 0 \Rightarrow x = 3$ or $x = 1$

17. Continuous everywhere. ($|x - 1| + \sin x$ defined for all x; limits exist and are equal to function values.)

19. Discontinuous only at $x = 0$

21. Discontinuous when 2x is an integer multiple of π, i.e., $2x = n\pi$, n an integer $\Rightarrow$ $x = \frac{n\pi}{2}$, n an integer, but continuous at all other x.

23. Discontinuous at odd integer multiples of $\frac{\pi}{2}$, i.e., $x = (2n-1)\frac{\pi}{2}$, n an integer, but continuous at all other x.

25. Discontinuous when $2x + 3 < 0$ or $x < -\frac{3}{2}$ $\Rightarrow$ continuous on the interval $\left[-\frac{3}{2}, \infty\right)$.

27. Continuous everywhere: $(2x-1)^{1/3}$ is defined for all x; limits exist and are equal to function values.

29. $\lim\limits_{x \to \pi} \sin(x - \sin x) = \sin(\pi - \sin \pi) = \sin(\pi - 0) = \sin \pi = 0$, and function continuous at $x = \pi$.

31. $\lim\limits_{y \to 1} \sec(y \sec^2 y - \tan^2 y - 1) = \lim\limits_{y \to 1} \sec(y \sec^2 y - \sec^2 y) = \lim\limits_{y \to 1} \sec((y-1)\sec^2 y) = \sec((1-1)\sec^2 1)$
    $= \sec 0 = 1$, , and function continuous at $y = 1$.

33. $\lim\limits_{t \to 0} \cos\left[\frac{\pi}{\sqrt{19 - 3\sec 2t}}\right] = \cos\left[\frac{\pi}{\sqrt{19 - 3\sec 0}}\right] = \cos\frac{\pi}{\sqrt{16}} = \cos\frac{\pi}{4} = \frac{\sqrt{2}}{2}$, and function continuous at $t = 0$.

35. $g(x) = \frac{x^2 - 9}{x - 3} = \frac{(x+3)(x-3)}{(x-3)} = x + 3, x \neq 3 \Rightarrow g(3) = \lim\limits_{x \to 3}(x+3) = 6$

37. $f(s) = \frac{s^3 - 1}{s^2 - 1} = \frac{(s^2 + s + 1)(s - 1)}{(s+1)(s-1)} = \frac{s^2 + s + 1}{s + 1}, s \neq 1 \Rightarrow f(1) = \lim\limits_{s \to 1}\left(\frac{s^2 + s + 1}{s + 1}\right) = \frac{3}{2}$

39. As defined, $\lim\limits_{x \to 3^-} f(x) = (3)^2 - 1 = 8$ and $\lim\limits_{x \to 3^+}(2a)(3) = 6a$. For f(x) to be continuous we must have
    $6a = 8 \Rightarrow a = \frac{4}{3}$.

41. The function can be extended: $f(0) \approx 2.3$.

43. The function cannot be extended to be continuous at $x = 0$. If $f(0) = 1$, it will be continuous from the right. Or if $f(0) = -1$, it will be continuous from the left.

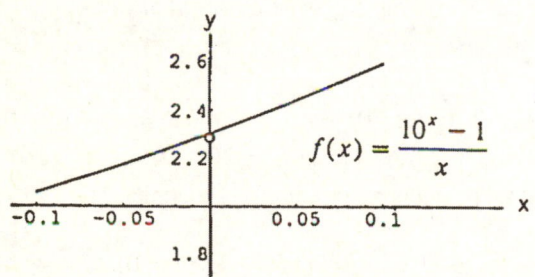

$$f(x) = \frac{10^x - 1}{x}$$

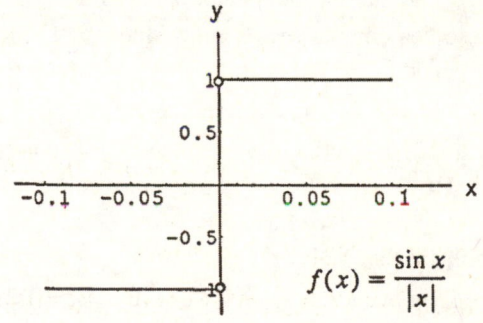

$$f(x) = \frac{\sin x}{|x|}$$

45. f(x) is continuous on $[0, 1]$ and $f(0) < 0$, $f(1) > 0$ $\Rightarrow$ by the Intermediate Value Theorem f(x) takes on every value between f(0) and f(1) $\Rightarrow$ the equation $f(x) = 0$ has at least one solution between $x = 0$ and $x = 1$.

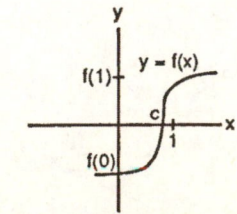

47. Let $f(x) = x^3 - 15x + 1$ which is continuous on $[-4, 4]$. Then $f(-4) = -3$, $f(-1) = 15$, $f(1) = -13$, and $f(4) = 5$. By the Intermediate Value Theorem, $f(x) = 0$ for some x in each of the intervals $-4 < x < -1$, $-1 < x < 1$, and $1 < x < 4$. That is, $x^3 - 15x + 1 = 0$ has three solutions in $[-4, 4]$. Since a polynomial of degree 3 can have at most 3 solutions, these are the only solutions.

49. Answers may vary. Note that f is continuous for every value of x.
    (a) $f(0) = 10$, $f(1) = 1^3 - 8(1) + 10 = 3$. Since $3 < \pi < 10$, by the Intermediate Value Theorem, there exists a c so that $0 < c < 1$ and $f(c) = \pi$.
    (b) $f(0) = 10$, $f(-4) = (-4)^3 - 8(-4) + 10 = -22$. Since $-22 < -\sqrt{3} < 10$, by the Intermediate Value Theorem, there exists a c so that $-4 < c < 0$ and $f(c) = -\sqrt{3}$.
    (c) $f(0) = 10$, $f(1000) = (1000)^3 - 8(1000) + 10 = 999,992,010$. Since $10 < 5,000,000 < 999,992,010$, by the Intermediate Value Theorem, there exists a c so that $0 < c < 1000$ and $f(c) = 5,000,000$.

51. Answers may vary. For example, $f(x) = \frac{\sin(x-2)}{x-2}$ is discontinuous at $x = 2$ because it is not defined there. However, the discontinuity can be removed because f has a limit (namely 1) as $x \to 2$.

53. (a) Suppose x_0 is rational $\Rightarrow f(x_0) = 1$. Choose $\epsilon = \frac{1}{2}$. For any $\delta > 0$ there is an irrational number x (actually infinitely many) in the interval $(x_0 - \delta, x_0 + \delta) \Rightarrow f(x) = 0$. Then $0 < |x - x_0| < \delta$ but $|f(x) - f(x_0)|$ $= 1 > \frac{1}{2} = \epsilon$, so $\lim_{x \to x_0} f(x)$ fails to exist $\Rightarrow$ f is discontinuous at x_0 rational.

   On the other hand, x_0 irrational $\Rightarrow f(x_0) = 0$ and there is a rational number x in $(x_0 - \delta, x_0 + \delta) \Rightarrow f(x)$ $= 1$. Again $\lim_{x \to x_0} f(x)$ fails to exist $\Rightarrow$ f is discontinuous at x_0 irrational. That is, f is discontinuous at every point.

   (b) f is neither right-continuous nor left-continuous at any point x_0 because in every interval $(x_0 - \delta, x_0)$ or $(x_0, x_0 + \delta)$ there exist both rational and irrational real numbers. Thus neither limits $\lim_{x \to x_0^-} f(x)$ and $\lim_{x \to x_0^+} f(x)$ exist by the same arguments used in part (a).

55. No. For instance, if $f(x) = 0$, $g(x) = \lceil x \rceil$, then $h(x) = 0 (\lceil x \rceil) = 0$ is continuous at $x = 0$ and $g(x)$ is not.

57. Yes, because of the Intermediate Value Theorem. If $f(a)$ and $f(b)$ did have different signs then f would have to equal zero at some point between a and b since f is continuous on $[a, b]$.

59. If $f(0) = 0$ or $f(1) = 1$, we are done (i.e., $c = 0$ or $c = 1$ in those cases). Then let $f(0) = a > 0$ and $f(1) = b < 1$ because $0 \le f(x) \le 1$. Define $g(x) = f(x) - x \Rightarrow$ g is continuous on $[0, 1]$. Moreover, $g(0) = f(0) - 0 = a > 0$ and $g(1) = f(1) - 1 = b - 1 < 0 \Rightarrow$ by the Intermediate Value Theorem there is a number c in $(0, 1)$ such that $g(c) = 0 \Rightarrow f(c) - c = 0$ or $f(c) = c$.

61. By Exercises 52 in Section 2.3, we have $\lim_{x \to c} f(x) = L \Leftrightarrow \lim_{h \to 0} f(c + h) = L$.

   Thus, $f(x)$ is continuous at $x = c \Leftrightarrow \lim_{x \to c} f(x) = f(c) \Leftrightarrow \lim_{h \to 0} f(c + h) = f(c)$.

63. $x \approx 1.8794, -1.5321, -0.3473$

65. $x \approx 1.7549$

67. $x \approx 3.5156$

69. $x \approx 0.7391$

2.7 TANGENTS AND DERIVATIVES

1. P_1: $m_1 = 1$, P_2: $m_2 = 5$

3. P_1: $m_1 = \frac{5}{2}$, P_2: $m_2 = -\frac{1}{2}$

5. $m = \lim\limits_{h \to 0} \frac{[4 - (-1+h)^2] - (4 - (-1)^2)}{h}$
   $= \lim\limits_{h \to 0} \frac{-(1 - 2h + h^2) + 1}{h} = \lim\limits_{h \to 0} \frac{h(2-h)}{h} = 2$;
   at $(-1, 3)$: $y = 3 + 2(x - (-1)) \Rightarrow y = 2x + 5$,
   tangent line

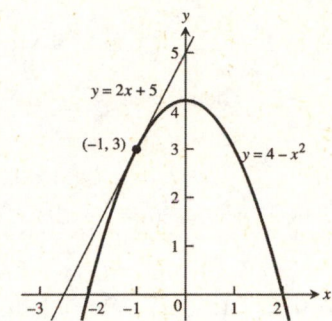

7. $m = \lim\limits_{h \to 0} \frac{2\sqrt{1+h} - 2\sqrt{1}}{h} = \lim\limits_{h \to 0} \frac{2\sqrt{1+h} - 2}{h} \cdot \frac{2\sqrt{1+h} + 2}{2\sqrt{1+h} + 2}$
   $= \lim\limits_{h \to 0} \frac{4(1+h) - 4}{2h\left(\sqrt{1+h} + 1\right)} = \lim\limits_{h \to 0} \frac{2}{\sqrt{1+h} + 1} = 1$;
   at $(1, 2)$: $y = 2 + 1(x - 1) \Rightarrow y = x + 1$, tangent line

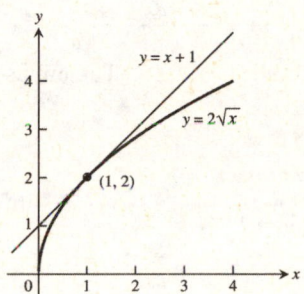

9. $m = \lim\limits_{h \to 0} \frac{(-2+h)^3 - (-2)^3}{h} = \lim\limits_{h \to 0} \frac{-8 + 12h - 6h^2 + h^3 + 8}{h}$
   $= \lim\limits_{h \to 0} (12 - 6h + h^2) = 12$;
   at $(-2, -8)$: $y = -8 + 12(x - (-2)) \Rightarrow y = 12x + 16$,
   tangent line

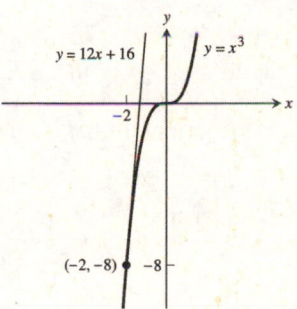

11. $m = \lim\limits_{h \to 0} \frac{[(2+h)^2 + 1] - 5}{h} = \lim\limits_{h \to 0} \frac{(5 + 4h + h^2) - 5}{h} = \lim\limits_{h \to 0} \frac{h(4+h)}{h} = 4$;
    at $(2, 5)$: $y - 5 = 4(x - 2)$, tangent line

13. $m = \lim\limits_{h \to 0} \frac{\frac{3+h}{(3+h)-2} - 3}{h} = \lim\limits_{h \to 0} \frac{(3+h) - 3(h+1)}{h(h+1)} = \lim\limits_{h \to 0} \frac{-2h}{h(h+1)} = -2$;
    at $(3, 3)$: $y - 3 = -2(x - 3)$, tangent line

15. $m = \lim\limits_{h \to 0} \frac{(2+h)^3 - 8}{h} = \lim\limits_{h \to 0} \frac{(8 + 12h + 6h^2 + h^3) - 8}{h} = \lim\limits_{h \to 0} \frac{h(12 + 6h + h^2)}{h} = 12$;
    at $(2, 8)$: $y - 8 = 12(t - 2)$, tangent line

17. $m = \lim\limits_{h \to 0} \frac{\sqrt{4+h} - 2}{h} = \lim\limits_{h \to 0} \frac{\sqrt{4+h} - 2}{h} \cdot \frac{\sqrt{4+h} + 2}{\sqrt{4+h} + 2} = \lim\limits_{h \to 0} \frac{(4+h) - 4}{h\left(\sqrt{4+h} + 2\right)} = \lim\limits_{h \to 0} \frac{h}{h\left(\sqrt{4+h} + 2\right)} = \frac{1}{\sqrt{4} + 2}$
    $= \frac{1}{4}$; at $(4, 2)$: $y - 2 = \frac{1}{4}(x - 4)$, tangent line

19. At $x = -1, y = 5 \Rightarrow m = \lim\limits_{h \to 0} \frac{5(-1+h)^2 - 5}{h} = \lim\limits_{h \to 0} \frac{5(1 - 2h + h^2) - 5}{h} = \lim\limits_{h \to 0} \frac{5h(-2+h)}{h} = -10$, slope

21. At $x = 3, y = \frac{1}{2} \Rightarrow m = \lim\limits_{h \to 0} \frac{\frac{1}{(3+h)-1} - \frac{1}{2}}{h} = \lim\limits_{h \to 0} \frac{2 - (2+h)}{2h(2+h)} = \lim\limits_{h \to 0} \frac{-h}{2h(2+h)} = -\frac{1}{4}$, slope

23. At a horizontal tangent the slope $m = 0 \Rightarrow 0 = m = \lim\limits_{h \to 0} \frac{[(x+h)^2 + 4(x+h) - 1] - (x^2 + 4x - 1)}{h}$

$= \lim\limits_{h \to 0} \frac{(x^2 + 2xh + h^2 + 4x + 4h - 1) - (x^2 + 4x - 1)}{h} = \lim\limits_{h \to 0} \frac{(2xh + h^2 + 4h)}{h} = \lim\limits_{h \to 0} (2x + h + 4) = 2x + 4;$

$2x + 4 = 0 \Rightarrow x = -2$. Then $f(-2) = 4 - 8 - 1 = -5 \Rightarrow (-2, -5)$ is the point on the graph where there is a horizontal tangent.

25. $-1 = m = \lim\limits_{h \to 0} \frac{\frac{1}{(x+h)-1} - \frac{1}{x-1}}{h} = \lim\limits_{h \to 0} \frac{(x-1) - (x+h-1)}{h(x-1)(x+h-1)} = \lim\limits_{h \to 0} \frac{-h}{h(x-1)(x+h-1)} = -\frac{1}{(x-1)^2}$

$\Rightarrow (x-1)^2 = 1 \Rightarrow x^2 - 2x = 0 \Rightarrow x(x-2) = 0 \Rightarrow x = 0$ or $x = 2$. If $x = 0$, then $y = -1$ and $m = -1$

$\Rightarrow y = -1 - (x - 0) = -(x+1)$. If $x = 2$, then $y = 1$ and $m = -1 \Rightarrow y = 1 - (x - 2) = -(x - 3)$.

27. $\lim\limits_{h \to 0} \frac{f(2+h) - f(2)}{h} = \lim\limits_{h \to 0} \frac{(100 - 4.9(2+h)^2) - (100 - 4.9(2)^2)}{h} = \lim\limits_{h \to 0} \frac{-4.9(4 + 4h + h^2) + 4.9(4)}{h}$

$= \lim\limits_{h \to 0} (-19.6 - 4.9h) = -19.6$. The minus sign indicates the object is falling <u>downward</u> at a speed of 19.6 m/sec.

29. $\lim\limits_{h \to 0} \frac{f(3+h) - f(3)}{h} = \lim\limits_{h \to 0} \frac{\pi(3+h)^2 - \pi(3)^2}{h} = \lim\limits_{h \to 0} \frac{\pi[9 + 6h + h^2 - 9]}{h} = \lim\limits_{h \to 0} \pi(6 + h) = 6\pi$

31. Slope at origin $= \lim\limits_{h \to 0} \frac{f(0+h) - f(0)}{h} = \lim\limits_{h \to 0} \frac{h^2 \sin\left(\frac{1}{h}\right)}{h} = \lim\limits_{h \to 0} h \sin\left(\frac{1}{h}\right) = 0 \Rightarrow$ yes, f(x) does have a tangent at the origin with slope 0.

33. $\lim\limits_{h \to 0^-} \frac{f(0+h) - f(0)}{h} = \lim\limits_{h \to 0^-} \frac{-1 - 0}{h} = \infty$, and $\lim\limits_{h \to 0^+} \frac{f(0+h) - f(0)}{h} = \lim\limits_{h \to 0^+} \frac{1 - 0}{h} = \infty$. Therefore,

$\lim\limits_{h \to 0} \frac{f(0+h) - f(0)}{h} = \infty \Rightarrow$ yes, the graph of f has a vertical tangent at the origin.

35. (a) The graph appears to have a cusp at $x = 0$.

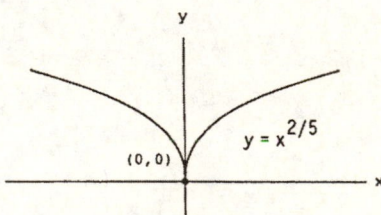

(b) $\lim\limits_{h \to 0^-} \frac{f(0+h) - f(0)}{h} = \lim\limits_{h \to 0^-} \frac{h^{2/5} - 0}{h} = \lim\limits_{h \to 0^-} \frac{1}{h^{3/5}} = -\infty$ and $\lim\limits_{h \to 0^+} \frac{1}{h^{3/5}} = \infty \Rightarrow$ limit does not exist

$\Rightarrow$ the graph of $y = x^{2/5}$ does not have a vertical tangent at $x = 0$.

37. (a) The graph appears to have a vertical tangent at $x = 0$.

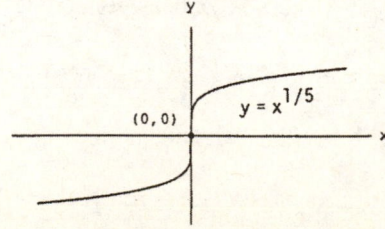

(b) $\lim\limits_{h \to 0} \frac{f(0+h) - f(0)}{h} = \lim\limits_{h \to 0} \frac{h^{1/5} - 0}{h} = \lim\limits_{h \to 0} \frac{1}{h^{4/5}} = \infty \Rightarrow y = x^{1/5}$ has a vertical tangent at $x = 0$.

39. (a) The graph appears to have a cusp at $x = 0$.

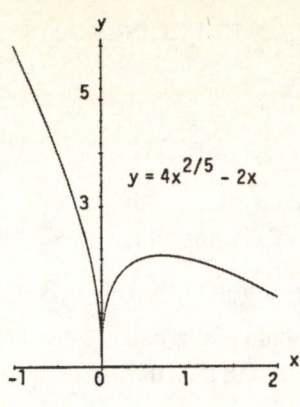

$y = 4x^{2/5} - 2x$

(b) $\lim\limits_{h \to 0^-} \dfrac{f(0+h) - f(0)}{h} = \lim\limits_{h \to 0^-} \dfrac{4h^{2/5} - 2h}{h} = \lim\limits_{h \to 0^-} \dfrac{4}{h^{3/5}} - 2 = -\infty$ and $\lim\limits_{h \to 0^+} \dfrac{4}{h^{3/5}} - 2 = \infty$

$\Rightarrow$ limit does not exist $\Rightarrow$ the graph of $y = 4x^{2/5} - 2x$ does not have a vertical tangent at $x = 0$.

41. (a) The graph appears to have a vertical tangent at $x = 1$ and a cusp at $x = 0$.

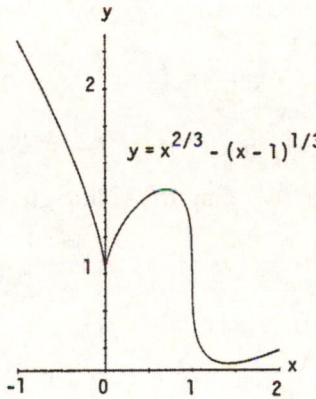

$y = x^{2/3} - (x - 1)^{1/3}$

(b) $x = 1$: $\lim\limits_{h \to 0} \dfrac{(1+h)^{2/3} - (1 + h - 1)^{1/3} - 1}{h} = \lim\limits_{h \to 0} \dfrac{(1+h)^{2/3} - h^{1/3} - 1}{h} = -\infty$

$\Rightarrow y = x^{2/3} - (x - 1)^{1/3}$ has a vertical tangent at $x = 1$;

$x = 0$: $\lim\limits_{h \to 0} \dfrac{f(0+h) - f(0)}{h} = \lim\limits_{h \to 0} \dfrac{h^{2/3} - (h - 1)^{1/3} - (-1)^{1/3}}{h} = \lim\limits_{h \to 0} \left[\dfrac{1}{h^{1/3}} - \dfrac{(h-1)^{1/3}}{h} + \dfrac{1}{h}\right]$

does not exist $\Rightarrow y = x^{2/3} - (x - 1)^{1/3}$ does not have a vertical tangent at $x = 0$.

43. (a) The graph appears to have a vertical tangent at $x = 0$.

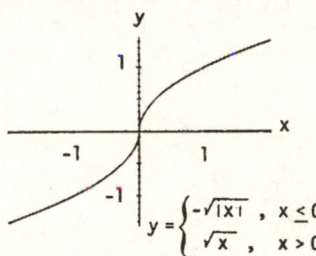

$y = \begin{cases} -\sqrt{|x|}, & x \le 0 \\ \sqrt{x}, & x > 0 \end{cases}$

(b) $\lim\limits_{h \to 0^+} \dfrac{f(0+h) - f(0)}{h} = \lim\limits_{x \to 0^+} \dfrac{\sqrt{h} - 0}{h} = \lim\limits_{h \to 0} \dfrac{1}{\sqrt{h}} = \infty$;

$\lim\limits_{h \to 0^-} \dfrac{f(0+h) - f(0)}{h} = \lim\limits_{h \to 0^-} \dfrac{-\sqrt{|h|} - 0}{h} = \lim\limits_{h \to 0^-} \dfrac{-\sqrt{|h|}}{-|h|} = \lim\limits_{h \to 0^-} \dfrac{1}{\sqrt{|h|}} = \infty$

$\Rightarrow y$ has a vertical tangent at $x = 0$.

CHAPTER 2 PRACTICE EXERCISES

1. At $x = -1$: $\lim\limits_{x \to -1^-} f(x) = \lim\limits_{x \to -1^+} f(x) = 1$

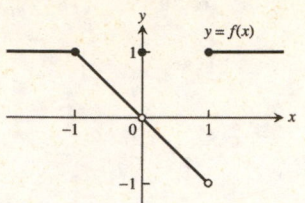

$\Rightarrow \lim\limits_{x \to -1} f(x) = 1 = f(-1)$

$\Rightarrow$ f is continuous at $x = -1$.

At $x = 0$: $\lim\limits_{x \to 0^-} f(x) = \lim\limits_{x \to 0^+} f(x) = 0 \Rightarrow \lim\limits_{x \to 0} f(x) = 0$.

But $f(0) = 1 \neq \lim\limits_{x \to 0} f(x)$

$\Rightarrow$ f is discontinuous at $x = 0$.

If we define $f(0) = 0$, then the discontinuity at $x = 0$ is removable.

At $x = 1$: $\lim\limits_{x \to 1^-} f(x) = -1$ and $\lim\limits_{x \to 1^+} f(x) = 1$

$\Rightarrow \lim\limits_{x \to 1} f(x)$ does not exist

$\Rightarrow$ f is discontinuous at $x = 1$.

3. (a) $\lim\limits_{t \to t_0} (3f(t)) = 3 \lim\limits_{t \to t_0} f(t) = 3(-7) = -21$

(b) $\lim\limits_{t \to t_0} (f(t))^2 = \left(\lim\limits_{t \to t_0} f(t) \right)^2 = (-7)^2 = 49$

(c) $\lim\limits_{t \to t_0} (f(t) \cdot g(t)) = \lim\limits_{t \to t_0} f(t) \cdot \lim\limits_{t \to t_0} g(t) = (-7)(0) = 0$

(d) $\lim\limits_{t \to t_0} \frac{f(t)}{g(t)-7} = \frac{\lim\limits_{t \to t_0} f(t)}{\lim\limits_{t \to t_0} (g(t) - 7)} = \frac{\lim\limits_{t \to t_0} f(t)}{\lim\limits_{t \to t_0} g(t) - \lim\limits_{t \to t_0} 7} = \frac{-7}{0-7} = 1$

(e) $\lim\limits_{t \to t_0} \cos(g(t)) = \cos\left(\lim\limits_{t \to t_0} g(t) \right) = \cos 0 = 1$

(f) $\lim\limits_{t \to t_0} |f(t)| = \left| \lim\limits_{t \to t_0} f(t) \right| = |-7| = 7$

(g) $\lim\limits_{t \to t_0} (f(t) + g(t)) = \lim\limits_{t \to t_0} f(t) + \lim\limits_{t \to t_0} g(t) = -7 + 0 = -7$

(h) $\lim\limits_{t \to t_0} \left(\frac{1}{f(t)} \right) = \frac{1}{\lim\limits_{t \to t_0} f(t)} = \frac{1}{-7} = -\frac{1}{7}$

5. Since $\lim\limits_{x \to 0} x = 0$ we must have that $\lim\limits_{x \to 0} (4 - g(x)) = 0$. Otherwise, if $\lim\limits_{x \to 0} (4 - g(x))$ is a finite positive number, we would have $\lim\limits_{x \to 0^-} \left[\frac{4-g(x)}{x} \right] = -\infty$ and $\lim\limits_{x \to 0^+} \left[\frac{4-g(x)}{x} \right] = \infty$ so the limit could not equal 1 as $x \to 0$. Similar reasoning holds if $\lim\limits_{x \to 0} (4 - g(x))$ is a finite negative number. We conclude that $\lim\limits_{x \to 0} g(x) = 4$.

7. (a) $\lim\limits_{x \to c} f(x) = \lim\limits_{x \to c} x^{1/3} = c^{1/3} = f(c)$ for every real number $c \Rightarrow$ f is continuous on $(-\infty, \infty)$.

(b) $\lim\limits_{x \to c} g(x) = \lim\limits_{x \to c} x^{3/4} = c^{3/4} = g(c)$ for every nonnegative real number $c \Rightarrow$ g is continuous on $[0, \infty)$.

(c) $\lim\limits_{x \to c} h(x) = \lim\limits_{x \to c} x^{-2/3} = \frac{1}{c^{2/3}} = h(c)$ for every nonzero real number $c \Rightarrow$ h is continuous on $(-\infty, 0)$ and $(-\infty, \infty)$.

(d) $\lim\limits_{x \to c} k(x) = \lim\limits_{x \to c} x^{-1/6} = \frac{1}{c^{1/6}} = k(c)$ for every positive real number $c \Rightarrow$ k is continuous on $(0, \infty)$

9. (a) $\lim\limits_{x \to 0} \frac{x^2 - 4x + 4}{x^3 + 5x^2 - 14x} = \lim\limits_{x \to 0} \frac{(x-2)(x-2)}{x(x+7)(x-2)} = \lim\limits_{x \to 0} \frac{x-2}{x(x+7)}$, $x \neq 2$; the limit does not exist because

$\lim\limits_{x \to 0^-} \frac{x-2}{x(x+7)} = \infty$ and $\lim\limits_{x \to 0^+} \frac{x-2}{x(x+7)} = -\infty$

(b) $\lim\limits_{x \to 2} \frac{x^2 - 4x + 4}{x^3 + 5x^2 - 14x} = \lim\limits_{x \to 2} \frac{(x-2)(x-2)}{x(x+7)(x-2)} = \lim\limits_{x \to 2} \frac{x-2}{x(x+7)}$, $x \neq 2$, and $\lim\limits_{x \to 2} \frac{x-2}{x(x+7)} = \frac{0}{2(9)} = 0$

11. $\lim\limits_{x \to 1} \frac{1-\sqrt{x}}{1-x} = \lim\limits_{x \to 1} \frac{1-\sqrt{x}}{(1-\sqrt{x})(1+\sqrt{x})} = \lim\limits_{x \to 1} \frac{1}{1+\sqrt{x}} = \frac{1}{2}$

13. $\lim\limits_{h \to 0} \frac{(x+h)^2 - x^2}{h} = \lim\limits_{h \to 0} \frac{(x^2 + 2hx + h^2) - x^2}{h} = \lim\limits_{h \to 0} (2x + h) = 2x$

15. $\lim\limits_{x \to 0} \frac{\frac{1}{2+x} - \frac{1}{2}}{x} = \lim\limits_{x \to 0} \frac{2-(2+x)}{2x(2+x)} = \lim\limits_{x \to 0} \frac{-1}{4+2x} = -\frac{1}{4}$

17. $\lim\limits_{x \to 0^+} [4\, g(x)]^{1/3} = 2 \Rightarrow \left[\lim\limits_{x \to 0^+} 4\, g(x)\right]^{1/3} = 2 \Rightarrow \lim\limits_{x \to 0^+} 4\, g(x) = 8$, since $2^3 = 8$. Then $\lim\limits_{x \to 0^+} g(x) = 2$.

19. $\lim\limits_{x \to 1} \frac{3x^2 + 1}{g(x)} = \infty \Rightarrow \lim\limits_{x \to 1} g(x) = 0$ since $\lim\limits_{x \to 1} (3x^2 + 1) = 4$

21. $\lim\limits_{x \to \infty} \frac{2x+3}{5x+7} = \lim\limits_{x \to \infty} \frac{2 + \frac{3}{x}}{5 + \frac{7}{x}} = \frac{2+0}{5+0} = \frac{2}{5}$

23. $\lim\limits_{x \to -\infty} \frac{x^2 - 4x + 8}{3x^3} = \lim\limits_{x \to -\infty} \left(\frac{1}{3x} - \frac{4}{3x^2} + \frac{8}{3x^3}\right) = 0 - 0 + 0 = 0$

25. $\lim\limits_{x \to -\infty} \frac{x^2 - 7x}{x+1} = \lim\limits_{x \to -\infty} \frac{x-7}{1 + \frac{1}{x}} = -\infty$

27. $\lim\limits_{x \to \infty} \frac{|\sin x|}{\lfloor x \rfloor} \leq \lim\limits_{x \to \infty} \frac{1}{\lfloor x \rfloor} = 0$ since int $x \to \infty$ as $x \to \infty \Rightarrow \lim\limits_{x \to \infty} \frac{|\sin x|}{\lfloor x \rfloor} = 0$.

29. $\lim\limits_{x \to \infty} \frac{x + \sin x + 2\sqrt{x}}{x + \sin x} = \lim\limits_{x \to \infty} \frac{1 + \frac{\sin x}{x} + \frac{2}{\sqrt{x}}}{1 + \frac{\sin x}{x}} = \frac{1+0+0}{1+0} = 1$

31. At $x = -1$: $\lim\limits_{x \to -1^-} f(x) = \lim\limits_{x \to -1^-} \frac{x(x^2 - 1)}{|x^2 - 1|}$

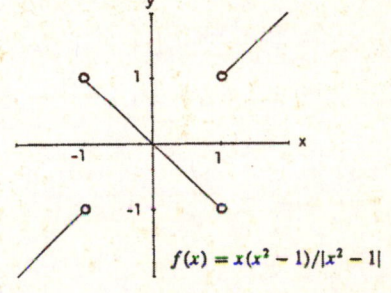

$f(x) = x(x^2 - 1)/|x^2 - 1|$

   $= \lim\limits_{x \to -1^-} \frac{x(x^2 - 1)}{x^2 - 1} = \lim\limits_{x \to -1^-} x = -1$, and

   $\lim\limits_{x \to -1^+} f(x) = \lim\limits_{x \to -1^+} \frac{x(x^2 - 1)}{|x^2 - 1|} = \lim\limits_{x \to -1^+} \frac{x(x^2 - 1)}{-(x^2 - 1)}$

   $= \lim\limits_{x \to -1} (-x) = -(-1) = 1$. Since

   $\lim\limits_{x \to -1^-} f(x) \neq \lim\limits_{x \to -1^+} f(x)$

   $\Rightarrow \lim\limits_{x \to -1} f(x)$ does not exist, the function f <u>cannot</u> be

   extended to a continuous function at $x = -1$.

   At $x = 1$: $\lim\limits_{x \to 1^-} f(x) = \lim\limits_{x \to 1^-} \frac{x(x^2 - 1)}{|x^2 - 1|} = \lim\limits_{x \to 1^-} \frac{x(x^2 - 1)}{-(x^2 - 1)} = \lim\limits_{x \to 1^-} (-x) = -1$, and

   $\lim\limits_{x \to 1^+} f(x) = \lim\limits_{x \to 1^+} \frac{x(x^2 - 1)}{|x^2 - 1|} = \lim\limits_{x \to 1^+} \frac{x(x^2 - 1)}{x^2 - 1} = \lim\limits_{x \to 1^+} x = 1$. Again $\lim\limits_{x \to 1} f(x)$ does not exist so f

   <u>cannot</u> be extended to a continuous function at $x = 1$ either.

33. Yes, f does have a continuous extension to $a = 1$:

   define $f(1) = \lim\limits_{x \to 1} \frac{x-1}{x - \sqrt[4]{x}} = \frac{4}{3}$.

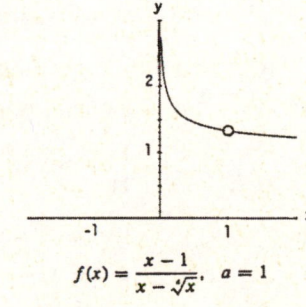

$f(x) = \dfrac{x-1}{x - \sqrt[4]{x}}, \quad a = 1$

35. From the graph we see that $\lim\limits_{t \to 0^-} h(t) \neq \lim\limits_{t \to 0^+} h(t)$

so h <u>cannot</u> be extended to a continuous function

at $a = 0$.

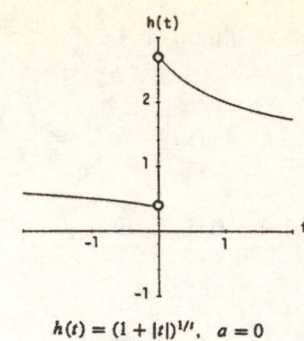

$h(t) = (1 + |t|)^{1/t}, \quad a = 0$

37. (a) $f(-1) = -1$ and $f(2) = 5 \Rightarrow$ f has a root between -1 and 2 by the Intermediate Value Theorem.

(b), (c) root is 1.32471795724

CHAPTER 2 ADDITIONAL AND ADVANCED EXERCISES

1. (a)

x	0.1	0.01	0.001	0.0001	0.00001
x^x	0.7943	0.9550	0.9931	0.9991	0.9999

Apparently, $\lim\limits_{x \to 0^+} x^x = 1$

(b)

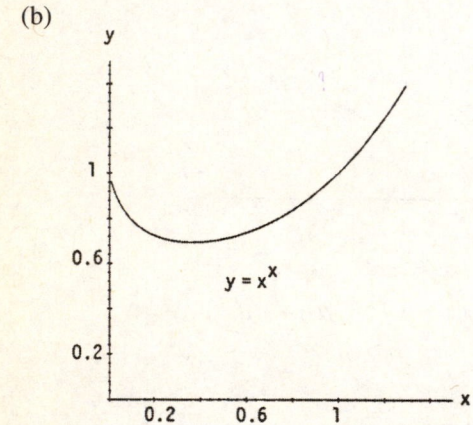

3. $\lim\limits_{v \to c^-} L = \lim\limits_{v \to c^-} L_0 \sqrt{1 - \frac{v^2}{c^2}} = L_0 \sqrt{1 - \frac{\lim\limits_{v \to c^-} v^2}{c^2}} = L_0 \sqrt{1 - \frac{c^2}{c^2}} = 0$

The left-hand limit was needed because the function L is undefined if $v > c$ (the rocket cannot move faster than the speed of light).

5. $|10 + (t - 70) \times 10^{-4} - 10| < 0.0005 \Rightarrow |(t - 70) \times 10^{-4}| < 0.0005 \Rightarrow -0.0005 < (t - 70) \times 10^{-4} < 0.0005$

$\Rightarrow -5 < t - 70 < 5 \Rightarrow 65° < t < 75° \Rightarrow$ Within 5° F.

7. Show $\lim\limits_{x \to 1} f(x) = \lim\limits_{x \to 1} (x^2 - 7) = -6 = f(1)$.

Step 1: $|(x^2 - 7) + 6| < \epsilon \Rightarrow -\epsilon < x^2 - 1 < \epsilon \Rightarrow 1 - \epsilon < x^2 < 1 + \epsilon \Rightarrow \sqrt{1 - \epsilon} < x < \sqrt{1 + \epsilon}$.

Step 2: $|x - 1| < \delta \Rightarrow -\delta < x - 1 < \delta \Rightarrow -\delta + 1 < x < \delta + 1$.

Then $-\delta + 1 = \sqrt{1 - \epsilon}$ or $\delta + 1 = \sqrt{1 + \epsilon}$. Choose $\delta = \min\left\{1 - \sqrt{1 - \epsilon}, \sqrt{1 + \epsilon} - 1\right\}$, then

$0 < |x - 1| < \delta \Rightarrow |(x^2 - 7) - 6| < \epsilon$ and $\lim\limits_{x \to 1} f(x) = -6$. By the continuity test, f(x) is continuous at $x = 1$.

9. Show $\lim\limits_{x \to 2} h(x) = \lim\limits_{x \to 2} \sqrt{2x - 3} = 1 = h(2)$.

Step 1: $\left|\sqrt{2x - 3} - 1\right| < \epsilon \Rightarrow -\epsilon < \sqrt{2x - 3} - 1 < \epsilon \Rightarrow 1 - \epsilon < \sqrt{2x - 3} < 1 + \epsilon \Rightarrow \frac{(1 - \epsilon)^2 + 3}{2} < x < \frac{(1 + \epsilon)^2 + 3}{2}$.

Step 2: $|x - 2| < \delta \Rightarrow -\delta < x - 2 < \delta$ or $-\delta + 2 < x < \delta + 2$.

Then $-\delta + 2 = \frac{(1-\epsilon)^2 + 3}{2} \Rightarrow \delta = 2 - \frac{(1-\epsilon)^2 + 3}{2} = \frac{1 - (1-\epsilon)^2}{2} = \epsilon - \frac{\epsilon^2}{2}$, or $\delta + 2 = \frac{(1+\epsilon)^2 + 3}{2}$

$\Rightarrow \delta = \frac{(1+\epsilon)^2 + 3}{2} - 2 = \frac{(1+\epsilon)^2 - 1}{2} = \epsilon + \frac{\epsilon^2}{2}$. Choose $\delta = \epsilon - \frac{\epsilon^2}{2}$, the smaller of the two values. Then,

$0 < |x - 2| < \delta \Rightarrow \left|\sqrt{2x - 3} - 1\right| < \epsilon$, so $\lim_{x \to 2} \sqrt{2x - 3} = 1$. By the continuity test, h(x) is continuous at $x = 2$.

11. Suppose L_1 and L_2 are two different limits. Without loss of generality assume $L_2 > L_1$. Let $\epsilon = \frac{1}{3}(L_2 - L_1)$.

Since $\lim_{x \to x_0} f(x) = L_1$ there is a $\delta_1 > 0$ such that $0 < |x - x_0| < \delta_1 \Rightarrow |f(x) - L_1| < \epsilon \Rightarrow -\epsilon < f(x) - L_1 < \epsilon$

$\Rightarrow -\frac{1}{3}(L_2 - L_1) + L_1 < f(x) < \frac{1}{3}(L_2 - L_1) + L_1 \Rightarrow 4L_1 - L_2 < 3f(x) < 2L_1 + L_2$. Likewise, $\lim_{x \to x_0} f(x) = L_2$

so there is a δ_2 such that $0 < |x - x_0| < \delta_2 \Rightarrow |f(x) - L_2| < \epsilon \Rightarrow -\epsilon < f(x) - L_2 < \epsilon$

$\Rightarrow -\frac{1}{3}(L_2 - L_1) + L_2 < f(x) < \frac{1}{3}(L_2 - L_1) + L_2 \Rightarrow 2L_2 + L_1 < 3f(x) < 4L_2 - L_1$

$\Rightarrow L_1 - 4L_2 < -3f(x) < -2L_2 - L_1$. If $\delta = \min\{\delta_1, \delta_2\}$ both inequalities must hold for $0 < |x - x_0| < \delta$:

$\left.\begin{array}{l} 4L_1 - L_2 < 3f(x) < 2L_1 + L_2 \\ L_1 - 4L_2 < -3f(x) < -2L_2 - L_1 \end{array}\right\} \Rightarrow 5(L_1 - L_2) < 0 < L_1 - L_2$. That is, $L_1 - L_2 < 0$ and $L_1 - L_2 > 0$,

a contradiction.

13. (a) Since $x \to 0^+, 0 < x^3 < x < 1 \Rightarrow (x^3 - x) \to 0^- \Rightarrow \lim_{x \to 0^+} f(x^3 - x) = \lim_{y \to 0^-} f(y) = B$ where $y = x^3 - x$.

(b) Since $x \to 0^-, -1 < x < x^3 < 0 \Rightarrow (x^3 - x) \to 0^+ \Rightarrow \lim_{x \to 0^-} f(x^3 - x) = \lim_{y \to 0^+} f(y) = A$ where $y = x^3 - x$.

(c) Since $x \to 0^+, 0 < x^4 < x^2 < 1 \Rightarrow (x^2 - x^4) \to 0^+ \Rightarrow \lim_{x \to 0^+} f(x^2 - x^4) = \lim_{y \to 0^+} f(y) = A$ where $y = x^2 - x^4$.

(d) Since $x \to 0^-, -1 < x < 0 \Rightarrow 0 < x^4 < x^2 < 1 \Rightarrow (x^2 - x^4) \to 0^+ \Rightarrow \lim_{x \to 0^+} f(x^2 - x^4) = A$ as in part (c).

15. Show $\lim_{x \to -1} f(x) = \lim_{x \to -1} \frac{x^2 - 1}{x + 1} = \lim_{x \to -1} \frac{(x+1)(x-1)}{(x+1)} = -2, x \neq -1$.

Define the continuous extension of f(x) as $F(x) = \begin{cases} \frac{x^2 - 1}{x+1}, & x \neq -1 \\ -2, & x = -1 \end{cases}$. We now prove the limit of f(x) as $x \to -1$

exists and has the correct value.

Step 1: $\left|\frac{x^2 - 1}{x+1} - (-2)\right| < \epsilon \Rightarrow -\epsilon < \frac{(x+1)(x-1)}{(x+1)} + 2 < \epsilon \Rightarrow -\epsilon < (x - 1) + 2 < \epsilon, x \neq -1 \Rightarrow -\epsilon - 1 < x < \epsilon - 1$.

Step 2: $|x - (-1)| < \delta \Rightarrow -\delta < x + 1 < \delta \Rightarrow -\delta - 1 < x < \delta - 1$.

Then $-\delta - 1 = -\epsilon - 1 \Rightarrow \delta = \epsilon$, or $\delta - 1 = \epsilon - 1 \Rightarrow \delta = \epsilon$. Choose $\delta = \epsilon$. Then $0 < |x - (-1)| < \delta$

$\Rightarrow \left|\frac{x^2 - 1}{x+1} - (-2)\right| < \epsilon \Rightarrow \lim_{x \to -1} F(x) = -2$. Since the conditions of the continuity test are met by F(x), then f(x) has a

continuous extension to F(x) at $x = -1$.

17. (a) Let $\epsilon > 0$ be given. If x is rational, then $f(x) = x \Rightarrow |f(x) - 0| = |x - 0| < \epsilon \Leftrightarrow |x - 0| < \epsilon$; i.e., choose

$\delta = \epsilon$. Then $|x - 0| < \delta \Rightarrow |f(x) - 0| < \epsilon$ for x rational. If x is irrational, then $f(x) = 0 \Rightarrow |f(x) - 0| < \epsilon$

$\Leftrightarrow 0 < \epsilon$ which is true no matter how close irrational x is to 0, so again we can choose $\delta = \epsilon$. In either case,

given $\epsilon > 0$ there is a $\delta = \epsilon > 0$ such that $0 < |x - 0| < \delta \Rightarrow |f(x) - 0| < \epsilon$. Therefore, f is continuous at

$x = 0$.

(b) Choose $x = c > 0$. Then within any interval $(c - \delta, c + \delta)$ there are both rational and irrational numbers.

If c is rational, pick $\epsilon = \frac{c}{2}$. No matter how small we choose $\delta > 0$ there is an irrational number x in

$(c - \delta, c + \delta) \Rightarrow |f(x) - f(c)| = |0 - c| = c > \frac{c}{2} = \epsilon$. That is, f is not continuous at any rational $c > 0$. On

the other hand, suppose c is irrational $\Rightarrow f(c) = 0$. Again pick $\epsilon = \frac{c}{2}$. No matter how small we choose $\delta > 0$

there is a rational number x in $(c - \delta, c + \delta)$ with $|x - c| < \frac{c}{2} = \epsilon \Leftrightarrow \frac{c}{2} < x < \frac{3c}{2}$. Then $|f(x) - f(c)| = |x - 0|$

$= |x| > \frac{c}{2} = \epsilon \Rightarrow$ f is not continuous at any irrational $c > 0$.

If $x = c < 0$, repeat the argument picking $\epsilon = \frac{|c|}{2} = \frac{-c}{2}$. Therefore f fails to be continuous at any

nonzero value $x = c$.

19. Yes. Let R be the radius of the equator (earth) and suppose at a fixed instant of time we label noon as the zero point, 0, on the equator $\Rightarrow 0 + \pi R$ represents the midnight point (at the same exact time). Suppose x_1 is a point on the equator "just after" noon $\Rightarrow x_1 + \pi R$ is simultaneously "just after" midnight. It seems reasonable that the temperature T at a point just after noon is hotter than it would be at the diametrically opposite point just after midnight: That is, $T(x_1) - T(x_1 + \pi R) > 0$. At exactly the same moment in time pick x_2 to be a point just before midnight $\Rightarrow x_2 + \pi R$ is just before noon. Then $T(x_2) - T(x_2 + \pi R) < 0$. Assuming the temperature function T is continuous along the equator (which is reasonable), the Intermediate Value Theorem says there is a point c between 0 (noon) and πR (simultaneously midnight) such that $T(c) - T(c + \pi R) = 0$; i.e., there is always a pair of antipodal points on the earth's equator where the temperatures are the same.

21. (a) At $x = 0$: $\displaystyle\lim_{a \to 0} r_+(a) = \lim_{a \to 0} \frac{-1+\sqrt{1+a}}{a} = \lim_{a \to 0} \left(\frac{-1+\sqrt{1+a}}{a}\right)\left(\frac{-1-\sqrt{1+a}}{-1-\sqrt{1+a}}\right)$

$\displaystyle = \lim_{a \to 0} \frac{1-(1+a)}{a\left(-1-\sqrt{1+a}\right)} = \frac{-1}{-1-\sqrt{1+0}} = \frac{1}{2}$

At $x = -1$: $\displaystyle\lim_{a \to -1^+} r_+(a) = \lim_{a \to -1^+} \frac{1-(1+a)}{a\left(-1-\sqrt{1+a}\right)} = \lim_{a \to -1} \frac{-a}{a\left(-1-\sqrt{1+a}\right)} = \frac{-1}{-1-\sqrt{0}} = 1$

(b) At $x = 0$: $\displaystyle\lim_{a \to 0^-} r_-(a) = \lim_{a \to 0^-} \frac{-1-\sqrt{1+a}}{a} = \lim_{a \to 0^-} \left(\frac{-1-\sqrt{1+a}}{a}\right)\left(\frac{-1+\sqrt{1+a}}{-1+\sqrt{1+a}}\right)$

$\displaystyle = \lim_{a \to 0^-} \frac{1-(1+a)}{a\left(-1+\sqrt{1+a}\right)} = \lim_{a \to 0^-} \frac{-a}{a\left(-1+\sqrt{1+a}\right)} = \lim_{a \to 0^-} \frac{-1}{-1+\sqrt{1+a}} = \infty$ (because the

denominator is always negative); $\displaystyle\lim_{a \to 0^+} r_-(a) = \lim_{a \to 0^+} \frac{-1}{-1+\sqrt{1+a}} = -\infty$ (because the denominator

is always positive). Therefore, $\displaystyle\lim_{a \to 0} r_-(a)$ does not exist.

At $x = -1$: $\displaystyle\lim_{a \to -1^+} r_-(a) = \lim_{a \to -1^+} \frac{-1-\sqrt{1+a}}{a} = \lim_{a \to -1^+} \frac{-1}{-1+\sqrt{1+a}} = 1$

(c)

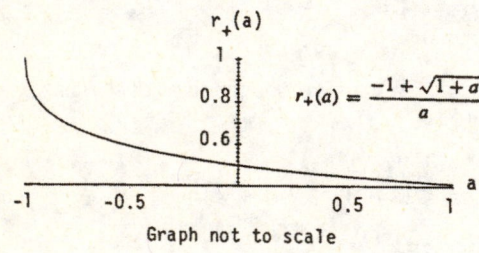

$r_+(a) = \dfrac{-1+\sqrt{1+a}}{a}$

Graph not to scale

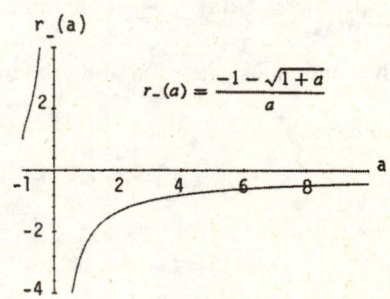

$r_-(a) = \dfrac{-1-\sqrt{1+a}}{a}$

(d)

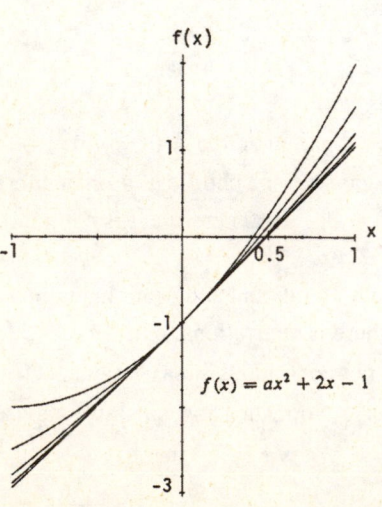

$f(x) = ax^2 + 2x - 1$

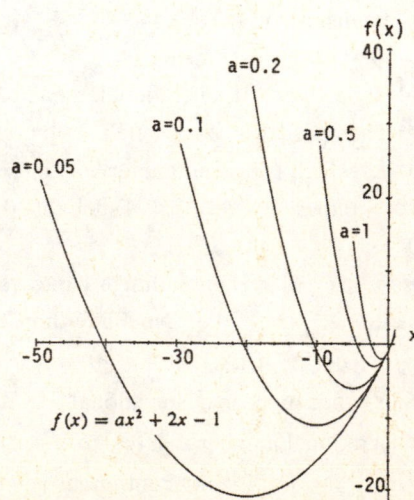

$f(x) = ax^2 + 2x - 1$

23. (a) The function f is bounded on D if $f(x) \geq M$ and $f(x) \leq N$ for all x in D. This means $M \leq f(x) \leq N$ for all x in D. Choose B to be max $\{|M|, |N|\}$. Then $|f(x)| \leq B$. On the other hand, if $|f(x)| \leq B$, then $-B \leq f(x) \leq B \Rightarrow f(x) \geq -B$ and $f(x) \leq B \Rightarrow f(x)$ is bounded on D with $N = B$ an upper bound and $M = -B$ a lower bound.

(b) Assume $f(x) \leq N$ for all x and that $L > N$. Let $\epsilon = \frac{L-N}{2}$. Since $\lim_{x \to x_0} f(x) = L$ there is a $\delta > 0$ such that

$$0 < |x - x_0| < \delta \Rightarrow |f(x) - L| < \epsilon \Leftrightarrow L - \epsilon < f(x) < L + \epsilon \Leftrightarrow L - \frac{L-N}{2} < f(x) < L + \frac{L-N}{2}$$

$\Leftrightarrow \frac{L+N}{2} < f(x) < \frac{3L-N}{2}$. But $L > N \Rightarrow \frac{L+N}{2} > N \Rightarrow N < f(x)$ contrary to the boundedness assumption $f(x) \leq N$. This contradiction proves $L \leq N$.

(c) Assume $M \leq f(x)$ for all x and that $L < M$. Let $\epsilon = \frac{M-L}{2}$. As in part (b), $0 < |x - x_0| < \delta$

$\Rightarrow L - \frac{M-L}{2} < f(x) < L + \frac{M-L}{2} \Leftrightarrow \frac{3L-M}{2} < f(x) < \frac{M+L}{2} < M$, a contradiction.

25. $\lim_{x \to 0} = \frac{\sin(1 - \cos x)}{x} = \lim_{x \to 0} \frac{\sin(1 - \cos x)}{1 - \cos x} \cdot \frac{1 - \cos x}{x} \cdot \frac{1 + \cos x}{1 + \cos x} = \lim_{x \to 0} \frac{\sin(1 - \cos x)}{1 - \cos x} \cdot \lim_{x \to 0} \frac{1 - \cos^2 x}{x(1 + \cos x)} = 1 \cdot \lim_{x \to 0} \frac{\sin^2 x}{x(1 + \cos x)}$

$= \lim_{x \to 0} \frac{\sin x}{x} \cdot \frac{\sin x}{1 + \cos x} = 1 \cdot \left(\frac{0}{2}\right) = 0.$

27. $\lim_{x \to 0} \frac{\sin(\sin x)}{x} = \lim_{x \to 0} \frac{\sin(\sin x)}{\sin x} \cdot \frac{\sin x}{x} = \lim_{x \to 0} \frac{\sin(\sin x)}{\sin x} \cdot \lim_{x \to 0} \frac{\sin x}{x} = 1 \cdot 1 = 1.$

29. $\lim_{x \to 2} \frac{\sin(x^2 - 4)}{x - 2} = \lim_{x \to 2} \frac{\sin(x^2 - 4)}{x^2 - 4} \cdot (x + 2) = \lim_{x \to 2} \frac{\sin(x^2 - 4)}{x^2 - 4} \cdot \lim_{x \to 2} (x + 2) = 1 \cdot 4 = 4$

NOTES:

CHAPTER 3 DIFFERENTIATION

3.1 THE DERIVATIVE AS A FUNCTION

1. Step 1: $f(x) = 4 - x^2$ and $f(x + h) = 4 - (x + h)^2$

   Step 2: $\frac{f(x+h) - f(x)}{h} = \frac{[4 - (x+h)^2] - (4 - x^2)}{h} = \frac{(4 - x^2 - 2xh - h^2) - 4 + x^2}{h} = \frac{-2xh - h^2}{h} = \frac{h(-2x - h)}{h}$
   $= -2x - h$

   Step 3: $f'(x) = \lim_{h \to 0} (-2x - h) = -2x; f'(-3) = 6, f'(0) = 0, f'(1) = -2$

3. Step 1: $g(t) = \frac{1}{t^2}$ and $g(t + h) = \frac{1}{(t+h)^2}$

   Step 2: $\frac{g(t+h) - g(t)}{h} = \frac{\frac{1}{(t+h)^2} - \frac{1}{t^2}}{h} = \frac{\left(\frac{t^2 - (t+h)^2}{(t+h)^2 \cdot t^2}\right)}{h} = \frac{t^2 - (t^2 + 2th + h^2)}{(t+h)^2 \cdot t^2 \cdot h} = \frac{-2th - h^2}{(t+h)^2 t^2 h}$
   $= \frac{h(-2t - h)}{(t+h)^2 t^2 h} = \frac{-2t - h}{(t+h)^2 t^2}$

   Step 3: $g'(t) = \lim_{h \to 0} \frac{-2t - h}{(t+h)^2 t^2} = \frac{-2t}{t^2 \cdot t^2} = \frac{-2}{t^3}; g'(-1) = 2, g'(2) = -\frac{1}{4}, g'\left(\sqrt{3}\right) = -\frac{2}{3\sqrt{3}}$

5. Step 1: $p(\theta) = \sqrt{3\theta}$ and $p(\theta + h) = \sqrt{3(\theta + h)}$

   Step 2: $\frac{p(\theta+h) - p(\theta)}{h} = \frac{\sqrt{3(\theta+h)} - \sqrt{3\theta}}{h} = \frac{\left(\sqrt{3\theta + 3h} - \sqrt{3\theta}\right)}{h} \cdot \frac{\left(\sqrt{3\theta + 3h} + \sqrt{3\theta}\right)}{\left(\sqrt{3\theta + 3h} + \sqrt{3\theta}\right)} = \frac{(3\theta + 3h) - 3\theta}{h\left(\sqrt{3\theta + 3h} + \sqrt{3\theta}\right)}$

   $= \frac{3h}{h\left(\sqrt{3\theta + 3h} + \sqrt{3\theta}\right)} = \frac{3}{\sqrt{3\theta + 3h} + \sqrt{3\theta}}$

   Step 3: $p'(\theta) = \lim_{h \to 0} \frac{3}{\sqrt{3\theta + 3h} + \sqrt{3\theta}} = \frac{3}{\sqrt{3\theta} + \sqrt{3\theta}} = \frac{3}{2\sqrt{3\theta}}; p'(1) = \frac{3}{2\sqrt{3}}, p'(3) = \frac{1}{2}, p'\left(\frac{2}{3}\right) = \frac{3}{2\sqrt{2}}$

7. $y = f(x) = 2x^3$ and $f(x + h) = 2(x + h)^3 \Rightarrow \frac{dy}{dx} = \lim_{h \to 0} \frac{2(x+h)^3 - 2x^3}{h} = \lim_{h \to 0} \frac{2(x^3 + 3x^2h + 3xh^2 + h^3) - 2x^3}{h}$

   $= \lim_{h \to 0} \frac{6x^2h + 6xh^2 + 2h^3}{h} = \lim_{h \to 0} \frac{h(6x^2 + 6xh + 2h^2)}{h} = \lim_{h \to 0} (6x^2 + 6xh + 2h^2) = 6x^2$

9. $s = r(t) = \frac{t}{2t+1}$ and $r(t + h) = \frac{t+h}{2(t+h)+1} \Rightarrow \frac{ds}{dt} = \lim_{h \to 0} \frac{\left(\frac{t+h}{2(t+h)+1}\right) - \left(\frac{t}{2t+1}\right)}{h}$

   $= \lim_{h \to 0} \frac{\left(\frac{(t+h)(2t+1) - t(2t+2h+1)}{(2t+2h+1)(2t+1)}\right)}{h} = \lim_{h \to 0} \frac{(t+h)(2t+1) - t(2t+2h+1)}{(2t+2h+1)(2t+1)h}$

   $= \lim_{h \to 0} \frac{2t^2 + t + 2ht + h - 2t^2 - 2ht - t}{(2t+2h+1)(2t+1)h} = \lim_{h \to 0} \frac{h}{(2t+2h+1)(2t+1)h} = \lim_{h \to 0} \frac{1}{(2t+2h+1)(2t+1)}$

   $= \frac{1}{(2t+1)(2t+1)} = \frac{1}{(2t+1)^2}$

11. $p = f(q) = \frac{1}{\sqrt{q+1}}$ and $f(q + h) = \frac{1}{\sqrt{(q+h)+1}} \Rightarrow \frac{dp}{dq} = \lim_{h \to 0} \frac{\left(\frac{1}{\sqrt{(q+h)+1}}\right) - \left(\frac{1}{\sqrt{q+1}}\right)}{h}$

   $= \lim_{h \to 0} \frac{\left(\frac{\sqrt{q+1} - \sqrt{q+h+1}}{\sqrt{q+h+1}\sqrt{q+1}}\right)}{h} = \lim_{h \to 0} \frac{\sqrt{q+1} - \sqrt{q+h+1}}{h\sqrt{q+h+1}\sqrt{q+1}}$

   $= \lim_{h \to 0} \frac{\left(\sqrt{q+1} - \sqrt{q+h+1}\right)}{h\sqrt{q+h+1}\sqrt{q+1}} \cdot \frac{\left(\sqrt{q+1} + \sqrt{q+h+1}\right)}{\left(\sqrt{q+1} + \sqrt{q+h+1}\right)} = \lim_{h \to 0} \frac{(q+1) - (q+h+1)}{h\sqrt{q+h+1}\sqrt{q+1}\left(\sqrt{q+1} + \sqrt{q+h+1}\right)}$

   $= \lim_{h \to 0} \frac{-h}{h\sqrt{q+h+1}\sqrt{q+1}\left(\sqrt{q+1} + \sqrt{q+h+1}\right)} = \lim_{h \to 0} \frac{-1}{\sqrt{q+h+1}\sqrt{q+1}\left(\sqrt{q+1} + \sqrt{q+h+1}\right)}$

   $= \frac{-1}{\sqrt{q+1}\sqrt{q+1}\left(\sqrt{q+1} + \sqrt{q+1}\right)} = \frac{-1}{2(q+1)\sqrt{q+1}}$

13. $f(x) = x + \frac{9}{x}$ and $f(x+h) = (x+h) + \frac{9}{(x+h)}$ $\Rightarrow$ $\frac{f(x+h)-f(x)}{h} = \frac{\left[(x+h)+\frac{9}{(x+h)}\right] - \left[x+\frac{9}{x}\right]}{h}$

$= \frac{x(x+h)^2 + 9x - x^2(x+h) - 9(x+h)}{x(x+h)h} = \frac{x^3 + 2x^2h + xh^2 + 9x - x^3 - x^2h - 9x - 9h}{x(x+h)h} = \frac{x^2h + xh^2 - 9h}{x(x+h)h}$

$= \frac{h(x^2 + xh - 9)}{x(x+h)h} = \frac{x^2 + xh - 9}{x(x+h)}$; $f'(x) = \lim\limits_{h \to 0} \frac{x^2 + xh - 9}{x(x+h)} = \frac{x^2 - 9}{x^2} = 1 - \frac{9}{x^2}$; $m = f'(-3) = 0$

15. $\frac{ds}{dt} = \lim\limits_{h \to 0} \frac{[(t+h)^3 - (t+h)^2] - (t^3 - t^2)}{h} = \lim\limits_{h \to 0} \frac{(t^3 + 3t^2h + 3th^2 + h^3) - (t^2 + 2th + h^2) - t^3 + t^2}{h}$

$= \lim\limits_{h \to 0} \frac{3t^2h + 3th^2 + h^3 - 2th - h^2}{h} = \lim\limits_{h \to 0} \frac{h(3t^2 + 3th + h^2 - 2t - h)}{h} = \lim\limits_{h \to 0} (3t^2 + 3th + h^2 - 2t - h)$

$= 3t^2 - 2t$; $m = \frac{ds}{dt}\Big|_{t=-1} = 5$

17. $f(x) = \frac{8}{\sqrt{x-2}}$ and $f(x+h) = \frac{8}{\sqrt{(x+h)-2}}$ $\Rightarrow$ $\frac{f(x+h)-f(x)}{h} = \frac{\frac{8}{\sqrt{(x+h)-2}} - \frac{8}{\sqrt{x-2}}}{h}$

$= \frac{8\left(\sqrt{x-2} - \sqrt{x+h-2}\right)}{h\sqrt{x+h-2}\sqrt{x-2}} \cdot \frac{\left(\sqrt{x-2} + \sqrt{x+h-2}\right)}{\left(\sqrt{x-2} + \sqrt{x+h-2}\right)} = \frac{8[(x-2) - (x+h-2)]}{h\sqrt{x+h-2}\sqrt{x-2}\left(\sqrt{x-2} + \sqrt{x+h-2}\right)}$

$= \frac{-8h}{h\sqrt{x+h-2}\sqrt{x-2}\left(\sqrt{x-2} + \sqrt{x+h-2}\right)}$ $\Rightarrow$ $f'(x) = \lim\limits_{h \to 0} \frac{-8}{\sqrt{x+h-2}\sqrt{x-2}\left(\sqrt{x-2} + \sqrt{x+h-2}\right)}$

$= \frac{-8}{\sqrt{x-2}\sqrt{x-2}\left(\sqrt{x-2} + \sqrt{x-2}\right)} = \frac{-4}{(x-2)\sqrt{x-2}}$; $m = f'(6) = \frac{-4}{4\sqrt{4}} = -\frac{1}{2}$ $\Rightarrow$ the equation of the tangent

line at $(6, 4)$ is $y - 4 = -\frac{1}{2}(x - 6) \Rightarrow y = -\frac{1}{2}x + 3 + 4 \Rightarrow y = -\frac{1}{2}x + 7$.

19. $s = f(t) = 1 - 3t^2$ and $f(t+h) = 1 - 3(t+h)^2 = 1 - 3t^2 - 6th - 3h^2$ $\Rightarrow$ $\frac{ds}{dt} = \lim\limits_{h \to 0} \frac{f(t+h)-f(t)}{h}$

$= \lim\limits_{h \to 0} \frac{(1 - 3t^2 - 6th - 3h^2) - (1 - 3t^2)}{h} = \lim\limits_{h \to 0} (-6t - 3h) = -6t$ $\Rightarrow$ $\frac{ds}{dt}\Big|_{t=-1} = 6$

21. $r = f(\theta) = \frac{2}{\sqrt{4-\theta}}$ and $f(\theta + h) = \frac{2}{\sqrt{4-(\theta+h)}}$ $\Rightarrow$ $\frac{dr}{d\theta} = \lim\limits_{h \to 0} \frac{f(\theta+h)-f(\theta)}{h} = \lim\limits_{h \to 0} \frac{\frac{2}{\sqrt{4-\theta-h}} - \frac{2}{\sqrt{4-\theta}}}{h}$

$= \lim\limits_{h \to 0} \frac{2\sqrt{4-\theta} - 2\sqrt{4-\theta-h}}{h\sqrt{4-\theta}\sqrt{4-\theta-h}} = \lim\limits_{h \to 0} \frac{2\sqrt{4-\theta} - 2\sqrt{4-\theta-h}}{h\sqrt{4-\theta}\sqrt{4-\theta-h}} \cdot \frac{\left(2\sqrt{4-\theta} + 2\sqrt{4-\theta-h}\right)}{\left(2\sqrt{4-\theta} + 2\sqrt{4-\theta-h}\right)}$

$= \lim\limits_{h \to 0} \frac{4(4-\theta) - 4(4-\theta-h)}{2h\sqrt{4-\theta}\sqrt{4-\theta-h}\left(\sqrt{4-\theta} + \sqrt{4-\theta-h}\right)} = \lim\limits_{h \to 0} \frac{2}{\sqrt{4-\theta}\sqrt{4-\theta-h}\left(\sqrt{4-\theta} + \sqrt{4-\theta-h}\right)}$

$= \frac{2}{(4-\theta)\left(2\sqrt{4-\theta}\right)} = \frac{1}{(4-\theta)\sqrt{4-\theta}}$ $\Rightarrow$ $\frac{dr}{d\theta}\Big|_{\theta=0} = \frac{1}{8}$

23. $f'(x) = \lim\limits_{z \to x} \frac{f(z)-f(x)}{z-x} = \lim\limits_{z \to x} \frac{\frac{1}{z+2} - \frac{1}{x+2}}{z-x} = \lim\limits_{z \to x} \frac{(x+2)-(z+2)}{(z-x)(z+2)(x+2)} = \lim\limits_{z \to x} \frac{x-z}{(z-x)(z+2)(x+2)} = \lim\limits_{z \to x} \frac{-1}{(z+2)(x+2)}$

$= \frac{-1}{(x+2)^2}$

25. $g'(x) = \lim\limits_{z \to x} \frac{g(z)-g(x)}{z-x} = \lim\limits_{z \to x} \frac{\frac{z}{z-1} - \frac{x}{x-1}}{z-x} = \lim\limits_{z \to x} \frac{z(x-1) - x(z-1)}{(z-x)(z-1)(x-1)} = \lim\limits_{z \to x} \frac{-z+x}{(z-x)(z-1)(x-1)} = \lim\limits_{z \to x} \frac{-1}{(z-1)(x-1)}$

$= \frac{-1}{(x-1)^2}$

27. Note that as x increases, the slope of the tangent line to the curve is first negative, then zero (when x = 0), then positive $\Rightarrow$ the slope is always increasing which matches (b).

29. $f_3(x)$ is an oscillating function like the cosine. Everywhere that the graph of f_3 has a horizontal tangent we expect f_3' to be zero, and (d) matches this condition.

31. (a) f' is not defined at x = 0, 1, 4. At these points, the left-hand and right-hand derivatives do not agree.

For example, $\lim\limits_{x \to 0^-} \frac{f(x)-f(0)}{x-0} = $ slope of line joining $(-4, 0)$ and $(0, 2) = \frac{1}{2}$ but $\lim\limits_{x \to 0^+} \frac{f(x)-f(0)}{x-0} = $ slope of line joining $(0, 2)$ and $(1, -2) = -4$. Since these values are not equal, $f'(0) = \lim\limits_{x \to 0} \frac{f(x) - f(0)}{x - 0}$ does not exist.

(b)

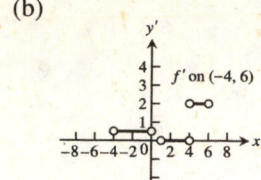

33.

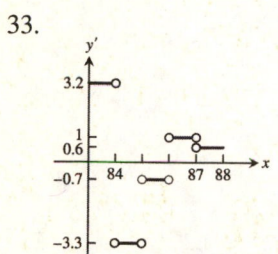

35. **Left-hand derivative:** For $h < 0$, $f(0 + h) = f(h) = h^2$ (using $y = x^2$ curve) $\Rightarrow \lim\limits_{h \to 0^-} \frac{f(0 + h) - f(0)}{h}$

$= \lim\limits_{h \to 0^-} \frac{h^2 - 0}{h} = \lim\limits_{h \to 0^-} h = 0$;

Right-hand derivative: For $h > 0$, $f(0 + h) = f(h) = h$ (using $y = x$ curve) $\Rightarrow \lim\limits_{h \to 0^+} \frac{f(0 + h) - f(0)}{h}$

$= \lim\limits_{h \to 0^+} \frac{h - 0}{h} = \lim\limits_{h \to 0^+} 1 = 1$;

Then $\lim\limits_{h \to 0^-} \frac{f(0 + h) - f(0)}{h} \neq \lim\limits_{h \to 0^+} \frac{f(0 + h) - f(0)}{h} \Rightarrow$ the derivative $f'(0)$ does not exist.

37. **Left-hand derivative:** When $h < 0$, $1 + h < 1 \Rightarrow f(1 + h) = \sqrt{1 + h} \Rightarrow \lim\limits_{h \to 0^-} \frac{f(1 + h) - f(1)}{h}$

$= \lim\limits_{h \to 0^-} \frac{\sqrt{1 + h} - 1}{h} = \lim\limits_{h \to 0^-} \frac{\left(\sqrt{1 + h} - 1\right)}{h} \cdot \frac{\left(\sqrt{1 + h} + 1\right)}{\left(\sqrt{1 + h} + 1\right)} = \lim\limits_{h \to 0^-} \frac{(1 + h) - 1}{h \left(\sqrt{1 + h} + 1\right)} = \lim\limits_{h \to 0^-} \frac{1}{\sqrt{1 + h} + 1} = \frac{1}{2}$;

Right-hand derivative: When $h > 0$, $1 + h > 1 \Rightarrow f(1 + h) = 2(1 + h) - 1 = 2h + 1 \Rightarrow \lim\limits_{h \to 0^+} \frac{f(1 + h) - f(1)}{h}$

$= \lim\limits_{h \to 0^+} \frac{(2h + 1) - 1}{h} = \lim\limits_{h \to 0^+} 2 = 2$;

Then $\lim\limits_{h \to 0^-} \frac{f(1 + h) - f(1)}{h} \neq \lim\limits_{h \to 0^+} \frac{f(1 + h) - f(1)}{h} \Rightarrow$ the derivative $f'(1)$ does not exist.

39. (a) The function is differentiable on its domain $-3 \le x \le 2$ (it is smooth)

    (b) none

    (c) none

41. (a) The function is differentiable on $-3 \le x < 0$ and $0 < x \le 3$

    (b) none

    (c) The function is neither continuous nor differentiable at $x = 0$ since $\lim\limits_{x \to 0^-} f(x) \neq \lim\limits_{x \to 0^+} f(x)$

43. (a) f is differentiable on $-1 \le x < 0$ and $0 < x \le 2$

    (b) f is continuous but not differentiable at $x = 0$: $\lim\limits_{x \to 0} f(x) = 0$ exists but there is a cusp at $x = 0$, so

       $f'(0) = \lim\limits_{h \to 0} \frac{f(0 + h) - f(0)}{h}$ does not exist

    (c) none

45. (a) $f'(x) = \lim\limits_{h \to 0} \frac{f(x+h) - f(x)}{h} = \lim\limits_{h \to 0} \frac{-(x+h)^2 - (-x^2)}{h} = \lim\limits_{h \to 0} \frac{-x^2 - 2xh - h^2 + x^2}{h} = \lim\limits_{h \to 0} (-2x - h) = -2x$

(b)

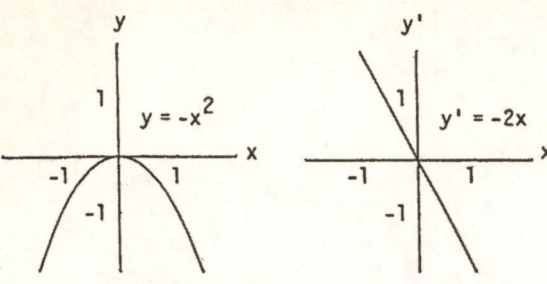

(c) $y' = -2x$ is positive for $x < 0$, y' is zero when $x = 0$, y' is negative when $x > 0$

(d) $y = -x^2$ is increasing for $-\infty < x < 0$ and decreasing for $0 < x < \infty$; the function is increasing on intervals where $y' > 0$ and decreasing on intervals where $y' < 0$

47. (a) Using the alternate formula for calculating derivatives: $f'(x) = \lim\limits_{z \to x} \frac{f(z) - f(x)}{z - x} = \lim\limits_{z \to x} \frac{\left(\frac{z^3}{3} - \frac{x^3}{3}\right)}{z - x}$

$= \lim\limits_{z \to x} \frac{z^3 - x^3}{3(z - x)} = \lim\limits_{z \to x} \frac{(z - x)(z^2 + zx + x^2)}{3(z - x)} = \lim\limits_{z \to x} \frac{z^2 + zx + x^2}{3} = x^2 \Rightarrow f'(x) = x^2$

(b)

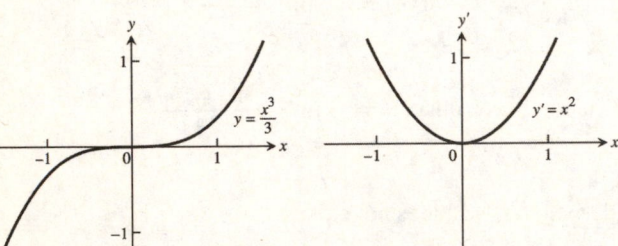

(c) y' is positive for all $x \ne 0$, and $y' = 0$ when $x = 0$; y' is never negative

(d) $y = \frac{x^3}{3}$ is increasing for all $x \ne 0$ (the graph is horizontal at $x = 0$) because y is increasing where $y' > 0$; y is never decreasing

49. $y' = \lim\limits_{x \to c} \frac{f(x) - f(c)}{x - c} = \lim\limits_{x \to c} \frac{x^3 - c^3}{x - c} = \lim\limits_{x \to c} \frac{(x-c)(x^2 + xc + c^2)}{x - c} = \lim\limits_{x \to c} (x^2 + xc + c^2) = 3c^2$.

The slope of the curve $y = x^3$ at $x = c$ is $y' = 3c^2$. Notice that $3c^2 \ge 0$ for all $c \Rightarrow y = x^3$ never has a negative slope.

51. $y' = \lim\limits_{h \to 0} \frac{(2(x + h)^2 - 13(x + h) + 5) - (2x^2 - 13x + 5)}{h} = \lim\limits_{h \to 0} \frac{2x^2 + 4xh + 2h^2 - 13x - 13h + 5 - 2x^2 + 13x - 5}{h}$

$= \lim\limits_{h \to 0} \frac{4xh + 2h^2 - 13h}{h} = \lim\limits_{h \to 0} (4x + 2h - 13) = 4x - 13$, slope at x. The slope is -1 when $4x - 13 = -1$

$\Rightarrow 4x = 12 \Rightarrow x = 3 \Rightarrow y = 2 \cdot 3^2 - 13 \cdot 3 + 5 = -16$. Thus the tangent line is $y + 16 = (-1)(x - 3)$

$\Rightarrow y = -x - 13$ and the point of tangency is $(3, -16)$.

53. No. Derivatives of functions have the intermediate value property. The function $f(x) = \lfloor x \rfloor$ satisfies $f(0) = 0$ and $f(1) = 1$ but does not take on the value $\frac{1}{2}$ anywhere in $[0, 1] \Rightarrow f$ does not have the intermediate value property. Thus f cannot be the derivative of any function on $[0, 1] \Rightarrow f$ cannot be the derivative of any function on $(-\infty, \infty)$.

55. Yes; the derivative of $-f$ is $-f'$ so that $f'(x_0)$ exists $\Rightarrow -f'(x_0)$ exists as well.

57. Yes, $\lim\limits_{t \to 0} \frac{g(t)}{h(t)}$ can exist but it need not equal zero. For example, let $g(t) = mt$ and $h(t) = t$. Then $g(0) = h(0)$

$= 0$, but $\lim\limits_{t \to 0} \frac{g(t)}{h(t)} = \lim\limits_{t \to 0} \frac{mt}{t} = \lim\limits_{t \to 0} m = m$, which need not be zero.

59. The graphs are shown below for h = 1, 0.5, 0.1. The function $y = \frac{1}{2\sqrt{x}}$ is the derivative of the function

$y = \sqrt{x}$ so that $\frac{1}{2\sqrt{x}} = \lim\limits_{h \to 0} \frac{\sqrt{x+h} - \sqrt{x}}{h}$. The graphs reveal that $y = \frac{\sqrt{x+h} - \sqrt{x}}{h}$ gets closer to $y = \frac{1}{2\sqrt{x}}$

as h gets smaller and smaller.

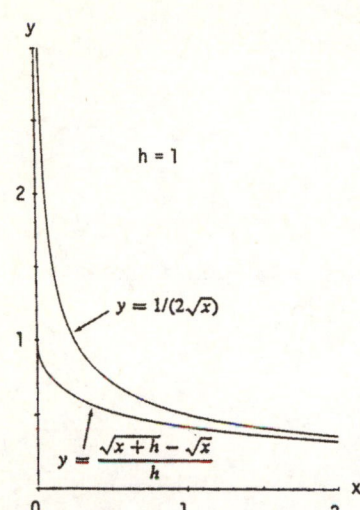 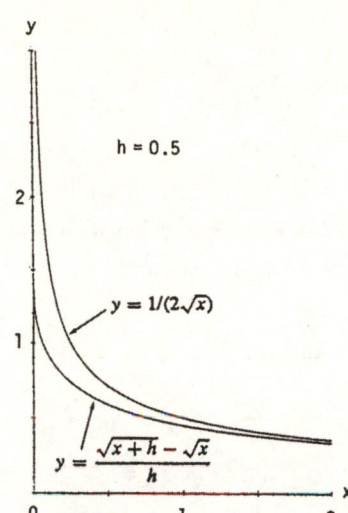 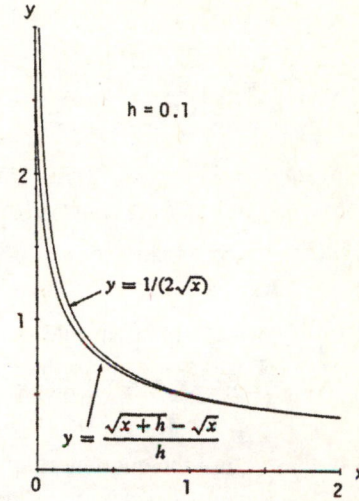

61. Weierstrass's nowhere differentiable continuous function.

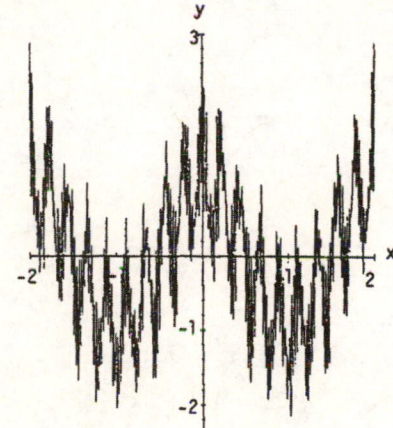

$$g(x) = \cos(\pi x) + \left(\frac{2}{3}\right)^1 \cos(9\pi x) + \left(\frac{2}{3}\right)^2 \cos(9^2\pi x) + \left(\frac{2}{3}\right)^3 \cos(9^3\pi x)$$

$$+ \cdots + \left(\frac{2}{3}\right)^7 \cos(9^7\pi x)$$

3.2 DIFFERENTIATION RULES

1. $y = -x^2 + 3 \Rightarrow \frac{dy}{dx} = \frac{d}{dx}(-x^2) + \frac{d}{dx}(3) = -2x + 0 = -2x \Rightarrow \frac{d^2y}{dx^2} = -2$

3. $s = 5t^3 - 3t^5 \Rightarrow \frac{ds}{dt} = \frac{d}{dt}(5t^3) - \frac{d}{dt}(3t^5) = 15t^2 - 15t^4 \Rightarrow \frac{d^2s}{dt^2} = \frac{d}{dt}(15t^2) - \frac{d}{dt}(15t^4) = 30t - 60t^3$

5. $y = \frac{4}{3}x^3 - x \Rightarrow \frac{dy}{dx} = 4x^2 - 1 \Rightarrow \frac{d^2y}{dx^2} = 8x$

7. $w = 3z^{-2} - z^{-1} \Rightarrow \frac{dw}{dz} = -6z^{-3} + z^{-2} = \frac{-6}{z^3} + \frac{1}{z^2} \Rightarrow \frac{d^2w}{dz^2} = 18z^{-4} - 2z^{-3} = \frac{18}{z^4} - \frac{2}{z^3}$

9. $y = 6x^2 - 10x - 5x^{-2} \Rightarrow \frac{dy}{dx} = 12x - 10 + 10x^{-3} = 12x - 10 + \frac{10}{x^3} \Rightarrow \frac{d^2y}{dx^2} = 12 - 0 - 30x^{-4} = 12 - \frac{30}{x^4}$

11. $r = \frac{1}{3}s^{-2} - \frac{5}{2}s^{-1} \Rightarrow \frac{dr}{ds} = -\frac{2}{3}s^{-3} + \frac{5}{2}s^{-2} = \frac{-2}{3s^3} + \frac{5}{2s^2} \Rightarrow \frac{d^2r}{ds^2} = 2s^{-4} - 5s^{-3} = \frac{2}{s^4} - \frac{5}{s^3}$

13. (a) $y = (3 - x^2)(x^3 - x + 1) \Rightarrow y' = (3 - x^2) \cdot \frac{d}{dx}(x^3 - x + 1) + (x^3 - x + 1) \cdot \frac{d}{dx}(3 - x^2)$

  $= (3 - x^2)(3x^2 - 1) + (x^3 - x + 1)(-2x) = -5x^4 + 12x^2 - 2x - 3$

  (b) $y = -x^5 + 4x^3 - x^2 - 3x + 3 \Rightarrow y' = -5x^4 + 12x^2 - 2x - 3$

15. (a) $y = (x^2 + 1)\left(x + 5 + \frac{1}{x}\right) \Rightarrow y' = (x^2 + 1) \cdot \frac{d}{dx}\left(x + 5 + \frac{1}{x}\right) + \left(x + 5 + \frac{1}{x}\right) \cdot \frac{d}{dx}(x^2 + 1)$

  $= (x^2 + 1)(1 - x^{-2}) + (x + 5 + x^{-1})(2x) = (x^2 - 1 + 1 - x^{-2}) + (2x^2 + 10x + 2) = 3x^2 + 10x + 2 - \frac{1}{x^2}$

  (b) $y = x^3 + 5x^2 + 2x + 5 + \frac{1}{x} \Rightarrow y' = 3x^2 + 10x + 2 - \frac{1}{x^2}$

17. $y = \frac{2x + 5}{3x - 2}$; use the quotient rule: $u = 2x + 5$ and $v = 3x - 2 \Rightarrow u' = 2$ and $v' = 3 \Rightarrow y' = \frac{vu' - uv'}{v^2}$

  $= \frac{(3x - 2)(2) - (2x + 5)(3)}{(3x - 2)^2} = \frac{6x - 4 - 6x - 15}{(3x - 2)^2} = \frac{-19}{(3x - 2)^2}$

19. $g(x) = \frac{x^2 - 4}{x + 0.5}$; use the quotient rule: $u = x^2 - 4$ and $v = x + 0.5 \Rightarrow u' = 2x$ and $v' = 1 \Rightarrow g'(x) = \frac{vu' - uv'}{v^2}$

  $= \frac{(x + 0.5)(2x) - (x^2 - 4)(1)}{(x + 0.5)^2} = \frac{2x^2 + x - x^2 + 4}{(x + 0.5)^2} = \frac{x^2 + x + 4}{(x + 0.5)^2}$

21. $v = (1 - t)(1 + t^2)^{-1} = \frac{1 - t}{1 + t^2} \Rightarrow \frac{dv}{dt} = \frac{(1 + t^2)(-1) - (1 - t)(2t)}{(1 + t^2)^2} = \frac{-1 - t^2 - 2t + 2t^2}{(1 + t^2)^2} = \frac{t^2 - 2t - 1}{(1 + t^2)^2}$

23. $f(s) = \frac{\sqrt{s} - 1}{\sqrt{s} + 1} \Rightarrow f'(s) = \frac{(\sqrt{s} + 1)\left(\frac{1}{2\sqrt{s}}\right) - (\sqrt{s} - 1)\left(\frac{1}{2\sqrt{s}}\right)}{(\sqrt{s} + 1)^2} = \frac{(\sqrt{s} + 1) - (\sqrt{s} - 1)}{2\sqrt{s}(\sqrt{s} + 1)^2} = \frac{1}{\sqrt{s}(\sqrt{s} + 1)^2}$

  NOTE: $\frac{d}{ds}(\sqrt{s}) = \frac{1}{2\sqrt{s}}$ from Example 2 in Section 2.1

25. $v = \frac{1 + x - 4\sqrt{x}}{x} \Rightarrow v' = \frac{x\left(1 - \frac{2}{\sqrt{x}}\right) - (1 + x - 4\sqrt{x})}{x^2} = \frac{2\sqrt{x} - 1}{x^2}$

27. $y = \frac{1}{(x^2 - 1)(x^2 + x + 1)}$; use the quotient rule: $u = 1$ and $v = (x^2 - 1)(x^2 + x + 1) \Rightarrow u' = 0$ and

  $v' = (x^2 - 1)(2x + 1) + (x^2 + x + 1)(2x) = 2x^3 + x^2 - 2x - 1 + 2x^3 + 2x^2 + 2x = 4x^3 + 3x^2 - 1$

  $\Rightarrow \frac{dy}{dx} = \frac{vu' - uv'}{v^2} = \frac{0 - 1(4x^3 + 3x^2 - 1)}{(x^2 - 1)^2(x^2 + x + 1)^2} = \frac{-4x^3 - 3x^2 + 1}{(x^2 - 1)^2(x^2 + x + 1)^2}$

29. $y = \frac{1}{2}x^4 - \frac{3}{2}x^2 - x \Rightarrow y' = 2x^3 - 3x - 1 \Rightarrow y'' = 6x^2 - 3 \Rightarrow y''' = 12x \Rightarrow y^{(4)} = 12 \Rightarrow y^{(n)} = 0$ for all $n \geq 5$

31. $y = \frac{x^3 + 7}{x} = x^2 + 7x^{-1} \Rightarrow \frac{dy}{dx} = 2x - 7x^{-2} = 2x - \frac{7}{x^2} \Rightarrow \frac{d^2y}{dx^2} = 2 + 14x^{-3} = 2 + \frac{14}{x^3}$

33. $r = \frac{(\theta - 1)(\theta^2 + \theta + 1)}{\theta^3} = \frac{\theta^3 - 1}{\theta^3} = 1 - \frac{1}{\theta^3} = 1 - \theta^{-3} \Rightarrow \frac{dr}{d\theta} = 0 + 3\theta^{-4} = 3\theta^{-4} = \frac{3}{\theta^4} \Rightarrow \frac{d^2r}{d\theta^2} = -12\theta^{-5} = \frac{-12}{\theta^5}$

35. $w = \left(\frac{1 + 3z}{3z}\right)(3 - z) = \left(\frac{1}{3}z^{-1} + 1\right)(3 - z) = z^{-1} - \frac{1}{3} + 3 - z = z^{-1} + \frac{8}{3} - z \Rightarrow \frac{dw}{dz} = -z^{-2} + 0 - 1 = -z^{-2} - 1$

  $= \frac{-1}{z^2} - 1 \Rightarrow \frac{d^2w}{dz^2} = 2z^{-3} - 0 = 2z^{-3} = \frac{2}{z^3}$

37. $p = \left(\frac{q^2 + 3}{12q}\right)\left(\frac{q^4 - 1}{q^3}\right) = \frac{q^6 - q^2 + 3q^4 - 3}{12q^4} = \frac{1}{12}q^2 - \frac{1}{12}q^{-2} + \frac{1}{4} - \frac{1}{4}q^{-4} \Rightarrow \frac{dp}{dq} = \frac{1}{6}q + \frac{1}{6}q^{-3} + q^{-5} = \frac{1}{6}q + \frac{1}{6q^3} + \frac{1}{q^5}$

  $\Rightarrow \frac{d^2p}{dq^2} = \frac{1}{6} - \frac{1}{2}q^{-4} - 5q^{-6} = \frac{1}{6} - \frac{1}{2q^4} - \frac{5}{q^6}$

39. $u(0) = 5$, $u'(0) = -3$, $v(0) = -1$, $v'(0) = 2$

(a) $\frac{d}{dx}(uv) = uv' + vu' \Rightarrow \frac{d}{dx}(uv)\big|_{x=0} = u(0)v'(0) + v(0)u'(0) = 5 \cdot 2 + (-1)(-3) = 13$

(b) $\frac{d}{dx}\left(\frac{u}{v}\right) = \frac{vu' - uv'}{v^2} \Rightarrow \frac{d}{dx}\left(\frac{u}{v}\right)\big|_{x=0} = \frac{v(0)u'(0) - u(0)v'(0)}{(v(0))^2} = \frac{(-1)(-3) - (5)(2)}{(-1)^2} = -7$

(c) $\frac{d}{dx}\left(\frac{v}{u}\right) = \frac{uv' - vu'}{u^2} \Rightarrow \frac{d}{dx}\left(\frac{v}{u}\right)\big|_{x=0} = \frac{u(0)v'(0) - v(0)u'(0)}{(u(0))^2} = \frac{(5)(2) - (-1)(-3)}{(5)^2} = \frac{7}{25}$

(d) $\frac{d}{dx}(7v - 2u) = 7v' - 2u' \Rightarrow \frac{d}{dx}(7v - 2u)\big|_{x=0} = 7v'(0) - 2u'(0) = 7 \cdot 2 - 2(-3) = 20$

41. $y = x^3 - 4x + 1$. Note that $(2, 1)$ is on the curve: $1 = 2^3 - 4(2) + 1$

(a) Slope of the tangent at (x, y) is $y' = 3x^2 - 4 \Rightarrow$ slope of the tangent at $(2, 1)$ is $y'(2) = 3(2)^2 - 4 = 8$. Thus the slope of the line perpendicular to the tangent at $(2, 1)$ is $-\frac{1}{8} \Rightarrow$ the equation of the line perpendicular to to the tangent line at $(2, 1)$ is $y - 1 = -\frac{1}{8}(x - 2)$ or $y = -\frac{x}{8} + \frac{5}{4}$.

(b) The slope of the curve at x is $m = 3x^2 - 4$ and the smallest value for m is -4 when $x = 0$ and $y = 1$.

(c) We want the slope of the curve to be $8 \Rightarrow y' = 8 \Rightarrow 3x^2 - 4 = 8 \Rightarrow 3x^2 = 12 \Rightarrow x^2 = 4 \Rightarrow x = \pm 2$. When $x = 2$, $y = 1$ and the tangent line has equation $y - 1 = 8(x - 2)$ or $y = 8x - 15$; when $x = -2$, $y = (-2)^3 - 4(-2) + 1 = 1$, and the tangent line has equation $y - 1 = 8(x + 2)$ or $y = 8x + 17$.

43. $y = \frac{4x}{x^2 + 1} \Rightarrow \frac{dy}{dx} = \frac{(x^2 + 1)(4) - (4x)(2x)}{(x^2 + 1)^2} = \frac{4x^2 + 4 - 8x^2}{(x^2 + 1)^2} = \frac{4(-x^2 + 1)}{(x^2 + 1)^2}$. When $x = 0$, $y = 0$ and $y' = \frac{4(0 + 1)}{1}$

$= 4$, so the tangent to the curve at $(0, 0)$ is the line $y = 4x$. When $x = 1$, $y = 2 \Rightarrow y' = 0$, so the tangent to the curve at $(1, 2)$ is the line $y = 2$.

45. $y = ax^2 + bx + c$ passes through $(0, 0) \Rightarrow 0 = a(0) + b(0) + c \Rightarrow c = 0$; $y = ax^2 + bx$ passes through $(1, 2)$ $\Rightarrow 2 = a + b$; $y' = 2ax + b$ and since the curve is tangent to $y = x$ at the origin, its slope is 1 at $x = 0$ $\Rightarrow y' = 1$ when $x = 0 \Rightarrow 1 = 2a(0) + b \Rightarrow b = 1$. Then $a + b = 2 \Rightarrow a = 1$. In summary $a = b = 1$ and $c = 0$ so the curve is $y = x^2 + x$.

47. (a) $y = x^3 - x \Rightarrow y' = 3x^2 - 1$. When $x = -1$, $y = 0$ and $y' = 2 \Rightarrow$ the tangent line to the curve at $(-1, 0)$ is $y = 2(x + 1)$ or $y = 2x + 2$.

(b)

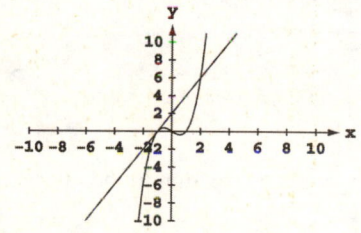

(c) $\left. \begin{array}{l} y = x^3 - x \\ y = 2x + 2 \end{array} \right\} \Rightarrow x^3 - x = 2x + 2 \Rightarrow x^3 - 3x - 2 = (x - 2)(x + 1)^2 = 0 \Rightarrow x = 2$ or $x = -1$. Since $y = 2(2) + 2 = 6$; the other intersection point is $(2, 6)$

49. $P(x) = a_n x^n + a_{n-1} x^{n-1} + \cdots + a_2 x^2 + a_1 x + a_0 \Rightarrow P'(x) = na_n x^{n-1} + (n - 1)a_{n-1} x^{n-2} + \cdots + 2a_2 x + a_1$

51. Let c be a constant $\Rightarrow \frac{dc}{dx} = 0 \Rightarrow \frac{d}{dx}(u \cdot c) = u \cdot \frac{dc}{dx} + c \cdot \frac{du}{dx} = u \cdot 0 + c\frac{du}{dx} = c\frac{du}{dx}$. Thus when one of the functions is a constant, the Product Rule is just the Constant Multiple Rule $\Rightarrow$ the Constant Multiple Rule is a special case of the Product Rule.

53. (a) $\frac{d}{dx}(uvw) = \frac{d}{dx}((uv) \cdot w) = (uv)\frac{dw}{dx} + w \cdot \frac{d}{dx}(uv) = uv\frac{dw}{dx} + w\left(u\frac{dv}{dx} + v\frac{du}{dx}\right) = uv\frac{dw}{dx} + wu\frac{dv}{dx} + wv\frac{du}{dx}$

$= uvw' + uv'w + u'vw$

(b) $\frac{d}{dx}(u_1u_2u_3u_4) = \frac{d}{dx}((u_1u_2u_3)u_4) = (u_1u_2u_3)\frac{du_4}{dx} + u_4\frac{d}{dx}(u_1u_2u_3) \Rightarrow \frac{d}{dx}(u_1u_2u_3u_4)$

$= u_1u_2u_3\frac{du_4}{dx} + u_4\left(u_1u_2\frac{du_3}{dx} + u_3u_1\frac{du_2}{dx} + u_3u_2\frac{du_1}{dx}\right)$ (using (a) above)

$\Rightarrow \frac{d}{dx}(u_1u_2u_3u_4) = u_1u_2u_3\frac{du_4}{dx} + u_1u_2u_4\frac{du_3}{dx} + u_1u_3u_4\frac{du_2}{dx} + u_2u_3u_4\frac{du_1}{dx}$

$= u_1u_2u_3u_4' + u_1u_2u_3'u_4 + u_1u_2'u_3u_4 + u_1'u_2u_3u_4$

(c) Generalizing (a) and (b) above, $\frac{d}{dx}(u_1 \cdots u_n) = u_1u_2\cdots u_{n-1}u_n' + u_1u_2\cdots u_{n-2}u_{n-1}'u_n + \ldots + u_1'u_2\cdots u_n$

55. $p = \frac{nRT}{V-nb} - \frac{an^2}{V^2}$. We are holding T constant, and a, b, n, R are also constant so their derivatives are zero

$\Rightarrow \frac{dP}{dV} = \frac{(V-nb)\cdot 0 - (nRT)(1)}{(V-nb)^2} - \frac{V^2(0) - (an^2)(2V)}{(V^2)^2} = \frac{-nRT}{(V-nb)^2} + \frac{2an^2}{V^3}$

3.3 THE DERIVATIVE AS A RATE OF CHANGE

1. $s = t^2 - 3t + 2, 0 \le t \le 2$

(a) displacement $= \Delta s = s(2) - s(0) = 0m - 2m = -2$ m, $v_{av} = \frac{\Delta s}{\Delta t} = \frac{-2}{2} = -1$ m/sec

(b) $v = \frac{ds}{dt} = 2t - 3 \Rightarrow |v(0)| = |-3| = 3$ m/sec and $|v(2)| = 1$ m/sec;

$a = \frac{d^2s}{dt^2} = 2 \Rightarrow a(0) = 2$ m/sec^2 and $a(2) = 2$ m/sec^2

(c) $v = 0 \Rightarrow 2t - 3 = 0 \Rightarrow t = \frac{3}{2}$. v is negative in the interval $0 < t < \frac{3}{2}$ and v is positive when $\frac{3}{2} < t < 2 \Rightarrow$ the body

changes direction at $t = \frac{3}{2}$.

3. $s = -t^3 + 3t^2 - 3t, 0 \le t \le 3$

(a) displacement $= \Delta s = s(3) - s(0) = -9$ m, $v_{av} = \frac{\Delta s}{\Delta t} = \frac{-9}{3} = -3$ m/sec

(b) $v = \frac{ds}{dt} = -3t^2 + 6t - 3 \Rightarrow |v(0)| = |-3| = 3$ m/sec and $|v(3)| = |-12| = 12$ m/sec; $a = \frac{d^2s}{dt^2} = -6t + 6$

$\Rightarrow a(0) = 6$ m/sec^2 and $a(3) = -12$ m/sec^2

(c) $v = 0 \Rightarrow -3t^2 + 6t - 3 = 0 \Rightarrow t^2 - 2t + 1 = 0 \Rightarrow (t-1)^2 = 0 \Rightarrow t = 1$. For all other values of t in the

interval the velocity v is negative (the graph of $v = -3t^2 + 6t - 3$ is a parabola with vertex at $t = 1$ which

opens downward $\Rightarrow$ the body never changes direction).

5. $s = \frac{25}{t^2} - \frac{5}{t}, 1 \le t \le 5$

(a) $\Delta s = s(5) - s(1) = -20$ m, $v_{av} = \frac{-20}{4} = -5$ m/sec

(b) $v = \frac{-50}{t^3} + \frac{5}{t^2} \Rightarrow |v(1)| = 45$ m/sec and $|v(5)| = \frac{1}{5}$ m/sec; $a = \frac{150}{t^4} - \frac{10}{t^3} \Rightarrow a(1) = 140$ m/sec^2 and

$a(5) = \frac{4}{25}$ m/sec^2

(c) $v = 0 \Rightarrow \frac{-50 + 5t}{t^3} = 0 \Rightarrow -50 + 5t = 0 \Rightarrow t = 10 \Rightarrow$ the body does not change direction in the interval

7. $s = t^3 - 6t^2 + 9t$ and let the positive direction be to the right on the s-axis.

(a) $v = 3t^2 - 12t + 9$ so that $v = 0 \Rightarrow t^2 - 4t + 3 = (t-3)(t-1) = 0 \Rightarrow t = 1$ or $3; a = 6t - 12 \Rightarrow a(1)$

$= -6$ m/sec^2 and $a(3) = 6$ m/sec^2. Thus the body is motionless but being accelerated left when $t = 1$, and

motionless but being accelerated right when $t = 3$.

(b) $a = 0 \Rightarrow 6t - 12 = 0 \Rightarrow t = 2$ with speed $|v(2)| = |12 - 24 + 9| = 3$ m/sec

(c) The body moves to the right or forward on $0 \le t < 1$, and to the left or backward on $1 < t < 2$. The

positions are $s(0) = 0, s(1) = 4$ and $s(2) = 2 \Rightarrow$ total distance $= |s(1) - s(0)| + |s(2) - s(1)| = |4| + |-2|$

$= 6$ m.

9. $s_m = 1.86t^2 \Rightarrow v_m = 3.72t$ and solving $3.72t = 27.8 \Rightarrow t \approx 7.5$ sec on Mars; $s_j = 11.44t^2 \Rightarrow v_j = 22.88t$ and

solving $22.88t = 27.8 \Rightarrow t \approx 1.2$ sec on Jupiter.

11. $s = 15t - \frac{1}{2}g_s t^2 \Rightarrow v = 15 - g_s t$ so that $v = 0 \Rightarrow 15 - g_s t = 0 \Rightarrow g_s = \frac{15}{t}$. Therefore $g_s = \frac{15}{20} = \frac{3}{4} = 0.75$ m/sec^2

13. (a) $s = 179 - 16t^2 \Rightarrow v = -32t \Rightarrow$ speed $= |v| = 32t$ ft/sec and $a = -32$ ft/sec^2

(b) $s = 0 \Rightarrow 179 - 16t^2 = 0 \Rightarrow t = \sqrt{\frac{179}{16}} \approx 3.3$ sec

(c) When $t = \sqrt{\frac{179}{16}}$, $v = -32\sqrt{\frac{179}{16}} = -8\sqrt{179} \approx -107.0$ ft/sec

15. (a) at 2 and 7 seconds   (b) between 3 and 6 seconds: $3 \le t \le 6$

(c)   (d)

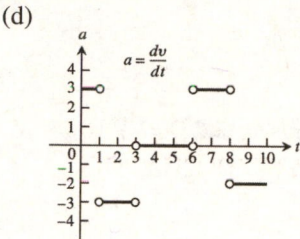

17. (a) 190 ft/sec   (b) 2 sec

(c) at 8 sec, 0 ft/sec   (d) 10.8 sec, 90 ft/sec

(e) From $t = 8$ until $t = 10.8$ sec, a total of 2.8 sec

(f) Greatest acceleration happens 2 sec after launch

(g) From $t = 2$ to $t = 10.8$ sec; during this period, $a = \frac{v(10.8)-v(2)}{10.8-2} \approx -32$ ft/sec^2

19. $s = 490t^2 \Rightarrow v = 980t \Rightarrow a = 980$

(a) Solving $160 = 490t^2 \Rightarrow t = \frac{4}{7}$ sec. The average velocity was $\frac{s(4/7)-s(0)}{4/7} = 280$ cm/sec.

(b) At the 160 cm mark the balls are falling at $v(4/7) = 560$ cm/sec. The acceleration at the 160 cm mark was 980 cm/sec^2.

(c) The light was flashing at a rate of $\frac{17}{4/7} = 29.75$ flashes per second.

21. C = position, A = velocity, and B = acceleration. Neither A nor C can be the derivative of B because B's derivative is constant. Graph C cannot be the derivative of A either, because A has some negative slopes while C has only positive values. So, C (being the derivative of neither A nor B) must be the graph of position. Curve C has both positive and negative slopes, so its derivative, the velocity, must be A and not B. That leaves B for acceleration.

23. (a) $c(100) = 11,000 \Rightarrow c_{av} = \frac{11,000}{100} = \$110$

(b) $c(x) = 2000 + 100x - .1x^2 \Rightarrow c'(x) = 100 - .2x$. Marginal cost $= c'(x) \Rightarrow$ the marginal cost of producing 100 machines is $c'(100) = \$80$

(c) The cost of producing the 101st machine is $c(101) - c(100) = 100 - \frac{201}{10} = \$79.90$

25. $b(t) = 10^6 + 10^4t - 10^3t^2 \Rightarrow b'(t) = 10^4 - (2)(10^3t) = 10^3(10 - 2t)$

(a) $b'(0) = 10^4$ bacteria/hr   (b) $b'(5) = 0$ bacteria/hr

(c) $b'(10) = -10^4$ bacteria/hr

27. (a) $y = 6\left(1 - \frac{t}{12}\right)^2 = 6\left(1 - \frac{t}{6} + \frac{t^2}{144}\right) \Rightarrow \frac{dy}{dt} = \frac{t}{12} - 1$

(b) The largest value of $\frac{dy}{dt}$ is 0 m/h when $t = 12$ and the fluid level is falling the slowest at that time. The smallest value of $\frac{dy}{dt}$ is -1 m/h, when $t = 0$, and the fluid level is falling the fastest at that time.

(c) In this situation, $\frac{dy}{dt} \leq 0 \Rightarrow$ the graph of y is always decreasing. As $\frac{dy}{dt}$ increases in value, the slope of the graph of y increases from -1 to 0 over the interval $0 \leq t \leq 12$.

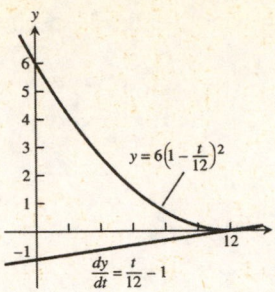

29. 200 km/hr $= 55\frac{5}{9}$ m/sec $= \frac{500}{9}$ m/sec, and D $= \frac{10}{9}t^2 \Rightarrow$ V $= \frac{20}{9}$ t. Thus V $= \frac{500}{9} \Rightarrow \frac{20}{9}t = \frac{500}{9} \Rightarrow t = 25$ sec. When $t = 25$, D $= \frac{10}{9}(25)^2 = \frac{6250}{9}$ m

31.

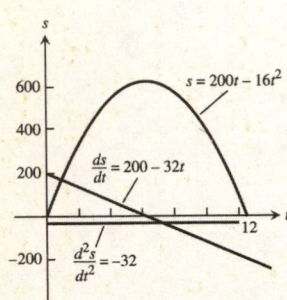

(a) $v = 0$ when $t = 6.25$ sec

(b) $v > 0$ when $0 \leq t < 6.25 \Rightarrow$ body moves up; $v < 0$ when $6.25 < t \leq 12.5 \Rightarrow$ body moves down

(c) body changes direction at $t = 6.25$ sec

(d) body speeds up on $(6.25, 12.5]$ and slows down on $[0, 6.25)$

(e) The body is moving fastest at the endpoints $t = 0$ and $t = 12.5$ when it is traveling 200 ft/sec. It's moving slowest at $t = 6.25$ when the speed is 0.

(f) When $t = 6.25$ the body is $s = 625$ m from the origin and farthest away.

33.

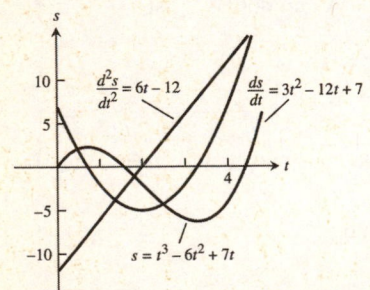

(a) $v = 0$ when $t = \frac{6 \pm \sqrt{15}}{3}$ sec

(b) $v < 0$ when $\frac{6 - \sqrt{15}}{3} < t < \frac{6 + \sqrt{15}}{3} \Rightarrow$ body moves left; $v > 0$ when $0 \leq t < \frac{6 - \sqrt{15}}{3}$ or $\frac{6 + \sqrt{15}}{3} < t \leq 4$
    $\Rightarrow$ body moves right

(c) body changes direction at $t = \frac{6 \pm \sqrt{15}}{3}$ sec

(d) body speeds up on $\left(\frac{6 - \sqrt{15}}{3}, 2 \right) \cup \left(\frac{6 + \sqrt{15}}{3}, 4 \right]$ and slows down on $\left[0, \frac{6 - \sqrt{15}}{3} \right) \cup \left(2, \frac{6 + \sqrt{15}}{3} \right)$.

(e) The body is moving fastest at $t = 0$ and $t = 4$ when it is moving 7 units/sec and slowest at $t = \frac{6 \pm \sqrt{15}}{3}$ sec

(f) When $t = \frac{6 + \sqrt{15}}{3}$ the body is at position $s \approx -6.303$ units and farthest from the origin.

35. (a) It takes 135 seconds.

   (b) Average speed $= \frac{\Delta F}{\Delta t} = \frac{5 - 0}{73 - 0} = \frac{5}{73} \approx 0.068$ furlongs/sec.

(c) Using a symmetric difference quotient, the horse's speed is approximately $\frac{\Delta F}{\Delta t} = \frac{4-2}{59-33} = \frac{2}{26} \approx 0.077$ furlongs/sec.

(d) The horse is running the fastest during the last furlong (between the 9th and 10th furlong markers). This furlong takes only 11 seconds to run, which is the least amount of time for a furlong.

(e) The horse accelerates the fastest during the first furlong (between markers 0 and 1).

3.4 DERIVATIVES OF TRIGONOMETRIC FUNCTIONS

1. $y = -10x + 3\cos x \;\Rightarrow\; \frac{dy}{dx} = -10 + 3\frac{d}{dx}(\cos x) = -10 - 3\sin x$

3. $y = \csc x - 4\sqrt{x} + 7 \;\Rightarrow\; \frac{dy}{dx} = -\csc x \cot x - \frac{4}{2\sqrt{x}} + 0 = -\csc x \cot x - \frac{2}{\sqrt{x}}$

5. $y = (\sec x + \tan x)(\sec x - \tan x) \;\Rightarrow\; \frac{dy}{dx} = (\sec x + \tan x)\frac{d}{dx}(\sec x - \tan x) + (\sec x - \tan x)\frac{d}{dx}(\sec x + \tan x)$

$= (\sec x + \tan x)(\sec x \tan x - \sec^2 x) + (\sec x - \tan x)(\sec x \tan x + \sec^2 x)$

$= (\sec^2 x \tan x + \sec x \tan^2 x - \sec^3 x - \sec^2 x \tan x) + (\sec^2 x \tan x - \sec x \tan^2 x + \sec^3 x - \tan x \sec^2 x) = 0.$

$\left(\text{Note also that } y = \sec^2 x - \tan^2 x = (\tan^2 x + 1) - \tan^2 x = 1 \;\Rightarrow\; \frac{dy}{dx} = 0.\right)$

7. $y = \frac{\cot x}{1 + \cot x} \;\Rightarrow\; \frac{dy}{dx} = \frac{(1 + \cot x)\frac{d}{dx}(\cot x) - (\cot x)\frac{d}{dx}(1 + \cot x)}{(1 + \cot x)^2} = \frac{(1 + \cot x)(-\csc^2 x) - (\cot x)(-\csc^2 x)}{(1 + \cot x)^2}$

$= \frac{-\csc^2 x - \csc^2 x \cot x + \csc^2 x \cot x}{(1 + \cot x)^2} = \frac{-\csc^2 x}{(1 + \cot x)^2}$

9. $y = \frac{4}{\cos x} + \frac{1}{\tan x} = 4\sec x + \cot x \;\Rightarrow\; \frac{dy}{dx} = 4\sec x \tan x - \csc^2 x$

11. $y = x^2 \sin x + 2x \cos x - 2\sin x \;\Rightarrow\; \frac{dy}{dx} = (x^2 \cos x + (\sin x)(2x)) + ((2x)(-\sin x) + (\cos x)(2)) - 2\cos x$

$= x^2 \cos x + 2x \sin x - 2x \sin x + 2\cos x - 2\cos x = x^2 \cos x$

13. $s = \tan t - t \;\Rightarrow\; \frac{ds}{dt} = \frac{d}{dt}(\tan t) - 1 = \sec^2 t - 1 = \tan^2 t$

15. $s = \frac{1 + \csc t}{1 - \csc t} \;\Rightarrow\; \frac{ds}{dt} = \frac{(1 - \csc t)(-\csc t \cot t) - (1 + \csc t)(\csc t \cot t)}{(1 - \csc t)^2}$

$= \frac{-\csc t \cot t + \csc^2 t \cot t - \csc t \cot t - \csc^2 t \cot t}{(1 - \csc t)^2} = \frac{-2\csc t \cot t}{(1 - \csc t)^2}$

17. $r = 4 - \theta^2 \sin \theta \;\Rightarrow\; \frac{dr}{d\theta} = -\left(\theta^2 \frac{d}{d\theta}(\sin \theta) + (\sin \theta)(2\theta)\right) = -\left(\theta^2 \cos \theta + 2\theta \sin \theta\right) = -\theta(\theta \cos \theta + 2\sin \theta)$

19. $r = \sec \theta \csc \theta \;\Rightarrow\; \frac{dr}{d\theta} = (\sec \theta)(-\csc \theta \cot \theta) + (\csc \theta)(\sec \theta \tan \theta)$

$= \left(\frac{-1}{\cos \theta}\right)\left(\frac{1}{\sin \theta}\right)\left(\frac{\cos \theta}{\sin \theta}\right) + \left(\frac{1}{\sin \theta}\right)\left(\frac{1}{\cos \theta}\right)\left(\frac{\sin \theta}{\cos \theta}\right) = \frac{-1}{\sin^2 \theta} + \frac{1}{\cos^2 \theta} = \sec^2 \theta - \csc^2 \theta$

21. $p = 5 + \frac{1}{\cot q} = 5 + \tan q \;\Rightarrow\; \frac{dp}{dq} = \sec^2 q$

23. $p = \frac{\sin q + \cos q}{\cos q} \;\Rightarrow\; \frac{dp}{dq} = \frac{(\cos q)(\cos q - \sin q) - (\sin q + \cos q)(-\sin q)}{\cos^2 q}$

$= \frac{\cos^2 q - \cos q \sin q + \sin^2 q + \cos q \sin q}{\cos^2 q} = \frac{1}{\cos^2 q} = \sec^2 q$

25. (a) $y = \csc x \;\Rightarrow\; y' = -\csc x \cot x \;\Rightarrow\; y'' = -((\csc x)(-\csc^2 x) + (\cot x)(-\csc x \cot x)) = \csc^3 x + \csc x \cot^2 x$

$= (\csc x)(\csc^2 x + \cot^2 x) = (\csc x)(\csc^2 x + \csc^2 x - 1) = 2\csc^3 x - \csc x$

(b) $y = \sec x \;\Rightarrow\; y' = \sec x \tan x \;\Rightarrow\; y'' = (\sec x)(\sec^2 x) + (\tan x)(\sec x \tan x) = \sec^3 x + \sec x \tan^2 x$

$= (\sec x)(\sec^2 x + \tan^2 x) = (\sec x)(\sec^2 x + \sec^2 x - 1) = 2\sec^3 x - \sec x$

27. $y = \sin x \Rightarrow y' = \cos x \Rightarrow$ slope of tangent at
    $x = -\pi$ is $y'(-\pi) = \cos(-\pi) = -1$; slope of
    tangent at $x = 0$ is $y'(0) = \cos(0) = 1$; and
    slope of tangent at $x = \frac{3\pi}{2}$ is $y'\left(\frac{3\pi}{2}\right) = \cos\frac{3\pi}{2}$
    $= 0$. The tangent at $(-\pi, 0)$ is $y - 0 = -1(x + \pi)$,
    or $y = -x - \pi$; the tangent at $(0, 0)$ is
    $y - 0 = 1(x - 0)$, or $y = x$; and the tangent at
    $\left(\frac{3\pi}{2}, -1\right)$ is $y = -1$.

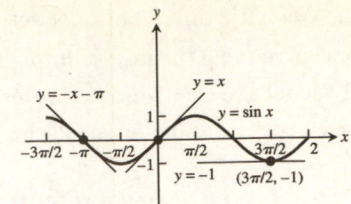

29. $y = \sec x \Rightarrow y' = \sec x \tan x \Rightarrow$ slope of tangent at
    $x = -\frac{\pi}{3}$ is $\sec\left(-\frac{\pi}{3}\right)\tan\left(-\frac{\pi}{3}\right) = -2\sqrt{3}$; slope of tangent
    at $x = \frac{\pi}{4}$ is $\sec\left(\frac{\pi}{4}\right)\tan\left(\frac{\pi}{4}\right) = \sqrt{2}$. The tangent at the point
    $\left(-\frac{\pi}{3}, \sec\left(-\frac{\pi}{3}\right)\right) = \left(-\frac{\pi}{3}, 2\right)$ is $y - 2 = -2\sqrt{3}\left(x + \frac{\pi}{3}\right)$;
    the tangent at the point $\left(\frac{\pi}{4}, \sec\left(\frac{\pi}{4}\right)\right) = \left(\frac{\pi}{4}, \sqrt{2}\right)$ is $y - \sqrt{2}$
    $= \sqrt{2}\left(x - \frac{\pi}{4}\right)$.

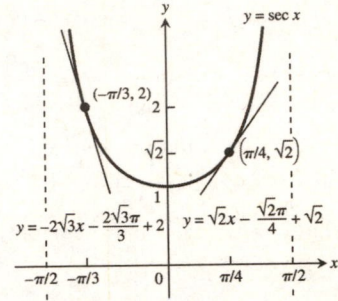

31. Yes, $y = x + \sin x \Rightarrow y' = 1 + \cos x$; horizontal tangent occurs where $1 + \cos x = 0 \Rightarrow \cos x = -1$
    $\Rightarrow x = \pi$

33. No, $y = x - \cot x \Rightarrow y' = 1 + \csc^2 x$; horizontal tangent occurs where $1 + \csc^2 x = 0 \Rightarrow \csc^2 x = -1$. But there
    are no x-values for which $\csc^2 x = -1$.

35. We want all points on the curve where the tangent
    line has slope 2. Thus, $y = \tan x \Rightarrow y' = \sec^2 x$ so
    that $y' = 2 \Rightarrow \sec^2 x = 2 \Rightarrow \sec x = \pm\sqrt{2}$
    $\Rightarrow x = \pm\frac{\pi}{4}$. Then the tangent line at $\left(\frac{\pi}{4}, 1\right)$ has
    equation $y - 1 = 2\left(x - \frac{\pi}{4}\right)$; the tangent line at
    $\left(-\frac{\pi}{4}, -1\right)$ has equation $y + 1 = 2\left(x + \frac{\pi}{4}\right)$.

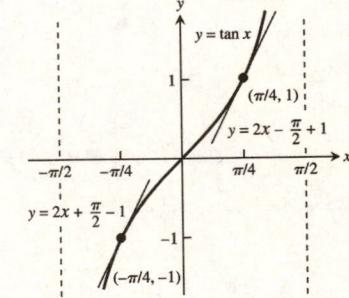

37. $y = 4 + \cot x - 2\csc x \Rightarrow y' = -\csc^2 x + 2\csc x \cot x = -\left(\frac{1}{\sin x}\right)\left(\frac{1 - 2\cos x}{\sin x}\right)$
    (a) When $x = \frac{\pi}{2}$, then $y' = -1$; the tangent line is $y = -x + \frac{\pi}{2} + 2$.
    (b) To find the location of the horizontal tangent set $y' = 0 \Rightarrow 1 - 2\cos x = 0 \Rightarrow x = \frac{\pi}{3}$ radians. When $x = \frac{\pi}{3}$,
        then $y = 4 - \sqrt{3}$ is the horizontal tangent.

39. $\lim\limits_{x \to 2} \sin\left(\frac{1}{x} - \frac{1}{2}\right) = \sin\left(\frac{1}{2} - \frac{1}{2}\right) = \sin 0 = 0$

41. $\lim\limits_{x \to 0} \sec\left[\cos x + \pi\tan\left(\frac{\pi}{4\sec x}\right) - 1\right] = \sec\left[\cos 0 + \pi\tan\left(\frac{\pi}{4\sec 0}\right) - 1\right] = \sec\left[1 + \pi\tan\left(\frac{\pi}{4}\right) - 1\right] = \sec\pi = -1$

43. $\lim\limits_{t \to 0} \tan\left(1 - \frac{\sin t}{t}\right) = \tan\left(1 - \lim\limits_{t \to 0}\frac{\sin t}{t}\right) = \tan(1 - 1) = 0$

45. $s = 2 - 2\sin t \Rightarrow v = \frac{ds}{dt} = -2\cos t \Rightarrow a = \frac{dv}{dt} = 2\sin t \Rightarrow j = \frac{da}{dt} = 2\cos t$. Therefore, velocity $= v\left(\frac{\pi}{4}\right)$
    $= -\sqrt{2}$ m/sec; speed $= \left|v\left(\frac{\pi}{4}\right)\right| = \sqrt{2}$ m/sec; acceleration $= a\left(\frac{\pi}{4}\right) = \sqrt{2}$ m/sec^2; jerk $= j\left(\frac{\pi}{4}\right) = \sqrt{2}$ m/sec^3.

47. $\lim\limits_{x \to 0} f(x) = \lim\limits_{x \to 0} \frac{\sin^2 3x}{x^2} = \lim\limits_{x \to 0} 9 \left(\frac{\sin 3x}{3x}\right) \left(\frac{\sin 3x}{3x}\right) = 9$ so that f is continuous at $x = 0 \Rightarrow \lim\limits_{x \to 0} f(x) = f(0)$
$\Rightarrow 9 = c$.

49. $\frac{d^{999}}{dx^{999}} (\cos x) = \sin x$ because $\frac{d^4}{dx^4} (\cos x) = \cos x \Rightarrow$ the derivative of cos x any number of times that is a
multiple of 4 is cos x. Thus, dividing 999 by 4 gives $999 = 249 \cdot 4 + 3 \Rightarrow \frac{d^{999}}{dx^{999}} (\cos x)$
$= \frac{d^3}{dx^3} \left[\frac{d^{249 \cdot 4}}{dx^{249 \cdot 4}} (\cos x) \right] = \frac{d^3}{dx^3} (\cos x) = \sin x$.

51.

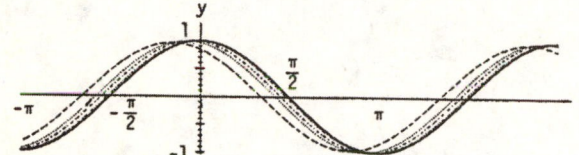 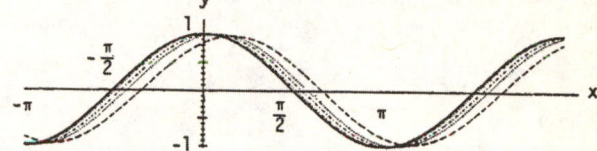

As h takes on the values of 1, 0.5, 0.3 and 0.1 the corresponding dashed curves of $y = \frac{\sin(x+h) - \sin x}{h}$ get
closer and closer to the black curve $y = \cos x$ because $\frac{d}{dx} (\sin x) = \lim\limits_{h \to 0} \frac{\sin(x+h) - \sin x}{h} = \cos x$. The same
is true as h takes on the values of $-1, -0.5, -0.3$ and -0.1.

53. (a)

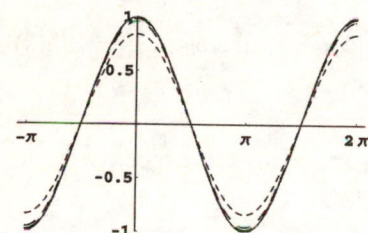

The dashed curves of $y = \frac{\sin(x+h) - \sin(x-h)}{2h}$ are closer to the black curve $y = \cos x$ than the corresponding dashed
curves in Exercise 51 illustrating that the centered difference quotient is a better approximation of the derivative of
this function.

(b)

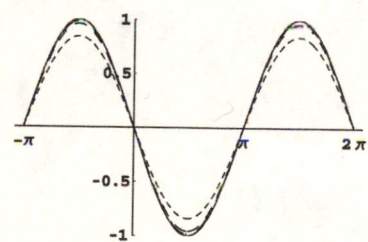

The dashed curves of $y = \frac{\cos(x+h) - \cos(x-h)}{2h}$ are closer to the black curve $y = -\sin x$ than the corresponding dashed
curves in Exercise 52 illustrating that the centered difference quotient is a better approximation of the derivative of
this function.

55. $y = \tan x \Rightarrow y' = \sec^2 x$, so the smallest value
$y' = \sec^2 x$ takes on is $y' = 1$ when $x = 0$;
y' has no maximum value since $\sec^2 x$ has no
largest value on $\left(-\frac{\pi}{2}, \frac{\pi}{2}\right)$; y' is never negative
since $\sec^2 x \geq 1$.

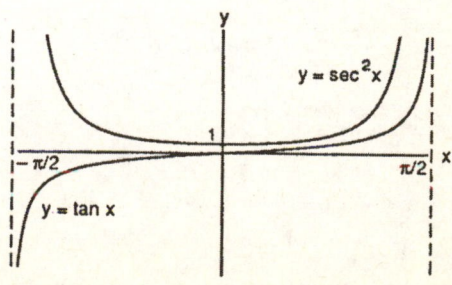

57. $y = \frac{\sin x}{x}$ appears to cross the y-axis at $y = 1$, since

$\lim\limits_{x \to 0} \frac{\sin x}{x} = 1$; $y = \frac{\sin 2x}{x}$ appears to cross the y-axis

at $y = 2$, since $\lim\limits_{x \to 0} \frac{\sin 2x}{x} = 2$; $y = \frac{\sin 4x}{x}$ appears to

cross the y-axis at $y = 4$, since $\lim\limits_{x \to 0} \frac{\sin 4x}{x} = 4$.

However, none of these graphs actually cross the y-axis

since $x = 0$ is not in the domain of the functions. Also,

$\lim\limits_{x \to 0} \frac{\sin 5x}{x} = 5$, $\lim\limits_{x \to 0} \frac{\sin(-3x)}{x} = -3$, and $\lim\limits_{x \to 0} \frac{\sin kx}{x}$

$= k \Rightarrow$ the graphs of $y = \frac{\sin 5x}{x}$, $y = \frac{\sin(-3x)}{x}$, and

$y = \frac{\sin kx}{x}$ approach 5, −3, and k, respectively, as

$x \to 0$. However, the graphs do not actually cross the

y-axis.

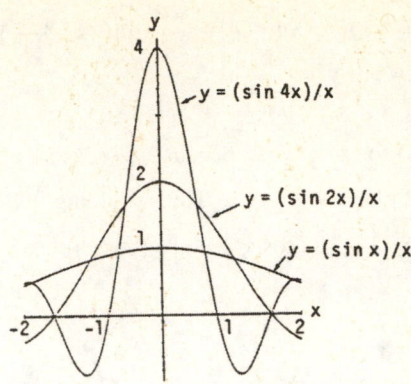

3.5 THE CHAIN RULE AND PARAMETRIC EQUATIONS

1. $f(u) = 6u - 9 \Rightarrow f'(u) = 6 \Rightarrow f'(g(x)) = 6$; $g(x) = \frac{1}{2}x^4 \Rightarrow g'(x) = 2x^3$; therefore $\frac{dy}{dx} = f'(g(x))g'(x)$

$= 6 \cdot 2x^3 = 12x^3$

3. $f(u) = \sin u \Rightarrow f'(u) = \cos u \Rightarrow f'(g(x)) = \cos(3x + 1)$; $g(x) = 3x + 1 \Rightarrow g'(x) = 3$; therefore $\frac{dy}{dx} = f'(g(x))g'(x)$

$= (\cos(3x + 1))(3) = 3\cos(3x + 1)$

5. $f(u) = \cos u \Rightarrow f'(u) = -\sin u \Rightarrow f'(g(x)) = -\sin(\sin x)$; $g(x) = \sin x \Rightarrow g'(x) = \cos x$; therefore

$\frac{dy}{dx} = f'(g(x))g'(x) = -(\sin(\sin x))\cos x$

7. $f(u) = \tan u \Rightarrow f'(u) = \sec^2 u \Rightarrow f'(g(x)) = \sec^2(10x - 5)$; $g(x) = 10x - 5 \Rightarrow g'(x) = 10$; therefore

$\frac{dy}{dx} = f'(g(x))g'(x) = (\sec^2(10x - 5))(10) = 10\sec^2(10x - 5)$

9. With $u = (2x + 1)$, $y = u^5$: $\frac{dy}{dx} = \frac{dy}{du}\frac{du}{dx} = 5u^4 \cdot 2 = 10(2x + 1)^4$

11. With $u = \left(1 - \frac{x}{7}\right)$, $y = u^{-7}$: $\frac{dy}{dx} = \frac{dy}{du}\frac{du}{dx} = -7u^{-8} \cdot \left(-\frac{1}{7}\right) = \left(1 - \frac{x}{7}\right)^{-8}$

13. With $u = \left(\frac{x^2}{8} + x - \frac{1}{x}\right)$, $y = u^4$: $\frac{dy}{dx} = \frac{dy}{du}\frac{du}{dx} = 4u^3 \cdot \left(\frac{x}{4} + 1 + \frac{1}{x^2}\right) = 4\left(\frac{x^2}{8} + x - \frac{1}{x}\right)^3\left(\frac{x}{4} + 1 + \frac{1}{x^2}\right)$

15. With $u = \tan x$, $y = \sec u$: $\frac{dy}{dx} = \frac{dy}{du}\frac{du}{dx} = (\sec u \tan u)(\sec^2 x) = (\sec(\tan x)\tan(\tan x))\sec^2 x$

17. With $u = \sin x$, $y = u^3$: $\frac{dy}{dx} = \frac{dy}{du}\frac{du}{dx} = 3u^2 \cos x = 3(\sin^2 x)(\cos x)$

19. $p = \sqrt{3 - t} = (3 - t)^{1/2} \Rightarrow \frac{dp}{dt} = \frac{1}{2}(3 - t)^{-1/2} \cdot \frac{d}{dt}(3 - t) = -\frac{1}{2}(3 - t)^{-1/2} = \frac{-1}{2\sqrt{3 - t}}$

21. $s = \frac{4}{3\pi}\sin 3t + \frac{4}{5\pi}\cos 5t \Rightarrow \frac{ds}{dt} = \frac{4}{3\pi}\cos 3t \cdot \frac{d}{dt}(3t) + \frac{4}{5\pi}(-\sin 5t) \cdot \frac{d}{dt}(5t) = \frac{4}{\pi}\cos 3t - \frac{4}{\pi}\sin 5t$

$= \frac{4}{\pi}(\cos 3t - \sin 5t)$

23. $r = (\csc\theta + \cot\theta)^{-1} \Rightarrow \frac{dr}{d\theta} = -(\csc\theta + \cot\theta)^{-2}\frac{d}{d\theta}(\csc\theta + \cot\theta) = \frac{\csc\theta\cot\theta + \csc^2\theta}{(\csc\theta + \cot\theta)^2} = \frac{\csc\theta(\cot\theta + \csc\theta)}{(\csc\theta + \cot\theta)^2}$

$= \frac{\csc\theta}{\csc\theta + \cot\theta}$

25. $y = x^2 \sin^4 x + x \cos^{-2} x \Rightarrow \frac{dy}{dx} = x^2 \frac{d}{dx}(\sin^4 x) + \sin^4 x \cdot \frac{d}{dx}(x^2) + x \frac{d}{dx}(\cos^{-2} x) + \cos^{-2} x \cdot \frac{d}{dx}(x)$

$= x^2 \left(4 \sin^3 x \frac{d}{dx}(\sin x)\right) + 2x \sin^4 x + x \left(-2 \cos^{-3} x \cdot \frac{d}{dx}(\cos x)\right) + \cos^{-2} x$

$= x^2 \left(4 \sin^3 x \cos x\right) + 2x \sin^4 x + x((-2 \cos^{-3} x)(-\sin x)) + \cos^{-2} x$

$= 4x^2 \sin^3 x \cos x + 2x \sin^4 x + 2x \sin x \cos^{-3} x + \cos^{-2} x$

27. $y = \frac{1}{21}(3x-2)^7 + \left(4 - \frac{1}{2x^2}\right)^{-1} \Rightarrow \frac{dy}{dx} = \frac{7}{21}(3x-2)^6 \cdot \frac{d}{dx}(3x-2) + (-1)\left(4 - \frac{1}{2x^2}\right)^{-2} \cdot \frac{d}{dx}\left(4 - \frac{1}{2x^2}\right)$

$= \frac{7}{21}(3x-2)^6 \cdot 3 + (-1)\left(4 - \frac{1}{2x^2}\right)^{-2}\left(\frac{1}{x^3}\right) = (3x-2)^6 - \frac{1}{x^3\left(4 - \frac{1}{2x^2}\right)^2}$

29. $y = (4x+3)^4(x+1)^{-3} \Rightarrow \frac{dy}{dx} = (4x+3)^4(-3)(x+1)^{-4} \cdot \frac{d}{dx}(x+1) + (x+1)^{-3}(4)(4x+3)^3 \cdot \frac{d}{dx}(4x+3)$

$= (4x+3)^4(-3)(x+1)^{-4}(1) + (x+1)^{-3}(4)(4x+3)^3(4) = -3(4x+3)^4(x+1)^{-4} + 16(4x+3)^3(x+1)^{-3}$

$= \frac{(4x+3)^3}{(x+1)^4}[-3(4x+3) + 16(x+1)] = \frac{(4x+3)^3(4x+7)}{(x+1)^4}$

31. $h(x) = x \tan\left(2\sqrt{x}\right) + 7 \Rightarrow h'(x) = x \frac{d}{dx}\left(\tan\left(2x^{1/2}\right)\right) + \tan\left(2x^{1/2}\right) \cdot \frac{d}{dx}(x) + 0$

$= x \sec^2\left(2x^{1/2}\right) \cdot \frac{d}{dx}\left(2x^{1/2}\right) + \tan\left(2x^{1/2}\right) = x \sec^2\left(2\sqrt{x}\right) \cdot \frac{1}{\sqrt{x}} + \tan\left(2\sqrt{x}\right) = \sqrt{x}\sec^2\left(2\sqrt{x}\right) + \tan\left(2\sqrt{x}\right)$

33. $f(\theta) = \left(\frac{\sin\theta}{1+\cos\theta}\right)^2 \Rightarrow f'(\theta) = 2\left(\frac{\sin\theta}{1+\cos\theta}\right) \cdot \frac{d}{d\theta}\left(\frac{\sin\theta}{1+\cos\theta}\right) = \frac{2\sin\theta}{1+\cos\theta} \cdot \frac{(1+\cos\theta)(\cos\theta) - (\sin\theta)(-\sin\theta)}{(1+\cos\theta)^2}$

$= \frac{(2\sin\theta)(\cos\theta + \cos^2\theta + \sin^2\theta)}{(1+\cos\theta)^3} = \frac{(2\sin\theta)(\cos\theta + 1)}{(1+\cos\theta)^3} = \frac{2\sin\theta}{(1+\cos\theta)^2}$

35. $r = \sin\left(\theta^2\right)\cos\left(2\theta\right) \Rightarrow \frac{dr}{d\theta} = \sin\left(\theta^2\right)(-\sin 2\theta)\frac{d}{d\theta}(2\theta) + \cos\left(2\theta\right)(\cos\left(\theta^2\right)) \cdot \frac{d}{d\theta}\left(\theta^2\right)$

$= \sin\left(\theta^2\right)(-\sin 2\theta)(2) + (\cos 2\theta)(\cos\left(\theta^2\right))(2\theta) = -2\sin\left(\theta^2\right)\sin\left(2\theta\right) + 2\theta\cos\left(2\theta\right)\cos\left(\theta^2\right)$

37. $q = \sin\left(\frac{t}{\sqrt{t+1}}\right) \Rightarrow \frac{dq}{dt} = \cos\left(\frac{t}{\sqrt{t+1}}\right) \cdot \frac{d}{dt}\left(\frac{t}{\sqrt{t+1}}\right) = \cos\left(\frac{t}{\sqrt{t+1}}\right) \cdot \frac{\sqrt{t+1}(1) - t \cdot \frac{d}{dt}\left(\sqrt{t+1}\right)}{\left(\sqrt{t+1}\right)^2}$

$= \cos\left(\frac{t}{\sqrt{t+1}}\right) \cdot \frac{\sqrt{t+1} - \frac{t}{2\sqrt{t+1}}}{t+1} = \cos\left(\frac{t}{\sqrt{t+1}}\right)\left(\frac{2(t+1) - t}{2(t+1)^{3/2}}\right) = \left(\frac{t+2}{2(t+1)^{3/2}}\right)\cos\left(\frac{t}{\sqrt{t+1}}\right)$

39. $y = \sin^2\left(\pi t - 2\right) \Rightarrow \frac{dy}{dt} = 2\sin\left(\pi t - 2\right) \cdot \frac{d}{dt}\sin\left(\pi t - 2\right) = 2\sin\left(\pi t - 2\right) \cdot \cos\left(\pi t - 2\right) \cdot \frac{d}{dt}\left(\pi t - 2\right)$

$= 2\pi\sin\left(\pi t - 2\right)\cos\left(\pi t - 2\right)$

41. $y = (1 + \cos 2t)^{-4} \Rightarrow \frac{dy}{dt} = -4(1+\cos 2t)^{-5} \cdot \frac{d}{dt}(1+\cos 2t) = -4(1+\cos 2t)^{-5}(-\sin 2t) \cdot \frac{d}{dt}(2t) = \frac{8\sin 2t}{(1+\cos 2t)^5}$

43. $y = \sin\left(\cos\left(2t - 5\right)\right) \Rightarrow \frac{dy}{dt} = \cos\left(\cos\left(2t-5\right)\right) \cdot \frac{d}{dt}\cos\left(2t-5\right) = \cos\left(\cos\left(2t-5\right)\right) \cdot (-\sin\left(2t-5\right)) \cdot \frac{d}{dt}(2t-5)$

$= -2\cos\left(\cos\left(2t-5\right)\right)(\sin\left(2t-5\right))$

45. $y = \left[1 + \tan^4\left(\frac{t}{12}\right)\right]^3 \Rightarrow \frac{dy}{dt} = 3\left[1 + \tan^4\left(\frac{t}{12}\right)\right]^2 \cdot \frac{d}{dt}\left[1 + \tan^4\left(\frac{t}{12}\right)\right] = 3\left[1 + \tan^4\left(\frac{t}{12}\right)\right]^2\left[4\tan^3\left(\frac{t}{12}\right) \cdot \frac{d}{dt}\tan\left(\frac{t}{12}\right)\right]$

$= 12\left[1 + \tan^4\left(\frac{t}{12}\right)\right]^2\left[\tan^3\left(\frac{t}{12}\right)\sec^2\left(\frac{t}{12}\right) \cdot \frac{1}{12}\right] = \left[1 + \tan^4\left(\frac{t}{12}\right)\right]^2\left[\tan^3\left(\frac{t}{12}\right)\sec^2\left(\frac{t}{12}\right)\right]$

47. $y = (1 + \cos\left(t^2\right))^{1/2} \Rightarrow \frac{dy}{dt} = \frac{1}{2}(1 + \cos\left(t^2\right))^{-1/2} \cdot \frac{d}{dt}(1 + \cos\left(t^2\right)) = \frac{1}{2}(1 + \cos\left(t^2\right))^{-1/2}\left(-\sin\left(t^2\right) \cdot \frac{d}{dt}\left(t^2\right)\right)$

$= -\frac{1}{2}(1 + \cos\left(t^2\right))^{-1/2}(\sin\left(t^2\right)) \cdot 2t = -\frac{t\sin\left(t^2\right)}{\sqrt{1 + \cos\left(t^2\right)}}$

49. $y = \left(1 + \frac{1}{x}\right)^3 \Rightarrow y' = 3\left(1 + \frac{1}{x}\right)^2\left(-\frac{1}{x^2}\right) = -\frac{3}{x^2}\left(1 + \frac{1}{x}\right)^2 \Rightarrow y'' = \left(-\frac{3}{x^2}\right) \cdot \frac{d}{dx}\left(1 + \frac{1}{x}\right)^2 - \left(1 + \frac{1}{x}\right)^2 \cdot \frac{d}{dx}\left(\frac{3}{x^2}\right)$

$= \left(-\frac{3}{x^2}\right)\left(2\left(1 + \frac{1}{x}\right)\left(-\frac{1}{x^2}\right)\right) + \left(\frac{6}{x^3}\right)\left(1 + \frac{1}{x}\right)^2 = \frac{6}{x^4}\left(1 + \frac{1}{x}\right) + \frac{6}{x^3}\left(1 + \frac{1}{x}\right)^2 = \frac{6}{x^3}\left(1 + \frac{1}{x}\right)\left(\frac{1}{x} + 1 + \frac{1}{x}\right)$

$= \frac{6}{x^3}\left(1 + \frac{1}{x}\right)\left(1 + \frac{2}{x}\right)$

51. $y = \frac{1}{9}\cot(3x-1) \Rightarrow y' = -\frac{1}{9}\csc^2(3x-1)(3) = -\frac{1}{3}\csc^2(3x-1) \Rightarrow y'' = \left(-\frac{2}{3}\right)\left(\csc(3x-1)\cdot\frac{d}{dx}\csc(3x-1)\right)$

$= -\frac{2}{3}\csc(3x-1)(-\csc(3x-1)\cot(3x-1)\cdot\frac{d}{dx}(3x-1)) = 2\csc^2(3x-1)\cot(3x-1)$

53. $g(x) = \sqrt{x} \Rightarrow g'(x) = \frac{1}{2\sqrt{x}} \Rightarrow g(1) = 1$ and $g'(1) = \frac{1}{2}$; $f(u) = u^5+1 \Rightarrow f'(u) = 5u^4 \Rightarrow f'(g(1)) = f'(1) = 5$;

therefore, $(f \circ g)'(1) = f'(g(1))\cdot g'(1) = 5\cdot\frac{1}{2} = \frac{5}{2}$

55. $g(x) = 5\sqrt{x} \Rightarrow g'(x) = \frac{5}{2\sqrt{x}} \Rightarrow g(1) = 5$ and $g'(1) = \frac{5}{2}$; $f(u) = \cot\left(\frac{\pi u}{10}\right) \Rightarrow f'(u) = -\csc^2\left(\frac{\pi u}{10}\right)\left(\frac{\pi}{10}\right)$

$= \frac{-\pi}{10}\csc^2\left(\frac{\pi u}{10}\right) \Rightarrow f'(g(1)) = f'(5) = -\frac{\pi}{10}\csc^2\left(\frac{\pi}{2}\right) = -\frac{\pi}{10}$; therefore, $(f \circ g)'(1) = f'(g(1))g'(1) = -\frac{\pi}{10}\cdot\frac{5}{2}$

$= -\frac{\pi}{4}$

57. $g(x) = 10x^2+x+1 \Rightarrow g'(x) = 20x+1 \Rightarrow g(0) = 1$ and $g'(0) = 1$; $f(u) = \frac{2u}{u^2+1} \Rightarrow f'(u) = \frac{(u^2+1)(2)-(2u)(2u)}{(u^2+1)^2}$

$= \frac{-2u^2+2}{(u^2+1)^2} \Rightarrow f'(g(0)) = f'(1) = 0$; therefore, $(f \circ g)'(0) = f'(g(0))g'(0) = 0\cdot 1 = 0$

59. (a) $y = 2f(x) \Rightarrow \frac{dy}{dx} = 2f'(x) \Rightarrow \left.\frac{dy}{dx}\right|_{x=2} = 2f'(2) = 2\left(\frac{1}{3}\right) = \frac{2}{3}$

(b) $y = f(x)+g(x) \Rightarrow \frac{dy}{dx} = f'(x)+g'(x) \Rightarrow \left.\frac{dy}{dx}\right|_{x=3} = f'(3)+g'(3) = 2\pi+5$

(c) $y = f(x)\cdot g(x) \Rightarrow \frac{dy}{dx} = f(x)g'(x)+g(x)f'(x) \Rightarrow \left.\frac{dy}{dx}\right|_{x=3} = f(3)g'(3)+g(3)f'(3) = 3\cdot 5+(-4)(2\pi) = 15-8\pi$

(d) $y = \frac{f(x)}{g(x)} \Rightarrow \frac{dy}{dx} = \frac{g(x)f'(x)-f(x)g'(x)}{[g(x)]^2} \Rightarrow \left.\frac{dy}{dx}\right|_{x=2} = \frac{g(2)f'(2)-f(2)g'(2)}{[g(2)]^2} = \frac{(2)\left(\frac{1}{3}\right)-(8)(-3)}{2^2} = \frac{37}{6}$

(e) $y = f(g(x)) \Rightarrow \frac{dy}{dx} = f'(g(x))g'(x) \Rightarrow \left.\frac{dy}{dx}\right|_{x=2} = f'(g(2))g'(2) = f'(2)(-3) = \frac{1}{3}(-3) = -1$

(f) $y = (f(x))^{1/2} \Rightarrow \frac{dy}{dx} = \frac{1}{2}(f(x))^{-1/2}\cdot f'(x) = \frac{f'(x)}{2\sqrt{f(x)}} \Rightarrow \left.\frac{dy}{dx}\right|_{x=2} = \frac{f'(2)}{2\sqrt{f(2)}} = \frac{\left(\frac{1}{3}\right)}{2\sqrt{8}} = \frac{1}{6\sqrt{8}} = \frac{1}{12\sqrt{2}} = \frac{\sqrt{2}}{24}$

(g) $y = (g(x))^{-2} \Rightarrow \frac{dy}{dx} = -2(g(x))^{-3}\cdot g'(x) \Rightarrow \left.\frac{dy}{dx}\right|_{x=3} = -2(g(3))^{-3}g'(3) = -2(-4)^{-3}\cdot 5 = \frac{5}{32}$

(h) $y = \left((f(x))^2+(g(x))^2\right)^{1/2} \Rightarrow \frac{dy}{dx} = \frac{1}{2}\left((f(x))^2+(g(x))^2\right)^{-1/2}(2f(x)\cdot f'(x)+2g(x)\cdot g'(x))$

$\Rightarrow \left.\frac{dy}{dx}\right|_{x=2} = \frac{1}{2}\left((f(2))^2+(g(2))^2\right)^{-1/2}(2f(2)f'(2)+2g(2)g'(2)) = \frac{1}{2}\left(8^2+2^2\right)^{-1/2}\left(2\cdot 8\cdot\frac{1}{3}+2\cdot 2\cdot(-3)\right)$

$= -\frac{5}{3\sqrt{17}}$

61. $\frac{ds}{dt} = \frac{ds}{d\theta}\cdot\frac{d\theta}{dt}$: $s = \cos\theta \Rightarrow \frac{ds}{d\theta} = -\sin\theta \Rightarrow \left.\frac{ds}{d\theta}\right|_{\theta=\frac{3\pi}{2}} = -\sin\left(\frac{3\pi}{2}\right) = 1$ so that $\frac{ds}{dt} = \frac{ds}{d\theta}\cdot\frac{d\theta}{dt} = 1\cdot 5 = 5$

63. With $y = x$, we should get $\frac{dy}{dx} = 1$ for both (a) and (b):

(a) $y = \frac{u}{5}+7 \Rightarrow \frac{dy}{du} = \frac{1}{5}$; $u = 5x-35 \Rightarrow \frac{du}{dx} = 5$; therefore, $\frac{dy}{dx} = \frac{dy}{du}\cdot\frac{du}{dx} = \frac{1}{5}\cdot 5 = 1$, as expected

(b) $y = 1+\frac{1}{u} \Rightarrow \frac{dy}{du} = -\frac{1}{u^2}$; $u = (x-1)^{-1} \Rightarrow \frac{du}{dx} = -(x-1)^{-2}(1) = \frac{-1}{(x-1)^2}$; therefore $\frac{dy}{dx} = \frac{dy}{du}\cdot\frac{du}{dx}$

$= \frac{-1}{u^2}\cdot\frac{-1}{(x-1)^2} = \frac{-1}{((x-1)^{-1})^2}\cdot\frac{-1}{(x-1)^2} = (x-1)^2\cdot\frac{1}{(x-1)^2} = 1$, again as expected

65. $y = 2\tan\left(\frac{\pi x}{4}\right) \Rightarrow \frac{dy}{dx} = \left(2\sec^2\frac{\pi x}{4}\right)\left(\frac{\pi}{4}\right) = \frac{\pi}{2}\sec^2\frac{\pi x}{4}$

(a) $\left.\frac{dy}{dx}\right|_{x=1} = \frac{\pi}{2}\sec^2\left(\frac{\pi}{4}\right) = \pi \Rightarrow$ slope of tangent is 2; thus, $y(1) = 2\tan\left(\frac{\pi}{4}\right) = 2$ and $y'(1) = \pi \Rightarrow$ tangent line is

given by $y-2 = \pi(x-1) \Rightarrow y = \pi x+2-\pi$

(b) $y' = \frac{\pi}{2}\sec^2\left(\frac{\pi x}{4}\right)$ and the smallest value the secant function can have in $-2 < x < 2$ is $1 \Rightarrow$ the minimum

value of y' is $\frac{\pi}{2}$ and that occurs when $\frac{\pi}{2} = \frac{\pi}{2}\sec^2\left(\frac{\pi x}{4}\right) \Rightarrow 1 = \sec^2\left(\frac{\pi x}{4}\right) \Rightarrow \pm 1 = \sec\left(\frac{\pi x}{4}\right) \Rightarrow x = 0$.

67. $x = \cos 2t, \ y = \sin 2t, \ 0 \le t \le \pi$

    $\Rightarrow \cos^2 2t + \sin^2 2t = 1 \ \Rightarrow \ x^2 + y^2 = 1$

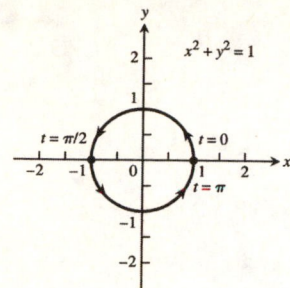

69. $x = 4 \cos t, \ y = 2 \sin t, \ 0 \le t \le 2\pi$

    $\Rightarrow \frac{16 \cos^2 t}{16} + \frac{4 \sin^2 t}{4} = 1 \ \Rightarrow \ \frac{x^2}{16} + \frac{y^2}{4} = 1$

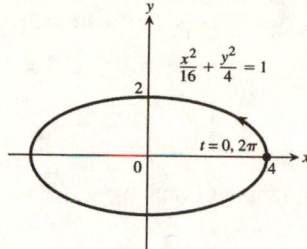

71. $x = 3t, \ y = 9t^2, \ -\infty < t < \infty \ \Rightarrow \ y = x^2$

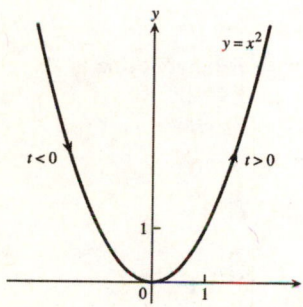

73. $x = 2t - 5, \ y = 4t - 7, \ -\infty < t < \infty$

    $\Rightarrow \ x + 5 = 2t \ \Rightarrow \ 2(x + 5) = 4t$

    $\Rightarrow \ y = 2(x + 5) - 7 \ \Rightarrow \ y = 2x + 3$

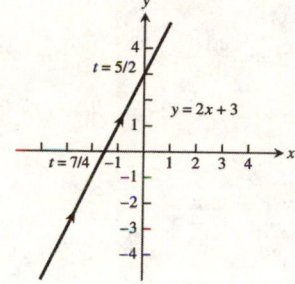

75. $x = t, \ y = \sqrt{1 - t^2}, \ -1 \le t \le 0$

    $\Rightarrow \ y = \sqrt{1 - x^2}$

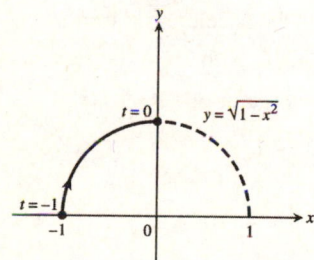

77. $x = \sec^2 t - 1, \ y = \tan t, \ -\frac{\pi}{2} < t < \frac{\pi}{2}$

    $\Rightarrow \ \sec^2 t - 1 = \tan^2 t \ \Rightarrow \ x = y^2$

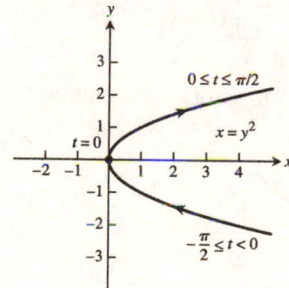

79. (a) $x = a \cos t, \ y = -a \sin t, \ 0 \le t \le 2\pi$

     (b) $x = a \cos t, \ y = a \sin t, \ 0 \le t \le 2\pi$

     (c) $x = a \cos t, \ y = -a \sin t, \ 0 \le t \le 4\pi$

     (d) $x = a \cos t, \ y = a \sin t, \ 0 \le t \le 4\pi$

81. Using $(-1, -3)$ we create the parametric equations $x = -1 + at$ and $y = -3 + bt$, representing a line which goes through $(-1, -3)$ at $t = 0$. We determine a and b so that the line goes through $(4, 1)$ when $t = 1$. Since $4 = -1 + a \Rightarrow a = 5$. Since $1 = -3 + b \Rightarrow b = 4$. Therefore, one possible parameterization is $x = -1 + 5t$, $y = -3 - 4t, \ 0 \le t \le 1$.

83. The lower half of the parabola is given by $x = y^2 + 1$ for $y \le 0$. Substituting t for y, we obtain one possible parameterization $x = t^2 + 1, \ y = t, \ t \le 0$.

85. For simplicity, we assume that x and y are linear functions of t and that the point(x, y) starts at $(2, 3)$ for $t = 0$ and passes through $(-1, -1)$ at $t = 1$. Then $x = f(t)$, where $f(0) = 2$ and $f(1) = -1$.

Since slope $= \frac{\Delta x}{\Delta t} = \frac{-1-2}{1-0} = -3$, $x = f(t) = -3t + 2 = 2 - 3t$. Also, $y = g(t)$, where $g(0) = 3$ and $g(1) = -1$.

Since slope $= \frac{\Delta y}{\Delta t} = \frac{-1-3}{1-0} = -4$. $y = g(t) = -4t + 3 = 3 - 4t$.

One possible parameterization is: $x = 2 - 3t$, $y = 3 - 4t$, $t \geq 0$.

87. $t = \frac{\pi}{4} \Rightarrow x = 2 \cos \frac{\pi}{4} = \sqrt{2}$, $y = 2 \sin \frac{\pi}{4} = \sqrt{2}$; $\frac{dx}{dt} = -2 \sin t$, $\frac{dy}{dt} = 2 \cos t \Rightarrow \frac{dy}{dx} = \frac{dy/dt}{dx/dt} = \frac{2 \cos t}{-2 \sin t} = -\cot t$

$\Rightarrow \frac{dy}{dx}\Big|_{t=\frac{\pi}{4}} = -\cot \frac{\pi}{4} = -1$; tangent line is $y - \sqrt{2} = -1\left(x - \sqrt{2}\right)$ or $y = -x + 2\sqrt{2}$; $\frac{dy'}{dt} = \csc^2 t$

$\Rightarrow \frac{d^2y}{dx^2} = \frac{dy'/dt}{dx/dt} = \frac{\csc^2 t}{-2 \sin t} = -\frac{1}{2 \sin^3 t} \Rightarrow \frac{d^2y}{dx^2}\Big|_{t=\frac{\pi}{4}} = -\sqrt{2}$

89. $t = \frac{1}{4} \Rightarrow x = \frac{1}{4}$, $y = \frac{1}{2}$; $\frac{dx}{dt} = 1$, $\frac{dy}{dt} = \frac{1}{2\sqrt{t}} \Rightarrow \frac{dy}{dx} = \frac{dy/dt}{dx/dt} = \frac{1}{2\sqrt{t}} \Rightarrow \frac{dy}{dx}\Big|_{t=\frac{1}{4}} = \frac{1}{2\sqrt{\frac{1}{4}}} = 1$; tangent line is

$y - \frac{1}{2} = 1 \cdot \left(x - \frac{1}{4}\right)$ or $y = x + \frac{1}{4}$; $\frac{dy'}{dt} = -\frac{1}{4} t^{-3/2} \Rightarrow \frac{d^2y}{dx^2} = \frac{dy'/dt}{dx/dt} = -\frac{1}{4} t^{-3/2} \Rightarrow \frac{d^2y}{dx^2}\Big|_{t=\frac{1}{4}} = -2$

91. $t = -1 \Rightarrow x = 5$, $y = 1$; $\frac{dx}{dt} = 4t$, $\frac{dy}{dt} = 4t^3 \Rightarrow \frac{dy}{dx} = \frac{dy/dt}{dx/dt} = \frac{4t^3}{4t} = t^2 \Rightarrow \frac{dy}{dx}\Big|_{t=-1} = (-1)^2 = 1$; tangent line is

$y - 1 = 1 \cdot (x - 5)$ or $y = x - 4$; $\frac{dy'}{dt} = 2t \Rightarrow \frac{d^2y}{dx^2} = \frac{dy'/dt}{dx/dt} = \frac{2t}{4t} = \frac{1}{2} \Rightarrow \frac{d^2y}{dx^2}\Big|_{t=-1} = \frac{1}{2}$

93. $t = \frac{\pi}{2} \Rightarrow x = \cos \frac{\pi}{2} = 0$, $y = 1 + \sin \frac{\pi}{2} = 2$; $\frac{dx}{dt} = -\sin t$, $\frac{dy}{dt} = \cos t \Rightarrow \frac{dy}{dx} = \frac{\cos t}{-\sin t} = -\cot t$

$\Rightarrow \frac{dy}{dx}\Big|_{t=\frac{\pi}{2}} = -\cot \frac{\pi}{2} = 0$; tangent line is $y = 2$; $\frac{dy'}{dt} = \csc^2 t \Rightarrow \frac{d^2y}{dx^2} = \frac{\csc^2 t}{-\sin t} = -\csc^3 t \Rightarrow \frac{d^2y}{dx^2}\Big|_{t=\frac{\pi}{2}} = -1$

95. $s = A \cos (2\pi bt) \Rightarrow v = \frac{ds}{dt} = -A \sin (2\pi bt)(2\pi b) = -2\pi bA \sin (2\pi bt)$. If we replace b with 2b to double the frequency, the velocity formula gives $v = -4\pi bA \sin (4\pi bt) \Rightarrow$ doubling the frequency causes the velocity to double. Also $v = -2\pi bA \sin (2\pi bt) \Rightarrow a = \frac{dv}{dt} = -4\pi^2 b^2 A \cos (2\pi bt)$. If we replace b with 2b in the acceleration formula, we get $a = -16\pi^2 b^2 A \cos (4\pi bt) \Rightarrow$ doubling the frequency causes the acceleration to quadruple. Finally, $a = -4\pi^2 b^2 A \cos (2\pi bt) \Rightarrow j = \frac{da}{dt} = 8\pi^3 b^3 A \sin (2\pi bt)$. If we replace b with 2b in the jerk formula, we get $j = 64\pi^3 b^3 A \sin (4\pi bt) \Rightarrow$ doubling the frequency multiplies the jerk by a factor of 8.

97. $s = (1 + 4t)^{1/2} \Rightarrow v = \frac{ds}{dt} = \frac{1}{2} (1 + 4t)^{-1/2}(4) = 2(1 + 4t)^{-1/2} \Rightarrow v(6) = 2(1 + 4 \cdot 6)^{-1/2} = \frac{2}{5}$ m/sec;

$v = 2(1 + 4t)^{-1/2} \Rightarrow a = \frac{dv}{dt} = -\frac{1}{2} \cdot 2(1 + 4t)^{-3/2}(4) = -4(1 + 4t)^{-3/2} \Rightarrow a(6) = -4(1 + 4 \cdot 6)^{-3/2} = -\frac{4}{125}$ m/sec^2

99. v proportional to $\frac{1}{\sqrt{s}} \Rightarrow v = \frac{k}{\sqrt{s}}$ for some constant $k \Rightarrow \frac{dv}{ds} = -\frac{k}{2s^{3/2}}$. Thus, $a = \frac{dv}{dt} = \frac{dv}{ds} \cdot \frac{ds}{dt} = \frac{dv}{ds} \cdot v$

$= -\frac{k}{2s^{3/2}} \cdot \frac{k}{\sqrt{s}} = -\frac{k^2}{2} \left(\frac{1}{s^2}\right) \Rightarrow$ acceleration is a constant times $\frac{1}{s^2}$ so a is inversely proportional to s^2.

101. $T = 2\pi \sqrt{\frac{L}{g}} \Rightarrow \frac{dT}{dL} = 2\pi \cdot \frac{1}{2\sqrt{\frac{L}{g}}} \cdot \frac{1}{g} = \frac{\pi}{g\sqrt{\frac{L}{g}}} = \frac{\pi}{\sqrt{gL}}$. Therefore, $\frac{dT}{du} = \frac{dT}{dL} \cdot \frac{dL}{du} = \frac{\pi}{\sqrt{gL}} \cdot kL = \frac{\pi k \sqrt{L}}{\sqrt{g}} = \frac{1}{2} \cdot 2\pi k \sqrt{\frac{L}{g}}$

$= \frac{kT}{2}$, as required.

103. The graph of $y = (f \circ g)(x)$ has a horizontal tangent at $x = 1$ provided that $(f \circ g)'(1) = 0 \Rightarrow f'(g(1))g'(1) = 0$
$\Rightarrow$ either $f'(g(1)) = 0$ or $g'(1) = 0$ (or both) $\Rightarrow$ either the graph of f has a horizontal tangent at $u = g(1)$, or the graph of g has a horizontal tangent at $x = 1$ (or both).

105. As $h \to 0$, the graph of $y = \frac{\sin 2(x+h) - \sin 2x}{h}$
approaches the graph of $y = 2 \cos 2x$ because

$$\lim_{h \to 0} \frac{\sin 2(x+h) - \sin 2x}{h} = \frac{d}{dx}(\sin 2x) = 2 \cos 2x.$$

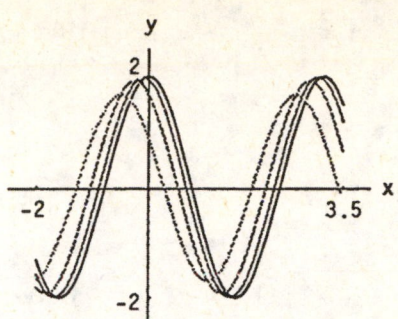

107. $\frac{dx}{dt} = \cos t$ and $\frac{dy}{dt} = 2 \cos 2t \Rightarrow \frac{dy}{dx} = \frac{dy/dt}{dx/dt} = \frac{2 \cos 2t}{\cos t} = \frac{2(2\cos^2 t - 1)}{\cos t}$; then $\frac{dy}{dx} = 0 \Rightarrow \frac{2(2\cos^2 t - 1)}{\cos t} = 0$

$\Rightarrow 2\cos^2 t - 1 = 0 \Rightarrow \cos t = \pm \frac{1}{\sqrt{2}} \Rightarrow t = \frac{\pi}{4}, \frac{3\pi}{4}, \frac{5\pi}{4}, \frac{7\pi}{4}$. In the 1st quadrant: $t = \frac{\pi}{4} \Rightarrow x = \sin \frac{\pi}{4} = \frac{\sqrt{2}}{2}$ and

$y = \sin 2 \left(\frac{\pi}{4}\right) = 1 \Rightarrow \left(\frac{\sqrt{2}}{2}, 1\right)$ is the point where the tangent line is horizontal. At the origin: $x = 0$ and $y = 0$

$\Rightarrow \sin t = 0 \Rightarrow t = 0$ or $t = \pi$ and $\sin 2t = 0 \Rightarrow t = 0, \frac{\pi}{2}, \pi, \frac{3\pi}{2}$; thus $t = 0$ and $t = \pi$ give the tangent lines at

the origin. Tangents at origin: $\left.\frac{dy}{dx}\right|_{t=0} = 2 \Rightarrow y = 2x$ and $\left.\frac{dy}{dx}\right|_{t=\pi} = -2 \Rightarrow y = -2x$

109. From the power rule, with $y = x^{1/4}$, we get $\frac{dy}{dx} = \frac{1}{4} x^{-3/4}$. From the chain rule, $y = \sqrt{\sqrt{\sqrt{x}}}$

$\Rightarrow \frac{dy}{dx} = \frac{1}{2\sqrt{\sqrt{x}}} \cdot \frac{d}{dx}\left(\sqrt{x}\right) = \frac{1}{2\sqrt{\sqrt{x}}} \cdot \frac{1}{2\sqrt{x}} = \frac{1}{4} x^{-3/4}$, in agreement.

111. (a)

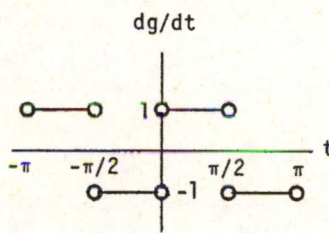

dg/dt

(b) $\frac{df}{dt} = 1.27324 \sin 2t + 0.42444 \sin 6t + 0.2546 \sin 10t + 0.18186 \sin 14t$

(c) The curve of $y = \frac{df}{dt}$ approximates $y = \frac{dg}{dt}$
the best when t is not $-\pi, -\frac{\pi}{2}, 0, \frac{\pi}{2}$, nor π.

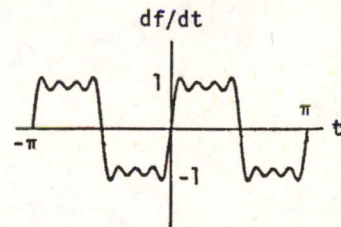

3.6 IMPLICIT DIFFERENTIATION

1. $y = x^{9/4} \Rightarrow \frac{dy}{dx} = \frac{9}{4} x^{5/4}$

3. $y = \sqrt[3]{2x} = (2x)^{1/3} \Rightarrow \frac{dy}{dx} = \frac{1}{3}(2x)^{-2/3} \cdot 2 = \frac{2^{1/3}}{3x^{2/3}}$

5. $y = 7\sqrt{x+6} = 7(x+6)^{1/2} \Rightarrow \frac{dy}{dx} = \frac{7}{2}(x+6)^{-1/2} = \frac{7}{2\sqrt{x+6}}$

7. $y = (2x+5)^{-1/2} \Rightarrow \frac{dy}{dx} = -\frac{1}{2}(2x+5)^{-3/2} \cdot 2 = -(2x+5)^{-3/2}$

9. $y = x\left(x^2 + 1\right)^{1/2} \Rightarrow y' = x \cdot \frac{1}{2}(x^2+1)^{-1/2}(2x) + (x^2+1)^{1/2} \cdot 1 = (x^2+1)^{-1/2}(x^2+x^2+1) = \frac{2x^2+1}{\sqrt{x^2+1}}$

11. $s = \sqrt[7]{t^2} = t^{2/7} \Rightarrow \frac{ds}{dt} = \frac{2}{7}t^{-5/7}$

13. $y = \sin\left((2t+5)^{-2/3}\right) \Rightarrow \frac{dy}{dt} = \cos\left((2t+5)^{-2/3}\right) \cdot \left(-\frac{2}{3}\right)(2t+5)^{-5/3} \cdot 2 = -\frac{4}{3}(2t+5)^{-5/3}\cos\left((2t+5)^{-2/3}\right)$

15. $f(x) = \sqrt{1 - \sqrt{x}} = \left(1 - x^{1/2}\right)^{1/2} \Rightarrow f'(x) = \frac{1}{2}\left(1 - x^{1/2}\right)^{-1/2}\left(-\frac{1}{2}x^{-1/2}\right) = \frac{-1}{4\left(\sqrt{1-\sqrt{x}}\right)\sqrt{x}} = \frac{-1}{4\sqrt{x\left(1-\sqrt{x}\right)}}$

17. $h(\theta) = \sqrt[3]{1 + \cos(2\theta)} = (1 + \cos 2\theta)^{1/3} \Rightarrow h'(\theta) = \frac{1}{3}(1 + \cos 2\theta)^{-2/3} \cdot (-\sin 2\theta) \cdot 2 = -\frac{2}{3}(\sin 2\theta)(1 + \cos 2\theta)^{-2/3}$

19. $x^2 y + xy^2 = 6$:

   Step 1: $\left(x^2 \frac{dy}{dx} + y \cdot 2x\right) + \left(x \cdot 2y\frac{dy}{dx} + y^2 \cdot 1\right) = 0$

   Step 2: $x^2 \frac{dy}{dx} + 2xy\frac{dy}{dx} = -2xy - y^2$

   Step 3: $\frac{dy}{dx}\left(x^2 + 2xy\right) = -2xy - y^2$

   Step 4: $\frac{dy}{dx} = \frac{-2xy - y^2}{x^2 + 2xy}$

21. $2xy + y^2 = x + y$:

   Step 1: $\left(2x\frac{dy}{dx} + 2y\right) + 2y\frac{dy}{dx} = 1 + \frac{dy}{dx}$

   Step 2: $2x\frac{dy}{dx} + 2y\frac{dy}{dx} - \frac{dy}{dx} = 1 - 2y$

   Step 3: $\frac{dy}{dx}(2x + 2y - 1) = 1 - 2y$

   Step 4: $\frac{dy}{dx} = \frac{1 - 2y}{2x + 2y - 1}$

23. $x^2(x - y)^2 = x^2 - y^2$:

   Step 1: $x^2\left[2(x - y)\left(1 - \frac{dy}{dx}\right)\right] + (x - y)^2(2x) = 2x - 2y\frac{dy}{dx}$

   Step 2: $-2x^2(x - y)\frac{dy}{dx} + 2y\frac{dy}{dx} = 2x - 2x^2(x - y) - 2x(x - y)^2$

   Step 3: $\frac{dy}{dx}\left[-2x^2(x - y) + 2y\right] = 2x\left[1 - x(x - y) - (x - y)^2\right]$

   Step 4: $\frac{dy}{dx} = \frac{2x\left[1 - x(x-y) - (x-y)^2\right]}{-2x^2(x-y) + 2y} = \frac{x\left[1 - x(x-y) - (x-y)^2\right]}{y - x^2(x-y)} = \frac{x\left(1 - x^2 + xy - x^2 + 2xy - y^2\right)}{x^2 y - x^3 + y}$

   $= \frac{x - 2x^3 + 3x^2 y - xy^2}{x^2 y - x^3 + y}$

25. $y^2 = \frac{x-1}{x+1} \Rightarrow 2y\frac{dy}{dx} = \frac{(x+1) - (x-1)}{(x+1)^2} = \frac{2}{(x+1)^2} \Rightarrow \frac{dy}{dx} = \frac{1}{y(x+1)^2}$

27. $x = \tan y \Rightarrow 1 = (\sec^2 y)\frac{dy}{dx} \Rightarrow \frac{dy}{dx} = \frac{1}{\sec^2 y} = \cos^2 y$

29. $x + \tan(xy) = 0 \Rightarrow 1 + [\sec^2(xy)]\left(y + x\frac{dy}{dx}\right) = 0 \Rightarrow x\sec^2(xy)\frac{dy}{dx} = -1 - y\sec^2(xy) \Rightarrow \frac{dy}{dx} = \frac{-1 - y\sec^2(xy)}{x\sec^2(xy)}$

   $= \frac{-1}{x\sec^2(xy)} - \frac{y}{x} = \frac{-\cos^2(xy)}{x} - \frac{y}{x} = \frac{-\cos^2(xy) - y}{x}$

31. $y\sin\left(\frac{1}{y}\right) = 1 - xy \Rightarrow y\left[\cos\left(\frac{1}{y}\right) \cdot (-1)\frac{1}{y^2} \cdot \frac{dy}{dx}\right] + \sin\left(\frac{1}{y}\right) \cdot \frac{dy}{dx} = -x\frac{dy}{dx} - y \Rightarrow$

   $\frac{dy}{dx}\left[-\frac{1}{y}\cos\left(\frac{1}{y}\right) + \sin\left(\frac{1}{y}\right) + x\right] = -y \Rightarrow \frac{dy}{dx} = \frac{-y}{-\frac{1}{y}\cos\left(\frac{1}{y}\right) + \sin\left(\frac{1}{y}\right) + x} = \frac{-y^2}{y\sin\left(\frac{1}{y}\right) - \cos\left(\frac{1}{y}\right) + xy}$

33. $\theta^{1/2} + r^{1/2} = 1 \;\Rightarrow\; \frac{1}{2}\theta^{-1/2} + \frac{1}{2}r^{-1/2} \cdot \frac{dr}{d\theta} = 0 \;\Rightarrow\; \frac{dr}{d\theta}\left[\frac{1}{2\sqrt{r}}\right] = \frac{-1}{2\sqrt{\theta}} \;\Rightarrow\; \frac{dr}{d\theta} = -\frac{2\sqrt{r}}{2\sqrt{\theta}} = -\frac{\sqrt{r}}{\sqrt{\theta}}$

35. $\sin(r\theta) = \frac{1}{2} \;\Rightarrow\; [\cos(r\theta)]\left(r + \theta\frac{dr}{d\theta}\right) = 0 \;\Rightarrow\; \frac{dr}{d\theta}[\theta\cos(r\theta)] = -r\cos(r\theta) \;\Rightarrow\; \frac{dr}{d\theta} = \frac{-r\cos(r\theta)}{\theta\cos(r\theta)} = -\frac{r}{\theta}$,

$\cos(r\theta) \neq 0$

37. $x^2 + y^2 = 1 \;\Rightarrow\; 2x + 2yy' = 0 \;\Rightarrow\; 2yy' = -2x \;\Rightarrow\; \frac{dy}{dx} = y' = -\frac{x}{y}$; now to find $\frac{d^2y}{dx^2}$, $\frac{d}{dx}(y') = \frac{d}{dx}\left(-\frac{x}{y}\right)$

$\Rightarrow\; y'' = \frac{y(-1)+xy'}{y^2} = \frac{-y+x\left(-\frac{x}{y}\right)}{y^2}$ since $y' = -\frac{x}{y} \;\Rightarrow\; \frac{d^2y}{dx^2} = y'' = \frac{-y^2-x^2}{y^3} = \frac{-y^2-(1-y^2)}{y^3} = \frac{-1}{y^3}$

39. $y^2 = x^2 + 2x \;\Rightarrow\; 2yy' = 2x + 2 \;\Rightarrow\; y' = \frac{2x+2}{2y} = \frac{x+1}{y}$; then $y'' = \frac{y-(x+1)y'}{y^2} = \frac{y-(x+1)\left(\frac{x+1}{y}\right)}{y^2}$

$\Rightarrow\; \frac{d^2y}{dx^2} = y'' = \frac{y^2-(x+1)^2}{y^3}$

41. $2\sqrt{y} = x - y \;\Rightarrow\; y^{-1/2}y' = 1 - y' \;\Rightarrow\; y'\left(y^{-1/2}+1\right) = 1 \;\Rightarrow\; \frac{dy}{dx} = y' = \frac{1}{y^{-1/2}+1} = \frac{\sqrt{y}}{\sqrt{y}+1}$; we can

differentiate the equation $y'\left(y^{-1/2}+1\right) = 1$ again to find y'': $y'\left(-\frac{1}{2}y^{-3/2}y'\right) + \left(y^{-1/2}+1\right)y'' = 0$

$\Rightarrow\; \left(y^{-1/2}+1\right)y'' = \frac{1}{2}[y']^2 y^{-3/2} \;\Rightarrow\; \frac{d^2y}{dx^2} = y'' = \frac{\frac{1}{2}\left(\frac{1}{y^{-1/2}+1}\right)^2 y^{-3/2}}{\left(y^{-1/2}+1\right)} = \frac{1}{2y^{3/2}\left(y^{-1/2}+1\right)^3} = \frac{1}{2\left(1+\sqrt{y}\right)^3}$

43. $x^3 + y^3 = 16 \;\Rightarrow\; 3x^2 + 3y^2y' = 0 \;\Rightarrow\; 3y^2y' = -3x^2 \;\Rightarrow\; y' = -\frac{x^2}{y^2}$; we differentiate $y^2y' = -x^2$ to find y'':

$y^2y'' + y'[2y\cdot y'] = -2x \;\Rightarrow\; y^2y'' = -2x - 2y[y']^2 \;\Rightarrow\; y'' = \frac{-2x-2y\left(-\frac{x^2}{y^2}\right)^2}{y^2} = \frac{-2x-\frac{2x^4}{y^3}}{y^2}$

$= \frac{-2xy^3-2x^4}{y^5} \;\Rightarrow\; \frac{d^2y}{dx^2}\Big|_{(2,2)} = \frac{-32-32}{32} = -2$

45. $y^2 + x^2 = y^4 - 2x$ at $(-2,1)$ and $(-2,-1) \;\Rightarrow\; 2y\frac{dy}{dx} + 2x = 4y^3\frac{dy}{dx} - 2 \;\Rightarrow\; 2y\frac{dy}{dx} - 4y^3\frac{dy}{dx} = -2 - 2x$

$\Rightarrow\; \frac{dy}{dx}(2y-4y^3) = -2-2x \;\Rightarrow\; \frac{dy}{dx} = \frac{x+1}{2y^3-y} \;\Rightarrow\; \frac{dy}{dx}\Big|_{(-2,1)} = -1$ and $\frac{dy}{dx}\Big|_{(-2,-1)} = 1$

47. $x^2 + xy - y^2 = 1 \;\Rightarrow\; 2x + y + xy' - 2yy' = 0 \;\Rightarrow\; (x-2y)y' = -2x - y \;\Rightarrow\; y' = \frac{2x+y}{2y-x}$;

(a) the slope of the tangent line $m = y'|_{(2,3)} = \frac{7}{4} \;\Rightarrow\;$ the tangent line is $y - 3 = \frac{7}{4}(x-2) \;\Rightarrow\; y = \frac{7}{4}x - \frac{1}{2}$

(b) the normal line is $y - 3 = -\frac{4}{7}(x-2) \;\Rightarrow\; y = -\frac{4}{7}x + \frac{29}{7}$

49. $x^2y^2 = 9 \;\Rightarrow\; 2xy^2 + 2x^2yy' = 0 \;\Rightarrow\; x^2yy' = -xy^2 \;\Rightarrow\; y' = -\frac{y}{x}$;

(a) the slope of the tangent line $m = y'|_{(-1,3)} = -\frac{y}{x}\Big|_{(-1,3)} = 3 \;\Rightarrow\;$ the tangent line is $y - 3 = 3(x+1)$

$\Rightarrow\; y = 3x + 6$

(b) the normal line is $y - 3 = -\frac{1}{3}(x+1) \;\Rightarrow\; y = -\frac{1}{3}x + \frac{8}{3}$

51. $6x^2 + 3xy + 2y^2 + 17y - 6 = 0 \;\Rightarrow\; 12x + 3y + 3xy' + 4yy' + 17y' = 0 \;\Rightarrow\; y'(3x+4y+17) = -12x - 3y$

$\Rightarrow\; y' = \frac{-12x-3y}{3x+4y+17}$;

(a) the slope of the tangent line $m = y'|_{(-1,0)} = \frac{-12x-3y}{3x+4y+17}\Big|_{(-1,0)} = \frac{6}{7} \;\Rightarrow\;$ the tangent line is $y - 0 = \frac{6}{7}(x+1)$

$\Rightarrow\; y = \frac{6}{7}x + \frac{6}{7}$

(b) the normal line is $y - 0 = -\frac{7}{6}(x+1) \;\Rightarrow\; y = -\frac{7}{6}x - \frac{7}{6}$

53. $2xy + \pi\sin y = 2\pi \;\Rightarrow\; 2xy' + 2y + \pi(\cos y)y' = 0 \;\Rightarrow\; y'(2x + \pi\cos y) = -2y \;\Rightarrow\; y' = \frac{-2y}{2x+\pi\cos y}$;

(a) the slope of the tangent line $m = y'|_{(1,\frac{\pi}{2})} = \frac{-2y}{2x+\pi\cos y}\Big|_{(1,\frac{\pi}{2})} = -\frac{\pi}{2} \;\Rightarrow\;$ the tangent line is

$y - \frac{\pi}{2} = -\frac{\pi}{2}(x - 1) \Rightarrow y = -\frac{\pi}{2}x + \pi$

(b) the normal line is $y - \frac{\pi}{2} = \frac{2}{\pi}(x - 1) \Rightarrow y = \frac{2}{\pi}x - \frac{2}{\pi} + \frac{\pi}{2}$

55. $y = 2\sin(\pi x - y) \Rightarrow y' = 2[\cos(\pi x - y)] \cdot (\pi - y') \Rightarrow y'[1 + 2\cos(\pi x - y)] = 2\pi\cos(\pi x - y)$

$\Rightarrow y' = \frac{2\pi\cos(\pi x - y)}{1 + 2\cos(\pi x - y)}$;

(a) the slope of the tangent line $m = y'|_{(1,0)} = \frac{2\pi\cos(\pi x - y)}{1 + 2\cos(\pi x - y)}\Big|_{(1,0)} = 2\pi \Rightarrow$ the tangent line is

$y - 0 = 2\pi(x - 1) \Rightarrow y = 2\pi x - 2\pi$

(b) the normal line is $y - 0 = -\frac{1}{2\pi}(x - 1) \Rightarrow y = -\frac{x}{2\pi} + \frac{1}{2\pi}$

57. Solving $x^2 + xy + y^2 = 7$ and $y = 0 \Rightarrow x^2 = 7 \Rightarrow x = \pm\sqrt{7} \Rightarrow \left(-\sqrt{7}, 0\right)$ and $\left(\sqrt{7}, 0\right)$ are the points where the

curve crosses the x-axis. Now $x^2 + xy + y^2 = 7 \Rightarrow 2x + y + xy' + 2yy' = 0 \Rightarrow (x + 2y)y' = -2x - y$

$\Rightarrow y' = -\frac{2x + y}{x + 2y} \Rightarrow m = -\frac{2x + y}{x + 2y} \Rightarrow$ the slope at $\left(-\sqrt{7}, 0\right)$ is $m = -\frac{-2\sqrt{7}}{-\sqrt{7}} = -2$ and the slope at $\left(\sqrt{7}, 0\right)$ is

$m = -\frac{2\sqrt{7}}{\sqrt{7}} = -2$. Since the slope is -2 in each case, the corresponding tangents must be parallel.

59. $y^4 = y^2 - x^2 \Rightarrow 4y^3y' = 2yy' - 2x \Rightarrow 2(2y^3 - y)y' = -2x \Rightarrow y' = \frac{x}{y - 2y^3}$; the slope of the tangent line at

$\left(\frac{\sqrt{3}}{4}, \frac{\sqrt{3}}{2}\right)$ is $\frac{x}{y - 2y^3}\Big|_{\left(\frac{\sqrt{3}}{4}, \frac{\sqrt{3}}{2}\right)} = \frac{\frac{\sqrt{3}}{4}}{\frac{\sqrt{3}}{2} - \frac{6\sqrt{3}}{8}} = \frac{\frac{1}{4}}{\frac{1}{2} - \frac{3}{4}} = \frac{1}{2 - 3} = -1$; the slope of the tangent line at $\left(\frac{\sqrt{3}}{4}, \frac{1}{2}\right)$

is $\frac{x}{y - 2y^3}\Big|_{\left(\frac{\sqrt{3}}{4}, \frac{1}{2}\right)} = \frac{\frac{\sqrt{3}}{4}}{\frac{1}{2} - \frac{2}{8}} = \frac{2\sqrt{3}}{4 - 2} = \sqrt{3}$

61. $y^4 - 4y^2 = x^4 - 9x^2 \Rightarrow 4y^3y' - 8yy' = 4x^3 - 18x \Rightarrow y'(4y^3 - 8y) = 4x^3 - 18x \Rightarrow y' = \frac{4x^3 - 18x}{4y^3 - 8y} = \frac{2x^3 - 9x}{2y^3 - 4y}$

$= \frac{x(2x^2 - 9)}{y(2y^2 - 4)} = m$; $(-3, 2)$: $m = \frac{(-3)(18 - 9)}{2(8 - 4)} = -\frac{27}{8}$; $(-3, -2)$: $m = \frac{27}{8}$; $(3, 2)$: $m = \frac{27}{8}$; $(3, -2)$: $m = -\frac{27}{8}$

63. $x^2 - 2tx + 2t^2 = 4 \Rightarrow 2x\frac{dx}{dt} - 2x - 2t\frac{dx}{dt} + 4t = 0 \Rightarrow (2x - 2t)\frac{dx}{dt} = 2x - 4t \Rightarrow \frac{dx}{dt} = \frac{2x - 4t}{2x - 2t} = \frac{x - 2t}{x - t}$;

$2y^3 - 3t^2 = 4 \Rightarrow 6y^2\frac{dy}{dt} - 6t = 0 \Rightarrow \frac{dy}{dt} = \frac{6t}{6y^2} = \frac{t}{y^2}$; thus $\frac{dy}{dx} = \frac{dy/dt}{dx/dt} = \frac{\left(\frac{t}{y^2}\right)}{\left(\frac{x - 2t}{x - t}\right)} = \frac{t(x - t)}{y^2(x - 2t)}$; $t = 2$

$\Rightarrow x^2 - 2(2)x + 2(2)^2 = 4 \Rightarrow x^2 - 4x + 4 = 0 \Rightarrow (x - 2)^2 = 0 \Rightarrow x = 2$; $t = 2 \Rightarrow 2y^3 - 3(2)^2 = 4$

$\Rightarrow 2y^3 = 16 \Rightarrow y^3 = 8 \Rightarrow y = 2$; therefore $\frac{dy}{dx}\Big|_{t=2} = \frac{2(2 - 2)}{(2)^2(2 - 2(2))} = 0$

65. $x + 2x^{3/2} = t^2 + t \Rightarrow \frac{dx}{dt} + 3x^{1/2}\frac{dx}{dt} = 2t + 1 \Rightarrow (1 + 3x^{1/2})\frac{dx}{dt} = 2t + 1 \Rightarrow \frac{dx}{dt} = \frac{2t + 1}{1 + 3x^{1/2}}$; $y\sqrt{t + 1} + 2t\sqrt{y} = 4$

$\Rightarrow \frac{dy}{dt}\sqrt{t + 1} + y\left(\frac{1}{2}\right)(t + 1)^{-1/2} + 2\sqrt{y} + 2t\left(\frac{1}{2}y^{-1/2}\right)\frac{dy}{dt} = 0 \Rightarrow \frac{dy}{dt}\sqrt{t + 1} + \frac{y}{2\sqrt{t + 1}} + 2\sqrt{y} + \left(\frac{t}{\sqrt{y}}\right)\frac{dy}{dt} = 0$

$\Rightarrow \left(\sqrt{t + 1} + \frac{t}{\sqrt{y}}\right)\frac{dy}{dt} = \frac{-y}{2\sqrt{t + 1}} - 2\sqrt{y} \Rightarrow \frac{dy}{dt} = \frac{\left(\frac{-y}{2\sqrt{t + 1}} - 2\sqrt{y}\right)}{\left(\sqrt{t + 1} + \frac{t}{\sqrt{y}}\right)} = \frac{-y\sqrt{y} - 4y\sqrt{t + 1}}{2\sqrt{y}(t + 1) + 2t\sqrt{t + 1}}$; thus

$\frac{dy}{dx} = \frac{dy/dt}{dx/dt} = \frac{\left(\frac{-y\sqrt{y} - 4y\sqrt{t + 1}}{2\sqrt{y}(t + 1) + 2t\sqrt{t + 1}}\right)}{\left(\frac{2t + 1}{1 + 3x^{1/2}}\right)}$; $t = 0 \Rightarrow x + 2x^{3/2} = 0 \Rightarrow x(1 + 2x^{1/2}) = 0 \Rightarrow x = 0$; $t = 0$

$\Rightarrow y\sqrt{0 + 1} + 2(0)\sqrt{y} = 4 \Rightarrow y = 4$; therefore $\frac{dy}{dx}\Big|_{t=0} = \frac{\left(\frac{-4\sqrt{4} - 4(4)\sqrt{0 + 1}}{2\sqrt{4}(0 + 1) + 2(0)\sqrt{0 + 1}}\right)}{\left(\frac{2(0) + 1}{1 + 3(0)^{1/2}}\right)} = -6$

67. (a) if $f(x) = \frac{3}{2}x^{2/3} - 3$, then $f'(x) = x^{-1/3}$ and $f''(x) = -\frac{1}{3}x^{-4/3}$ so the claim $f''(x) = x^{-1/3}$ is false

(b) if $f(x) = \frac{9}{10}x^{5/3} - 7$, then $f'(x) = \frac{3}{2}x^{2/3}$ and $f''(x) = x^{-1/3}$ is true

(c) $f''(x) = x^{-1/3} \Rightarrow f'''(x) = -\frac{1}{3}x^{-4/3}$ is true

(d) if $f'(x) = \frac{3}{2}x^{2/3} + 6$, then $f''(x) = x^{-1/3}$ is true

69. $x^2 + 2xy - 3y^2 = 0 \Rightarrow 2x + 2xy' + 2y - 6yy' = 0 \Rightarrow y'(2x - 6y) = -2x - 2y \Rightarrow y' = \frac{x+y}{3y-x} \Rightarrow$ the slope of the

tangent line $m = y'|_{(1,1)} = \frac{x+y}{3y-x}\Big|_{(1,1)} = 1 \Rightarrow$ the equation of the normal line at $(1,1)$ is $y - 1 = -1(x - 1)$

$\Rightarrow y = -x + 2$. To find where the normal line intersects the curve we substitute into its equation:

$x^2 + 2x(2 - x) - 3(2 - x)^2 = 0 \Rightarrow x^2 + 4x - 2x^2 - 3\left(4 - 4x + x^2\right) = 0 \Rightarrow -4x^2 + 16x - 12 = 0$

$\Rightarrow x^2 - 4x + 3 = 0 \Rightarrow (x - 3)(x - 1) = 0 \Rightarrow x = 3$ and $y = -x + 2 = -1$. Therefore, the normal to the curve

at $(1,1)$ intersects the curve at the point $(3, -1)$. Note that it also intersects the curve at $(1,1)$.

71. $y^2 = x \Rightarrow \frac{dy}{dx} = \frac{1}{2y}$. If a normal is drawn from $(a, 0)$ to (x_1, y_1) on the curve its slope satisfies $\frac{y_1 - 0}{x_1 - a} = -2y_1$

$\Rightarrow y_1 = -2y_1(x_1 - a)$ or $a = x_1 + \frac{1}{2}$. Since $x_1 \geq 0$ on the curve, we must have that $a \geq \frac{1}{2}$. By symmetry, the

two points on the parabola are $\left(x_1, \sqrt{x_1}\right)$ and $\left(x_1, -\sqrt{x_1}\right)$. For the normal to be perpendicular,

$\left(\frac{\sqrt{x_1}}{x_1 - a}\right)\left(\frac{\sqrt{x_1}}{a - x_1}\right) = -1 \Rightarrow \frac{x_1}{(a - x_1)^2} = 1 \Rightarrow x_1 = (a - x_1)^2 \Rightarrow x_1 = \left(x_1 + \frac{1}{2} - x_1\right)^2 \Rightarrow x_1 = \frac{1}{4}$ and $y_1 = \pm\frac{1}{2}$.

Therefore, $\left(\frac{1}{4}, \pm\frac{1}{2}\right)$ and $a = \frac{3}{4}$.

73. $xy^3 + x^2 y = 6 \Rightarrow x\left(3y^2 \frac{dy}{dx}\right) + y^3 + x^2 \frac{dy}{dx} + 2xy = 0 \Rightarrow \frac{dy}{dx}\left(3xy^2 + x^2\right) = -y^3 - 2xy \Rightarrow \frac{dy}{dx} = \frac{-y^3 - 2xy}{3xy^2 + x^2}$

$= -\frac{y^3 + 2xy}{3xy^2 + x^2}$; also, $xy^3 + x^2 y = 6 \Rightarrow x\left(3y^2\right) + y^3 \frac{dx}{dy} + x^2 + y\left(2x \frac{dx}{dy}\right) = 0 \Rightarrow \frac{dx}{dy}\left(y^3 + 2xy\right) = -3xy^2 - x^2$

$\Rightarrow \frac{dx}{dy} = -\frac{3xy^2 + x^2}{y^3 + 2xy}$; thus $\frac{dx}{dy}$ appears to equal $\frac{1}{\frac{dy}{dx}}$. The two different treatments view the graphs as functions

symmetric across the line $y = x$, so their slopes are reciprocals of one another at the corresponding points
(a, b) and (b, a).

75. $x^4 + 4y^2 = 1$:

(a) $y^2 = \frac{1 - x^4}{4} \Rightarrow y = \pm\frac{1}{2}\sqrt{1 - x^4}$

$\Rightarrow \frac{dy}{dx} = \pm\frac{1}{4}\left(1 - x^4\right)^{-1/2}\left(-4x^3\right) = \frac{\pm x^3}{(1 - x^4)^{1/2}}$;

differentiating implicitly, we find, $4x^3 + 8y \frac{dy}{dx} = 0$

$\Rightarrow \frac{dy}{dx} = \frac{-4x^3}{8y} = \frac{-4x^3}{8\left(\pm\frac{1}{2}\sqrt{1 - x^4}\right)} = \frac{\pm x^3}{(1 - x^4)^{1/2}}$.

(b)

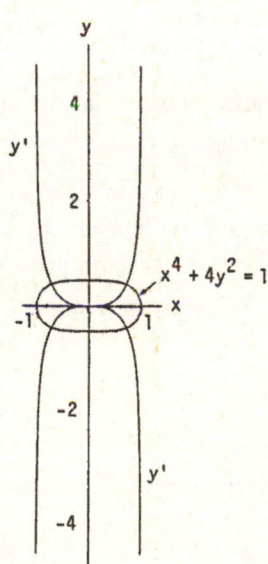

3.7 RELATED RATES

1. $A = \pi r^2 \Rightarrow \frac{dA}{dt} = 2\pi r \frac{dr}{dt}$

3. (a) $V = \pi r^2 h \Rightarrow \frac{dV}{dt} = \pi r^2 \frac{dh}{dt}$ 　　　　(b) $V = \pi r^2 h \Rightarrow \frac{dV}{dt} = 2\pi r h \frac{dr}{dt}$

　　(c) $V = \pi r^2 h \Rightarrow \frac{dV}{dt} = \pi r^2 \frac{dh}{dt} + 2\pi r h \frac{dr}{dt}$

5. (a) $\frac{dV}{dt} = 1$ volt/sec 　　　　　　(b) $\frac{dI}{dt} = -\frac{1}{3}$ amp/sec

　　(c) $\frac{dV}{dt} = R\left(\frac{dI}{dt}\right) + I\left(\frac{dR}{dt}\right) \Rightarrow \frac{dR}{dt} = \frac{1}{I}\left(\frac{dV}{dt} - R\frac{dI}{dt}\right) \Rightarrow \frac{dR}{dt} = \frac{1}{I}\left(\frac{dV}{dt} - \frac{V}{I}\frac{dI}{dt}\right)$

(d) $\frac{dR}{dt} = \frac{1}{2}\left[1 - \frac{12}{2}\left(-\frac{1}{3}\right)\right] = \left(\frac{1}{2}\right)(3) = \frac{3}{2}$ ohms/sec, R is increasing

7. (a) $s = \sqrt{x^2 + y^2} = (x^2 + y^2)^{1/2} \Rightarrow \frac{ds}{dt} = \frac{x}{\sqrt{x^2+y^2}}\frac{dx}{dt}$

    (b). $s = \sqrt{x^2 + y^2} = (x^2 + y^2)^{1/2} \Rightarrow \frac{ds}{dt} = \frac{x}{\sqrt{x^2+y^2}}\frac{dx}{dt} + \frac{y}{\sqrt{x^2+y^2}}\frac{dy}{dt}$

    (c) $s = \sqrt{x^2 + y^2} \Rightarrow s^2 = x^2 + y^2 \Rightarrow 2s\frac{ds}{dt} = 2x\frac{dx}{dt} + 2y\frac{dy}{dt} \Rightarrow 2s\cdot 0 = 2x\frac{dx}{dt} + 2y\frac{dy}{dt} \Rightarrow \frac{dx}{dt} = -\frac{y}{x}\frac{dy}{dt}$

9. (a) $A = \frac{1}{2}ab\sin\theta \Rightarrow \frac{dA}{dt} = \frac{1}{2}ab\cos\theta\frac{d\theta}{dt}$ (b) $A = \frac{1}{2}ab\sin\theta \Rightarrow \frac{dA}{dt} = \frac{1}{2}ab\cos\theta\frac{d\theta}{dt} + \frac{1}{2}b\sin\theta\frac{da}{dt}$

    (c) $A = \frac{1}{2}ab\sin\theta \Rightarrow \frac{dA}{dt} = \frac{1}{2}ab\cos\theta\frac{d\theta}{dt} + \frac{1}{2}b\sin\theta\frac{da}{dt} + \frac{1}{2}a\sin\theta\frac{db}{dt}$

11. Given $\frac{d\ell}{dt} = -2$ cm/sec, $\frac{dw}{dt} = 2$ cm/sec, $\ell = 12$ cm and $w = 5$ cm.

    (a) $A = \ell w \Rightarrow \frac{dA}{dt} = \ell\frac{dw}{dt} + w\frac{d\ell}{dt} \Rightarrow \frac{dA}{dt} = 12(2) + 5(-2) = 14$ cm^2/sec, increasing

    (b) $P = 2\ell + 2w \Rightarrow \frac{dP}{dt} = 2\frac{d\ell}{dt} + 2\frac{dw}{dt} = 2(-2) + 2(2) = 0$ cm/sec, constant

    (c) $D = \sqrt{w^2 + \ell^2} = (w^2 + \ell^2)^{1/2} \Rightarrow \frac{dD}{dt} = \frac{1}{2}(w^2+\ell^2)^{-1/2}\left(2w\frac{dw}{dt} + 2\ell\frac{d\ell}{dt}\right) \Rightarrow \frac{dD}{dt} = \frac{w\frac{dw}{dt} + \ell\frac{d\ell}{dt}}{\sqrt{w^2+\ell^2}}$

      $= \frac{(5)(2) + (12)(-2)}{\sqrt{25+144}} = -\frac{14}{13}$ cm/sec, decreasing

13. Given: $\frac{dx}{dt} = 5$ ft/sec, the ladder is 13 ft long, and $x = 12$, $y = 5$ at the instant of time

    (a) Since $x^2 + y^2 = 169 \Rightarrow \frac{dy}{dt} = -\frac{x}{y}\frac{dx}{dt} = -\left(\frac{12}{5}\right)(5) = -12$ ft/sec, the ladder is sliding down the wall

    (b) The area of the triangle formed by the ladder and walls is $A = \frac{1}{2}xy \Rightarrow \frac{dA}{dt} = \left(\frac{1}{2}\right)\left(x\frac{dy}{dt} + y\frac{dx}{dt}\right)$. The area

      is changing at $\frac{1}{2}[12(-12) + 5(5)] = -\frac{119}{2} = -59.5$ ft^2/sec.

    (c) $\cos\theta = \frac{x}{13} \Rightarrow -\sin\theta\frac{d\theta}{dt} = \frac{1}{13}\cdot\frac{dx}{dt} \Rightarrow \frac{d\theta}{dt} = -\frac{1}{13\sin\theta}\cdot\frac{dx}{dt} = -\left(\frac{1}{5}\right)(5) = -1$ rad/sec

15. Let s represent the distance between the girl and the kite and x represents the horizontal distance between the girl and kite $\Rightarrow s^2 = (300)^2 + x^2 \Rightarrow \frac{ds}{dt} = \frac{x}{s}\frac{dx}{dt} = \frac{400(25)}{500} = 20$ ft/sec.

17. $V = \frac{1}{3}\pi r^2 h$, $h = \frac{3}{8}(2r) = \frac{3r}{4} \Rightarrow r = \frac{4h}{3} \Rightarrow V = \frac{1}{3}\pi\left(\frac{4h}{3}\right)^2 h = \frac{16\pi h^3}{27} \Rightarrow \frac{dV}{dt} = \frac{16\pi h^2}{9}\frac{dh}{dt}$

    (a) $\frac{dh}{dt}\big|_{h=4} = \left(\frac{9}{16\pi 4^2}\right)(10) = \frac{90}{256\pi} \approx 0.1119$ m/sec $= 11.19$ cm/sec

    (b) $r = \frac{4h}{3} \Rightarrow \frac{dr}{dt} = \frac{4}{3}\frac{dh}{dt} = \frac{4}{3}\left(\frac{90}{256\pi}\right) = \frac{15}{32\pi} \approx 0.1492$ m/sec $= 14.92$ cm/sec

19. (a) $V = \frac{\pi}{3}y^2(3R - y) \Rightarrow \frac{dV}{dt} = \frac{\pi}{3}[2y(3R - y) + y^2(-1)]\frac{dy}{dt} \Rightarrow \frac{dy}{dt} = \left[\frac{\pi}{3}(6Ry - 3y^2)\right]^{-1}\frac{dV}{dt} \Rightarrow$ at $R = 13$ and

      $y = 8$ we have $\frac{dy}{dt} = \frac{1}{144\pi}(-6) = \frac{-1}{24\pi}$ m/min

    (b) The hemisphere is on the circle $r^2 + (13 - y)^2 = 169 \Rightarrow r = \sqrt{26y - y^2}$ m

    (c) $r = (26y - y^2)^{1/2} \Rightarrow \frac{dr}{dt} = \frac{1}{2}(26y - y^2)^{-1/2}(26 - 2y)\frac{dy}{dt} \Rightarrow \frac{dr}{dt} = \frac{13 - y}{\sqrt{26y - y^2}}\frac{dy}{dt} \Rightarrow \frac{dr}{dt}\big|_{y=8} = \frac{13 - 8}{\sqrt{26\cdot 8 - 64}}\left(\frac{-1}{24\pi}\right)$

      $= \frac{-5}{288\pi}$ m/min

21. If $V = \frac{4}{3}\pi r^3$, $r = 5$, and $\frac{dV}{dt} = 100\pi$ ft^3/min, then $\frac{dV}{dt} = 4\pi r^2\frac{dr}{dt} \Rightarrow \frac{dr}{dt} = 1$ ft/min. Then $S = 4\pi r^2 \Rightarrow \frac{dS}{dt}$

    $= 8\pi r\frac{dr}{dt} = 8\pi(5)(1) = 40\pi$ ft^2/min, the rate at which the surface area is increasing.

23. Let s represent the distance between the bicycle and balloon, h the height of the balloon and x the horizontal distance between the balloon and the bicycle. The relationship between the variables is $s^2 = h^2 + x^2$

    $\Rightarrow \frac{ds}{dt} = \frac{1}{s}\left(h\frac{dh}{dt} + x\frac{dx}{dt}\right) \Rightarrow \frac{ds}{dt} = \frac{1}{85}[68(1) + 51(17)] = 11$ ft/sec.

25. $y = QD^{-1} \Rightarrow \frac{dy}{dt} = D^{-1}\frac{dQ}{dt} - QD^{-2}\frac{dD}{dt} = \frac{1}{41}(0) - \frac{233}{(41)^2}(-2) = \frac{466}{1681}$ L/min $\Rightarrow$ increasing about 0.2772 L/min

27. Let P(x, y) represent a point on the curve $y = x^2$ and θ the angle of inclination of a line containing P and the origin. Consequently, $\tan \theta = \frac{y}{x} \Rightarrow \tan \theta = \frac{x^2}{x} = x \Rightarrow \sec^2 \theta \frac{d\theta}{dt} = \frac{dx}{dt} \Rightarrow \frac{d\theta}{dt} = \cos^2 \theta \frac{dx}{dt}$. Since $\frac{dx}{dt} = 10$ m/sec and $\cos^2 \theta|_{x=3} = \frac{x^2}{y^2 + x^2} = \frac{3^2}{9^2 + 3^2} = \frac{1}{10}$, we have $\frac{d\theta}{dt}\big|_{x=3} = 1$ rad/sec.

29. The distance from the origin is $s = \sqrt{x^2 + y^2}$ and we wish to find $\frac{ds}{dt}\big|_{(5,12)}$

$$= \frac{1}{2}(x^2 + y^2)^{-1/2}\left(2x\frac{dx}{dt} + 2y\frac{dy}{dt}\right)\bigg|_{(5,12)} = \frac{(5)(-1) + (12)(-5)}{\sqrt{25 + 144}} = -5 \text{ m/sec}$$

31. Let $s = 16t^2$ represent the distance the ball has fallen, h the distance between the ball and the ground, and I the distance between the shadow and the point directly beneath the ball. Accordingly, $s + h = 50$ and since the triangle LOQ and triangle PRQ are similar we have

$I = \frac{30h}{50-h} \Rightarrow h = 50 - 16t^2$ and $I = \frac{30(50 - 16t^2)}{50 - (50 - 16t^2)}$

$= \frac{1500}{16t^2} - 30 \Rightarrow \frac{dI}{dt} = -\frac{1500}{8t^3} \Rightarrow \frac{dI}{dt}\big|_{t=\frac{1}{2}} = -1500$ ft/sec.

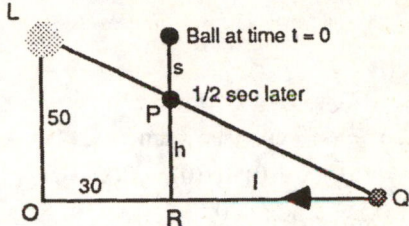

33. The volume of the ice is $V = \frac{4}{3}\pi r^3 - \frac{4}{3}\pi 4^3 \Rightarrow \frac{dV}{dt} = 4\pi r^2 \frac{dr}{dt} \Rightarrow \frac{dr}{dt}\big|_{r=6} = \frac{-5}{72\pi}$ in./min when $\frac{dV}{dt} = -10$ in^3/min, the thickness of the ice is decreasing at $\frac{5}{72\pi}$ in/min. The surface area is $S = 4\pi r^2 \Rightarrow \frac{dS}{dt} = 8\pi r \frac{dr}{dt} \Rightarrow \frac{dS}{dt}\big|_{r=6} = 48\pi\left(\frac{-5}{72\pi}\right)$ $= -\frac{10}{3}$ in^2/min, the outer surface area of the ice is decreasing at $\frac{10}{3}$ in^2/min.

35. When x represents the length of the shadow, then $\tan \theta = \frac{80}{x} \Rightarrow \sec^2 \theta \frac{d\theta}{dt} = -\frac{80}{x^2}\frac{dx}{dt} \Rightarrow \frac{dx}{dt} = \frac{-x^2 \sec^2 \theta}{80}\frac{d\theta}{dt}$.

We are given that $\frac{d\theta}{dt} = 0.27° = \frac{3\pi}{2000}$ rad/min. At $x = 60$, $\cos \theta = \frac{3}{5} \Rightarrow$

$$\left|\frac{dx}{dt}\right| = \left|\frac{-x^2 \sec^2 \theta}{80}\frac{d\theta}{dt}\right|\bigg|_{\left(\frac{d\theta}{dt} = \frac{3\pi}{2000}\text{ and } \sec \theta = \frac{5}{3}\right)} = \frac{3\pi}{16} \text{ ft/min} \approx 0.589 \text{ ft/min} \approx 7.1 \text{ in./min.}$$

37. Let x represent distance of the player from second base and s the distance to third base. Then $\frac{dx}{dt} = -16$ ft/sec

   (a) $s^2 = x^2 + 8100 \Rightarrow 2s\frac{ds}{dt} = 2x\frac{dx}{dt} \Rightarrow \frac{ds}{dt} = \frac{x}{s}\frac{dx}{dt}$. When the player is 30 ft from first base, $x = 60$
   $\Rightarrow s = 30\sqrt{13}$ and $\frac{ds}{dt} = \frac{60}{30\sqrt{13}}(-16) = \frac{-32}{\sqrt{13}} \approx -8.875$ ft/sec

   (b) $\cos \theta_1 = \frac{90}{s} \Rightarrow -\sin \theta_1 \frac{d\theta_1}{dt} = -\frac{90}{s^2} \cdot \frac{ds}{dt} \Rightarrow \frac{d\theta_1}{dt} = \frac{90}{s^2 \sin \theta_1} \cdot \frac{ds}{dt} = \frac{90}{sx} \cdot \frac{ds}{dt}$. Therefore, $x = 60$ and $s = 30\sqrt{13}$
   $\Rightarrow \frac{d\theta_1}{dt} = \frac{90}{(30\sqrt{13})(60)} \cdot \left(\frac{-32}{\sqrt{13}}\right) = \frac{-8}{65}$ rad/sec; $\sin \theta_2 = \frac{90}{s} \Rightarrow \cos \theta_2 \frac{d\theta_2}{dt} = -\frac{90}{s^2} \cdot \frac{ds}{dt} \Rightarrow \frac{d\theta_2}{dt} = \frac{-90}{s^2 \cos \theta_2} \cdot \frac{ds}{dt}$
   $= \frac{-90}{sx} \cdot \frac{ds}{dt}$. Therefore, $x = 60$ and $s = 30\sqrt{13} \Rightarrow \frac{d\theta_2}{dt} = \frac{8}{65}$ rad/sec.

   (c) $\frac{d\theta_1}{dt} = \frac{90}{s^2 \sin \theta_1} \cdot \frac{ds}{dt} = \frac{90}{\left(s^2 \cdot \frac{x}{s}\right)} \cdot \left(\frac{x}{s}\right) \cdot \left(\frac{dx}{dt}\right) = \left(\frac{90}{s^2}\right)\left(\frac{dx}{dt}\right) = \left(\frac{90}{x^2 + 8100}\right)\frac{dx}{dt} \Rightarrow \lim_{x \to 0} \frac{d\theta_1}{dt}$
   $= \lim_{x \to 0}\left(\frac{90}{x^2 + 8100}\right)(-15) = -\frac{1}{6}$ rad/sec; $\frac{d\theta_2}{dt} = \frac{-90}{s^2 \cos \theta_2} \cdot \frac{ds}{dt} = \left(\frac{-90}{s^2 \cdot \frac{x}{s}}\right)\left(\frac{x}{s}\right)\left(\frac{dx}{dt}\right) = \left(\frac{-90}{s^2}\right)\left(\frac{dx}{dt}\right)$
   $= \left(\frac{-90}{x^2 + 8100}\right)\frac{dx}{dt} \Rightarrow \lim_{x \to 0} \frac{d\theta_2}{dt} = \frac{1}{6}$ rad/sec

3.8 LINEARIZATION AND DIFFERENTIALS

1. $f(x) = x^3 - 2x + 3 \Rightarrow f'(x) = 3x^2 - 2 \Rightarrow L(x) = f'(2)(x - 2) + f(2) = 10(x - 2) + 7 \Rightarrow L(x) = 10x - 13$ at $x = 2$

3. $f(x) = x + \frac{1}{x} \Rightarrow f'(x) = 1 - x^{-2} \Rightarrow L(x) = f(1) + f'(1)(x - 1) = 2 + 0(x - 1) = 2$

5. $f(x) = x^2 + 2x \Rightarrow f'(x) = 2x + 2 \Rightarrow L(x) = f'(0)(x - 0) + f(0) = 2(x - 0) + 0 \Rightarrow L(x) = 2x$ at $x = 0$

7. $f(x) = 2x^2 + 4x - 3 \Rightarrow f'(x) = 4x + 4 \Rightarrow L(x) = f'(-1)(x + 1) + f(-1) = 0(x + 1) + (-5) \Rightarrow L(x) = -5$ at $x = -1$

9. $f(x) = \sqrt[3]{x} = x^{1/3} \Rightarrow f'(x) = \left(\frac{1}{3}\right)x^{-2/3} \Rightarrow L(x) = f'(8)(x - 8) + f(8) = \frac{1}{12}(x - 8) + 2 \Rightarrow L(x) = \frac{1}{12}x + \frac{4}{3}$ at $x = 8$

11. $f(x) = \sin x \Rightarrow f'(x) = \cos x$

    (a) $L(x) = f'(0)(x - 0) + f(0) = 1(x - 0) + 0$

    $\Rightarrow L(x) = x$ at $x = 0$

    (b) $L(x) = f'(\pi)(x - \pi) + f(\pi) = (-1)(x - \pi) + 0$

    $\Rightarrow L(x) = \pi - x$ at $x = \pi$

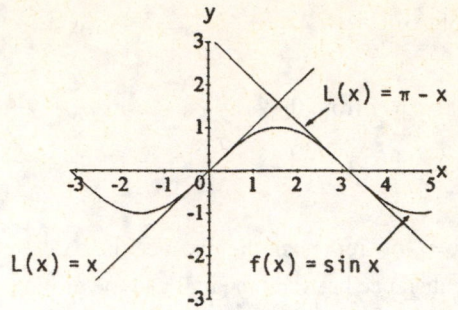

13. $f(x) = \sec x \Rightarrow f'(x) = \sec x \tan x$

    (a) $L(x) = f'(0)(x - 0) + f(0) = 0(x - 0) + 1$

    $\Rightarrow L(x) = 1$ at $x = 0$

    (b) $L(x) = f'\left(-\frac{\pi}{3}\right)\left(x + \frac{\pi}{3}\right) + f\left(-\frac{\pi}{3}\right)$

    $= -2\sqrt{3}\left(x + \frac{\pi}{3}\right) + 2 \Rightarrow L(x) = 2 - 2\sqrt{3}\left(x + \frac{\pi}{3}\right)$

    at $x = -\frac{\pi}{3}$

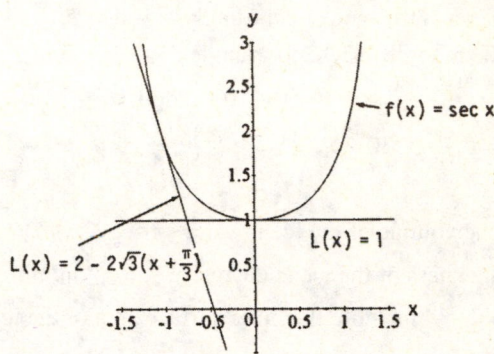

15. $f'(x) = k(1 + x)^{k-1}$. We have $f(0) = 1$ and $f'(0) = k$. $L(x) = f(0) + f'(0)(x - 0) = 1 + k(x - 0) = 1 + kx$

17. (a) $(1.0002)^{50} = (1 + 0.0002)^{50} \approx 1 + 50(0.0002) = 1 + .01 = 1.01$

    (b) $\sqrt[3]{1.009} = (1 + 0.009)^{1/3} \approx 1 + \left(\frac{1}{3}\right)(0.009) = 1 + 0.003 = 1.003$

19. $y = x^3 - 3\sqrt{x} = x^3 - 3x^{1/2} \Rightarrow dy = \left(3x^2 - \frac{3}{2}x^{-1/2}\right)dx \Rightarrow dy = \left(3x^2 - \frac{3}{2\sqrt{x}}\right)dx$

21. $y = \frac{2x}{1+x^2} \Rightarrow dy = \left(\frac{(2)(1+x^2) - (2x)(2x)}{(1+x^2)^2}\right)dx = \frac{2 - 2x^2}{(1+x^2)^2}dx$

23. $2y^{3/2} + xy - x = 0 \Rightarrow 3y^{1/2}dy + y\,dx + x\,dy - dx = 0 \Rightarrow \left(3y^{1/2} + x\right)dy = (1 - y)dx \Rightarrow dy = \frac{1-y}{3\sqrt{y}+x}dx$

25. $y = \sin\left(5\sqrt{x}\right) = \sin\left(5x^{1/2}\right) \Rightarrow dy = \left(\cos\left(5x^{1/2}\right)\right)\left(\frac{5}{2}x^{-1/2}\right)dx \Rightarrow dy = \frac{5\cos\left(5\sqrt{x}\right)}{2\sqrt{x}}dx$

27. $y = 4\tan\left(\frac{x^3}{3}\right) \Rightarrow dy = 4\left(\sec^2\left(\frac{x^3}{3}\right)\right)(x^2)dx \Rightarrow dy = 4x^2\sec^2\left(\frac{x^3}{3}\right)dx$

29. $y = 3\csc\left(1 - 2\sqrt{x}\right) = 3\csc\left(1 - 2x^{1/2}\right) \Rightarrow dy = 3\left(-\csc\left(1 - 2x^{1/2}\right)\right)\cot\left(1 - 2x^{1/2}\right)\left(-x^{-1/2}\right)dx$

    $\Rightarrow dy = \frac{3}{\sqrt{x}}\csc\left(1 - 2\sqrt{x}\right)\cot\left(1 - 2\sqrt{x}\right)dx$

31. $f(x) = x^2 + 2x, x_0 = 1, dx = 0.1 \Rightarrow f'(x) = 2x + 2$

    (a) $\Delta f = f(x_0 + dx) - f(x_0) = f(1.1) - f(1) = 3.41 - 3 = 0.41$

    (b) $df = f'(x_0)dx = [2(1) + 2](0.1) = 0.4$

    (c) $|\Delta f - df| = |0.41 - 0.4| = 0.01$

33. $f(x) = x^3 - x$, $x_0 = 1$, $dx = 0.1 \Rightarrow f'(x) = 3x^2 - 1$

   (a) $\Delta f = f(x_0 + dx) - f(x_0) = f(1.1) - f(1) = .231$

   (b) $df = f'(x_0)\, dx = [3(1)^2 - 1](.1) = .2$

   (c) $|\Delta f - df| = |.231 - .2| = .031$

35. $f(x) = x^{-1}$, $x_0 = 0.5$, $dx = 0.1 \Rightarrow f'(x) = -x^{-2}$

   (a) $\Delta f = f(x_0 + dx) - f(x_0) = f(.6) - f(.5) = -\frac{1}{3}$

   (b) $df = f'(x_0)\, dx = (-4)\left(\frac{1}{10}\right) = -\frac{2}{5}$

   (c) $|\Delta f - df| = \left|-\frac{1}{3} + \frac{2}{5}\right| = \frac{1}{15}$

37. $V = \frac{4}{3}\pi r^3 \Rightarrow dV = 4\pi r_0^2\, dr$

39. $S = 6x^2 \Rightarrow dS = 12x_0\, dx$

41. $V = \pi r^2 h$, height constant $\Rightarrow dV = 2\pi r_0 h\, dr$

43. Given $r = 2$ m, $dr = .02$ m

   (a) $A = \pi r^2 \Rightarrow dA = 2\pi r\, dr = 2\pi(2)(.02) = .08\pi$ m^2

   (b) $\left(\frac{.08\pi}{4\pi}\right)(100\%) = 2\%$

45. The volume of a cylinder is $V = \pi r^2 h$. When h is held fixed, we have $\frac{dV}{dr} = 2\pi rh$, and so $dV = 2\pi rh\, dr$. For $h = 30$ in., $r = 6$ in., and $dr = 0.5$ in., the volume of the material in the shell is approximately $dV = 2\pi rh\, dr = 2\pi(6)(30)(0.5) = 180\pi \approx 565.5$ in^3.

47. $V = \pi h^3 \Rightarrow dV = 3\pi h^2\, dh$; recall that $\Delta V \approx dV$. Then $|\Delta V| \le (1\%)(V) = \frac{(1)(\pi h^3)}{100} \Rightarrow |dV| \le \frac{(1)(\pi h^3)}{100}$

   $\Rightarrow |3\pi h^2\, dh| \le \frac{(1)(\pi h^3)}{100} \Rightarrow |dh| \le \frac{1}{300} h = \left(\frac{1}{3}\%\right) h$. Therefore the greatest tolerated error in the measurement of h is $\frac{1}{3}\%$.

49. $V = \pi r^2 h$, h is constant $\Rightarrow dV = 2\pi rh\, dr$; recall that $\Delta V \approx dV$. We want $|\Delta V| \le \frac{1}{1000} V \Rightarrow |dV| \le \frac{\pi r^2 h}{1000}$

   $\Rightarrow |2\pi rh\, dr| \le \frac{\pi r^2 h}{1000} \Rightarrow |dr| \le \frac{r}{2000} = (.05\%)r \Rightarrow$ a $.05\%$ variation in the radius can be tolerated.

51. $W = a + \frac{b}{g} = a + bg^{-1} \Rightarrow dW = -bg^{-2}\, dg = -\frac{b\, dg}{g^2} \Rightarrow \frac{dW_{moon}}{dW_{earth}} = \frac{\left(-\frac{b\, dg}{(5.2)^2}\right)}{\left(-\frac{b\, dg}{(32)^2}\right)} = \left(\frac{32}{5.2}\right)^2 = 37.87$, so a change of

   gravity on the moon has about 38 times the effect that a change of the same magnitude has on Earth.

53. The error in measurement $dx = (1\%)(10) = 0.1$ cm; $V = x^3 \Rightarrow dV = 3x^2\, dx = 3(10)^2(0.1) = 30$ cm$^3 \Rightarrow$ the percentage error in the volume calculation is $\left(\frac{30}{1000}\right)(100\%) = 3\%$

55. Given $D = 100$ cm, $dD = 1$ cm, $V = \frac{4}{3}\pi\left(\frac{D}{2}\right)^3 = \frac{\pi D^3}{6} \Rightarrow dV = \frac{\pi}{2}D^2\, dD = \frac{\pi}{2}(100)^2(1) = \frac{10^4\pi}{2}$. Then $\frac{dV}{V}(100\%)$

   $= \left[\frac{\frac{10^4\pi}{2}}{\frac{10^6\pi}{6}}\right](10^2\%) = \left[\frac{\frac{10^6\pi}{2}}{\frac{10^6\pi}{6}}\right]\% = 3\%$

57. A 5% error in measuring $t \Rightarrow dt = (5\%)t = \frac{t}{20}$. Then $s = 16t^2 \Rightarrow ds = 32t\, dt = 32t\left(\frac{t}{20}\right) = \frac{32t^2}{20} = \frac{16t^2}{10} = \left(\frac{1}{10}\right)s$

   $= (10\%)s \Rightarrow$ a 10% error in the calculation of s.

59. $\lim\limits_{x \to 0} \frac{\sqrt{1+x}}{1 + \frac{x}{2}} = \frac{\sqrt{1+0}}{1 + \frac{0}{2}} = 1$

61. $E(x) = f(x) - g(x) \Rightarrow E(x) = f(x) - m(x - a) - c$. Then $E(a) = 0 \Rightarrow f(a) - m(a - a) - c = 0 \Rightarrow c = f(a)$. Next
we calculate m: $\lim\limits_{x \to a} \frac{E(x)}{x-a} = 0 \Rightarrow \lim\limits_{x \to a} \frac{f(x) - m(x-a) - c}{x-a} = 0 \Rightarrow \lim\limits_{x \to a} \left[\frac{f(x) - f(a)}{x-a} - m\right] = 0$ (since $c = f(a)$)
$\Rightarrow f'(a) - m = 0 \Rightarrow m = f'(a)$. Therefore, $g(x) = m(x - a) + c = f'(a)(x - a) + f(a)$ is the linear approximation,
as claimed.

63. (a) $x = 1$

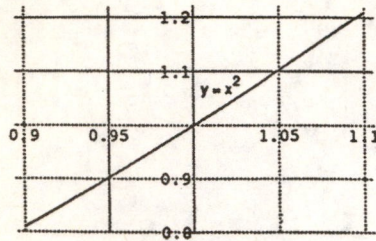

(b) $x = 1$; $m = 2.5$, $e^1 \approx 2.7$ $x = 0$; $m = 1$, $e^0 = 1$ $x = -1$; $m = 0.3$, $e^{-1} \approx 0.4$

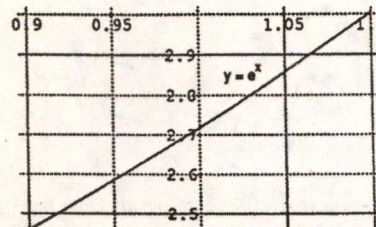 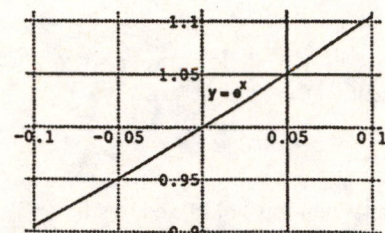 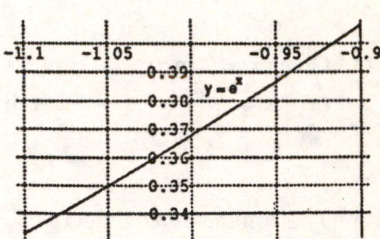

65. Find $|v|$ when $m = 1.01m_0$. $m = \frac{m_0}{\sqrt{1 - \frac{v^2}{c^2}}} \Rightarrow m\sqrt{1 - \frac{v^2}{c^2}} = m_0 \Rightarrow \sqrt{1 - \frac{v^2}{c^2}} = \frac{m_0}{m} \Rightarrow 1 - \frac{v^2}{c^2} = \frac{m_0^2}{m^2} \Rightarrow v^2 = c^2\left(1 - \frac{m_0^2}{m^2}\right)$
$\Rightarrow |v| = c\sqrt{1 - \frac{m_0^2}{m^2}} \Rightarrow dv = c \cdot \frac{1}{2}\left(1 - \frac{m_0^2}{m^2}\right)^{-1/2}\left(\frac{2m_0^2}{m^3}\right)dm$, $dm = 0.01m_0 \Rightarrow dv = \frac{c\,m_0^2}{m^3\sqrt{1 - \frac{m_0^2}{m^2}}}\left(\frac{m_0}{100}\right)$. $m = \frac{101}{100}m_0$,
$dv = \frac{c \cdot m_0^2}{\frac{101^3}{100^3}m_0^3\sqrt{1 - \frac{m_0^2}{\frac{101^2}{100^2}m_0^2}}}\left(\frac{m_0}{100}\right) = \frac{1000}{101^3\sqrt{1 - \frac{100^2}{101^2}}} \approx 0.69c$. Body at rest $\Rightarrow v_0 = 0$ and $v = v_0 + dv$
$\Rightarrow v = 0.69c$.

CHAPTER 3 PRACTICE EXERCISES

1. $y = x^5 - 0.125x^2 + 0.25x \Rightarrow \frac{dy}{dx} = 5x^4 - 0.25x + 0.25$

3. $y = x^3 - 3(x^2 + \pi^2) \Rightarrow \frac{dy}{dx} = 3x^2 - 3(2x + 0) = 3x^2 - 6x = 3x(x - 2)$

5. $y = (x + 1)^2(x^2 + 2x) \Rightarrow \frac{dy}{dx} = (x + 1)^2(2x + 2) + (x^2 + 2x)(2(x + 1)) = 2(x + 1)[(x + 1)^2 + x(x + 2)]$
   $= 2(x + 1)(2x^2 + 4x + 1)$

7. $y = (\theta^2 + \sec\theta + 1)^3 \Rightarrow \frac{dy}{d\theta} = 3(\theta^2 + \sec\theta + 1)^2(2\theta + \sec\theta\tan\theta)$

9. $s = \frac{\sqrt{t}}{1 + \sqrt{t}} \Rightarrow \frac{ds}{dt} = \frac{(1 + \sqrt{t}) \cdot \frac{1}{2\sqrt{t}} - \sqrt{t}\left(\frac{1}{2\sqrt{t}}\right)}{(1 + \sqrt{t})^2} = \frac{(1 + \sqrt{t}) - \sqrt{t}}{2\sqrt{t}(1 + \sqrt{t})^2} = \frac{1}{2\sqrt{t}(1 + \sqrt{t})^2}$

11. $y = 2\tan^2 x - \sec^2 x \Rightarrow \frac{dy}{dx} = (4\tan x)(\sec^2 x) - (2\sec x)(\sec x\tan x) = 2\sec^2 x\tan x$

13. $s = \cos^4 (1 - 2t) \Rightarrow \frac{ds}{dt} = 4 \cos^3 (1 - 2t)(-\sin (1 - 2t))(-2) = 8 \cos^3 (1 - 2t) \sin (1 - 2t)$

15. $s = (\sec t + \tan t)^5 \Rightarrow \frac{ds}{dt} = 5(\sec t + \tan t)^4 (\sec t \tan t + \sec^2 t) = 5(\sec t)(\sec t + \tan t)^5$

17. $r = \sqrt{2\theta \sin \theta} = (2\theta \sin \theta)^{1/2} \Rightarrow \frac{dr}{d\theta} = \frac{1}{2} (2\theta \sin \theta)^{-1/2}(2\theta \cos \theta + 2 \sin \theta) = \frac{\theta \cos \theta + \sin \theta}{\sqrt{2\theta \sin \theta}}$

19. $r = \sin \sqrt{2\theta} = \sin (2\theta)^{1/2} \Rightarrow \frac{dr}{d\theta} = \cos (2\theta)^{1/2} \left(\frac{1}{2} (2\theta)^{-1/2}(2)\right) = \frac{\cos \sqrt{2\theta}}{\sqrt{2\theta}}$

21. $y = \frac{1}{2} x^2 \csc \frac{2}{x} \Rightarrow \frac{dy}{dx} = \frac{1}{2} x^2 \left(-\csc \frac{2}{x} \cot \frac{2}{x}\right) \left(\frac{-2}{x^2}\right) + \left(\csc \frac{2}{x}\right) \left(\frac{1}{2} \cdot 2x\right) = \csc \frac{2}{x} \cot \frac{2}{x} + x \csc \frac{2}{x}$

23. $y = x^{-1/2} \sec (2x)^2 \Rightarrow \frac{dy}{dx} = x^{-1/2} \sec (2x)^2 \tan (2x)^2(2(2x) \cdot 2) + \sec (2x)^2 \left(-\frac{1}{2} x^{-3/2}\right)$

$= 8x^{1/2} \sec (2x)^2 \tan (2x)^2 - \frac{1}{2} x^{-3/2} \sec (2x)^2 = \frac{1}{2} x^{1/2} \sec (2x)^2 \left[16 \tan (2x)^2 - x^{-2}\right]$ or $\frac{1}{2x^{3/2}} \sec(2x)^2 \left[16x^2 \tan(2x)^2 - 1\right]$

25. $y = 5 \cot x^2 \Rightarrow \frac{dy}{dx} = 5 \left(-\csc^2 x^2\right) (2x) = -10x \csc^2 (x^2)$

27. $y = x^2 \sin^2 (2x^2) \Rightarrow \frac{dy}{dx} = x^2 (2 \sin (2x^2)) (\cos (2x^2)) (4x) + \sin^2 (2x^2) (2x) = 8x^3 \sin (2x^2) \cos (2x^2) + 2x \sin^2 (2x^2)$

29. $s = \left(\frac{4t}{t+1}\right)^{-2} \Rightarrow \frac{ds}{dt} = -2 \left(\frac{4t}{t+1}\right)^{-3} \left(\frac{(t+1)(4) - (4t)(1)}{(t+1)^2}\right) = -2 \left(\frac{4t}{t+1}\right)^{-3} \frac{4}{(t+1)^2} = -\frac{(t+1)}{8t^3}$

31. $y = \left(\frac{\sqrt{x}}{x+1}\right)^2 \Rightarrow \frac{dy}{dx} = 2 \left(\frac{\sqrt{x}}{x+1}\right) \cdot \frac{(x+1)\left(\frac{1}{2\sqrt{x}}\right) - (\sqrt{x})(1)}{(x+1)^2} = \frac{(x+1) - 2x}{(x+1)^3} = \frac{1-x}{(x+1)^3}$

33. $y = \sqrt{\frac{x^2 + x}{x^2}} = \left(1 + \frac{1}{x}\right)^{1/2} \Rightarrow \frac{dy}{dx} = \frac{1}{2} \left(1 + \frac{1}{x}\right)^{-1/2} \left(-\frac{1}{x^2}\right) = -\frac{1}{2x^2 \sqrt{1 + \frac{1}{x}}}$

35. $r = \left(\frac{\sin \theta}{\cos \theta - 1}\right)^2 \Rightarrow \frac{dr}{d\theta} = 2 \left(\frac{\sin \theta}{\cos \theta - 1}\right) \left[\frac{(\cos \theta - 1)(\cos \theta) - (\sin \theta)(-\sin \theta)}{(\cos \theta - 1)^2}\right]$

$= 2 \left(\frac{\sin \theta}{\cos \theta - 1}\right) \left(\frac{\cos^2 \theta - \cos \theta + \sin^2 \theta}{(\cos \theta - 1)^2}\right) = \frac{(2 \sin \theta)(1 - \cos \theta)}{(\cos \theta - 1)^3} = \frac{-2 \sin \theta}{(\cos \theta - 1)^2}$

37. $y = (2x + 1) \sqrt{2x + 1} = (2x + 1)^{3/2} \Rightarrow \frac{dy}{dx} = \frac{3}{2} (2x + 1)^{1/2}(2) = 3\sqrt{2x + 1}$

39. $y = 3 \left(5x^2 + \sin 2x\right)^{-3/2} \Rightarrow \frac{dy}{dx} = 3 \left(-\frac{3}{2}\right) \left(5x^2 + \sin 2x\right)^{-5/2}[10x + (\cos 2x)(2)] = \frac{-9(5x + \cos 2x)}{(5x^2 + \sin 2x)^{5/2}}$

41. $xy + 2x + 3y = 1 \Rightarrow (xy' + y) + 2 + 3y' = 0 \Rightarrow xy' + 3y' = -2 - y \Rightarrow y'(x + 3) = -2 - y \Rightarrow y' = -\frac{y+2}{x+3}$

43. $x^3 + 4xy - 3y^{4/3} = 2x \Rightarrow 3x^2 + \left(4x \frac{dy}{dx} + 4y\right) - 4y^{1/3} \frac{dy}{dx} = 2 \Rightarrow 4x \frac{dy}{dx} - 4y^{1/3} \frac{dy}{dx} = 2 - 3x^2 - 4y$

$\Rightarrow \frac{dy}{dx} \left(4x - 4y^{1/3}\right) = 2 - 3x^2 - 4y \Rightarrow \frac{dy}{dx} = \frac{2 - 3x^2 - 4y}{4x - 4y^{1/3}}$

45. $(xy)^{1/2} = 1 \Rightarrow \frac{1}{2} (xy)^{-1/2} \left(x \frac{dy}{dx} + y\right) = 0 \Rightarrow x^{1/2} y^{-1/2} \frac{dy}{dx} = -x^{-1/2} y^{1/2} \Rightarrow \frac{dy}{dx} = -x^{-1} y \Rightarrow \frac{dy}{dx} = -\frac{y}{x}$

47. $y^2 = \frac{x}{x+1} \Rightarrow 2y \frac{dy}{dx} = \frac{(x+1)(1) - (x)(1)}{(x+1)^2} \Rightarrow \frac{dy}{dx} = \frac{1}{2y(x+1)^2}$

49. $p^3 + 4pq - 3q^2 = 2 \Rightarrow 3p^2 \frac{dp}{dq} + 4 \left(p + q \frac{dp}{dq}\right) - 6q = 0 \Rightarrow 3p^2 \frac{dp}{dq} + 4q \frac{dp}{dq} = 6q - 4p \Rightarrow \frac{dp}{dq} (3p^2 + 4q) = 6q - 4p$

$\Rightarrow \frac{dp}{dq} = \frac{6q - 4p}{3p^2 + 4q}$

51. $r \cos 2s + \sin^2 s = \pi \Rightarrow r(-\sin 2s)(2) + (\cos 2s)\left(\frac{dr}{ds}\right) + 2 \sin s \cos s = 0 \Rightarrow \frac{dr}{ds}(\cos 2s) = 2r \sin 2s - 2 \sin s \cos s$

$\Rightarrow \frac{dr}{ds} = \frac{2r \sin 2s - \sin 2s}{\cos 2s} = \frac{(2r-1)(\sin 2s)}{\cos 2s} = (2r-1)(\tan 2s)$

53. (a) $x^3 + y^3 = 1 \Rightarrow 3x^2 + 3y^2 \frac{dy}{dx} = 0 \Rightarrow \frac{dy}{dx} = -\frac{x^2}{y^2} \Rightarrow \frac{d^2y}{dx^2} = \frac{y^2(-2x) - (-x^2)\left(2y\frac{dy}{dx}\right)}{y^4}$

$\Rightarrow \frac{d^2y}{dx^2} = \frac{-2xy^2 + (2yx^2)\left(-\frac{x^2}{y^2}\right)}{y^4} = \frac{-2xy^2 - \frac{2x^4}{y}}{y^4} = \frac{-2xy^3 - 2x^4}{y^5}$

(b) $y^2 = 1 - \frac{2}{x} \Rightarrow 2y \frac{dy}{dx} = \frac{2}{x^2} \Rightarrow \frac{dy}{dx} = \frac{1}{yx^2} \Rightarrow \frac{dy}{dx} = (yx^2)^{-1} \Rightarrow \frac{d^2y}{dx^2} = -(yx^2)^{-2}\left[y(2x) + x^2\frac{dy}{dx}\right]$

$\Rightarrow \frac{d^2y}{dx^2} = \frac{-2xy - x^2\left(\frac{1}{yx^2}\right)}{y^2x^4} = \frac{-2xy^2 - 1}{y^3x^4}$

55. (a) Let $h(x) = 6f(x) - g(x) \Rightarrow h'(x) = 6f'(x) - g'(x) \Rightarrow h'(1) = 6f'(1) - g'(1) = 6\left(\frac{1}{2}\right) - (-4) = 7$

(b) Let $h(x) = f(x)g^2(x) \Rightarrow h'(x) = f(x)\left(2g(x)\right)g'(x) + g^2(x)f'(x) \Rightarrow h'(0) = 2f(0)g(0)g'(0) + g^2(0)f'(0)$

$= 2(1)(1)\left(\frac{1}{2}\right) + (1)^2(-3) = -2$

(c) Let $h(x) = \frac{f(x)}{g(x)+1} \Rightarrow h'(x) = \frac{(g(x)+1)f'(x) - f(x)g'(x)}{(g(x)+1)^2} \Rightarrow h'(1) = \frac{(g(1)+1)f'(1) - f(1)g'(1)}{(g(1)+1)^2}$

$= \frac{(5+1)\left(\frac{1}{2}\right) - 3(-4)}{(5+1)^2} = \frac{5}{12}$

(d) Let $h(x) = f(g(x)) \Rightarrow h'(x) = f'(g(x))g'(x) \Rightarrow h'(0) = f'(g(0))g'(0) = f'(1)\left(\frac{1}{2}\right) = \left(\frac{1}{2}\right)\left(\frac{1}{2}\right) = \frac{1}{4}$

(e) Let $h(x) = g(f(x)) \Rightarrow h'(x) = g'(f(x))f'(x) \Rightarrow h'(0) = g'(f(0))f'(0) = g'(1)f'(0) = (-4)(-3) = 12$

(f) Let $h(x) = (x + f(x))^{3/2} \Rightarrow h'(x) = \frac{3}{2}(x + f(x))^{1/2}(1 + f'(x)) \Rightarrow h'(1) = \frac{3}{2}(1 + f(1))^{1/2}(1 + f'(1))$

$= \frac{3}{2}(1 + 3)^{1/2}\left(1 + \frac{1}{2}\right) = \frac{9}{2}$

(g) Let $h(x) = f(x + g(x)) \Rightarrow h'(x) = f'(x + g(x))(1 + g'(x)) \Rightarrow h'(0) = f'(g(0))(1 + g'(0))$

$= f'(1)\left(1 + \frac{1}{2}\right) = \left(\frac{1}{2}\right)\left(\frac{3}{2}\right) = \frac{3}{4}$

57. $x = t^2 + \pi \Rightarrow \frac{dx}{dt} = 2t; \ y = 3 \sin 2x \Rightarrow \frac{dy}{dx} = 3(\cos 2x)(2) = 6 \cos 2x = 6 \cos(2t^2 + 2\pi) = 6 \cos(2t^2)$; thus,

$\frac{dy}{dt} = \frac{dy}{dx} \cdot \frac{dx}{dt} = 6 \cos(2t^2) \cdot 2t \Rightarrow \frac{dy}{dt}\Big|_{t=0} = 6 \cos(0) \cdot 0 = 0$

59. $r = 8 \sin\left(s + \frac{\pi}{6}\right) \Rightarrow \frac{dr}{ds} = 8 \cos\left(s + \frac{\pi}{6}\right); \ w = \sin\left(\sqrt{r} - 2\right) \Rightarrow \frac{dw}{dr} = \cos\left(\sqrt{r} - 2\right)\left(\frac{1}{2\sqrt{r}}\right)$

$= \frac{\cos\sqrt{8 \sin\left(s + \frac{\pi}{6}\right)} - 2}{2\sqrt{8 \sin\left(s + \frac{\pi}{6}\right)}}$; thus, $\frac{dw}{ds} = \frac{dw}{dr} \cdot \frac{dr}{ds} = \frac{\cos\left(\sqrt{8 \sin\left(s + \frac{\pi}{6}\right)} - 2\right)}{2\sqrt{8 \sin\left(s + \frac{\pi}{6}\right)}} \cdot \left[8 \cos\left(s + \frac{\pi}{6}\right)\right]$

$\Rightarrow \frac{dw}{ds}\Big|_{s=0} = \frac{\cos\left(\sqrt{8 \sin\left(\frac{\pi}{6}\right)} - 2\right) \cdot 8 \cos\left(\frac{\pi}{6}\right)}{2\sqrt{8 \sin\left(\frac{\pi}{6}\right)}} = \frac{(\cos 0)(8)\left(\frac{\sqrt{3}}{2}\right)}{2\sqrt{4}} = \sqrt{3}$

61. $y^3 + y = 2 \cos x \Rightarrow 3y^2 \frac{dy}{dx} + \frac{dy}{dx} = -2 \sin x \Rightarrow \frac{dy}{dx}(3y^2 + 1) = -2 \sin x \Rightarrow \frac{dy}{dx} = \frac{-2 \sin x}{3y^2 + 1} \Rightarrow \frac{dy}{dx}\Big|_{(0,1)}$

$= \frac{-2 \sin(0)}{3+1} = 0; \ \frac{d^2y}{dx^2} = \frac{(3y^2 + 1)(-2 \cos x) - (-2 \sin x)\left(6y\frac{dy}{dx}\right)}{(3y^2 + 1)^2}$

$\Rightarrow \frac{d^2y}{dx^2}\Big|_{(0,1)} = \frac{(3+1)(-2 \cos 0) - (-2 \sin 0)(6\cdot0)}{(3+1)^2} = -\frac{1}{2}$

63. $f(t) = \frac{1}{2t+1}$ and $f(t + h) = \frac{1}{2(t+h)+1} \Rightarrow \frac{f(t+h) - f(t)}{h} = \frac{\frac{1}{2(t+h)+1} - \frac{1}{2t+1}}{h} = \frac{2t+1 - (2t+2h+1)}{(2t+2h+1)(2t+1)h}$

$= \frac{-2h}{(2t+2h+1)(2t+1)h} = \frac{-2}{(2t+2h+1)(2t+1)} \Rightarrow f'(t) = \lim_{h \to 0} \frac{f(t+h) - f(t)}{h} = \lim_{h \to 0} \frac{-2}{(2t+2h+1)(2t+1)}$

$= \frac{-2}{(2t+1)^2}$

65. (a)

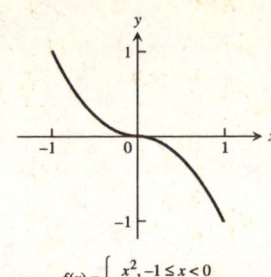

$$f(x) = \begin{cases} x^2, & -1 \le x < 0 \\ -x^2, & 0 \le x < 1 \end{cases}$$

(b) $\lim\limits_{x \to 0^-} f(x) = \lim\limits_{x \to 0^-} x^2 = 0$ and $\lim\limits_{x \to 0^+} f(x) = \lim\limits_{x \to 0^+} -x^2 = 0 \Rightarrow \lim\limits_{x \to 0} f(x) = 0$. Since $\lim\limits_{x \to 0} f(x) = 0 = f(0)$ it

follows that f is continuous at $x = 0$.

(c) $\lim\limits_{x \to 0^-} f'(x) = \lim\limits_{x \to 0^-} (2x) = 0$ and $\lim\limits_{x \to 0^+} f'(x) = \lim\limits_{x \to 0^+} (-2x) = 0 \Rightarrow \lim\limits_{x \to 0} f'(x) = 0$. Since this limit exists, it

follows that f is differentiable at $x = 0$.

67. (a)

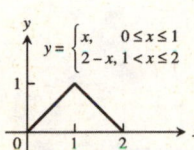

$$y = \begin{cases} x, & 0 \le x \le 1 \\ 2 - x, & 1 < x \le 2 \end{cases}$$

(b) $\lim\limits_{x \to 1^-} f(x) = \lim\limits_{x \to 1^-} x = 1$ and $\lim\limits_{x \to 1^+} f(x) = \lim\limits_{x \to 1^+} (2 - x) = 1 \Rightarrow \lim\limits_{x \to 1} f(x) = 1$. Since $\lim\limits_{x \to 1} f(x) = 1 = f(1)$, it

follows that f is continuous at $x = 1$.

(c) $\lim\limits_{x \to 1^-} f'(x) = \lim\limits_{x \to 1^-} 1 = 1$ and $\lim\limits_{x \to 1^+} f'(x) = \lim\limits_{x \to 1^+} -1 = -1 \Rightarrow \lim\limits_{x \to 1^-} f'(x) \neq \lim\limits_{x \to 1^+} f'(x)$, so $\lim\limits_{x \to 1} f'(x)$ does

not exist $\Rightarrow$ f is not differentiable at $x = 1$.

69. $y = \frac{x}{2} + \frac{1}{2x - 4} = \frac{1}{2}x + (2x - 4)^{-1} \Rightarrow \frac{dy}{dx} = \frac{1}{2} - 2(2x - 4)^{-2}$; the slope of the tangent is $-\frac{3}{2} \Rightarrow -\frac{3}{2}$

$= \frac{1}{2} - 2(2x - 4)^{-2} \Rightarrow -2 = -2(2x - 4)^{-2} \Rightarrow 1 = \frac{1}{(2x - 4)^2} \Rightarrow (2x - 4)^2 = 1 \Rightarrow 4x^2 - 16x + 16 = 1$

$\Rightarrow 4x^2 - 16x + 15 = 0 \Rightarrow (2x - 5)(2x - 3) = 0 \Rightarrow x = \frac{5}{2}$ or $x = \frac{3}{2} \Rightarrow \left(\frac{5}{2}, \frac{9}{4}\right)$ and $\left(\frac{3}{2}, -\frac{1}{4}\right)$ are points on the

curve where the slope is $-\frac{3}{2}$.

71. $y = 2x^3 - 3x^2 - 12x + 20 \Rightarrow \frac{dy}{dx} = 6x^2 - 6x - 12$; the tangent is parallel to the x-axis when $\frac{dy}{dx} = 0$

$\Rightarrow 6x^2 - 6x - 12 = 0 \Rightarrow x^2 - x - 2 = 0 \Rightarrow (x - 2)(x + 1) = 0 \Rightarrow x = 2$ or $x = -1 \Rightarrow (2, 0)$ and $(-1, 27)$ are

points on the curve where the tangent is parallel to the x-axis.

73. $y = 2x^3 - 3x^2 - 12x + 20 \Rightarrow \frac{dy}{dx} = 6x^2 - 6x - 12$

(a) The tangent is perpendicular to the line $y = 1 - \frac{x}{24}$ when $\frac{dy}{dx} = -\left(\frac{1}{-\left(\frac{1}{24}\right)}\right) = 24$; $6x^2 - 6x - 12 = 24$

$\Rightarrow x^2 - x - 2 = 4 \Rightarrow x^2 - x - 6 = 0 \Rightarrow (x - 3)(x + 2) = 0 \Rightarrow x = -2$ or $x = 3 \Rightarrow (-2, 16)$ and $(3, 11)$ are

points where the tangent is perpendicular to $y = 1 - \frac{x}{24}$.

(b) The tangent is parallel to the line $y = \sqrt{2} - 12x$ when $\frac{dy}{dx} = -12 \Rightarrow 6x^2 - 6x - 12 = -12 \Rightarrow x^2 - x = 0$

$\Rightarrow x(x - 1) = 0 \Rightarrow x = 0$ or $x = 1 \Rightarrow (0, 20)$ and $(1, 7)$ are points where the tangent is parallel to

$y = \sqrt{2} - 12x$.

75. $y = \tan x, -\frac{\pi}{2} < x < \frac{\pi}{2} \Rightarrow \frac{dy}{dx} = \sec^2 x$; now the slope
   of $y = -\frac{x}{2}$ is $-\frac{1}{2} \Rightarrow$ the normal line is parallel to
   $y = -\frac{x}{2}$ when $\frac{dy}{dx} = 2$. Thus, $\sec^2 x = 2 \Rightarrow \frac{1}{\cos^2 x} = 2$
   $\Rightarrow \cos^2 x = \frac{1}{2} \Rightarrow \cos x = \frac{\pm 1}{\sqrt{2}} \Rightarrow x = -\frac{\pi}{4}$ and $x = \frac{\pi}{4}$
   for $-\frac{\pi}{2} < x < \frac{\pi}{2} \Rightarrow \left(-\frac{\pi}{4}, -1\right)$ and $\left(\frac{\pi}{4}, 1\right)$ are points
   where the normal is parallel to $y = -\frac{x}{2}$.

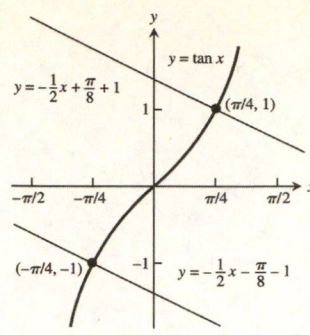

77. $y = x^2 + C \Rightarrow \frac{dy}{dx} = 2x$ and $y = x \Rightarrow \frac{dy}{dx} = 1$; the parabola is tangent to $y = x$ when $2x = 1 \Rightarrow x = \frac{1}{2} \Rightarrow y = \frac{1}{2}$;
   thus, $\frac{1}{2} = \left(\frac{1}{2}\right)^2 + C \Rightarrow C = \frac{1}{4}$

79. The line through $(0, 3)$ and $(5, -2)$ has slope $m = \frac{3-(-2)}{0-5} = -1 \Rightarrow$ the line through $(0, 3)$ and $(5, -2)$ is
   $y = -x + 3$; $y = \frac{c}{x+1} \Rightarrow \frac{dy}{dx} = \frac{-c}{(x+1)^2}$, so the curve is tangent to $y = -x + 3 \Rightarrow \frac{dy}{dx} = -1 = \frac{-c}{(x+1)^2}$
   $\Rightarrow (x+1)^2 = c, x \neq -1$. Moreover, $y = \frac{c}{x+1}$ intersects $y = -x + 3 \Rightarrow \frac{c}{x+1} = -x + 3, x \neq -1$
   $\Rightarrow c = (x+1)(-x+3), x \neq -1$. Thus $c = c \Rightarrow (x+1)^2 = (x+1)(-x+3) \Rightarrow (x+1)[x+1-(-x+3)]$
   $= 0, x \neq -1 \Rightarrow (x+1)(2x-2) = 0 \Rightarrow x = 1$ (since $x \neq -1$) $\Rightarrow c = 4$.

81. $x^2 + 2y^2 = 9 \Rightarrow 2x + 4y\frac{dy}{dx} = 0 \Rightarrow \frac{dy}{dx} = -\frac{x}{2y} \Rightarrow \frac{dy}{dx}\Big|_{(1,2)} = -\frac{1}{4} \Rightarrow$ the tangent line is $y = 2 - \frac{1}{4}(x-1)$
   $= -\frac{1}{4}x + \frac{9}{4}$ and the normal line is $y = 2 + 4(x-1) = 4x - 2$.

83. $xy + 2x - 5y = 2 \Rightarrow \left(x\frac{dy}{dx} + y\right) + 2 - 5\frac{dy}{dx} = 0 \Rightarrow \frac{dy}{dx}(x-5) = -y - 2 \Rightarrow \frac{dy}{dx} = \frac{-y-2}{x-5} \Rightarrow \frac{dy}{dx}\Big|_{(3,2)} = 2$
   $\Rightarrow$ the tangent line is $y = 2 + 2(x-3) = 2x - 4$ and the normal line is $y = 2 + \frac{-1}{2}(x-3) = -\frac{1}{2}x + \frac{7}{2}$.

85. $x + \sqrt{xy} = 6 \Rightarrow 1 + \frac{1}{2\sqrt{xy}}\left(x\frac{dy}{dx} + y\right) = 0 \Rightarrow x\frac{dy}{dx} + y = -2\sqrt{xy} \Rightarrow \frac{dy}{dx} = \frac{-2\sqrt{xy}-y}{x} \Rightarrow \frac{dy}{dx}\Big|_{(4,1)} = \frac{-5}{4}$
   $\Rightarrow$ the tangent line is $y = 1 - \frac{5}{4}(x-4) = -\frac{5}{4}x + 6$ and the normal line is $y = 1 + \frac{4}{5}(x-4) = \frac{4}{5}x - \frac{11}{5}$.

87. $x^3y^3 + y^2 = x + y \Rightarrow \left[x^3\left(3y^2\frac{dy}{dx}\right) + y^3(3x^2)\right] + 2y\frac{dy}{dx} = 1 + \frac{dy}{dx} \Rightarrow 3x^3y^2\frac{dy}{dx} + 2y\frac{dy}{dx} - \frac{dy}{dx} = 1 - 3x^2y^3$
   $\Rightarrow \frac{dy}{dx}(3x^3y^2 + 2y - 1) = 1 - 3x^2y^3 \Rightarrow \frac{dy}{dx} = \frac{1-3x^2y^3}{3x^3y^2+2y-1} \Rightarrow \frac{dy}{dx}\Big|_{(1,1)} = -\frac{2}{4}$, but $\frac{dy}{dx}\Big|_{(1,-1)}$ is undefined.
   Therefore, the curve has slope $-\frac{1}{2}$ at $(1, 1)$ but the slope is undefined at $(1, -1)$.

89. $x = \frac{1}{2}\tan t, y = \frac{1}{2}\sec t \Rightarrow \frac{dy}{dx} = \frac{dy/dt}{dx/dt} = \frac{\frac{1}{2}\sec t \tan t}{\frac{1}{2}\sec^2 t} = \frac{\tan t}{\sec t} = \sin t \Rightarrow \frac{dy}{dx}\Big|_{t=\pi/3} = \sin\frac{\pi}{3} = \frac{\sqrt{3}}{2}$; $t = \frac{\pi}{3}$
   $\Rightarrow x = \frac{1}{2}\tan\frac{\pi}{3} = \frac{\sqrt{3}}{2}$ and $y = \frac{1}{2}\sec\frac{\pi}{3} = 1 \Rightarrow y = \frac{\sqrt{3}}{2}x + \frac{1}{4}$; $\frac{d^2y}{dx^2} = \frac{dy'/dt}{dx/dt} = \frac{\cos t}{\frac{1}{2}\sec^2 t} = 2\cos^3 t \Rightarrow \frac{d^2y}{dx^2}\Big|_{t=\pi/3}$
   $= 2\cos^3\left(\frac{\pi}{3}\right) = \frac{1}{4}$

91. B = graph of f, A = graph of f'. Curve B cannot be the derivative of A because A has only negative slopes
   while some of B's values are positive.

93.

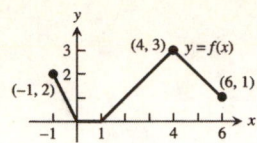

95. (a) $0, 0$ (b) largest 1700, smallest about 1400

97. $\lim\limits_{x \to 0} \dfrac{\sin x}{2x^2 - x} = \lim\limits_{x \to 0} \left[\left(\dfrac{\sin x}{x}\right) \cdot \dfrac{1}{(2x-1)} \right] = (1)\left(\dfrac{1}{-1}\right) = -1$

99. $\lim\limits_{r \to 0} \dfrac{\sin r}{\tan 2r} = \lim\limits_{r \to 0} \left(\dfrac{\sin r}{r} \cdot \dfrac{2r}{\tan 2r} \cdot \dfrac{1}{2} \right) = \left(\dfrac{1}{2}\right)(1) \lim\limits_{r \to 0} \dfrac{\cos 2r}{\left(\frac{\sin 2r}{2r}\right)} = \left(\dfrac{1}{2}\right)(1)\left(\dfrac{1}{1}\right) = \dfrac{1}{2}$

101. $\lim\limits_{\theta \to \left(\frac{\pi}{2}\right)^-} \dfrac{4\tan^2\theta + \tan\theta + 1}{\tan^2\theta + 5} = \lim\limits_{\theta \to \left(\frac{\pi}{2}\right)^-} \dfrac{\left(4 + \frac{1}{\tan\theta} + \frac{1}{\tan^2\theta}\right)}{\left(1 + \frac{5}{\tan^2\theta}\right)} = \dfrac{(4+0+0)}{(1+0)} = 4$

103. $\lim\limits_{x \to 0} \dfrac{x \sin x}{2 - 2\cos x} = \lim\limits_{x \to 0} \dfrac{x \sin x}{2(1 - \cos x)} = \lim\limits_{x \to 0} \dfrac{x \sin x}{2\left(2\sin^2\left(\frac{x}{2}\right)\right)} = \lim\limits_{x \to 0} \left[\dfrac{\frac{x}{2} \cdot \frac{x}{2}}{\sin^2\left(\frac{x}{2}\right)} \cdot \dfrac{\sin x}{x} \right]$

$= \lim\limits_{x \to 0} \left[\dfrac{\left(\frac{x}{2}\right)}{\sin\left(\frac{x}{2}\right)} \cdot \dfrac{\left(\frac{x}{2}\right)}{\sin\left(\frac{x}{2}\right)} \cdot \dfrac{\sin x}{x} \right] = (1)(1)(1) = 1$

105. $\lim\limits_{x \to 0} \dfrac{\tan x}{x} = \lim\limits_{x \to 0} \left(\dfrac{1}{\cos x} \cdot \dfrac{\sin x}{x} \right) = 1$; let $\theta = \tan x \Rightarrow \theta \to 0$ as $x \to 0 \Rightarrow \lim\limits_{x \to 0} g(x) = \lim\limits_{x \to 0} \dfrac{\tan(\tan x)}{\tan x}$

$= \lim\limits_{\theta \to 0} \dfrac{\tan\theta}{\theta} = 1$. Therefore, to make g continuous at the origin, define $g(0) = 1$.

107. (a) $S = 2\pi r^2 + 2\pi rh$ and h constant $\Rightarrow \dfrac{dS}{dt} = 4\pi r \dfrac{dr}{dt} + 2\pi h \dfrac{dr}{dt} = (4\pi r + 2\pi h) \dfrac{dr}{dt}$

(b) $S = 2\pi r^2 + 2\pi rh$ and r constant $\Rightarrow \dfrac{dS}{dt} = 2\pi r \dfrac{dh}{dt}$

(c) $S = 2\pi r^2 + 2\pi rh \Rightarrow \dfrac{dS}{dt} = 4\pi r \dfrac{dr}{dt} + 2\pi \left(r \dfrac{dh}{dt} + h \dfrac{dr}{dt} \right) = (4\pi r + 2\pi h) \dfrac{dr}{dt} + 2\pi r \dfrac{dh}{dt}$

(d) S constant $\Rightarrow \dfrac{dS}{dt} = 0 \Rightarrow 0 = (4\pi r + 2\pi h) \dfrac{dr}{dt} + 2\pi r \dfrac{dh}{dt} \Rightarrow (2r + h) \dfrac{dr}{dt} = -r \dfrac{dh}{dt} \Rightarrow \dfrac{dr}{dt} = \dfrac{-r}{2r+h} \dfrac{dh}{dt}$

109. $A = \pi r^2 \Rightarrow \dfrac{dA}{dt} = 2\pi r \dfrac{dr}{dt}$; so $r = 10$ and $\dfrac{dr}{dt} = -\dfrac{2}{\pi}$ m/sec $\Rightarrow \dfrac{dA}{dt} = (2\pi)(10)\left(-\dfrac{2}{\pi}\right) = -40$ m^2/sec

111. $\dfrac{dR_1}{dt} = -1$ ohm/sec, $\dfrac{dR_2}{dt} = 0.5$ ohm/sec; and $\dfrac{1}{R} = \dfrac{1}{R_1} + \dfrac{1}{R_2} \Rightarrow \dfrac{-1}{R^2} \dfrac{dR}{dt} = \dfrac{-1}{R_1^2} \dfrac{dR_1}{dt} - \dfrac{1}{R_2^2} \dfrac{dR_2}{dt}$. Also,

$R_1 = 75$ ohms and $R_2 = 50$ ohms $\Rightarrow \dfrac{1}{R} = \dfrac{1}{75} + \dfrac{1}{50} \Rightarrow R = 30$ ohms. Therefore, from the derivative equation,

$\dfrac{-1}{(30)^2} \dfrac{dR}{dt} = \dfrac{-1}{(75)^2}(-1) - \dfrac{1}{(50)^2}(0.5) = \left(\dfrac{1}{5625} - \dfrac{1}{5000} \right) \Rightarrow \dfrac{dR}{dt} = (-900)\left(\dfrac{5000 - 5625}{5625 \cdot 5000} \right) = \dfrac{9(625)}{50(5625)} = \dfrac{1}{50}$

$= 0.02$ ohm/sec.

113. Given $\dfrac{dx}{dt} = 10$ m/sec and $\dfrac{dy}{dt} = 5$ m/sec, let D be the distance from the origin $\Rightarrow D^2 = x^2 + y^2 \Rightarrow 2D \dfrac{dD}{dt}$

$= 2x \dfrac{dx}{dt} + 2y \dfrac{dy}{dt} \Rightarrow D \dfrac{dD}{dt} = x \dfrac{dx}{dt} + y \dfrac{dy}{dt}$. When $(x, y) = (3, -4)$, $D = \sqrt{3^2 + (-4)^2} = 5$ and

$5 \dfrac{dD}{dt} = (5)(10) + (12)(5) \Rightarrow \dfrac{dD}{dt} = \dfrac{110}{5} = 22$. Therefore, the particle is moving <u>away from</u> the origin at 22 m/sec

(because the distance D is increasing).

115. (a) From the diagram we have $\dfrac{10}{h} = \dfrac{4}{r} \Rightarrow r = \dfrac{2}{5} h$.

(b) $V = \dfrac{1}{3}\pi r^2 h = \dfrac{1}{3}\pi \left(\dfrac{2}{5} h\right)^2 h = \dfrac{4\pi h^3}{75} \Rightarrow \dfrac{dV}{dt} = \dfrac{4\pi h^2}{25} \dfrac{dh}{dt}$, so $\dfrac{dV}{dt} = -5$ and $h = 6 \Rightarrow \dfrac{dh}{dt} = -\dfrac{125}{144\pi}$ ft/min.

117. (a) From the sketch in the text, $\dfrac{d\theta}{dt} = -0.6$ rad/sec and $x = \tan\theta$. Also $x = \tan\theta \Rightarrow \dfrac{dx}{dt} = \sec^2\theta \dfrac{d\theta}{dt}$; at

point A, $x = 0 \Rightarrow \theta = 0 \Rightarrow \dfrac{dx}{dt} = (\sec^2 0)(-0.6) = -0.6$. Therefore the speed of the light is $0.6 = \dfrac{3}{5}$ km/sec

when it reaches point A.

(b) $\frac{(3/5)\,\text{rad}}{\text{sec}} \cdot \frac{1\,\text{rev}}{2\pi\,\text{rad}} \cdot \frac{60\,\text{sec}}{\text{min}} = \frac{18}{\pi}$ revs/min

119. (a) If $f(x) = \tan x$ and $x = -\frac{\pi}{4}$, then $f'(x) = \sec^2 x$,
$f\left(-\frac{\pi}{4}\right) = -1$ and $f'\left(-\frac{\pi}{4}\right) = 2$. The linearization of
$f(x)$ is $L(x) = 2\left(x + \frac{\pi}{4}\right) + (-1) = 2x + \frac{\pi-2}{2}$.

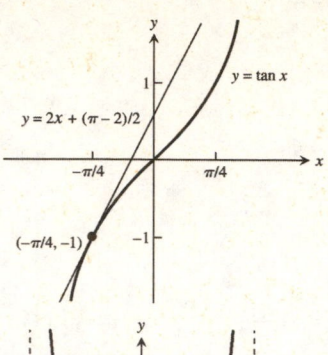

(b) If $f(x) = \sec x$ and $x = -\frac{\pi}{4}$, then $f'(x) = \sec x \tan x$,
$f\left(-\frac{\pi}{4}\right) = \sqrt{2}$ and $f'\left(-\frac{\pi}{4}\right) = -\sqrt{2}$. The
linearization of $f(x)$ is $L(x) = -\sqrt{2}\left(x + \frac{\pi}{4}\right) + \sqrt{2}$
$= -\sqrt{2}x + \frac{\sqrt{2}(4-\pi)}{4}$.

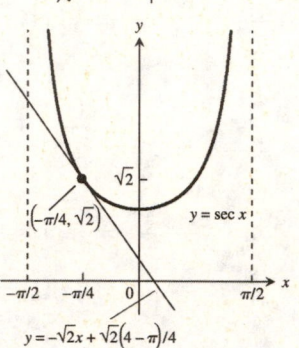

121. $f(x) = \sqrt{x+1} + \sin x - 0.5 = (x+1)^{1/2} + \sin x - 0.5 \Rightarrow f'(x) = \left(\frac{1}{2}\right)(x+1)^{-1/2} + \cos x$
$\Rightarrow L(x) = f'(0)(x - 0) + f(0) = 1.5(x - 0) + 0.5 \Rightarrow L(x) = 1.5x + 0.5$, the linearization of $f(x)$.

123. $S = \pi r\sqrt{r^2 + h^2}$, r constant $\Rightarrow dS = \pi r \cdot \frac{1}{2}(r^2 + h^2)^{-1/2} 2h\,dh = \frac{\pi r h}{\sqrt{r^2 + h^2}}\,dh$. Height changes from h_0 to $h_0 + dh$
$\Rightarrow dS = \frac{\pi r h_0 (dh)}{\sqrt{r^2 + h_0^2}}$

125. $C = 2\pi r \Rightarrow r = \frac{C}{2\pi}$, $S = 4\pi r^2 = \frac{C^2}{\pi}$, and $V = \frac{4}{3}\pi r^3 = \frac{C^3}{6\pi^2}$. It also follows that $dr = \frac{1}{2\pi}\,dC$, $dS = \frac{2C}{\pi}\,dC$ and
$dV = \frac{C^2}{2\pi^2}\,dC$. Recall that $C = 10$ cm and $dC = 0.4$ cm.
(a) $dr = \frac{0.4}{2\pi} = \frac{0.2}{\pi}$ cm $\Rightarrow \left(\frac{dr}{r}\right)(100\%) = \left(\frac{0.2}{\pi}\right)\left(\frac{2\pi}{10}\right)(100\%) = (.04)(100\%) = 4\%$
(b) $dS = \frac{20}{\pi}(0.4) = \frac{8}{\pi}$ cm $\Rightarrow \left(\frac{dS}{S}\right)(100\%) = \left(\frac{8}{\pi}\right)\left(\frac{\pi}{100}\right)(100\%) = 8\%$
(c) $dV = \frac{10^2}{2\pi^2}(0.4) = \frac{20}{\pi^2}$ cm $\Rightarrow \left(\frac{dV}{V}\right)(100\%) = \left(\frac{20}{\pi^2}\right)\left(\frac{6\pi^2}{1000}\right)(100\%) = 12\%$

CHAPTER 3 ADDITIONAL AND ADVANCED EXERCISES

1. (a) $\sin 2\theta = 2\sin\theta\cos\theta \Rightarrow \frac{d}{d\theta}(\sin 2\theta) = \frac{d}{d\theta}(2\sin\theta\cos\theta) \Rightarrow 2\cos 2\theta = 2[(\sin\theta)(-\sin\theta) + (\cos\theta)(\cos\theta)]$
$\Rightarrow \cos 2\theta = \cos^2\theta - \sin^2\theta$
(b) $\cos 2\theta = \cos^2\theta - \sin^2\theta \Rightarrow \frac{d}{d\theta}(\cos 2\theta) = \frac{d}{d\theta}(\cos^2\theta - \sin^2\theta) \Rightarrow -2\sin 2\theta = (2\cos\theta)(-\sin\theta) - (2\sin\theta)(\cos\theta)$
$\Rightarrow \sin 2\theta = \cos\theta\sin\theta + \sin\theta\cos\theta \Rightarrow \sin 2\theta = 2\sin\theta\cos\theta$

3. (a) $f(x) = \cos x \Rightarrow f'(x) = -\sin x \Rightarrow f''(x) = -\cos x$, and $g(x) = a + bx + cx^2 \Rightarrow g'(x) = b + 2cx \Rightarrow g''(x) = 2c$;
also, $f(0) = g(0) \Rightarrow \cos(0) = a \Rightarrow a = 1$; $f'(0) = g'(0) \Rightarrow -\sin(0) = b \Rightarrow b = 0$; $f''(0) = g''(0)$
$\Rightarrow -\cos(0) = 2c \Rightarrow c = -\frac{1}{2}$. Therefore, $g(x) = 1 - \frac{1}{2}x^2$.
(b) $f(x) = \sin(x + a) \Rightarrow f'(x) = \cos(x + a)$, and $g(x) = b\sin x + c\cos x \Rightarrow g'(x) = b\cos x - c\sin x$; also,
$f(0) = g(0) \Rightarrow \sin(a) = b\sin(0) + c\cos(0) \Rightarrow c = \sin a$; $f'(0) = g'(0) \Rightarrow \cos(a) = b\cos(0) - c\sin(0)$
$\Rightarrow b = \cos a$. Therefore, $g(x) = \sin x \cos a + \cos x \sin a$.

(c) When $f(x) = \cos x$, $f'''(x) = \sin x$ and $f^{(4)}(x) = \cos x$; when $g(x) = 1 - \frac{1}{2}x^2$, $g'''(x) = 0$ and $g^{(4)}(x) = 0$.
Thus $f'''(0) = 0 = g'''(0)$ so the third derivatives agree at $x = 0$. However, the fourth derivatives do not
agree since $f^{(4)}(0) = 1$ but $g^{(4)}(0) = 0$. In case (b), when $f(x) = \sin(x + a)$ and $g(x)$
$= \sin x \cos a + \cos x \sin a$, notice that $f(x) = g(x)$ for all x, not just $x = 0$. Since this is an identity, we
have $f^{(n)}(x) = g^{(n)}(x)$ for any x and any positive integer n.

5. If the circle $(x - h)^2 + (y - k)^2 = a^2$ and $y = x^2 + 1$ are tangent at $(1, 2)$, then the slope of this tangent is
$m = 2x|_{(1,2)} = 2$ and the tangent line is $y = 2x$. The line containing (h, k) and $(1, 2)$ is perpendicular to
$y = 2x \Rightarrow \frac{k-2}{h-1} = -\frac{1}{2} \Rightarrow h = 5 - 2k \Rightarrow$ the location of the center is $(5 - 2k, k)$. Also, $(x - h)^2 + (y - k)^2 = a^2$
$\Rightarrow x - h + (y - k)y' = 0 \Rightarrow 1 + (y')^2 + (y - k)y'' = 0 \Rightarrow y'' = \frac{1+(y')^2}{k-y}$. At the point $(1, 2)$ we know
$y' = 2$ from the tangent line and that $y'' = 2$ from the parabola. Since the second derivatives are equal at $(1, 2)$
we obtain $2 = \frac{1+(2)^2}{k-2} \Rightarrow k = \frac{9}{2}$. Then $h = 5 - 2k = -4 \Rightarrow$ the circle is $(x + 4)^2 + \left(y - \frac{9}{2}\right)^2 = a^2$. Since $(1, 2)$
lies on the circle we have that $a = \frac{5\sqrt{5}}{2}$.

7. (a) $y = uv \Rightarrow \frac{dy}{dt} = \frac{du}{dt}v + u\frac{dv}{dt} = (0.04u)v + u(0.05v) = 0.09uv = 0.09y \Rightarrow$ the rate of growth of the total production is
9% per year.

(b) If $\frac{du}{dt} = -0.02u$ and $\frac{dv}{dt} = 0.03v$, then $\frac{dy}{dt} = (-0.02u)v + (0.03v)u = 0.01uv = 0.01y$, increasing at 1% per
year.

9. Answers will vary. Here is one possibility.

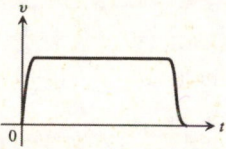

11. (a) $s(t) = 64t - 16t^2 \Rightarrow v(t) = \frac{ds}{dt} = 64 - 32t = 32(2 - t)$. The maximum height is reached when $v(t) = 0$
$\Rightarrow t = 2$ sec. The velocity when it leaves the hand is $v(0) = 64$ ft/sec.

(b) $s(t) = 64t - 2.6t^2 \Rightarrow v(t) = \frac{ds}{dt} = 64 - 5.2t$. The maximum height is reached when $v(t) = 0 \Rightarrow t \approx 12.31$ sec.
The maximum height is about $s(12.31) = 393.85$ ft.

13. $m(v^2 - v_0^2) = k(x_0^2 - x^2) \Rightarrow m\left(2v\frac{dv}{dt}\right) = k\left(-2x\frac{dx}{dt}\right) \Rightarrow m\frac{dv}{dt} = k\left(-\frac{2x}{2v}\right)\frac{dx}{dt} \Rightarrow m\frac{dv}{dt} = -kx\left(\frac{1}{v}\right)\frac{dx}{dt}$. Then
substituting $\frac{dx}{dt} = v \Rightarrow m\frac{dv}{dt} = -kx$, as claimed.

15. (a) To be continuous at $x = \pi$ requires that $\lim\limits_{x \to \pi^-} \sin x = \lim\limits_{x \to \pi^+} (mx + b) \Rightarrow 0 = m\pi + b \Rightarrow m = -\frac{b}{\pi}$;

(b) If $y' = \begin{cases} \cos x, & x < \pi \\ m, & x \geq \pi \end{cases}$ is differentiable at $x = \pi$, then $\lim\limits_{x \to \pi^-} \cos x = m \Rightarrow m = -1$ and $b = \pi$.

17. (a) For all a, b and for all $x \neq 2$, f is differentiable at x. Next, f differentiable at $x = 2 \Rightarrow$ f continuous at $x = 2$
$\Rightarrow \lim\limits_{x \to 2^-} f(x) = f(2) \Rightarrow 2a = 4a - 2b + 3 \Rightarrow 2a - 2b + 3 = 0$. Also, f differentiable at $x \neq 2$
$\Rightarrow f'(x) = \begin{cases} a, & x < 2 \\ 2ax - b, & x > 2 \end{cases}$. In order that $f'(2)$ exist we must have $a = 2a(2) - b \Rightarrow a = 4a - b \Rightarrow 3a = b$.
Then $2a - 2b + 3 = 0$ and $3a = b \Rightarrow a = \frac{3}{4}$ and $b = \frac{9}{4}$.

(b) For $x < 2$, the graph of f is a straight line having a slope of $\frac{3}{4}$ and passing through the origin; for $x \geq 2$, the graph of f
is a parabola. At $x = 2$, the value of the y-coordinate on the parabola is $\frac{3}{2}$ which matches the y-coordinate of the point
on the straight line at $x = 2$. In addition, the slope of the parabola at the match up point is $\frac{3}{4}$ which is equal to the
slope of the straight line. Therefore, since the graph is differentiable at the match up point, the graph is smooth there.

19. f odd $\Rightarrow$ f($-$x) $=$ $-$f(x) $\Rightarrow$ $\frac{d}{dx}$ (f($-$x)) $=$ $\frac{d}{dx}$ ($-$f(x)) $\Rightarrow$ f$'$($-$x)($-$1) $=$ $-$f$'$(x) $\Rightarrow$ f$'$($-$x) $=$ f$'$(x) $\Rightarrow$ f$'$ is even.

21. Let h(x) $=$ (fg)(x) $=$ f(x) g(x) $\Rightarrow$ h$'$(x) $=$ $\lim\limits_{x \to x_0}$ $\frac{h(x) - h(x_0)}{x - x_0}$ $=$ $\lim\limits_{x \to x_0}$ $\frac{f(x) g(x) - f(x_0) g(x_0)}{x - x_0}$

$=$ $\lim\limits_{x \to x_0}$ $\frac{f(x) g(x) - f(x) g(x_0) + f(x) g(x_0) - f(x_0) g(x_0)}{x - x_0}$ $=$ $\lim\limits_{x \to x_0}$ $\left[f(x) \left[\frac{g(x) - g(x_0)}{x - x_0} \right] \right]$ $+$ $\lim\limits_{x \to x_0}$ $\left[g(x_0) \left[\frac{f(x) - f(x_0)}{x - x_0} \right] \right]$

$=$ f(x_0) $\lim\limits_{x \to x_0}$ $\left[\frac{g(x) - g(x_0)}{x - x_0} \right]$ $+$ g(x_0) f$'$(x_0) $=$ 0 $\cdot$ $\lim\limits_{x \to x_0}$ $\left[\frac{g(x) - g(x_0)}{x - x_0} \right]$ $+$ g(x_0) f$'$(x_0) $=$ g(x_0) f$'$(x_0), if g is

continuous at x_0. Therefore (fg)(x) is differentiable at x_0 if f(x_0) $=$ 0, and (fg)$'$ (x_0) $=$ g(x_0) f$'$(x_0).

23. If f(x) $=$ x and g(x) $=$ x sin $\left(\frac{1}{x} \right)$, then x^2 sin $\left(\frac{1}{x} \right)$ is differentiable at x $=$ 0 because f$'$(0) $=$ 1, f(0) $=$ 0 and

$\lim\limits_{x \to 0}$ x sin $\left(\frac{1}{x} \right)$ $=$ $\lim\limits_{x \to 0}$ $\frac{\sin \left(\frac{1}{x} \right)}{\frac{1}{x}}$ $=$ $\lim\limits_{t \to \infty}$ $\frac{\sin t}{t}$ $=$ 0 (so g is continuous at x $=$ 0). In fact, from Exercise 21,

h$'$(0) $=$ g(0) f$'$(0) $=$ 0. However, for x $\neq$ 0, h$'$(x) $=$ $\left[x^2 \cos \left(\frac{1}{x} \right) \right] \left(-\frac{1}{x^2} \right)$ $+$ 2x sin $\left(\frac{1}{x} \right)$. But

$\lim\limits_{x \to 0}$ h$'$(x) $=$ $\lim\limits_{x \to 0}$ $\left[-\cos \left(\frac{1}{x} \right) + 2x \sin \left(\frac{1}{x} \right) \right]$ does not exist because cos $\left(\frac{1}{x} \right)$ has no limit as x $\to$ 0. Therefore,

the derivative is not continuous at x $=$ 0 because it has no limit there.

25. Step 1:   The formula holds for n $=$ 2 (a single product) since y $=$ $u_1 u_2$ $\Rightarrow$ $\frac{dy}{dx}$ $=$ $\frac{du_1}{dx}$ u_2 $+$ u_1 $\frac{du_2}{dx}$.

   Step 2:   Assume the formula holds for n $=$ k:

$$y = u_1 u_2 \cdots u_k \Rightarrow \frac{dy}{dx} = \frac{du_1}{dx} u_2 u_3 \cdots u_k + u_1 \frac{du_2}{dx} u_3 \cdots u_k + \ldots + u_1 u_2 \cdots u_{k-1} \frac{du_k}{dx} .$$

If y $=$ $u_1 u_2 \cdots u_k u_{k+1}$ $=$ ($u_1 u_2 \cdots u_k$) u_{k+1}, then $\frac{dy}{dx}$ $=$ $\frac{d(u_1 u_2 \cdots u_k)}{dx}$ u_{k+1} $+$ $u_1 u_2 \cdots u_k$ $\frac{du_{k+1}}{dx}$

$=$ $\left(\frac{du_1}{dx} u_2 u_3 \cdots u_k + u_1 \frac{du_2}{dx} u_3 \cdots u_k + \cdots + u_1 u_2 \cdots u_{k-1} \frac{du_k}{dx} \right)$ u_{k+1} $+$ $u_1 u_2 \cdots u_k$ $\frac{du_{k+1}}{dx}$

$=$ $\frac{du_1}{dx}$ $u_2 u_3 \cdots u_{k+1}$ $+$ u_1 $\frac{du_2}{dx}$ $u_3 \cdots u_{k+1}$ $+$ $\cdots$ $+$ $u_1 u_2 \cdots u_{k-1}$ $\frac{du_k}{dx}$ u_{k+1} $+$ $u_1 u_2 \cdots u_k$ $\frac{du_{k+1}}{dx}$.

Thus the original formula holds for n $=$ (k+1) whenever it holds for n $=$ k.

27. (a) T^2 $=$ $\frac{4\pi^2 L}{g}$ $\Rightarrow$ L $=$ $\frac{T^2 g}{4\pi^2}$ $\Rightarrow$ L $=$ $\frac{(1 \text{ sec}^2)(32.2 \text{ ft/sec}^2)}{4\pi^2}$ $\Rightarrow$ L $\approx$ 0.8156 ft

   (b) T^2 $=$ $\frac{4\pi^2 L}{g}$ $\Rightarrow$ T $=$ $\frac{2\pi}{\sqrt{g}}$ $\sqrt{L}$; dT $=$ $\frac{2\pi}{\sqrt{g}}$ $\cdot$ $\frac{1}{2\sqrt{L}}$dL $=$ $\frac{\pi}{\sqrt{Lg}}$dL; dT $=$ $\frac{\pi}{\sqrt{(0.8156 \text{ ft})(32.2 \text{ ft/sec}^2)}}$(0.01 ft) $\approx$ 0.00613 sec.

   (c)   Since there are 86,400 sec in a day, we have (0.00613 sec)(86,400 sec/day) $\approx$ 529.6 sec/day, or 8.83 min/day; the

   clock will lose about 8.83 min/day.

CHAPTER 4 APPLICATIONS OF DERIVATIVES

4.1 EXTREME VALUES OF FUNCTIONS

1. An absolute minimum at $x = c_2$, an absolute maximum at $x = b$. Theorem 1 guarantees the existence of such extreme values because h is continuous on [a, b].

3. No absolute minimum. An absolute maximum at $x = c$. Since the function's domain is an open interval, the function does not satisfy the hypotheses of Theorem 1 and need not have absolute extreme values.

5. An absolute minimum at $x = a$ and an absolute maximum at $x = c$. Note that $y = g(x)$ is not continuous but still has extrema. When the hypothesis of Theorem 1 is satisfied then extrema are guaranteed, but when the hypothesis is not satisfied, absolute extrema may or may not occur.

7. Local minimum at $(-1, 0)$, local maximum at $(1, 0)$

9. Maximum at $(0, 5)$. Note that there is no minimum since the endpoint $(2, 0)$ is excluded from the graph.

11. Graph (c), since this the only graph that has positive slope at c.

13. Graph (d), since this is the only graph representing a funtion that is differentiable at b but not at a.

15. $f(x) = \frac{2}{3}x - 5 \Rightarrow f'(x) = \frac{2}{3} \Rightarrow$ no critical points;
$f(-2) = -\frac{19}{3}$, $f(3) = -3 \Rightarrow$ the absolute maximum is -3 at $x = 3$ and the absolute minimum is $-\frac{19}{3}$ at $x = -2$

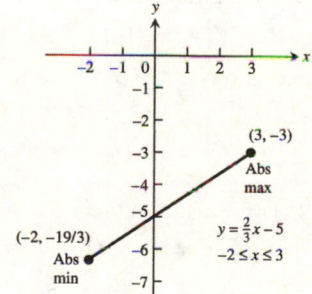

17. $f(x) = x^2 - 1 \Rightarrow f'(x) = 2x \Rightarrow$ a critical point at $x = 0$; $f(-1) = 0$, $f(0) = -1$, $f(2) = 3 \Rightarrow$ the absolute maximum is 3 at $x = 2$ and the absolute minimum is -1 at $x = 0$

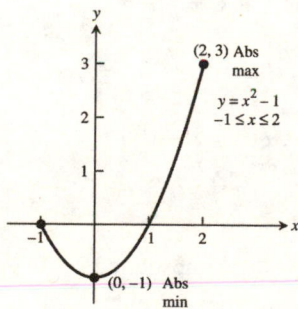

19. $F(x) = -\frac{1}{x^2} = -x^{-2} \Rightarrow F'(x) = 2x^{-3} = \frac{2}{x^3}$, however

x = 0 is not a critical point since 0 is not in the domain;

$F(0.5) = -4, F(2) = -0.25 \Rightarrow$ the absolute maximum is

-0.25 at x = 2 and the absolute minimum is -4 at

x = 0.5

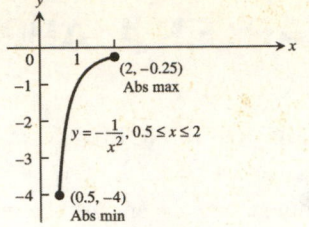

21. $h(x) = \sqrt[3]{x} = x^{1/3} \Rightarrow h'(x) = \frac{1}{3}x^{-2/3} \Rightarrow$ a critical point

at x = 0; h(-1) = -1, h(0) = 0, h(8) = 2 $\Rightarrow$ the absolute

maximum is 2 at x = 8 and the absolute minimum is -1

at x = -1

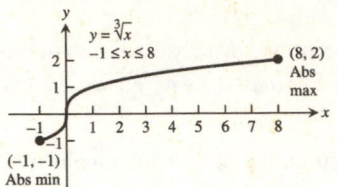

23. $g(x) = \sqrt{4 - x^2} = (4 - x^2)^{1/2}$

$\Rightarrow g'(x) = \frac{1}{2}(4 - x^2)^{-1/2}(-2x) = \frac{-x}{\sqrt{4 - x^2}}$

$\Rightarrow$ critical points at x = -2 and x = 0, but not at x = 2

because 2 is not in the domain; g(-2) = 0, g(0) = 2,

$g(1) = \sqrt{3} \Rightarrow$ the absolute maximum is 2 at x = 0 and the

absolute minimum is 0 at x = -2

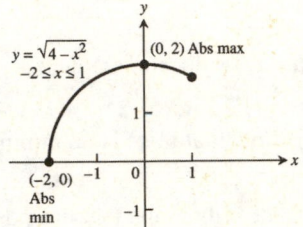

25. $f(\theta) = \sin \theta \Rightarrow f'(\theta) = \cos \theta \Rightarrow \theta = \frac{\pi}{2}$ is a critical point,

but $\theta = \frac{-\pi}{2}$ is not a critical point because $\frac{-\pi}{2}$ is not interior to

the domain; $f\left(\frac{-\pi}{2}\right) = -1, f\left(\frac{\pi}{2}\right) = 1, f\left(\frac{5\pi}{6}\right) = \frac{1}{2}$

$\Rightarrow$ the absolute maximum is 1 at $\theta = \frac{\pi}{2}$ and the absolute

minimum is -1 at $\theta = \frac{-\pi}{2}$

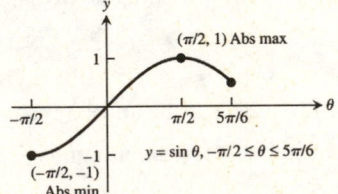

27. $g(x) = \csc x \Rightarrow g'(x) = -(\csc x)(\cot x) \Rightarrow$ a critical point

at $x = \frac{\pi}{2}$; $g\left(\frac{\pi}{3}\right) = \frac{2}{\sqrt{3}}, g\left(\frac{\pi}{2}\right) = 1, g\left(\frac{2\pi}{3}\right) = \frac{2}{\sqrt{3}} \Rightarrow$ the

absolute maximum is $\frac{2}{\sqrt{3}}$ at $x = \frac{\pi}{3}$ and $x = \frac{2\pi}{3}$, and the

absolute minimum is 1 at $x = \frac{\pi}{2}$

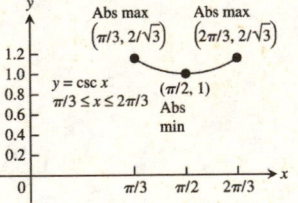

29. $f(t) = 2 - |t| = 2 - \sqrt{t^2} = 2 - (t^2)^{1/2}$

$\Rightarrow f'(t) = -\frac{1}{2}(t^2)^{-1/2}(2t) = -\frac{t}{\sqrt{t^2}} = -\frac{t}{|t|}$

$\Rightarrow$ a critical point at t = 0; f(-1) = 1,

f(0) = 2, f(3) = -1 $\Rightarrow$ the absolute maximum is 2 at t = 0

and the absolute minimum is -1 at t = 3

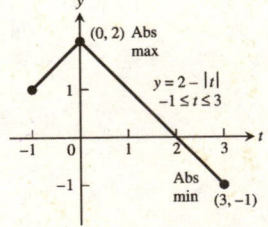

31. $f(x) = x^{4/3} \Rightarrow f'(x) = \frac{4}{3}x^{1/3} \Rightarrow$ a critical point at x = 0; f(-1) = 1, f(0) = 0, f(8) = 16 $\Rightarrow$ the absolute

maximum is 16 at x = 8 and the absolute minimum is 0 at x = 0

33. $g(\theta) = \theta^{3/5} \Rightarrow g'(\theta) = \frac{3}{5}\theta^{-2/5} \Rightarrow$ a critical point at $\theta = 0$; g(-32) = -8, g(0) = 0, g(1) = 1 $\Rightarrow$ the absolute

maximum is 1 at $\theta = 1$ and the absolute minimum is -8 at $\theta = -32$

35. Minimum value is 1 at x = 2.

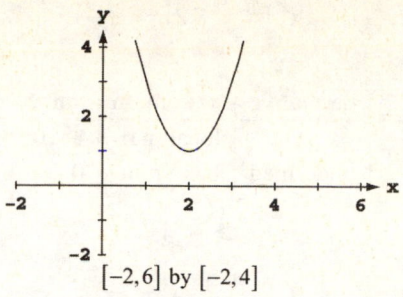

$[-2, 6]$ by $[-2, 4]$

37. To find the exact values, note that that $y' = 3x^2 + 2x - 8$
     $= (3x - 4)(x + 2)$, which is zero when $x = -2$ or $x = \frac{4}{3}$.
     Local maximum at $(-2, 17)$; local minimum at $\left(\frac{4}{3}, -\frac{41}{27}\right)$

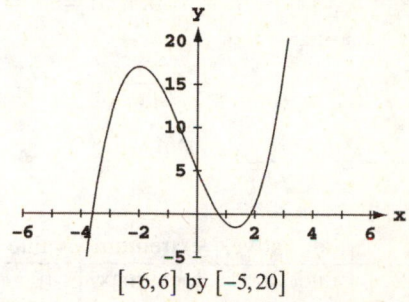

$[-6, 6]$ by $[-5, 20]$

39. Minimum value is 0 when $x = -1$ or $x = 1$.

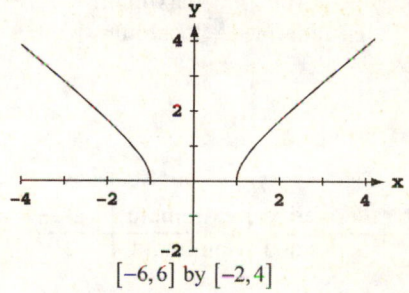

$[-6, 6]$ by $[-2, 4]$

41. The actual graph of the function has asymptotes at $x = \pm 1$,
     so there are no extrema near these values. (This is an
     example of grapher failure.) There is a local minimum at
     $(0, 1)$.

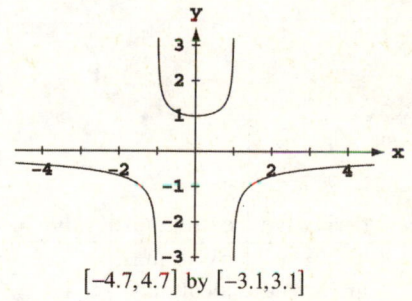

$[-4.7, 4.7]$ by $[-3.1, 3.1]$

43. Maximum value is $\frac{1}{2}$ at $x = 1$;
     minimum value is $-\frac{1}{2}$ as $x = -1$.

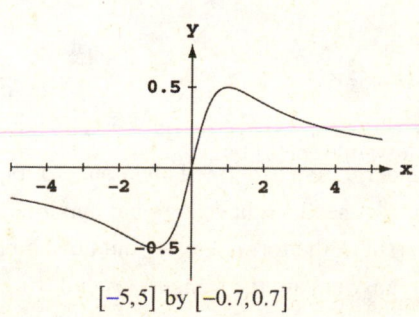

$[-5, 5]$ by $[-0.7, 0.7]$

45. $y' = x^{2/3}(1) + \frac{2}{3}x^{-1/3}(x+2) = \frac{5x+4}{3\sqrt[3]{x}}$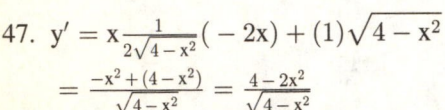

crit. pt.	derivative	extremum	value
$x = -\frac{4}{5}$	0	local max	$\frac{12}{25}10^{1/3} = 1.034$
$x = 0$	undefined	local min	0

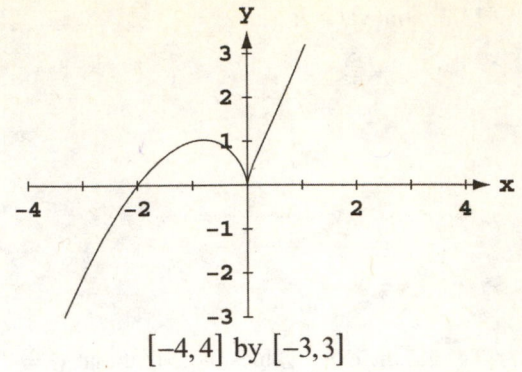

$[-4,4]$ by $[-3,3]$

47. $y' = x\frac{1}{2\sqrt{4-x^2}}(-2x) + (1)\sqrt{4-x^2}$

$= \frac{-x^2 + (4-x^2)}{\sqrt{4-x^2}} = \frac{4-2x^2}{\sqrt{4-x^2}}$

crit. pt.	derivative	extremum	value
$x = -2$	undefined	local max	0
$x = -\sqrt{2}$	0	minimum	-2
$x = \sqrt{2}$	0	maximum	2
$x = 2$	undefined	local min	0

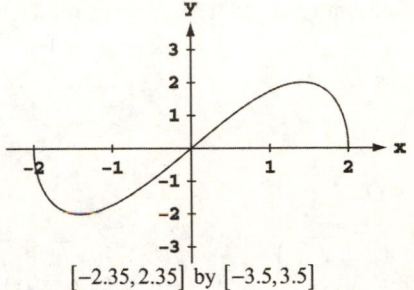

$[-2.35, 2.35]$ by $[-3.5, 3.5]$

49. $y' = \begin{cases} -2, & x < 1 \\ 1, & x > 1 \end{cases}$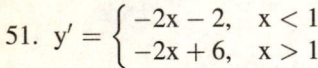

crit. pt.	derivative	extremum	value
$x = 1$	undefined	minimum	2

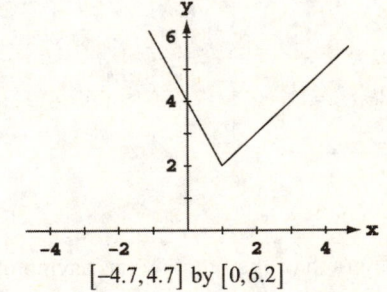

$[-4.7, 4.7]$ by $[0, 6.2]$

51. $y' = \begin{cases} -2x - 2, & x < 1 \\ -2x + 6, & x > 1 \end{cases}$

crit. pt.	derivative	extremum	value
$x = -1$	0	maximum	5
$x = 1$	undefined	local min	1
$x = 3$	0	maximum	5

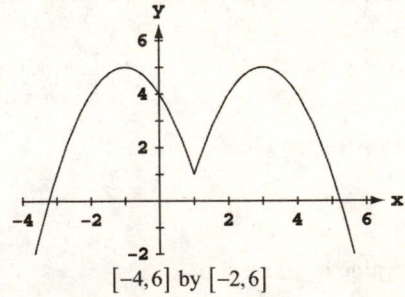

$[-4, 6]$ by $[-2, 6]$

53. (a) No, since $f'(x) = \frac{2}{3}(x-2)^{-1/3}$, which is undefined at $x = 2$.

(b) The derivative is defined and nonzero for all $x \neq 2$. Also, $f(2) = 0$ and $f(x) > 0$ for all $x \neq 2$.

(c) No, $f(x)$ need not have a global maximum because its domain is all real numbers. Any restriction of f to a closed interval of the form [a, b] would have both a maximum value and minimum value on the interval.

(d) The answers are the same as (a) and (b) with 2 replaced by a.

55.

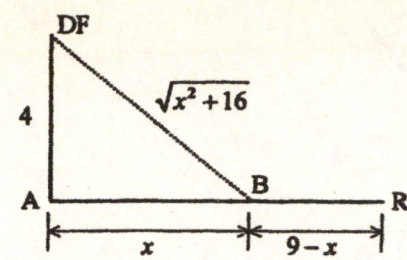

(a) The construction cost is $C(x) = 0.3\sqrt{16 + x^2} + 0.2(9 - x)$ million dollars, where $0 \leq x \leq 9$ miles. The following is a graph of $C(x)$.

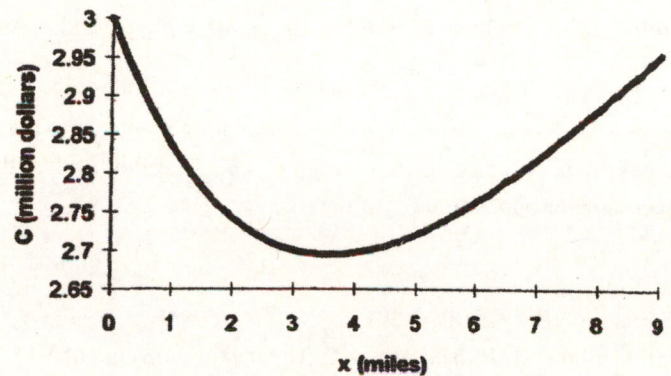

Solving $C'(x) = \frac{0.3x}{\sqrt{16 + x^2}} - 0.2 = 0$ gives $x = \pm \frac{8\sqrt{5}}{5} \approx \pm 3.58$ miles, but only $x = 3.58$ miles is a critical point is

the specified domain. Evaluating the costs at the critical and endpoints gives $C(0) = \$3$ million, $C\left(\frac{8\sqrt{5}}{5}\right) \approx \$2.694$

million, and $C(9) \approx \$2.955$ million. Therefore, to minimize the cost of construction, the pipeline should be placed

from the docking facility to point B, 3.58 miles along the shore from point A, and then along the shore from B to the

refinery.

(b) If the per mile cost of underwater construction is p, then $C(x) = p\sqrt{16 + x^2} + 0.2(9 - x)$ and

$C'(x) = \frac{0.3x}{\sqrt{16 + x^2}} - 0.2 = 0$ gives $x_c = \frac{0.8}{\sqrt{p^2 - 0.04}}$, which minimizes the construction cost provided $x_c \leq 9$. The value

of p that gives $x_c = 9$ miles is 0.218864. Consequently, if the underwater construction costs $218,864 per mile or less,

then running the pipeline along a straight line directly from the docking facility to the refinery will minimize the cost

of construction.

In theory, p would have to be infinite to justify running the pipe directly from the docking facility to point A (i.e., for

x_c to be zero). For all values of $p > 0.218864$ there is always an $x_c \in (0, 9)$ that will give a minimum value for C.

This is proved by looking at $C''(x_c) = \frac{16p}{(16 + x_c^2)^{3/2}}$ which is always positive for $p > 0$.

57.

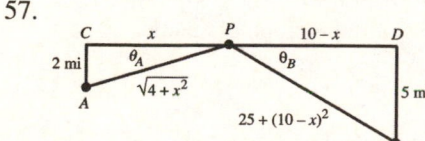

The length of pipeline is $L(x) = \sqrt{4 + x^2} + \sqrt{25 + (10 - x)^2}$ for $0 \leq x \leq 10$. The following is a graph of $L(x)$.

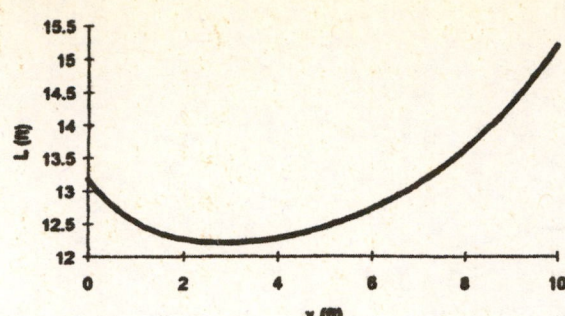

Setting the derivative of $L(x)$ equal to zero gives $L'(x) = \frac{x}{\sqrt{4+x^2}} - \frac{(10-x)}{\sqrt{25+(10-x)^2}} = 0$. Note that $\frac{x}{\sqrt{4+x^2}} = \cos\theta_A$ and

$\frac{10-x}{\sqrt{25+(10-x)^2}} = \cos\theta_B$, therefore, $L'(x) = 0$ when $\cos\theta_A = \cos\theta_B$, or $\theta_A = \theta_B$ and $\triangle ACP$ is similar to $\triangle BDP$. Use

simple proportions to determine x as follows: $\frac{x}{2} = \frac{10-x}{5} \Rightarrow x = \frac{20}{7} \approx 2.857$ miles along the coast from town A to town B.
If the two towns were on opposite sides of the river, the obvious solution would be to place the pump station on a straight
line (the shortest distance) between two towns, again forcing $\theta_A = \theta_B$. The shortest length of pipe is the same regardless of
whether the towns are on thee same or opposite sides of the river.

59. (a) $V(x) = 160x - 52x^2 + 4x^3$
    $V'(x) = 160 - 104x + 12x^2 = 4(x-2)(3x-20)$
    The only critical point in the interval $(0, 5)$ is at $x = 2$. The maximum value of $V(x)$ is 144 at $x = 2$.
    (b) The largest possible volume of the box is 144 cubic units, and it occurs when $x = 2$ units.

61. Let x represent the length of the base and $\sqrt{25-x^2}$ the height of the triangle. The area of the triangle is represented by
    $A(x) = \frac{x}{2}\sqrt{25-x^2}$ where $0 \leq x \leq 5$. Consequently, solving $A'(x) = 0 \Rightarrow \frac{25-2x^2}{2\sqrt{25-x^2}} = 0 \Rightarrow x = \frac{5}{\sqrt{2}}$. Since
    $A(0) = A(5) = 0$, $A(x)$ is maximized at $x = \frac{5}{\sqrt{2}}$. The largest possible area is $A\left(\frac{5}{\sqrt{2}}\right) = \frac{25}{4}$ cm^2.

63. $s = -\frac{1}{2}gt^2 + v_0t + s_0 \Rightarrow \frac{ds}{dt} = -gt + v_0 = 0 \Rightarrow t = \frac{v_0}{g}$. Now $s(t) = s_0 \Leftrightarrow t\left(-\frac{gt}{2} + v_0\right) = 0 \Leftrightarrow t = 0$ or $t = \frac{2v_0}{g}$.
    Thus $s\left(\frac{v_0}{g}\right) = -\frac{1}{2}g\left(\frac{v_0}{g}\right)^2 + v_0\left(\frac{v_0}{g}\right) + s_0 = \frac{v_0^2}{2g} + s_0 > s_0$ is the <u>maximum</u> height over the interval $0 \leq t \leq \frac{2v_0}{g}$.

65. Yes, since $f(x) = |x| = \sqrt{x^2} = (x^2)^{1/2} \Rightarrow f'(x) = \frac{1}{2}(x^2)^{-1/2}(2x) = \frac{x}{(x^2)^{1/2}} = \frac{x}{|x|}$ is not defined at $x = 0$. Thus it
    is not required that f' be zero at a local extreme point since f' may be undefined there.

67. If $g(c)$ is a local minimum value of g, then $g(x) \geq g(c)$ for all x in some open interval (a, b) containing c. Since
    g is odd, $g(-x) = -g(x) \leq -g(c) = g(-c)$ for all $-x$ in the open interval $(-b, -a)$ containing $-c$. That is, g
    assumes a local maximum at the point $-c$. This is also clear from the graph of g because the graph of an odd
    function is symmetric about the origin.

69. (a) $f'(x) = 3ax^2 + 2bx + c$ is a quadratic, so it can have 0, 1, or 2 zeros, which would be the critical points of f. The
    function $f(x) = x^3 - 3x$ has two critical points at $x = -1$ and $x = 1$. The function $f(x) = x^3 - 1$ has one critical point
    at $x = 0$. The function $f(x) = x^3 + x$ has no critical points.

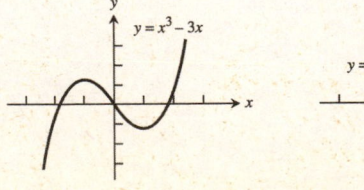
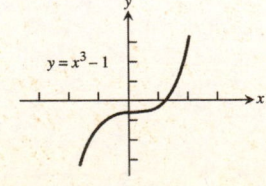
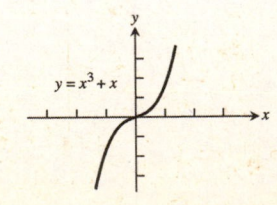

(b) The function can have either two local extreme values or no extreme values. (If there is only one critical point, the cubic function has no extreme values.)

71. Maximum value is 11 at x = 5;
minimum value is 5 on the interval [−3, 2];
local maximum at (−5, 9)

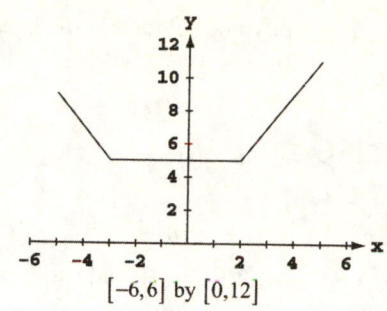

$[-6,6]$ by $[0,12]$

73. Maximum value is 5 on the interval $[3, \infty)$;
minimum value is −5 on the interval $(-\infty, -2]$.

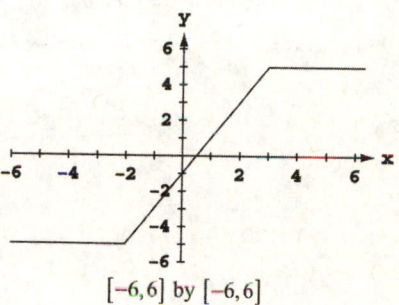

$[-6,6]$ by $[-6,6]$

4.2 THE MEAN VALUE THEOREM

1. When $f(x) = x^2 + 2x - 1$ for $0 \le x \le 1$, then $\frac{f(1)-f(0)}{1-0} = f'(c) \Rightarrow 3 = 2c + 2 \Rightarrow c = \frac{1}{2}$.

3. When $f(x) = x + \frac{1}{x}$ for $\frac{1}{2} \le x \le 2$, then $\frac{f(2)-f(1/2)}{2-1/2} = f'(c) \Rightarrow 0 = 1 - \frac{1}{c^2} \Rightarrow c = 1$.

5. Does not; $f(x)$ is not differentiable at $x = 0$ in $(-1, 8)$.

7. Does; $f(x)$ is continuous for every point of $[0, 1]$ and differentiable for every point in $(0, 1)$.

9. Since $f(x)$ is not continuous on $0 \le x \le 1$, Rolle's Theorem does not apply: $\lim\limits_{x \to 1^-} f(x) = \lim\limits_{x \to 1^-} x = 1$
$\ne 0 = f(1)$.

11. (a) i

ii

iii

iv

(b) Let r_1 and r_2 be zeros of the polynomial $P(x) = x^n + a_{n-1}x^{n-1} + \ldots + a_1x + a_0$, then $P(r_1) = P(r_2) = 0$.
Since polynomials are everywhere continuous and differentiable, by Rolle's Theorem $P'(r) = 0$ for some r
between r_1 and r_2, where $P'(x) = nx^{n-1} + (n - 1)a_{n-1}x^{n-2} + \ldots + a_1$.

13. Since f'' exists throughout $[a, b]$ the derivative function f' is continuous there. If f' has more than one zero in
$[a, b]$, say $f'(r_1) = f'(r_2) = 0$ for $r_1 \ne r_2$, then by Rolle's Theorem there is a c between r_1 and r_2 such that
$f''(c) = 0$, contrary to $f'' > 0$ throughout $[a, b]$. Therefore f' has at most one zero in $[a, b]$. The same argument
holds if $f'' < 0$ throughout $[a, b]$.

15. With $f(-2) = 11 > 0$ and $f(-1) = -1 < 0$ we conclude from the Intermediate Value Theorem that
$f(x) = x^4 + 3x + 1$ has at least one zero between -2 and -1. Then $-2 < x < -1 \Rightarrow -8 < x^3 < -1$
$\Rightarrow -32 < 4x^3 < -4 \Rightarrow -29 < 4x^3 + 3 < -1 \Rightarrow f'(x) < 0$ for $-2 < x < -1 \Rightarrow f(x)$ is decreasing on $[-2, -1]$
$\Rightarrow f(x) = 0$ has exactly one solution in the interval $(-2, -1)$.

17. $g(t) = \sqrt{t} + \sqrt{t+1} - 4 \Rightarrow g'(t) = \frac{1}{2\sqrt{t}} + \frac{1}{2\sqrt{t+1}} > 0 \Rightarrow g(t)$ is increasing for t in $(0, \infty)$; $g(3) = \sqrt{3} - 2 < 0$
and $g(15) = \sqrt{15} > 0 \Rightarrow g(t)$ has exactly one zero in $(0, \infty)$.

19. $r(\theta) = \theta + \sin^2\left(\frac{\theta}{3}\right) - 8 \Rightarrow r'(\theta) = 1 + \frac{2}{3}\sin\left(\frac{\theta}{3}\right)\cos\left(\frac{\theta}{3}\right) = 1 + \frac{1}{3}\sin\left(\frac{2\theta}{3}\right) > 0$ on $(-\infty, \infty) \Rightarrow r(\theta)$ is
increasing on $(-\infty, \infty)$; $r(0) = -8$ and $r(8) = \sin^2\left(\frac{8}{3}\right) > 0 \Rightarrow r(\theta)$ has exactly one zero in $(-\infty, \infty)$.

21. $r(\theta) = \sec\theta - \frac{1}{\theta^3} + 5 \Rightarrow r'(\theta) = (\sec\theta)(\tan\theta) + \frac{3}{\theta^4} > 0$ on $\left(0, \frac{\pi}{2}\right) \Rightarrow r(\theta)$ is increasing on $\left(0, \frac{\pi}{2}\right)$;
$r(0.1) \approx -994$ and $r(1.57) \approx 1260.5 \Rightarrow r(\theta)$ has exactly one zero in $\left(0, \frac{\pi}{2}\right)$.

23. By Corollary 1, $f'(x) = 0$ for all $x \Rightarrow f(x) = C$, where C is a constant. Since $f(-1) = 3$ we have $C = 3$
$\Rightarrow f(x) = 3$ for all x.

25. $g(x) = x^2 \Rightarrow g'(x) = 2x = f'(x)$ for all x. By Corollary 2, $f(x) = g(x) + C$.
(a) $f(0) = 0 \Rightarrow 0 = g(0) + C = 0 + C \Rightarrow C = 0 \Rightarrow f(x) = x^2 \Rightarrow f(2) = 4$
(b) $f(1) = 0 \Rightarrow 0 = g(1) + C = 1 + C \Rightarrow C = -1 \Rightarrow f(x) = x^2 - 1 \Rightarrow f(2) = 3$
(c) $f(-2) = 3 \Rightarrow 3 = g(-2) + C \Rightarrow 3 = 4 + C \Rightarrow C = -1 \Rightarrow f(x) = x^2 - 1 \Rightarrow f(2) = 3$

27. (a) $y = \frac{x^2}{2} + C$ (b) $y = \frac{x^3}{3} + C$ (c) $y = \frac{x^4}{4} + C$

29. (a) $y' = -x^{-2} \Rightarrow y = \frac{1}{x} + C$ (b) $y = x + \frac{1}{x} + C$ (c) $y = 5x - \frac{1}{x} + C$

31. (a) $y = -\frac{1}{2}\cos 2t + C$ (b) $y = 2\sin\frac{t}{2} + C$
(c) $y = -\frac{1}{2}\cos 2t + 2\sin\frac{t}{2} + C$

33. $f(x) = x^2 - x + C$; $0 = f(0) = 0^2 - 0 + C \Rightarrow C = 0 \Rightarrow f(x) = x^2 - x$

35. $r(\theta) = 8\theta + \cot\theta + C$; $0 = r\left(\frac{\pi}{4}\right) = 8\left(\frac{\pi}{4}\right) + \cot\left(\frac{\pi}{4}\right) + C \Rightarrow 0 = 2\pi + 1 + C \Rightarrow C = -2\pi - 1$
$\Rightarrow r(\theta) = 8\theta + \cot\theta - 2\pi - 1$

37. $v = \frac{ds}{dt} = 9.8t + 5 \Rightarrow s = 4.9t^2 + 5t + C$; at $s = 10$ and $t = 0$ we have $C = 10 \Rightarrow s = 4.9t^2 + 5t + 10$

39. $v = \frac{ds}{dt} = \sin(\pi t) \Rightarrow s = -\frac{1}{\pi}\cos(\pi t) + C$; at $s = 0$ and $t = 0$ we have $C = \frac{1}{\pi} \Rightarrow s = \frac{1 - \cos(\pi t)}{\pi}$

41. $a = 32 \Rightarrow v = 32t + C_1$; at $v = 20$ and $t = 0$ we have $C_1 = 20 \Rightarrow v = 32t + 20 \Rightarrow s = 16t^2 + 20t + C_2$; at $s = 5$ and
$t = 0$ we have $C_2 = 5 \Rightarrow s = 16t^2 + 20t + 5$

43. $a = -4\sin(2t) \Rightarrow v = 2\cos(2t) + C_1$; at $v = 2$ and $t = 0$ we have $C_1 = 0 \Rightarrow v = 2\cos(2t) \Rightarrow s = \sin(2t) + C_2$; at $s = -3$
and $t = 0$ we have $C_2 = -3 \Rightarrow s = \sin(2t) - 3$

45. If $T(t)$ is the temperature of the thermometer at time t, then $T(0) = -19°$ C and $T(14) = 100°$ C. From the
Mean Value Theorem there exists a $0 < t_0 < 14$ such that $\frac{T(14) - T(0)}{14 - 0} = 8.5°$ C/sec $= T'(t_0)$, the rate at which
the temperature was changing at $t = t_0$ as measured by the rising mercury on the thermometer.

47. Because its average speed was approximately 7.667 knots, and by the Mean Value Theorem, it must have been going that speed at least once during the trip.

49. Let d(t) represent the distance the automobile traveled in time t. The average speed over $0 \le t \le 2$ is $\frac{d(2) - d(0)}{2 - 0}$. The Mean Value Theorem says that for some $0 < t_0 < 2$, $d'(t_0) = \frac{d(2) - d(0)}{2 - 0}$. The value $d'(t_0)$ is the speed of the automobile at time t_0 (which is read on the speedometer).

51. The conclusion of the Mean Value Theorem yields $\frac{\frac{1}{b} - \frac{1}{a}}{b - a} = -\frac{1}{c^2} \Rightarrow c^2 \left(\frac{a - b}{ab} \right) = a - b \Rightarrow c = \sqrt{ab}$.

53. $f'(x) = [\cos x \sin(x + 2) + \sin x \cos(x + 2)] - 2\sin(x + 1)\cos(x + 1) = \sin(x + x + 2) - \sin 2(x + 1)$
$= \sin(2x + 2) - \sin(2x + 2) = 0$. Therefore, the function has the constant value $f(0) = -\sin^2 1 \approx -0.7081$ which explains why the graph is a horizontal line.

55. $f(x)$ must be zero at least once between a and b by the Intermediate Value Theorem. Now suppose that $f(x)$ is zero twice between a and b. Then by the Mean Value Theorem, $f'(x)$ would have to be zero at least once between the two zeros of $f(x)$, but this can't be true since we are given that $f'(x) \ne 0$ on this interval. Therefore, $f(x)$ is zero once and only once between a and b.

57. Yes. By Corollary 2 we have $f(x) = g(x) + c$ since $f'(x) = g'(x)$. If the graphs start at the same point $x = a$, then $f(a) = g(a) \Rightarrow c = 0 \Rightarrow f(x) = g(x)$.

59. By the Mean Value Theorem we have $\frac{f(b) - f(a)}{b - a} = f'(c)$ for some point c between a and b. Since $b - a > 0$ and $f(b) < f(a)$, we have $f(b) - f(a) < 0 \Rightarrow f'(c) < 0$.

61. $f'(x) = (1 + x^4 \cos x)^{-1} \Rightarrow f''(x) = -(1 + x^4 \cos x)^{-2} (4x^3 \cos x - x^4 \sin x)$
$= -x^3 (1 + x^4 \cos x)^{-2}(4 \cos x - x \sin x) < 0$ for $0 \le x \le 0.1 \Rightarrow f'(x)$ is decreasing when $0 \le x \le 0.1$
$\Rightarrow$ min $f' \approx 0.9999$ and max $f' = 1$. Now we have $0.9999 \le \frac{f(0.1) - 1}{0.1} \le 1 \Rightarrow 0.09999 \le f(0.1) - 1 \le 0.1$
$\Rightarrow 1.09999 \le f(0.1) \le 1.1$.

63. (a) Suppose $x < 1$, then by the Mean Value Theorem $\frac{f(x) - f(1)}{x - 1} < 0 \Rightarrow f(x) > f(1)$. Suppose $x > 1$, then by the Mean Value Theorem $\frac{f(x) - f(1)}{x - 1} > 0 \Rightarrow f(x) > f(1)$. Therefore $f(x) \ge 1$ for all x since $f(1) = 1$.

(b) Yes. From part (a), $\lim\limits_{x \to 1^-} \frac{f(x) - f(1)}{x - 1} \le 0$ and $\lim\limits_{x \to 1^+} \frac{f(x) - f(1)}{x - 1} \ge 0$. Since $f'(1)$ exists, these two one-sided limits are equal and have the value $f'(1) \Rightarrow f'(1) \le 0$ and $f'(1) \ge 0 \Rightarrow f'(1) = 0$.

4.3 MONOTONIC FUNCTIONS AND THE FIRST DERIVATIVE TEST

1. (a) $f'(x) = x(x - 1) \Rightarrow$ critical points at 0 and 1
   (b) $f' = +++ | --- | +++ \Rightarrow$ increasing on $(-\infty, 0)$ and $(1, \infty)$, decreasing on $(0, 1)$
              0   1
   (c) Local maximum at $x = 0$ and a local minimum at $x = 1$

3. (a) $f'(x) = (x - 1)^2(x + 2) \Rightarrow$ critical points at -2 and 1
   (b) $f' = --- | +++ | +++ \Rightarrow$ increasing on $(-2, 1)$ and $(1, \infty)$, decreasing on $(-\infty, -2)$
             -2 1
   (c) No local maximum and a local minimum at $x = -2$

5. (a) $f'(x) = (x-1)(x+2)(x-3) \Rightarrow$ critical points at -2, 1 and 3
    (b) $f' = --- \mid_{-2} +++ \mid_{1} --- \mid_{3} +++ \Rightarrow$ increasing on $(-2, 1)$ and $(3, \infty)$, decreasing on $(-\infty, -2)$ and $(1, 3)$
    (c) Local maximum at $x = 1$, local minima at $x = -2$ and $x = 3$

7. (a) $f'(x) = x^{-1/3}(x+2) \Rightarrow$ critical points at -2 and 0
    (b) $f' = +++ \mid_{-2} ---)(+++ \Rightarrow$ increasing on $(-\infty, -2)$ and $(0, \infty)$, decreasing on $(-2, 0)$
    (c) Local maximum at $x = -2$, local minimum at $x = 0$

9. (a) $g(t) = -t^2 - 3t + 3 \Rightarrow g'(t) = -2t - 3 \Rightarrow$ a critical point at $t = -\frac{3}{2}$; $g' = +++ \mid_{-3/2} ---$, increasing on $\left(-\infty, -\frac{3}{2}\right)$, decreasing on $\left(-\frac{3}{2}, \infty\right)$
    (b) local maximum value of $g\left(-\frac{3}{2}\right) = \frac{21}{4}$ at $t = -\frac{3}{2}$
    (c) absolute maximum is $\frac{21}{4}$ at $t = -\frac{3}{2}$
    (d)

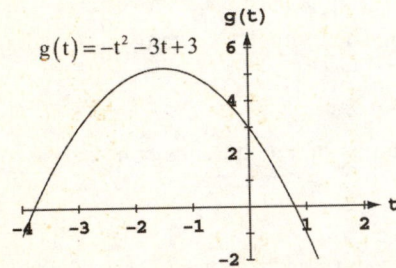

11. (a) $h(x) = -x^3 + 2x^2 \Rightarrow h'(x) = -3x^2 + 4x = x(4 - 3x) \Rightarrow$ critical points at $x = 0, \frac{4}{3}$
    $\Rightarrow h' = --- \mid_{0} +++ \mid_{4/3} ---$, increasing on $\left(0, \frac{4}{3}\right)$, decreasing on $(-\infty, 0)$ and $\left(\frac{4}{3}, \infty\right)$
    (b) local maximum value of $h\left(\frac{4}{3}\right) = \frac{32}{27}$ at $x = \frac{4}{3}$; local minimum value of $h(0) = 0$ at $x = 0$
    (c) no absolute extrema
    (d)

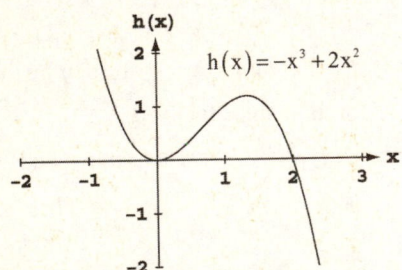

13. (a) $f(\theta) = 3\theta^2 - 4\theta^3 \Rightarrow f'(\theta) = 6\theta - 12\theta^2 = 6\theta(1 - 2\theta) \Rightarrow$ critical points at $\theta = 0, \frac{1}{2} \Rightarrow f' = --- \mid_{0} +++ \mid_{1/2} ---$, increasing on $\left(0, \frac{1}{2}\right)$, decreasing on $(-\infty, 0)$ and $\left(\frac{1}{2}, \infty\right)$
    (b) a local maximum is $f\left(\frac{1}{2}\right) = \frac{1}{4}$ at $\theta = \frac{1}{2}$, a local minimum is $f(0) = 0$ at $\theta = 0$
    (c) no absolute extrema

(d)

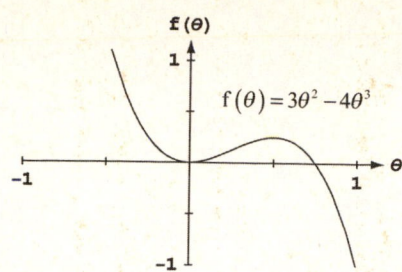

$$f(\theta) = 3\theta^2 - 4\theta^3$$

15. (a) $f(r) = 3r^3 + 16r \Rightarrow f'(r) = 9r^2 + 16 \Rightarrow$ no critical points $\Rightarrow f' = +++++$, increasing on $(-\infty, \infty)$, never decreasing

(b) no local extrema

(c) no absolute extrema

(d)

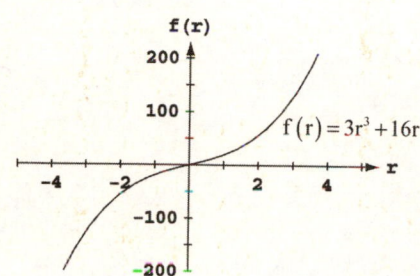

$$f(r) = 3r^3 + 16r$$

17. (a) $f(x) = x^4 - 8x^2 + 16 \Rightarrow f'(x) = 4x^3 - 16x = 4x(x + 2)(x - 2) \Rightarrow$ critical points at $x = 0$ and $x = \pm 2$
$\Rightarrow f' = --- \mid +++ \mid --- \mid +++$, increasing on $(-2, 0)$ and $(2, \infty)$, decreasing on $(-\infty, -2)$ and $(0, 2)$
          -202

(b) a local maximum is $f(0) = 16$ at $x = 0$, local minima are $f(\pm 2) = 0$ at $x = \pm 2$

(c) no absolute maximum; absolute minimum is 0 at $x = \pm 2$

(d)

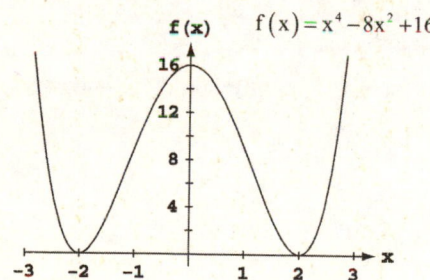

$$f(x) = x^4 - 8x^2 + 16$$

19. (a) $H(t) = \frac{3}{2}t^4 - t^6 \Rightarrow H'(t) = 6t^3 - 6t^5 = 6t^3(1 + t)(1 - t) \Rightarrow$ critical points at $t = 0, \pm 1$
$\Rightarrow H' = +++ \mid --- \mid +++ \mid ---$, increasing on $(-\infty, -1)$ and $(0, 1)$, decreasing on $(-1, 0)$ and $(1, \infty)$
          -101

(b) the local maxima are $H(-1) = \frac{1}{2}$ at $t = -1$ and $H(1) = \frac{1}{2}$ at $t = 1$, the local minimum is $H(0) = 0$ at $t = 0$

(c) absolute maximum is $\frac{1}{2}$ at $t = \pm 1$; no absolute minimum

(d)

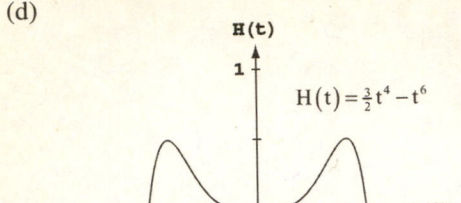

21. (a) $g(x) = x\sqrt{8 - x^2} = x(8 - x^2)^{1/2} \Rightarrow g'(x) = (8 - x^2)^{1/2} + x\left(\frac{1}{2}\right)(8 - x^2)^{-1/2}(-2x) = \dfrac{2(2 - x)(2 + x)}{\sqrt{\left(2\sqrt{2} - x\right)\left(2\sqrt{2} + x\right)}}$

$\Rightarrow$ critical points at $x = \pm 2, \pm 2\sqrt{2} \Rightarrow g' = (\ \underset{-2\sqrt{2}}{\quad} ---\ |\ \underset{-2}{+++}\ |\ \underset{2}{---}\ \underset{2\sqrt{2}}{)}$, increasing on $(-2, 2)$, decreasing on

$\left(-2\sqrt{2}, -2\right)$ and $\left(2, 2\sqrt{2}\right)$

(b) local maxima are $g(2) = 4$ at $x = 2$ and $g\left(-2\sqrt{2}\right) = 0$ at $x = -2\sqrt{2}$, local minima are $g(-2) = -4$ at

$x = -2$ and $g\left(2\sqrt{2}\right) = 0$ at $x = 2\sqrt{2}$

(c) absolute maximum is 4 at $x = 2$; absolute minimum is -4 at $x = -2$

(d)

$g(x) = x\sqrt{8-x^2}$

23. (a) $f(x) = \frac{x^2 - 3}{x - 2} \Rightarrow f'(x) = \frac{2x(x - 2) - (x^2 - 3)(1)}{(x - 2)^2} = \frac{(x - 3)(x - 1)}{(x - 2)^2} \Rightarrow$ critical points at $x = 1, 3$

$\Rightarrow f' = +++ \ |\ \underset{1}{---} \ )(\ \underset{2}{---} \ |\ \underset{3}{+++}$, increasing on $(-\infty, 1)$ and $(3, \infty)$, decreasing on $(1, 2)$ and $(2, 3)$,

discontinuous at $x = 2$

(b) a local maximum is $f(1) = 2$ at $x = 1$, a local minimum is $f(3) = 6$ at $x = 3$

(c) no absolute extrema

(d)

$f(x) = \dfrac{x^2 - 3}{x - 2}$

25. (a) $f(x) = x^{1/3}(x + 8) = x^{4/3} + 8x^{1/3} \Rightarrow f'(x) = \frac{4}{3}x^{1/3} + \frac{8}{3}x^{-2/3} = \frac{4(x + 2)}{3x^{2/3}} \Rightarrow$ critical points at $x = 0, -2$

$\Rightarrow f' = --- \ |\ \underset{-2}{+++} \ )(\underset{0}{+++}$, increasing on $(-2, 0) \cup (0, \infty)$, decreasing on $(-\infty, -2)$

(b) no local maximum, a local minimum is $f(-2) = -6\sqrt[3]{2} \approx -7.56$ at $x = -2$

(c) no absolute maximum; absolute minimum is $-6\sqrt[3]{2}$ at $x = -2$

(d)

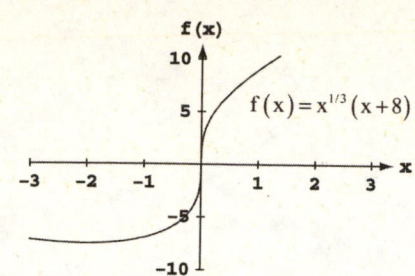

27. (a) $h(x) = x^{1/3}(x^2 - 4) = x^{7/3} - 4x^{1/3} \Rightarrow h'(x) = \frac{7}{3}x^{4/3} - \frac{4}{3}x^{-2/3} = \frac{(\sqrt{7}x + 2)(\sqrt{7}x - 2)}{3\sqrt[3]{x^2}} \Rightarrow$ critical points at

$x = 0, \frac{\pm 2}{\sqrt{7}} \Rightarrow h' = +++ \ | \ \underset{-2/\sqrt{7}}{} \ --- \ )(--- \ | \ \underset{0}{} \ +++ \ \underset{2/\sqrt{7}}{} ,$ increasing on $\left(-\infty, \frac{-2}{\sqrt{7}}\right)$ and $\left(\frac{2}{\sqrt{7}}, \infty\right)$, decreasing on

$\left(\frac{-2}{\sqrt{7}}, 0\right)$ and $\left(0, \frac{2}{\sqrt{7}}\right)$

(b) local maximum is $h\left(\frac{-2}{\sqrt{7}}\right) = \frac{24\sqrt[3]{2}}{7^{7/6}} \approx 3.12$ at $x = \frac{-2}{\sqrt{7}}$, the local minimum is $h\left(\frac{2}{\sqrt{7}}\right) = -\frac{24\sqrt[3]{2}}{7^{7/6}} \approx -3.12$

(c) no absolute extrema

(d)

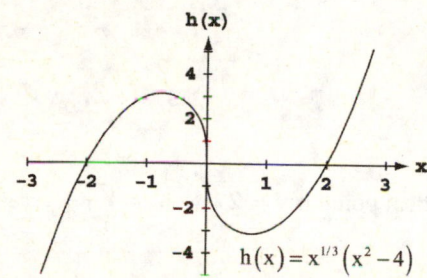

29. (a) $f(x) = 2x - x^2 \Rightarrow f'(x) = 2 - 2x = 2(1 - x) \Rightarrow$ a critical point at $x = 1 \Rightarrow f' = +++ \ | \ \underset{1}{} \ --- \ ] \ \underset{2}{}$ and $f(1) = 1$,

$f(2) = 0 \Rightarrow$ a local maximum is 1 at $x = 1$, a local minimum is 0 at $x = 2$

(b) absolute maximum is 1 at $x = 1$; no absolute minimum

(c)

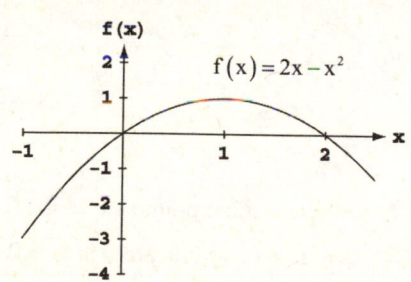

31. (a) $g(x) = x^2 - 4x + 4 \Rightarrow g'(x) = 2x - 4 = 2(x - 2) \Rightarrow$ a critical point at $x = 2 \Rightarrow g' = [\ \underset{1}{} \ --- \ | \ \underset{2}{} \ +++$ and

$g(1) = 1, g(2) = 0 \Rightarrow$ a local maximum is 1 at $x = 1$, a local minimum is $g(2) = 0$ at $x = 2$

(b) no absolute maximum; absolute minimum is 0 at $x = 2$

(c)

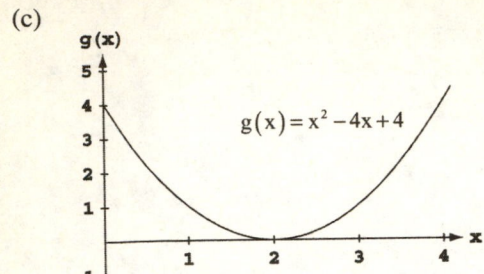

33. (a) $f(t) = 12t - t^3 \Rightarrow f'(t) = 12 - 3t^2 = 3(2+t)(2-t) \Rightarrow$ critical points at $t = \pm 2 \Rightarrow f' = [\underset{-3}{} \; --- \underset{-2}{|} \; +++ \underset{2}{|} \; ---$

and $f(-3) = -9, f(-2) = -16, f(2) = 16 \Rightarrow$ local maxima are -9 at $t = -3$ and 16 at $t = -2$, a local

minimum is -16 at $t = -2$

(b) absolute maximum is 16 at $t = 2$; no absolute minimum

(c)

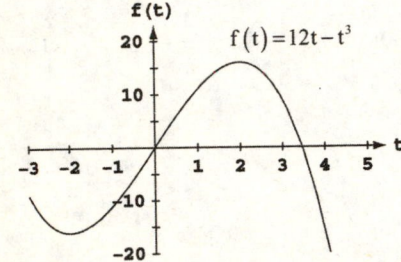

35. (a) $h(x) = \frac{x^3}{3} - 2x^2 + 4x \Rightarrow h'(x) = x^2 - 4x + 4 = (x-2)^2 \Rightarrow$ a critical point at $x = 2 \Rightarrow h' = [\underset{0}{} \; +++ \underset{2}{|} \; +++$ and

$h(0) = 0 \Rightarrow$ no local maximum, a local minimum is 0 at $x = 0$

(b) no absolute maximum; absolute minimum is 0 at $x = 0$

(c)

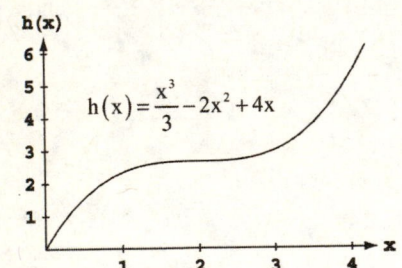

37. (a) $f(x) = \frac{x}{2} - 2\sin\left(\frac{x}{2}\right) \Rightarrow f'(x) = \frac{1}{2} - \cos\left(\frac{x}{2}\right), f'(x) = 0 \Rightarrow \cos\left(\frac{x}{2}\right) = \frac{1}{2} \Rightarrow$ a critical point at $x = \frac{2\pi}{3}$

$\Rightarrow f' = [\underset{0}{} \; --- \underset{2\pi/3}{|} \; +++ \underset{2\pi}{]}$ and $f(0) = 0, f\left(\frac{2\pi}{3}\right) = \frac{\pi}{3} - \sqrt{3}, f(2\pi) = \pi \Rightarrow$ local maxima are 0 at $x = 0$ and π

at $x = 2\pi$, a local minimum is $\frac{\pi}{3} - \sqrt{3}$ at $x = \frac{2\pi}{3}$

(b) The graph of f rises when $f' > 0$, falls when $f' < 0$, and has a local minimum value at the point where f' changes from negative to positive.

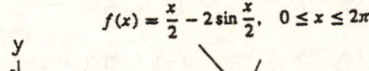
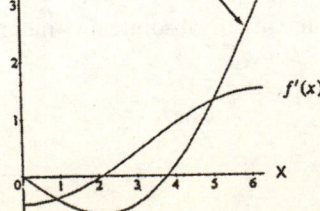

39. (a) $f(x) = \csc^2 x - 2 \cot x \Rightarrow f'(x) = 2(\csc x)(-\csc x)(\cot x) - 2(-\csc^2 x) = -2(\csc^2 x)(\cot x - 1) \Rightarrow$ a critical

point at $x = \frac{\pi}{4} \Rightarrow f' = (\underset{0}{---} | \underset{\pi/4}{+++})_{\pi}$ and $f\left(\frac{\pi}{4}\right) = 0 \Rightarrow$ no local maximum, a local minimum is 0 at $x = \frac{\pi}{4}$

(b) The graph of f rises when $f' > 0$, falls when $f' < 0$, and has a local minimum value at the point where $f' = 0$ and the values of f' change from negative to positive. The graph of f steepens as $f'(x) \to \pm\infty$.

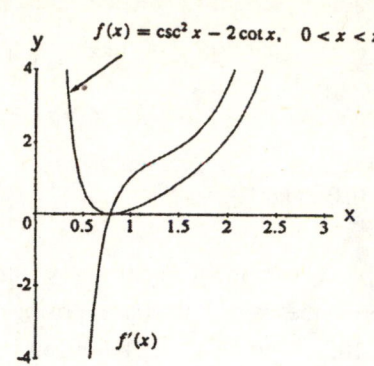

41. $h(\theta) = 3 \cos\left(\frac{\theta}{2}\right) \Rightarrow h'(\theta) = -\frac{3}{2} \sin\left(\frac{\theta}{2}\right) \Rightarrow h' = [\underset{0}{---}]_{2\pi}$, $(0, 3)$ and $(2\pi, -3) \Rightarrow$ a local maximum is 3 at $\theta = 0$,

a local minimum is -3 at $\theta = 2\pi$

43. (a)

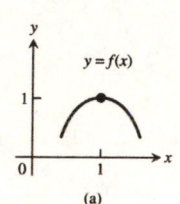

(a)

(b)

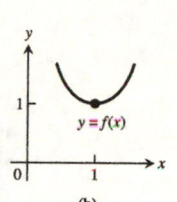

(b)

(c)

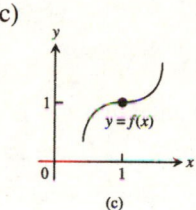

(c)

(d)

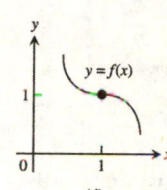

(d)

45. (a)

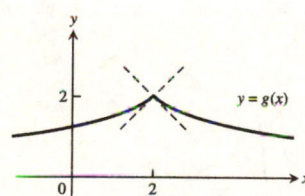

(b)

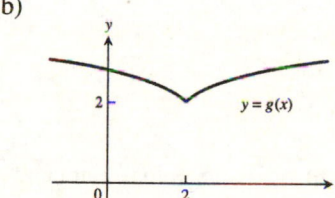

47. $f(x) = x^3 - 3x + 2 \Rightarrow f'(x) = 3x^2 - 3 = 3(x - 1)(x + 1) \Rightarrow f' = +++ | \underset{-1}{---} | \underset{1}{+++} \Rightarrow$ rising for $x = c = 2$ since

$f'(x) > 0$ for $x = c = 2$.

4.4 CONCAVITY AND CURVE SKETCHING

1. $y = \frac{x^3}{3} - \frac{x^2}{2} - 2x + \frac{1}{3} \Rightarrow y' = x^2 - x - 2 = (x - 2)(x + 1) \Rightarrow y'' = 2x - 1 = 2\left(x - \frac{1}{2}\right)$. The graph is rising on $(-\infty, -1)$ and $(2, \infty)$, falling on $(-1, 2)$, concave up on $\left(\frac{1}{2}, \infty\right)$ and concave down on $\left(-\infty, \frac{1}{2}\right)$. Consequently, a local maximum is $\frac{3}{2}$ at $x = -1$, a local minimum is -3 at $x = 2$, and $\left(\frac{1}{2}, -\frac{3}{4}\right)$ is a point of inflection.

3. $y = \frac{3}{4}(x^2 - 1)^{2/3} \Rightarrow y' = \left(\frac{3}{4}\right)\left(\frac{2}{3}\right)(x^2 - 1)^{-1/3}(2x) = x(x^2 - 1)^{-1/3}, y' = \underset{-1}{---}) (\underset{0}{+++} | \underset{1}{---})(+++$

$\Rightarrow$ the graph is rising on $(-1, 0)$ and $(1, \infty)$, falling on $(-\infty, -1)$ and $(0, 1) \Rightarrow$ a local maximum is $\frac{3}{4}$ at $x = 0$, local

minima are 0 at $x = \pm 1$; $y'' = (x^2 - 1)^{-1/3} + (x)\left(-\frac{1}{3}\right)(x^2 - 1)^{-4/3}(2x) = \frac{x^2 - 3}{3\sqrt[3]{(x^2 - 1)^4}}$,

$y'' = +++ | \underset{-\sqrt{3}}{---}) (\underset{-1}{---})(\underset{1}{---} | \underset{\sqrt{3}}{+++} \Rightarrow$ the graph is concave up on $\left(-\infty, -\sqrt{3}\right)$ and $\left(\sqrt{3}, \infty\right)$, concave

down on $\left(-\sqrt{3}, \sqrt{3}\right) \Rightarrow$ points of inflection at $\left(\pm\sqrt{3}, \frac{3\sqrt[3]{4}}{4}\right)$

5. $y = x + \sin 2x \Rightarrow y' = 1 + 2\cos 2x$, $y' = [--- | +++ | ---]$ $\Rightarrow$ the graph is rising on $\left(-\frac{\pi}{3}, \frac{\pi}{3}\right)$, falling
    $$-2\pi/3 \quad -\pi/3 \quad \pi/3 \quad 2\pi/3$$

    on $\left(-\frac{2\pi}{3}, -\frac{\pi}{3}\right)$ and $\left(\frac{\pi}{3}, \frac{2\pi}{3}\right)$ $\Rightarrow$ local maxima are $-\frac{2\pi}{3} + \frac{\sqrt{3}}{2}$ at $x = -\frac{2\pi}{3}$ and $\frac{\pi}{3} + \frac{\sqrt{3}}{2}$ at $x = \frac{\pi}{3}$, local minima are

    $-\frac{\pi}{3} - \frac{\sqrt{3}}{2}$ at $x = -\frac{\pi}{3}$ and $\frac{2\pi}{3} - \frac{\sqrt{3}}{2}$ at $x = \frac{2\pi}{3}$; $y'' = -4\sin 2x$, $y'' = [--- | +++ | --- | +++]$ $\Rightarrow$ the
    $$-2\pi/3 \quad -\pi/2 \quad 0 \quad \pi/2 \quad 2\pi/3$$

    graph is concave up on $\left(-\frac{\pi}{2}, 0\right)$ and $\left(\frac{\pi}{2}, \frac{2\pi}{3}\right)$, concave down on $\left(-\frac{2\pi}{3}, -\frac{\pi}{2}\right)$ and $\left(0, \frac{\pi}{2}\right)$ $\Rightarrow$ points of inflection at

    $\left(-\frac{\pi}{2}, -\frac{\pi}{2}\right)$, $(0,0)$, and $\left(\frac{\pi}{2}, \frac{\pi}{2}\right)$

7. If $x \geq 0$, $\sin |x| = \sin x$ and if $x < 0$, $\sin |x| = \sin(-x)$
    $= -\sin x$. From the sketch the graph is rising on
    $\left(-\frac{3\pi}{2}, -\frac{\pi}{2}\right)$, $\left(0, \frac{\pi}{2}\right)$ and $\left(\frac{3\pi}{2}, 2\pi\right)$, falling on $\left(-2\pi, -\frac{3\pi}{2}\right)$,
    $\left(-\frac{\pi}{2}, 0\right)$ and $\left(\frac{\pi}{2}, \frac{3\pi}{2}\right)$; local minima are -1 at $x = \pm\frac{3\pi}{2}$
    and 0 at $x = 0$; local maxima are 1 at $x = \pm\frac{\pi}{2}$ and
    0 at $x = \pm 2\pi$; concave up on $(-2\pi, -\pi)$ and $(\pi, 2\pi)$, and
    concave down on $(-\pi, 0)$ and $(0, \pi)$ $\Rightarrow$ points of inflection
    are $(-\pi, 0)$ and $(\pi, 0)$

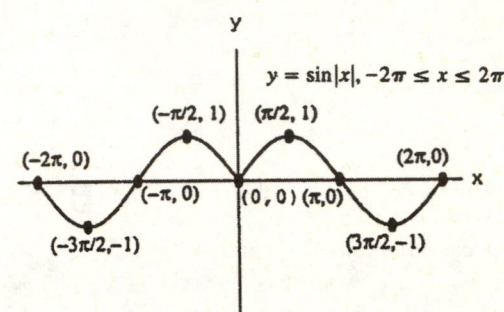

9. When $y = x^2 - 4x + 3$, then $y' = 2x - 4 = 2(x - 2)$ and
    $y'' = 2$. The curve rises on $(2, \infty)$ and falls on $(-\infty, 2)$.
    At $x = 2$ there is a minimum. Since $y'' > 0$, the curve is
    concave up for all x.

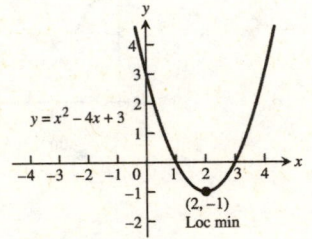

11. When $y = x^3 - 3x + 3$, then $y' = 3x^2 - 3 = 3(x - 1)(x + 1)$
    and $y'' = 6x$. The curve rises on $(-\infty, -1) \cup (1, \infty)$ and
    falls on $(-1, 1)$. At $x = -1$ there is a local maximum and at
    $x = 1$ a local minimum. The curve is concave down on
    $(-\infty, 0)$ and concave up on $(0, \infty)$. There is a point of
    inflection at $x = 0$.

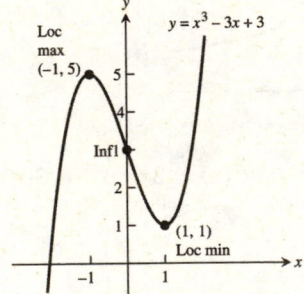

13. When $y = -2x^3 + 6x^2 - 3$, then $y' = -6x^2 + 12x$
    $= -6x(x - 2)$ and $y'' = -12x + 12 = -12(x - 1)$. The
    curve rises on $(0, 2)$ and falls on $(-\infty, 0)$ and $(2, \infty)$.
    At $x = 0$ there is a local minimum and at $x = 2$ a local
    maximum. The curve is concave up on $(-\infty, 1)$ and
    concave down on $(1, \infty)$. At $x = 1$ there is a point of
    inflection.

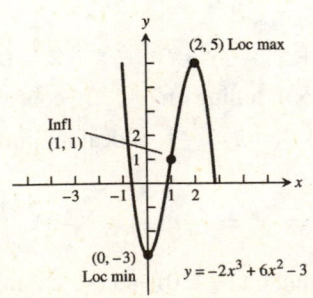

15. When $y = (x - 2)^3 + 1$, then $y' = 3(x - 2)^2$ and $y'' = 6(x - 2)$. The curve never falls and there are no local extrema. The curve is concave down on $(-\infty, 2)$ and concave up on $(2, \infty)$. At $x = 2$ there is a point of inflection.

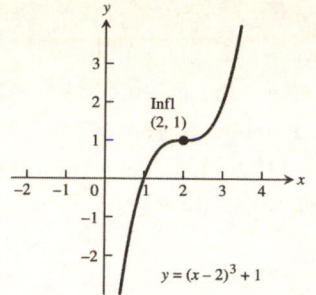

17. When $y = x^4 - 2x^2$, then $y' = 4x^3 - 4x = 4x(x + 1)(x - 1)$ and $y'' = 12x^2 - 4 = 12\left(x + \frac{1}{\sqrt{3}}\right)\left(x - \frac{1}{\sqrt{3}}\right)$. The curve rises on $(-1, 0)$ and $(1, \infty)$ and falls on $(-\infty, -1)$ and $(0, 1)$. At $x = \pm 1$ there are local minima and at $x = 0$ a local maximum. The curve is concave up on $\left(-\infty, -\frac{1}{\sqrt{3}}\right)$ and $\left(\frac{1}{\sqrt{3}}, \infty\right)$ and concave down on $\left(-\frac{1}{\sqrt{3}}, \frac{1}{\sqrt{3}}\right)$. At $x = \frac{\pm 1}{\sqrt{3}}$ there are points of inflection.

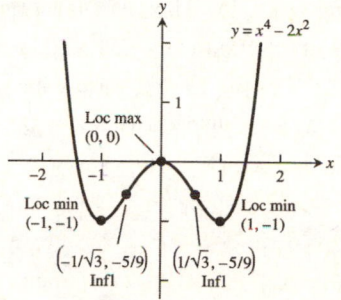

19. When $y = 4x^3 - x^4$, then $y' = 12x^2 - 4x^3 = 4x^2(3 - x)$ and $y'' = 24x - 12x^2 = 12x(2 - x)$. The curve rises on $(-\infty, 3)$ and falls on $(3, \infty)$. At $x = 3$ there is a local maximum, but there is no local minimum. The graph is concave up on $(0, 2)$ and concave down on $(-\infty, 0)$ and $(2, \infty)$. There are inflection points at $x = 0$ and $x = 2$.

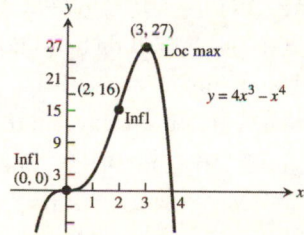

21. When $y = x^5 - 5x^4$, then $y' = 5x^4 - 20x^3 = 5x^3(x - 4)$ and $y'' = 20x^3 - 60x^2 = 20x^2(x - 3)$. The curve rises on $(-\infty, 0)$ and $(4, \infty)$, and falls on $(0, 4)$. There is a local maximum at $x = 0$, and a local minimum at $x = 4$. The curve is concave down on $(-\infty, 3)$ and concave up on $(3, \infty)$. At $x = 3$ there is a point of inflection.

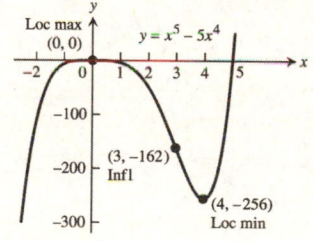

23. When $y = x + \sin x$, then $y' = 1 + \cos x$ and $y'' = -\sin x$. The curve rises on $(0, 2\pi)$. At $x = 0$ there is a local and absolute minimum and at $x = 2\pi$ there is a local and absolute maximum. The curve is concave down on $(0, \pi)$ and concave up on $(\pi, 2\pi)$. At $x = \pi$ there is a point of inflection.

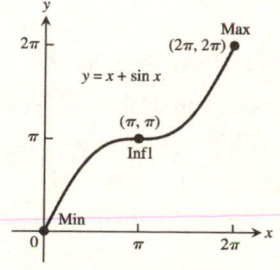

25. When $y = x^{1/5}$, then $y' = \frac{1}{5} x^{-4/5}$ and $y'' = -\frac{4}{25} x^{-9/5}$. The curve rises on $(-\infty, \infty)$ and there are no extrema. The curve is concave up on $(-\infty, 0)$ and concave down on $(0, \infty)$. At $x = 0$ there is a point of inflection.

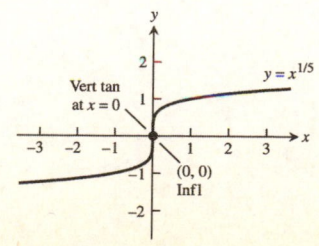

27. When $y = x^{2/5}$, then $y' = \frac{2}{5} x^{-3/5}$ and $y'' = -\frac{6}{25} x^{-8/5}$. The curve is rising on $(0, \infty)$ and falling on $(-\infty, 0)$. At $x = 0$ there is a local and absolute minimum. There is no local or absolute maximum. The curve is concave down on $(-\infty, 0)$ and $(0, \infty)$. There are no points of inflection, but a cusp exists at $x = 0$.

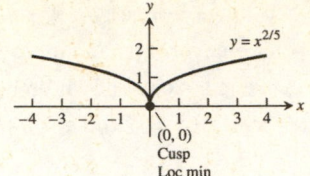

29. When $y = 2x - 3x^{2/3}$, then $y' = 2 - 2x^{-1/3}$ and $y'' = \frac{2}{3} x^{-4/3}$. The curve is rising on $(-\infty, 0)$ and $(1, \infty)$, and falling on $(0, 1)$. There is a local maximum at $x = 0$ and a local minimum at $x = 1$. The curve is concave up on $(-\infty, 0)$ and $(0, \infty)$. There are no points of inflection, but a cusp exists at $x = 0$.

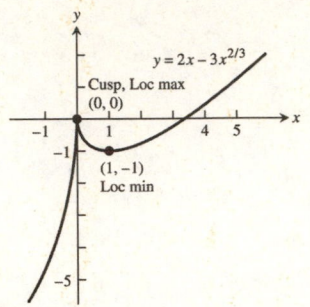

31. When $y = x^{2/3}\left(\frac{5}{2} - x\right) = \frac{5}{2} x^{2/3} - x^{5/3}$, then $y' = \frac{5}{3} x^{-1/3} - \frac{5}{3} x^{2/3} = \frac{5}{3} x^{-1/3}(1 - x)$ and $y'' = -\frac{5}{9} x^{-4/3} - \frac{10}{9} x^{-1/3} = -\frac{5}{9} x^{-4/3}(1 + 2x)$. The curve is rising on $(0, 1)$ and falling on $(-\infty, 0)$ and $(1, \infty)$. There is a local minimum at $x = 0$ and a local maximum at $x = 1$. The curve is concave up on $\left(-\infty, -\frac{1}{2}\right)$ and concave down on $\left(-\frac{1}{2}, 0\right)$ and $(0, \infty)$. There is a point of inflection at $x = -\frac{1}{2}$ and a cusp at $x = 0$.

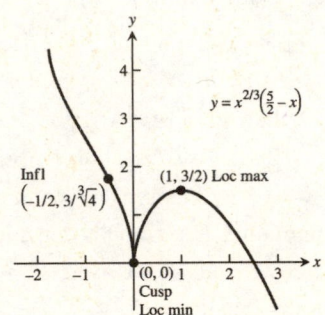

33. When $y = x\sqrt{8 - x^2} = x\left(8 - x^2\right)^{1/2}$, then
$y' = \left(8 - x^2\right)^{1/2} + (x)\left(\frac{1}{2}\right)\left(8 - x^2\right)^{-1/2}(-2x)$
$= \left(8 - x^2\right)^{-1/2}\left(8 - 2x^2\right) = \frac{2(2 - x)(2 + x)}{\sqrt{\left(2\sqrt{2} + x\right)\left(2\sqrt{2} - x\right)}}$ and
$y'' = \left(-\frac{1}{2}\right)\left(8 - x^2\right)^{-\frac{3}{2}}(-2x)\left(8 - 2x^2\right) + \left(8 - x^2\right)^{-\frac{1}{2}}(-4x)$
$= \frac{2x\left(x^2 - 12\right)}{\sqrt{\left(8 - x^2\right)^3}}$. The curve is rising on $(-2, 2)$, and falling on $\left(-2\sqrt{2}, -2\right)$ and $\left(2, 2\sqrt{2}\right)$. There are local minima $x = -2$ and $x = 2\sqrt{2}$, and local maxima at $x = -2\sqrt{2}$ and $x = 2$. The curve is concave up on $\left(-2\sqrt{2}, 0\right)$ and concave down on $\left(0, 2\sqrt{2}\right)$. There is a point of inflection at $x = 0$.

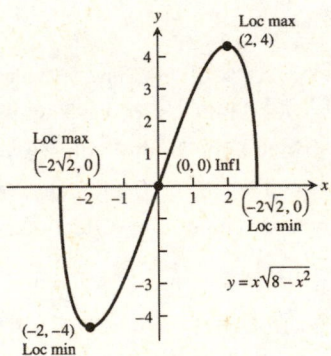

35. When $y = \frac{x^2-3}{x-2}$, then $y' = \frac{2x(x-2)-(x^2-3)(1)}{(x-2)^2}$

$= \frac{(x-3)(x-1)}{(x-2)^2}$ and

$y'' = \frac{(2x-4)(x-2)^2-(x^2-4x+3)2(x-2)}{(x-2)^4} = \frac{2}{(x-2)^3}$.

The curve is rising on $(-\infty, 1)$ and $(3, \infty)$, and falling on $(1, 2)$ and $(2, 3)$. There is a local maximum at $x = 1$ and a local minimum at $x = 3$. The curve is concave down on $(-\infty, 2)$ and concave up on $(2, \infty)$. There are no points of inflection because $x = 2$ is not in the domain.

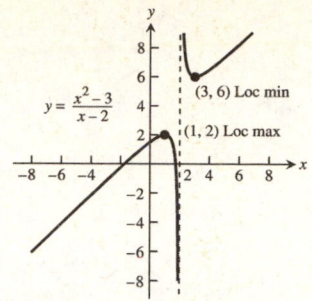

37. When $y = |x^2 - 1| = \begin{cases} x^2 - 1, & |x| \geq 1 \\ 1 - x^2, & |x| < 1 \end{cases}$, then

$y' = \begin{cases} 2x, & |x| > 1 \\ -2x, & |x| < 1 \end{cases}$ and $y'' = \begin{cases} 2, & |x| > 1 \\ -2, & |x| < 1 \end{cases}$. The

curve rises on $(-1, 0)$ and $(1, \infty)$ and falls on $(-\infty, -1)$ and $(0, 1)$. There is a local maximum at $x = 0$ and local minima at $x = \pm 1$. The curve is concave up on $(-\infty, -1)$ and $(1, \infty)$, and concave down on $(-1, 1)$. There are no points of inflection because y is not differentiable at $x = \pm 1$ (so there is no tangent line at those points).

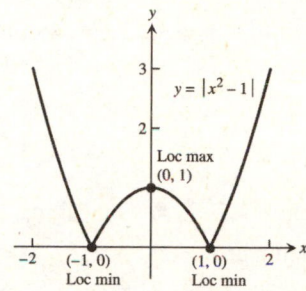

39. When $y = \sqrt{|x|} = \begin{cases} \sqrt{x}, & x \geq 0 \\ \sqrt{-x}, & x < 0 \end{cases}$, then

$y' = \begin{cases} \frac{1}{2\sqrt{x}}, & x > 0 \\ \frac{-1}{2\sqrt{-x}}, & x < 0 \end{cases}$ and $y'' = \begin{cases} \frac{-x^{-3/2}}{4}, & x > 0 \\ \frac{-(-x)^{-3/2}}{4}, & x < 0 \end{cases}$.

Since $\lim\limits_{x \to 0^-} y' = -\infty$ and $\lim\limits_{x \to 0^+} y' = \infty$ there is a cusp at $x = 0$. There is a local minimum at $x = 0$, but no local maximum. The curve is concave down on $(-\infty, 0)$ and $(0, \infty)$. There are no points of inflection.

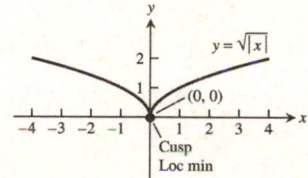

41. $y' = 2 + x - x^2 = (1 + x)(2 - x)$, $y' = ---\,|\,+++\,|\,---$

$\qquad\qquad\qquad\qquad\qquad\qquad\qquad\quad -1 \qquad 2$

$\Rightarrow$ rising on $(-1, 2)$, falling on $(-\infty, -1)$ and $(2, \infty)$

$\Rightarrow$ there is a local maximum at $x = 2$ and a local minimum at $x = -1$; $y'' = 1 - 2x$, $y'' = +++\,|\,---$

$\qquad\qquad\qquad\qquad\qquad\qquad\qquad\quad 1/2$

$\Rightarrow$ concave up on $\left(-\infty, \frac{1}{2}\right)$, concave down on $\left(\frac{1}{2}, \infty\right)$

$\Rightarrow$ a point of inflection at $x = \frac{1}{2}$

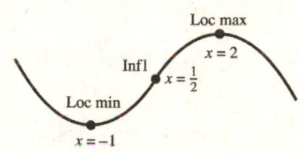

43. $y' = x(x - 3)^2$, $y' = ---\,|\,+++\,|\,+++ \Rightarrow$ rising on

$\qquad\qquad\qquad\qquad\qquad\quad 0 \qquad 3$

$(0, \infty)$, falling on $(-\infty, 0) \Rightarrow$ no local maximum, but there is a local minimum at $x = 0$; $y'' = (x - 3)^2 + x(2)(x - 3)$

$= 3(x - 3)(x - 1)$, $y'' = +++\,|\,---\,|\,+++ \Rightarrow$ concave

$\qquad\qquad\qquad\qquad\qquad\qquad\quad 1 \qquad 3$

up on $(-\infty, 1)$ and $(3, \infty)$, concave down on $(1, 3) \Rightarrow$ points of inflection at $x = 1$ and $x = 3$

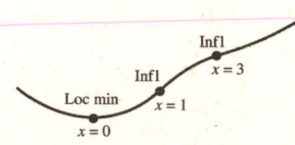

45. $y' = x(x^2 - 12) = x\left(x - 2\sqrt{3}\right)\left(x + 2\sqrt{3}\right),$

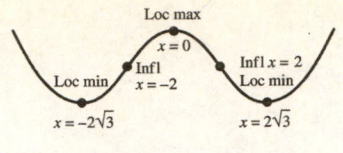

$y' = --- \mid \underset{0}{+++} \mid \underset{2\sqrt{3}}{---} \mid +++ \Rightarrow$ rising on

$\left(-2\sqrt{3}, 0\right)$ and $\left(2\sqrt{3}, \infty\right)$, falling on $\left(-\infty, -2\sqrt{3}\right)$

and $\left(0, 2\sqrt{3}\right) \Rightarrow$ a local maximum at $x = 0$, local minima

at $x = \pm 2\sqrt{3}$; $y'' = (1)(x^2 - 12) + (x)(2x)$

$= 3(x - 2)(x + 2),\ y'' = \underset{-2}{+++ \mid} \underset{2}{--- \mid} +++$

$\Rightarrow$ concave up on $(-\infty, -2)$ and $(2, \infty)$, concave down on

$(-2, 2) \Rightarrow$ points of inflection at $x = \pm 2$

47. $y' = (8x - 5x^2)(4 - x)^2 = x(8 - 5x)(4 - x)^2,$

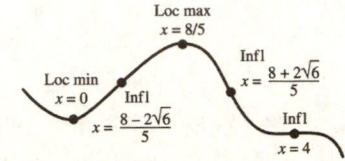

$y' = --- \mid \underset{0}{+++} \mid \underset{8/5}{---} \mid \underset{4}{---} \Rightarrow$ rising on $\left(0, \frac{8}{5}\right),$

falling on $(-\infty, 0)$ and $\left(\frac{8}{5}, \infty\right) \Rightarrow$ a local maximum at

$x = \frac{8}{5}$, a local minimum at $x = 0$;

$y'' = (8 - 10x)(4 - x)^2 + (8x - 5x^2)(2)(4 - x)(-1)$

$= 4(4 - x)(5x^2 - 16x + 8),$

$y'' = +++ \mid \underset{\frac{8-2\sqrt{6}}{5}}{} \; --- \; \mid \underset{\frac{8+2\sqrt{6}}{5}}{} \; +++ \mid \underset{4}{---} \Rightarrow$ concave up

on $\left(-\infty, \frac{8-2\sqrt{6}}{5}\right)$ and $\left(\frac{8+2\sqrt{6}}{5}, 4\right)$, concave down on

$\left(\frac{8-2\sqrt{6}}{5}, \frac{8+2\sqrt{6}}{5}\right)$ and $(4, \infty) \Rightarrow$ points of inflection at

$x = \frac{8 \pm 2\sqrt{6}}{5}$ and $x = 4$

49. $y' = \sec^2 x,\ y' = (\ \underset{-\pi/2 \quad \pi/2}{+++}\ ) \Rightarrow$ rising on $\left(-\frac{\pi}{2}, \frac{\pi}{2}\right),$

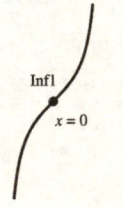

never falling $\Rightarrow$ no local extrema;

$y'' = 2(\sec x)(\sec x)(\tan x) = 2(\sec^2 x)(\tan x),$

$y'' = (\ \underset{-\pi/2}{---} \mid \underset{0 \quad \pi/2}{+++}\ ) \Rightarrow$ concave up on $\left(0, \frac{\pi}{2}\right),$

concave down on $\left(-\frac{\pi}{2}, 0\right)$, 0 is a opoint of inflection.

51. $y' = \cot \frac{\theta}{2},\ y' = (\underset{0}{+++} \mid \underset{\pi}{---}) \Rightarrow$ rising on $(0, \pi),$

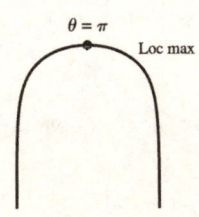

falling on $(\pi, 2\pi) \Rightarrow$ a local maximum at $\theta = \pi$, no local

minimum; $y'' = -\frac{1}{2}\csc^2 \frac{\theta}{2},\ y'' = (\underset{0 \quad 2\pi}{---}) \Rightarrow$ never

concave up, concave down on $(0, 2\pi) \Rightarrow$ no points of

inflection

53. $y' = \tan^2\theta - 1 = (\tan\theta - 1)(\tan\theta + 1)$,

$y' = (\quad +++ | \quad --- | \ +++) \quad \Rightarrow$ rising on
$\quad -\pi/2 \quad -\pi/4 \quad \pi/4 \quad \pi/2$

$\left(-\frac{\pi}{2}, -\frac{\pi}{4}\right)$ and $\left(\frac{\pi}{4}, \frac{\pi}{2}\right)$, falling on $\left(-\frac{\pi}{4}, \frac{\pi}{4}\right)$

$\Rightarrow$ a local maximum at $\theta = -\frac{\pi}{4}$, a local minimum at $\theta = \frac{\pi}{4}$;

$y'' = 2\tan\theta \sec^2\theta$, $y'' = (\quad --- | +++)$
$\qquad\qquad\qquad\quad -\pi/2 \quad 0 \quad \pi/2$

$\Rightarrow$ concave up on $\left(0, \frac{\pi}{2}\right)$, concave down on $\left(-\frac{\pi}{2}, 0\right)$

$\Rightarrow$ a point of inflection at $\theta = 0$

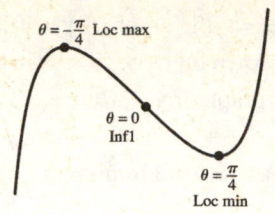

55. $y' = \cos t$, $y' = [+++ | \quad --- | \quad +++] \quad \Rightarrow$ rising on
$\qquad\qquad\quad 0 \quad \pi/2 \quad 3\pi/2 \quad 2\pi$

$\left(0, \frac{\pi}{2}\right)$ and $\left(\frac{3\pi}{2}, 2\pi\right)$, falling on $\left(\frac{\pi}{2}, \frac{3\pi}{2}\right) \Rightarrow$ local maxima at

$t = \frac{\pi}{2}$ and $t = 2\pi$, local minima at $t = 0$ and $t = \frac{3\pi}{2}$;

$y'' = -\sin t$, $y'' = [\quad --- | +++]$
$\qquad\qquad\qquad 0 \quad \pi \quad 2\pi$

$\Rightarrow$ concave up on $(\pi, 2\pi)$, concave down

on $(0, \pi) \Rightarrow$ a point of inflection at $t = \pi$

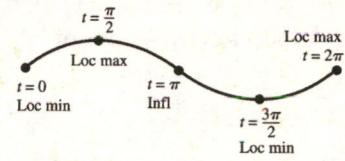

57. $y' = (x + 1)^{-2/3}$, $y' = +++) \ (+++ \Rightarrow$ rising on
$\qquad\qquad\qquad\qquad\quad -1$

$(-\infty, \infty)$, never falling $\Rightarrow$ no local extrema;

$y'' = -\frac{2}{3}(x + 1)^{-5/3}$, $y'' = +++) \ (---$
$\qquad\qquad\qquad\qquad\qquad\quad -1$

$\Rightarrow$ concave up on $(-\infty, -1)$, concave down on $(-1, \infty)$

$\Rightarrow$ a point of inflection and vertical tangent at $x = -1$

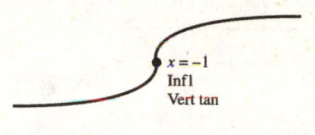

59. $y' = x^{-2/3}(x - 1)$, $y' = ---) (--- | +++ \Rightarrow$ rising on
$\qquad\qquad\qquad\qquad\qquad 0 \quad 1$

$(1, \infty)$, falling on $(-\infty, 1) \Rightarrow$ no local maximum, but a

local minimum at $x = 1$; $y'' = \frac{1}{3}x^{-2/3} + \frac{2}{3}x^{-5/3}$

$= \frac{1}{3}x^{-5/3}(x + 2)$, $y'' = +++ | ---) (+++$
$\qquad\qquad\qquad\qquad\qquad -2 \quad 0$

$\Rightarrow$ concave up on $(-\infty, -2)$ and $(0, \infty)$, concave down on

$(-2, 0) \Rightarrow$ points of inflection at $x = -2$ and $x = 0$, and a

vertical tangent at $x = 0$

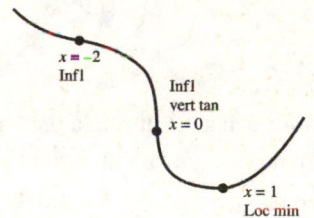

61. $y' = \begin{cases} -2x, & x \le 0 \\ 2x, & x > 0 \end{cases}$, $y' = +++ | +++ \Rightarrow$ rising on
$\qquad\qquad\qquad\qquad\qquad\qquad\qquad\qquad 0$

$(-\infty, \infty) \Rightarrow$ no local extrema; $y'' = \begin{cases} -2, & x < 0 \\ 2, & x > 0 \end{cases}$,

$y'' = ---) (+++ \Rightarrow$ concave up on $(0, \infty)$, concave
$\qquad\quad 0$

down on $(-\infty, 0) \Rightarrow$ a point of inflection at $x = 0$

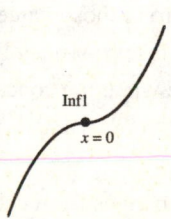

63. The graph of $y = f''(x) \Rightarrow$ the graph of $y = f(x)$ is concave
up on $(0, \infty)$, concave down on $(-\infty, 0) \Rightarrow$ a point of
inflection at $x = 0$; the graph of $y = f'(x)$
$\Rightarrow y' = +++ \mid --- \mid +++ \Rightarrow$ the graph $y = f(x)$ has
both a local maximum and a local minimum

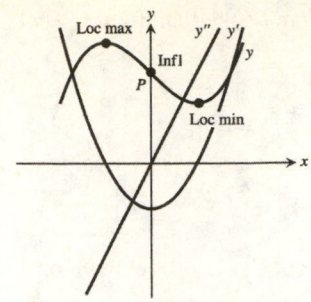

65. The graph of $y = f''(x) \Rightarrow y'' = --- \mid +++ \mid ---$
$\Rightarrow$ the graph of $y = f(x)$ has two points of inflection, the
graph of $y = f'(x) \Rightarrow y' = --- \mid +++ \Rightarrow$ the graph of
$y = f(x)$ has a local minimum

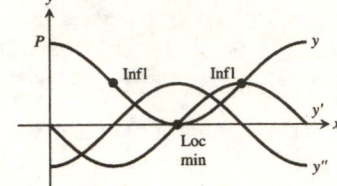

67.

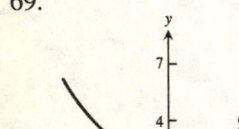

Point	y'	y''
P	−	+
Q	+	0
R	+	−
S	0	−
T	−	−

69.

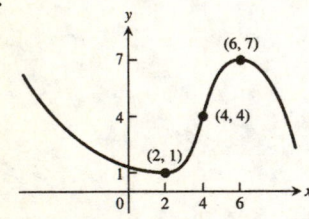

71. Graphs printed in color can shift during a press run, so your values may differ somewhat from those given here.
   (a) The body is moving away from the origin when |displacement| is increasing as t increases, $0 < t < 2$ and
       $6 < t < 9.5$; the body is moving toward the origin when |displacement| is decreasing as t increases, $2 < t < 6$
       and $9.5 < t < 15$.
   (b) The velocity will be zero when the slope of the tangent line for $y = s(t)$ is horizontal. The velocity is zero
       when t is approximately 2, 6, or 9.5 sec.
   (c) The acceleration will be zero at those values of t where the curve $y = s(t)$ has points of inflection. The
       acceleration is zero when t is approximately 4, 7.5, or 12.5 sec.
   (d) The acceleration is positive when the concavity is up, $4 < t < 7.5$ and $12.5 < t < 15$; the acceleration is
       negative when the concavity is down, $0 < t < 4$ and $7.5 < t < 12.5$.

73. The marginal cost is $\frac{dc}{dx}$ which changes from decreasing to increasing when its derivative $\frac{d^2c}{dx^2}$ is zero. This is a
   point of inflection of the cost curve and occurs when the production level x is approximately 60 thousand units.

75. When $y' = (x - 1)^2(x - 2)$, then $y'' = 2(x - 1)(x - 2) + (x - 1)^2$. The curve falls on $(-\infty, 2)$ and rises on
   $(2, \infty)$. At $x = 2$ there is a local minimum. There is no local maximum. The curve is concave upward on $(-\infty, 1)$ and
   $\left(\frac{5}{3}, \infty\right)$, and concave downward on $\left(1, \frac{5}{3}\right)$. At $x = 1$ or $x = \frac{5}{3}$ there are inflection points.

77. The graph must be concave down for x > 0 because
    $f''(x) = -\frac{1}{x^2} < 0.$

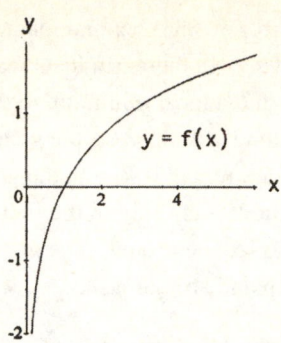

79. The curve will have a point of inflection at x = 1 if 1 is a solution of $y'' = 0$; $y = x^3 + bx^2 + cx + d$
    $\Rightarrow y' = 3x^2 + 2bx + c \Rightarrow y'' = 6x + 2b$ and $6(1) + 2b = 0 \Rightarrow b = -3.$

81. (a) $f(x) = ax^2 + bx + c = a\left(x^2 + \frac{b}{a}x\right) + c = a\left(x^2 + \frac{b}{a}x + \frac{b^2}{4a^2}\right) - \frac{b^2}{4a} + c = a\left(x + \frac{b}{2a}\right)^2 - \frac{b^2 - 4ac}{4a}$ a parabola

    whose vertex is at $x = -\frac{b}{2a} \Rightarrow$ the coordinates of the vertex are $\left(-\frac{b}{2a}, -\frac{b^2 - 4ac}{4a}\right)$

    (b) The second derivative, $f''(x) = 2a$, describes concavity $\Rightarrow$ when a > 0 the parabola is concave up and
    when a < 0 the parabola is concave down.

83. A quadratic curve never has an inflection point. If $y = ax^2 + bx + c$ where $a \neq 0$, then $y' = 2ax + b$ and
    $y'' = 2a.$ Since 2a is a constant, it is not possible for y'' to change signs.

85. If $y = x^5 - 5x^4 - 240$, then $y' = 5x^3(x - 4)$ and
    $y'' = 20x^2(x - 3).$ The zeros of y' are extrema, and
    there is a point of inflection at x = 3.

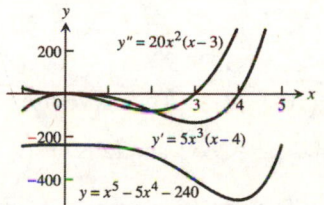

87. If $y = \frac{4}{5}x^5 + 16x^2 - 25$, then $y' = 4x(x^3 + 8)$ and
    $y'' = 16(x^3 + 2).$ The zeros of y' and y'' are
    extrema and points of inflection, respectively.

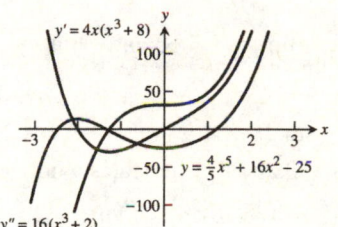

89. The graph of f falls where $f' < 0$, rises where $f' > 0$,
    and has horizontal tangents where $f' = 0.$ It has local
    minima at points where f' changes from negative to
    positive and local maxima where f' changes from
    positive to negative. The graph of f is concave down
    where $f'' < 0$ and concave up where $f'' > 0.$ It has an
    inflection point each time f'' changes sign, provided a tangent
    line exists there.

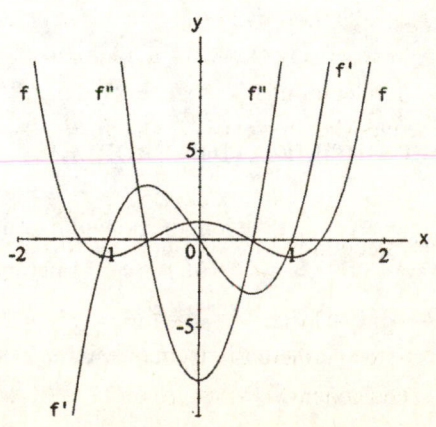

91. (a) It appears to control the number and magnitude of the local extrema. If $k < 0$, there is a local maximum to the left of the origin and a local minimum to the right. The larger the magnitude of k (k < 0), the greater the magnitude of the extrema. If $k > 0$, the graph has only positive slopes and lies entirely in the first and third quadrants with no local extrema. The graph becomes increasingly steep and straight as $k \to \infty$.

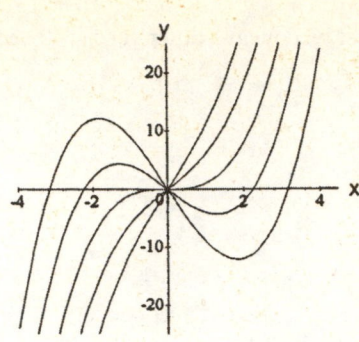

(b) $f'(x) = 3x^2 + k \Rightarrow$ the discriminant $0^2 - 4(3)(k) = -12k$ is positive for $k < 0$, zero for $k = 0$, and negative for $k > 0$; f' has two zeros $x = \pm\sqrt{-\frac{k}{3}}$ when $k < 0$, one zero $x = 0$ when $k = 0$ and no real zeros when $k > 0$; the sign of k controls the number of local extrema.

(c) As $k \to \infty$, $f'(x) \to \infty$ and the graph becomes increasingly steep and straight. As $k \to -\infty$, the crest of the graph (local maximum) in the second quadrant becomes increasingly high and the trough (local minimum) in the fourth quadrant becomes increasingly deep.

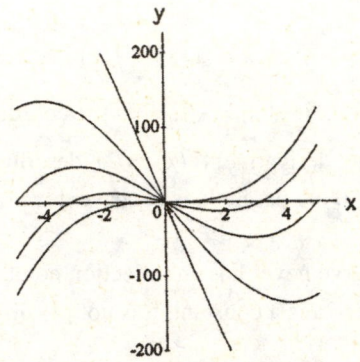

93. (a) If $y = x^{2/3}\left(x^2 - 2\right)$, then $y' = \frac{4}{3}x^{-1/3}\left(2x^2 - 1\right)$ and $y'' = \frac{4}{9}x^{-4/3}\left(10x^2 + 1\right)$. The curve rises on $\left(-\frac{1}{\sqrt{2}}, 0\right)$ and $\left(\frac{1}{\sqrt{2}}, \infty\right)$ and falls on $\left(-\infty, -\frac{1}{\sqrt{2}}\right)$ and $\left(0, \frac{1}{\sqrt{2}}\right)$. The curve is concave up on $(-\infty, 0)$ and $(0, \infty)$.

(b) A cusp since $\lim\limits_{x \to 0^-} y' = \infty$ and $\lim\limits_{x \to 0^+} y' = -\infty$.

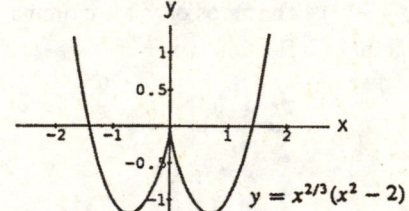

$y = x^{2/3}(x^2 - 2)$

95. Yes: $y = x^2 + 3\sin 2x \Rightarrow y' = 2x + 6\cos 2x$. The graph of y' is zero near -3 and this indicates a horizontal tangent near $x = -3$.

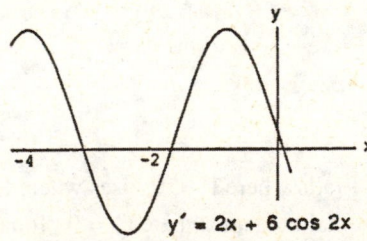

$y' = 2x + 6\cos 2x$

4.5 APPLIED OPTIMIZATION PROBLEMS

1. Let ℓ and w represent the length and width of the rectangle, respectively. With an area of 16 in.^{2}, we have that $(\ell)(w) = 16 \Rightarrow w = 16\ell^{-1} \Rightarrow$ the perimeter is $P = 2\ell + 2w = 2\ell + 32\ell^{-1}$ and $P'(\ell) = 2 - \frac{32}{\ell^2} = \frac{2(\ell^2 - 16)}{\ell^2}$. Solving $P'(\ell) = 0 \Rightarrow \frac{2(\ell + 4)(\ell - 4)}{\ell^2} = 0 \Rightarrow \ell = -4, 4$. Since $\ell > 0$ for the length of a rectangle, ℓ must be 4 and $w = 4 \Rightarrow$ the perimeter is 16 in., a minimum since $P''(\ell) = \frac{16}{\ell^3} > 0$.

3. (a) The line containing point P also contains the points $(0, 1)$ and $(1, 0) \Rightarrow$ the line containing P is $y = 1 - x$ $\Rightarrow$ a general point on that line is $(x, 1 - x)$.

(b) The area $A(x) = 2x(1 - x)$, where $0 \le x \le 1$.

(c) When $A(x) = 2x - 2x^2$, then $A'(x) = 0 \Rightarrow 2 - 4x = 0 \Rightarrow x = \frac{1}{2}$. Since $A(0) = 0$ and $A(1) = 0$, we conclude that $A\left(\frac{1}{2}\right) = \frac{1}{2}$ sq units is the largest area. The dimensions are 1 unit by $\frac{1}{2}$ unit.

5. The volume of the box is $V(x) = x(15 - 2x)(8 - 2x)$
    $= 120x - 46x^2 + 4x^3$, where $0 \le x \le 4$. Solving $V'(x) = 0$
    $\Rightarrow 120 - 92x + 12x^2 = 4(6 - x)(5 - 3x) = 0 \Rightarrow x = \frac{5}{3}$
    or 6, but 6 is not in the domain. Since $V(0) = V(4) = 0$,
    $V\left(\frac{5}{3}\right) = \frac{2450}{27} \approx 91$ in^3 must be the maximum volume of
    the box with dimensions $\frac{14}{3} \times \frac{35}{3} \times \frac{5}{3}$ inches.

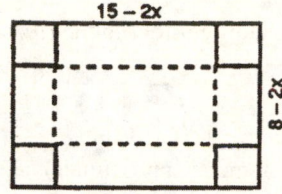

7. The area is $A(x) = x(800 - 2x)$, where $0 \le x \le 400$.
    Solving $A'(x) = 800 - 4x = 0 \Rightarrow x = 200$. With
    $A(0) = A(400) = 0$, the maximum area is
    $A(200) = 80{,}000$ m^2. The dimensions are 200 m by 400 m.

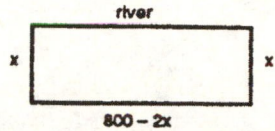

9. (a) We minimize the weight $= tS$ where S is the surface area, and t is the thickness of the steel walls of the tank. The surace area is $S = x^2 + 4xy$ where x is the length of a side of the square base of the tank, and y is its depth. The volume of the tank must be 500ft^3 $\Rightarrow y = \frac{500}{x^2}$. Therefore, the weight of the tank is $w(x) = t\left(x^2 + \frac{2000}{x}\right)$. Treating the thickness as a constant gives $w'(x) = t\left(2x - \frac{2000}{x^2}\right)$ for x.0. The critical value is at $x = 10$. Since $w''(10) = t\left(2 + \frac{4000}{10^3}\right) > 0$, there is a minimum at $x = 10$. Therefore, the optimum dimensions of the tank are 10 ft on the base edges and 5 ft deep.

    (b) Minimizing the surface area of the tank minimizes its weight for a given wall thickness. The thickness of the steel walls would likely be determined by other considerations such as structural requirements.

11. The area of the printing is $(y - 4)(x - 8) = 50$.
     Consequently, $y = \left(\frac{50}{x - 8}\right) + 4$. The area of the paper is
     $A(x) = x\left(\frac{50}{x - 8} + 4\right)$, where $8 < x$. Then
     $A'(x) = \left(\frac{50}{x - 8} + 4\right) - x\left(\frac{50}{(x - 8)^2}\right) = \frac{4(x - 8)^2 - 400}{(x - 8)^2} = 0$
     $\Rightarrow$ the critical points are -2 and 18, but -2 is not in the
     domain. Thus $A''(18) > 0 \Rightarrow$ at $x = 18$ we have
     a minimum. Therefore the dimensions 18 by 9 inches
     minimize the amount of paper.

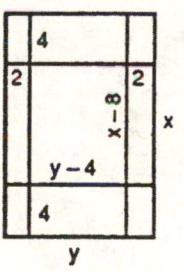

13. The area of the triangle is $A(\theta) = \frac{ab \sin \theta}{2}$, where $0 < \theta < \pi$.
     Solving $A'(\theta) = 0 \Rightarrow \frac{ab \cos \theta}{2} = 0 \Rightarrow \theta = \frac{\pi}{2}$. Since $A''(\theta)$
     $= -\frac{ab \sin \theta}{2} \Rightarrow A''\left(\frac{\pi}{2}\right) < 0$, there is a maximum at $\theta = \frac{\pi}{2}$.

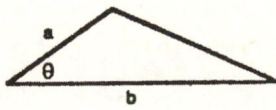

15. With a volume of 1000 cm and $V = \pi r^2 h$, then $h = \frac{1000}{\pi r^2}$. The amount of aluminum used per can is
     $A = 8r^2 + 2\pi rh = 8r^2 + \frac{2000}{r}$. Then $A'(r) = 16r - \frac{2000}{r^2} = 0 \Rightarrow \frac{8r^3 - 1000}{r^2} = 0 \Rightarrow$ the critical points are 0 and 5,
     but $r = 0$ results in no can. Since $A''(r) = 16 + \frac{1000}{r^3} > 0$ we have a minimum at $r = 5 \Rightarrow h = \frac{40}{\pi}$ and h:r = 8:π.

17. (a) The" sides" of the suitcase will measure $24 - 2x$ in. by $18 - 2x$ in. and will be 2x in. apart, so the volume formula is
     $V(x) = 2x(24 - 2x)(18 - 2x) = 8x^3 - 168x^2 + 862x$.

(b) We require x > 0, 2x < 18, and 2x < 24. Combining these requirements, the domain is the interval (0, 9).

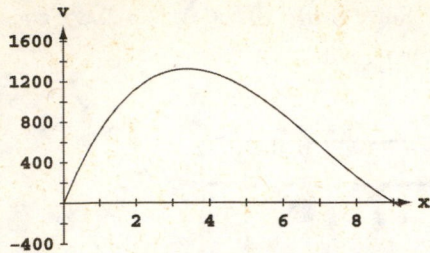

(c) The maximum volume is approximately 1309.95 in.^{3} when x $\approx$ 3.39 in.

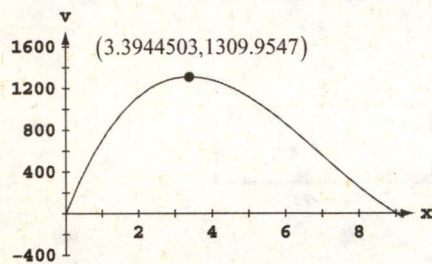

(d) $V'(x) = 24x^2 - 336x + 864 = 24(x^2 - 14x + 36)$. The critical point is at $x = \dfrac{14 \pm \sqrt{(-14)^2 - 4(1)(36)}}{2(1)} = \dfrac{14 \pm \sqrt{52}}{2}$

$= 7 \pm \sqrt{13}$, that is, x $\approx$ 3.39 or x $\approx$ 10.61. We discard the larger value because it is not in the domain. Since $V''(x) = 24(2x - 14)$ which is negative when x $\approx$ 3.39, the critical point corresponds to the maximum volume. The maximum value occurs at $x = 7 - \sqrt{13} \approx 3.39$, which confirms the results in (c).

(e) $8x^3 - 168x^2 + 862x = 1120 \Rightarrow 8(x^3 - 21x^2 + 108x - 140) = 0 \Rightarrow 8(x - 2)(x - 5)(x - 14) = 0$. Since 14 is not in the fomain, the possible values of x are x = 2 in. or x = 5 in.

(f) The dimensions of the resulting box are 2x in., (24 − 2x) in., and (18 − 2x). Each of these measurements must be positive, so that gives the domain of (0, 9).

19. Let the radius of the cylinder be r cm, $0 < r < 10$. Then the height is $2\sqrt{100 - r^2}$ and the volume is

$V(r) = 2\pi r^2 \sqrt{100 - r^2}$ cm^3. Then, $V'(r) = 2\pi r^2 \left(\dfrac{1}{\sqrt{100 - r^2}}\right)(-2r) + \left(2\pi \sqrt{100 - r^2}\right)(2r)$

$= \dfrac{-2\pi r^3 + 4\pi r(100 - r^2)}{\sqrt{100 - r^2}} = \dfrac{2\pi r(200 - 3r^2)}{\sqrt{100 - r^2}}$. The critical point for $0 < r < 10$ occurs at $r = \sqrt{\dfrac{200}{3}} = 10\sqrt{\dfrac{2}{3}}$. Since $V'(r) > 0$ for

$0 < r < 10\sqrt{\dfrac{2}{3}}$ and $V'(r) < 0$ for $10\sqrt{\dfrac{2}{3}} < r < 10$, the critical point corresponds to the maximum volume. The

dimensions are $r = 10\sqrt{\dfrac{2}{3}} \approx 8.16$ cm and $h = \dfrac{20}{\sqrt{3}} \approx 11.55$ cm, and the volume is $\dfrac{4000\pi}{3\sqrt{3}} \approx 2418.40$ cm^3.

21. (a) From the diagram we have 3h + 2w = 108 and

$V = h^2 w \Rightarrow V(h) = h^2 \left(54 - \frac{3}{2}h\right) = 54h^2 - \frac{3}{2}h^3$.

Then $V'(h) = 108h - \frac{9}{2}h^2 = \frac{9}{2}h(24 - h) = 0$

$\Rightarrow$ h = 0 or h = 24, but h = 0 results in no box. Since

$V''(h) = 108 - 9h < 0$ at h = 24, we have a maximum

volume at h = 24 and $w = 54 - \frac{3}{2}h = 18$.

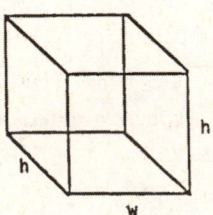

(b)

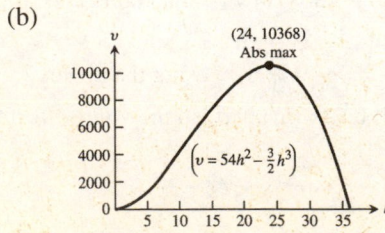

23. The fixed volume is $V = \pi r^2 h + \frac{2}{3}\pi r^3 \Rightarrow h = \frac{V}{\pi r^2} - \frac{2r}{3}$, where h is the height of the cylinder and r is the radius of the hemisphere. To minimize the cost we must minimize surface area of the cylinder added to twice the surface area of the hemisphere. Thus, we minimize $C = 2\pi r h + 4\pi r^2 = 2\pi r\left(\frac{V}{\pi r^2} - \frac{2r}{3}\right) + 4\pi r^2 = \frac{2V}{r} + \frac{8}{3}\pi r^2$.

Then $\frac{dC}{dr} = -\frac{2V}{r^2} + \frac{16}{3}\pi r = 0 \Rightarrow V = \frac{8}{3}\pi r^3 \Rightarrow r = \left(\frac{3V}{8\pi}\right)^{1/3}$. From the volume equation, $h = \frac{V}{\pi r^2} - \frac{2r}{3}$

$= \frac{4V^{1/3}}{\pi^{1/3}\cdot 3^{2/3}} - \frac{2\cdot 3^{1/3}\cdot V^{1/3}}{3\cdot 2\cdot \pi^{1/3}} = \frac{3^{1/3}\cdot 2\cdot 4\cdot V^{1/3} - 2\cdot 3^{1/3}\cdot V^{1/3}}{3\cdot 2\cdot \pi^{1/3}} = \left(\frac{3V}{\pi}\right)^{1/3}$. Since $\frac{d^2 C}{dr^2} = \frac{4V}{r^3} + \frac{16}{3}\pi > 0$, these

dimensions do minimize the cost.

25. (a) From the diagram we have: $\overline{AP} = x$, $\overline{RA} = \sqrt{L - x^2}$,

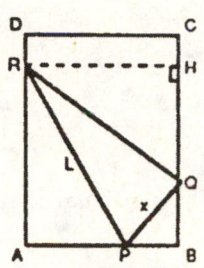

$\overline{PB} = 8.5 - x$, $\overline{CH} = \overline{DR} = 11 - \overline{RA} = 11 - \sqrt{L - x^2}$,

$\overline{QB} = \sqrt{x^2 - (8.5 - x)^2}$, $\overline{HQ} = 11 - \overline{CH} - \overline{QB}$

$= 11 - \left[11 - \sqrt{L - x^2} + \sqrt{x^2 - (8.5 - x)^2}\right]$

$= \sqrt{L - x^2} - \sqrt{x^2 - (8.5 - x)^2}$, $\overline{RQ}^2 = \overline{RH}^2 + \overline{HQ}^2$

$= (8.5)^2 + \left(\sqrt{L - x^2} - \sqrt{x^2 - (8.5 - x)^2}\right)^2$. It

follows that $\overline{RP}^2 = \overline{PQ}^2 + \overline{RQ}^2 \Rightarrow L^2 = x^2 + \left(\sqrt{L^2 - x^2} - \sqrt{x^2 - (x - 8.5)^2}\right)^2 + (8.5)^2$

$\Rightarrow L^2 = x^2 + L^2 - x^2 - 2\sqrt{L^2 - x^2}\sqrt{17x - (8.5)^2} + 17x - (8.5)^2 + (8.5)^2$

$\Rightarrow 17^2 x^2 = 4\left(L^2 - x^2\right)\left(17x - (8.5)^2\right) \Rightarrow L^2 = x^2 + \frac{17^2 x^2}{4[17x - (8.5)^2]} = \frac{17x^3}{17x - (8.5)^2}$

$= \frac{17x^3}{17x - \left(\frac{17}{2}\right)^2} = \frac{4x^3}{4x - 17} = \frac{2x^3}{2x - 8.5}$.

(b) If $f(x) = \frac{4x^3}{4x - 17}$ is minimized, then L^2 is minimized. Now $f'(x) = \frac{4x^2(8x - 51)}{(4x - 17)^2} \Rightarrow f'(x) < 0$ when $x < \frac{51}{8}$ and $f'(x) > 0$ when $x > \frac{51}{8}$. Thus L^2 is minimized when $x = \frac{51}{8}$.

(c) When $x = \frac{51}{8}$, then $L \approx 11.0$ in.

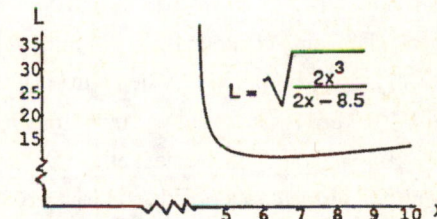

27. Note that $h^2 + r^2 = 3$ and so $r = \sqrt{3 - h^2}$. Then the volume is given by $V = \frac{\pi}{3}r^2 h = \frac{\pi}{3}(3 - h^2)h = \pi h - \frac{\pi}{3}h^3$ for $0 < h < \sqrt{3}$, and so $\frac{dV}{dh} = \pi - \pi r^2 = \pi(1 - r^2)$. The critical point (for $h > 0$) occurs at $h = 1$. Since $\frac{dV}{dh} > 0$ for $0 < h < 1$, and $\frac{dV}{dh} < 0$ for $1 < h < \sqrt{3}$, the critical point corresponds to the maximum volume. The cone of greatest volume has radius $\sqrt{2}$ m, height 1m, and volume $\frac{2\pi}{3}$ m^3.

29. If $f(x) = x^2 + \frac{a}{x}$, then $f'(x) = 2x - ax^{-2}$ and $f''(x) = 2 + 2ax^{-3}$. The critical points are 0 and $\sqrt[3]{\frac{a}{2}}$, but $x \neq 0$. Now $f''\left(\sqrt[3]{\frac{a}{2}}\right) = 6 > 0 \Rightarrow$ at $x = \sqrt[3]{\frac{a}{2}}$ there is a local minimum. However, no local maximum exists for any a.

31. (a) $s(t) = -16t^2 + 96t + 112 \Rightarrow v(t) = s'(t) = -32t + 96$. At $t = 0$, the velocity is $v(0) = 96$ ft/sec.

(b) The maximum height ocurs when $v(t) = 0$, when $t = 3$. The maximum height is $s(3) = 256$ ft and it occurs at $t = 3$ sec.

(c) Note that $s(t) = -16t^2 + 96t + 112 = -16(t + 1)(t - 7)$, so $s = 0$ at $t = -1$ or $t = 7$. Choosing the positive value of t, the velocity when $s = 0$ is $v(7) = -128$ ft/sec.

33. $\frac{8}{x} = \frac{h}{x+27} \Rightarrow h = 8 + \frac{216}{x}$ and $L(x) = \sqrt{h^2 + (x+27)^2}$

$= \sqrt{\left(8 + \frac{216}{x}\right)^2 + (x+27)^2}$ when $x \geq 0$. Note that $L(x)$ is

minimized when $f(x) = \left(8 + \frac{216}{x}\right)^2 + (x+27)^2$ is

minimized. If $f'(x) = 0$, then

$2\left(8 + \frac{216}{x}\right)\left(-\frac{216}{x^2}\right) + 2(x+27) = 0$

$\Rightarrow (x+27)\left(1 - \frac{1728}{x^3}\right) = 0 \Rightarrow x = -27$ (not acceptable

since distance is never negative or $x = 12$. Then $L(12) = \sqrt{2197} \approx 46.87$ ft.

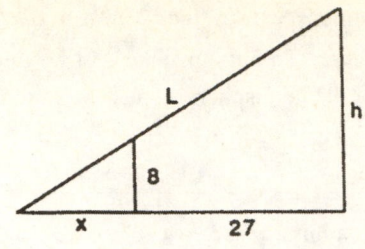

35. (a) From the situation we have $w^2 = 144 - d^2$. The stiffness of the beam is $S = kwd^3 = kd^3\left(144 - d^2\right)^{1/2}$,

where $0 \leq d \leq 12$. Also, $S'(d) = \frac{4kd^2\left(108 - d^2\right)}{\sqrt{144 - d^2}} \Rightarrow$ critical points at 0, 12, and $6\sqrt{3}$. Both $d = 0$ and

$d = 12$ cause $S = 0$. The maximum occurs at $d = 6\sqrt{3}$. The dimensions are 6 by $6\sqrt{3}$ inches.

(b)

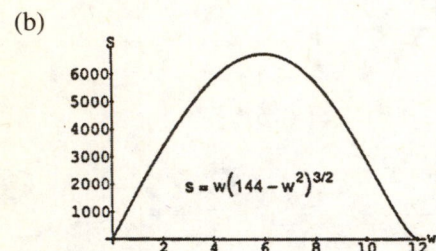

(c)

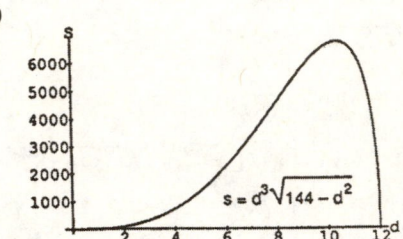

Both graphs indicate the same maximum value and are consistent with each other. The changing of k has
no effect.

37. (a) $s = 10\cos(\pi t) \Rightarrow v = -10\pi\sin(\pi t) \Rightarrow$ speed $= |10\pi\sin(\pi t)| = 10\pi|\sin(\pi t)| \Rightarrow$ the maximum speed is

$10\pi \approx 31.42$ cm/sec since the maximum value of $|\sin(\pi t)|$ is 1; the cart is moving the fastest at $t = 0.5$ sec,

1.5 sec, 2.5 sec and 3.5 sec when $|\sin(\pi t)|$ is 1. At these times the distance is $s = 10\cos\left(\frac{\pi}{2}\right) = 0$ cm and

$a = -10\pi^2\cos(\pi t) \Rightarrow |a| = 10\pi^2|\cos(\pi t)| \Rightarrow |a| = 0$ cm/sec^2.

(b) $|a| = 10\pi^2|\cos(\pi t)|$ is greatest at $t = 0.0$ sec, 1.0 sec, 2.0 sec, 3.0 sec and 4.0 sec, and at these times the

magnitude of the cart's position is $|s| = 10$ cm from the rest position and the speed is 0 cm/sec.

39. (a) $s = \sqrt{(12 - 12t)^2 + (8t)^2} = \left((12 - 12t)^2 + 64t^2\right)^{1/2}$

(b) $\frac{ds}{dt} = \frac{1}{2}\left((12 - 12t)^2 + 64t^2\right)^{-1/2}[2(12 - 12t)(-12) + 128t] = \frac{208t - 144}{\sqrt{(12 - 12t)^2 + 64t^2}}$

$\Rightarrow \frac{ds}{dt}\Big|_{t=0} = -12$ knots and $\frac{ds}{dt}\Big|_{t=1} = 8$ knots

(c) The graph indicates that the ships did not see
each other because $s(t) > 5$ for all values of t.

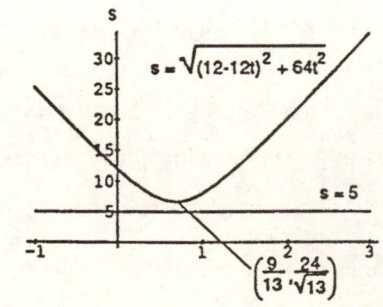

(d) The graph supports the conclusions in parts (b) and (c).

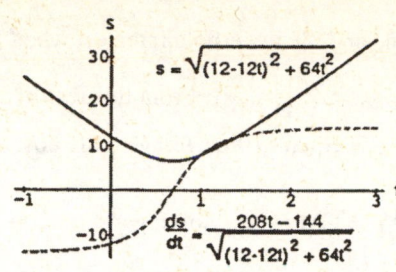

(e) $\displaystyle\lim_{t \to \infty} \frac{ds}{dt} = \sqrt{\lim_{t \to \infty} \frac{(208t - 144)^2}{144(1-t)^2 + 64t^2}} = \sqrt{\lim_{t \to \infty} \frac{\left(208 - \frac{144}{t}\right)^2}{144\left(\frac{1}{t} - 1\right)^2 + 64}} = \sqrt{\frac{208^2}{144 + 64}} = \sqrt{208} = 4\sqrt{13}$

which equals the square root of the sums of the squares of the individual speeds.

41. If $v = kax - kx^2$, then $v' = ka - 2kx$ and $v'' = -2k$, so $v' = 0 \Rightarrow x = \frac{a}{2}$. At $x = \frac{a}{2}$ there is a maximum since $v''\left(\frac{a}{2}\right) = -2k < 0$. The maximum value of v is $\frac{ka^2}{4}$.

43. The profit is $p = nx - nc = n(x - c) = \left[a(x - c)^{-1} + b(100 - x)\right](x - c) = a + b(100 - x)(x - c)$
     $= a + (bc + 100b)x - 100bc - bx^2$. Then $p'(x) = bc + 100b - 2bx$ and $p''(x) = -2b$. Solving $p'(x) = 0 \Rightarrow$
     $x = \frac{c}{2} + 50$. At $x = \frac{c}{2} + 50$ there is a maximum profit since $p''(x) = -2b < 0$ for all x.

45. (a) $A(q) = kmq^{-1} + cm + \frac{h}{2}q$, where $q > 0 \Rightarrow A'(q) = -kmq^{-2} + \frac{h}{2} = \frac{hq^2 - 2km}{2q^2}$ and $A''(q) = 2kmq^{-3}$. The
          critical points are $-\sqrt{\frac{2km}{h}}$, 0, and $\sqrt{\frac{2km}{h}}$, but only $\sqrt{\frac{2km}{h}}$ is in the domain. Then $A''\left(\sqrt{\frac{2km}{h}}\right) > 0 \Rightarrow$ at
          $q = \sqrt{\frac{2km}{h}}$ there is a minimum average weekly cost.

     (b) $A(q) = \frac{(k + bq)m}{q} + cm + \frac{h}{2}q = kmq^{-1} + bm + cm + \frac{h}{2}q$, where $q > 0 \Rightarrow A'(q) = 0$ at $q = \sqrt{\frac{2km}{h}}$ as in (a).
          Also $A''(q) = 2kmq^{-3} > 0$ so the most economical quantity to order is still $q = \sqrt{\frac{2km}{h}}$ which minimizes
          the average weekly cost.

47. The profit $p(x) = r(x) - c(x) = 6x - (x^3 - 6x^2 + 15x) = -x^3 + 6x^2 - 9x$, where $x \geq 0$. Then
     $p'(x) = -3x^2 + 12x - 9 = -3(x - 3)(x - 1)$ and $p''(x) = -6x + 12$. The critical points are 1 and 3. Thus
     $p''(1) = 6 > 0 \Rightarrow$ at $x = 1$ there is a local minimum, and $p''(3) = -6 < 0 \Rightarrow$ at $x = 3$ there is a local maximum.
     But $p(3) = 0 \Rightarrow$ the best you can do is break even.

49. (a) The artisan should order px units of material in order to have enough until the next delivery.

     (b) The average number of units in storage until the next delivery is $\frac{px}{2}$ and so the cost of storing then is $s\left(\frac{px}{2}\right)$ per
          day, and the total cost for x days is $\left(\frac{px}{2}\right)sx$. When added to the delivery cost, the total cost for delivery and storage
          for each cycle is: cost per cycle $= d + \frac{px}{2}sx$.

     (c) The average cost per day for storage and delivery of materials is: average cost per day $= \frac{\left(d + \frac{ps}{2}x^2\right)}{x} + \frac{d}{x} + \frac{ps}{2}x$.
          To minimize the average cost per day, set the derivative equal to zero. $\frac{d}{dx}\left(d(x)^{-1} + \frac{ps}{2}x\right) = -d(x)^{-2} + \frac{ps}{2} = 0$
          $\Rightarrow x = \pm\sqrt{\frac{2d}{ps}}$. Only the positive root makes sense in this context so that $x^* = \sqrt{\frac{2d}{ps}}$. To verify that x^* gives a
          minimum, check the second derivative $\left[\frac{d}{dx}\left(-d(x)^{-2} + \frac{ps}{2}\right)\right]\Bigg|_{\sqrt{\frac{2d}{ps}}} = \frac{2d}{x^3}\Bigg|_{\sqrt{\frac{2d}{ps}}} = \frac{2d}{\left(\sqrt{\frac{2d}{ps}}\right)^3} > 0 \Rightarrow$ a minimum.

          The amount to deliver is $px^* = \sqrt{\frac{2pd}{s}}$.

(d) The line and the hyperbola intersect when $\frac{d}{x} = \frac{ps}{2}x$. Solving for x gives $x_{intersection} = \pm\sqrt{\frac{2d}{ps}}$. For $x > 0$,

$x_{intersection} = \sqrt{\frac{2d}{ps}} = x^*$. From this result, the average cost per day is minimized when the average daily cost of delivery is equal to the average daily cost of storage.

51. We have $\frac{dR}{dM} = CM - M^2$. Solving $\frac{d^2R}{dM^2} = C - 2M = 0 \Rightarrow M = \frac{C}{2}$. Also, $\frac{d^3R}{dM^3} = -2 < 0 \Rightarrow$ at $M = \frac{C}{2}$ there is a maximum.

53. If $x > 0$, then $(x - 1)^2 \geq 0 \Rightarrow x^2 + 1 \geq 2x \Rightarrow \frac{x^2+1}{x} \geq 2$. In particular if a, b, c and d are positive integers, then $\left(\frac{a^2+1}{a}\right)\left(\frac{b^2+1}{b}\right)\left(\frac{c^2+1}{c}\right)\left(\frac{d^2+1}{d}\right) \geq 16$.

55. At $x = c$, the tangents to the curves are parallel. Justification: The vertical distance between the curves is $D(x) = f(x) - g(x)$, so $D'(x) = f'(x) - g'(x)$. The maximum value of D will occur at a point c where $D' = 0$. At such a point, $f'(c) - g'(c) = 0$, or $f'(c) = g'(c)$.

57. (a) If $y = \cot x - \sqrt{2}\csc x$ where $0 < x < \pi$, then $y' = (\csc x)\left(\sqrt{2}\cot x - \csc x\right)$. Solving $y' = 0$

$\Rightarrow \cos x = \frac{1}{\sqrt{2}} \Rightarrow x = \frac{\pi}{4}$. For $0 < x < \frac{\pi}{4}$ we have $y' > 0$, and $y' < 0$ when $\frac{\pi}{4} < x < \pi$. Therefore, at $x = \frac{\pi}{4}$ there is a maximum value of $y = -1$.

(b)

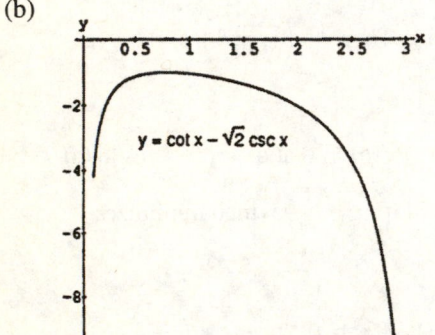

$y = \cot x - \sqrt{2}\csc x$

The graph confirms the findings in (a).

59. (a) The square of the distance is $D(x) = \left(x - \frac{3}{2}\right)^2 + \left(\sqrt{x} + 0\right)^2 = x^2 - 2x + \frac{9}{4}$, so $D'(x) = 2x - 2$ and the critical point occurs at $x = 1$. Since $D'(x) < 0$ for $x < 1$ and $D'(x) > 0$ for $x > 1$, the critical point corresponds to the minimum distance. The minimum distance is $\sqrt{D(1)} = \frac{\sqrt{5}}{2}$.

(b)

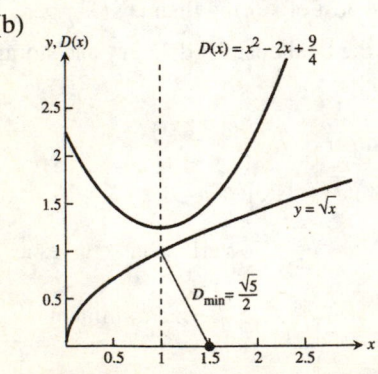

$D(x) = x^2 - 2x + \frac{9}{4}$

$y = \sqrt{x}$

$D_{min} = \frac{\sqrt{5}}{2}$

The minimum distance is from the point $\left(\frac{3}{2}, 0\right)$ to the point $(1, 1)$ on the graph of $y = \sqrt{x}$, and this occurs at the value $x = 1$ where $D(x)$, the distance squared, has its minimum value.

61. (a) The base radius of the cone is $r = \frac{2\pi a - x}{2\pi}$ and so the height is $h = \sqrt{a^2 - r^2} = \sqrt{a^2 - \left(\frac{2\pi a - x}{2\pi}\right)^2}$. Therefore,

$$V(x) = \frac{\pi}{3}r^2 h = \frac{\pi}{3}\left(\frac{2\pi a - x}{2\pi}\right)^2 \sqrt{a^2 - \left(\frac{2\pi a - x}{2\pi}\right)^2}.$$

(b) To simplify the calculations, we shall consider the volume as a function of r: volume $= f(r) = \frac{\pi}{3}r^2\sqrt{a^2 - r^2}$, where

$$0 < r < a. \ f'(r) = \frac{\pi}{3}\frac{d}{dr}\left(r^2\sqrt{a^2 - r^2}\right) = \frac{\pi}{3}\left[r^2 \cdot \frac{1}{2\sqrt{a^2 - r^2}}(-2r) + \left(\sqrt{a^2 - r^2}\right)(2r)\right] = \frac{\pi}{3}\left[\frac{-r^3 + 2r(a^2 - r^2)}{\sqrt{a^2 - r^2}}\right]$$

$$= \frac{\pi}{3}\left[\frac{2a^2 r - 3r^3}{\sqrt{a^2 - r^2}}\right] = \frac{\pi r(2a^2 - 3r^2)}{3\sqrt{a^2 - r^2}}.$$ The critical point occurs when $r^2 = \frac{2a^2}{3}$, which gives $r = a\sqrt{\frac{2}{3}} = \frac{a\sqrt{6}}{3}$. Then

$$h = \sqrt{a^2 - r^2} = \sqrt{a^2 - \frac{2a^2}{3}} = \sqrt{\frac{a^2}{3}} = \frac{a\sqrt{3}}{3}.$$ Using $r = \frac{a\sqrt{6}}{3}$ and $h = \frac{a\sqrt{3}}{3}$, we may now find the values of r and h for the given values of a.

When $a = 4$: $r = \frac{4\sqrt{6}}{3}, h = \frac{4\sqrt{3}}{3}$;

When $a = 5$: $r = \frac{5\sqrt{6}}{3}, h = \frac{5\sqrt{3}}{3}$;

When $a = 6$: $r = 2\sqrt{6}, h = 2\sqrt{3}$;

When $a = 8$: $r = \frac{8\sqrt{6}}{3}, h = \frac{8\sqrt{3}}{3}$;

(c) Since $r = \frac{a\sqrt{6}}{3}$ and $h = \frac{a\sqrt{3}}{3}$, the relationship is $\frac{r}{h} = \sqrt{2}$.

4.6 INDETERMINATE FORMS AND L'HÔPITAL'S RULE

1. l'Hôpital: $\lim\limits_{x \to 2} \frac{x - 2}{x^2 - 4} = \frac{1}{2x}\Big|_{x=2} = \frac{1}{4}$ or $\lim\limits_{x \to 2} \frac{x - 2}{x^2 - 4} = \lim\limits_{x \to 2} \frac{x - 2}{(x - 2)(x + 2)} = \lim\limits_{x \to 2} \frac{1}{x + 2} = \frac{1}{4}$

3. l'Hôpital: $\lim\limits_{x \to \infty} \frac{5x^2 - 3x}{7x^2 + 1} = \lim\limits_{x \to \infty} \frac{10x - 3}{14x} = \lim\limits_{x \to \infty} \frac{10}{14} = \frac{5}{7}$ or $\lim\limits_{x \to \infty} \frac{5x^2 - 3x}{7x^2 + 1} = \lim\limits_{x \to \infty} \frac{5 - \frac{3}{x}}{7 + \frac{1}{x}} = \frac{5}{7}$

5. l'Hôpital: $\lim\limits_{x \to 0} \frac{1 - \cos x}{x^2} = \lim\limits_{x \to 0} \frac{\sin x}{2x} = \lim\limits_{x \to 0} \frac{\cos x}{2} = \frac{1}{2}$ or $\lim\limits_{x \to 0} \frac{1 - \cos x}{x^2} = \lim\limits_{x \to 0} \left[\frac{(1 - \cos x)}{x^2}\left(\frac{1 + \cos x}{1 + \cos x}\right)\right]$

$$= \lim\limits_{x \to 0} \frac{\sin^2 x}{x^2(1 + \cos x)} = \lim\limits_{x \to 0} \left[\left(\frac{\sin x}{x}\right)\left(\frac{\sin x}{x}\right)\left(\frac{1}{1 + \cos x}\right)\right] = \frac{1}{2}$$

7. $\lim\limits_{t \to 0} \frac{\sin t^2}{t} = \lim\limits_{t \to 0} \frac{2t \cos t^2}{1} = 0$

9. $\lim\limits_{\theta \to \pi} \frac{\sin \theta}{\pi - \theta} = \lim\limits_{\theta \to \pi} \frac{\cos \theta}{-1} = \frac{-1}{-1} = 1$

11. $\lim\limits_{x \to \pi/4} \frac{\sin x - \cos x}{x - \frac{\pi}{4}} = \lim\limits_{x \to \pi/4} \frac{\cos x + \sin x}{1} = \frac{\sqrt{2}}{2} + \frac{\sqrt{2}}{2} = \sqrt{2}$

13. $\lim\limits_{x \to \pi/2} -\left(x - \frac{\pi}{2}\right)\tan x = \lim\limits_{x \to \pi/2} \frac{-\left(x - \frac{\pi}{2}\right)\sin x}{\cos x} = \lim\limits_{x \to \pi/2} \frac{\left(\frac{\pi}{2} - x\right)\cos x + \sin x(-1)}{-\sin x} = \frac{-1}{-1} = 1$

15. $\lim\limits_{x \to 1} \frac{2x^2 - (3x + 1)\sqrt{x} + 2}{x - 1} = \lim\limits_{x \to 1} \frac{2x^2 - 3x^{3/2} - x^{1/2} + 2}{x - 1} = \lim\limits_{x \to 1} \frac{4x - \frac{9}{2}x^{1/2} - \frac{1}{2\sqrt{x}}}{1} = -1$

17. $\lim\limits_{x \to 0} \frac{\sqrt{a(a + x)} - a}{x} = \lim\limits_{x \to 0} \frac{a}{2\sqrt{a^2 + ax}} = \frac{a}{2\sqrt{a^2}} = \frac{1}{2}$, where $a > 0$.

19. $\lim\limits_{x \to 0} \frac{x(\cos x - 1)}{\sin x - x} = \lim\limits_{x \to 0} \frac{-x\sin x + \cos x - 1}{\cos x - 1} = \lim\limits_{x \to 0} \frac{-x\cos x - 2\sin x}{-\sin x} = \lim\limits_{x \to 0} \frac{x\cos x + 2\sin x}{\sin x} = \lim\limits_{x \to 0} \frac{-x\sin x + 3\cos x}{\cos x} = \frac{3}{1} = 3$

21. $\lim\limits_{r \to 1} \frac{a(r^n - 1)}{r - 1} = \lim\limits_{r \to 1} \frac{a(n \cdot r^{n-1})}{1} = an \lim\limits_{r \to 1} r^{n-1} = an$, where n is a positive integer.

23. $\lim\limits_{x \to \infty} \left(x - \sqrt{x^2 + x}\right) = \lim\limits_{x \to \infty} \left(x - \sqrt{x^2 + x}\right)\left(\frac{x + \sqrt{x^2 + x}}{x + \sqrt{x^2 + x}}\right) = \lim\limits_{x \to \infty} \frac{x^2 - (x^2 + x)}{x + \sqrt{x^2 + x}} = \lim\limits_{x \to \infty} \frac{-\frac{x}{x}}{\frac{x}{x} + \frac{\sqrt{x^2 + x}}{\sqrt{x^2}}}$

$= \lim\limits_{x \to \infty} \frac{-1}{1 + \sqrt{1 + \frac{1}{x}}} = -\frac{1}{2} \left(\begin{array}{l} \text{l'Hopital's rule} \\ \text{is unnecessary} \end{array}\right)$

25. $\lim\limits_{x \to \pm\infty} \frac{3x - 5}{2x^2 - x + 2} = \lim\limits_{x \to \pm\infty} \frac{3}{4x - 1} = 0$

27. $\lim\limits_{x \to \infty} \frac{\sqrt{9x + 1}}{\sqrt{x + 1}} = \sqrt{\lim\limits_{x \to \infty} \frac{9x + 1}{x + 1}} = \sqrt{\lim\limits_{x \to \infty} \frac{9}{1}} = \sqrt{9} = 3$

29. $\lim\limits_{x \to \pi/2^-} \frac{\sec x}{\tan x} = \lim\limits_{x \to \pi/2^-} \left(\frac{1}{\cos x}\right)\left(\frac{\cos x}{\sin x}\right) = \lim\limits_{x \to \pi/2^-} \frac{1}{\sin x} = 1$

31. Part (b) is correct because part (a) is neither in the $\frac{0}{0}$ nor $\frac{\infty}{\infty}$ form and so l'Hôpital's rule may not be used.

33. If f(x) is to be continuous at x = 0, then $\lim\limits_{x \to 0} f(x) = f(0) \Rightarrow c = f(0) = \lim\limits_{x \to 0} \frac{9x - 3\sin 3x}{5x^3} = \lim\limits_{x \to 0} \frac{9 - 9\cos 3x}{15x^2}$

$= \lim\limits_{x \to 0} \frac{27 \sin 3x}{30x} = \lim\limits_{x \to 0} \frac{81 \cos 3x}{30} = \frac{27}{10}$.

35. The graph indicates a limit near -1. The limit leads to the

indeterminate form $\frac{0}{0}$: $\lim\limits_{x \to 1} \frac{2x^2 - (3x + 1)\sqrt{x} + 2}{x - 1}$

$= \lim\limits_{x \to 1} \frac{2x^2 - 3x^{3/2} - x^{1/2} + 2}{x - 1} = \lim\limits_{x \to 1} \frac{4x - \frac{9}{2}x^{1/2} - \frac{1}{2}x^{-1/2}}{1}$

$= \frac{4 - \frac{9}{2} - \frac{1}{2}}{1} = \frac{4 - 5}{1} = -1$

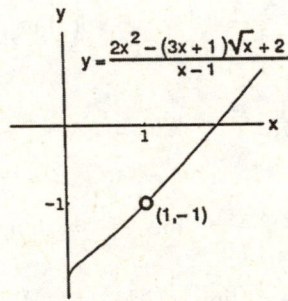

37. Graphing $f(x) = \frac{1 - \cos x^6}{x^{12}}$ on th window $[-1, 1]$ by $[-0.5, 1]$ it appears that $\lim\limits_{x \to 0} f(x) = 0$. However, we see that if we let

$u = x^6$, then $\lim\limits_{x \to 0} f(x) = \lim\limits_{u \to 0} \frac{1 - \cos u}{u^2} = \lim\limits_{u \to 0} \frac{\sin u}{2u} = \lim\limits_{u \to 0} \frac{\cos u}{2} = \frac{1}{2}$.

39. (a) By similar triangles, $\frac{PA}{AB} = \frac{CE}{EB}$ where E is the point on $\overleftrightarrow{AB}$ such that $\overleftrightarrow{CE} \perp \overleftrightarrow{AB}$:

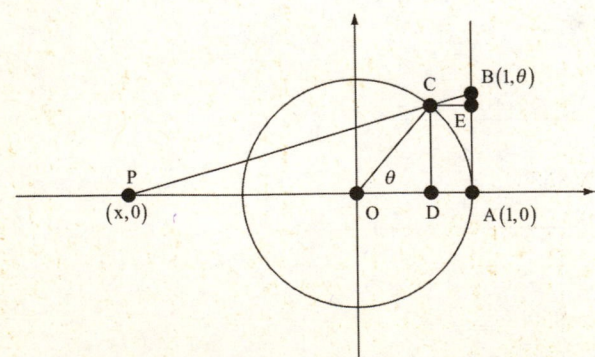

Thus $\frac{1 - x}{\theta} = \frac{1 - \cos\theta}{\theta - \sin\theta}$, since the coordinates of C are $(\cos\theta, \sin\theta)$. Hence, $1 - x = \frac{\theta(1 - \cos\theta)}{\theta - \sin\theta}$.

(b) $\lim\limits_{\theta \to 0} (1 - x) = \lim\limits_{\theta \to 0} \frac{\theta(1 - \cos\theta)}{\theta - \sin\theta} = \lim\limits_{\theta \to 0} \frac{\theta \sin\theta + 1 - \cos\theta}{1 - \cos\theta} = \lim\limits_{\theta \to 0} \frac{\theta \cos\theta + \sin\theta + \sin\theta}{\sin\theta} = \lim\limits_{\theta \to 0} \frac{\theta \cos\theta + 2\sin\theta}{\sin\theta}$

$= \lim\limits_{\theta \to 0} \frac{\theta(-\sin\theta) + \cos\theta + 2\cos\theta}{\cos\theta} = \lim\limits_{\theta \to 0} \frac{-\theta \sin\theta + 3\cos\theta}{\cos\theta} = \frac{0 + 3}{1} = 3$

(c) We have that $\lim\limits_{\theta \to \infty} \left[(1-x)-(1-\cos\theta)\right] = \lim\limits_{\theta \to \infty}\left[\frac{\theta(1-\cos\theta)}{\theta-\sin\theta}-(1-\cos\theta)\right] = \lim\limits_{\theta \to \infty}(1-\cos\theta)\left[\frac{\theta}{\theta-\sin\theta}-1\right]$

As $\theta \to \infty$, $(1-\cos\theta)$ oscillates between 0 and 2, and so it is bounded. Since $\lim\limits_{\theta \to \infty}\left(\frac{\theta}{\theta-\sin\theta}-1\right)=1-1=0$,

$\lim\limits_{\theta \to \infty}(1-\cos\theta)\left[\frac{\theta}{\theta-\sin\theta}-1\right]=0$. Geometrically, this means that as $\theta \to \infty$, the distance between points P and D approaches 0.

4.7 NEWTON'S METHOD

1. $y=x^2+x-1 \Rightarrow y'=2x+1 \Rightarrow x_{n+1}=x_n-\frac{x_n^2+x_n-1}{2x_n+1}$; $x_0=1 \Rightarrow x_1=1-\frac{1+1-1}{2+1}=\frac{2}{3}$

$\Rightarrow x_2=\frac{2}{3}-\frac{\frac{4}{9}+\frac{2}{3}-1}{\frac{4}{3}+1} \Rightarrow x_2=\frac{2}{3}-\frac{4+6-9}{12+9}=\frac{2}{3}-\frac{1}{21}=\frac{13}{21}\approx .61905$; $x_0=-1 \Rightarrow x_1=1-\frac{1-1-1}{-2+1}=-2$

$\Rightarrow x_2=-2-\frac{4-2-1}{-4+1}=-\frac{5}{3}\approx -1.66667$

3. $y=x^4+x-3 \Rightarrow y'=4x^3+1 \Rightarrow x_{n+1}=x_n-\frac{x_n^4+x_n-3}{4x_n^3+1}$; $x_0=1 \Rightarrow x_1=1-\frac{1+1-3}{4+1}=\frac{6}{5}$

$\Rightarrow x_2=\frac{6}{5}-\frac{\frac{1296}{625}+\frac{6}{5}-3}{\frac{864}{125}+1}=\frac{6}{5}-\frac{1296+750-1875}{4320+625}=\frac{6}{5}-\frac{171}{4945}=\frac{5763}{4945}\approx 1.16542$; $x_0=-1 \Rightarrow x_1=-1-\frac{1-1-3}{-4+1}$

$=-2 \Rightarrow x_2=-2-\frac{16-2-3}{-32+1}=-2+\frac{11}{31}=-\frac{51}{31}\approx -1.64516$

5. $y=x^4-2 \Rightarrow y'=4x^3 \Rightarrow x_{n+1}=x_n-\frac{x_n^4-2}{4x_n^3}$; $x_0=1 \Rightarrow x_1=1-\frac{1-2}{4}=\frac{5}{4} \Rightarrow x_2=\frac{5}{4}-\frac{\frac{625}{256}-2}{\frac{125}{16}}=\frac{5}{4}-\frac{625-512}{2000}$

$=\frac{5}{4}-\frac{113}{2000}=\frac{2500-113}{2000}=\frac{2387}{2000}\approx 1.1935$

7. $f(x_0)=0$ and $f'(x_0)\neq 0 \Rightarrow x_{n+1}=x_n-\frac{f(x_n)}{f'(x_n)}$ gives $x_1=x_0 \Rightarrow x_2=x_0 \Rightarrow x_n=x_0$ for all $n \geq 0$. That is, all of the approximations in Newton's method will be the root of $f(x)=0$.

9. If $x_0=h>0 \Rightarrow x_1=x_0-\frac{f(x_0)}{f'(x_0)}=h-\frac{f(h)}{f'(h)}$

$=h-\frac{\sqrt{h}}{\left(\frac{1}{2\sqrt{h}}\right)}=h-\left(\sqrt{h}\right)\left(2\sqrt{h}\right)=-h$;

if $x_0=-h<0 \Rightarrow x_1=x_0-\frac{f(x_0)}{f'(x_0)}=-h-\frac{f(-h)}{f'(-h)}$

$=-h-\frac{\sqrt{h}}{\left(\frac{-1}{2\sqrt{h}}\right)}=-h+\left(\sqrt{h}\right)\left(2\sqrt{h}\right)=h.$

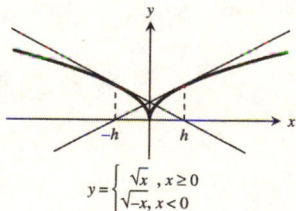

$y=\begin{cases}\sqrt{x}, & x\geq 0 \\ \sqrt{-x}, & x<0\end{cases}$

11. i) is equivalent to solving $x^3-3x-1=0$.
ii) is equivalent to solving $x^3-3x-1=0$.
iii) is equivalent to solving $x^3-3x-1=0$.
iv) is equivalent to solving $x^3-3x-1=0$.
All four equations are equivalent.

13. For $x_0=-0.3$, the procedure converges to the root $-0.32218535....$

(a)

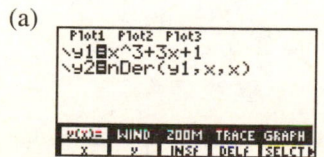

(b)

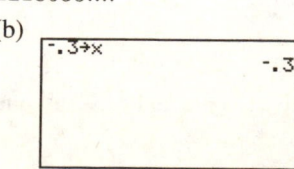

(c)

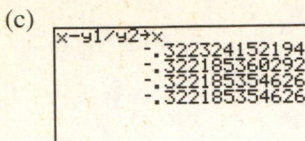

(d) Values for x will vary. One possible choice is $x_0 = 0.1$.

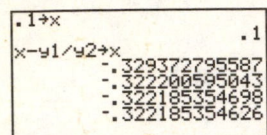

(e) Values for x will vary.

15. $f(x) = \tan x - 2x \Rightarrow f'(x) = \sec^2 x - 2 \Rightarrow x_{n+1} = x_n - \frac{\tan(x_n) - 2x_n}{\sec^2(x_n)}$; $x_0 = 1 \Rightarrow x_1 = 12920445$

$\Rightarrow x_2 = 1.155327774 \Rightarrow x_{16} = x_{17} = 1.165561185$

17. (a) The graph of $f(x) = \sin 3x - 0.99 + x^2$ in the window
$-2 \le x \le 2, -2 \le y \le 3$ suggests three roots.
However, when you zoom in on the x-axis near $x = 1.2$,
you can see that the graph lies above the axis there.
There are only two roots, one near $x = -1$, the other
near $x = 0.4$.

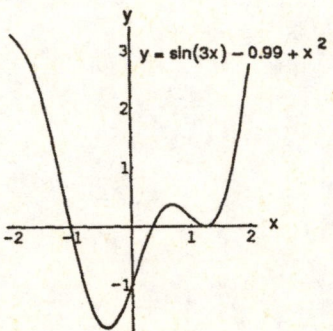

(b) $f(x) = \sin 3x - 0.99 + x^2 \Rightarrow f'(x) = 3 \cos 3x + 2x$

$\Rightarrow x_{n+1} = x_n - \frac{\sin(3x_n) - 0.99 + x_n^2}{3 \cos(3x_n) + 2x_n}$ and the solutions
are approximately 0.35003501505249 and
-1.0261731615301

19. $f(x) = 2x^4 - 4x^2 + 1 \Rightarrow f'(x) = 8x^3 - 8x \Rightarrow x_{n+1} = x_n - \frac{2x_n^4 - 4x_n^2 + 1}{8x_n^3 - 8x_n}$; if $x_0 = -2$, then $x_6 = -1.30656296$; if

$x_0 = -0.5$, then $x_3 = -0.5411961$; the roots are approximately ± 0.5411961 and ± 1.30656296 because $f(x)$ is
an even function.

21. From the graph we let $x_0 = 0.5$ and $f(x) = \cos x - 2x$

$\Rightarrow x_{n+1} = x_n - \frac{\cos(x_n) - 2x_n}{-\sin(x_n) - 2} \Rightarrow x_1 = .45063$

$\Rightarrow x_2 = .45018 \Rightarrow$ at $x \approx 0.45$ we have $\cos x = 2x$.

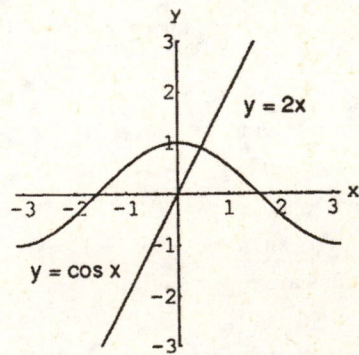

23. If $f(x) = x^3 + 2x - 4$, then $f(1) = -1 < 0$ and $f(2) = 8 > 0 \Rightarrow$ by the Intermediate Value Theorem the equation

$x^3 + 2x - 4 = 0$ has a solution between 1 and 2. Consequently, $f'(x) = 3x^2 + 2$ and $x_{n+1} = x_n - \frac{x_n^3 + 2x_n - 4}{3x_n^2 + 2}$.

Then $x_0 = 1 \Rightarrow x_1 = 1.2 \Rightarrow x_2 = 1.17975 \Rightarrow x_3 = 1.179509 \Rightarrow x_4 = 1.1795090 \Rightarrow$ the root is approximately
1.17951.

25. $f(x) = 4x^4 - 4x^2 \Rightarrow f'(x) = 16x^3 - 8x \Rightarrow x_{i+1} = x_i - \frac{f(x_i)}{f'(x_i)} = x_i - \frac{x_i^3 - x_i}{4x_i^2 - 2}$. Iterations are performed using the procedure in problem 13 in this section.

   (a) For $x_0 = 2$ or $x_0 = -0.8$, $x_i \to -1$ as i gets large.

   (b) For $x_0 = -0.5$ or $x_0 = 0.25$, $x_i \to 0$ as i gets large.

   (c) For $x_0 = 0.8$ or $x_0 = 2$, $x_i \to 1$ as i gets large.

   (d) (If your calculator has a CAS, put it in exact mode, otherwise approximate the radicals with a decimal value.)

     For $x_0 = -\frac{\sqrt{21}}{7}$ or $x_0 = -\frac{\sqrt{21}}{7}$, Newton's method does not converge. The values of x_i alternate between $x_0 = -\frac{\sqrt{21}}{7}$ or $x_0 = -\frac{\sqrt{21}}{7}$ as i increases.

27. $f(x) = (x-1)^{40} \Rightarrow f'(x) = 40(x-1)^{39} \Rightarrow x_{n+1} = x_n - \frac{(x_n-1)^{40}}{40(x_n-1)^{39}} = \frac{39x_n + 1}{40}$. With $x_0 = 2$, our computer gave $x_{87} = x_{88} = x_{89} = \cdots = x_{200} = 1.11051$, coming within 0.11051 of the root $x = 1$.

29. $f(x) = x^3 + 3.6x^2 - 36.4 \Rightarrow f'(x) = 3x^2 + 7.2x \Rightarrow x_{n+1} = x_n - \frac{x_n^3 + 3.6x_n^2 - 36.4}{3x_n^2 + 7.2x_n}$; $x_0 = 2 \Rightarrow x_1 = 2.5\overline{303}$
$\Rightarrow x_2 = 2.45418225 \Rightarrow x_3 = 2.45238021 \Rightarrow x_4 = 2.45237921$ which is 2.45 to two decimal places. Recall that $x = 10^4 [H_3O^+] \Rightarrow [H_3O^+] = (x)(10^{-4}) = (2.45)(10^{-4}) = 0.000245$

4.8 ANTIDERIVATIVES

1. (a) x^2 (b) $\frac{x^3}{3}$ (c) $\frac{x^3}{3} - x^2 + x$

3. (a) x^{-3} (b) $-\frac{x^{-3}}{3}$ (c) $-\frac{x^{-3}}{3} + x^2 + 3x$

5. (a) $\frac{-1}{x}$ (b) $\frac{-5}{x}$ (c) $2x + \frac{5}{x}$

7. (a) $\sqrt{x^3}$ (b) $\sqrt{x}$ (c) $\frac{2}{3}\sqrt{x^3} + 2\sqrt{x}$

9. (a) $x^{2/3}$ (b) $x^{1/3}$ (c) $x^{-1/3}$

11. (a) $\cos(\pi x)$ (b) $-3\cos x$ (c) $\frac{-\cos(\pi x)}{\pi} + \cos(3x)$

13. (a) $\tan x$ (b) $2\tan\left(\frac{x}{3}\right)$ (c) $-\frac{2}{3}\tan\left(\frac{3x}{2}\right)$

15. (a) $-\csc x$ (b) $\frac{1}{5}\csc(5x)$ (c) $2\csc\left(\frac{\pi x}{2}\right)$

17. $\int (x+1)\,dx = \frac{x^2}{2} + x + C$

19. $\int \left(3t^2 + \frac{t}{2}\right) dt = t^3 + \frac{t^2}{4} + C$

21. $\int (2x^3 - 5x + 7)\,dx = \frac{1}{2}x^4 - \frac{5}{2}x^2 + 7x + C$

23. $\int \left(\frac{1}{x^2} - x^2 - \frac{1}{3}\right) dx = \int \left(x^{-2} - x^2 - \frac{1}{3}\right) dx = \frac{x^{-1}}{-1} - \frac{x^3}{3} - \frac{1}{3}x + C = -\frac{1}{x} - \frac{x^3}{3} - \frac{x}{3} + C$

25. $\int x^{-1/3}\,dx = \frac{x^{2/3}}{\frac{2}{3}} + C = \frac{3}{2}x^{2/3} + C$

27. $\int \left(\sqrt{x} + \sqrt[3]{x} \right) dx = \int \left(x^{1/2} + x^{1/3} \right) dx = \frac{x^{3/2}}{\frac{3}{2}} + \frac{x^{4/3}}{\frac{4}{3}} + C = \frac{2}{3} x^{3/2} + \frac{3}{4} x^{4/3} + C$

29. $\int \left(8y - \frac{2}{y^{1/4}} \right) dy = \int \left(8y - 2y^{-1/4} \right) dy = \frac{8y^2}{2} - 2 \left(\frac{y^{3/4}}{\frac{3}{4}} \right) + C = 4y^2 - \frac{8}{3} y^{3/4} + C$

31. $\int 2x \left(1 - x^{-3} \right) dx = \int \left(2x - 2x^{-2} \right) dx = \frac{2x^2}{2} - 2 \left(\frac{x^{-1}}{-1} \right) + C = x^2 + \frac{2}{x} + C$

33. $\int \frac{t\sqrt{t} + \sqrt{t}}{t^2} dt = \int \left(\frac{t^{3/2}}{t^2} + \frac{t^{1/2}}{t^2} \right) dt = \int \left(t^{-1/2} + t^{-3/2} \right) dt = \frac{t^{1/2}}{\frac{1}{2}} + \left(\frac{t^{-1/2}}{-\frac{1}{2}} \right) + C = 2\sqrt{t} - \frac{2}{\sqrt{t}} + C$

35. $\int -2 \cos t \, dt = -2 \sin t + C$

37. $\int 7 \sin \frac{\theta}{3} \, d\theta = -21 \cos \frac{\theta}{3} + C$

39. $\int -3 \csc^2 x \, dx = 3 \cot x + C$

41. $\int \frac{\csc \theta \cot \theta}{2} \, d\theta = -\frac{1}{2} \csc \theta + C$

43. $\int \left(4 \sec x \tan x - 2 \sec^2 x \right) dx = 4 \sec x - 2 \tan x + C$

45. $\int \left(\sin 2x - \csc^2 x \right) dx = -\frac{1}{2} \cos 2x + \cot x + C$

47. $\int \frac{1 + \cos 4t}{2} \, dt = \int \left(\frac{1}{2} + \frac{1}{2} \cos 4t \right) dt = \frac{1}{2} t + \frac{1}{2} \left(\frac{\sin 4t}{4} \right) + C = \frac{t}{2} + \frac{\sin 4t}{8} + C$

49. $\int \left(1 + \tan^2 \theta \right) d\theta = \int \sec^2 \theta \, d\theta = \tan \theta + C$

51. $\int \cot^2 x \, dx = \int \left(\csc^2 x - 1 \right) dx = -\cot x - x + C$

53. $\int \cos \theta \left(\tan \theta + \sec \theta \right) d\theta = \int \left(\sin \theta + 1 \right) d\theta = -\cos \theta + \theta + C$

55. $\frac{d}{dx} \left(\frac{(7x - 2)^4}{28} + C \right) = \frac{4(7x - 2)^3(7)}{28} = (7x - 2)^3$

57. $\frac{d}{dx} \left(\frac{1}{5} \tan (5x - 1) + C \right) = \frac{1}{5} \left(\sec^2 (5x - 1) \right) (5) = \sec^2 (5x - 1)$

59. $\frac{d}{dx} \left(\frac{-1}{x+1} + C \right) = (-1)(-1)(x + 1)^{-2} = \frac{1}{(x + 1)^2}$

61. (a) Wrong: $\frac{d}{dx} \left(\frac{x^2}{2} \sin x + C \right) = \frac{2x}{2} \sin x + \frac{x^2}{2} \cos x = x \sin x + \frac{x^2}{2} \cos x \neq x \sin x$

    (b) Wrong: $\frac{d}{dx} (-x \cos x + C) = -\cos x + x \sin x \neq x \sin x$

    (c) Right: $\frac{d}{dx} (-x \cos x + \sin x + C) = -\cos x + x \sin x + \cos x = x \sin x$

63. (a) Wrong: $\frac{d}{dx} \left(\frac{(2x + 1)^3}{3} + C \right) = \frac{3(2x + 1)^2(2)}{3} = 2(2x + 1)^2 \neq (2x + 1)^2$

    (b) Wrong: $\frac{d}{dx} \left((2x + 1)^3 + C \right) = 3(2x + 1)^2(2) = 6(2x + 1)^2 \neq 3(2x + 1)^2$

    (c) Right: $\frac{d}{dx} \left((2x + 1)^3 + C \right) = 6(2x + 1)^2$

65. Graph (b), because $\frac{dy}{dx} = 2x \Rightarrow y = x^2 + C$. Then $y(1) = 4 \Rightarrow C = 3$.

67. $\frac{dy}{dx} = 2x - 7 \Rightarrow y = x^2 - 7x + C$; at $x = 2$ and $y = 0$ we have $0 = 2^2 - 7(2) + C \Rightarrow C = 10 \Rightarrow y = x^2 - 7x + 10$

69. $\frac{dy}{dx} = \frac{1}{x^2} + x = x^{-2} + x \Rightarrow y = -x^{-1} + \frac{x^2}{2} + C$; at $x = 2$ and $y = 1$ we have $1 = -2^{-1} + \frac{2^2}{2} + C \Rightarrow C = -\frac{1}{2}$
$\Rightarrow y = -x^{-1} + \frac{x^2}{2} - \frac{1}{2}$ or $y = -\frac{1}{x} + \frac{x^2}{2} - \frac{1}{2}$

71. $\frac{dy}{dx} = 3x^{-2/3} \Rightarrow y = \frac{3x^{1/3}}{\frac{1}{3}} + C = 9$; at $x = 9x^{1/3} + C$; at $x = -1$ and $y = -5$ we have $-5 = 9(-1)^{1/3} + C \Rightarrow C = 4$
$\Rightarrow y = 9x^{1/3} + 4$

73. $\frac{ds}{dt} = 1 + \cos t \Rightarrow s = t + \sin t + C$; at $t = 0$ and $s = 4$ we have $4 = 0 + \sin 0 + C \Rightarrow C = 4 \Rightarrow s = t + \sin t + 4$

75. $\frac{dr}{d\theta} = -\pi \sin \pi\theta \Rightarrow r = \cos(\pi\theta) + C$; at $r = 0$ and $\theta = 0$ we have $0 = \cos(\pi 0) + C \Rightarrow C = -1 \Rightarrow r = \cos(\pi\theta) - 1$

77. $\frac{dv}{dt} = \frac{1}{2} \sec t \tan t \Rightarrow v = \frac{1}{2} \sec t + C$; at $v = 1$ and $t = 0$ we have $1 = \frac{1}{2} \sec(0) + C \Rightarrow C = \frac{1}{2} \Rightarrow v = \frac{1}{2} \sec t + \frac{1}{2}$

79. $\frac{d^2y}{dx^2} = 2 - 6x \Rightarrow \frac{dy}{dx} = 2x - 3x^2 + C_1$; at $\frac{dy}{dx} = 4$ and $x = 0$ we have $4 = 2(0) - 3(0)^2 + C_1 \Rightarrow C_1 = 4$
$\Rightarrow \frac{dy}{dx} = 2x - 3x^2 + 4 \Rightarrow y = x^2 - x^3 + 4x + C_2$; at $y = 1$ and $x = 0$ we have $1 = 0^2 - 0^3 + 4(0) + C_2 \Rightarrow C_2 = 1$
$\Rightarrow y = x^2 - x^3 + 4x + 1$

81. $\frac{d^2r}{dt^2} = \frac{2}{t^3} = 2t^{-3} \Rightarrow \frac{dr}{dt} = -t^{-2} + C_1$; at $\frac{dr}{dt} = 1$ and $t = 1$ we have $1 = -(1)^{-2} + C_1 \Rightarrow C_1 = 2 \Rightarrow \frac{dr}{dt} = -t^{-2} + 2$
$\Rightarrow r = t^{-1} + 2t + C_2$; at $r = 1$ and $t = 1$ we have $1 = 1^{-1} + 2(1) + C_2 \Rightarrow C_2 = -2 \Rightarrow r = t^{-1} + 2t - 2$ or
$r = \frac{1}{t} + 2t - 2$

83. $\frac{d^3y}{dx^3} = 6 \Rightarrow \frac{d^2y}{dx^2} = 6x + C_1$; at $\frac{d^2y}{dx^2} = -8$ and $x = 0$ we have $-8 = 6(0) + C_1 \Rightarrow C_1 = -8 \Rightarrow \frac{d^2y}{dx^2} = 6x - 8$
$\Rightarrow \frac{dy}{dx} = 3x^2 - 8x + C_2$; at $\frac{dy}{dx} = 0$ and $x = 0$ we have $0 = 3(0)^2 - 8(0) + C_2 \Rightarrow C_2 = 0 \Rightarrow \frac{dy}{dx} = 3x^2 - 8x$
$\Rightarrow y = x^3 - 4x^2 + C_3$; at $y = 5$ and $x = 0$ we have $5 = 0^3 - 4(0)^2 + C_3 \Rightarrow C_3 = 5 \Rightarrow y = x^3 - 4x^2 + 5$

85. $y^{(4)} = -\sin t + \cos t \Rightarrow y''' = \cos t + \sin t + C_1$; at $y''' = 7$ and $t = 0$ we have $7 = \cos(0) + \sin(0) + C_1$
$\Rightarrow C_1 = 6 \Rightarrow y''' = \cos t + \sin t + 6 \Rightarrow y'' = \sin t - \cos t + 6t + C_2$; at $y'' = -1$ and $t = 0$ we have
$-1 = \sin(0) - \cos(0) + 6(0) + C_2 \Rightarrow C_2 = 0 \Rightarrow y'' = \sin t - \cos t + 6t \Rightarrow y' = -\cos t - \sin t + 3t^2 + C_3$;
at $y' = -1$ and $t = 0$ we have $-1 = -\cos(0) - \sin(0) + 3(0)^2 + C_3 \Rightarrow C_3 = 0 \Rightarrow y' = -\cos t - \sin t + 3t^2$
$\Rightarrow y = -\sin t + \cos t + t^3 + C_4$; at $y = 0$ and $t = 0$ we have $0 = -\sin(0) + \cos(0) + 0^3 + C_4 \Rightarrow C_4 = -1$
$\Rightarrow y = -\sin t + \cos t + t^3 - 1$

87. $m = y' = 3\sqrt{x} = 3x^{1/2} \Rightarrow y = 2x^{3/2} + C$; at $(9, 4)$ we have $4 = 2(9)^{3/2} + C \Rightarrow C = -50 \Rightarrow y = 2x^{3/2} - 50$

89. $\frac{dy}{dx} = 1 - \frac{4}{3}x^{1/3} \Rightarrow y = \int \left(1 - \frac{4}{3}x^{1/3}\right) dx = x - x^{4/3} + C$; at $(1, 0.5)$ on the curve we have $0.5 = 1 - 1^{4/3} + C$
$\Rightarrow C = 0.5 \Rightarrow y = x - x^{4/3} + \frac{1}{2}$

91. $\frac{dy}{dx} = \sin x - \cos x \Rightarrow y = \int (\sin x - \cos x) dx = -\cos x - \sin x + C$; at $(-\pi, -1)$ on the curve we have
$-1 = -\cos(-\pi) - \sin(-\pi) + C \Rightarrow C = -2 \Rightarrow y = -\cos x - \sin x - 2$

93. (a) $\frac{ds}{dt} = 9.8t - 3 \Rightarrow s = 4.9t^2 - 3t + C$; (i) at $s = 5$ and $t = 0$ we have $C = 5 \Rightarrow s = 4.9t^2 - 3t + 5$;
displacement $= s(3) - s(1) = ((4.9)(9) - 9 + 5) - (4.9 - 3 + 5) = 33.2$ units; (ii) at $s = -2$ and $t = 0$ we have
$C = -2 \Rightarrow s = 4.9t^2 - 3t - 2$; displacement $= s(3) - s(1) = ((4.9)(9) - 9 - 2) - (4.9 - 3 - 2) = 33.2$ units;

(iii) at $s = s_0$ and $t = 0$ we have $C = s_0 \Rightarrow s = 4.9t^2 - 3t + s_0$; displacement $= s(3) - s(1)$
$= ((4.9)(9) - 9 + s_0) - (4.9 - 3 + s_0) = 33.2$ units

(b) True. Given an antiderivative f(t) of the velocity function, we know that the body's position function is
$s = f(t) + C$ for some constant C. Therefore, the displacement from $t = a$ to $t = b$ is $(f(b) + C) - (f(a) + C)$
$= f(b) - f(a)$. Thus we can find the displacement from any antiderivative f as the numerical difference
$f(b) - f(a)$ without knowing the exact values of C and s.

95. Step 1: $\frac{d^2s}{dt^2} = -k \Rightarrow \frac{ds}{dt} = -kt + C_1$; at $\frac{ds}{dt} = 88$ and $t = 0$ we have $C_1 = 88 \Rightarrow \frac{ds}{dt} = -kt + 88 \Rightarrow$
$s = -k\left(\frac{t^2}{2}\right) + 88t + C_2$; at $s = 0$ and $t = 0$ we have $C_2 = 0 \Rightarrow s = -\frac{kt^2}{2} + 88t$

Step 2: $\frac{ds}{dt} = 0 \Rightarrow 0 = -kt + 88 \Rightarrow t = \frac{88}{k}$

Step 3: $242 = \frac{-k\left(\frac{88}{k}\right)^2}{2} + 88\left(\frac{88}{k}\right) \Rightarrow 242 = -\frac{(88)^2}{2k} + \frac{(88)^2}{k} \Rightarrow 242 = \frac{(88)^2}{2k} \Rightarrow k = 16$

97. (a) $v = \int a \, dt = \int \left(15t^{1/2} - 3t^{-1/2}\right) dt = 10t^{3/2} - 6t^{1/2} + C$; $\frac{ds}{dt}(1) = 4 \Rightarrow 4 = 10(1)^{3/2} - 6(1)^{1/2} + C \Rightarrow C = 0$
$\Rightarrow v = 10t^{3/2} - 6t^{1/2}$

(b) $s = \int v \, dt = \int \left(10t^{3/2} - 6t^{1/2}\right) dt = 4t^{5/2} - 4t^{3/2} + C$; $s(1) = 0 \Rightarrow 0 = 4(1)^{5/2} - 4(1)^{3/2} + C \Rightarrow C = 0$
$\Rightarrow s = 4t^{5/2} - 4t^{3/2}$

99. $\frac{d^2s}{dt^2} = a \Rightarrow \frac{ds}{dt} = \int a \, dt = at + C$; $\frac{ds}{dt} = v_0$ when $t = 0 \Rightarrow C = v_0 \Rightarrow \frac{ds}{dt} = at + v_0 \Rightarrow s = \frac{at^2}{2} + v_0 t + C_1$; $s = s_0$
when $t = 0 \Rightarrow s_0 = \frac{a(0)^2}{2} + v_0(0) + C_1 \Rightarrow C_1 = s_0 \Rightarrow s = \frac{at^2}{2} + v_0 t + s_0$

101. (a) $\int f(x) \, dx = 1 - \sqrt{x} + C_1 = -\sqrt{x} + C$ (b) $\int g(x) \, dx = x + 2 + C_1 = x + C$

(c) $\int -f(x) \, dx = -\left(1 - \sqrt{x}\right) + C_1 = \sqrt{x} + C$ (d) $\int -g(x) \, dx = -(x + 2) + C_1 = -x + C$

(e) $\int [f(x) + g(x)] \, dx = \left(1 - \sqrt{x}\right) + (x + 2) + C_1 = x - \sqrt{x} + C$

(f) $\int [f(x) - g(x)] \, dx = \left(1 - \sqrt{x}\right) - (x + 2) + C_1 = -x - \sqrt{x} + C$

CHAPTER 4 PRACTICE EXERCISES

1. No, since $f(x) = x^3 + 2x + \tan x \Rightarrow f'(x) = 3x^2 + 2 + \sec^2 x > 0 \Rightarrow f(x)$ is always increasing on its domain

3. No absolute minimum because $\lim\limits_{x \to \infty} (7 + x)(11 - 3x)^{1/3} = -\infty$. Next $f'(x) =$
$(11 - 3x)^{1/3} - (7 + x)(11 - 3x)^{-2/3} = \frac{(11 - 3x) - (7 + x)}{(11 - 3x)^{2/3}} = \frac{4(1 - x)}{(11 - 3x)^{2/3}} \Rightarrow x = 1$ and $x = \frac{11}{3}$ are critical points.
Since $f' > 0$ if $x < 1$ and $f' < 0$ if $x > 1$, $f(1) = 16$ is the absolute maximum.

5. Yes, because at each point of $[0, 1)$ except $x = 0$, the function's value is a local minimum value as well as a
local maximum value. At $x = 0$ the function's value, 0, is not a local minimum value because each open
interval around $x = 0$ on the x-axis contains points to the left of 0 where f equals -1.

7. No, because the interval $0 < x < 1$ fails to be closed. The Extreme Value Theorem says that if the function is
continuous throughout a finite closed interval $a \le x \le b$ then the existence of absolute extrema is guaranteed on
that interval.

9. (a) There appear to be local minima at x = -1.75
and 1.8. Points of inflection are indicated at
approximately x = 0 and x = ± 1.

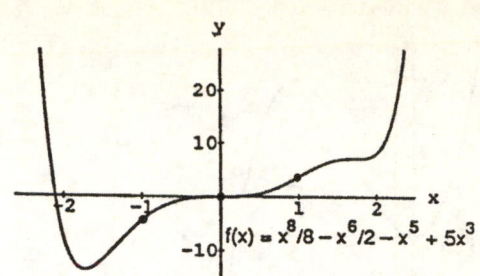

(b) $f'(x) = x^7 - 3x^5 - 5x^4 + 15x^2 = x^2(x^2 - 3)(x^3 - 5)$. The pattern $y' = \; --- \; | \; +++ \; | \; +++ \; | \; --- \; | \; +++$
$\qquad\qquad\qquad\qquad\qquad\qquad\qquad\qquad\qquad -\sqrt{3} \quad\; 0 \quad\; \sqrt[3]{5} \quad\; \sqrt{3}$

indicates a local maximum at x = $\sqrt[3]{5}$ and local minima at x = $\pm\sqrt{3}$.

(c)

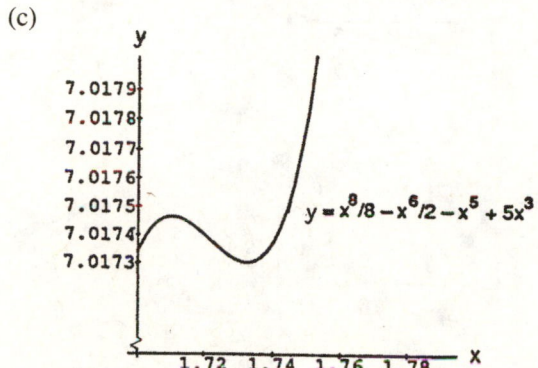

11. (a) $g(t) = \sin^2 t - 3t \Rightarrow g'(t) = 2\sin t \cos t - 3 = \sin(2t) - 3 \Rightarrow g' < 0 \Rightarrow g(t)$ is always falling and hence must
decrease on every interval in its domain.

(b) One, since $\sin^2 t - 3t - 5 = 0$ and $\sin^2 t - 3t = 5$ have the same solutions: $f(t) = \sin^2 t - 3t - 5$ has the same
derivative as g(t) in part (a) and is always decreasing with $f(-3) > 0$ and $f(0) < 0$. The Intermediate Value
Theorem guarantees the continuous function f has a root in $[-3, 0]$.

13. (a) $f(x) = x^4 + 2x^2 - 2 \Rightarrow f'(x) = 4x^3 + 4x$. Since $f(0) = -2 < 0$, $f(1) = 1 > 0$ and $f'(x) \geq 0$ for $0 \leq x \leq 1$, we
may conclude from the Intermediate Value Theorem that f(x) has exactly one solution when $0 \leq x \leq 1$.

(b) $x^2 = \frac{-2 \pm \sqrt{4+8}}{2} > 0 \Rightarrow x^2 = \sqrt{3} - 1$ and $x \geq 0 \Rightarrow x \approx \sqrt{.7320508076} \approx .8555996772$

15. Let V(t) represent the volume of the water in the reservoir at time t, in minutes, let $V(0) = a_0$ be the initial
amount and $V(1440) = a_0 + (1400)(43,560)(7.48)$ gallons be the amount of water contained in the reservoir
after the rain, where 24 hr = 1440 min. Assume that V(t) is continuous on $[0, 1440]$ and differentiable on
$(0, 1440)$. The Mean Value Theorem says that for some t_0 in $(0, 1440)$ we have $V'(t_0) = \frac{V(1440) - V(0)}{1440 - 0}$
$= \frac{a_0 + (1400)(43,560)(7.48) - a_0}{1440} = \frac{456,160,320 \text{ gal}}{1440 \text{ min}} = 316,778$ gal/min. Therefore at t_0 the reservoir's volume
was increasing at a rate in excess of 225,000 gal/min.

17. No, $\frac{x}{x+1} = 1 + \frac{-1}{x+1} \Rightarrow \frac{x}{x+1}$ differs from $\frac{-1}{x+1}$ by the constant 1. Both functions have the same derivative
$\frac{d}{dx}\left(\frac{x}{x+1}\right) = \frac{(x+1) - x(1)}{(x+1)^2} = \frac{1}{(x+1)^2} = \frac{d}{dx}\left(\frac{-1}{x+1}\right)$.

19. The global minimum value of $\frac{1}{2}$ occurs at x = 2.

21. (a) t = 0, 6, 12   (b) t = 3, 9   (c) $6 < t < 12$ (d) $0 < t < 6$, $12 < t < 14$

23.

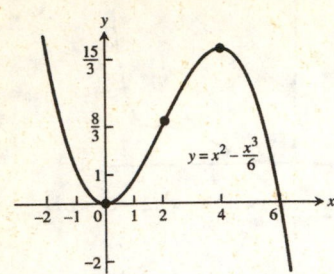

25.

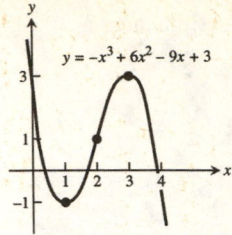

27.

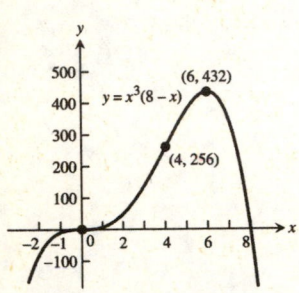

29.

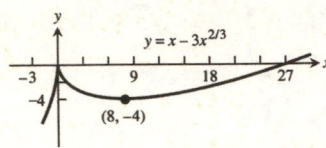

31.

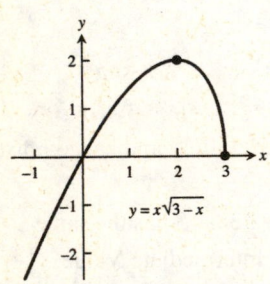

33. (a) $y' = 16 - x^2 \Rightarrow y' = --- \mid +++ \mid --- \Rightarrow$ the curve is rising on $(-4, 4)$, falling on $(-\infty, -4)$ and $(4, \infty)$
$\qquad\qquad\qquad\qquad\qquad\quad -4 \qquad 4$

$\Rightarrow$ a local maximum at $x = 4$ and a local minimum at $x = -4$; $y'' = -2x \Rightarrow y'' = +++ \mid --- \Rightarrow$ the curve
$\qquad\qquad\qquad\qquad\qquad\qquad\qquad\qquad\qquad\qquad\qquad\qquad\qquad\qquad\qquad\quad 0$

is concave up on $(-\infty, 0)$, concave down on $(0, \infty) \Rightarrow$ a point of inflection at $x = 0$

(b)

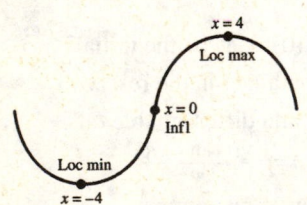

35. (a) $y' = 6x(x + 1)(x - 2) = 6x^3 - 6x^2 - 12x \Rightarrow y' = --- \mid +++ \mid --- \mid +++ \Rightarrow$ the graph is rising on $(-1, 0)$
$\qquad\qquad\qquad\qquad\qquad\qquad\qquad\qquad\qquad\qquad\qquad\qquad\quad -1 \qquad 0 \qquad 2$

and $(2, \infty)$, falling on $(-\infty, -1)$ and $(0, 2) \Rightarrow$ a local maximum at $x = 0$, local minima at $x = -1$ and

$x = 2$; $y'' = 18x^2 - 12x - 12 = 6(3x^2 - 2x - 2) = 6\left(x - \frac{1 - \sqrt{7}}{3}\right)\left(x - \frac{1 + \sqrt{7}}{3}\right) \Rightarrow$

$y'' = +++ \mid --- \mid +++ \Rightarrow$ the curve is concave up on $\left(-\infty, \frac{1 - \sqrt{7}}{3}\right)$ and $\left(\frac{1 + \sqrt{7}}{3}, \infty\right)$, concave down
$\qquad\quad \frac{1 - \sqrt{7}}{3} \qquad \frac{1 + \sqrt{7}}{3}$

on $\left(\frac{1 - \sqrt{7}}{3}, \frac{1 + \sqrt{7}}{3}\right) \Rightarrow$ points of inflection at $x = \frac{1 \pm \sqrt{7}}{3}$

(b)

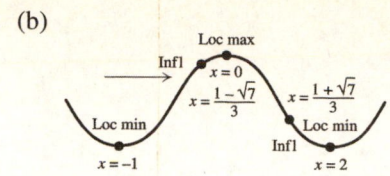

37. (a) $y' = x^4 - 2x^2 = x^2(x^2 - 2) \Rightarrow y' = {+}{+}{+} \underset{-\sqrt{2}}{|} \; {-}{-}{-} \underset{0}{|} {-}{-}{-} \underset{\sqrt{2}}{|} \; {+}{+}{+} \Rightarrow$ the curve is rising on $\left(-\infty, -\sqrt{2}\right)$ and

$\left(\sqrt{2}, \infty\right)$, falling on $\left(-\sqrt{2}, \sqrt{2}\right) \Rightarrow$ a local maximum at $x = -\sqrt{2}$ and a local minimum at $x = \sqrt{2}$;

$y'' = 4x^3 - 4x = 4x(x-1)(x+1) \Rightarrow y'' = {-}{-}{-} \underset{-1}{|} {+}{+}{+} \underset{0}{|} {-}{-}{-} \underset{1}{|} {+}{+}{+} \Rightarrow$ concave up on $(-1, 0)$ and $(1, \infty)$,

concave down on $(-\infty, -1)$ and $(0, 1) \Rightarrow$ points of inflection at $x = 0$ and $x = \pm 1$

(b)

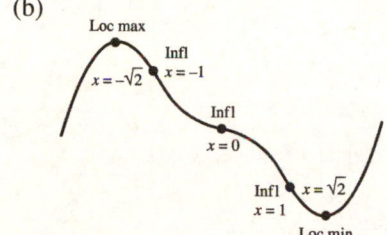

39. The values of the first derivative indicate that the curve is rising on $(0, \infty)$ and falling on $(-\infty, 0)$. The slope of the curve approaches $-\infty$ as $x \to 0^-$, and approaches ∞ as $x \to 0^+$ and $x \to 1$. The curve should therefore have a cusp and local minimum at $x = 0$, and a vertical tangent at $x = 1$.

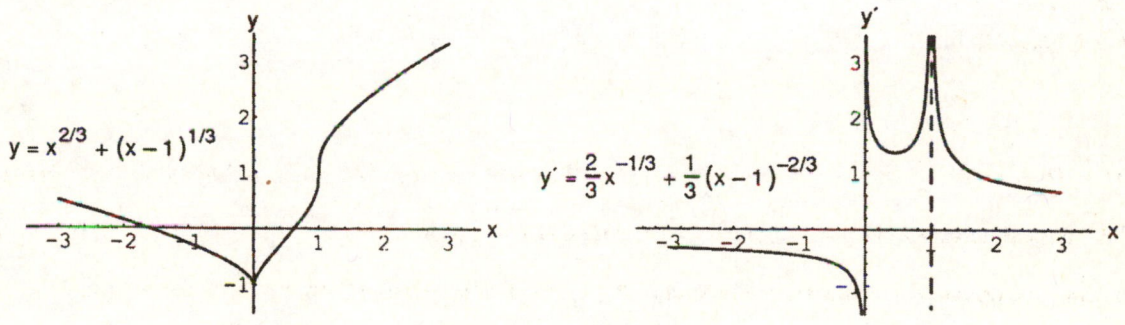

41. The values of the first derivative indicate that the curve is always rising. The slope of the curve approaches ∞ as $x \to 0$ and as $x \to 1$, indicating vertical tangents at both $x = 0$ and $x = 1$.

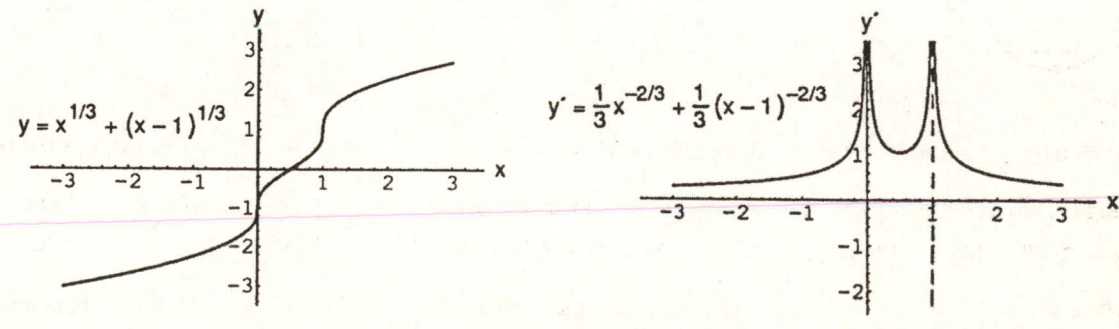

43. $y = \frac{x+1}{x-3} = 1 + \frac{4}{x-3}$

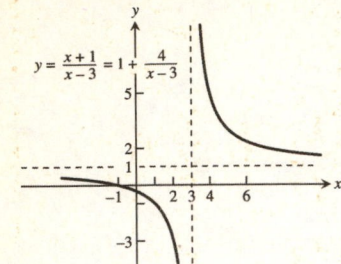

45. $y = \frac{x^2+1}{x} = x + \frac{1}{x}$

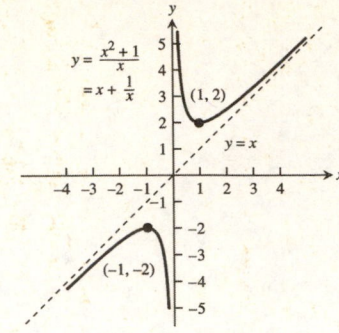

47. $y = \frac{x^3+2}{2x} = \frac{x^2}{2} + \frac{1}{x}$

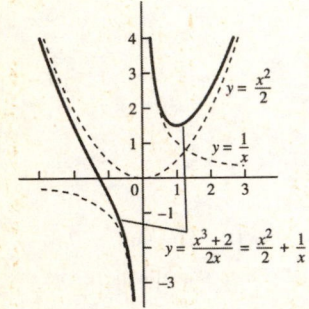

49. $y = \frac{x^2-4}{x^2-3} = 1 - \frac{1}{x^2-3}$

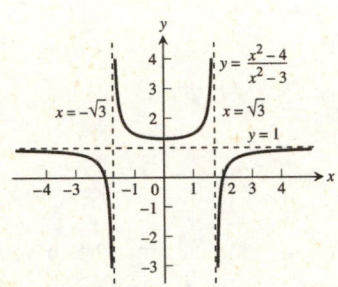

51. $\lim\limits_{x \to 1} \frac{x^2 + 3x - 4}{x - 1} = \lim\limits_{x \to 1} \frac{2x + 3}{1} = 5$

53. $\lim\limits_{x \to \pi} \frac{\tan x}{x} = \frac{\tan \pi}{\pi} = 0$

55. $\lim\limits_{x \to 0} \frac{\sin^2 x}{\tan(x^2)} = \lim\limits_{x \to 0} \frac{2\sin x \cdot \cos x}{2x \sec^2(x^2)} = \lim\limits_{x \to 0} \frac{\sin(2x)}{2x \sec^2(x^2)} = \lim\limits_{x \to 0} \frac{2\cos(2x)}{2x\,(2\sec^2(x^2)\tan(x^2)\cdot 2x) + 2\sec^2(x^2)} = \frac{2}{0 + 2 \cdot 1} = 1$

57. $\lim\limits_{x \to \pi/2^-} \sec(7x)\cos(3x) = \lim\limits_{x \to \pi/2^-} \frac{\cos(3x)}{\cos(7x)} = \lim\limits_{x \to \pi/2^-} \frac{-3\sin(3x)}{-7\sin(7x)} = \frac{3}{7}$

59. $\lim\limits_{x \to 0} (\csc x - \cot x) = \lim\limits_{x \to 0} \frac{1 - \cos x}{\sin x} = \lim\limits_{x \to 0} \frac{\sin x}{\cos x} = \frac{0}{1} = 0$

61. $\lim\limits_{x \to \infty} \left(\sqrt{x^2 + x + 1} - \sqrt{x^2 - x} \right) = \lim\limits_{x \to \infty} \left(\sqrt{x^2 + x + 1} - \sqrt{x^2 - x} \right) \cdot \frac{\sqrt{x^2 + x + 1} + \sqrt{x^2 - x}}{\sqrt{x^2 + x + 1} + \sqrt{x^2 - x}}$

$= \lim\limits_{x \to \infty} \frac{2x + 1}{\sqrt{x^2 + x + 1} + \sqrt{x^2 - x}}.$

Notice that $x = \sqrt{x^2}$ for $x > 0$ so this is equivalent to

$= \lim\limits_{x \to \infty} \frac{\frac{2x+1}{x}}{\sqrt{\frac{x^2+x+1}{x^2}} + \sqrt{\frac{x^2-x}{x^2}}} = \lim\limits_{x \to \infty} \frac{2 + \frac{1}{x}}{\sqrt{1 + \frac{1}{x} + \frac{1}{x^2}} + \sqrt{1 - \frac{1}{x}}} = \frac{2}{\sqrt{1} + \sqrt{1}} = 1$

63. (a) Maximize $f(x) = \sqrt{x} - \sqrt{36 - x} = x^{1/2} - (36 - x)^{1/2}$ where $0 \le x \le 36$

$\Rightarrow f'(x) = \frac{1}{2}x^{-1/2} - \frac{1}{2}(36 - x)^{-1/2}(-1) = \frac{\sqrt{36-x} + \sqrt{x}}{2\sqrt{x}\sqrt{36-x}} \Rightarrow$ derivative fails to exist at 0 and 36; $f(0) = -6$,

and $f(36) = 6 \Rightarrow$ the numbers are 0 and 36

(b) Maximize $g(x) = \sqrt{x} + \sqrt{36 - x} = x^{1/2} + (36 - x)^{1/2}$ where $0 \le x \le 36$

$\Rightarrow g'(x) = \frac{1}{2}x^{-1/2} + \frac{1}{2}(36 - x)^{-1/2}(-1) = \frac{\sqrt{36-x} - \sqrt{x}}{2\sqrt{x}\sqrt{36-x}} \Rightarrow$ critical points at 0, 18 and 36; $g(0) = 6$,

$g(18) = 2\sqrt{18} = 6\sqrt{2}$ and $g(36) = 6 \Rightarrow$ the numbers are 18 and 18

65. $A(x) = \frac{1}{2}(2x)(27 - x^2)$ for $0 \le x \le \sqrt{27}$

$\Rightarrow A'(x) = 3(3 + x)(3 - x)$ and $A''(x) = -6x$.

The critical points are -3 and 3, but -3 is not in the

domain. Since $A''(3) = -18 < 0$ and $A\left(\sqrt{27}\right) = 0$,

the maximum occurs at $x = 3 \Rightarrow$ the largest area is

$A(3) = 54$ sq units.

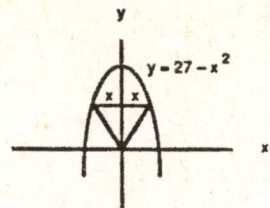

67. From the diagram we have $\left(\frac{h}{2}\right)^2 + r^2 = \left(\sqrt{3}\right)^2$

$\Rightarrow r^2 = \frac{12 - h^2}{4}$. The volume of the cylinder is

$V = \pi r^2 h = \pi \left(\frac{12 - h^2}{4}\right) h = \frac{\pi}{4}(12h - h^3)$, where

$0 \le h \le 2\sqrt{3}$. Then $V'(h) = \frac{3\pi}{4}(2 + h)(2 - h)$

$\Rightarrow$ the critical points are -2 and 2, but -2 is not in

the domain. At $h = 2$ there is a maximum since

$V''(2) = -3\pi < 0$. The dimensions of the largest

cylinder are radius $= \sqrt{2}$ and height $= 2$.

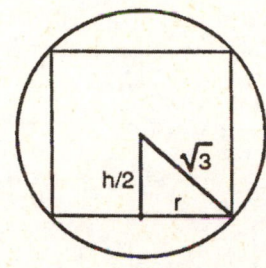

69. The profit $P = 2px + py = 2px + p\left(\frac{40 - 10x}{5 - x}\right)$, where p is the profit on grade B tires and $0 \le x \le 4$. Thus

$P'(x) = \frac{2p}{(5-x)^2}(x^2 - 10x + 20) \Rightarrow$ the critical points are $\left(5 - \sqrt{5}\right)$, 5, and $\left(5 + \sqrt{5}\right)$, but only $\left(5 - \sqrt{5}\right)$ is in

the domain. Now $P'(x) > 0$ for $0 < x < \left(5 - \sqrt{5}\right)$ and $P'(x) < 0$ for $\left(5 - \sqrt{5}\right) < x < 4 \Rightarrow$ at $x = \left(5 - \sqrt{5}\right)$ there

is a local maximum. Also $P(0) = 8p$, $P\left(5 - \sqrt{5}\right) = 4p\left(5 - \sqrt{5}\right) \approx 11p$, and $P(4) = 8p \Rightarrow$ at $x = \left(5 - \sqrt{5}\right)$ there

is an absolute maximum. The maximum occurs when $x = \left(5 - \sqrt{5}\right)$ and $y = 2\left(5 - \sqrt{5}\right)$, the units are

hundreds of tires, i.e., $x \approx 276$ tires and $y \approx 553$ tires.

71. The dimensions will be x in. by $10 - 2x$ in. by $16 - 2x$ in., so $V(x) = x(10 - 2x)(16 - 2x) = 4x^3 - 52x^2 + 160x$ for

$0 < x < 5$. Then $V'(x) = 12x^2 - 104x + 160 = 4(x - 2)(3x - 20)$, so the critical point in the correct domain is $x = 2$.

This critical point corresponds to the maximum possible volume because $V'(x) > 0$ for $0 < x < 2$ and $V'(x) < 0$ for

$2 < x < 5$. The box of largest volume has a height of 2 in. and a base measuring 6 in. by 12 in., and its volume is 144 in.³

Graphical support:

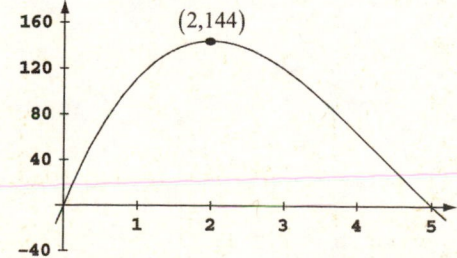

73. $g(x) = 3x - x^3 + 4 \Rightarrow g(2) = 2 > 0$ and $g(3) = -14 < 0 \Rightarrow g(x) = 0$ in the interval $[2, 3]$ by the Intermediate

Value Theorem. Then $g'(x) = 3 - 3x^2 \Rightarrow x_{n+1} = x_n - \frac{3x_n - x_n^3 + 4}{3 - 3x_n^2}$; $x_0 = 2 \Rightarrow x_1 = 2.2\overline{2} \Rightarrow x_2 = 2.196215$, and

so forth to $x_5 = 2.195823345$.

75. $\int (x^3 + 5x - 7)\,dx = \frac{x^4}{4} + \frac{5x^2}{2} - 7x + C$

77. $\int \left(3\sqrt{t} + \frac{4}{t^2}\right) dt = \int \left(3t^{1/2} + 4t^{-2}\right) dt = \frac{3t^{3/2}}{\left(\frac{3}{2}\right)} + \frac{4t^{-1}}{-1} + C = 2t^{3/2} - \frac{4}{t} + C$

79. Let $u = r + 5 \Rightarrow du = dr$

$\int \frac{dr}{(r+5)^2} = \int \frac{du}{u^2} = \int u^{-2}\,du = \frac{u^{-1}}{-1} + C = -u^{-1} + C = -\frac{1}{(r+5)} + C$

81. Let $u = \theta^2 + 1 \Rightarrow du = 2\theta\,d\theta \Rightarrow \frac{1}{2}\,du = \theta\,d\theta$

$\int 3\theta\sqrt{\theta^2 + 1}\,d\theta = \int \sqrt{u}\left(\frac{3}{2}\,du\right) = \frac{3}{2}\int u^{1/2}\,du = \frac{3}{2}\left(\frac{u^{3/2}}{\frac{3}{2}}\right) + C = u^{3/2} + C = \left(\theta^2 + 1\right)^{3/2} + C$

83. Let $u = 1 + x^4 \Rightarrow du = 4x^3\,dx \Rightarrow \frac{1}{4}\,du = x^3\,dx$

$\int x^3 \left(1 + x^4\right)^{-1/4} dx = \int u^{-1/4}\left(\frac{1}{4}\,du\right) = \frac{1}{4}\int u^{-1/4}\,du = \frac{1}{4}\left(\frac{u^{3/4}}{\frac{3}{4}}\right) + C = \frac{1}{3}u^{3/4} + C = \frac{1}{3}\left(1 + x^4\right)^{3/4} + C$

85. Let $u = \frac{s}{10} \Rightarrow du = \frac{1}{10}\,ds \Rightarrow 10\,du = ds$

$\int \sec^2 \frac{s}{10}\,ds = \int \left(\sec^2 u\right)(10\,du) = 10\int \sec^2 u\,du = 10\tan u + C = 10\tan \frac{s}{10} + C$

87. Let $u = \sqrt{2}\,\theta \Rightarrow du = \sqrt{2}\,d\theta \Rightarrow \frac{1}{\sqrt{2}}\,du = d\theta$

$\int \csc \sqrt{2}\theta \cot \sqrt{2}\theta\,d\theta = \int (\csc u \cot u)\left(\frac{1}{\sqrt{2}}\,du\right) = \frac{1}{\sqrt{2}}(-\csc u) + C = -\frac{1}{\sqrt{2}}\csc \sqrt{2}\theta + C$

89. Let $u = \frac{x}{4} \Rightarrow du = \frac{1}{4}\,dx \Rightarrow 4\,du = dx$

$\int \sin^2 \frac{x}{4}\,dx = \int (\sin^2 u)(4\,du) = \int 4\left(\frac{1 - \cos 2u}{2}\right) du = 2\int (1 - \cos 2u)\,du = 2\left(u - \frac{\sin 2u}{2}\right) + C$

$= 2u - \sin 2u + C = 2\left(\frac{x}{4}\right) - \sin 2\left(\frac{x}{4}\right) + C = \frac{x}{2} - \sin \frac{x}{2} + C$

91. $y = \int \frac{x^2 + 1}{x^2}\,dx = \int \left(1 + x^{-2}\right) dx = x - x^{-1} + C = x - \frac{1}{x} + C;\ y = -1$ when $x = 1 \Rightarrow 1 - \frac{1}{1} + C = -1$

$\Rightarrow C = -1 \Rightarrow y = x - \frac{1}{x} - 1$

93. $\frac{dr}{dt} = \int \left(15\sqrt{t} + \frac{3}{\sqrt{t}}\right) dt = \int \left(15t^{1/2} + 3t^{-1/2}\right) dt = 10t^{3/2} + 6t^{1/2} + C;\ \frac{dr}{dt} = 8$ when $t = 1$

$\Rightarrow 10(1)^{3/2} + 6(1)^{1/2} + C = 8 \Rightarrow C = -8.$ Thus $\frac{dr}{dt} = 10t^{3/2} + 6t^{1/2} - 8 \Rightarrow r = \int \left(10t^{3/2} + 6t^{1/2} - 8\right) dt$

$= 4t^{5/2} + 4t^{3/2} - 8t + C;\ r = 0$ when $t = 1 \Rightarrow 4(1)^{5/2} + 4(1)^{3/2} - 8(1) + C_1 = 0 \Rightarrow C_1 = 0.$ Therefore,

$r = 4t^{5/2} + 4t^{3/2} - 8t$

CHAPTER 4 ADDITIONAL AND ADVANCED EXERCISES

1. If M and m are the maximum and minimum values, respectively, then $m \leq f(x) \leq M$ for all $x \in I$. If $m = M$ then f is constant on I.

3. On an open interval the extreme values of a continuous function (if any) must occur at an interior critical point. On a half-open interval the extreme values of a continuous function may be at a critical point or at the closed endpoint. Extreme values occur only where $f' = 0$, f' does not exist, or at the endpoints of the interval. Thus the extreme points will not be at the ends of an open interval.

5. (a) If $y' = 6(x + 1)(x - 2)^2$, then $y' < 0$ for $x < -1$ and $y' > 0$ for $x > -1$. The sign pattern is
$f' = --- \mid +++ \mid +++ \Rightarrow$ f has a local minimum at $x = -1$. Also $y'' = 6(x - 2)^2 + 12(x + 1)(x - 2)$
$\quad \quad \quad \quad \quad _{-1} \quad \quad _{2}$
$= 6(x - 2)(3x) \Rightarrow y'' > 0$ for $x < 0$ or $x > 2$, while $y'' < 0$ for $0 < x < 2$. Therefore f has points of inflection
at $x = 0$ and $x = 2$. There is no local maximum.

(b) If $y' = 6x(x + 1)(x - 2)$, then $y' < 0$ for $x < -1$ and $0 < x < 2$; $y' > 0$ for $-1 < x < 0$ and $x > 2$. The sign
sign pattern is $y' = --- \mid +++ \mid --- \mid +++$. Therefore f has a local maximum at $x = 0$ and
$\quad \quad \quad \quad \quad \quad \quad \quad _{-1} \quad \quad _{0} \quad \quad _{2}$
local minima at $x = -1$ and $x = 2$. Also, $y'' = 18 \left[x - \left(\frac{1 - \sqrt{7}}{3} \right) \right] \left[x - \left(\frac{1 + \sqrt{7}}{3} \right) \right]$, so $y'' < 0$ for
$\frac{1 - \sqrt{7}}{3} < x < \frac{1 + \sqrt{7}}{3}$ and $y'' > 0$ for all other $x \Rightarrow$ f has points of inflection at $x = \frac{1 \pm \sqrt{7}}{3}$.

7. If f is continuous on $[a, c)$ and $f'(x) \le 0$ on $[a, c)$, then by the Mean Value Theorem for all $x \in [a, c)$ we have
$\frac{f(c) - f(x)}{c - x} \le 0 \Rightarrow f(c) - f(x) \le 0 \Rightarrow f(x) \ge f(c)$. Also if f is continuous on $(c, b]$ and $f'(x) \ge 0$ on $(c, b]$, then for
all $x \in (c, b]$ we have $\frac{f(x) - f(c)}{x - c} \ge 0 \Rightarrow f(x) - f(c) \ge 0 \Rightarrow f(x) \ge f(c)$. Therefore $f(x) \ge f(c)$ for all $x \in [a, b]$.

9. No. Corollary 1 requires that $f'(x) = 0$ for <u>all</u> x in some interval I, not $f'(x) = 0$ at a single point in I.

11. From (ii), $f(-1) = \frac{-1 + a}{b - c + 2} = 0 \Rightarrow a = 1$; from (iii), either $1 = \lim_{x \to \infty} f(x)$ or $1 = \lim_{x \to -\infty} f(x)$. In either case,
$\lim_{x \to \pm\infty} f(x) = \lim_{x \to \pm\infty} \frac{x + 1}{bx^2 + cx + 2} = \lim_{x \to \pm\infty} \frac{1 + \frac{1}{x}}{bx + c + \frac{2}{x}} = 1 \Rightarrow b = 0$ and $c = 1$. For if $b = 1$, then
$\lim_{x \to \pm\infty} \frac{1 + \frac{1}{x}}{x + c + \frac{2}{x}} = 0$ and if $c = 0$, then $\lim_{x \to \pm\infty} \frac{1 + \frac{1}{x}}{bx + \frac{2}{x}} = \lim_{x \to \pm\infty} \frac{1 + \frac{1}{x}}{\frac{2}{x}} = \pm\infty$. Thus $a = 1$, $b = 0$, and $c = 1$.

13. The area of the $\triangle ABC$ is $A(x) = \frac{1}{2}(2)\sqrt{1 - x^2} = (1 - x^2)^{1/2}$,
where $0 \le x \le 1$. Thus $A'(x) = \frac{-x}{\sqrt{1 - x^2}} \Rightarrow 0$ and ± 1 are
critical points. Also $A(\pm 1) = 0$ so $A(0) = 1$ is the
maximum. When $x = 0$ the $\triangle ABC$ is isosceles since
$AC = BC = \sqrt{2}$.

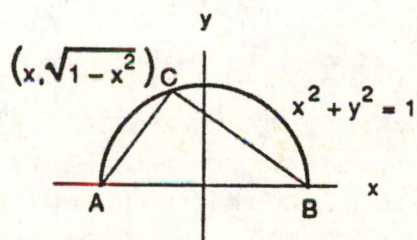

15. The time it would take the water to hit the ground from height y is $\sqrt{\frac{2y}{g}}$, where g is the acceleration of
gravity. The product of time and exit velocity (rate) yields the distance the water travels:
$D(y) = \sqrt{\frac{2y}{g}}\sqrt{64(h - y)} = 8\sqrt{\frac{2}{g}}(hy - y^2)^{1/2}, 0 \le y \le h \Rightarrow D'(y) = -4\sqrt{\frac{2}{g}}(hy - y^2)^{-1/2}(h - 2y) \Rightarrow 0, \frac{h}{2}$ and h
are critical points. Now $D(0) = 0$, $D\left(\frac{h}{2}\right) = \frac{8h}{\sqrt{g}}$ and $D(h) = 0 \Rightarrow$ the best place to drill the hole is at $y = \frac{h}{2}$.

17. The surface area of the cylinder is $S = 2\pi r^2 + 2\pi rh$. From
the diagram we have $\frac{r}{R} = \frac{H - h}{H} \Rightarrow h = \frac{RH - rH}{R}$ and
$S(r) = 2\pi r(r + h) = 2\pi r\left(r + H - r\frac{H}{R}\right)$
$= 2\pi\left(1 - \frac{H}{R}\right)r^2 + 2\pi Hr$, where $0 \le r \le R$.
Case 1: $H < R \Rightarrow S(r)$ is a quadratic equation containing
    the origin and concave upward $\Rightarrow S(r)$ is maximum at
    $r = R$.
Case 2: $H = R \Rightarrow S(r)$ is a linear equation containing the
    origin with a positive slope $\Rightarrow S(r)$ is maximum at
    $r = R$.

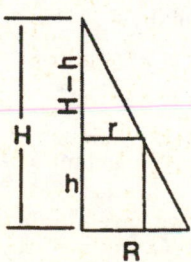

Case 3: $H > R \Rightarrow S(r)$ is a quadratic equation containing the origin and concave downward. Then

$\frac{dS}{dr} = 4\pi\left(1 - \frac{H}{R}\right)r + 2\pi H$ and $\frac{dS}{dr} = 0 \Rightarrow 4\pi\left(1 - \frac{H}{R}\right)r + 2\pi H = 0 \Rightarrow r = \frac{RH}{2(H-R)}$. For simplification

we let $r^* = \frac{RH}{2(H-R)}$.

(a) If $R < H < 2R$, then $0 > H - 2R \Rightarrow H > 2(H - R) \Rightarrow \frac{RH}{2(H-R)} > R$ which is impossible.

(b) If $H = 2R$, then $r^* = \frac{2R^2}{2R} = R \Rightarrow S(r)$ is maximum at $r = R$.

(c) If $H > 2R$, then $2R + H < 2H \Rightarrow H < 2(H - R) \Rightarrow \frac{H}{2(H-R)} < 1 \Rightarrow \frac{RH}{2(H-R)} < R \Rightarrow r^* < R$. Therefore,

$S(r)$ is a maximum at $r = r^* = \frac{RH}{2(H-R)}$.

Conclusion: If $H \in (0, R]$ or $H = 2R$, then the maximum surface area is at $r = R$. If $H \in (R, 2R)$, then $r > R$

which is not possible. If $H \in (2R, \infty)$, then the maximum is at $r = r^* = \frac{RH}{2(H-R)}$.

19. (a) $\lim\limits_{x \to 0} \frac{2\sin(5x)}{3x} = \lim\limits_{x \to 0} \frac{2\sin(5x)}{\frac{3}{5}(5x)} = \lim\limits_{x \to 0} \frac{10}{3}\frac{\sin(5x)}{(5x)} = \frac{10}{3} \cdot 1 = \frac{10}{3}$

(b) $\lim\limits_{x \to 0} \sin(5x)\cot(3x) = \lim\limits_{x \to 0} \frac{\sin(5x)\cos(3x)}{\sin(3x)} = \lim\limits_{x \to 0} \frac{-3\sin(5x)\sin(3x) + 5\cos(5x)\cos(3x)}{3\cos(3x)} = \frac{5}{3}$

(c) $\lim\limits_{x \to 0} x\csc^2\sqrt{2x} = \lim\limits_{x \to 0} \frac{x}{\sin^2\sqrt{2x}} = \lim\limits_{x \to 0} \frac{1}{\frac{2\sin\sqrt{2x}\cos\sqrt{2x}}{\sqrt{2x}}} = \lim\limits_{x \to 0} \frac{\sqrt{2x}}{\sin(2\sqrt{2x})} = \lim\limits_{x \to 0} \frac{\frac{1}{\sqrt{2x}}}{\cos(2\sqrt{2x})\frac{2}{\sqrt{2x}}}$

$= \lim\limits_{x \to 0} \frac{1}{\cos(2\sqrt{2x})\cdot 2} = \frac{1}{2}$

(d) $\lim\limits_{x \to \pi/2} (\sec x - \tan x) = \lim\limits_{x \to \pi/2} \frac{1 - \sin x}{\cos x} = \lim\limits_{x \to \pi/2} \frac{-\cos x}{-\sin x} = 0.$

(e) $\lim\limits_{x \to 0} \frac{x - \sin x}{x - \tan x} = \lim\limits_{x \to 0} \frac{1 - \cos x}{1 - \sec^2 x} = \lim\limits_{x \to 0} \frac{1 - \cos x}{-\tan^2 x} = \lim\limits_{x \to 0} \frac{\cos x - 1}{\tan^2 x} = \lim\limits_{x \to 0} \frac{-\sin x}{2\tan x \sec^2 x} = \lim\limits_{x \to 0} \frac{-\sin x}{\frac{2\sin x}{\cos^3 x}} =$

$= \lim\limits_{x \to 0} \frac{\cos^3 x}{-2} = -\frac{1}{2}$

(f) $\lim\limits_{x \to 0} \frac{\sin(x^2)}{x\sin x} = \lim\limits_{x \to 0} \frac{2x\cos(x^2)}{x\cos x + \sin x} = \lim\limits_{x \to 0} \frac{-(2x^2)\sin(x^2) + 2\cos(x^2)}{-x\sin x + 2\cos x} = \frac{2}{2} = 1$

(g) $\lim\limits_{x \to 0} \frac{\sec x - 1}{x^2} = \lim\limits_{x \to 0} \frac{\sec x \tan x}{2x} = \lim\limits_{x \to 0} \frac{\sec^3 x + \tan^2 x \sec x}{2} = \frac{1 + 0}{2} = \frac{1}{2}$

(h) $\lim\limits_{x \to 2} \frac{x^3 - 8}{x^2 - 4} = \lim\limits_{x \to 2} \frac{(x - 2)(x^2 + 2x + 4)}{(x - 2)(x + 2)} = \lim\limits_{x \to 2} \frac{x^2 + 2x + 4}{x + 2} = \frac{4 + 4 + 4}{4} = 3$

21. (a) The profit function is $P(x) = (c - ex)x - (a + bx) = -ex^2 + (c - b)x - a$. $P'(x) = -2ex + c - b = 0$

$\Rightarrow x = \frac{c-b}{2e}$. $P''(x) = -2e < 0$ if $e > 0$ so that the profit function is maximized at $x = \frac{c-b}{2e}$.

(b) The price therefore that corresponds to a production level yeilding a maximum profit is

$p\Big|_{x=\frac{c-b}{2e}} = c - e\left(\frac{c-b}{2e}\right) = \frac{c+b}{2}$ dollars.

(c) The weekly profit at this production level is $P(x) = -e\left(\frac{c-b}{2e}\right)^2 + (c - b)\left(\frac{c-b}{2e}\right) - a = \frac{(c-b)^2}{4e} - a.$

(d) The tax increases cost to the new profit function is $F(x) = (c - ex)x - (a + bx + tx) = -ex^2 + (c - b - t)x - a$.

Now $F'(x) = -2ex + c - b - t = 0$ when $x = \frac{t+b-c}{-2e} = \frac{c-b-t}{2e}$. Since $F''(x) = -2e < 0$ if $e > 0$, F is maximized

when $x = \frac{c-b-t}{2e}$ units per week. Thus the price per unit is $p = c - e\left(\frac{c-b-t}{2e}\right) = \frac{c+b+t}{2}$ dollars. Thus, such a tax

increases the cost per unit by $\frac{c+b+t}{2} - \frac{c+b}{2} = \frac{t}{2}$ dollars if units are priced to maximize profit.

23. $x_1 = x_0 - \frac{f(x_0)}{f'(x_0)} = x_0 - \frac{x_0^q - a}{qx_0^{q-1}} = \frac{qx_0^q - x_0^q - a}{qx_0^{q-1}} = \frac{x_0^q(q-1) - a}{qx_0^{q-1}} = x_0\left(\frac{q-1}{q}\right) + \frac{a}{x_0^{q-1}}\left(\frac{1}{q}\right)$ so that x_1 is a weighted average of x_0

and $\frac{a}{x_0^{q-1}}$ with weights $m_0 = \frac{q-1}{q}$ and $m_1 = \frac{1}{q}$.

In the case where $x_0 = \frac{a}{x_0^{q-1}}$ we have $x_0^q = a$ and $x_1 = \frac{a}{x_0^{q-1}}\left(\frac{q-1}{q}\right) + \frac{a}{x_0^{q-1}}\left(\frac{1}{q}\right) = \frac{a}{x_0^{q-1}}\left(\frac{q-1}{q} + \frac{1}{q}\right) = \frac{a}{x_0^{q-1}}.$

25. (a) $a(t) = s''(t) = -k$ $(k > 0) \Rightarrow s'(t) = -kt + C_1$, where $s'(0) = 88 \Rightarrow C_1 = 88 \Rightarrow s'(t) = -kt + 88$. So

$s(t) = \frac{-kt^2}{2} + 88t + C_2$ where $s(0) = 0 \Rightarrow C_2 = 0$ so $s(t) = \frac{-kt^2}{2} + 88t$. Now $s(t) = 100$ when

$\frac{-kt^2}{2} + 88t = 100$. Solving for t we obtain $t = \frac{88 \pm \sqrt{88^2 - 200k}}{k}$. At such t we want $s'(t) = 0$, thus

$-k\left(\frac{88+\sqrt{88^2-200k}}{k}\right)+88=0$ or $-k\left(\frac{88-\sqrt{88^2-200k}}{k}\right)+88=0$. In either case we obtain $88^2-200k=0$

so that $k=\frac{88^2}{200}\approx 38.72$ ft/sec^2.

(b) The initial condition that $s'(0)=44$ ft/sec implies that $s'(t)=-kt+44$ and $s(t)=\frac{-kt^2}{2}+44t$ where k is as above.

The car is stopped at a time t such that $s'(t)=-kt+44=0\Rightarrow t=\frac{44}{k}$. At this time the car has traveled a distance

$s\left(\frac{44}{k}\right)=\frac{-k}{2}\left(\frac{44}{k}\right)^2+44\left(\frac{44}{k}\right)=\frac{44^2}{2k}=\frac{968}{k}=968\left(\frac{200}{88^2}\right)=25$ feet. Thus halving the initial velocity quarters

stopping distance.

27. Yes. The curve $y=x$ satisfies all three conditions since $\frac{dy}{dx}=1$ everywhere, when $x=0$, $y=0$, and $\frac{d^2y}{dx^2}=0$ everywhere.

29. $s''(t)=a=-t^2\Rightarrow v=s'(t)=\frac{-t^3}{3}+C$. We seek $v_0=s'(0)=C$. We know that $s(t^*)=b$ for some t^* and s is at a

maximum for this t^*. Since $s(t)=\frac{-t^4}{12}+Ct+k$ and $s(0)=0$ we have that $s(t)=\frac{-t^4}{12}+Ct$ and also $s'(t^*)=0$ so that

$t^*=(3C)^{1/3}$. So $\frac{[-(3C)^{1/3}]^4}{12}+C(3C)^{1/3}=b\Rightarrow(3C)^{1/3}\left(C-\frac{3C}{12}\right)=b\Rightarrow(3C)^{1/3}\left(\frac{3C}{4}\right)=b\Rightarrow 3^{1/3}C^{4/3}=\frac{4b}{3}$

$\Rightarrow C=\frac{(4b)^{3/4}}{3}$. Thus $v_0=s'(0)=\frac{(4b)^{3/4}}{3}=\frac{2\sqrt{2}}{3}b^{3/4}$.

31. The graph of $f(x)=ax^2+bx+c$ with $a>0$ is a parabola opening upwards. Thus $f(x)\geq 0$ for all x if $f(x)=0$ for at most

one real value of x. The solutions to $f(x)=0$ are, by the quadratic equation $\frac{-2b\pm\sqrt{(2b)^2-4ac}}{2a}$. Thus we require

$(2b)^2-4ac\leq 0\Rightarrow b^2-ac\leq 0$.

NOTES:

CHAPTER 5 INTEGRATION

5.1 ESTIMATING WITH FINITE SUMS

1. $f(x) = x^2$

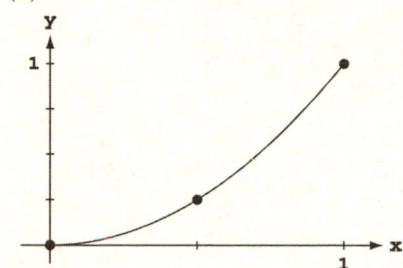

Since f is increasing on $[0, 1]$, we use left endpoints to obtain lower sums and right endpoints to obtain upper sums.

(a) $\triangle x = \frac{1-0}{2} = \frac{1}{2}$ and $x_i = i\triangle x = \frac{i}{2} \Rightarrow$ a lower sum is $\sum_{i=0}^{1} \left(\frac{i}{2}\right)^2 \cdot \frac{1}{2} = \frac{1}{2}\left(0^2 + \left(\frac{1}{2}\right)^2\right) = \frac{1}{8}$

(b) $\triangle x = \frac{1-0}{4} = \frac{1}{4}$ and $x_i = i\triangle x = \frac{i}{4} \Rightarrow$ a lower sum is $\sum_{i=0}^{3} \left(\frac{i}{4}\right)^2 \cdot \frac{1}{4} = \frac{1}{4}\left(0^2 + \left(\frac{1}{4}\right)^2 + \left(\frac{1}{2}\right)^2 + \left(\frac{3}{4}\right)^2\right) = \frac{1}{4} \cdot \frac{7}{8} = \frac{7}{32}$

(c) $\triangle x = \frac{1-0}{2} = \frac{1}{2}$ and $x_i = i\triangle x = \frac{i}{2} \Rightarrow$ an upper sum is $\sum_{i=1}^{2} \left(\frac{i}{2}\right)^2 \cdot \frac{1}{2} = \frac{1}{2}\left(\left(\frac{1}{2}\right)^2 + 1^2\right) = \frac{5}{8}$

(d) $\triangle x = \frac{1-0}{4} = \frac{1}{4}$ and $x_i = i\triangle x = \frac{i}{4} \Rightarrow$ an upper sum is $\sum_{i=1}^{4} \left(\frac{i}{4}\right)^2 \cdot \frac{1}{4} = \frac{1}{4}\left(\left(\frac{1}{4}\right)^2 + \left(\frac{1}{2}\right)^2 + \left(\frac{3}{4}\right)^2 + 1^2\right) = \frac{1}{4} \cdot \left(\frac{30}{16}\right) = \frac{15}{32}$

3. $f(x) = \frac{1}{x}$

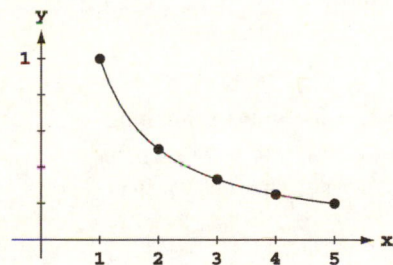

Since f is decreasing on $[0, 1]$, we use left endpoints to obtain upper sums and right endpoints to obtain lower sums.

(a) $\triangle x = \frac{5-1}{2} = 2$ and $x_i = 1 + i\triangle x = 1 + 2i \Rightarrow$ a lower sum is $\sum_{i=1}^{2} \frac{1}{x_i} \cdot 2 = 2\left(\frac{1}{3} + \frac{1}{5}\right) = \frac{16}{15}$

(b) $\triangle x = \frac{5-1}{4} = 1$ and $x_i = 1 + i\triangle x = 1 + i \Rightarrow$ a lower sum is $\sum_{i=1}^{4} \frac{1}{x_i} \cdot 1 = 1\left(\frac{1}{2} + \frac{1}{3} + \frac{1}{4} + \frac{1}{5}\right) = \frac{77}{60}$

(c) $\triangle x = \frac{5-1}{2} = 2$ and $x_i = 1 + i\triangle x = 1 + 2i \Rightarrow$ an upper sum is $\sum_{i=0}^{1} \frac{1}{x_i} \cdot 2 = 2\left(1 + \frac{1}{3}\right) = \frac{8}{3}$

(d) $\triangle x = \frac{5-1}{4} = 1$ and $x_i = 1 + i\triangle x = 1 + i \Rightarrow$ an upper sum is $\sum_{i=0}^{3} \frac{1}{x_i} \cdot 1 = 1\left(1 + \frac{1}{2} + \frac{1}{3} + \frac{1}{4}\right) = \frac{25}{12}$

5. $f(x) = x^2$

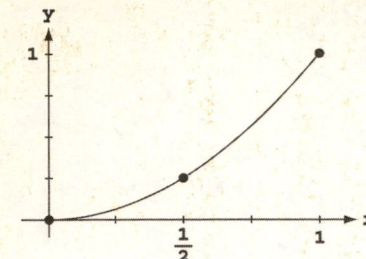

Using 2 rectangles $\Rightarrow \triangle x = \frac{1-0}{2} = \frac{1}{2} \Rightarrow \frac{1}{2}\left(f\left(\frac{1}{4}\right) + f\left(\frac{3}{4}\right)\right)$

$= \frac{1}{2}\left(\left(\frac{1}{4}\right)^2 + \left(\frac{3}{4}\right)^2\right) = \frac{10}{32} = \frac{5}{16}$

Using 4 rectangles $\Rightarrow \triangle x = \frac{1-0}{4} = \frac{1}{4}$

$\Rightarrow \frac{1}{4}\left(f\left(\frac{1}{8}\right) + f\left(\frac{3}{8}\right) + f\left(\frac{5}{8}\right) + f\left(\frac{7}{8}\right)\right)$

$= \frac{1}{4}\left(\left(\frac{1}{8}\right)^2 + \left(\frac{3}{8}\right)^2 + \left(\frac{5}{8}\right)^2 + \left(\frac{7}{8}\right)^2\right) = \frac{21}{64}$

7. $f(x) = \frac{1}{x}$

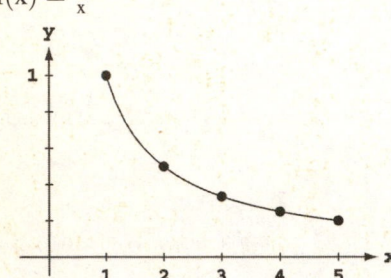

Using 2 rectangles $\Rightarrow \triangle x = \frac{5-1}{2} = 2 \Rightarrow 2(f(2) + f(4))$

$= 2\left(\frac{1}{2} + \frac{1}{4}\right) = \frac{3}{2}$

Using 4 rectangles $\Rightarrow \triangle x = \frac{5-1}{4} = 1$

$\Rightarrow 1\left(f\left(\frac{3}{2}\right) + f\left(\frac{5}{2}\right) + f\left(\frac{7}{2}\right) + f\left(\frac{9}{2}\right)\right)$

$= 1\left(\frac{2}{3} + \frac{2}{5} + \frac{2}{7} + \frac{2}{9}\right) = \frac{1488}{3 \cdot 5 \cdot 7 \cdot 9} = \frac{496}{5 \cdot 7 \cdot 9} = \frac{496}{315}$

9. (a) $D \approx (0)(1) + (12)(1) + (22)(1) + (10)(1) + (5)(1) + (13)(1) + (11)(1) + (6)(1) + (2)(1) + (6)(1) = 87$ inches
   (b) $D \approx (12)(1) + (22)(1) + (10)(1) + (5)(1) + (13)(1) + (11)(1) + (6)(1) + (2)(1) + (6)(1) + (0)(1) = 87$ inches

11. (a) $D \approx (0)(10) + (44)(10) + (15)(10) + (35)(10) + (30)(10) + (44)(10) + (35)(10) + (15)(10) + (22)(10)$
       $+ (35)(10) + (44)(10) + (30)(10) = 3490$ feet ≈ 0.66 miles
    (b) $D \approx (44)(10) + (15)(10) + (35)(10) + (30)(10) + (44)(10) + (35)(10) + (15)(10) + (22)(10) + (35)(10)$
       $+ (44)(10) + (30)(10) + (35)(10) = 3840$ feet ≈ 0.73 miles

13. (a) Because the acceleration is decreasing, an upper estimate is obtained using left end-points in summing
       acceleration $\cdot \Delta t$. Thus, $\Delta t = 1$ and speed $\approx [32.00 + 19.41 + 11.77 + 7.14 + 4.33](1) = 74.65$ ft/sec
    (b) Using right end-points we obtain a lower estimate: speed $\approx [19.41 + 11.77 + 7.14 + 4.33 + 2.63](1)$
       $= 45.28$ ft/sec
    (c) Upper estimates for the speed at each second are:

t	0	1	2	3	4	5
v	0	32.00	51.41	63.18	70.32	74.65

Thus, the distance fallen when t = 3 seconds is s $\approx [32.00 + 51.41 + 63.18](1) = 146.59$ ft.

15. Partition $[0, 2]$ into the four subintervals $[0, 0.5]$, $[0.5, 1]$, $[1, 1.5]$, and $[1.5, 2]$. The midpoints of these
    subintervals are $m_1 = 0.25$, $m_2 = 0.75$, $m_3 = 1.25$, and $m_4 = 1.75$. The heights of the four approximating
    rectangles are $f(m_1) = (0.25)^3 = \frac{1}{64}$, $f(m_2) = (0.75)^3 = \frac{27}{64}$, $f(m_3) = (1.25)^3 = \frac{125}{64}$, and $f(m_4) = (1.75)^3 = \frac{343}{64}$
    Notice that the average value is approximated by $\frac{1}{2}\left[\left(\frac{1}{4}\right)^3\left(\frac{1}{2}\right) + \left(\frac{3}{4}\right)^3\left(\frac{1}{2}\right) + \left(\frac{5}{4}\right)^3\left(\frac{1}{2}\right) + \left(\frac{7}{4}\right)^3\left(\frac{1}{2}\right)\right] = \frac{31}{16}$

$= \frac{1}{\text{length of } [0,2]} \cdot \left[\begin{array}{c}\text{approximate area under}\\\text{curve } f(x) = x^3\end{array}\right]$. We use this observation in solving the next several exercises.

17. Partition $[0, 2]$ into the four subintervals $[0, 0.5]$, $[0.5, 1]$, $[1, 1.5]$, and $[1.5, 2]$. The midpoints of the subintervals
    are $m_1 = 0.25$, $m_2 = 0.75$, $m_3 = 1.25$, and $m_4 = 1.75$. The heights of the four approximating rectangles are

$f(m_1) = \frac{1}{2} + \sin^2 \frac{\pi}{4} = \frac{1}{2} + \frac{1}{2} = 1$, $f(m_2) = \frac{1}{2} + \sin^2 \frac{3\pi}{4} = \frac{1}{2} + \frac{1}{2} = 1$, $f(m_3) = \frac{1}{2} + \sin^2 \frac{5\pi}{4} = \frac{1}{2} + \left(-\frac{1}{\sqrt{2}}\right)^2$

$= \frac{1}{2} + \frac{1}{2} = 1$, and $f(m_4) = \frac{1}{2} + \sin^2 \frac{7\pi}{4} = \frac{1}{2} + \left(-\frac{1}{\sqrt{2}}\right)^2 = 1$. The width of each rectangle is $\Delta x = \frac{1}{2}$. Thus,

Area $\approx (1 + 1 + 1 + 1)\left(\frac{1}{2}\right) = 2 \Rightarrow$ average value $\approx \frac{\text{area}}{\text{length of }[0,2]} = \frac{2}{2} = 1.$

19. Since the leakage is increasing, an upper estimate uses right endpoints and a lower estimate uses left endpoints:

   (a) upper estimate $= (70)(1) + (97)(1) + (136)(1) + (190)(1) + (265)(1) = 758$ gal,
       lower estimate $= (50)(1) + (70)(1) + (97)(1) + (136)(1) + (190)(1) = 543$ gal.

   (b) upper estimate $= (70 + 97 + 136 + 190 + 265 + 369 + 516 + 720) = 2363$ gal,
       lower estimate $= (50 + 70 + 97 + 136 + 190 + 265 + 369 + 516) = 1693$ gal.

   (c) worst case: $2363 + 720t = 25{,}000 \Rightarrow t \approx 31.4$ hrs;
       best case: $1693 + 720t = 25{,}000 \Rightarrow t \approx 32.4$ hrs

21. (a) The diagonal of the square has length 2, so the side length is $\sqrt{2}$. Area $= \left(\sqrt{2}\right)^2 = 2$

   (b) Think of the octagon as a collection of 16 right triangles with a hypotenuse of length 1 and an acute angle measuring $\frac{2\pi}{16} = \frac{\pi}{8}$.
       Area $= 16\left(\frac{1}{2}\right)\left(\sin\frac{\pi}{8}\right)\left(\cos\frac{\pi}{8}\right) = 4\sin\frac{\pi}{4} = 2\sqrt{2} \approx 2.828$

   (c) Think of the 16-gon as a collection of 32 right triangles with a hypotenuse of length 1 and an acute angle measuring $\frac{2\pi}{32} = \frac{\pi}{16}$.
       Area $= 32\left(\frac{1}{2}\right)\left(\sin\frac{\pi}{16}\right)\left(\cos\frac{\pi}{16}\right) = 8\sin\frac{\pi}{8} = 2\sqrt{2} \approx 3.061$

   (d) Each area is less than the area of the circle, π. As n increases, the area approaches π.

5.2 SIGMA NOTATION AND LIMITS OF FINITE SUMS

1. $\displaystyle\sum_{k=1}^{2} \frac{6k}{k+1} = \frac{6(1)}{1+1} + \frac{6(2)}{2+1} = \frac{6}{2} + \frac{12}{3} = 7$

3. $\displaystyle\sum_{k=1}^{4} \cos k\pi = \cos(1\pi) + \cos(2\pi) + \cos(3\pi) + \cos(4\pi) = -1 + 1 - 1 + 1 = 0$

5. $\displaystyle\sum_{k=1}^{3} (-1)^{k+1} \sin\frac{\pi}{k} = (-1)^{1+1} \sin\frac{\pi}{1} + (-1)^{2+1} \sin\frac{\pi}{2} + (-1)^{3+1} \sin\frac{\pi}{3} = 0 - 1 + \frac{\sqrt{3}}{2} = \frac{\sqrt{3}-2}{2}$

7. (a) $\displaystyle\sum_{k=1}^{6} 2^{k-1} = 2^{1-1} + 2^{2-1} + 2^{3-1} + 2^{4-1} + 2^{5-1} + 2^{6-1} = 1 + 2 + 4 + 8 + 16 + 32$

   (b) $\displaystyle\sum_{k=0}^{5} 2^k = 2^0 + 2^1 + 2^2 + 2^3 + 2^4 + 2^5 = 1 + 2 + 4 + 8 + 16 + 32$

   (c) $\displaystyle\sum_{k=-1}^{4} 2^{k+1} = 2^{-1+1} + 2^{0+1} + 2^{1+1} + 2^{2+1} + 2^{3+1} + 2^{4+1} = 1 + 2 + 4 + 8 + 16 + 32$

   All of them represent $1 + 2 + 4 + 8 + 16 + 32$

9. (a) $\displaystyle\sum_{k=2}^{4} \frac{(-1)^{k-1}}{k-1} = \frac{(-1)^{2-1}}{2-1} + \frac{(-1)^{3-1}}{3-1} + \frac{(-1)^{4-1}}{4-1} = -1 + \frac{1}{2} - \frac{1}{3}$

   (b) $\displaystyle\sum_{k=0}^{2} \frac{(-1)^k}{k+1} = \frac{(-1)^0}{0+1} + \frac{(-1)^1}{1+1} + \frac{(-1)^2}{2+1} = 1 - \frac{1}{2} + \frac{1}{3}$

   (c) $\displaystyle\sum_{k=-1}^{1} \frac{(-1)^k}{k+2} = \frac{(-1)^{-1}}{-1+2} + \frac{(-1)^0}{0+2} + \frac{(-1)^1}{1+2} = -1 + \frac{1}{2} - \frac{1}{3}$

   (a) and (c) are equivalent; (b) is not equivalent to the other two.

11. $\displaystyle\sum_{k=1}^{6} k$

13. $\displaystyle\sum_{k=1}^{4} \frac{1}{2^k}$

15. $\displaystyle\sum_{k=1}^{5} (-1)^{k+1} \frac{1}{k}$

17. (a) $\displaystyle\sum_{k=1}^{n} 3a_k = 3 \sum_{k=1}^{n} a_k = 3(-5) = -15$

(b) $\displaystyle\sum_{k=1}^{n} \frac{b_k}{6} = \frac{1}{6} \sum_{k=1}^{n} b_k = \frac{1}{6}(6) = 1$

(c) $\displaystyle\sum_{k=1}^{n} (a_k + b_k) = \sum_{k=1}^{n} a_k + \sum_{k=1}^{n} b_k = -5 + 6 = 1$

(d) $\displaystyle\sum_{k=1}^{n} (a_k - b_k) = \sum_{k=1}^{n} a_k - \sum_{k=1}^{n} b_k = -5 - 6 = -11$

(e) $\displaystyle\sum_{k=1}^{n} (b_k - 2a_k) = \sum_{k=1}^{n} b_k - 2 \sum_{k=1}^{n} a_k = 6 - 2(-5) = 16$

19. (a) $\displaystyle\sum_{k=1}^{10} k = \frac{10(10+1)}{2} = 55$ 　　　　 (b) $\displaystyle\sum_{k=1}^{10} k^2 = \frac{10(10+1)(2(10)+1)}{6} = 385$

(c) $\displaystyle\sum_{k=1}^{10} k^3 = \left[\frac{10(10+1)}{2} \right]^2 = 55^2 = 3025$

21. $\displaystyle\sum_{k=1}^{7} -2k = -2 \sum_{k=1}^{7} k = -2 \left(\frac{7(7+1)}{2} \right) = -56$

23. $\displaystyle\sum_{k=1}^{6} (3 - k^2) = \sum_{k=1}^{6} 3 - \sum_{k=1}^{6} k^2 = 3(6) - \frac{6(6+1)(2(6)+1)}{6} = -73$

25. $\displaystyle\sum_{k=1}^{5} k(3k + 5) = \sum_{k=1}^{5} (3k^2 + 5k) = 3 \sum_{k=1}^{5} k^2 + 5 \sum_{k=1}^{5} k = 3 \left(\frac{5(5+1)(2(5)+1)}{6} \right) + 5 \left(\frac{5(5+1)}{2} \right) = 240$

27. $\displaystyle\sum_{k=1}^{5} \frac{k^3}{225} + \left(\sum_{k=1}^{5} k \right)^3 = \frac{1}{225} \sum_{k=1}^{5} k^3 + \left(\sum_{k=1}^{5} k \right)^3 = \frac{1}{225} \left(\frac{5(5+1)}{2} \right)^2 + \left(\frac{5(5+1)}{2} \right)^3 = 3376$

29. (a) 　　　　　　　　　　 (b) 　　　　　　　　　　 (c)

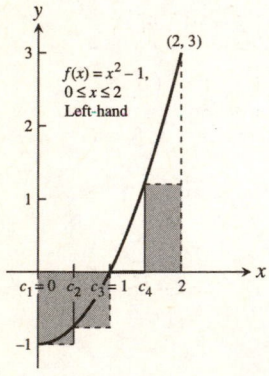 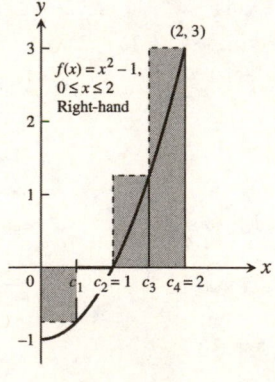 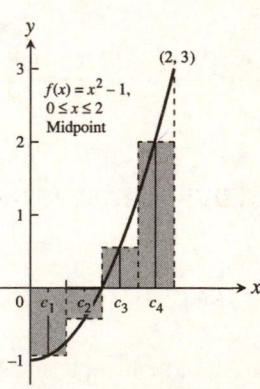

31. (a) 　　　　　　　　　　 (b) 　　　　　　　　　　 (c)

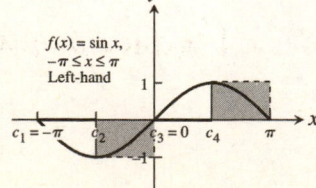 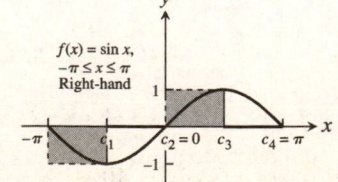 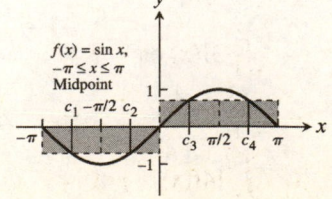

33. $|x_1 - x_0| = |1.2 - 0| = 1.2$, $|x_2 - x_1| = |1.5 - 1.2| = 0.3$, $|x_3 - x_2| = |2.3 - 1.5| = 0.8$, $|x_4 - x_3| = |2.6 - 2.3| = 0.3$, and $|x_5 - x_4| = |3 - 2.6| = 0.4$; the largest is $\|P\| = 1.2$.

35. $f(x) = 1 - x^2$

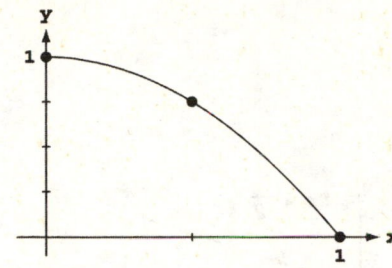

Since f is decreasing on $[0, 1]$ we use left endpoints to obtain upper sums. $\triangle x = \frac{1-0}{n} = \frac{1}{n}$ and $x_i = i\triangle x = \frac{i}{n}$. So an upper sum

is $\sum_{i=0}^{n-1}(1 - x_i^2)\frac{1}{n} = \frac{1}{n}\sum_{i=0}^{n-1}\left(1 - \left(\frac{i}{n}\right)^2\right) = \frac{1}{n^3}\sum_{i=0}^{n-1}(n^2 - i^2)$

$= \frac{n^3}{n^3} - \frac{1}{n^3}\sum_{i=0}^{n}i^2 = 1 - \frac{(n-1)n(2(n-1)+1)}{6n^3} = 1 - \frac{2n^3 - 3n^2 + n}{6n^3}$

$= 1 - \frac{2 - \frac{3}{n} + \frac{1}{n^2}}{6}$. Thus,

$\lim_{n\to\infty}\sum_{i=0}^{n-1}(1 - x_i^2)\frac{1}{n} = \lim_{n\to\infty}\left(1 - \frac{2 - \frac{3}{n} + \frac{1}{n^2}}{6}\right) = 1 - \frac{1}{3} = \frac{2}{3}$

37. $f(x) = x^2 + 1$

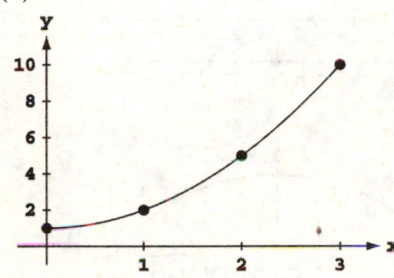

Since f is increasing on $[0, 3]$ we use right endpoints to obtain upper sums. $\triangle x = \frac{3-0}{n} = \frac{3}{n}$ and $x_i = i\triangle x = \frac{3i}{n}$. So an upper

sum is $\sum_{i=1}^{n}(x_i^2 + 1)\frac{3}{n} = \sum_{i=1}^{n}\left(\left(\frac{3i}{n}\right)^2 + 1\right)\frac{3}{n} = \frac{3}{n}\sum_{i=1}^{n}\left(\frac{9i^2}{n^2} + 1\right)$

$= \frac{27}{n}\sum_{i=1}^{n}i^2 + \frac{3}{n}\cdot n = \frac{27}{n^3}\left(\frac{n(n+1)(2n+1)}{6}\right) + 3$

$= \frac{9(2n^3 + 3n^2 + n)}{2n^3} + 3 = \frac{18 + \frac{27}{n} + \frac{9}{n^2}}{2} + 3$. Thus,

$\lim_{n\to\infty}\sum_{i=1}^{n}(x_i^2 + 1)\frac{3}{n} = \lim_{n\to\infty}\left(\frac{18 + \frac{27}{n} + \frac{9}{n^2}}{2} + 3\right) = 9 + 3 = 12$.

39. $f(x) = x + x^2 = x(1 + x)$

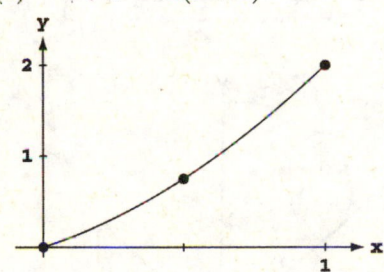

Since f is increasing on $[0, 1]$ we use right endpoints to obtain upper sums. $\triangle x = \frac{1-0}{n} = \frac{1}{n}$ and $x_i = i\triangle x = \frac{i}{n}$. So an upper sum

is $\sum_{i=1}^{n}(x_i + x_i^2)\frac{1}{n} = \sum_{i=1}^{n}\left(\frac{i}{n} + \left(\frac{i}{n}\right)^2\right)\frac{1}{n} = \frac{1}{n^2}\sum_{i=1}^{n}i + \frac{1}{n^3}\sum_{i=1}^{n}i^2$

$= \frac{1}{n^2}\left(\frac{n(n+1)}{2}\right) + \frac{1}{n^3}\left(\frac{n(n+1)(2n+1)}{6}\right) = \frac{n^2 + n}{2n^2} + \frac{2n^3 + 3n^2 + n}{6n^3}$

$= \frac{1 + \frac{1}{n}}{2} + \frac{2 + \frac{3}{n} + \frac{1}{n^2}}{6}$. Thus, $\lim_{n\to\infty}\sum_{i=1}^{n}(x_i + x_i^2)\frac{1}{n}$

$= \lim_{n\to\infty}\left[\left(\frac{1 + \frac{1}{n}}{2}\right) + \left(\frac{2 + \frac{3}{n} + \frac{1}{n^2}}{6}\right)\right] = \frac{1}{2} + \frac{2}{6} = \frac{5}{6}$.

5.3 THE DEFINITE INTEGRAL

1. $\int_0^2 x^2\, dx$

3. $\int_{-7}^5 (x^2 - 3x)\, dx$

5. $\int_2^3 \frac{1}{1-x}\, dx$

7. $\int_{-\pi/4}^0 (\sec x)\, dx$

9. (a) $\int_2^2 g(x)\, dx = 0$

(b) $\int_5^1 g(x)\, dx = -\int_1^5 g(x)\, dx = -8$

(c) $\int_1^2 3f(x)\, dx = 3\int_1^2 f(x)\, dx = 3(-4) = -12$

(d) $\int_2^5 f(x)\, dx = \int_1^5 f(x)\, dx - \int_1^2 f(x)\, dx = 6 - (-4) = 10$

(e) $\int_1^5 [f(x) - g(x)]\, dx = \int_1^5 f(x)\, dx - \int_1^5 g(x)\, dx = 6 - 8 = -2$

(f) $\int_1^5 [4f(x) - g(x)]\, dx = 4\int_1^5 f(x)\, dx - \int_1^5 g(x)\, dx = 4(6) - 8 = 16$

11. (a) $\int_1^2 f(u)\,du = \int_1^2 f(x)\,dx = 5$

(b) $\int_1^2 \sqrt{3}\,f(z)\,dz = \sqrt{3}\int_1^2 f(z)\,dz = 5\sqrt{3}$

(c) $\int_2^1 f(t)\,dt = -\int_1^2 f(t)\,dt = -5$

(d) $\int_1^2 [-f(x)]\,dx = -\int_1^2 f(x)\,dx = -5$

13. (a) $\int_3^4 f(z)\,dz = \int_0^4 f(z)\,dz - \int_0^3 f(z)\,dz = 7 - 3 = 4$

(b) $\int_4^3 f(t)\,dt = -\int_3^4 f(t)\,dt = -4$

15. The area of the trapezoid is $A = \frac{1}{2}(B + b)h$

$= \frac{1}{2}(5 + 2)(6) = 21 \Rightarrow \int_{-2}^4 \left(\frac{x}{2} + 3\right)\,dx$

$= 21$ square units

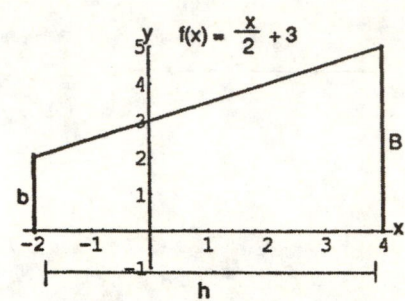

17. The area of the semicircle is $A = \frac{1}{2}\pi r^2 = \frac{1}{2}\pi(3)^2$

$= \frac{9}{2}\pi \Rightarrow \int_{-3}^3 \sqrt{9 - x^2}\,dx = \frac{9}{2}\pi$ square units

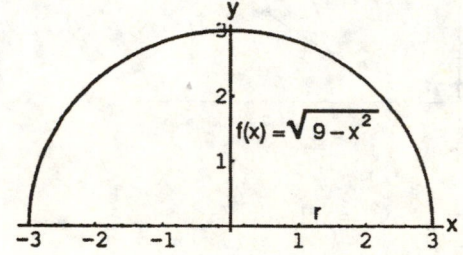

19. The area of the triangle on the left is $A = \frac{1}{2}bh = \frac{1}{2}(2)(2)$

$= 2$. The area of the triangle on the right is $A = \frac{1}{2}bh$

$= \frac{1}{2}(1)(1) = \frac{1}{2}$. Then, the total area is 2.5

$\Rightarrow \int_{-2}^1 |x|\,dx = 2.5$ square units

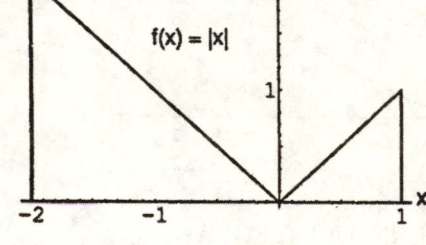

21. The area of the triangular peak is $A = \frac{1}{2}bh = \frac{1}{2}(2)(1) = 1$.
The area of the rectangular base is $S = \ell w = (2)(1) = 2$.

Then the total area is $3 \Rightarrow \int_{-1}^1 (2 - |x|)\,dx = 3$ square units

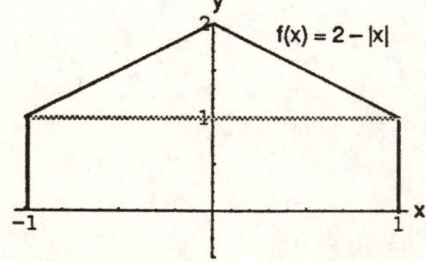

23. $\int_0^b \frac{x}{2} \, dx = \frac{1}{2}(b)(\frac{b}{2}) = \frac{b^2}{4}$

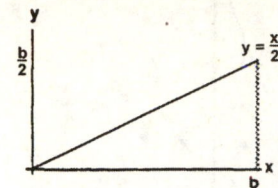

25. $\int_a^b 2s \, ds = \frac{1}{2}b(2b) - \frac{1}{2}a(2a) = b^2 - a^2$

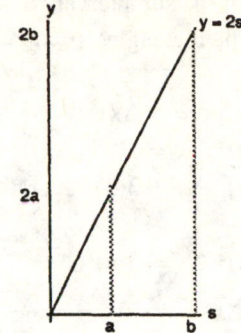

27. $\int_1^{\sqrt{2}} x \, dx = \frac{\left(\sqrt{2}\right)^2}{2} - \frac{(1)^2}{2} = \frac{1}{2}$

29. $\int_\pi^{2\pi} \theta \, d\theta = \frac{(2\pi)^2}{2} - \frac{\pi^2}{2} = \frac{3\pi^2}{2}$

31. $\int_0^{\sqrt[3]{7}} x^2 \, dx = \frac{\left(\sqrt[3]{7}\right)^3}{3} = \frac{7}{3}$

33. $\int_0^{1/2} t^2 \, dt = \frac{\left(\frac{1}{2}\right)^3}{3} = \frac{1}{24}$

35. $\int_a^{2a} x \, dx = \frac{(2a)^2}{2} - \frac{a^2}{2} = \frac{3a^2}{2}$

37. $\int_0^{\sqrt[3]{b}} x^2 \, dx = \frac{\left(\sqrt[3]{b}\right)^3}{3} = \frac{b}{3}$

39. $\int_3^1 7 \, dx = 7(1 - 3) = -14$

41. $\int_0^2 5x \, dx = 5 \int_0^2 x \, dx = 5 \left[\frac{2^2}{2} - \frac{0^2}{2}\right] = 10$

43. $\int_0^2 (2t - 3) \, dt = 2 \int_1^1 t \, dt - \int_0^2 3 \, dt = 2\left[\frac{2^2}{2} - \frac{0^2}{2}\right] - 3(2 - 0) = 4 - 6 = -2$

45. $\int_2^1 \left(1 + \frac{z}{2}\right) dz = \int_2^1 1 \, dz + \int_2^1 \frac{z}{2} \, dz = \int_2^1 1 \, dz - \frac{1}{2}\int_1^2 z \, dz = 1[1 - 2] - \frac{1}{2}\left[\frac{2^2}{2} - \frac{1^2}{2}\right] = -1 - \frac{1}{2}\left(\frac{3}{2}\right) = -\frac{7}{4}$

47. $\int_1^2 3u^2 \, du = 3 \int_1^2 u^2 \, du = 3\left[\int_0^2 u^2 \, du - \int_0^1 u^2 \, du\right] = 3\left(\left[\frac{2^3}{3} - \frac{0^3}{3}\right] - \left[\frac{1^3}{3} - \frac{0^3}{3}\right]\right) = 3\left[\frac{2^3}{3} - \frac{1^3}{3}\right] = 3\left(\frac{7}{3}\right) = 7$

49. $\int_0^2 (3x^2 + x - 5) \, dx = 3 \int_0^2 x^2 \, dx + \int_0^2 x \, dx - \int_0^2 5 \, dx = 3\left[\frac{2^3}{3} - \frac{0^3}{3}\right] + \left[\frac{2^2}{2} - \frac{0^2}{2}\right] - 5[2 - 0] = (8 + 2) - 10 = 0$

51. Let $\Delta x = \frac{b-0}{n} = \frac{b}{n}$ and let $x_0 = 0$, $x_1 = \Delta x$,

$x_2 = 2\Delta x, \ldots, x_{n-1} = (n-1)\Delta x$, $x_n = n\Delta x = b$.

Let the c_k's be the right end-points of the subintervals

$\Rightarrow c_1 = x_1$, $c_2 = x_2$, and so on. The rectangles

defined have areas:

$f(c_1)\,\Delta x = f(\Delta x)\,\Delta x = 3(\Delta x)^2\,\Delta x = 3(\Delta x)^3$

$f(c_2)\,\Delta x = f(2\Delta x)\,\Delta x = 3(2\Delta x)^2\,\Delta x = 3(2)^2(\Delta x)^3$

$f(c_3)\,\Delta x = f(3\Delta x)\,\Delta x = 3(3\Delta x)^2\,\Delta x = 3(3)^2(\Delta x)^3$

$\vdots$

$f(c_n)\,\Delta x = f(n\Delta x)\,\Delta x = 3(n\Delta x)^2\,\Delta x = 3(n)^2(\Delta x)^3$

Then $S_n = \sum\limits_{k=1}^{n} f(c_k)\,\Delta x = \sum\limits_{k=1}^{n} 3k^2(\Delta x)^3$

$= 3(\Delta x)^3 \sum\limits_{k=1}^{n} k^2 = 3\left(\frac{b^3}{n^3}\right)\left(\frac{n(n+1)(2n+1)}{6}\right)$

$= \frac{b^3}{2}\left(2 + \frac{3}{n} + \frac{1}{n^2}\right) \Rightarrow \int_0^b 3x^2\,dx = \lim\limits_{n \to \infty} \frac{b^3}{2}\left(2 + \frac{3}{n} + \frac{1}{n^2}\right) = b^3.$

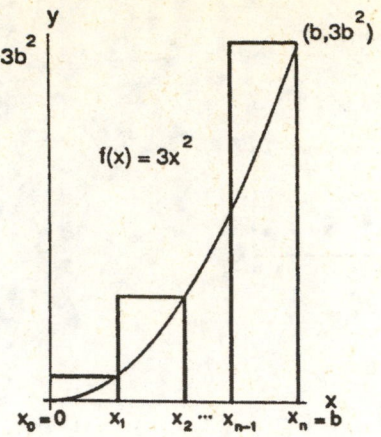

53. Let $\Delta x = \frac{b-0}{n} = \frac{b}{n}$ and let $x_0 = 0$, $x_1 = \Delta x$,

$x_2 = 2\Delta x, \ldots, x_{n-1} = (n-1)\Delta x$, $x_n = n\Delta x = b$.

Let the c_k's be the right end-points of the subintervals

$\Rightarrow c_1 = x_1$, $c_2 = x_2$, and so on. The rectangles

defined have areas:

$f(c_1)\,\Delta x = f(\Delta x)\,\Delta x = 2(\Delta x)(\Delta x) = 2(\Delta x)^2$

$f(c_2)\,\Delta x = f(2\Delta x)\,\Delta x = 2(2\Delta x)(\Delta x) = 2(2)(\Delta x)^2$

$f(c_3)\,\Delta x = f(3\Delta x)\,\Delta x = 2(3\Delta x)(\Delta x) = 2(3)(\Delta x)^2$

$\vdots$

$f(c_n)\,\Delta x = f(n\Delta x)\,\Delta x = 2(n\Delta x)(\Delta x) = 2(n)(\Delta x)^2$

Then $S_n = \sum\limits_{k=1}^{n} f(c_k)\,\Delta x = \sum\limits_{k=1}^{n} 2k(\Delta x)^2$

$= 2(\Delta x)^2 \sum\limits_{k=1}^{n} k = 2\left(\frac{b^2}{n^2}\right)\left(\frac{n(n+1)}{2}\right)$

$= b^2\left(1 + \frac{1}{n}\right) \Rightarrow \int_0^b 2x\,dx = \lim\limits_{n \to \infty} b^2\left(1 + \frac{1}{n}\right) = b^2.$

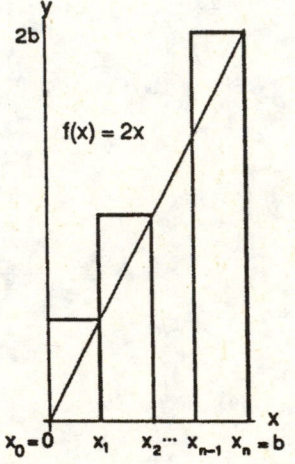

55. $\text{av}(f) = \left(\frac{1}{\sqrt{3}-0}\right)\int_0^{\sqrt{3}} (x^2 - 1)\,dx$

$= \frac{1}{\sqrt{3}}\int_0^{\sqrt{3}} x^2\,dx - \frac{1}{\sqrt{3}}\int_0^{\sqrt{3}} 1\,dx$

$= \frac{1}{\sqrt{3}}\left(\frac{\left(\sqrt{3}\right)^3}{3}\right) - \frac{1}{\sqrt{3}}\left(\sqrt{3} - 0\right) = 1 - 1 = 0.$

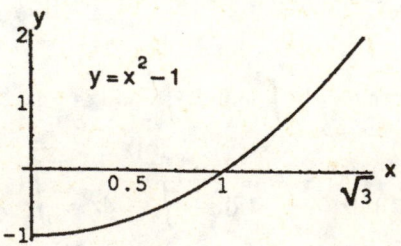

57. $\text{av}(f) = \left(\frac{1}{1-0}\right)\int_0^1 (-3x^2 - 1)\,dx =$

$= -3\int_0^1 x^2\,dx - \int_0^1 1\,dx = -3\left(\frac{1^3}{3}\right) - (1 - 0)$

$= -2.$

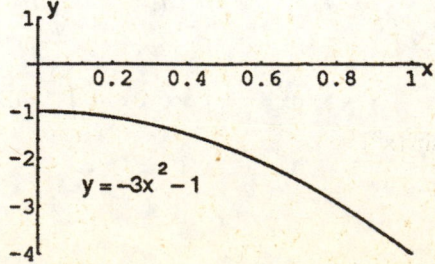

59. $\text{av}(f) = \left(\frac{1}{3-0}\right) \int_0^3 (t-1)^2 \, dt$

$= \frac{1}{3} \int_0^3 t^2 \, dt - \frac{2}{3} \int_0^3 t \, dt + \frac{1}{3} \int_0^3 1 \, dt$

$= \frac{1}{3} \left(\frac{3^3}{3}\right) - \frac{2}{3} \left(\frac{3^2}{2} - \frac{0^2}{2}\right) + \frac{1}{3}(3-0) = 1.$

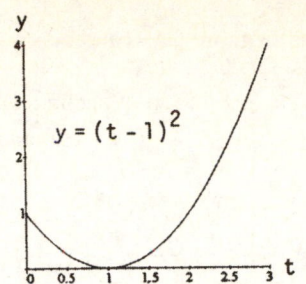

61. (a) $\text{av}(g) = \left(\frac{1}{1-(-1)}\right) \int_{-1}^1 (|x|-1) \, dx$

$= \frac{1}{2} \int_{-1}^0 (-x-1) \, dx + \frac{1}{2} \int_0^1 (x-1) \, dx$

$= -\frac{1}{2} \int_{-1}^0 x \, dx - \frac{1}{2} \int_{-1}^0 1 \, dx + \frac{1}{2} \int_0^1 x \, dx - \frac{1}{2} \int_0^1 1 \, dx$

$= -\frac{1}{2} \left(\frac{0^2}{2} - \frac{(-1)^2}{2}\right) - \frac{1}{2}(0-(-1)) + \frac{1}{2}\left(\frac{1^2}{2} - \frac{0^2}{2}\right) - \frac{1}{2}(1-0)$

$= -\frac{1}{2}.$

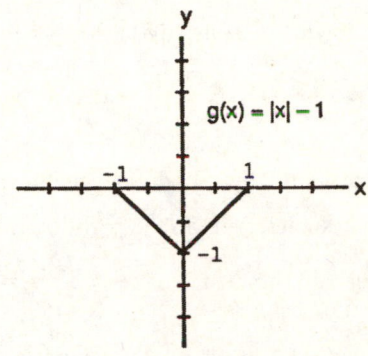

(b) $\text{av}(g) = \left(\frac{1}{3-1}\right) \int_1^3 (|x|-1) \, dx = \frac{1}{2} \int_1^3 (x-1) \, dx$

$= \frac{1}{2} \int_1^3 x \, dx - \frac{1}{2} \int_1^3 1 \, dx = \frac{1}{2} \left(\frac{3^2}{2} - \frac{1^2}{2}\right) - \frac{1}{2}(3-1)$

$= 1.$

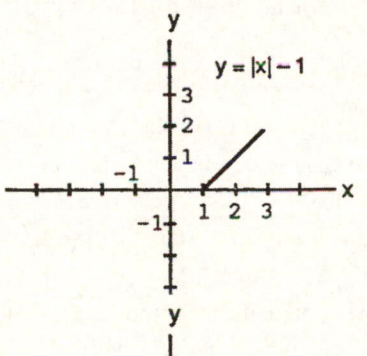

(c) $\text{av}(g) = \left(\frac{1}{3-(-1)}\right) \int_{-1}^3 (|x|-1) \, dx$

$= \frac{1}{4} \int_{-1}^1 (|x|-1) \, dx + \frac{1}{4} \int_1^3 (|x|-1) \, dx$

$= \frac{1}{4}(-1+2) = \frac{1}{4}$ (see parts (a) and (b) above).

63. To find where $x - x^2 \geq 0$, let $x - x^2 = 0 \Rightarrow x(1-x) = 0 \Rightarrow x = 0$ or $x = 1$. If $0 < x < 1$, then $0 < x - x^2 \Rightarrow a = 0$ and $b = 1$ maximize the integral.

65. $f(x) = \frac{1}{1+x^2}$ is decreasing on $[0, 1] \Rightarrow$ maximum value of f occurs at $0 \Rightarrow$ max $f = f(0) = 1$; minimum value of f

occurs at $1 \Rightarrow$ min $f = f(1) = \frac{1}{1+1^2} = \frac{1}{2}$. Therefore, $(1-0)$ min $f \leq \int_0^1 \frac{1}{1+x^2} \, dx \leq (1-0)$ max f

$\Rightarrow \frac{1}{2} \leq \int_0^1 \frac{1}{1+x^2} \, dx \leq 1$. That is, an upper bound $= 1$ and a lower bound $= \frac{1}{2}$.

67. $-1 \leq \sin(x^2) \leq 1$ for all $x \Rightarrow (1-0)(-1) \leq \int_0^1 \sin(x^2) \, dx \leq (1-0)(1)$ or $\int_0^1 \sin x^2 \, dx \leq 1 \Rightarrow \int_0^1 \sin x^2 \, dx$ cannot equal 2.

69. If $f(x) \geq 0$ on $[a, b]$, then min $f \geq 0$ and max $f \geq 0$ on $[a, b]$. Now, $(b-a)\min f \leq \int_a^b f(x)\, dx \leq (b-a)\max f$.

Then $b \geq a \Rightarrow b - a \geq 0 \Rightarrow (b-a)\min f \geq 0 \Rightarrow \int_a^b f(x)\, dx \geq 0$.

71. $\sin x \leq x$ for $x \geq 0 \Rightarrow \sin x - x \leq 0$ for $x \geq 0 \Rightarrow \int_0^1 (\sin x - x)\, dx \leq 0$ (see Exercise 70) $\Rightarrow \int_0^1 \sin x\, dx - \int_0^1 x\, dx$

$\leq 0 \Rightarrow \int_0^1 \sin x\, dx \leq \int_0^1 x\, dx \Rightarrow \int_0^1 \sin x\, dx \leq \left(\frac{1^2}{2} - \frac{0^2}{2}\right) \Rightarrow \int_0^1 \sin x\, dx \leq \frac{1}{2}$. Thus an upper bound is $\frac{1}{2}$.

73. Yes, for the following reasons: $\text{av}(f) = \frac{1}{b-a}\int_a^b f(x)\, dx$ is a constant K. Thus $\int_a^b \text{av}(f)\, dx = \int_a^b K\, dx$

$= K(b-a) \Rightarrow \int_a^b \text{av}(f)\, dx = (b-a)K = (b-a)\cdot\frac{1}{b-a}\int_a^b f(x)\, dx = \int_a^b f(x)\, dx$.

75. Consider the partition P that subdivides the interval $[a, b]$ into n subintervals of width $\triangle x = \frac{b-a}{n}$ and let c_k be the right endpoint of each subinterval. So the partition is $P = \{a, a + \frac{b-a}{n}, a + \frac{2(b-a)}{n}, \ldots, a + \frac{n(b-a)}{n}\}$ and $c_k = a + \frac{k(b-a)}{n}$.

We get the Riemann sum $\sum_{k=1}^n f(c_k)\triangle x = \sum_{k=1}^n c \cdot \frac{b-a}{n} = \frac{c(b-a)}{n}\sum_{k=1}^n 1 = \frac{c(b-a)}{n}\cdot n = c(b-a)$. As $n \to \infty$ and $\|P\| \to 0$

this expression remains $c(b-a)$. Thus, $\int_a^b c\, dx = c(b-a)$.

77. (a) $U = \max_1 \triangle x + \max_2 \triangle x + \ldots + \max_n \triangle x$ where $\max_1 = f(x_1)$, $\max_2 = f(x_2)$, $\ldots$, $\max_n = f(x_n)$ since f is
increasing on $[a, b]$; $L = \min_1 \triangle x + \min_2 \triangle x + \ldots + \min_n \triangle x$ where $\min_1 = f(x_0)$, $\min_2 = f(x_1)$, $\ldots$,
$\min_n = f(x_{n-1})$ since f is increasing on $[a, b]$. Therefore
$U - L = (\max_1 - \min_1)\triangle x + (\max_2 - \min_2)\triangle x + \ldots + (\max_n - \min_n)\triangle x$
$= (f(x_1) - f(x_0))\triangle x + (f(x_2) - f(x_1))\triangle x + \ldots + (f(x_n) - f(x_{n-1}))\triangle x = (f(x_n) - f(x_0))\triangle x = (f(b) - f(a))\triangle x$.

(b) $U = \max_1 \triangle x_1 + \max_2 \triangle x_2 + \ldots + \max_n \triangle x_n$ where $\max_1 = f(x_1)$, $\max_2 = f(x_2)$, $\ldots$, $\max_n = f(x_n)$ since f
is increasing on $[a, b]$; $L = \min_1 \triangle x_1 + \min_2 \triangle x_2 + \ldots + \min_n \triangle x_n$ where
$\min_1 = f(x_0)$, $\min_2 = f(x_1)$, $\ldots$, $\min_n = f(x_{n-1})$ since f is increasing on $[a, b]$. Therefore
$U - L = (\max_1 - \min_1)\triangle x_1 + (\max_2 - \min_2)\triangle x_2 + \ldots + (\max_n - \min_n)\triangle x_n$
$= (f(x_1) - f(x_0))\triangle x_1 + (f(x_2) - f(x_1))\triangle x_2 + \ldots + (f(x_n) - f(x_{n-1}))\triangle x_n$
$\leq (f(x_1) - f(x_0))\triangle x_{\max} + (f(x_2) - f(x_1))\triangle x_{\max} + \ldots + (f(x_n) - f(x_{n-1}))\triangle x_{\max}$. Then
$U - L \leq (f(x_n) - f(x_0))\triangle x_{\max} = (f(b) - f(a))\triangle x_{\max} = |f(b) - f(a)|\triangle x_{\max}$ since $f(b) \geq f(a)$. Thus
$\lim_{\|P\| \to 0} (U - L) = \lim_{\|P\| \to 0} (f(b) - f(a))\triangle x_{\max} = 0$, since $\triangle x_{\max} = \|P\|$.

79. (a) Partition $\left[0, \frac{\pi}{2}\right]$ into n subintervals, each of length $\triangle x = \frac{\pi}{2n}$ with points $x_0 = 0$, $x_1 = \triangle x$,
$x_2 = 2\triangle x, \ldots, x_n = n\triangle x = \frac{\pi}{2}$. Since $\sin x$ is increasing on $\left[0, \frac{\pi}{2}\right]$, the upper sum U is the sum of the areas
of the circumscribed rectangles of areas $f(x_1)\triangle x = (\sin \triangle x)\triangle x$, $f(x_2)\triangle x = (\sin 2\triangle x)\triangle x, \ldots, f(x_n)\triangle x$

$= (\sin n\triangle x)\triangle x$. Then $U = (\sin \triangle x + \sin 2\triangle x + \ldots + \sin n\triangle x)\triangle x = \left[\dfrac{\cos \frac{\triangle x}{2} - \cos\left(\left(n + \frac{1}{2}\right)\triangle x\right)}{2 \sin \frac{\triangle x}{2}}\right]\triangle x$

$= \left[\dfrac{\cos \frac{\pi}{4n} - \cos\left(\left(n + \frac{1}{2}\right)\frac{\pi}{2n}\right)}{2 \sin \frac{\pi}{4n}}\right]\left(\frac{\pi}{2n}\right) = \dfrac{\pi\left(\cos \frac{\pi}{4n} - \cos\left(\frac{\pi}{2} + \frac{\pi}{4n}\right)\right)}{4n \sin \frac{\pi}{4n}} = \dfrac{\cos \frac{\pi}{4n} - \cos\left(\frac{\pi}{2} + \frac{\pi}{4n}\right)}{\left(\frac{\sin \frac{\pi}{4n}}{\frac{\pi}{4n}}\right)}$

(b) The area is $\int_0^{\pi/2} \sin x\, dx = \lim_{n \to \infty} \dfrac{\cos \frac{\pi}{4n} - \cos\left(\frac{\pi}{2} + \frac{\pi}{4n}\right)}{\left(\frac{\sin \frac{\pi}{4n}}{\frac{\pi}{4n}}\right)} = \dfrac{1 - \cos \frac{\pi}{2}}{1} = 1$.

81. By Exercise 80, $U - L = \sum_{i=1}^n \triangle x_i \cdot M_i - \sum_{i=1}^n \triangle x_i \cdot m_i$ where $M_i = \max\{f(x)$ on the ith subinterval$\}$ and

$m_i = \min\{f(x)$ on the ith subinterval$\}$. Thus $U - L = \sum_{i=1}^n (M_i - m_i)\triangle x_i < \sum_{i=1}^n \epsilon \cdot \triangle x_i$ provided $\triangle x_i < \delta$ for each

$i = 1, \ldots, n$. Since $\sum_{i=1}^{n} \epsilon \cdot \triangle x_i = \epsilon \sum_{i=1}^{n} \triangle x_i = \epsilon(b - a)$ the result, $U - L < \epsilon(b - a)$ follows.

5.4 THE FUNDAMENTAL THEOREM OF CALCULUS

1. $\int_{-2}^{0} (2x + 5) \, dx = [x^2 + 5x]_{-2}^{0} = (0^2 + 5(0)) - ((-2)^2 + 5(-2)) = 6$

3. $\int_{0}^{4} \left(3x - \frac{x^3}{4}\right) dx = \left[\frac{3x^2}{2} - \frac{x^4}{16}\right]_{0}^{4} = \left(\frac{3(4)^2}{2} - \frac{4^4}{16}\right) - \left(\frac{3(0)^2}{2} - \frac{(0)^4}{16}\right) = 8$

5. $\int_{0}^{1} (x^2 + \sqrt{x}) \, dx = \left[\frac{x^3}{3} + \frac{2}{3}x^{3/2}\right]_{0}^{1} = \left(\frac{1}{3} + \frac{2}{3}\right) - 0 = 1$

7. $\int_{1}^{32} x^{-6/5} \, dx = \left[-5x^{-1/5}\right]_{1}^{32} = \left(-\frac{5}{2}\right) - (-5) = \frac{5}{2}$

9. $\int_{0}^{\pi} \sin x \, dx = [-\cos x]_{0}^{\pi} = (-\cos \pi) - (-\cos 0) = -(-1) - (-1) = 2$

11. $\int_{0}^{\pi/3} 2 \sec^2 x \, dx = [2 \tan x]_{0}^{\pi/3} = \left(2 \tan \left(\frac{\pi}{3}\right)\right) - (2 \tan 0) = 2\sqrt{3} - 0 = 2\sqrt{3}$

13. $\int_{\pi/4}^{3\pi/4} \csc \theta \cot \theta \, d\theta = [-\csc \theta]_{\pi/4}^{3\pi/4} = \left(-\csc \left(\frac{3\pi}{4}\right)\right) - \left(-\csc \left(\frac{\pi}{4}\right)\right) = -\sqrt{2} - \left(-\sqrt{2}\right) = 0$

15. $\int_{\pi/2}^{0} \frac{1 + \cos 2t}{2} \, dt = \int_{\pi/2}^{0} \left(\frac{1}{2} + \frac{1}{2} \cos 2t\right) dt = \left[\frac{1}{2}t + \frac{1}{4} \sin 2t\right]_{\pi/2}^{0} = \left(\frac{1}{2}(0) + \frac{1}{4} \sin 2(0)\right) - \left(\frac{1}{2}\left(\frac{\pi}{2}\right) + \frac{1}{4} \sin 2\left(\frac{\pi}{2}\right)\right)$
$= -\frac{\pi}{4}$

17. $\int_{-\pi/2}^{\pi/2} (8y^2 + \sin y) \, dy = \left[\frac{8y^3}{3} - \cos y\right]_{-\pi/2}^{\pi/2} = \left(\frac{8 \left(\frac{\pi}{2}\right)^3}{3} - \cos \frac{\pi}{2}\right) - \left(\frac{8 \left(-\frac{\pi}{2}\right)^3}{3} - \cos \left(-\frac{\pi}{2}\right)\right) = \frac{2\pi^3}{3}$

19. $\int_{1}^{-1} (r + 1)^2 \, dr = \int_{1}^{-1} (r^2 + 2r + 1) \, dr = \left[\frac{r^3}{3} + r^2 + r\right]_{1}^{-1} = \left(\frac{(-1)^3}{3} + (-1)^2 + (-1)\right) - \left(\frac{1^3}{3} + 1^2 + 1\right) = -\frac{8}{3}$

21. $\int_{\sqrt{2}}^{1} \left(\frac{u^7}{2} - \frac{1}{u^5}\right) du = \int_{\sqrt{2}}^{1} \left(\frac{u^7}{2} - u^{-5}\right) du = \left[\frac{u^8}{16} + \frac{1}{4u^4}\right]_{\sqrt{2}}^{1} = \left(\frac{1^8}{16} + \frac{1}{4(1)^4}\right) - \left(\frac{(\sqrt{2})^8}{16} + \frac{1}{4(\sqrt{2})^4}\right) = -\frac{3}{4}$

23. $\int_{1}^{\sqrt{2}} \frac{s^2 + \sqrt{s}}{s^2} \, ds = \int_{1}^{\sqrt{2}} (1 + s^{-3/2}) \, ds = \left[s - \frac{2}{\sqrt{s}}\right]_{1}^{\sqrt{2}} = \left(\sqrt{2} - \frac{2}{\sqrt{\sqrt{2}}}\right) - \left(1 - \frac{2}{\sqrt{1}}\right) = \sqrt{2} - 2^{3/4} + 1$
$= \sqrt{2} - \sqrt[4]{8} + 1$

25. $\int_{-4}^{4} |x| \, dx = \int_{-4}^{0} |x| \, dx + \int_{0}^{4} |x| \, dx = -\int_{-4}^{0} x \, dx + \int_{0}^{4} x \, dx = \left[-\frac{x^2}{2}\right]_{-4}^{0} + \left[\frac{x^2}{2}\right]_{0}^{4} = \left(-\frac{0^2}{2} + \frac{(-4)^2}{2}\right) + \left(\frac{4^2}{2} - \frac{0^2}{2}\right)$
$= 16$

27. (a) $\int_{0}^{\sqrt{x}} \cos t \, dt = [\sin t]_{0}^{\sqrt{x}} = \sin \sqrt{x} - \sin 0 = \sin \sqrt{x} \Rightarrow \frac{d}{dx} \left(\int_{0}^{\sqrt{x}} \cos t \, dt\right) = \frac{d}{dx} (\sin \sqrt{x}) = \cos \sqrt{x} \left(\frac{1}{2} x^{-1/2}\right)$
$= \frac{\cos \sqrt{x}}{2\sqrt{x}}$

(b) $\frac{d}{dx}\left(\int_0^{\sqrt{x}} \cos t \, dt\right) = (\cos \sqrt{x})\left(\frac{d}{dx}\left(\sqrt{x}\right)\right) = (\cos \sqrt{x})\left(\frac{1}{2}x^{-1/2}\right) = \frac{\cos \sqrt{x}}{2\sqrt{x}}$

29. (a) $\int_0^{t^4} \sqrt{u}\, du = \int_0^{t^4} u^{1/2}\, du = \left[\frac{2}{3}u^{3/2}\right]_0^{t^4} = \frac{2}{3}(t^4)^{3/2} - 0 = \frac{2}{3}t^6 \Rightarrow \frac{d}{dt}\left(\int_0^{t^4} \sqrt{u}\, du\right) = \frac{d}{dt}\left(\frac{2}{3}t^6\right) = 4t^5$

   (b) $\frac{d}{dt}\left(\int_0^{t^4} \sqrt{u}\, du\right) = \sqrt{t^4}\left(\frac{d}{dt}(t^4)\right) = t^2(4t^3) = 4t^5$

31. $y = \int_0^x \sqrt{1+t^2}\, dt \Rightarrow \frac{dy}{dx} = \sqrt{1+x^2}$

33. $y = \int_{\sqrt{x}}^0 \sin t^2\, dt = -\int_0^{\sqrt{x}} \sin t^2\, dt \Rightarrow \frac{dy}{dx} = -\left(\sin\left(\sqrt{x}\right)^2\right)\left(\frac{d}{dx}\left(\sqrt{x}\right)\right) = -(\sin x)\left(\frac{1}{2}x^{-1/2}\right) = -\frac{\sin x}{2\sqrt{x}}$

35. $y = \int_0^{\sin x} \frac{dt}{\sqrt{1-t^2}}, |x| < \frac{\pi}{2} \Rightarrow \frac{dy}{dx} = \frac{1}{\sqrt{1-\sin^2 x}}\left(\frac{d}{dx}(\sin x)\right) = \frac{1}{\sqrt{\cos^2 x}}(\cos x) = \frac{\cos x}{|\cos x|} = \frac{\cos x}{\cos x} = 1$ since $|x| < \frac{\pi}{2}$

37. $-x^2 - 2x = 0 \Rightarrow -x(x+2) = 0 \Rightarrow x = 0$ or $x = -2$; Area

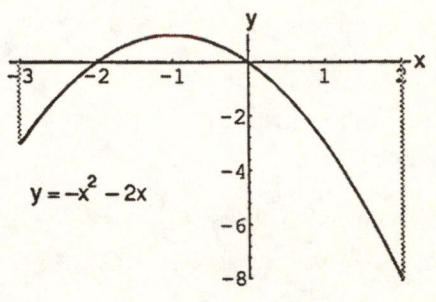

$= -\int_{-3}^{-2}(-x^2 - 2x)dx + \int_{-2}^{0}(-x^2 - 2x)dx - \int_0^2 (-x^2 - 2x)dx$

$= -\left[-\frac{x^3}{3} - x^2\right]_{-3}^{-2} + \left[-\frac{x^3}{3} - x^2\right]_{-2}^{0} - \left[-\frac{x^3}{3} - x^2\right]_0^2$

$= -\left(\left(-\frac{(-2)^3}{3} - (-2)^2\right) - \left(-\frac{(-3)^3}{3} - (-3)^2\right)\right)$

$+ \left(\left(-\frac{0^3}{3} - 0^2\right) - \left(-\frac{(-2)^3}{3} - (-2)^2\right)\right)$

$- \left(\left(-\frac{2^3}{3} - 2^2\right) - \left(-\frac{0^3}{3} - 0^2\right)\right) = \frac{28}{3}$

39. $x^3 - 3x^2 + 2x = 0 \Rightarrow x(x^2 - 3x + 2) = 0$

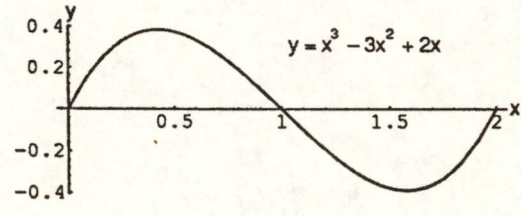

$\Rightarrow x(x-2)(x-1) = 0 \Rightarrow x = 0, 1,$ or 2;

Area $= \int_0^1 (x^3 - 3x^2 + 2x)dx - \int_1^2 (x^3 - 3x^2 + 2x)dx$

$= \left[\frac{x^4}{4} - x^3 + x^2\right]_0^1 - \left[\frac{x^4}{4} - x^3 + x^2\right]_1^2$

$= \left(\frac{1^4}{4} - 1^3 + 1^2\right) - \left(\frac{0^4}{4} - 0^3 + 0^2\right)$

$- \left[\left(\frac{2^4}{4} - 2^3 + 2^2\right) - \left(\frac{1^4}{4} - 1^3 + 1^2\right)\right] = \frac{1}{2}$

41. $x^{1/3} = 0 \Rightarrow x = 0$; Area $= -\int_{-1}^0 x^{1/3}\, dx + \int_0^8 x^{1/3}\, dx$

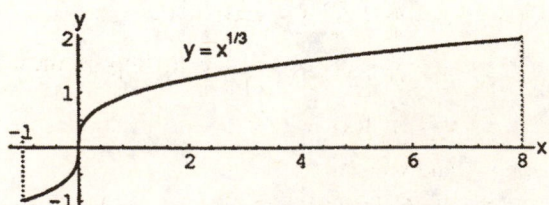

$= \left[-\frac{3}{4}x^{4/3}\right]_{-1}^0 + \left[\frac{3}{4}x^{4/3}\right]_0^8$

$= \left(-\frac{3}{4}(0)^{4/3}\right) - \left(-\frac{3}{4}(-1)^{4/3}\right) + \left(\frac{3}{4}(8)^{4/3}\right) - \left(\frac{3}{4}(0)^{4/3}\right)$

$= \frac{51}{4}$

43. The area of the rectangle bounded by the lines $y = 2$, $y = 0$, $x = \pi$, and $x = 0$ is 2π. The area under the curve $y = 1 + \cos x$ on $[0, \pi]$ is $\int_0^\pi (1 + \cos x)\, dx = [x + \sin x]_0^\pi = (\pi + \sin \pi) - (0 + \sin 0) = \pi$. Therefore the area of the shaded region is $2\pi - \pi = \pi$.

45. On $\left[-\frac{\pi}{4}, 0\right]$: The area of the rectangle bounded by the lines $y = \sqrt{2}$, $y = 0$, $\theta = 0$, and $\theta = -\frac{\pi}{4}$ is $\sqrt{2}\left(\frac{\pi}{4}\right)$

    $= \frac{\pi\sqrt{2}}{4}$. The area between the curve $y = \sec\theta\tan\theta$ and $y = 0$ is $-\int_{-\pi/4}^{0} \sec\theta\tan\theta\, d\theta = [-\sec\theta]_{-\pi/4}^{0}$

    $= (-\sec 0) - \left(-\sec\left(-\frac{\pi}{4}\right)\right) = \sqrt{2} - 1$. Therefore the area of the shaded region on $\left[-\frac{\pi}{4}, 0\right]$ is $\frac{\pi\sqrt{2}}{4} + \left(\sqrt{2} - 1\right)$.

    On $\left[0, \frac{\pi}{4}\right]$: The area of the rectangle bounded by $\theta = \frac{\pi}{4}$, $\theta = 0$, $y = \sqrt{2}$, and $y = 0$ is $\sqrt{2}\left(\frac{\pi}{4}\right) = \frac{\pi\sqrt{2}}{4}$. The area

    under the curve $y = \sec\theta\tan\theta$ is $\int_{0}^{\pi/4} \sec\theta\tan\theta\, d\theta = [\sec\theta]_{0}^{\pi/4} = \sec\frac{\pi}{4} - \sec 0 = \sqrt{2} - 1$. Therefore the area

    of the shaded region on $\left[0, \frac{\pi}{4}\right]$ is $\frac{\pi\sqrt{2}}{4} - \left(\sqrt{2} - 1\right)$. Thus, the area of the total shaded region is

    $\left(\frac{\pi\sqrt{2}}{4} + \sqrt{2} - 1\right) + \left(\frac{\pi\sqrt{2}}{4} - \sqrt{2} + 1\right) = \frac{\pi\sqrt{2}}{2}$.

47. $y = \int_{\pi}^{x} \frac{1}{t}\, dt - 3 \Rightarrow \frac{dy}{dx} = \frac{1}{x}$ and $y(\pi) = \int_{\pi}^{\pi} \frac{1}{t}\, dt - 3 = 0 - 3 = -3 \Rightarrow$ (d) is a solution to this problem.

49. $y = \int_{0}^{x} \sec t\, dt + 4 \Rightarrow \frac{dy}{dx} = \sec x$ and $y(0) = \int_{0}^{0} \sec t\, dt + 4 = 0 + 4 = 4 \Rightarrow$ (b) is a solution to this problem.

51. $y = \int_{2}^{x} \sec t\, dt + 3$

53. $s = \int_{t_0}^{t} f(x)\, dx + s_0$

55. Area $= \int_{-b/2}^{b/2} \left(h - \left(\frac{4h}{b^2}\right)x^2\right) dx = \left[hx - \frac{4hx^3}{3b^2}\right]_{-b/2}^{b/2}$

    $= \left(h\left(\frac{b}{2}\right) - \frac{4h\left(\frac{b}{2}\right)^3}{3b^2}\right) - \left(h\left(-\frac{b}{2}\right) - \frac{4h\left(-\frac{b}{2}\right)^3}{3b^2}\right)$

    $= \left(\frac{bh}{2} - \frac{bh}{6}\right) - \left(-\frac{bh}{2} + \frac{bh}{6}\right) = bh - \frac{bh}{3} = \frac{2}{3} bh$

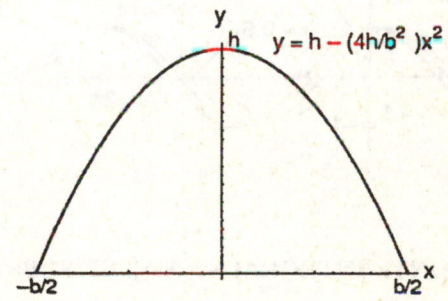

57. $\frac{dc}{dx} = \frac{1}{2\sqrt{x}} = \frac{1}{2} x^{-1/2} \Rightarrow c = \int_{0}^{x} \frac{1}{2} t^{-1/2} dt = \left[t^{1/2}\right]_{0}^{x} = \sqrt{x}$

    $c(100) - c(1) = \sqrt{100} - \sqrt{1} = \$9.00$

59. (a) $v = \frac{ds}{dt} = \frac{d}{dt}\int_{0}^{t} f(x)\, dx = f(t) \Rightarrow v(5) = f(5) = 2$ m/sec

    (b) $a = \frac{df}{dt}$ is negative since the slope of the tangent line at $t = 5$ is negative

    (c) $s = \int_{0}^{3} f(x)\, dx = \frac{1}{2}(3)(3) = \frac{9}{2}$ m since the integral is the area of the triangle formed by $y = f(x)$, the x-axis,
    and $x = 3$

    (d) $t = 6$ since from $t = 6$ to $t = 9$, the region lies below the x-axis

    (e) At $t = 4$ and $t = 7$, since there are horizontal tangents there

    (f) Toward the origin between $t = 6$ and $t = 9$ since the velocity is negative on this interval. Away from the
    origin between $t = 0$ and $t = 6$ since the velocity is positive there.

    (g) Right or positive side, because the integral of f from 0 to 9 is positive, there being more area above the
    x-axis than below it.

61. $k > 0 \Rightarrow$ one arch of $y = \sin kx$ will occur over the interval $\left[0, \frac{\pi}{k}\right] \Rightarrow$ the area $= \int_0^{\pi/k} \sin kx \, dx = \left[-\frac{1}{k} \cos kx\right]_0^{\pi/k}$

$= -\frac{1}{k} \cos \left(k\left(\frac{\pi}{k}\right)\right) - \left(-\frac{1}{k} \cos(0)\right) = \frac{2}{k}$

63. $\int_1^x f(t) \, dt = x^2 - 2x + 1 \Rightarrow f(x) = \frac{d}{dx} \int_1^x f(t) \, dt = \frac{d}{dx}(x^2 - 2x + 1) = 2x - 2$

65. $f(x) = 2 - \int_2^{x+1} \frac{9}{1+t} \, dt \Rightarrow f'(x) = -\frac{9}{1+(x+1)} = \frac{-9}{x+2} \Rightarrow f'(1) = -3; f(1) = 2 - \int_2^{1+1} \frac{9}{1+t} \, dt = 2 - 0 = 2;$

$L(x) = -3(x - 1) + f(1) = -3(x - 1) + 2 = -3x + 5$

67. (a) True: since f is continuous, g is differentiable by Part 1 of the Fundamental Theorem of Calculus.
    (b) True: g is continuous because it is differentiable.
    (c) True, since $g'(1) = f(1) = 0$.
    (d) False, since $g''(1) = f'(1) > 0$.
    (e) True, since $g'(1) = 0$ and $g''(1) = f'(1) > 0$.
    (f) False: $g''(x) = f'(x) > 0$, so g'' never changes sign.
    (g) True, since $g'(1) = f(1) = 0$ and $g'(x) = f(x)$ is an increasing function of x (because $f'(x) > 0$).

69.

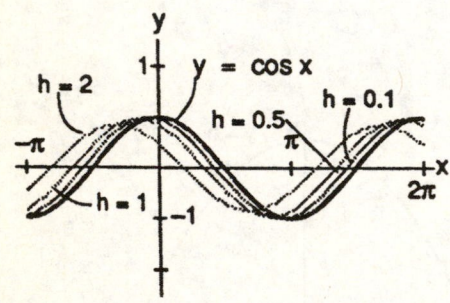

5.5 INDEFINTE INTEGRALS AND THE SUBSTITUTION RULE

1. Let $u = 3x \Rightarrow du = 3 \, dx \Rightarrow \frac{1}{3} \, du = dx$

    $\int \sin 3x \, dx = \int \frac{1}{3} \sin u \, du = -\frac{1}{3} \cos u + C = -\frac{1}{3} \cos 3x + C$

3. Let $u = 2t \Rightarrow du = 2 \, dt \Rightarrow \frac{1}{2} \, du = dt$

    $\int \sec 2t \tan 2t \, dt = \int \frac{1}{2} \sec u \tan u \, du = \frac{1}{2} \sec u + C = \frac{1}{2} \sec 2t + C$

5. Let $u = 7x - 2 \Rightarrow du = 7 \, dx \Rightarrow \frac{1}{7} \, du = dx$

    $\int 28(7x - 2)^{-5} \, dx = \int \frac{1}{7}(28)u^{-5} \, du = \int 4u^{-5} \, du = -u^{-4} + C = -(7x - 2)^{-4} + C$

7. Let $u = 1 - r^3 \Rightarrow du = -3r^2 \, dr \Rightarrow -3 \, du = 9r^2 \, dr$

    $\int \frac{9r^2 \, dr}{\sqrt{1 - r^3}} = \int -3u^{-1/2} \, du = -3(2)u^{1/2} + C = -6(1 - r^3)^{1/2} + C$

9. Let $u = x^{3/2} - 1 \Rightarrow du = \frac{3}{2} x^{1/2} \, dx \Rightarrow \frac{2}{3} \, du = \sqrt{x} \, dx$

    $\int \sqrt{x} \sin^2 (x^{3/2} - 1) \, dx = \int \frac{2}{3} \sin^2 u \, du = \frac{2}{3} \left(\frac{u}{2} - \frac{1}{4} \sin 2u\right) + C = \frac{1}{3}(x^{3/2} - 1) - \frac{1}{6} \sin(2x^{3/2} - 2) + C$

11. (a) Let $u = \cot 2\theta \Rightarrow du = -2\csc^2 2\theta \, d\theta \Rightarrow -\frac{1}{2} du = \csc^2 2\theta \, d\theta$

$$\int \csc^2 2\theta \cot 2\theta \, d\theta = -\int \frac{1}{2} u \, du = -\frac{1}{2}\left(\frac{u^2}{2}\right) + C = -\frac{u^2}{4} + C = -\frac{1}{4}\cot^2 2\theta + C$$

(b) Let $u = \csc 2\theta \Rightarrow du = -2\csc 2\theta \cot 2\theta \, d\theta \Rightarrow -\frac{1}{2} du = \csc 2\theta \cot 2\theta \, d\theta$

$$\int \csc^2 2\theta \cot 2\theta \, d\theta = \int -\frac{1}{2} u \, du = -\frac{1}{2}\left(\frac{u^2}{2}\right) + C = -\frac{u^2}{4} + C = -\frac{1}{4}\csc^2 2\theta + C$$

13. Let $u = 3 - 2s \Rightarrow du = -2\,ds \Rightarrow -\frac{1}{2} du = ds$

$$\int \sqrt{3 - 2s}\, ds = \int \sqrt{u}\left(-\frac{1}{2} du\right) = -\frac{1}{2}\int u^{1/2} \, du = \left(-\frac{1}{2}\right)\left(\frac{2}{3}u^{3/2}\right) + C = -\frac{1}{3}(3 - 2s)^{3/2} + C$$

15. Let $u = 5s + 4 \Rightarrow du = 5\,ds \Rightarrow \frac{1}{5} du = ds$

$$\int \frac{1}{\sqrt{5s+4}}\, ds = \int \frac{1}{\sqrt{u}}\left(\frac{1}{5} du\right) = \frac{1}{5}\int u^{-1/2} \, du = \left(\frac{1}{5}\right)\left(2u^{1/2}\right) + C = \frac{2}{5}\sqrt{5s+4} + C$$

17. Let $u = 1 - \theta^2 \Rightarrow du = -2\theta \, d\theta \Rightarrow -\frac{1}{2} du = \theta \, d\theta$

$$\int \theta \sqrt[4]{1 - \theta^2}\, d\theta = \int \sqrt[4]{u}\left(-\frac{1}{2} du\right) = -\frac{1}{2}\int u^{1/4} \, du = \left(-\frac{1}{2}\right)\left(\frac{4}{5}u^{5/4}\right) + C = -\frac{2}{5}(1 - \theta^2)^{5/4} + C$$

19. Let $u = 7 - 3y^2 \Rightarrow du = -6y \, dy \Rightarrow -\frac{1}{2} du = 3y \, dy$

$$\int 3y\sqrt{7 - 3y^2}\, dy = \int \sqrt{u}\left(-\frac{1}{2} du\right) = -\frac{1}{2}\int u^{1/2} \, du = \left(-\frac{1}{2}\right)\left(\frac{2}{3}u^{3/2}\right) + C = -\frac{1}{3}(7 - 3y^2)^{3/2} + C$$

21. Let $u = 1 + \sqrt{x} \Rightarrow du = \frac{1}{2\sqrt{x}} dx \Rightarrow 2\,du = \frac{1}{\sqrt{x}} dx$

$$\int \frac{1}{\sqrt{x}\left(1 + \sqrt{x}\right)^2}\, dx = \int \frac{2\,du}{u^2} = -\frac{2}{u} + C = \frac{-2}{1 + \sqrt{x}} + C$$

23. Let $u = 3z + 4 \Rightarrow du = 3\,dz \Rightarrow \frac{1}{3} du = dz$

$$\int \cos(3z + 4)\, dz = \int (\cos u)\left(\frac{1}{3} du\right) = \frac{1}{3}\int \cos u \, du = \frac{1}{3}\sin u + C = \frac{1}{3}\sin(3z + 4) + C$$

25. Let $u = 3x + 2 \Rightarrow du = 3\,dx \Rightarrow \frac{1}{3} du = dx$

$$\int \sec^2(3x + 2)\, dx = \int (\sec^2 u)\left(\frac{1}{3} du\right) = \frac{1}{3}\int \sec^2 u \, du = \frac{1}{3}\tan u + C = \frac{1}{3}\tan(3x + 2) + C$$

27. Let $u = \sin\left(\frac{x}{3}\right) \Rightarrow du = \frac{1}{3}\cos\left(\frac{x}{3}\right) dx \Rightarrow 3\,du = \cos\left(\frac{x}{3}\right) dx$

$$\int \sin^5\left(\frac{x}{3}\right)\cos\left(\frac{x}{3}\right) dx = \int u^5 (3\,du) = 3\left(\frac{1}{6}u^6\right) + C = \frac{1}{2}\sin^6\left(\frac{x}{3}\right) + C$$

29. Let $u = \frac{r^3}{18} - 1 \Rightarrow du = \frac{r^2}{6} dr \Rightarrow 6\,du = r^2 \, dr$

$$\int r^2 \left(\frac{r^3}{18} - 1\right)^5 dr = \int u^5 (6\,du) = 6\int u^5 \, du = 6\left(\frac{u^6}{6}\right) + C = \left(\frac{r^3}{18} - 1\right)^6 + C$$

31. Let $u = x^{3/2} + 1 \Rightarrow du = \frac{3}{2}x^{1/2} dx \Rightarrow \frac{2}{3} du = x^{1/2} dx$

$$\int x^{1/2} \sin\left(x^{3/2} + 1\right) dx = \int (\sin u)\left(\frac{2}{3} du\right) = \frac{2}{3}\int \sin u \, du = \frac{2}{3}(-\cos u) + C = -\frac{2}{3}\cos\left(x^{3/2} + 1\right) + C$$

33. Let $u = \sec\left(v + \frac{\pi}{2}\right) \Rightarrow du = \sec\left(v + \frac{\pi}{2}\right)\tan\left(v + \frac{\pi}{2}\right) dv$

$$\int \sec\left(v + \frac{\pi}{2}\right)\tan\left(v + \frac{\pi}{2}\right) dv = \int du = u + C = \sec\left(v + \frac{\pi}{2}\right) + C$$

35. Let $u = \cos(2t + 1) \Rightarrow du = -2\sin(2t + 1) \, dt \Rightarrow -\frac{1}{2} du = \sin(2t + 1) \, dt$

$$\int \frac{\sin(2t + 1)}{\cos^2(2t + 1)} \, dt = \int -\frac{1}{2}\frac{du}{u^2} = \frac{1}{2u} + C = \frac{1}{2\cos(2t + 1)} + C$$

37. Let $u = \cot y \Rightarrow du = -\csc^2 y\, dy \Rightarrow -du = \csc^2 y\, dy$

$\int \sqrt{\cot y}\, \csc^2 y\, dy = \int \sqrt{u}\,(-du) = -\int u^{1/2}\, du = -\frac{2}{3} u^{3/2} + C = -\frac{2}{3} (\cot y)^{3/2} + C = -\frac{2}{3} \left(\cot^3 y\right)^{1/2} + C$

39. Let $u = \frac{1}{t} - 1 = t^{-1} - 1 \Rightarrow du = -t^{-2}\, dt \Rightarrow -du = \frac{1}{t^2}\, dt$

$\int \frac{1}{t^2} \cos\left(\frac{1}{t} - 1\right) dt = \int (\cos u)(-du) = -\int \cos u\, du = -\sin u + C = -\sin\left(\frac{1}{t} - 1\right) + C$

41. Let $u = \sin \frac{1}{\theta} \Rightarrow du = \left(\cos \frac{1}{\theta}\right)\left(-\frac{1}{\theta^2}\right) d\theta \Rightarrow -du = \frac{1}{\theta^2} \cos \frac{1}{\theta}\, d\theta$

$\int \frac{1}{\theta^2} \sin \frac{1}{\theta} \cos \frac{1}{\theta}\, d\theta = \int -u\, du = -\frac{1}{2} u^2 + C = -\frac{1}{2} \sin^2 \frac{1}{\theta} + C$

43. Let $u = s^3 + 2s^2 - 5s + 5 \Rightarrow du = (3s^2 + 4s - 5)\, ds$

$\int (s^3 + 2s^2 - 5s + 5)(3s^2 + 4s - 5)\, ds = \int u\, du = \frac{u^2}{2} + C = \frac{(s^3 + 2s^2 - 5s + 5)^2}{2} + C$

45. Let $u = 1 + t^4 \Rightarrow du = 4t^3\, dt \Rightarrow \frac{1}{4}\, du = t^3\, dt$

$\int t^3 (1 + t^4)^3\, dt = \int u^3 \left(\frac{1}{4}\, du\right) = \frac{1}{4}\left(\frac{1}{4} u^4\right) + C = \frac{1}{16}(1 + t^4)^4 + C$

47. Let $u = x^2 + 1$. Then $du = 2x\,dx$ and $\frac{1}{2} du = x\,dx$ and $x^2 = u - 1$. Thus $\int x^3 \sqrt{x^2 + 1}\, dx = \int (u - 1) \frac{1}{2} \sqrt{u}\, du$

$= \frac{1}{2} \int \left(u^{3/2} - u^{1/2}\right) du = \frac{1}{2}\left[\frac{2}{5} u^{5/2} - \frac{2}{3} u^{3/2}\right] + C = \frac{1}{5} u^{5/2} - \frac{1}{3} u^{3/2} + C = \frac{1}{5}(x^2 + 1)^{5/2} - \frac{1}{3}(x^2 + 1)^{3/2} + C$

49. (a) Let $u = \tan x \Rightarrow du = \sec^2 x\, dx;\ v = u^3 \Rightarrow dv = 3u^2\, du \Rightarrow 6\, dv = 18u^2\, du;\ w = 2 + v \Rightarrow dw = dv$

$\int \frac{18 \tan^2 x \sec^2 x}{(2 + \tan^3 x)^2}\, dx = \int \frac{18u^2}{(2 + u^3)^2}\, du = \int \frac{6\, dv}{(2 + v)^2} = \int \frac{6\, dw}{w^2} = 6 \int w^{-2}\, dw = -6w^{-1} + C = -\frac{6}{2 + v} + C$

$= -\frac{6}{2 + u^3} + C = -\frac{6}{2 + \tan^3 x} + C$

(b) Let $u = \tan^3 x \Rightarrow du = 3 \tan^2 x \sec^2 x\, dx \Rightarrow 6\, du = 18 \tan^2 x \sec^2 x\, dx;\ v = 2 + u \Rightarrow dv = du$

$\int \frac{18 \tan^2 x \sec^2 x}{(2 + \tan^3 x)^2}\, dx = \int \frac{6\, du}{(2 + u)^2} = \int \frac{6\, dv}{v^2} = -\frac{6}{v} + C = -\frac{6}{2 + u} + C = -\frac{6}{2 + \tan^3 x} + C$

(c) Let $u = 2 + \tan^3 x \Rightarrow du = 3 \tan^2 x.\sec^2 x\, dx \Rightarrow 6\, du = 18 \tan^2 x \sec^2 x\, dx$

$\int \frac{18 \tan^2 x \sec^2 x}{(2 + \tan^3 x)^2}\, dx = \int \frac{6\, du}{u^2} = -\frac{6}{u} + C = -\frac{6}{2 + \tan^3 x} + C$

51. Let $u = 3(2r - 1)^2 + 6 \Rightarrow du = 6(2r - 1)(2)\, dr \Rightarrow \frac{1}{12}\, du = (2r - 1)\, dr;\ v = \sqrt{u} \Rightarrow dv = \frac{1}{2\sqrt{u}}\, du \Rightarrow \frac{1}{6}\, dv$

$= \frac{1}{12\sqrt{u}}\, du$

$\int \frac{(2r - 1) \cos \sqrt{3(2r - 1)^2 + 6}}{\sqrt{3(2r - 1)^2 + 6}}\, dr = \int \left(\frac{\cos \sqrt{u}}{\sqrt{u}}\right)\left(\frac{1}{12}\, du\right) = \int (\cos v)\left(\frac{1}{6}\, dv\right) = \frac{1}{6} \sin v + C = \frac{1}{6} \sin \sqrt{u} + C$

$= \frac{1}{6} \sin \sqrt{3(2r - 1)^2 + 6} + C$

53. Let $u = 3t^2 - 1 \Rightarrow du = 6t\, dt \Rightarrow 2\, du = 12t\, dt$

$s = \int 12t (3t^2 - 1)^3\, dt = \int u^3 (2\, du) = 2 \left(\frac{1}{4} u^4\right) + C = \frac{1}{2} u^4 + C = \frac{1}{2}(3t^2 - 1)^4 + C;$

$s = 3$ when $t = 1 \Rightarrow 3 = \frac{1}{2}(3 - 1)^4 + C \Rightarrow 3 = 8 + C \Rightarrow C = -5 \Rightarrow s = \frac{1}{2}(3t^2 - 1)^4 - 5$

55. Let $u = t + \frac{\pi}{12} \Rightarrow du = dt$

$s = \int 8 \sin^2 \left(t + \frac{\pi}{12}\right) dt = \int 8 \sin^2 u\, du = 8 \left(\frac{u}{2} - \frac{1}{4} \sin 2u\right) + C = 4 \left(t + \frac{\pi}{12}\right) - 2 \sin \left(2t + \frac{\pi}{6}\right) + C;$

$s = 8$ when $t = 0 \Rightarrow 8 = 4\left(\frac{\pi}{12}\right) - 2 \sin \left(\frac{\pi}{6}\right) + C \Rightarrow C = 8 - \frac{\pi}{3} + 1 = 9 - \frac{\pi}{3}$

$\Rightarrow s = 4\left(t + \frac{\pi}{12}\right) - 2 \sin \left(2t + \frac{\pi}{6}\right) + 9 - \frac{\pi}{3} = 4t - 2 \sin \left(2t + \frac{\pi}{6}\right) + 9$

57. Let $u = 2t - \frac{\pi}{2} \Rightarrow du = 2\,dt \Rightarrow -2\,du = -4\,dt$

$\frac{ds}{dt} = \int -4 \sin\left(2t - \frac{\pi}{2}\right) dt = \int (\sin u)(-2\,du) = 2\cos u + C_1 = 2\cos\left(2t - \frac{\pi}{2}\right) + C_1;$

at $t = 0$ and $\frac{ds}{dt} = 100$ we have $100 = 2\cos\left(-\frac{\pi}{2}\right) + C_1 \Rightarrow C_1 = 100 \Rightarrow \frac{ds}{dt} = 2\cos\left(2t - \frac{\pi}{2}\right) + 100$

$\Rightarrow s = \int \left(2\cos\left(2t - \frac{\pi}{2}\right) + 100\right) dt = \int (\cos u + 50)\,du = \sin u + 50u + C_2 = \sin\left(2t - \frac{\pi}{2}\right) + 50\left(2t - \frac{\pi}{2}\right) + C_2;$

at $t = 0$ and $s = 0$ we have $0 = \sin\left(-\frac{\pi}{2}\right) + 50\left(-\frac{\pi}{2}\right) + C_2 \Rightarrow C_2 = 1 + 25\pi$

$\Rightarrow s = \sin\left(2t - \frac{\pi}{2}\right) + 100t - 25\pi + (1 + 25\pi) \Rightarrow s = \sin\left(2t - \frac{\pi}{2}\right) + 100t + 1$

59. Let $u = 2t \Rightarrow du = 2\,dt \Rightarrow 3\,du = 6\,dt$

$s = \int 6 \sin 2t\,dt = \int (\sin u)(3\,du) = -3\cos u + C = -3\cos 2t + C;$

at $t = 0$ and $s = 0$ we have $0 = -3\cos 0 + C \Rightarrow C = 3 \Rightarrow s = 3 - 3\cos 2t \Rightarrow s\left(\frac{\pi}{2}\right) = 3 - 3\cos(\pi) = 6$ m

61. All three integrations are correct. In each case, the derivative of the function on the right is the integrand on the left, and each formula has an arbitrary constant for generating the remaining antiderivatives. Moreover, $\sin^2 x + C_1 = 1 - \cos^2 x + C_1 \Rightarrow C_2 = 1 + C_1$; also $-\cos^2 x + C_2 = -\frac{\cos 2x}{2} - \frac{1}{2} + C_2 \Rightarrow C_3 = C_2 - \frac{1}{2} = C_1 + \frac{1}{2}$.

63. (a) $\left(\frac{1}{\frac{1}{60} - 0}\right) \int_0^{1/60} V_{max} \sin 120\pi t\,dt = 60\left[-V_{max}\left(\frac{1}{120\pi}\right)\cos(120\pi t)\right]_0^{1/60} = -\frac{V_{max}}{2\pi}[\cos 2\pi - \cos 0]$

$= -\frac{V_{max}}{2\pi}[1 - 1] = 0$

(b) $V_{max} = \sqrt{2}\,V_{rms} = \sqrt{2}\,(240) \approx 339$ volts

(c) $\int_0^{1/60} (V_{max})^2 \sin^2 120\pi t\,dt = (V_{max})^2 \int_0^{1/60} \left(\frac{1 - \cos 240\pi t}{2}\right) dt = \frac{(V_{max})^2}{2} \int_0^{1/60} (1 - \cos 240\pi t)\,dt$

$= \frac{(V_{max})^2}{2}\left[t - \left(\frac{1}{240\pi}\right)\sin 240\pi t\right]_0^{1/60} = \frac{(V_{max})^2}{2}\left[\left(\frac{1}{60} - \left(\frac{1}{240\pi}\right)\sin(4\pi)\right) - \left(0 - \left(\frac{1}{240\pi}\right)\sin(0)\right)\right] = \frac{(V_{max})^2}{120}$

5.6 SUBSTITUTION AND AREA BETWEEN CURVES

1. (a) Let $u = y + 1 \Rightarrow du = dy$; $y = 0 \Rightarrow u = 1$, $y = 3 \Rightarrow u = 4$

$\int_0^3 \sqrt{y + 1}\,dy = \int_1^4 u^{1/2}\,du = \left[\frac{2}{3} u^{3/2}\right]_1^4 = \left(\frac{2}{3}\right)(4)^{3/2} - \left(\frac{2}{3}\right)(1)^{3/2} = \left(\frac{2}{3}\right)(8) - \left(\frac{2}{3}\right)(1) = \frac{14}{3}$

(b) Use the same substitution for u as in part (a); $y = -1 \Rightarrow u = 0$, $y = 0 \Rightarrow u = 1$

$\int_{-1}^0 \sqrt{y + 1}\,dy = \int_0^1 u^{1/2}\,du = \left[\frac{2}{3} u^{3/2}\right]_0^1 = \left(\frac{2}{3}\right)(1)^{3/2} - 0 = \frac{2}{3}$

3. (a) Let $u = \tan x \Rightarrow du = \sec^2 x\,dx$; $x = 0 \Rightarrow u = 0$, $x = \frac{\pi}{4} \Rightarrow u = 1$

$\int_0^{\pi/4} \tan x \sec^2 x\,dx = \int_0^1 u\,du = \left[\frac{u^2}{2}\right]_0^1 = \frac{1^2}{2} - 0 = \frac{1}{2}$

(b) Use the same substitution as in part (a); $x = -\frac{\pi}{4} \Rightarrow u = -1$, $x = 0 \Rightarrow u = 0$

$\int_{-\pi/4}^0 \tan x \sec^2 x\,dx = \int_{-1}^0 u\,du = \left[\frac{u^2}{2}\right]_{-1}^0 = 0 - \frac{1}{2} = -\frac{1}{2}$

5. (a) $u = 1 + t^4 \Rightarrow du = 4t^3\,dt \Rightarrow \frac{1}{4}\,du = t^3\,dt$; $t = 0 \Rightarrow u = 1$, $t = 1 \Rightarrow u = 2$

$\int_0^1 t^3 \left(1 + t^4\right)^3 dt = \int_1^2 \frac{1}{4} u^3\,du = \left[\frac{u^4}{16}\right]_1^2 = \frac{2^4}{16} - \frac{1^4}{16} = \frac{15}{16}$

(b) Use the same substitution as in part (a); $t = -1 \Rightarrow u = 2$, $t = 1 \Rightarrow u = 2$

$\int_{-1}^1 t^3 \left(1 + t^4\right)^3 dt = \int_2^2 \frac{1}{4} u^3\,du = 0$

7. (a) Let $u = 4 + r^2 \Rightarrow du = 2r\, dr \Rightarrow \frac{1}{2}\, du = r\, dr; r = -1 \Rightarrow u = 5, r = 1 \Rightarrow u = 5$

$$\int_{-1}^{1} \frac{5r}{(4+r^2)^2}\, dr = 5 \int_{5}^{5} \frac{1}{2} u^{-2}\, du = 0$$

(b) Use the same substitution as in part (a); $r = 0 \Rightarrow u = 4, r = 1 \Rightarrow u = 5$

$$\int_{0}^{1} \frac{5r}{(4+r^2)^2}\, dr = 5 \int_{4}^{5} \frac{1}{2} u^{-2}\, du = 5\left[-\frac{1}{2} u^{-1}\right]_{4}^{5} = 5\left(-\frac{1}{2}(5)^{-1}\right) - 5\left(-\frac{1}{2}(4)^{-1}\right) = \frac{1}{8}$$

9. (a) Let $u = x^2 + 1 \Rightarrow du = 2x\, dx \Rightarrow 2\, du = 4x\, dx; x = 0 \Rightarrow u = 1, x = \sqrt{3} \Rightarrow u = 4$

$$\int_{0}^{\sqrt{3}} \frac{4x}{\sqrt{x^2+1}}\, dx = \int_{1}^{4} \frac{2}{\sqrt{u}}\, du = \int_{1}^{4} 2u^{-1/2}\, du = \left[4u^{1/2}\right]_{1}^{4} = 4(4)^{1/2} - 4(1)^{1/2} = 4$$

(b) Use the same substitution as in part (a); $x = -\sqrt{3} \Rightarrow u = 4, x = \sqrt{3} \Rightarrow u = 4$

$$\int_{-\sqrt{3}}^{\sqrt{3}} \frac{4x}{\sqrt{x^2+1}}\, dx = \int_{4}^{4} \frac{2}{\sqrt{u}}\, du = 0$$

11. (a) Let $u = 1 - \cos 3t \Rightarrow du = 3 \sin 3t\, dt \Rightarrow \frac{1}{3}\, du = \sin 3t\, dt; t = 0 \Rightarrow u = 0, t = \frac{\pi}{6} \Rightarrow u = 1 - \cos\frac{\pi}{2} = 1$

$$\int_{0}^{\pi/6} (1 - \cos 3t) \sin 3t\, dt = \int_{0}^{1} \frac{1}{3} u\, du = \left[\frac{1}{3}\left(\frac{u^2}{2}\right)\right]_{0}^{1} = \frac{1}{6}(1)^2 - \frac{1}{6}(0)^2 = \frac{1}{6}$$

(b) Use the same substitution as in part (a); $t = \frac{\pi}{6} \Rightarrow u = 1, t = \frac{\pi}{3} \Rightarrow u = 1 - \cos \pi = 2$

$$\int_{\pi/6}^{\pi/3} (1 - \cos 3t) \sin 3t\, dt = \int_{1}^{2} \frac{1}{3} u\, du = \left[\frac{1}{3}\left(\frac{u^2}{2}\right)\right]_{1}^{2} = \frac{1}{6}(2)^2 - \frac{1}{6}(1)^2 = \frac{1}{2}$$

13. (a) Let $u = 4 + 3 \sin z \Rightarrow du = 3 \cos z\, dz \Rightarrow \frac{1}{3}\, du = \cos z\, dz; z = 0 \Rightarrow u = 4, z = 2\pi \Rightarrow u = 4$

$$\int_{0}^{2\pi} \frac{\cos z}{\sqrt{4+3\sin z}}\, dz = \int_{4}^{4} \frac{1}{\sqrt{u}}\left(\frac{1}{3}\, du\right) = 0$$

(b) Use the same substitution as in part (a); $z = -\pi \Rightarrow u = 4 + 3 \sin(-\pi) = 4, z = \pi \Rightarrow u = 4$

$$\int_{-\pi}^{\pi} \frac{\cos z}{\sqrt{4+3\sin z}}\, dz = \int_{4}^{4} \frac{1}{\sqrt{u}}\left(\frac{1}{3}\, du\right) = 0$$

15. Let $u = t^5 + 2t \Rightarrow du = (5t^4 + 2)\, dt; t = 0 \Rightarrow u = 0, t = 1 \Rightarrow u = 3$

$$\int_{0}^{1} \sqrt{t^5 + 2t}\,(5t^4 + 2)\, dt = \int_{0}^{3} u^{1/2}\, du = \left[\frac{2}{3} u^{3/2}\right]_{0}^{3} = \frac{2}{3}(3)^{3/2} - \frac{2}{3}(0)^{3/2} = 2\sqrt{3}$$

17. Let $u = \cos 2\theta \Rightarrow du = -2 \sin 2\theta\, d\theta \Rightarrow -\frac{1}{2}\, du = \sin 2\theta\, d\theta; \theta = 0 \Rightarrow u = 1, \theta = \frac{\pi}{6} \Rightarrow u = \cos 2\left(\frac{\pi}{6}\right) = \frac{1}{2}$

$$\int_{0}^{\pi/6} \cos^{-3} 2\theta \sin 2\theta\, d\theta = \int_{1}^{1/2} u^{-3}\left(-\frac{1}{2}\, du\right) = -\frac{1}{2} \int_{1}^{1/2} u^{-3}\, du = \left[-\frac{1}{2}\left(\frac{u^{-2}}{-2}\right)\right]_{1}^{1/2} = \frac{1}{4\left(\frac{1}{2}\right)^2} - \frac{1}{4(1)^2} = \frac{3}{4}$$

19. Let $u = 5 - 4 \cos t \Rightarrow du = 4 \sin t\, dt \Rightarrow \frac{1}{4}\, du = \sin t\, dt; t = 0 \Rightarrow u = 5 - 4 \cos 0 = 1, t = \pi \Rightarrow$
    $u = 5 - 4 \cos \pi = 9$

$$\int_{0}^{\pi} 5(5 - 4 \cos t)^{1/4} \sin t\, dt = \int_{1}^{9} 5u^{1/4}\left(\frac{1}{4}\, du\right) = \frac{5}{4} \int_{1}^{9} u^{1/4}\, du = \left[\frac{5}{4}\left(\frac{4}{5} u^{5/4}\right)\right]_{1}^{9} = 9^{5/4} - 1 = 3^{5/2} - 1$$

21. Let $u = 4y - y^2 + 4y^3 + 1 \Rightarrow du = (4 - 2y + 12y^2)\, dy; y = 0 \Rightarrow u = 1, y = 1 \Rightarrow u = 4(1) - (1)^2 + 4(1)^3 + 1 = 8$

$$\int_{0}^{1} (4y - y^2 + 4y^3 + 1)^{-2/3} (12y^2 - 2y + 4)\, dy = \int_{1}^{8} u^{-2/3}\, du = \left[3u^{1/3}\right]_{1}^{8} = 3(8)^{1/3} - 3(1)^{1/3} = 3$$

23. Let $u = \theta^{3/2} \Rightarrow du = \frac{3}{2} \theta^{1/2}\, d\theta \Rightarrow \frac{2}{3}\, du = \sqrt{\theta}\, d\theta; \theta = 0 \Rightarrow u = 0, \theta = \sqrt[3]{\pi^2} \Rightarrow u = \pi$

$$\int_{0}^{\sqrt[3]{\pi^2}} \sqrt{\theta} \cos^2\left(\theta^{3/2}\right) d\theta = \int_{0}^{\pi} \cos^2 u\left(\frac{2}{3}\, du\right) = \left[\frac{2}{3}\left(\frac{u}{2} + \frac{1}{4} \sin 2u\right)\right]_{0}^{\pi} = \frac{2}{3}\left(\frac{\pi}{2} + \frac{1}{4} \sin 2\pi\right) - \frac{2}{3}(0) = \frac{\pi}{3}$$

25. Let $u = 4 - x^2 \Rightarrow du = -2x\,dx \Rightarrow -\frac{1}{2}du = x\,dx;\ x = -2 \Rightarrow u = 0,\ x = 0 \Rightarrow u = 4,\ x = 2 \Rightarrow u = 0$

$A = -\int_{-2}^{0} x\sqrt{4-x^2}\,dx + \int_{0}^{2} x\sqrt{4-x^2}\,dx = -\int_{0}^{4} -\frac{1}{2}u^{1/2}\,du + \int_{4}^{0} -\frac{1}{2}u^{1/2}\,du = 2\int_{0}^{4}\frac{1}{2}u^{1/2}\,du = \int_{0}^{4}u^{1/2}\,du$

$= \left[\frac{2}{3}u^{3/2}\right]_{0}^{4} = \frac{2}{3}(4)^{3/2} - \frac{2}{3}(0)^{3/2} = \frac{16}{3}$

27. Let $u = 1 + \cos x \Rightarrow du = -\sin x\,dx \Rightarrow -du = \sin x\,dx;\ x = -\pi \Rightarrow u = 1 + \cos(-\pi) = 0,\ x = 0$
    $\Rightarrow u = 1 + \cos 0 = 2$

$A = -\int_{-\pi}^{0} 3(\sin x)\sqrt{1+\cos x}\,dx = -\int_{0}^{2} 3u^{1/2}(-du) = 3\int_{0}^{2}u^{1/2}\,du = \left[2u^{3/2}\right]_{0}^{2} = 2(2)^{3/2} - 2(0)^{3/2} = 2^{5/2}$

29. For the sketch given, $a = 0,\ b = \pi;\ f(x) - g(x) = 1 - \cos^2 x = \sin^2 x = \frac{1-\cos 2x}{2}$;

$A = \int_{0}^{\pi}\frac{(1-\cos 2x)}{2}\,dx = \frac{1}{2}\int_{0}^{\pi}(1-\cos 2x)\,dx = \frac{1}{2}\left[x - \frac{\sin 2x}{2}\right]_{0}^{\pi} = \frac{1}{2}[(\pi - 0) - (0 - 0)] = \frac{\pi}{2}$

31. For the sketch given, $a = -2,\ b = 2;\ f(x) - g(x) = 2x^2 - (x^4 - 2x^2) = 4x^2 - x^4$;

$A = \int_{-2}^{2}(4x^2 - x^4)\,dx = \left[\frac{4x^3}{3} - \frac{x^5}{5}\right]_{-2}^{2} = \left(\frac{32}{3} - \frac{32}{5}\right) - \left[-\frac{32}{3} - \left(-\frac{32}{5}\right)\right] = \frac{64}{3} - \frac{64}{5} = \frac{320-192}{15} = \frac{128}{15}$

33. For the sketch given, $c = 0,\ d = 1;\ f(y) - g(y) = (12y^2 - 12y^3) - (2y^2 - 2y) = 10y^2 - 12y^3 + 2y$;

$A = \int_{0}^{1}(10y^2 - 12y^3 + 2y)\,dy = \int_{0}^{1}10y^2\,dy - \int_{0}^{1}12y^3\,dy + \int_{0}^{1}2y\,dy = \left[\frac{10}{3}y^3\right]_{0}^{1} - \left[\frac{12}{4}y^4\right]_{0}^{1} + \left[\frac{2}{2}y^2\right]_{0}^{1}$

$= \left(\frac{10}{3} - 0\right) - (3 - 0) + (1 - 0) = \frac{4}{3}$

35. We want the area between the line $y = 1,\ 0 \le x \le 2$, and the curve $y = \frac{x^2}{4}$, *minus* the area of a triangle
    (formed by $y = x$ and $y = 1$) with base 1 and height 1. Thus, $A = \int_{0}^{2}\left(1 - \frac{x^2}{4}\right)dx - \frac{1}{2}(1)(1) = \left[x - \frac{x^3}{12}\right]_{0}^{2} - \frac{1}{2}$

$= \left(2 - \frac{8}{12}\right) - \frac{1}{2} = 2 - \frac{2}{3} - \frac{1}{2} = \frac{5}{6}$

37. AREA $= A1 + A2$
    A1: For the sketch given, $a = -3$ and we find b by solving the equations $y = x^2 - 4$ and $y = -x^2 - 2x$
    simultaneously for x: $x^2 - 4 = -x^2 - 2x \Rightarrow 2x^2 + 2x - 4 = 0 \Rightarrow 2(x+2)(x-1) \Rightarrow x = -2$ or $x = 1$ so
    $b = -2$: $f(x) - g(x) = (x^2 - 4) - (-x^2 - 2x) = 2x^2 + 2x - 4 \Rightarrow A1 = \int_{-3}^{-2}(2x^2 + 2x - 4)\,dx$
    $= \left[\frac{2x^3}{3} + \frac{2x^2}{2} - 4x\right]_{-3}^{-2} = \left(-\frac{16}{3} + 4 + 8\right) - (-18 + 9 + 12) = 9 - \frac{16}{3} = \frac{11}{3}$;

    A2: For the sketch given, $a = -2$ and $b = 1$: $f(x) - g(x) = (-x^2 - 2x) - (x^2 - 4) = -2x^2 - 2x + 4$
    $\Rightarrow A2 = -\int_{-2}^{1}(2x^2 + 2x - 4)\,dx = -\left[\frac{2x^3}{3} + x^2 - 4x\right]_{-2}^{1} = -\left(\frac{2}{3} + 1 - 4\right) + \left(-\frac{16}{3} + 4 + 8\right)$
    $= -\frac{2}{3} - 1 + 4 - \frac{16}{3} + 4 + 8 = 9$;
    Therefore, AREA $= A1 + A2 = \frac{11}{3} + 9 = \frac{38}{3}$

39. AREA $= A1 + A2 + A3$
    A1: For the sketch given, $a = -2$ and $b = -1$: $f(x) - g(x) = (-x + 2) - (4 - x^2) = x^2 - x - 2$
    $\Rightarrow A1 = \int_{-2}^{-1}(x^2 - x - 2)\,dx = \left[\frac{x^3}{3} - \frac{x^2}{2} - 2x\right]_{-2}^{-1} = \left(-\frac{1}{3} - \frac{1}{2} + 2\right) - \left(-\frac{8}{3} - \frac{4}{2} + 4\right) = \frac{7}{3} - \frac{1}{2} = \frac{14-3}{6} = \frac{11}{6}$;

    A2: For the sketch given, $a = -1$ and $b = 2$: $f(x) - g(x) = (4 - x^2) - (-x + 2) = -(x^2 - x - 2)$
    $\Rightarrow A2 = -\int_{-1}^{2}(x^2 - x - 2)\,dx = -\left[\frac{x^3}{3} - \frac{x^2}{2} - 2x\right]_{-1}^{2} = -\left(\frac{8}{3} - \frac{4}{2} - 4\right) + \left(-\frac{1}{3} - \frac{1}{2} + 2\right) = -3 + 8 - \frac{1}{2} = \frac{9}{2}$;

    A3: For the sketch given, $a = 2$ and $b = 3$: $f(x) - g(x) = (-x + 2) - (4 - x^2) = x^2 - x - 2$
    $\Rightarrow A3 = \int_{2}^{3}(x^2 - x - 2)\,dx = \left[\frac{x^3}{3} - \frac{x^2}{2} - 2x\right]_{2}^{3} = \left(\frac{27}{3} - \frac{9}{2} - 6\right) - \left(\frac{8}{3} - \frac{4}{2} - 4\right) = 9 - \frac{9}{2} - \frac{8}{3}$;

Therefore, $\text{AREA} = A1 + A2 + A3 = \frac{11}{6} + \frac{9}{2} + \left(9 - \frac{9}{2} - \frac{8}{3}\right) = 9 - \frac{5}{6} = \frac{49}{6}$

41. $a = -2, b = 2;$

$f(x) - g(x) = 2 - (x^2 - 2) = 4 - x^2$

$\Rightarrow A = \int_{-2}^{2} (4 - x^2)\,dx = \left[4x - \frac{x^3}{3}\right]_{-2}^{2} = \left(8 - \frac{8}{3}\right) - \left(-8 + \frac{8}{3}\right)$

$= 2 \cdot \left(\frac{24}{3} - \frac{8}{3}\right) = \frac{32}{3}$

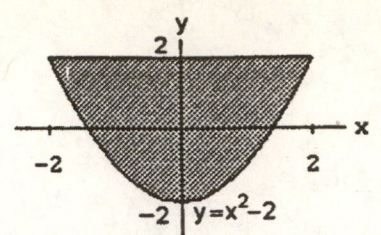

43. $a = 0, b = 2;$

$f(x) - g(x) = 8x - x^4 \Rightarrow A = \int_{0}^{2} (8x - x^4)\,dx$

$= \left[\frac{8x^2}{2} - \frac{x^5}{5}\right]_{0}^{2} = 16 - \frac{32}{5} = \frac{80 - 32}{5} = \frac{48}{5}$

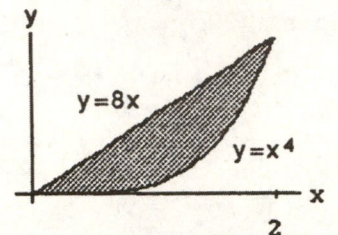

45. Limits of integration: $x^2 = -x^2 + 4x \Rightarrow 2x^2 - 4x = 0$

$\Rightarrow 2x(x - 2) = 0 \Rightarrow a = 0$ and $b = 2;$

$f(x) - g(x) = (-x^2 + 4x) - x^2 = -2x^2 + 4x$

$\Rightarrow A = \int_{0}^{2} (-2x^2 + 4x)\,dx = \left[\frac{-2x^3}{3} + \frac{4x^2}{2}\right]_{0}^{2}$

$= -\frac{16}{3} + \frac{16}{2} = \frac{-32 + 48}{6} = \frac{8}{3}$

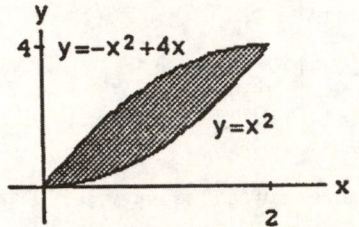

47. Limits of integration: $x^4 - 4x^2 + 4 = x^2$

$\Rightarrow x^4 - 5x^2 + 4 = 0 \Rightarrow (x^2 - 4)(x^2 - 1) = 0$

$\Rightarrow (x + 2)(x - 2)(x + 1)(x - 1) = 0 \Rightarrow x = -2, -1, 1, 2;$

$f(x) - g(x) = (x^4 - 4x^2 + 4) - x^2 = x^4 - 5x^2 + 4$ and

$g(x) - f(x) = x^2 - (x^4 - 4x^2 + 4) = -x^4 + 5x^2 - 4$

$\Rightarrow A = \int_{-2}^{-1} (-x^4 + 5x^2 - 4)\,dx + \int_{-1}^{1} (x^4 - 5x^2 + 4)\,dx$

$+ \int_{1}^{2} (-x^4 + 5x^2 - 4)\,dx$

$= \left[-\frac{x^5}{5} + \frac{5x^3}{3} - 4x\right]_{-2}^{-1} + \left[\frac{x^5}{5} - \frac{5x^3}{3} + 4x\right]_{-1}^{1} + \left[\frac{-x^5}{5} + \frac{5x^3}{3} - 4x\right]_{1}^{2}$

$= \left(\frac{1}{5} - \frac{5}{3} + 4\right) - \left(\frac{32}{5} - \frac{40}{3} + 8\right) + \left(\frac{1}{5} - \frac{5}{3} + 4\right) - \left(-\frac{1}{5} + \frac{5}{3} - 4\right) + \left(-\frac{32}{5} + \frac{40}{3} - 8\right) - \left(-\frac{1}{5} + \frac{5}{3} - 4\right)$

$= -\frac{60}{5} + \frac{60}{3} = \frac{300 - 180}{15} = 8$

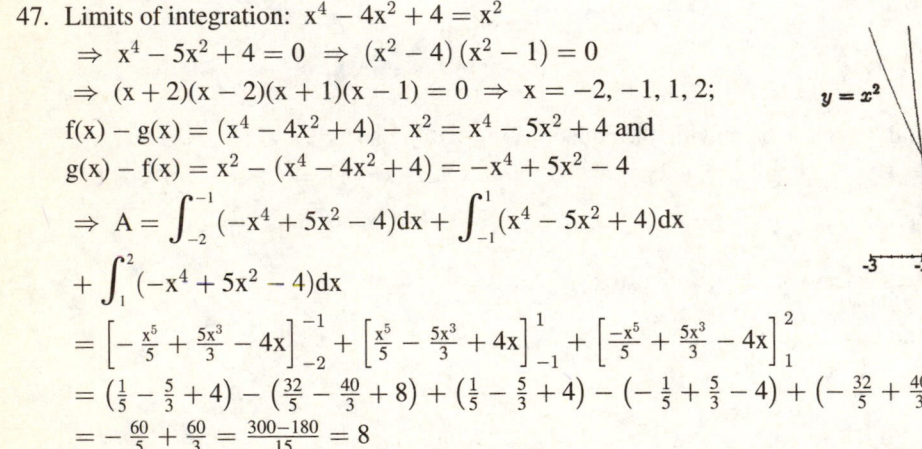
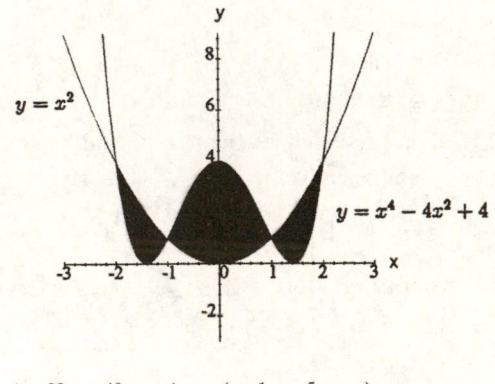

49. Limits of integration: $y = \sqrt{|x|} = \begin{cases} \sqrt{-x}, & x \le 0 \\ \sqrt{x}, & x \ge 0 \end{cases}$ and

$5y = x + 6$ or $y = \frac{x}{5} + \frac{6}{5};$ for $x \le 0$: $\sqrt{-x} = \frac{x}{5} + \frac{6}{5}$

$\Rightarrow 5\sqrt{-x} = x + 6 \Rightarrow 25(-x) = x^2 + 12x + 36$

$\Rightarrow x^2 + 37x + 36 = 0 \Rightarrow (x + 1)(x + 36) = 0$

$\Rightarrow x = -1, -36$ (but $x = -36$ is not a solution);

for $x \ge 0$: $5\sqrt{x} = x + 6 \Rightarrow 25x = x^2 + 12x + 36$

$\Rightarrow x^2 - 13x + 36 = 0 \Rightarrow (x - 4)(x - 9) = 0$

$\Rightarrow x = 4, 9;$ there are three intersection points and

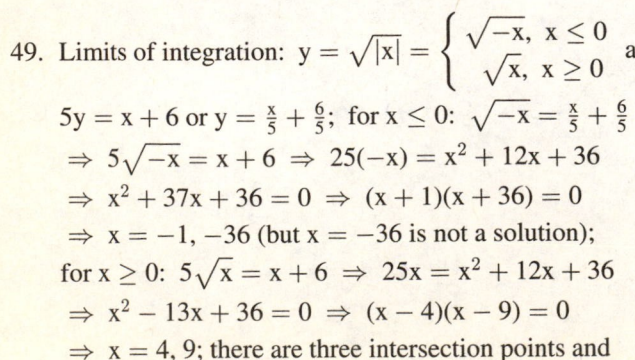
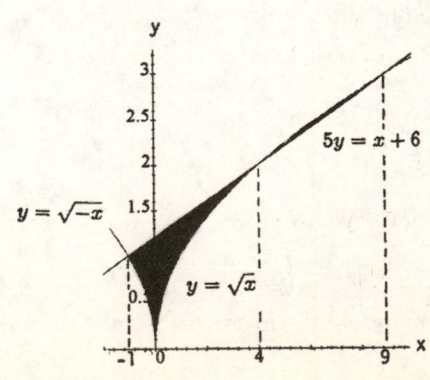

$$A = \int_{-1}^{0} \left(\tfrac{x+6}{5} - \sqrt{-x}\right) dx + \int_{0}^{4} \left(\tfrac{x+6}{5} - \sqrt{x}\right) dx + \int_{4}^{9} \left(\sqrt{x} - \tfrac{x+6}{5}\right) dx$$

$$= \left[\tfrac{(x+6)^2}{10} + \tfrac{2}{3}(-x)^{3/2}\right]_{-1}^{0} + \left[\tfrac{(x+6)^2}{10} - \tfrac{2}{3} x^{3/2}\right]_{0}^{4} + \left[\tfrac{2}{3} x^{3/2} - \tfrac{(x+6)^2}{10}\right]_{4}^{9}$$

$$= \left(\tfrac{36}{10} - \tfrac{25}{10} - \tfrac{2}{3}\right) + \left(\tfrac{100}{10} - \tfrac{2}{3} \cdot 4^{3/2} - \tfrac{36}{10} + 0\right) + \left(\tfrac{2}{3} \cdot 9^{3/2} - \tfrac{225}{10} - \tfrac{2}{3} \cdot 4^{3/2} + \tfrac{100}{10}\right) = -\tfrac{50}{10} + \tfrac{20}{3} = \tfrac{5}{3}$$

51. Limits of integration: $c = 0$ and $d = 3$;

$f(y) - g(y) = 2y^2 - 0 = 2y^2$

$\Rightarrow A = \int_{0}^{3} 2y^2 \, dy = \left[\tfrac{2y^3}{3}\right]_{0}^{3} = 2 \cdot 9 = 18$

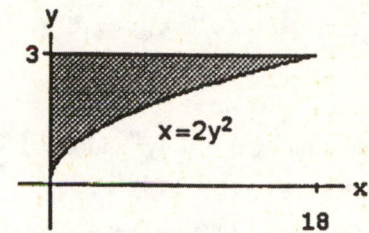

53. Limits of integration: $4x = y^2 - 4$ and $4x = 16 + y$

$\Rightarrow y^2 - 4 = 16 + y \Rightarrow y^2 - y - 20 = 0 \Rightarrow$

$(y - 5)(y + 4) = 0 \Rightarrow c = -4$ and $d = 5$;

$f(y) - g(y) = \left(\tfrac{16+y}{4}\right) - \left(\tfrac{y^2-4}{4}\right) = \tfrac{-y^2+y+20}{4}$

$\Rightarrow A = \tfrac{1}{4} \int_{-4}^{5} (-y^2 + y + 20) \, dy$

$= \tfrac{1}{4} \left[-\tfrac{y^3}{3} + \tfrac{y^2}{2} + 20y\right]_{-4}^{5}$

$= \tfrac{1}{4} \left(-\tfrac{125}{3} + \tfrac{25}{2} + 100\right) - \tfrac{1}{4} \left(\tfrac{64}{3} + \tfrac{16}{2} - 80\right)$

$= \tfrac{1}{4} \left(-\tfrac{189}{3} + \tfrac{9}{2} + 180\right) = \tfrac{243}{8}$

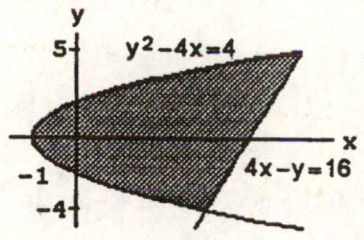

55. Limits of integration: $x = -y^2$ and $x = 2 - 3y^2$

$\Rightarrow -y^2 = 2 - 3y^2 \Rightarrow 2y^2 - 2 = 0$

$\Rightarrow 2(y - 1)(y + 1) = 0 \Rightarrow c = -1$ and $d = 1$;

$f(y) - g(y) = (2 - 3y^2) - (-y^2) = 2 - 2y^2 = 2(1 - y^2)$

$\Rightarrow A = 2 \int_{-1}^{1} (1 - y^2) \, dy = 2 \left[y - \tfrac{y^3}{3}\right]_{-1}^{1}$

$= 2 \left(1 - \tfrac{1}{3}\right) - 2 \left(-1 + \tfrac{1}{3}\right) = 4 \left(\tfrac{2}{3}\right) = \tfrac{8}{3}$

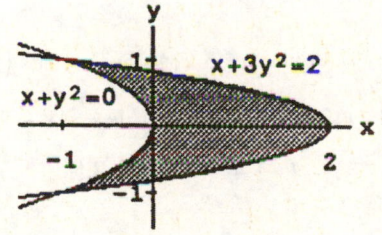

57. Limits of integration: $x = y^2 - 1$ and $x = |y| \sqrt{1 - y^2}$

$\Rightarrow y^2 - 1 = |y| \sqrt{1 - y^2} \Rightarrow y^4 - 2y^2 + 1 = y^2 (1 - y^2)$

$\Rightarrow y^4 - 2y^2 + 1 = y^2 - y^4 \Rightarrow 2y^4 - 3y^2 + 1 = 0$

$\Rightarrow (2y^2 - 1)(y^2 - 1) = 0 \Rightarrow 2y^2 - 1 = 0$ or $y^2 - 1 = 0$

$\Rightarrow y^2 = \tfrac{1}{2}$ or $y^2 = 1 \Rightarrow y = \pm \tfrac{\sqrt{2}}{2}$ or $y = \pm 1$.

Substitution shows that $\tfrac{\pm\sqrt{2}}{2}$ are not solutions $\Rightarrow y = \pm 1$;

for $-1 \le y \le 0$, $f(x) - g(x) = -y \sqrt{1 - y^2} - (y^2 - 1)$

$= 1 - y^2 - y (1 - y^2)^{1/2}$, and by symmetry of the graph,

$A = 2 \int_{-1}^{0} \left[1 - y^2 - y (1 - y^2)^{1/2}\right] dy$

$= 2 \int_{-1}^{0} (1 - y^2) \, dy - 2 \int_{-1}^{0} y (1 - y^2)^{1/2} \, dy$

$= 2 \left[y - \tfrac{y^3}{3}\right]_{-1}^{0} + 2 \left(\tfrac{1}{2}\right) \left[\tfrac{2(1-y^2)^{3/2}}{3}\right]_{-1}^{0} = 2 \left[(0 - 0) - (-1 + \tfrac{1}{3})\right] + \left(\tfrac{2}{3} - 0\right) = 2$

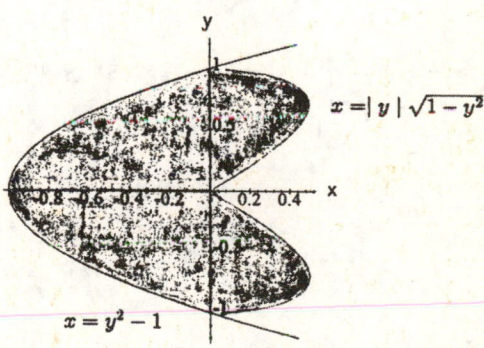

59. Limits of integration: $y = -4x^2 + 4$ and $y = x^4 - 1$

$\Rightarrow x^4 - 1 = -4x^2 + 4 \Rightarrow x^4 + 4x^2 - 5 = 0$

$\Rightarrow (x^2 + 5)(x - 1)(x + 1) = 0 \Rightarrow a = -1$ and $b = 1$;

$f(x) - g(x) = -4x^2 + 4 - x^4 + 1 = -4x^2 - x^4 + 5$

$\Rightarrow A = \int_{-1}^{1}(-4x^2 - x^4 + 5)\,dx = \left[-\frac{4x^3}{3} - \frac{x^5}{5} + 5x\right]_{-1}^{1}$

$= \left(-\frac{4}{3} - \frac{1}{5} + 5\right) - \left(\frac{4}{3} + \frac{1}{5} - 5\right) = 2\left(-\frac{4}{3} - \frac{1}{5} + 5\right) = \frac{104}{15}$

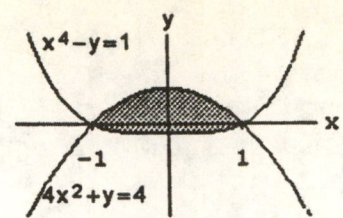

61. Limits of integration: $x = 4 - 4y^2$ and $x = 1 - y^4$

$\Rightarrow 4 - 4y^2 = 1 - y^4 \Rightarrow y^4 - 4y^2 + 3 = 0$

$\Rightarrow \left(y - \sqrt{3}\right)\left(y + \sqrt{3}\right)(y - 1)(y + 1) = 0 \Rightarrow c = -1$

and $d = 1$ since $x \geq 0$; $f(y) - g(y) = (4 - 4y^2) - (1 - y^4)$

$= 3 - 4y^2 + y^4 \Rightarrow A = \int_{-1}^{1}(3 - 4y^2 + y^4)\,dy$

$= \left[3y - \frac{4y^3}{3} + \frac{y^5}{5}\right]_{-1}^{1} = 2\left(3 - \frac{4}{3} + \frac{1}{5}\right) = \frac{56}{15}$

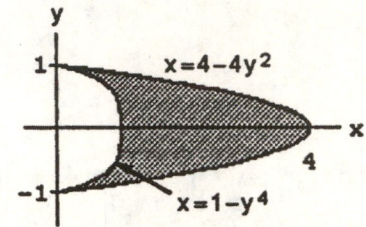

63. $a = 0$, $b = \pi$; $f(x) - g(x) = 2\sin x - \sin 2x$

$\Rightarrow A = \int_{0}^{\pi}(2\sin x - \sin 2x)\,dx = \left[-2\cos x + \frac{\cos 2x}{2}\right]_{0}^{\pi}$

$= \left[-2(-1) + \frac{1}{2}\right] - \left(-2\cdot 1 + \frac{1}{2}\right) = 4$

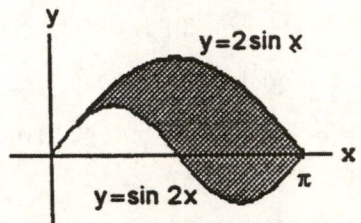

65. $a = -1$, $b = 1$; $f(x) - g(x) = (1 - x^2) - \cos\left(\frac{\pi x}{2}\right)$

$\Rightarrow A = \int_{-1}^{1}\left[1 - x^2 - \cos\left(\frac{\pi x}{2}\right)\right]\,dx = \left[x - \frac{x^3}{3} - \frac{2}{\pi}\sin\left(\frac{\pi x}{2}\right)\right]_{-1}^{1}$

$= \left(1 - \frac{1}{3} - \frac{2}{\pi}\right) - \left(-1 + \frac{1}{3} + \frac{2}{\pi}\right) = 2\left(\frac{2}{3} - \frac{2}{\pi}\right) = \frac{4}{3} - \frac{4}{\pi}$

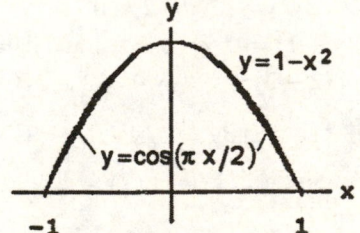

67. $a = -\frac{\pi}{4}$, $b = \frac{\pi}{4}$; $f(x) - g(x) = \sec^2 x - \tan^2 x$

$\Rightarrow A = \int_{-\pi/4}^{\pi/4}(\sec^2 x - \tan^2 x)\,dx$

$= \int_{-\pi/4}^{\pi/4}[\sec^2 x - (\sec^2 x - 1)]\,dx$

$= \int_{-\pi/4}^{\pi/4} 1\cdot dx = [x]_{-\pi/4}^{\pi/4} = \frac{\pi}{4} - \left(-\frac{\pi}{4}\right) = \frac{\pi}{2}$

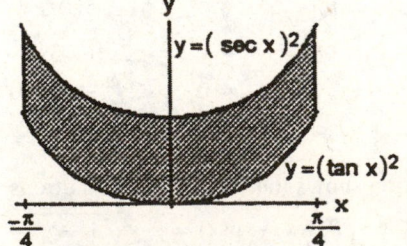

69. $c = 0$, $d = \frac{\pi}{2}$; $f(y) - g(y) = 3\sin y\sqrt{\cos y} - 0 = 3\sin y\sqrt{\cos y}$

$\Rightarrow A = 3\int_{0}^{\pi/2}\sin y\sqrt{\cos y}\,dy = -3\left[\frac{2}{3}(\cos y)^{3/2}\right]_{0}^{\pi/2}$

$= -2(0 - 1) = 2$

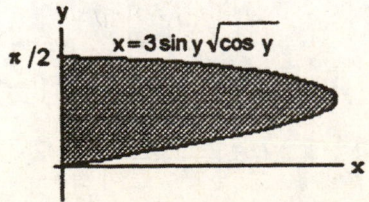

71. $A = A_1 + A_2$

Limits of integration: $x = y^3$ and $x = y \Rightarrow y = y^3$

$\Rightarrow y^3 - y = 0 \Rightarrow y(y-1)(y+1) = 0 \Rightarrow c_1 = -1, d_1 = 0$

and $c_2 = 0, d_2 = 1$; $f_1(y) - g_1(y) = y^3 - y$ and

$f_2(y) - g_2(y) = y - y^3 \Rightarrow$ by symmetry about the origin,

$A_1 + A_2 = 2A_2 \Rightarrow A = 2\int_0^1 (y - y^3)\, dy = 2\left[\frac{y^2}{2} - \frac{y^4}{4}\right]_0^1$

$= 2\left(\frac{1}{2} - \frac{1}{4}\right) = \frac{1}{2}$

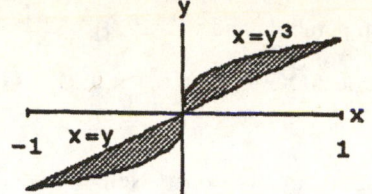

73. $A = A_1 + A_2$

Limits of integration: $y = x$ and $y = \frac{1}{x^2} \Rightarrow x = \frac{1}{x^2}, x \neq 0$

$\Rightarrow x^3 = 1 \Rightarrow x = 1$, $f_1(x) - g_1(x) = x - 0 = x$

$\Rightarrow A_1 = \int_0^1 x\, dx = \left[\frac{x^2}{2}\right]_0^1 = \frac{1}{2}$; $f_2(x) - g_2(x) = \frac{1}{x^2} - 0$

$= x^{-2} \Rightarrow A_2 = \int_1^2 x^{-2}\, dx = \left[\frac{-1}{x}\right]_1^2 = -\frac{1}{2} + 1 = \frac{1}{2}$;

$A = A_1 + A_2 = \frac{1}{2} + \frac{1}{2} = 1$

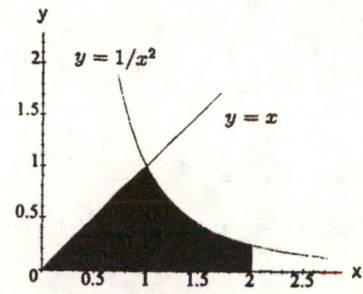

75. (a) The coordinates of the points of intersection of the
line and parabola are $c = x^2 \Rightarrow x = \pm\sqrt{c}$ and $y = c$

(b) $f(y) - g(y) = \sqrt{y} - (-\sqrt{y}) = 2\sqrt{y} \Rightarrow$ the area of the

lower section is, $A_L = \int_0^c [f(y) - g(y)]\, dy$

$= 2\int_0^c \sqrt{y}\, dy = 2\left[\frac{2}{3} y^{3/2}\right]_0^c = \frac{4}{3} c^{3/2}$. The area of the

entire shaded region can be found by setting $c = 4$:

$A = \left(\frac{4}{3}\right) 4^{3/2} = \frac{4 \cdot 8}{3} = \frac{32}{3}$. Since we want c to divide the

region into subsections of equal area we have $A = 2A_L$

$\Rightarrow \frac{32}{3} = 2\left(\frac{4}{3} c^{3/2}\right) \Rightarrow c = 4^{2/3}$

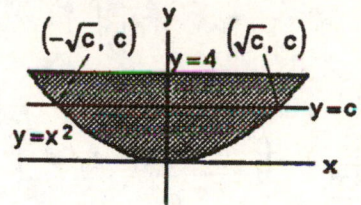

(c) $f(x) - g(x) = c - x^2 \Rightarrow A_L = \int_{-\sqrt{c}}^{\sqrt{c}} [f(x) - g(x)]\, dx = \int_{-\sqrt{c}}^{\sqrt{c}} (c - x^2)\, dx = \left[cx - \frac{x^3}{3}\right]_{-\sqrt{c}}^{\sqrt{c}} = 2\left[c^{3/2} - \frac{c^{3/2}}{3}\right]$

$= \frac{4}{3} c^{3/2}$. Again, the area of the whole shaded region can be found by setting $c = 4 \Rightarrow A = \frac{32}{3}$. From the

condition $A = 2A_L$, we get $\frac{4}{3} c^{3/2} = \frac{32}{3} \Rightarrow c = 4^{2/3}$ as in part (b).

77. Limits of integration: $y = 1 + \sqrt{x}$ and $y = \frac{2}{\sqrt{x}}$

$\Rightarrow 1 + \sqrt{x} = \frac{2}{\sqrt{x}}, x \neq 0 \Rightarrow \sqrt{x} + x = 2 \Rightarrow x = (2 - x)^2$

$\Rightarrow x = 4 - 4x + x^2 \Rightarrow x^2 - 5x + 4 = 0$

$\Rightarrow (x - 4)(x - 1) = 0 \Rightarrow x = 1, 4$ (but $x = 4$ does not

satisfy the equation); $y = \frac{2}{\sqrt{x}}$ and $y = \frac{x}{4} \Rightarrow \frac{2}{\sqrt{x}} = \frac{x}{4}$

$\Rightarrow 8 = x\sqrt{x} \Rightarrow 64 = x^3 \Rightarrow x = 4$.

Therefore, AREA $= A_1 + A_2$: $f_1(x) - g_1(x) = \left(1 + x^{1/2}\right) - \frac{x}{4}$

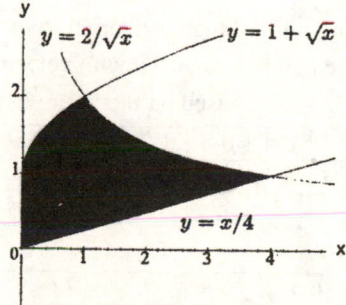

$\Rightarrow A_1 = \int_0^1 \left(1 + x^{1/2} - \frac{x}{4}\right) dx = \left[x + \frac{2}{3} x^{3/2} - \frac{x^2}{8}\right]_0^1$

$= \left(1 + \frac{2}{3} - \frac{1}{8}\right) - 0 = \frac{37}{24}$; $f_2(x) - g_2(x) = 2x^{-1/2} - \frac{x}{4}$

$\Rightarrow A_2 = \int_1^4 \left(2x^{-1/2} - \frac{x}{4}\right) dx = \left[4x^{1/2} - \frac{x^2}{8}\right]_1^4 = \left(4 \cdot 2 - \frac{16}{8}\right) - \left(4 - \frac{1}{8}\right) = 4 - \frac{15}{8} = \frac{17}{8}$;

Therefore, AREA $= A_1 + A_2 = \frac{37}{24} + \frac{17}{8} = \frac{37 + 51}{24} = \frac{88}{24} = \frac{11}{3}$

79. Area between parabola and $y = a^2$: $A = 2\int_0^a (a^2 - x^2)\, dx = 2\left[a^2x - \frac{1}{3}x^3\right]_0^a = 2\left(a^3 - \frac{a^3}{3}\right) - 0 = \frac{4a^3}{3}$;

   Area of triangle AOC: $\frac{1}{2}(2a)(a^2) = a^3$; limit of ratio $= \lim\limits_{a \to 0^+} \frac{a^3}{\left(\frac{4a^3}{3}\right)} = \frac{3}{4}$ which is independent of a.

81. Neither one; they are both zero. Neither integral takes into account the changes in the formulas for the region's upper and lower bounding curves at $x = 0$. The area of the shaded region is actually

   $A = \int_{-1}^0 [-x - (x)]\, dx + \int_0^1 [x - (-x)]\, dx = \int_{-1}^0 -2x\, dx + \int_0^1 2x\, dx = 2.$

83. Let $u = 2x \Rightarrow du = 2\, dx \Rightarrow \frac{1}{2}\, du = dx$; $x = 1 \Rightarrow u = 2$, $x = 3 \Rightarrow u = 6$

   $\int_1^3 \frac{\sin 2x}{x}\, dx = \int_2^6 \frac{\sin u}{\left(\frac{u}{2}\right)}\left(\frac{1}{2}\, du\right) = \int_2^6 \frac{\sin u}{u}\, du = [F(u)]_2^6 = F(6) - F(2)$

85. (a) Let $u = -x \Rightarrow du = -dx$; $x = -1 \Rightarrow u = 1$, $x = 0 \Rightarrow u = 0$

   f odd $\Rightarrow f(-x) = -f(x)$. Then $\int_{-1}^0 f(x)\, dx = \int_1^0 f(-u)\,(-du) = \int_1^0 -f(u)\,(-du) = \int_1^0 f(u)\, du = -\int_0^1 f(u)\, du$
   $= -3$

   (b) Let $u = -x \Rightarrow du = -dx$; $x = -1 \Rightarrow u = 1$, $x = 0 \Rightarrow u = 0$

   f even $\Rightarrow f(-x) = f(x)$. Then $\int_{-1}^0 f(x)\, dx = \int_1^0 f(-u)\,(-du) = -\int_1^0 f(u)\, du = \int_0^1 f(u)\, du = 3$

87. Let $u = a - x \Rightarrow du = -dx$; $x = 0 \Rightarrow u = a$, $x = a \Rightarrow u = 0$

   $I = \int_0^a \frac{f(x)\, dx}{f(x) + f(a-x)} = \int_a^0 \frac{f(a-u)}{f(a-u) + f(u)}\,(-du) = \int_0^a \frac{f(a-u)\, du}{f(u) + f(a-u)} = \int_0^a \frac{f(a-x)\, dx}{f(x) + f(a-x)}$

   $\Rightarrow I + I = \int_0^a \frac{f(x)\, dx}{f(x) + f(a-x)} + \int_0^a \frac{f(a-x)\, dx}{f(x) + f(a-x)} = \int_0^a \frac{f(x) + f(a-x)}{f(x) + f(a-x)}\, dx = \int_0^a dx = [x]_0^a = a - 0 = a.$

   Therefore, $2I = a \Rightarrow I = \frac{a}{2}$.

89. Let $u = x + c \Rightarrow du = dx$; $x = a - c \Rightarrow u = a$, $x = b - c \Rightarrow u = b$

   $\int_{a-c}^{b-c} f(x + c)\, dx = \int_a^b f(u)\, du = \int_a^b f(x)\, dx$

CHAPTER 5 PRACTICE EXERCISES

1. (a) Each time subinterval is of length $\Delta t = 0.4$ sec. The distance traveled over each subinterval, using the midpoint rule, is $\Delta h = \frac{1}{2}(v_i + v_{i+1})\Delta t$, where v_i is the velocity at the left endpoint and v_{i+1} the velocity at the right endpoint of the subinterval. We then add Δh to the height attained so far at the left endpoint v_i to arrive at the height associated with velocity v_{i+1} at the right endpoint. Using this methodology we build the following table based on the figure in the text:

t (sec)	0	0.4	0.8	1.2	1.6	2.0	2.4	2.8	3.2	3.6	4.0	4.4	4.8	5.2	5.6	6.0
v (fps)	0	10	25	55	100	190	180	165	150	140	130	115	105	90	76	65
h (ft)	0	2	9	25	56	114	188	257	320	378	432	481	525	564	592	620.2

t (sec)	6.4	6.8	7.2	7.6	8.0
v (fps)	50	37	25	12	0
h (ft)	643.2	660.6	672	679.4	681.8

   NOTE: Your table values may vary slightly from ours depending on the v-values you read from the graph. Remember that some shifting of the graph occurs in the printing process.

   The total height attained is about 680 ft.

(b) The graph is based on the table in part (a).

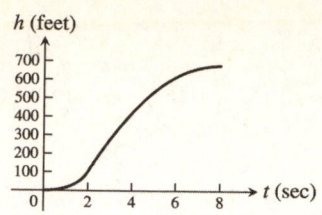

3. (a) $\sum_{k=1}^{10} \frac{a_k}{4} = \frac{1}{4} \sum_{k=1}^{10} a_k = \frac{1}{4}(-2) = -\frac{1}{2}$.

(b) $\sum_{k=1}^{10} (b_k - 3a_k) = \sum_{k=1}^{10} b_k - 3 \sum_{k=1}^{10} a_k = 25 - 3(-2) = 31$

(c) $\sum_{k=1}^{10} (a_k + b_k - 1) = \sum_{k=1}^{10} a_k + \sum_{k=1}^{10} b_k - \sum_{k=1}^{10} 1 = -2 + 25 - (1)(10) = 13$

(d) $\sum_{k=1}^{10} \left(\frac{5}{2} - b_k\right) = \sum_{k=1}^{10} \frac{5}{2} - \sum_{k=1}^{10} b_k = \frac{5}{2}(10) - 25 = 0$

5. Let $u = 2x - 1 \Rightarrow du = 2\,dx \Rightarrow \frac{1}{2}\,du = dx;\ x = 1 \Rightarrow u = 1,\ x = 5 \Rightarrow u = 9$

$\int_1^5 (2x - 1)^{-1/2}\,dx = \int_1^9 u^{-1/2} \left(\frac{1}{2}\,du\right) = \left[u^{1/2}\right]_1^9 = 3 - 1 = 2$

7. Let $u = \frac{x}{2} \Rightarrow 2\,du = dx;\ x = -\pi \Rightarrow u = -\frac{\pi}{2},\ x = 0 \Rightarrow u = 0$

$\int_{-\pi}^0 \cos\left(\frac{x}{2}\right) dx = \int_{-\pi/2}^0 (\cos u)(2\,du) = [2 \sin u]_{-\pi/2}^0 = 2 \sin 0 - 2 \sin\left(-\frac{\pi}{2}\right) = 2(0 - (-1)) = 2$

9. (a) $\int_{-2}^2 f(x)\,dx = \frac{1}{3} \int_{-2}^2 3\,f(x)\,dx = \frac{1}{3}(12) = 4$

(b) $\int_2^5 f(x)\,dx = \int_{-2}^5 f(x)\,dx - \int_{-2}^2 f(x)\,dx = 6 - 4 = 2$

(c) $\int_5^{-2} g(x)\,dx = -\int_{-2}^5 g(x)\,dx = -2$

(d) $\int_{-2}^5 (-\pi\,g(x))\,dx = -\pi \int_{-2}^5 g(x)\,dx = -\pi(2) = -2\pi$

(e) $\int_{-2}^5 \left(\frac{f(x) + g(x)}{5}\right) dx = \frac{1}{5} \int_{-2}^5 f(x)\,dx + \frac{1}{5} \int_{-2}^5 g(x)\,dx = \frac{1}{5}(6) + \frac{1}{5}(2) = \frac{8}{5}$

11. $x^2 - 4x + 3 = 0 \Rightarrow (x - 3)(x - 1) = 0 \Rightarrow x = 3$ or $x = 1$;

Area $= \int_0^1 (x^2 - 4x + 3)\,dx - \int_1^3 (x^2 - 4x + 3)\,dx$

$= \left[\frac{x^3}{3} - 2x^2 + 3x\right]_0^1 - \left[\frac{x^3}{3} - 2x^2 + 3x\right]_1^3$

$= \left[\left(\frac{1^3}{3} - 2(1)^2 + 3(1)\right) - 0\right]$

$\quad - \left[\left(\frac{3^3}{3} - 2(3)^2 + 3(3)\right) - \left(\frac{1^3}{3} - 2(1)^2 + 3(1)\right)\right]$

$= \left(\frac{1}{3} + 1\right) - \left[0 - \left(\frac{1}{3} + 1\right)\right] = \frac{8}{3}$

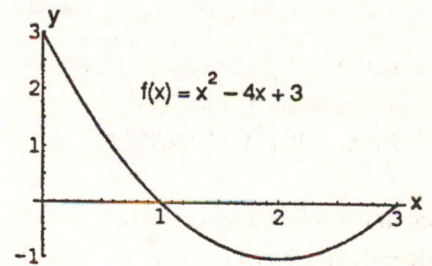

13. $5 - 5x^{2/3} = 0 \Rightarrow 1 - x^{2/3} = 0 \Rightarrow x = \pm 1$;

Area $= \int_{-1}^1 \left(5 - 5x^{2/3}\right) dx - \int_1^8 \left(5 - 5x^{2/3}\right) dx$

$= \left[5x - 3x^{5/3}\right]_{-1}^1 - \left[5x - 3x^{5/3}\right]_1^8$

$= \left[(5(1) - 3(1)^{5/3}) - (5(-1) - 3(-1)^{5/3})\right]$

$\quad - \left[(5(8) - 3(8)^{5/3}) - (5(1) - 3(1)^{5/3})\right]$

$= [2 - (-2)] - [(40 - 96) - 2] = 62$

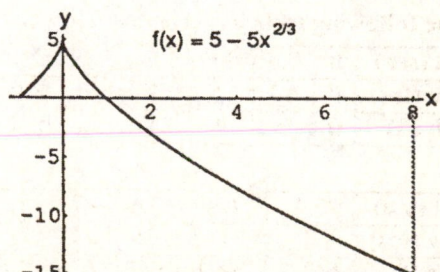

15. $f(x) = x, g(x) = \frac{1}{x^2}, a = 1, b = 2 \Rightarrow A = \int_a^b [f(x) - g(x)]\, dx$

$= \int_1^2 \left(x - \frac{1}{x^2}\right) dx = \left[\frac{x^2}{2} + \frac{1}{x}\right]_1^2 = \left(\frac{4}{2} + \frac{1}{2}\right) - \left(\frac{1}{2} + 1\right) = 1$

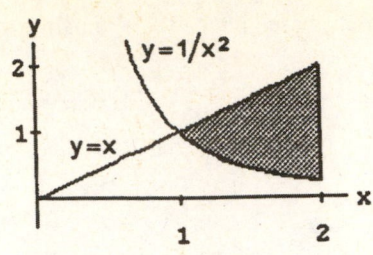

17. $f(x) = \left(1 - \sqrt{x}\right)^2, g(x) = 0, a = 0, b = 1 \Rightarrow A = \int_a^b [f(x) - g(x)]\, dx = \int_0^1 \left(1 - \sqrt{x}\right)^2 dx = \int_0^1 \left(1 - 2\sqrt{x} + x\right) dx$

$= \int_0^1 \left(1 - 2x^{1/2} + x\right) dx = \left[x - \frac{4}{3} x^{3/2} + \frac{x^2}{2}\right]_0^1 = 1 - \frac{4}{3} + \frac{1}{2} = \frac{1}{6}(6 - 8 + 3) = \frac{1}{6}$

19. $f(y) = 2y^2, g(y) = 0, c = 0, d = 3$

$\Rightarrow A = \int_c^d [f(y) - g(y)]\, dy = \int_0^3 (2y^2 - 0)\, dy$

$= 2\int_0^3 y^2\, dy = \frac{2}{3} [y^3]_0^3 = 18$

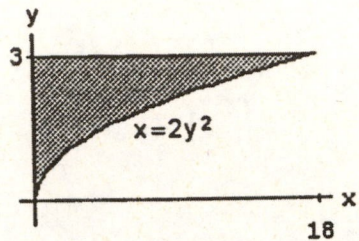

21. Let us find the intersection points: $\frac{y^2}{4} = \frac{y+2}{4}$

$\Rightarrow y^2 - y - 2 = 0 \Rightarrow (y - 2)(y + 1) = 0 \Rightarrow y = -1$

or $y = 2 \Rightarrow c = -1, d = 2; f(y) = \frac{y+2}{4}, g(y) = \frac{y^2}{4}$

$\Rightarrow A = \int_c^d [f(y) - g(y)]\, dy = \int_{-1}^2 \left(\frac{y+2}{4} - \frac{y^2}{4}\right) dy$

$= \frac{1}{4}\int_{-1}^2 (y + 2 - y^2)\, dy = \frac{1}{4}\left[\frac{y^2}{2} + 2y - \frac{y^3}{3}\right]_{-1}^2$

$= \frac{1}{4}\left[\left(\frac{4}{2} + 4 - \frac{8}{3}\right) - \left(\frac{1}{2} - 2 + \frac{1}{3}\right)\right] = \frac{9}{8}$

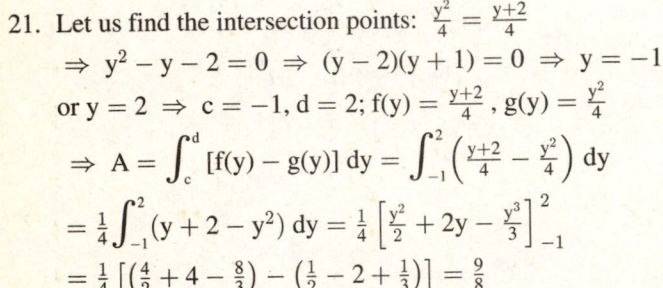

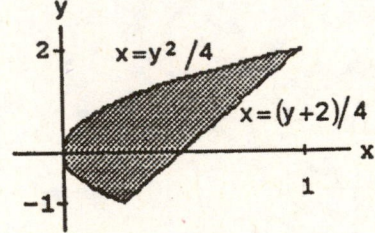

23. $f(x) = x, g(x) = \sin x, a = 0, b = \frac{\pi}{4}$

$\Rightarrow A = \int_a^b [f(x) - g(x)]\, dx = \int_0^{\pi/4} (x - \sin x)\, dx$

$= \left[\frac{x^2}{2} + \cos x\right]_0^{\pi/4} = \left(\frac{\pi^2}{32} + \frac{\sqrt{2}}{2}\right) - 1$

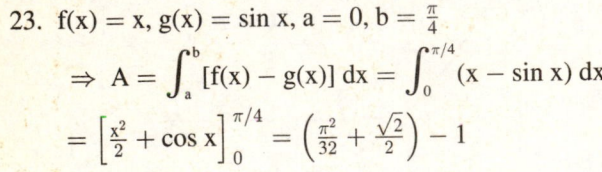

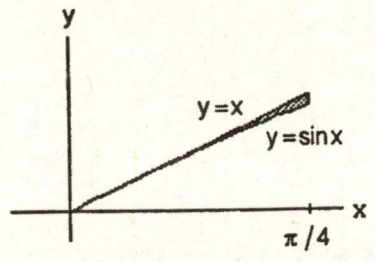

25. $a = 0, b = \pi, f(x) - g(x) = 2\sin x - \sin 2x$

$\Rightarrow A = \int_0^\pi (2\sin x - \sin 2x)\, dx = \left[-2\cos x + \frac{\cos 2x}{2}\right]_0^\pi$

$= \left[-2 \cdot (-1) + \frac{1}{2}\right] - \left(-2 \cdot 1 + \frac{1}{2}\right) = 4$

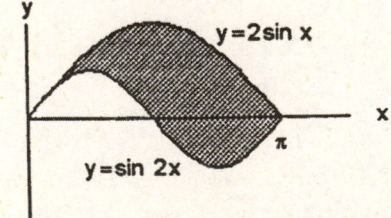

27. $f(y) = \sqrt{y}$, $g(y) = 2 - y$, $c = 1$, $d = 2$

$\Rightarrow A = \int_c^d [f(y) - g(y)]\, dy = \int_1^2 [\sqrt{y} - (2 - y)]\, dy$

$= \int_1^2 (\sqrt{y} - 2 + y)\, dy = \left[\frac{2}{3} y^{3/2} - 2y + \frac{y^2}{2}\right]_1^2$

$= \left(\frac{4}{3}\sqrt{2} - 4 + 2\right) - \left(\frac{2}{3} - 2 + \frac{1}{2}\right) = \frac{4}{3}\sqrt{2} - \frac{7}{6} = \frac{8\sqrt{2}-7}{6}$

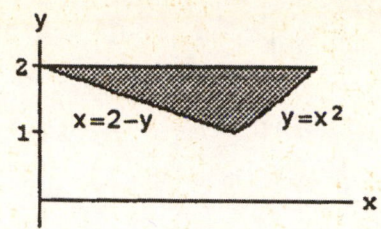

29. $f(x) = x^3 - 3x^2 = x^2(x - 3) \Rightarrow f'(x) = 3x^2 - 6x = 3x(x - 2) \Rightarrow f' = {+}{+}{+} \mid {-}{-}{-}{-} \mid {+}{+}{+}$
         0   2

$\Rightarrow f(0) = 0$ is a maximum and $f(2) = -4$ is a minimum. $A = -\int_0^3 (x^3 - 3x^2)\, dx = -\left[\frac{x^4}{4} - x^3\right]_0^3$

$= -\left(\frac{81}{4} - 27\right) = \frac{27}{4}$

31. The area above the x-axis is $A_1 = \int_0^1 (y^{2/3} - y)\, dy$

$= \left[\frac{3y^{5/3}}{5} - \frac{y^2}{2}\right]_0^1 = \frac{1}{10}$; the area below the x-axis is

$A_2 = \int_{-1}^0 (y^{2/3} - y)\, dy = \left[\frac{3y^{5/3}}{5} - \frac{y^2}{2}\right]_{-1}^0 = \frac{11}{10}$

$\Rightarrow$ the total area is $A_1 + A_2 = \frac{6}{5}$

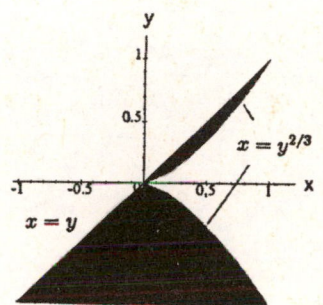

33. $y = x^2 + \int_1^x \frac{1}{t}\, dt \Rightarrow \frac{dy}{dx} = 2x + \frac{1}{x} \Rightarrow \frac{d^2y}{dx^2} = 2 - \frac{1}{x^2}$; $y(1) = 1 + \int_1^1 \frac{1}{t}\, dt = 1$ and $y'(1) = 2 + 1 = 3$

35. $y = \int_5^x \frac{\sin t}{t}\, dt - 3 \Rightarrow \frac{dy}{dx} = \frac{\sin x}{x}$; $x = 5 \Rightarrow y = \int_5^5 \frac{\sin t}{t}\, dt - 3 = -3$

37. Let $u = \cos x \Rightarrow du = -\sin x\, dx \Rightarrow -du = \sin x\, dx$

$\int 2(\cos x)^{-1/2} \sin x\, dx = \int 2u^{-1/2}(-du) = -2\int u^{-1/2}\, du = -2\left(\frac{u^{1/2}}{\frac{1}{2}}\right) + C = -4u^{1/2} + C$

$= -4(\cos x)^{1/2} + C$

39. Let $u = 2\theta + 1 \Rightarrow du = 2\, d\theta \Rightarrow \frac{1}{2}\, du = d\theta$

$\int [2\theta + 1 + 2\cos(2\theta + 1)]\, d\theta = \int (u + 2\cos u)\left(\frac{1}{2}\, du\right) = \frac{u^2}{4} + \sin u + C_1 = \frac{(2\theta+1)^2}{4} + \sin(2\theta + 1) + C_1$

$= \theta^2 + \theta + \sin(2\theta + 1) + C$, where $C = C_1 + \frac{1}{4}$ is still an arbitrary constant

41. $\int \left(t - \frac{2}{t}\right)\left(t + \frac{2}{t}\right)\, dt = \int \left(t^2 - \frac{4}{t^2}\right)\, dt = \int (t^2 - 4t^{-2})\, dt = \frac{t^3}{3} - 4\left(\frac{t^{-1}}{-1}\right) + C = \frac{t^3}{3} + \frac{4}{t} + C$

43. Let $u = 2t^{3/2} \Rightarrow du = 3\sqrt{t}\, dt \Rightarrow \frac{1}{3}\, du = \sqrt{t}\, dt$

$\int \sqrt{t}\sin\left(2t^{3/2}\right)dt = \frac{1}{3}\int \sin u\, du = -\frac{1}{3}\cos u + C = -\frac{1}{3}\cos\left(2t^{3/2}\right) + C$

45. $\int_{-1}^1 (3x^2 - 4x + 7)\, dx = [x^3 - 2x^2 + 7x]_{-1}^1 = [1^3 - 2(1)^2 + 7(1)] - [(-1)^3 - 2(-1)^2 + 7(-1)] = 6 - (-10) = 16$

47. $\int_1^2 \frac{4}{v^2}\, dv = \int_1^2 4v^{-2}\, dv = [-4v^{-1}]_1^2 = \left(\frac{-4}{2}\right) - \left(\frac{-4}{1}\right) = 2$

49. $\int_1^4 \frac{dt}{t\sqrt{t}} = \int_1^4 \frac{dt}{t^{3/2}} = \int_1^4 t^{-3/2}\,dt = \left[-2t^{-1/2}\right]_1^4 = \frac{-2}{\sqrt{4}} - \frac{(-2)}{\sqrt{1}} = 1$

51. Let $u = 2x + 1 \Rightarrow du = 2\,dx \Rightarrow 18\,du = 36\,dx;\ x = 0 \Rightarrow u = 1, x = 1 \Rightarrow u = 3$

$\int_0^1 \frac{36\,dx}{(2x+1)^3} = \int_1^3 18u^{-3}\,du = \left[\frac{18u^{-2}}{-2}\right]_1^3 = \left[\frac{-9}{u^2}\right]_1^3 = \left(\frac{-9}{3^2}\right) - \left(\frac{-9}{1^2}\right) = 8$

53. Let $u = 1 - x^{2/3} \Rightarrow du = -\frac{2}{3}x^{-1/3}\,dx \Rightarrow -\frac{3}{2}\,du = x^{-1/3}\,dx;\ x = \frac{1}{8} \Rightarrow u = 1 - \left(\frac{1}{8}\right)^{2/3} = \frac{3}{4},$

$x = 1 \Rightarrow u = 1 - 1^{2/3} = 0$

$\int_{1/8}^1 x^{-1/3}\left(1 - x^{2/3}\right)^{3/2}\,dx = \int_{3/4}^0 u^{3/2}\left(-\frac{3}{2}\,du\right) = \left[\left(-\frac{3}{2}\right)\left(\frac{u^{5/2}}{\frac{5}{2}}\right)\right]_{3/4}^0 = \left[-\frac{3}{5}u^{5/2}\right]_{3/4}^0 = -\frac{3}{5}(0)^{5/2} - \left(-\frac{3}{5}\right)\left(\frac{3}{4}\right)^{5/2}$

$= \frac{27\sqrt{3}}{160}$

55. Let $u = 5r \Rightarrow du = 5\,dr \Rightarrow \frac{1}{5}\,du = dr;\ r = 0 \Rightarrow u = 0, r = \pi \Rightarrow u = 5\pi$

$\int_0^\pi \sin^2 5r\,dr = \int_0^{5\pi} (\sin^2 u)\left(\frac{1}{5}\,du\right) = \frac{1}{5}\left[\frac{u}{2} - \frac{\sin 2u}{4}\right]_0^{5\pi} = \left(\frac{\pi}{2} - \frac{\sin 10\pi}{20}\right) - \left(0 - \frac{\sin 0}{20}\right) = \frac{\pi}{2}$

57. $\int_0^{\pi/3} \sec^2\theta\,d\theta = [\tan\theta]_0^{\pi/3} = \tan\frac{\pi}{3} - \tan 0 = \sqrt{3}$

59. Let $u = \frac{x}{6} \Rightarrow du = \frac{1}{6}\,dx \Rightarrow 6\,du = dx;\ x = \pi \Rightarrow u = \frac{\pi}{6}, x = 3\pi \Rightarrow u = \frac{\pi}{2}$

$\int_\pi^{3\pi} \cot^2\frac{x}{6}\,dx = \int_{\pi/6}^{\pi/2} 6\cot^2 u\,du = 6\int_{\pi/6}^{\pi/2}(\csc^2 u - 1)\,du = [6(-\cot u - u)]_{\pi/6}^{\pi/2} = 6\left(-\cot\frac{\pi}{2} - \frac{\pi}{2}\right) - 6\left(-\cot\frac{\pi}{6} - \frac{\pi}{6}\right)$

$= 6\sqrt{3} - 2\pi$

61. $\int_{-\pi/3}^0 \sec x \tan x\,dx = [\sec x]_{-\pi/3}^0 = \sec 0 - \sec\left(-\frac{\pi}{3}\right) = 1 - 2 = -1$

63. Let $u = \sin x \Rightarrow du = \cos x\,dx;\ x = 0 \Rightarrow u = 0, x = \frac{\pi}{2} \Rightarrow u = 1$

$\int_0^{\pi/2} 5(\sin x)^{3/2}\cos x\,dx = \int_0^1 5u^{3/2}\,du = \left[5\left(\frac{2}{5}\right)u^{5/2}\right]_0^1 = \left[2u^{5/2}\right]_0^1 = 2(1)^{5/2} - 2(0)^{5/2} = 2$

65. Let $u = \sin 3x \Rightarrow du = 3\cos 3x\,dx \Rightarrow \frac{1}{3}\,du = \cos 3x\,dx;\ x = -\frac{\pi}{2} \Rightarrow u = \sin\left(-\frac{3\pi}{2}\right) = 1, x = \frac{\pi}{2} \Rightarrow u = \sin\left(\frac{3\pi}{2}\right)$

$= -1$

$\int_{-\pi/2}^{\pi/2} 15\sin^4 3x\cos 3x\,dx = \int_1^{-1} 15u^4\left(\frac{1}{3}\,du\right) = \int_1^{-1} 5u^4\,du = [u^5]_1^{-1} = (-1)^5 - (1)^5 = -2$

67. Let $u = 1 + 3\sin^2 x \Rightarrow du = 6\sin x\cos x\,dx \Rightarrow \frac{1}{2}\,du = 3\sin x\cos x\,dx;\ x = 0 \Rightarrow u = 1, x = \frac{\pi}{2}$

$\Rightarrow u = 1 + 3\sin^2\frac{\pi}{2} = 4$

$\int_0^{\pi/2} \frac{3\sin x\cos x}{\sqrt{1 + 3\sin^2 x}}\,dx = \int_1^4 \frac{1}{\sqrt{u}}\left(\frac{1}{2}\,du\right) = \int_1^4 \frac{1}{2}u^{-1/2}\,du = \left[\frac{1}{2}\left(\frac{u^{1/2}}{\frac{1}{2}}\right)\right]_1^4 = \left[u^{1/2}\right]_1^4 = 4^{1/2} - 1^{1/2} = 1$

69. Let $u = \sec\theta \Rightarrow du = \sec\theta\tan\theta\,d\theta;\ \theta = 0 \Rightarrow u = \sec 0 = 1, \theta = \frac{\pi}{3} \Rightarrow u = \sec\frac{\pi}{3} = 2$

$\int_0^{\pi/3} \frac{\tan\theta}{\sqrt{2\sec\theta}}\,d\theta = \int_0^{\pi/3} \frac{\sec\theta\tan\theta}{\sec\theta\sqrt{2\sec\theta}}\,d\theta = \int_0^{\pi/3} \frac{\sec\theta\tan\theta}{\sqrt{2}(\sec\theta)^{3/2}}\,d\theta = \int_1^2 \frac{1}{\sqrt{2}u^{3/2}}\,du = \frac{1}{\sqrt{2}}\int_1^2 u^{-3/2}\,du$

$= \frac{1}{\sqrt{2}}\left[\frac{u^{-1/2}}{\left(-\frac{1}{2}\right)}\right]_1^2 = \left[-\frac{2}{\sqrt{2u}}\right]_1^2 = -\frac{2}{\sqrt{2(2)}} - \left(-\frac{2}{\sqrt{2(1)}}\right) = \sqrt{2} - 1$

71. (a) $\text{av}(f) = \frac{1}{1-(-1)} \int_{-1}^{1} (mx + b)\, dx = \frac{1}{2} \left[\frac{mx^2}{2} + bx \right]_{-1}^{1} = \frac{1}{2} \left[\left(\frac{m(1)^2}{2} + b(1) \right) - \left(\frac{m(-1)^2}{2} + b(-1) \right) \right] = \frac{1}{2}(2b) = b$

    (b) $\text{av}(f) = \frac{1}{k-(-k)} \int_{-k}^{k} (mx + b)\, dx = \frac{1}{2k} \left[\frac{mx^2}{2} + bx \right]_{-k}^{k} = \frac{1}{2k} \left[\left(\frac{m(k)^2}{2} + b(k) \right) - \left(\frac{m(-k)^2}{2} + b(-k) \right) \right]$

        $= \frac{1}{2k}(2bk) = b$

73. $f'_{av} = \frac{1}{b-a} \int_{a}^{b} \sqrt{a}xf'(x)\, dx = \frac{1}{b-a} [f(x)]_a^b = \frac{1}{b-a} [f(b) - f(a)] = \frac{f(b) - f(a)}{b-a}$ so the average value of f' over $[a, b]$ is the

    slope of the secant line joining the points $(a, f(a))$ and $(b, f(b))$, which is the average rate of change of f over $[a, b]$.

75. We want to evaluate

$$\frac{1}{365-0} \int_0^{365} f(x)\, dx = \frac{1}{365} \int_0^{365} \left(37\sin\left[\frac{2\pi}{365}(x - 101) \right] + 25 \right) dx = \frac{37}{365} \int_0^{365} \sin\left[\frac{2\pi}{365}(x - 101) \right] dx + \frac{25}{365} \int_0^{365} dx$$

    Notice that the period of $y = \sin\left[\frac{2\pi}{365}(x - 101) \right]$ is $\frac{2\pi}{\frac{2\pi}{365}} = 365$ and that we are integrating this function over an iterval of

    length 365. Thus the value of $\frac{37}{365} \int_0^{365} \sin\left[\frac{2\pi}{365}(x - 101) \right] dx + \frac{25}{365} \int_0^{365} dx$ is $\frac{37}{365} \cdot 0 + \frac{25}{365} \cdot 365 = 25$.

77. $\frac{dy}{dx} = \sqrt{2 + \cos^3 x}$

79. $\frac{dy}{dx} = \frac{d}{dx} \left(-\int_1^x \frac{6}{3+t^4} dt \right) = -\frac{6}{3+x^4}$

81. Yes. The function f, being differentiable on $[a, b]$, is then continuous on $[a, b]$. The Fundamental Theorem of

    Calculus says that every continuous function on $[a, b]$ is the derivative of a function on $[a, b]$.

83. $y = \int_1^x \sqrt{1 + t^2}\, dt = -\int_1^x \sqrt{1 + t^2}\, dt \Rightarrow \frac{dy}{dx} = \frac{d}{dx} \left[-\int_1^x \sqrt{1 + t^2}\, dt \right] = -\frac{d}{dx} \left[\int_1^x \sqrt{1 + t^2}\, dt \right] = -\sqrt{1 + x^2}$

85. We estimate the area A using midpoints of the vertical intervals, and we will estimate the width of the parking lot on each

    interval by averaging the widths at top and bottom. This gives the estimate

    $A \approx 15 \cdot \left(\frac{0+36}{2} + \frac{36+54}{2} + \frac{54+51}{2} + \frac{51+49.5}{2} + \frac{49.5+54}{2} + \frac{54+64.4}{2} + \frac{64.4+67.5}{2} + \frac{67.5+42}{2} \right)$

    $A \approx 5961$ ft^2. The cost is Area $\cdot$ ($2.10/ft^2) $\approx$ (5961 ft^2)($2.10/ft^2) = $12,518.10 $\Rightarrow$ the job cannot be done for $11,000.

87. $\text{av}(I) = \frac{1}{30} \int_0^{30} (1200 - 40t)\, dt = \frac{1}{30} [1200t - 20t^2]_0^{30} = \frac{1}{30} [((1200(30) - 20(30)^2) - (1200(0) - 20(0)^2)]$

    $= \frac{1}{30}(18,000) = 600$; Average Daily Holding Cost = (600)($0.03) = $18

89. $\text{av}(I) = \frac{1}{30} \int_0^{30} \left(450 - \frac{t^2}{2} \right) dt = \frac{1}{30} \left[450t - \frac{t^3}{6} \right]_0^{30} = \frac{1}{30} \left[450(30) - \frac{30^3}{6} - 0 \right] = 300$; Average Daily Holding Cost

    $= (300)($0.02) = $6

CHAPTER 5 ADDITIONAL AND ADVANCED EXERCISES

1. (a) Yes, because $\int_0^1 f(x)\, dx = \frac{1}{7} \int_0^1 7f(x)\, dx = \frac{1}{7}(7) = 1$

    (b) No. For example, $\int_0^1 8x\, dx = [4x^2]_0^1 = 4$, but $\int_0^1 \sqrt{8x}\, dx = \left[2\sqrt{2} \left(\frac{x^{3/2}}{\frac{3}{2}} \right) \right]_0^1 = \frac{4\sqrt{2}}{3} (1^{3/2} - 0^{3/2})$

        $= \frac{4\sqrt{2}}{3} \neq \sqrt{4}$

3. $y = \frac{1}{a} \int_0^x f(t) \sin a(x-t)\, dt = \frac{1}{a} \int_0^x f(t) \sin ax \cos at\, dt - \frac{1}{a} \int_0^x f(t) \cos ax \sin at\, dt$

$= \frac{\sin ax}{a} \int_0^x f(t) \cos at\, dt - \frac{\cos ax}{a} \int_0^x f(t) \sin at\, dt \Rightarrow \frac{dy}{dx} = \cos ax \left(\int_0^x f(t) \cos at\, dt \right)$

$+ \frac{\sin ax}{a} \left(\frac{d}{dx} \int_0^x f(t) \cos at\, dt \right) + \sin ax \int_0^x f(t) \sin at\, dt - \frac{\cos ax}{a} \left(\frac{d}{dx} \int_0^x f(t) \sin at\, dt \right)$

$= \cos ax \int_0^x f(t) \cos at\, dt + \frac{\sin ax}{a} (f(x) \cos ax) + \sin ax \int_0^x f(t) \sin at\, dt - \frac{\cos ax}{a} (f(x) \sin ax)$

$\Rightarrow \frac{dy}{dx} = \cos ax \int_0^x f(t) \cos at\, dt + \sin ax \int_0^x f(t) \sin at\, dt.$ Next,

$\frac{d^2y}{dx^2} = -a \sin ax \int_0^x f(t) \cos at\, dt + (\cos ax) \left(\frac{d}{dx} \int_0^x f(t) \cos at\, dt \right) + a \cos ax \int_0^x f(t) \sin at\, dt$

$+ (\sin ax) \left(\frac{d}{dx} \int_0^x f(t) \sin at\, dt \right) = -a \sin ax \int_0^x f(t) \cos at\, dt + (\cos ax) f(x) \cos ax$

$+ a \cos ax \int_0^x f(t) \sin at\, dt + (\sin ax) f(x) \sin ax = -a \sin ax \int_0^x f(t) \cos at\, dt + a \cos ax \int_0^x f(t) \sin at\, dt + f(x).$

Therefore, $y'' + a^2 y = a \cos ax \int_0^x f(t) \sin at\, dt - a \sin ax \int_0^x f(t) \cos at\, dt + f(x)$

$+ a^2 \left(\frac{\sin ax}{a} \int_0^x f(t) \cos at\, dt - \frac{\cos ax}{a} \int_0^x f(t) \sin at\, dt \right) = f(x).$ Note also that $y'(0) = y(0) = 0.$

5. (a) $\int_0^{x^2} f(t)\, dt = x \cos \pi x \Rightarrow \frac{d}{dx} \int_0^{x^2} f(t)\, dt = \cos \pi x - \pi x \sin \pi x \Rightarrow f(x^2)(2x) = \cos \pi x - \pi x \sin \pi x$

$\Rightarrow f(x^2) = \frac{\cos \pi x - \pi x \sin \pi x}{2x}.$ Thus, $x = 2 \Rightarrow f(4) = \frac{\cos 2\pi - 2\pi \sin 2\pi}{4} = \frac{1}{4}$

(b) $\int_0^{f(x)} t^2\, dt = \left[\frac{t^3}{3} \right]_0^{f(x)} = \frac{1}{3}(f(x))^3 \Rightarrow \frac{1}{3}(f(x))^3 = x \cos \pi x \Rightarrow (f(x))^3 = 3x \cos \pi x \Rightarrow f(x) = \sqrt[3]{3x \cos \pi x}$

$\Rightarrow f(4) = \sqrt[3]{3(4) \cos 4\pi} = \sqrt[3]{12}$

7. $\int_1^b f(x)\, dx = \sqrt{b^2 + 1} - \sqrt{2} \Rightarrow f(b) = \frac{d}{db} \int_1^b f(x)\, dx = \frac{1}{2}(b^2+1)^{-1/2}(2b) = \frac{b}{\sqrt{b^2+1}} \Rightarrow f(x) = \frac{x}{\sqrt{x^2+1}}$

9. $\frac{dy}{dx} = 3x^2 + 2 \Rightarrow y = \int (3x^2 + 2)\, dx = x^3 + 2x + C.$ Then $(1, -1)$ on the curve $\Rightarrow 1^3 + 2(1) + C = -1 \Rightarrow C = -4$

$\Rightarrow y = x^3 + 2x - 4$

11. $\int_{-8}^3 f(x)\, dx = \int_{-8}^0 x^{2/3}\, dx + \int_0^3 -4\, dx$

$= \left[\frac{3}{5} x^{5/3} \right]_{-8}^0 + [-4x]_0^3$

$= \left(0 - \frac{3}{5}(-8)^{5/3} \right) + (-4(3) - 0) = \frac{96}{5} - 12$

$= \frac{36}{5}$

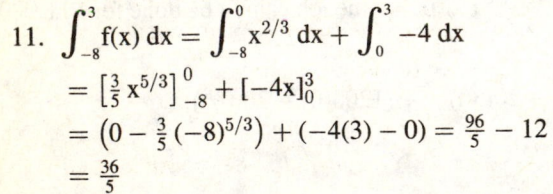
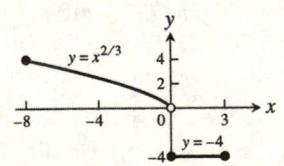

13. $\int_0^2 g(t)\, dt = \int_0^1 t\, dt + \int_1^2 \sin \pi t\, dt$

$= \left[\frac{t^2}{2} \right]_0^1 + \left[-\frac{1}{\pi} \cos \pi t \right]_1^2$

$= \left(\frac{1}{2} - 0 \right) + \left[-\frac{1}{\pi} \cos 2\pi - \left(-\frac{1}{\pi} \cos \pi \right) \right]$

$= \frac{1}{2} - \frac{2}{\pi}$

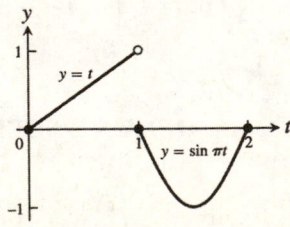

15. $\int_{-2}^{2} f(x)\,dx = \int_{-2}^{-1} dx + \int_{-1}^{1} (1-x^2)\,dx + \int_{1}^{2} 2\,dx$

$= [x]_{-2}^{-1} + \left[x - \frac{x^3}{3}\right]_{-1}^{1} + [2x]_{1}^{2}$

$= (-1-(-2)) + \left[\left(1-\frac{1^3}{3}\right) - \left(-1 - \frac{(-1)^3}{3}\right)\right] + \left[2(2) - 2(1)\right]$

$= 1 + \frac{2}{3} - \left(-\frac{2}{3}\right) + 4 - 2 = \frac{13}{3}$

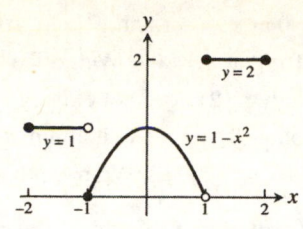

17. Ave. value $= \frac{1}{b-a}\int_{a}^{b} f(x)\,dx = \frac{1}{2-0}\int_{0}^{2} f(x)\,dx = \frac{1}{2}\left[\int_{0}^{1} x\,dx + \int_{1}^{2} (x-1)\,dx\right] = \frac{1}{2}\left[\frac{x^2}{2}\right]_{0}^{1} + \frac{1}{2}\left[\frac{x^2}{2} - x\right]_{1}^{2}$

$= \frac{1}{2}\left[\left(\frac{1^2}{2} - 0\right) + \left(\frac{2^2}{2} - 2\right) - \left(\frac{1^2}{2} - 1\right)\right] = \frac{1}{2}$

19. $f(x) = \int_{1/x}^{x} \frac{1}{t}\,dt \Rightarrow f'(x) = \frac{1}{x}\left(\frac{dx}{dx}\right) - \left(\frac{1}{\frac{1}{x}}\right)\left(\frac{d}{dx}\left(\frac{1}{x}\right)\right) = \frac{1}{x} - x\left(-\frac{1}{x^2}\right) = \frac{1}{x} + \frac{1}{x} = \frac{2}{x}$

21. $g(y) = \int_{\sqrt{y}}^{2\sqrt{y}} \sin t^2\,dt \Rightarrow g'(y) = \left(\sin\left(2\sqrt{y}\right)^2\right)\left(\frac{d}{dy}\left(2\sqrt{y}\right)\right) - \left(\sin\left(\sqrt{y}\right)^2\right)\left(\frac{d}{dy}\left(\sqrt{y}\right)\right) = \frac{\sin 4y}{\sqrt{y}} - \frac{\sin y}{2\sqrt{y}}$

23. Let $f(x) = x^5$ on $[0, 1]$. Partition $[0, 1]$ into n subintervals with $\Delta x = \frac{1-0}{n} = \frac{1}{n}$. Then $\frac{1}{n}, \frac{2}{n}, \ldots, \frac{n}{n}$ are the

right-hand endpoints of the subintervals. Since f is increasing on $[0, 1]$, $U = \sum_{j=1}^{\infty} \left(\frac{j}{n}\right)^5 \left(\frac{1}{n}\right)$ is the upper sum for

$f(x) = x^5$ on $[0, 1] \Rightarrow \lim_{n\to\infty} \sum_{j=1}^{\infty} \left(\frac{j}{n}\right)^5 \left(\frac{1}{n}\right) = \lim_{n\to\infty} \frac{1}{n}\left[\left(\frac{1}{n}\right)^5 + \left(\frac{2}{n}\right)^5 + \cdots + \left(\frac{n}{n}\right)^5\right] = \lim_{n\to\infty} \left[\frac{1^5 + 2^5 + \ldots + n^5}{n^6}\right]$

$= \int_{0}^{1} x^5\,dx = \left[\frac{x^6}{6}\right]_{0}^{1} = \frac{1}{6}$

25. Let $y = f(x)$ on $[0, 1]$. Partition $[0, 1]$ into n subintervals with $\Delta x = \frac{1-0}{n} = \frac{1}{n}$. Then $\frac{1}{n}, \frac{2}{n}, \ldots, \frac{n}{n}$ are the

right-hand endpoints of the subintervals. Since f is continuous on $[0, 1]$, $\sum_{j=1}^{\infty} f\left(\frac{j}{n}\right)\left(\frac{1}{n}\right)$ is a Riemann sum of

$y = f(x)$ on $[0, 1] \Rightarrow \lim_{n\to\infty} \sum_{j=1}^{\infty} f\left(\frac{j}{n}\right)\left(\frac{1}{n}\right) = \lim_{n\to\infty} \frac{1}{n}\left[f\left(\frac{1}{n}\right) + f\left(\frac{2}{n}\right) + \cdots + f\left(\frac{n}{n}\right)\right] = \int_{0}^{1} f(x)\,dx$

27. (a) Let the polygon be inscribed in a circle of radius r. If we draw a radius from the center of the circle (and

the polygon) to each vertex of the polygon, we have n isosceles triangles formed (the equal sides are equal

to r, the radius of the circle) and a vertex angle of θ_n where $\theta_n = \frac{2\pi}{n}$. The area of each triangle is

$A_n = \frac{1}{2}r^2 \sin\theta_n \Rightarrow$ the area of the polygon is $A = nA_n = \frac{nr^2}{2}\sin\theta_n = \frac{nr^2}{2}\sin\frac{2\pi}{n}$.

(b) $\lim_{n\to\infty} A = \lim_{n\to\infty} \frac{nr^2}{2}\sin\frac{2\pi}{n} = \lim_{n\to\infty} \frac{n\pi r^2}{2\pi}\sin\frac{2\pi}{n} = \lim_{n\to\infty} (\pi r^2)\frac{\sin\left(\frac{2\pi}{n}\right)}{\left(\frac{2\pi}{n}\right)} = (\pi r^2)\lim_{2\pi/n\to 0} \frac{\sin\left(\frac{2\pi}{n}\right)}{\left(\frac{2\pi}{n}\right)} = \pi r^2$

29. (a) $g(1) = \int_{1}^{1} f(t)\,dt = 0$

(b) $g(3) = \int_{1}^{3} f(t)\,dt = -\frac{1}{2}(2)(1) = -1$

(c) $g(-1) = \int_{1}^{-1} f(t)\,dt = -\int_{-1}^{1} f(t)\,dt = -\frac{1}{4}(\pi 2^2) = -\pi$

(d) $g'(x) = f(x) = 0 \Rightarrow x = -3, 1, 3$ and the sign chart for $g'(x) = f(x)$ is $\begin{array}{c} | \;+++ | \;--- | \;+++. \\ -3 \quad\;\; 1 \quad\;\; 3 \end{array}$ So g has a

relative maximum at $x = 1$.

(e) $g'(-1) = f(-1) = 2$ is the slope and $g(-1) = \int_{1}^{-1} f(t)\,dt = -\pi$, by (c). Thus the equation is $y + \pi = 2(x+1)$

$y = 2x + 2 - \pi$.

(f) $g''(x) = f'(x) = 0$ at $x = -1$ and $g''(x) = f'(x)$ is negative on $(-3, -1)$ and positive on $(-1, 1)$ so there is an inflection point for g at $x = -1$. We notice that $g''(x) = f'(x) < 0$ for x on $(-1, 2)$ and $g''(x) = f'(x) > 0$ for x on $(2, 4)$, even though $g''(2)$ does not exist, g has a tangent line at $x = 2$, so there is an inflection point at $x = 2$.

(g) g is continuous on $[-3, 4]$ and so it attains its absolute maximum and minimum values on this interval. We saw in (d) that $g'(x) = 0 \Rightarrow x = -3, 1, 3$. We have that

$$g(-3) = \int_1^{-3} f(t)\, dt = -\int_{-3}^1 f(t)\, dt = -\frac{\pi 2^2}{2} = -2\pi$$

$$g(1) = \int_1^1 f(t)\, dt = 0$$

$$g(3) = \int_1^3 f(t)\, dt = -1$$

$$g(4) = \int_1^4 f(t)\, dt = -1 + \frac{1}{2} \cdot 1 \cdot 1 = -\frac{1}{2}$$

Thus, the absolute minimum is -2π and the absolute maximum is 0. Thus, the range is $[-2\pi, 0]$.

CHAPTER 6 APPLICATIONS OF DEFINITE INTEGRALS

6.1 VOLUMES BY SLICING AND ROTATION ABOUT AN AXIS

1. (a) $A = \pi(\text{radius})^2$ and radius $= \sqrt{1 - x^2} \Rightarrow A(x) = \pi(1 - x^2)$

   (b) $A = \text{width} \cdot \text{height}$, width $=$ height $= 2\sqrt{1 - x^2} \Rightarrow A(x) = 4(1 - x^2)$

   (c) $A = (\text{side})^2$ and diagonal $= \sqrt{2}(\text{side}) \Rightarrow A = \frac{(\text{diagonal})^2}{2}$; diagonal $= 2\sqrt{1 - x^2} \Rightarrow A(x) = 2(1 - x^2)$

   (d) $A = \frac{\sqrt{3}}{4}(\text{side})^2$ and side $= 2\sqrt{1 - x^2} \Rightarrow A(x) = \sqrt{3}(1 - x^2)$

3. $A(x) = \frac{(\text{diagonal})^2}{2} = \frac{\left(\sqrt{x} - (-\sqrt{x})\right)^2}{2} = 2x$ (see Exercise 1c); $a = 0, b = 4$;

   $V = \int_a^b A(x)\,dx = \int_0^4 2x\,dx = [x^2]_0^4 = 16$

5. $A(x) = (\text{edge})^2 = \left[\sqrt{1 - x^2} - \left(-\sqrt{1 - x^2}\right)\right]^2 = \left(2\sqrt{1 - x^2}\right)^2 = 4(1 - x^2)$; $a = -1, b = 1$;

   $V = \int_a^b A(x)\,dx = \int_{-1}^1 4(1 - x^2)\,dx = 4\left[x - \frac{x^3}{3}\right]_{-1}^1 = 8\left(1 - \frac{1}{3}\right) = \frac{16}{3}$

7. (a) STEP 1) $A(x) = \frac{1}{2}(\text{side}) \cdot (\text{side}) \cdot \left(\sin \frac{\pi}{3}\right) = \frac{1}{2} \cdot \left(2\sqrt{\sin x}\right) \cdot \left(2\sqrt{\sin x}\right)\left(\sin \frac{\pi}{3}\right) = \sqrt{3} \sin x$

   STEP 2) $a = 0, b = \pi$

   STEP 3) $V = \int_a^b A(x)\,dx = \sqrt{3}\int_0^\pi \sin x\,dx = \left[-\sqrt{3}\cos x\right]_0^\pi = \sqrt{3}(1 + 1) = 2\sqrt{3}$

   (b) STEP 1) $A(x) = (\text{side})^2 = \left(2\sqrt{\sin x}\right)\left(2\sqrt{\sin x}\right) = 4\sin x$

   STEP 2) $a = 0, b = \pi$

   STEP 3) $V = \int_a^b A(x)\,dx = \int_0^\pi 4\sin x\,dx = [-4\cos x]_0^\pi = 8$

9. $A(y) = \frac{\pi}{4}(\text{diameter})^2 = \frac{\pi}{4}\left(\sqrt{5y^2} - 0\right)^2 = \frac{5\pi}{4}y^4$;

   $c = 0, d = 2; V = \int_c^d A(y)\,dy = \int_0^2 \frac{5\pi}{4}y^4\,dy$

   $= \left[\left(\frac{5\pi}{4}\right)\left(\frac{y^5}{5}\right)\right]_0^2 = \frac{\pi}{4}(2^5 - 0) = 8\pi$

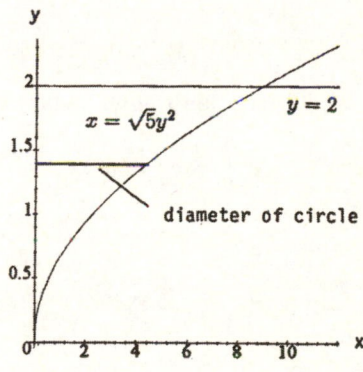

11. (a) It follows from Cavalieri's Principle that the volume of a column is the same as the volume of a right prism with a square base of side length s and altitude h. Thus, STEP 1) $A(x) = (\text{side length})^2 = s^2$;

   STEP 2) $a = 0, b = h$; STEP 3) $V = \int_a^b A(x)\,dx = \int_0^h s^2\,dx = s^2 h$

   (b) From Cavalieri's Principle we conclude that the volume of the column is the same as the volume of the prism described above, regardless of the number of turns $\Rightarrow V = s^2 h$

13. $R(x) = y = 1 - \frac{x}{2} \Rightarrow V = \int_0^2 \pi[R(x)]^2\,dx = \pi\int_0^2 \left(1 - \frac{x}{2}\right)^2\,dx = \pi\int_0^2 \left(1 - x + \frac{x^2}{4}\right)\,dx = \pi\left[x - \frac{x^2}{2} + \frac{x^3}{12}\right]_0^2$

   $= \pi\left(2 - \frac{4}{2} + \frac{8}{12}\right) = \frac{2\pi}{3}$

15. $R(x) = \tan\left(\frac{\pi}{4} y\right)$; $u = \frac{\pi}{4} y \Rightarrow du = \frac{\pi}{4} dy \Rightarrow 4\, du = \pi\, dy$; $y = 0 \Rightarrow u = 0$, $y = 1 \Rightarrow u = \frac{\pi}{4}$;

$V = \int_0^1 \pi[R(y)]^2\, dy = \pi \int_0^1 \left[\tan\left(\frac{\pi}{4} y\right)\right]^2 dy = 4 \int_0^{\pi/4} \tan^2 u\, du = 4 \int_0^{\pi/4} (-1 + \sec^2 u)\, du = 4[-u + \tan u]_0^{\pi/4}$

$= 4\left(-\frac{\pi}{4} + 1 - 0\right) = 4 - \pi$

17. $R(x) = x^2 \Rightarrow V = \int_0^2 \pi[R(x)]^2\, dx = \pi \int_0^2 (x^2)^2\, dx$

$= \pi \int_0^2 x^4\, dx = \pi \left[\frac{x^5}{5}\right]_0^2 = \frac{32\pi}{5}$

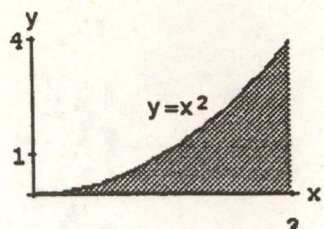

19. $R(x) = \sqrt{9 - x^2} \Rightarrow V = \int_{-3}^3 \pi[R(x)]^2\, dx = \pi \int_{-3}^3 (9 - x^2)\, dx$

$= \pi \left[9x - \frac{x^3}{3}\right]_{-3}^3 = 2\pi \left[9(3) - \frac{27}{3}\right] = 2 \cdot \pi \cdot 18 = 36\pi$

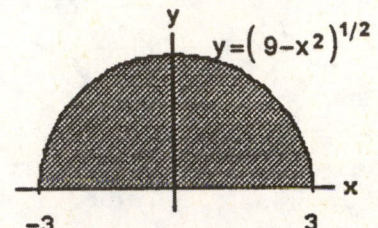

21. $R(x) = \sqrt{\cos x} \Rightarrow V = \int_0^{\pi/2} \pi[R(x)]^2\, dx = \pi \int_0^{\pi/2} \cos x\, dx$

$= \pi[\sin x]_0^{\pi/2} = \pi(1 - 0) = \pi$

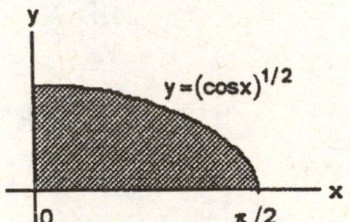

23. $R(x) = \sqrt{2} - \sec x \tan x \Rightarrow V = \int_0^{\pi/4} \pi[R(x)]^2\, dx$

$= \pi \int_0^{\pi/4} \left(\sqrt{2} - \sec x \tan x\right)^2 dx$

$= \pi \int_0^{\pi/4} \left(2 - 2\sqrt{2}\sec x \tan x + \sec^2 x \tan^2 x\right) dx$

$= \pi \left(\int_0^{\pi/4} 2\, dx - 2\sqrt{2} \int_0^{\pi/4} \sec x \tan x\, dx + \int_0^{\pi/4} (\tan x)^2 \sec^2 x\, dx\right)$

$= \pi \left([2x]_0^{\pi/4} - 2\sqrt{2}\,[\sec x]_0^{\pi/4} + \left[\frac{\tan^3 x}{3}\right]_0^{\pi/4}\right)$

$= \pi \left[\left(\frac{\pi}{2} - 0\right) - 2\sqrt{2}\left(\sqrt{2} - 1\right) + \frac{1}{3}\left(1^3 - 0\right)\right] = \pi \left(\frac{\pi}{2} + 2\sqrt{2} - \frac{11}{3}\right)$

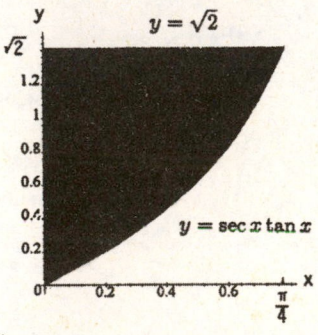

25. $R(y) = \sqrt{5} \cdot y^2 \Rightarrow V = \int_{-1}^1 \pi[R(y)]^2\, dy = \pi \int_{-1}^1 5y^4\, dy$

$= \pi[y^5]_{-1}^1 = \pi[1 - (-1)] = 2\pi$

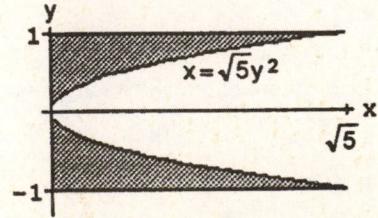

27. $R(y) = \sqrt{2 \sin 2y} \Rightarrow V = \int_0^{\pi/2} \pi [R(y)]^2 \, dy$

$= \pi \int_0^{\pi/2} 2 \sin 2y \, dy = \pi \left[-\cos 2y \right]_0^{\pi/2}$

$= \pi [1 - (-1)] = 2\pi$

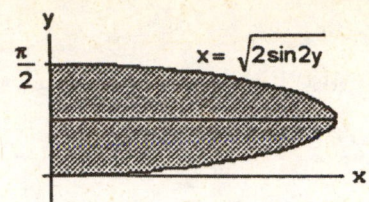

29. $R(y) = \frac{2}{y+1} \Rightarrow V = \int_0^3 \pi [R(y)]^2 \, dy = 4\pi \int_0^3 \frac{1}{(y+1)^2} \, dy$

$= 4\pi \left[\frac{-1}{y+1} \right]_0^3 = 4\pi \left[-\frac{1}{4} - (-1) \right] = 3\pi$

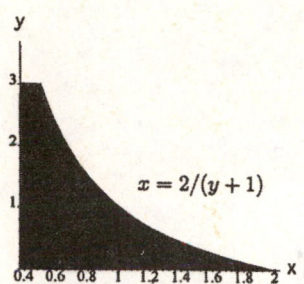

31. For the sketch given, $a = -\frac{\pi}{2}$, $b = \frac{\pi}{2}$; $R(x) = 1$, $r(x) = \sqrt{\cos x}$; $V = \int_a^b \pi \left([R(x)]^2 - [r(x)]^2 \right) dx$

$= \int_{-\pi/2}^{\pi/2} \pi(1 - \cos x) \, dx = 2\pi \int_0^{\pi/2} (1 - \cos x) \, dx = 2\pi [x - \sin x]_0^{\pi/2} = 2\pi \left(\frac{\pi}{2} - 1 \right) = \pi^2 - 2\pi$

33. $r(x) = x$ and $R(x) = 1 \Rightarrow V = \int_0^1 \pi \left([R(x)]^2 - [r(x)]^2 \right) dx$

$= \int_0^1 \pi (1 - x^2) \, dx = \pi \left[x - \frac{x^3}{3} \right]_0^1 = \pi \left[\left(1 - \frac{1}{3} \right) - 0 \right] = \frac{2\pi}{3}$

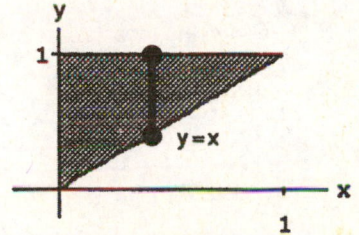

35. $r(x) = x^2 + 1$ and $R(x) = x + 3$

$\Rightarrow V = \int_{-1}^2 \pi \left([R(x)]^2 - [r(x)]^2 \right) dx$

$= \pi \int_{-1}^2 \left[(x+3)^2 - (x^2+1)^2 \right] dx$

$= \pi \int_{-1}^2 \left[(x^2 + 6x + 9) - (x^4 + 2x^2 + 1) \right] dx$

$= \pi \int_{-1}^2 (-x^4 - x^2 + 6x + 8) \, dx$

$= \pi \left[-\frac{x^5}{5} - \frac{x^3}{3} + \frac{6x^2}{2} + 8x \right]_{-1}^2$

$= \pi \left[\left(-\frac{32}{5} - \frac{8}{3} + \frac{24}{2} + 16 \right) - \left(\frac{1}{5} + \frac{1}{3} + \frac{6}{2} - 8 \right) \right] = \pi \left(-\frac{33}{5} - 3 + 28 - 3 + 8 \right) = \pi \left(\frac{5 \cdot 30 - 33}{5} \right) = \frac{117\pi}{5}$

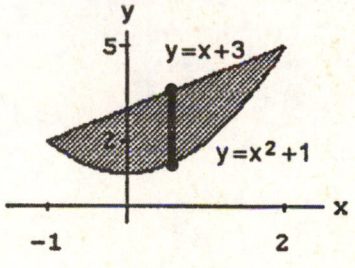

37. $r(x) = \sec x$ and $R(x) = \sqrt{2}$

$\Rightarrow V = \int_{-\pi/4}^{\pi/4} \pi \left([R(x)]^2 - [r(x)]^2 \right) dx$

$= \pi \int_{-\pi/4}^{\pi/4} (2 - \sec^2 x) \, dx = \pi [2x - \tan x]_{-\pi/4}^{\pi/4}$

$= \pi \left[\left(\frac{\pi}{2} - 1 \right) - \left(-\frac{\pi}{2} + 1 \right) \right] = \pi(\pi - 2)$

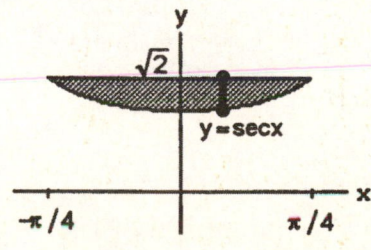

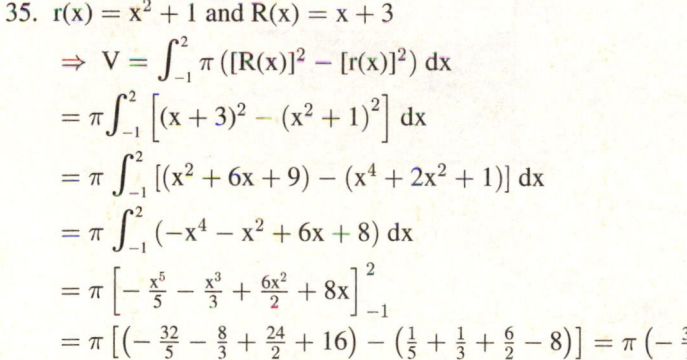

39. $r(y) = 1$ and $R(y) = 1 + y$

$\Rightarrow V = \int_0^1 \pi \left([R(y)]^2 - [r(y)]^2\right) dy$

$= \pi \int_0^1 [(1+y)^2 - 1] \, dy = \pi \int_0^1 (1 + 2y + y^2 - 1) \, dy$

$= \pi \int_0^1 (2y + y^2) \, dy = \pi \left[y^2 + \frac{y^3}{3}\right]_0^1 = \pi \left(1 + \frac{1}{3}\right) = \frac{4\pi}{3}$

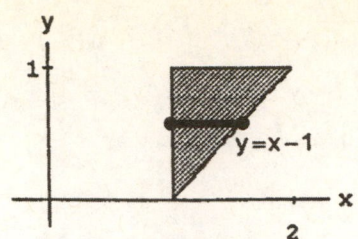

41. $R(y) = 2$ and $r(y) = \sqrt{y}$

$\Rightarrow V = \int_0^4 \pi \left([R(y)]^2 - [r(y)]^2\right) dy$

$= \pi \int_0^4 (4 - y) \, dy = \pi \left[4y - \frac{y^2}{2}\right]_0^4 = \pi(16 - 8) = 8\pi$

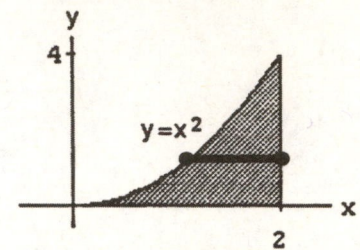

43. $R(y) = 2$ and $r(y) = 1 + \sqrt{y}$

$\Rightarrow V = \int_0^1 \pi \left([R(y)]^2 - [r(y)]^2\right) dy$

$= \pi \int_0^1 \left[4 - \left(1 + \sqrt{y}\right)^2\right] dy$

$= \pi \int_0^1 \left(4 - 1 - 2\sqrt{y} - y\right) dy$

$= \pi \int_0^1 \left(3 - 2\sqrt{y} - y\right) dy$

$= \pi \left[3y - \frac{4}{3} y^{3/2} - \frac{y^2}{2}\right]_0^1$

$= \pi \left(3 - \frac{4}{3} - \frac{1}{2}\right) = \pi \left(\frac{18-8-3}{6}\right) = \frac{7\pi}{6}$

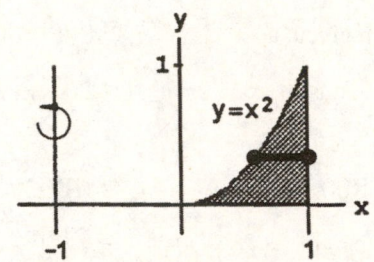

45. (a) $r(x) = \sqrt{x}$ and $R(x) = 2$

$\Rightarrow V = \int_0^4 \pi \left([R(x)]^2 - [r(x)]^2\right) dx$

$= \pi \int_0^4 (4 - x) \, dx = \pi \left[4x - \frac{x^2}{2}\right]_0^4 = \pi(16 - 8) = 8\pi$

(b) $r(y) = 0$ and $R(y) = y^2$

$\Rightarrow V = \int_0^2 \pi \left([R(y)]^2 - [r(y)]^2\right) dy$

$= \pi \int_0^2 y^4 \, dy = \pi \left[\frac{y^5}{5}\right]_0^2 = \frac{32\pi}{5}$

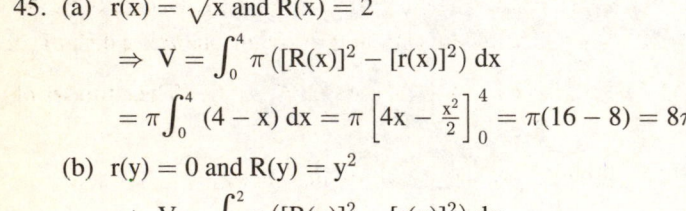
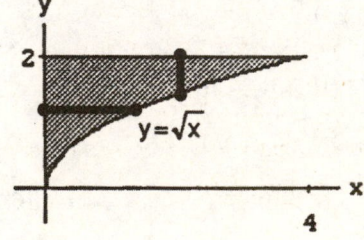

(c) $r(x) = 0$ and $R(x) = 2 - \sqrt{x}$ $\Rightarrow V = \int_0^4 \pi \left([R(x)]^2 - [r(x)]^2\right) dx = \pi \int_0^4 \left(2 - \sqrt{x}\right)^2 dx$

$= \pi \int_0^4 \left(4 - 4\sqrt{x} + x\right) dx = \pi \left[4x - \frac{8x^{3/2}}{3} + \frac{x^2}{2}\right]_0^4 = \pi \left(16 - \frac{64}{3} + \frac{16}{2}\right) = \frac{8\pi}{3}$

(d) $r(y) = 4 - y^2$ and $R(y) = 4$ $\Rightarrow V = \int_0^2 \pi \left([R(y)]^2 - [r(y)]^2\right) dy = \pi \int_0^2 \left[16 - \left(4 - y^2\right)^2\right] dy$

$= \pi \int_0^2 \left(16 - 16 + 8y^2 - y^4\right) dy = \pi \int_0^2 \left(8y^2 - y^4\right) dy = \pi \left[\frac{8}{3} y^3 - \frac{y^5}{5}\right]_0^2 = \pi \left(\frac{64}{3} - \frac{32}{5}\right) = \frac{224\pi}{15}$

47. (a) $r(x) = 0$ and $R(x) = 1 - x^2$

$\Rightarrow V = \int_{-1}^1 \pi \left([R(x)]^2 - [r(x)]^2\right) dx$

$= \pi \int_{-1}^1 \left(1 - x^2\right)^2 dx = \pi \int_{-1}^1 \left(1 - 2x^2 + x^4\right) dx$

$= \pi \left[x - \frac{2x^3}{3} + \frac{x^5}{5}\right]_{-1}^1 = 2\pi \left(1 - \frac{2}{3} + \frac{1}{5}\right)$

$= 2\pi \left(\frac{15-10+3}{15}\right) = \frac{16\pi}{15}$

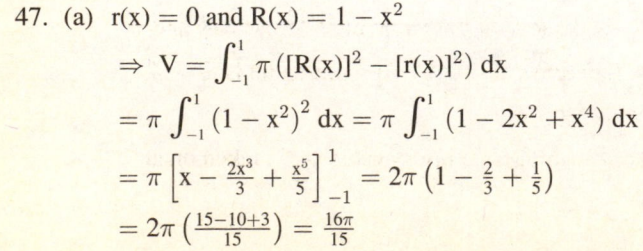
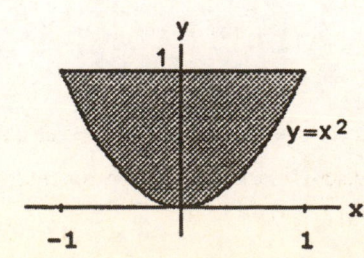

(b) $r(x) = 1$ and $R(x) = 2 - x^2 \Rightarrow V = \int_{-1}^{1} \pi \left([R(x)]^2 - [r(x)]^2\right) dx = \pi \int_{-1}^{1} \left[(2 - x^2)^2 - 1\right] dx$

$= \pi \int_{-1}^{1} (4 - 4x^2 + x^4 - 1)\, dx = \pi \int_{-1}^{1} (3 - 4x^2 + x^4)\, dx = \pi \left[3x - \frac{4}{3} x^3 + \frac{x^5}{5}\right]_{-1}^{1} = 2\pi \left(3 - \frac{4}{3} + \frac{1}{5}\right)$

$= \frac{2\pi}{15} (45 - 20 + 3) = \frac{56\pi}{15}$

(c) $r(x) = 1 + x^2$ and $R(x) = 2 \Rightarrow V = \int_{-1}^{1} \pi \left([R(x)]^2 - [r(x)]^2\right) dx = \pi \int_{-1}^{1} \left[4 - (1 + x^2)^2\right] dx$

$= \pi \int_{-1}^{1} (4 - 1 - 2x^2 - x^4)\, dx = \pi \int_{-1}^{1} (3 - 2x^2 - x^4)\, dx = \pi \left[3x - \frac{2}{3} x^3 - \frac{x^5}{5}\right]_{-1}^{1} = 2\pi \left(3 - \frac{2}{3} - \frac{1}{5}\right)$

$= \frac{2\pi}{15} (45 - 10 - 3) = \frac{64\pi}{15}$

49. $R(y) = b + \sqrt{a^2 - y^2}$ and $r(y) = b - \sqrt{a^2 - y^2}$

$\Rightarrow V = \int_{-a}^{a} \pi \left([R(y)]^2 - [r(y)]^2\right) dy$

$= \pi \int_{-a}^{a} \left[\left(b + \sqrt{a^2 - y^2}\right)^2 - \left(b - \sqrt{a^2 - y^2}\right)^2\right] dy$

$= \pi \int_{-a}^{a} 4b\sqrt{a^2 - y^2}\, dy = 4b\pi \int_{-a}^{a} \sqrt{a^2 - y^2}\, dy$

$= 4b\pi \cdot \text{area of semicircle of radius } a = 4b\pi \cdot \frac{\pi a^2}{2} = 2a^2 b\pi^2$

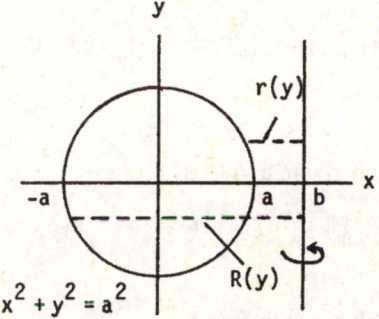

51. (a) $R(y) = \sqrt{a^2 - y^2} \Rightarrow V = \pi \int_{-a}^{h-a} (a^2 - y^2)\, dy = \pi \left[a^2 y - \frac{y^3}{3}\right]_{-a}^{h-a} = \pi \left[a^2 h - a^3 - \frac{(h-a)^3}{3} - \left(-a^3 + \frac{a^3}{3}\right)\right]$

$= \pi \left[a^2 h - \frac{1}{3} (h^3 - 3h^2 a + 3ha^2 - a^3) - \frac{a^3}{3}\right] = \pi \left(a^2 h - \frac{h^3}{3} + h^2 a - ha^2\right) = \frac{\pi h^2 (3a - h)}{3}$

(b) Given $\frac{dV}{dt} = 0.2$ m^3/sec and $a = 5$ m, find $\frac{dh}{dt}\big|_{h=4}$. From part (a), $V(h) = \frac{\pi h^2 (15 - h)}{3} = 5\pi h^2 - \frac{\pi h^3}{3}$

$\Rightarrow \frac{dV}{dh} = 10\pi h - \pi h^2 \Rightarrow \frac{dV}{dt} = \frac{dV}{dh} \cdot \frac{dh}{dt} = \pi h(10 - h) \frac{dh}{dt} \Rightarrow \frac{dh}{dt}\big|_{h=4} = \frac{0.2}{4\pi(10 - 4)} = \frac{1}{(20\pi)(6)} = \frac{1}{120\pi}$ m/sec.

53. The cross section of a solid right circular cylinder with a cone removed is a disk with radius R from which a disk of radius h has been removed. Thus its area is $A_1 = \pi R^2 - \pi h^2 = \pi (R^2 - h^2)$. The cross section of the hemisphere is a disk of radius $\sqrt{R^2 - h^2}$. Therefore its area is $A_2 = \pi \left(\sqrt{R^2 - h^2}\right)^2 = \pi (R^2 - h^2)$. We can see that $A_1 = A_2$. The altitudes of both solids are R. Applying Cavalieri's Principle we find

Volume of Hemisphere = (Volume of Cylinder) − (Volume of Cone) = $(\pi R^2) R - \frac{1}{3} \pi (R^2) R = \frac{2}{3} \pi R^3$.

55. $R(y) = \sqrt{256 - y^2} \Rightarrow V = \int_{-16}^{-7} \pi [R(y)]^2\, dy = \pi \int_{-16}^{-7} (256 - y^2)\, dy = \pi \left[256y - \frac{y^3}{3}\right]_{-16}^{-7}$

$= \pi \left[(256)(-7) + \frac{7^3}{3} - \left((256)(-16) + \frac{16^3}{3}\right)\right] = \pi \left(\frac{7^3}{3} + 256(16 - 7) - \frac{16^3}{3}\right) = 1053\pi$ cm$^3 \approx 3308$ cm^3

57. (a) $R(x) = |c - \sin x|$, so $V = \pi \int_0^\pi [R(x)]^2\, dx = \pi \int_0^\pi (c - \sin x)^2\, dx = \pi \int_0^\pi (c^2 - 2c \sin x + \sin^2 x)\, dx$

$= \pi \int_0^\pi \left(c^2 - 2c \sin x + \frac{1 - \cos 2x}{2}\right) dx = \pi \int_0^\pi \left(c^2 + \frac{1}{2} - 2c \sin x - \frac{\cos 2x}{2}\right) dx$

$= \pi \left[\left(c^2 + \frac{1}{2}\right) x + 2c \cos x - \frac{\sin 2x}{4}\right]_0^\pi = \pi \left[\left(c^2 \pi + \frac{\pi}{2} - 2c - 0\right) - (0 + 2c - 0)\right] = \pi \left(c^2 \pi + \frac{\pi}{2} - 4c\right)$. Let

$V(c) = \pi \left(c^2 \pi + \frac{\pi}{2} - 4c\right)$. We find the extreme values of $V(c)$: $\frac{dV}{dc} = \pi(2c\pi - 4) = 0 \Rightarrow c = \frac{2}{\pi}$ is a critical

point, and $V\left(\frac{2}{\pi}\right) = \pi \left(\frac{4}{\pi} + \frac{\pi}{2} - \frac{8}{\pi}\right) = \pi \left(\frac{\pi}{2} - \frac{4}{\pi}\right) = \frac{\pi^2}{2} - 4$; Evaluate V at the endpoints: $V(0) = \frac{\pi^2}{2}$ and

$V(1) = \pi \left(\frac{3}{2}\pi - 4\right) = \frac{\pi^2}{2} - (4 - \pi)\pi$. Now we see that the function's absolute minimum value is $\frac{\pi^2}{2} - 4$,

taken on at the critical point $c = \frac{2}{\pi}$. (See also the accompanying graph.)

(b) From the discussion in part (a) we conclude that the function's absolute maximum value is $\frac{\pi^2}{2}$, taken on at the endpoint $c = 0$.

(c) The graph of the solid's volume as a function of c for $0 \leq c \leq 1$ is given at the right. As c moves away from $[0, 1]$ the volume of the solid increases without bound. If we approximate the solid as a set of solid disks, we can see that the radius of a typical disk increases without bounds as c moves away from $[0, 1]$.

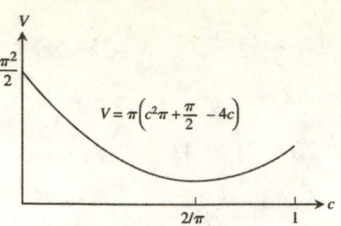

6.2 VOLUME BY CYLINDRICAL SHELLS

1. For the sketch given, $a = 0, b = 2$;
$$V = \int_a^b 2\pi \left(\begin{smallmatrix} \text{shell} \\ \text{radius} \end{smallmatrix}\right) \left(\begin{smallmatrix} \text{shell} \\ \text{height} \end{smallmatrix}\right) dx = \int_0^2 2\pi x \left(1 + \tfrac{x^2}{4}\right) dx = 2\pi \int_0^2 \left(x + \tfrac{x^3}{4}\right) dx = 2\pi \left[\tfrac{x^2}{2} + \tfrac{x^4}{16}\right]_0^2 = 2\pi \left(\tfrac{4}{2} + \tfrac{16}{16}\right)$$
$$= 2\pi \cdot 3 = 6\pi$$

3. For the sketch given, $c = 0, d = \sqrt{2}$;
$$V = \int_c^d 2\pi \left(\begin{smallmatrix} \text{shell} \\ \text{radius} \end{smallmatrix}\right) \left(\begin{smallmatrix} \text{shell} \\ \text{height} \end{smallmatrix}\right) dy = \int_0^{\sqrt{2}} 2\pi y \cdot (y^2) \, dy = 2\pi \int_0^{\sqrt{2}} y^3 \, dy = 2\pi \left[\tfrac{y^4}{4}\right]_0^{\sqrt{2}} = 2\pi$$

5. For the sketch given, $a = 0, b = \sqrt{3}$;
$$V = \int_a^b 2\pi \left(\begin{smallmatrix} \text{shell} \\ \text{radius} \end{smallmatrix}\right) \left(\begin{smallmatrix} \text{shell} \\ \text{height} \end{smallmatrix}\right) dx = \int_0^{\sqrt{3}} 2\pi x \cdot \left(\sqrt{x^2 + 1}\right) dx;$$
$$\left[u = x^2 + 1 \Rightarrow du = 2x \, dx; \, x = 0 \Rightarrow u = 1, \, x = \sqrt{3} \Rightarrow u = 4\right]$$
$$\rightarrow V = \pi \int_1^4 u^{1/2} \, du = \pi \left[\tfrac{2}{3} u^{3/2}\right]_1^4 = \tfrac{2\pi}{3} \left(4^{3/2} - 1\right) = \left(\tfrac{2\pi}{3}\right)(8 - 1) = \tfrac{14\pi}{3}$$

7. $a = 0, b = 2$;
$$V = \int_a^b 2\pi \left(\begin{smallmatrix} \text{shell} \\ \text{radius} \end{smallmatrix}\right) \left(\begin{smallmatrix} \text{shell} \\ \text{height} \end{smallmatrix}\right) dx = \int_0^2 2\pi x \left[x - \left(-\tfrac{x}{2}\right)\right] dx$$
$$= \int_0^2 2\pi x^2 \cdot \tfrac{3}{2} \, dx = \pi \int_0^2 3x^2 \, dx = \pi \left[x^3\right]_0^2 = 8\pi$$

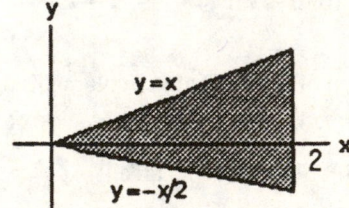

9. $a = 0, b = 1$;
$$V = \int_a^b 2\pi \left(\begin{smallmatrix} \text{shell} \\ \text{radius} \end{smallmatrix}\right) \left(\begin{smallmatrix} \text{shell} \\ \text{height} \end{smallmatrix}\right) dx = \int_0^1 2\pi x \left[(2 - x) - x^2\right] dx$$
$$= 2\pi \int_0^1 \left(2x - x^2 - x^3\right) dx = 2\pi \left[x^2 - \tfrac{x^3}{3} - \tfrac{x^4}{4}\right]_0^1$$
$$= 2\pi \left(1 - \tfrac{1}{3} - \tfrac{1}{4}\right) = 2\pi \left(\tfrac{12 - 4 - 3}{12}\right) = \tfrac{10\pi}{12} = \tfrac{5\pi}{6}$$

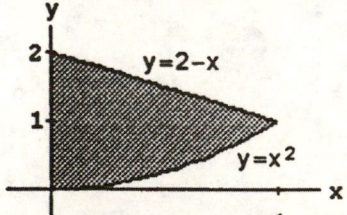

11. $a = 0, b = 1$;
$$V = \int_a^b 2\pi \left(\begin{smallmatrix} \text{shell} \\ \text{radius} \end{smallmatrix}\right) \left(\begin{smallmatrix} \text{shell} \\ \text{height} \end{smallmatrix}\right) dx = \int_0^1 2\pi x \left[\sqrt{x} - (2x - 1)\right] dx$$
$$= 2\pi \int_0^1 \left(x^{3/2} - 2x^2 + x\right) dx = 2\pi \left[\tfrac{2}{5} x^{5/2} - \tfrac{2}{3} x^3 + \tfrac{1}{2} x^2\right]_0^1$$
$$= 2\pi \left(\tfrac{2}{5} - \tfrac{2}{3} + \tfrac{1}{2}\right) = 2\pi \left(\tfrac{12 - 20 + 15}{30}\right) = \tfrac{7\pi}{15}$$

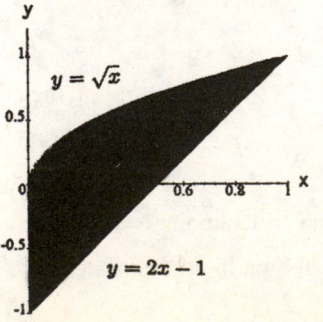

13. (a) $xf(x) = \begin{cases} x \cdot \frac{\sin x}{x}, & 0 < x \le \pi \\ x, & x = 0 \end{cases} \Rightarrow xf(x) = \begin{cases} \sin x, & 0 < x \le \pi \\ 0, & x = 0 \end{cases}$; since $\sin 0 = 0$ we have

$xf(x) = \begin{cases} \sin x, & 0 < x \le \pi \\ \sin x, & x = 0 \end{cases} \Rightarrow xf(x) = \sin x, 0 \le x \le \pi$

(b) $V = \int_a^b 2\pi \binom{\text{shell}}{\text{radius}} \binom{\text{shell}}{\text{height}} dx = \int_0^\pi 2\pi x \cdot f(x)\, dx$ and $x \cdot f(x) = \sin x, 0 \le x \le \pi$ by part (a)

$\Rightarrow V = 2\pi \int_0^\pi \sin x\, dx = 2\pi[-\cos x]_0^\pi = 2\pi(-\cos \pi + \cos 0) = 4\pi$

15. $c = 0, d = 2;$

$V = \int_c^d 2\pi \binom{\text{shell}}{\text{radius}} \binom{\text{shell}}{\text{height}} dy = \int_0^2 2\pi y \left[\sqrt{y} - (-y)\right] dy$

$= 2\pi \int_0^2 (y^{3/2} + y^2)\, dy = 2\pi \left[\frac{2y^{5/2}}{5} + \frac{y^3}{3}\right]_0^2$

$= 2\pi \left[\frac{2}{5}\left(\sqrt{2}\right)^5 + \frac{2^3}{3}\right] = 2\pi \left(\frac{8\sqrt{2}}{5} + \frac{8}{3}\right) = 16\pi \left(\frac{\sqrt{2}}{5} + \frac{1}{3}\right)$

$= \frac{16\pi}{15}\left(3\sqrt{2} + 5\right)$

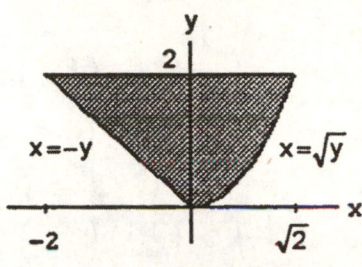

17. $c = 0, d = 2;$

$V = \int_c^d 2\pi \binom{\text{shell}}{\text{radius}} \binom{\text{shell}}{\text{height}} dy = \int_0^2 2\pi y\, (2y - y^2)dy$

$= 2\pi \int_0^2 (2y^2 - y^3)\, dy = 2\pi \left[\frac{2y^3}{3} - \frac{y^4}{4}\right]_0^2 = 2\pi \left(\frac{16}{3} - \frac{16}{4}\right)$

$= 32\pi \left(\frac{1}{3} - \frac{1}{4}\right) = \frac{32\pi}{12} = \frac{8\pi}{3}$

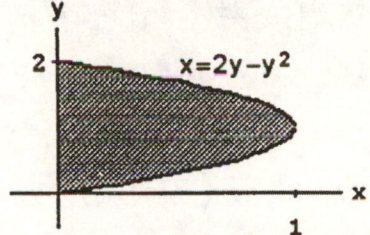

19. $c = 0, d = 1;$

$V = \int_c^d 2\pi \binom{\text{shell}}{\text{radius}} \binom{\text{shell}}{\text{height}} dy = 2\pi \int_0^1 y[y - (-y)]dy$

$= 2\pi \int_0^1 2y^2\, dy = \frac{4\pi}{3} [y^3]_0^1 = \frac{4\pi}{3}$

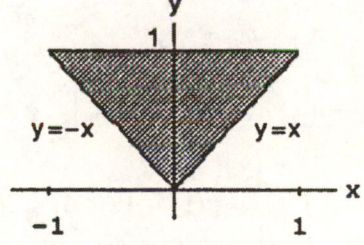

21. $c = 0, d = 2;$

$V = \int_c^d 2\pi \binom{\text{shell}}{\text{radius}} \binom{\text{shell}}{\text{height}} dy = \int_0^2 2\pi y\, [(2 + y) - y^2]\, dy$

$= 2\pi \int_0^2 (2y + y^2 - y^3)\, dy = 2\pi \left[y^2 + \frac{y^3}{3} - \frac{y^4}{4}\right]_0^2$

$= 2\pi \left(4 + \frac{8}{3} - \frac{16}{4}\right) = \frac{\pi}{6}(48 + 32 - 48) = \frac{16\pi}{3}$

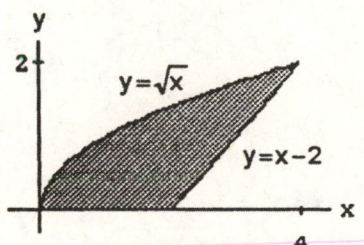

23. (a) $V = \int_c^d 2\pi \binom{\text{shell}}{\text{radius}} \binom{\text{shell}}{\text{height}} dy = \int_0^1 2\pi y \cdot 12\, (y^2 - y^3)\, dy = 24\pi \int_0^1 (y^3 - y^4)\, dy = 24\pi \left[\frac{y^4}{4} - \frac{y^5}{5}\right]_0^1$

$= 24\pi \left(\frac{1}{4} - \frac{1}{5}\right) = \frac{24\pi}{20} = \frac{6\pi}{5}$

(b) $V = \int_c^d 2\pi \binom{\text{shell}}{\text{radius}} \binom{\text{shell}}{\text{height}} dy = \int_0^1 2\pi(1 - y)\,[12\, (y^2 - y^3)]\, dy = 24\pi \int_0^1 (1 - y)\, (y^2 - y^3)\, dy$

$= 24\pi \int_0^1 (y^2 - 2y^3 + y^4)\, dy = 24\pi \left[\frac{y^3}{3} - \frac{y^4}{2} + \frac{y^5}{5}\right]_0^1 = 24\pi \left(\frac{1}{3} - \frac{1}{2} + \frac{1}{5}\right) = 24\pi \left(\frac{1}{30}\right) = \frac{4\pi}{5}$

(c) $V = \int_c^d 2\pi \left(\begin{smallmatrix}\text{shell}\\\text{radius}\end{smallmatrix}\right) \left(\begin{smallmatrix}\text{shell}\\\text{height}\end{smallmatrix}\right) dy = \int_0^1 2\pi \left(\frac{8}{5} - y\right) [12 (y^2 - y^3)] dy = 24\pi \int_0^1 \left(\frac{8}{5} - y\right) (y^2 - y^3) dy$

$= 24\pi \int_0^1 \left(\frac{8}{5} y^2 - \frac{13}{5} y^3 + y^4\right) dy = 24\pi \left[\frac{8}{15} y^3 - \frac{13}{20} y^4 + \frac{y^5}{5}\right]_0^1 = 24\pi \left(\frac{8}{15} - \frac{13}{20} + \frac{1}{5}\right) = \frac{24\pi}{60} (32 - 39 + 12)$

$= \frac{24\pi}{12} = 2\pi$

(d) $V = \int_c^d 2\pi \left(\begin{smallmatrix}\text{shell}\\\text{radius}\end{smallmatrix}\right) \left(\begin{smallmatrix}\text{shell}\\\text{height}\end{smallmatrix}\right) dy = \int_0^1 2\pi \left(y + \frac{2}{5}\right) [12 (y^2 - y^3)] dy = 24\pi \int_0^1 \left(y + \frac{2}{5}\right) (y^2 - y^3) dy$

$= 24\pi \int_0^1 \left(y^3 - y^4 + \frac{2}{5} y^2 - \frac{2}{5} y^3\right) dy = 24\pi \int_0^1 \left(\frac{2}{5} y^2 + \frac{3}{5} y^3 - y^4\right) dy = 24\pi \left[\frac{2}{15} y^3 + \frac{3}{20} y^4 - \frac{y^5}{5}\right]_0^1$

$= 24\pi \left(\frac{2}{15} + \frac{3}{20} - \frac{1}{5}\right) = \frac{24\pi}{60} (8 + 9 - 12) = \frac{24\pi}{12} = 2\pi$

25. (a) About x-axis: $V = \int_c^d 2\pi \left(\begin{smallmatrix}\text{shell}\\\text{radius}\end{smallmatrix}\right) \left(\begin{smallmatrix}\text{shell}\\\text{height}\end{smallmatrix}\right) dy$

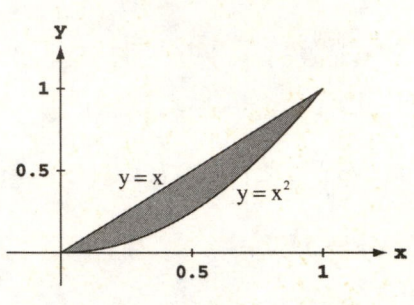

$= \int_0^1 2\pi y \left(\sqrt{y} - y\right) dy = 2\pi \int_0^1 \left(y^{3/2} - y^2\right) dy$

$= 2\pi \left[\frac{2}{5} y^{5/2} - \frac{1}{3} y^3\right]_0^1 = 2\pi \left(\frac{2}{5} - \frac{1}{3}\right) = \frac{2\pi}{15}$

About y-axis: $V = \int_a^b 2\pi \left(\begin{smallmatrix}\text{shell}\\\text{radius}\end{smallmatrix}\right) \left(\begin{smallmatrix}\text{shell}\\\text{height}\end{smallmatrix}\right) dx$

$= \int_0^1 2\pi x (x - x^2) dx = 2\pi \int_0^1 (x^2 - x^3) dx$

$= 2\pi \left[\frac{x^3}{3} - \frac{x^4}{4}\right]_0^1 = 2\pi \left(\frac{1}{3} - \frac{1}{4}\right) = \frac{\pi}{6}$

(b) About x-axis: $R(x) = x$ and $r(x) = x^2 \Rightarrow V = \int_a^b \pi \left[R(x)^2 - r(x)^2\right] dx = \int_0^1 \pi [x^2 - x^4] dx$

$= \pi \left[\frac{x^3}{3} - \frac{x^5}{5}\right]_0^1 = \pi \left(\frac{1}{3} - \frac{1}{5}\right) = \frac{2\pi}{15}$

About y-axis: $R(y) = \sqrt{y}$ and $r(y) = y \Rightarrow V = \int_c^d \pi \left[R(y)^2 - r(y)^2\right] dy = \int_0^1 \pi [y - y^2] dy$

$= \pi \left[\frac{y^2}{2} - \frac{y^3}{3}\right]_0^1 = \pi \left(\frac{1}{2} - \frac{1}{3}\right) = \frac{\pi}{6}$

27. (a) $V = \int_c^d 2\pi \left(\begin{smallmatrix}\text{shell}\\\text{radius}\end{smallmatrix}\right) \left(\begin{smallmatrix}\text{shell}\\\text{height}\end{smallmatrix}\right) dy = \int_1^2 2\pi y (y - 1) dy$

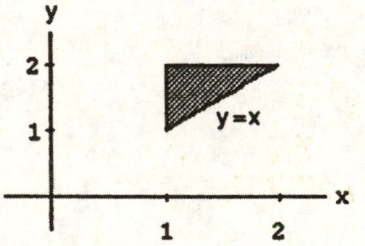

$= 2\pi \int_1^2 (y^2 - y) dy = 2\pi \left[\frac{y^3}{3} - \frac{y^2}{2}\right]_1^2$

$= 2\pi \left[\left(\frac{8}{3} - \frac{4}{2}\right) - \left(\frac{1}{3} - \frac{1}{2}\right)\right]$

$= 2\pi \left(\frac{7}{3} - 2 + \frac{1}{2}\right) = \frac{\pi}{3} (14 - 12 + 3) = \frac{5\pi}{3}$

(b) $V = \int_a^b 2\pi \left(\begin{smallmatrix}\text{shell}\\\text{radius}\end{smallmatrix}\right) \left(\begin{smallmatrix}\text{shell}\\\text{height}\end{smallmatrix}\right) dx$

$= \int_1^2 2\pi x (2 - x) dx = 2\pi \int_1^2 (2x - x^2) dx = 2\pi \left[x^2 - \frac{x^3}{3}\right]_1^2 = 2\pi \left[\left(4 - \frac{8}{3}\right) - \left(1 - \frac{1}{3}\right)\right]$

$= 2\pi \left[\left(\frac{12-8}{3}\right) - \left(\frac{3-1}{3}\right)\right] = 2\pi \left(\frac{4}{3} - \frac{2}{3}\right) = \frac{4\pi}{3}$

(c) $V = \int_a^b 2\pi \left(\begin{smallmatrix}\text{shell}\\\text{radius}\end{smallmatrix}\right) \left(\begin{smallmatrix}\text{shell}\\\text{height}\end{smallmatrix}\right) dx = \int_1^2 2\pi \left(\frac{10}{3} - x\right) (2 - x) dx = 2\pi \int_1^2 \left(\frac{20}{3} - \frac{16}{3} x + x^2\right) dx$

$= 2\pi \left[\frac{20}{3} x - \frac{8}{3} x^2 + \frac{1}{3} x^3\right]_1^2 = 2\pi \left[\left(\frac{40}{3} - \frac{32}{3} + \frac{8}{3}\right) - \left(\frac{20}{3} - \frac{8}{3} + \frac{1}{3}\right)\right] = 2\pi \left(\frac{3}{3}\right) = 2\pi$

(d) $V = \int_c^d 2\pi \left(\begin{smallmatrix}\text{shell}\\\text{radius}\end{smallmatrix}\right) \left(\begin{smallmatrix}\text{shell}\\\text{height}\end{smallmatrix}\right) dy = \int_1^2 2\pi (y - 1)(y - 1) dy = 2\pi \int_1^2 (y - 1)^2 = 2\pi \left[\frac{(y-1)^3}{3}\right]_1^2 = \frac{2\pi}{3}$

29. (a) $V = \int_c^d 2\pi \left(\begin{smallmatrix}\text{shell}\\\text{radius}\end{smallmatrix}\right) \left(\begin{smallmatrix}\text{shell}\\\text{height}\end{smallmatrix}\right) dy = \int_0^1 2\pi y (y - y^3) dy$

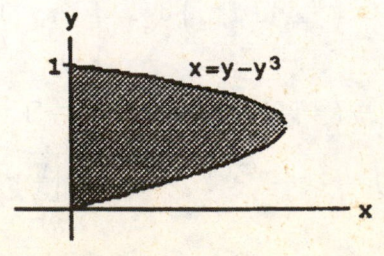

$= \int_0^1 2\pi (y^2 - y^4) dy = 2\pi \left[\frac{y^3}{3} - \frac{y^5}{5}\right]_0^1 = 2\pi \left(\frac{1}{3} - \frac{1}{5}\right)$

$= \frac{4\pi}{15}$

(b) $V = \int_c^d 2\pi \left(\begin{smallmatrix}\text{shell}\\\text{radius}\end{smallmatrix}\right) \left(\begin{smallmatrix}\text{shell}\\\text{height}\end{smallmatrix}\right) dy$

$= \int_0^1 2\pi (1 - y) (y - y^3) dy$

$$= 2\pi \int_0^1 (y - y^2 - y^3 + y^4)\, dy = 2\pi \left[\frac{y^2}{2} - \frac{y^3}{3} - \frac{y^4}{4} + \frac{y^5}{5}\right]_0^1 = 2\pi \left(\frac{1}{2} - \frac{1}{3} - \frac{1}{4} + \frac{1}{5}\right) = \frac{2\pi}{60}(30 - 20 - 15 + 12) = \frac{7\pi}{30}$$

31. (a) $V = \int_c^d 2\pi \left(\begin{smallmatrix} \text{shell} \\ \text{radius}\end{smallmatrix}\right)\left(\begin{smallmatrix} \text{shell} \\ \text{height}\end{smallmatrix}\right) dy = \int_0^2 2\pi y \left(\sqrt{8y} - y^2\right) dy$

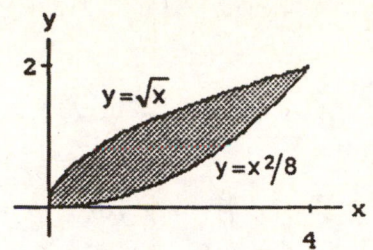

$$= 2\pi \int_0^2 \left(2\sqrt{2}\, y^{3/2} - y^3\right) dy = 2\pi \left[\frac{4\sqrt{2}}{5} y^{5/2} - \frac{y^4}{4}\right]_0^2$$

$$= 2\pi \left(\frac{4\sqrt{2}\cdot\left(\sqrt{2}\right)^5}{5} - \frac{2^4}{4}\right) = 2\pi \left(\frac{4\cdot 2^3}{5} - \frac{4\cdot 4}{4}\right)$$

$$= 2\pi\cdot 4 \left(\frac{8}{5} - 1\right) = \frac{8\pi}{5}(8 - 5) = \frac{24\pi}{5}$$

(b) $V = \int_a^b 2\pi \left(\begin{smallmatrix} \text{shell} \\ \text{radius}\end{smallmatrix}\right)\left(\begin{smallmatrix} \text{shell} \\ \text{height}\end{smallmatrix}\right) dx = \int_0^4 2\pi x \left(\sqrt{x} - \frac{x^2}{8}\right) dx = 2\pi \int_0^4 \left(x^{3/2} - \frac{x^3}{8}\right) dx = 2\pi \left[\frac{2}{5} x^{5/2} - \frac{x^4}{32}\right]_0^4$

$$= 2\pi \left(\frac{2\cdot 2^5}{5} - \frac{4^4}{32}\right) = 2\pi \left(\frac{2^6}{5} - \frac{2^8}{32}\right) = \frac{\pi\cdot 2^7}{160}(32 - 20) = \frac{\pi\cdot 2^9\cdot 3}{160} = \frac{\pi\cdot 2^4\cdot 3}{5} = \frac{48\pi}{5}$$

33. (a) $V = \int_a^b \pi \left[R^2(x) - r^2(x)\right] dx = \pi \int_{1/16}^1 \left(x^{-1/2} - 1\right) dx$

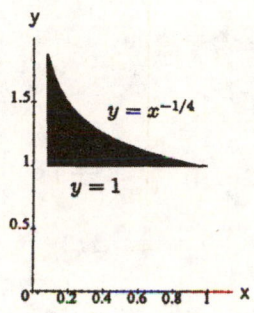

$$= \pi \left[2x^{1/2} - x\right]_{1/16}^1 = \pi \left[(2 - 1) - \left(2\cdot\frac{1}{4} - \frac{1}{16}\right)\right]$$

$$= \pi \left(1 - \frac{7}{16}\right) = \frac{9\pi}{16}$$

(b) $V = \int_a^b 2\pi \left(\begin{smallmatrix} \text{shell} \\ \text{radius}\end{smallmatrix}\right)\left(\begin{smallmatrix} \text{shell} \\ \text{height}\end{smallmatrix}\right) dy = \int_1^2 2\pi y \left(\frac{1}{y^4} - \frac{1}{16}\right) dy$

$$= 2\pi \int_1^2 \left(y^{-3} - \frac{y}{16}\right) dy = 2\pi \left[-\frac{1}{2} y^{-2} - \frac{y^2}{32}\right]_1^2$$

$$= 2\pi \left[\left(-\frac{1}{8} - \frac{1}{8}\right) - \left(-\frac{1}{2} - \frac{1}{32}\right)\right] = 2\pi \left(\frac{1}{4} + \frac{1}{32}\right)$$

$$= \frac{2\pi}{32}(8 + 1) = \frac{9\pi}{16}$$

35. (a) *Disk*: $V = V_1 - V_2$

$V_1 = \int_{a_1}^{b_1} \pi[R_1(x)]^2\, dx$ and $V_2 = \int_{a_2}^{b_2} \pi[R_2(x)]^2$ with $R_1(x) = \sqrt{\frac{x+2}{3}}$ and $R_2(x) = \sqrt{x}$,

$a_1 = -2, b_1 = 1; a_2 = 0, b_2 = 1 \Rightarrow$ two integrals are required

(b) *Washer*: $V = V_1 - V_2$

$V_1 = \int_{a_1}^{b_1} \pi \left([R_1(x)]^2 - [r_1(x)]^2\right) dx$ with $R_1(x) = \sqrt{\frac{x+2}{3}}$ and $r_1(x) = 0$; $a_1 = -2$ and $b_1 = 0$;

$V_2 = \int_{a_2}^{b_2} \pi \left([R_2(x)]^2 - [r_2(x)]^2\right) dx$ with $R_2(x) = \sqrt{\frac{x+2}{3}}$ and $r_2(x) = \sqrt{x}$; $a_2 = 0$ and $b_2 = 1$

$\Rightarrow$ two integrals are required

(c) *Shell*: $V = \int_c^d 2\pi \left(\begin{smallmatrix} \text{shell} \\ \text{radius}\end{smallmatrix}\right)\left(\begin{smallmatrix} \text{shell} \\ \text{height}\end{smallmatrix}\right) dy = \int_c^d 2\pi y \left(\begin{smallmatrix} \text{shell} \\ \text{height}\end{smallmatrix}\right) dy$ where shell height $= y^2 - (3y^2 - 2) = 2 - 2y^2$;

$c = 0$ and $d = 1$. Only *one* integral is required. It is, therefore preferable to use the *shell* method.
However, whichever method you use, you will get $V = \pi$.

6.3 LENGTHS OF PLANE CURVES

1. $\frac{dx}{dt} = -1$ and $\frac{dy}{dt} = 3 \Rightarrow \sqrt{\left(\frac{dx}{dt}\right)^2 + \left(\frac{dy}{dt}\right)^2} = \sqrt{(-1)^2 + (3)^2} = \sqrt{10}$

$\Rightarrow$ Length $= \int_{-2/3}^1 \sqrt{10}\, dt = \sqrt{10}\, [t]_{-2/3}^1 = \sqrt{10} - \left(-\frac{2}{3}\sqrt{10}\right) = \frac{5\sqrt{10}}{3}$

3. $\frac{dx}{dt} = 3t^2$ and $\frac{dy}{dt} = 3t \Rightarrow \sqrt{\left(\frac{dx}{dt}\right)^2 + \left(\frac{dy}{dt}\right)^2} = \sqrt{(3t^2)^2 + (3t)^2} = \sqrt{9t^4 + 9t^2} = 3t\sqrt{t^2 + 1}$ $\left(\text{since } t \geq 0 \text{ on } \left[0, \sqrt{3}\right]\right)$

$\Rightarrow$ Length $= \int_0^{\sqrt{3}} 3t\sqrt{t^2 + 1}\, dt;$ $\left[u = t^2 + 1 \Rightarrow \frac{3}{2}\, du = 3t\, dt; t = 0 \Rightarrow u = 1, t = \sqrt{3} \Rightarrow u = 4\right]$

$\rightarrow \int_1^4 \frac{3}{2} u^{1/2}\, du = \left[u^{3/2}\right]_1^4 = (8 - 1) = 7$

5. $\frac{dx}{dt} = (2t+3)^{1/2}$ and $\frac{dy}{dt} = 1+t \Rightarrow \sqrt{\left(\frac{dx}{dt}\right)^2 + \left(\frac{dy}{dt}\right)^2} = \sqrt{(2t+3)+(1+t)^2} = \sqrt{t^2+4t+4} = |t+2| = t+2$

since $0 \le t \le 3 \Rightarrow$ Length $= \int_0^3 (t+2)\,dt = \left[\frac{t^2}{2} + 2t\right]_0^3 = \frac{21}{2}$

7. $\frac{dy}{dx} = \frac{1}{3} \cdot \frac{3}{2}(x^2+2)^{1/2} \cdot 2x = \sqrt{(x^2+2)} \cdot x$

$\Rightarrow L = \int_0^3 \sqrt{1+(x^2+2)x^2}\,dx = \int_0^3 \sqrt{1+2x^2+x^4}\,dx$

$= \int_0^3 \sqrt{(1+x^2)^2}\,dx = \int_0^3 (1+x^2)\,dx = \left[x + \frac{x^3}{3}\right]_0^3$

$= 3 + \frac{27}{3} = 12$

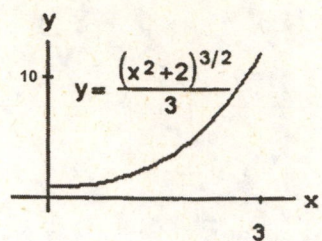

9. $\frac{dx}{dy} = y^2 - \frac{1}{4y^2} \Rightarrow \left(\frac{dx}{dy}\right)^2 = y^4 - \frac{1}{2} + \frac{1}{16y^4}$

$\Rightarrow L = \int_1^3 \sqrt{1+y^4-\frac{1}{2}+\frac{1}{16y^4}}\,dy$

$= \int_1^3 \sqrt{y^4+\frac{1}{2}+\frac{1}{16y^4}}\,dy$

$= \int_1^3 \sqrt{\left(y^2+\frac{1}{4y^2}\right)^2}\,dy = \int_1^3 \left(y^2+\frac{1}{4y^2}\right)\,dy$

$= \left[\frac{y^3}{3} - \frac{y^{-1}}{4}\right]_1^3 = \left(\frac{27}{3} - \frac{1}{12}\right) - \left(\frac{1}{3} - \frac{1}{4}\right) = 9 - \frac{1}{12} - \frac{1}{3} + \frac{1}{4} = 9 + \frac{(-1-4+3)}{12} = 9 + \frac{(-2)}{12} = \frac{53}{6}$

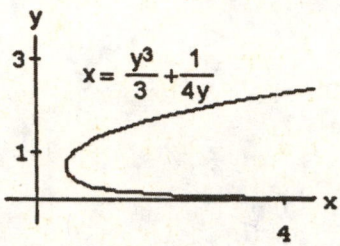

11. $\frac{dx}{dy} = y^3 - \frac{1}{4y^3} \Rightarrow \left(\frac{dx}{dy}\right)^2 = y^6 - \frac{1}{2} + \frac{1}{16y^6}$

$\Rightarrow L = \int_1^2 \sqrt{1+y^6-\frac{1}{2}+\frac{1}{16y^6}}\,dy$

$= \int_1^2 \sqrt{y^6+\frac{1}{2}+\frac{1}{16y^6}}\,dy = \int_1^2 \sqrt{\left(y^3+\frac{y^{-3}}{4}\right)^2}\,dy$

$= \int_1^2 \left(y^3+\frac{y^{-3}}{4}\right)\,dy = \left[\frac{y^4}{4} - \frac{y^{-2}}{8}\right]_1^2$

$= \left(\frac{16}{4} - \frac{1}{(16)(2)}\right) - \left(\frac{1}{4} - \frac{1}{8}\right) = 4 - \frac{1}{32} - \frac{1}{4} + \frac{1}{8} = \frac{128-1-8+4}{32} = \frac{123}{32}$

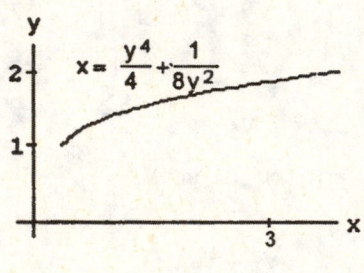

13. $\frac{dy}{dx} = x^{1/3} - \frac{1}{4}x^{-1/3} \Rightarrow \left(\frac{dy}{dx}\right)^2 = x^{2/3} - \frac{1}{2} + \frac{x^{-2/3}}{16}$

$\Rightarrow L = \int_1^8 \sqrt{1+x^{2/3}-\frac{1}{2}+\frac{x^{-2/3}}{16}}\,dx$

$= \int_1^8 \sqrt{x^{2/3}+\frac{1}{2}+\frac{x^{-2/3}}{16}}\,dx$

$= \int_1^8 \sqrt{\left(x^{1/3}+\frac{1}{4}x^{-1/3}\right)^2}\,dx = \int_1^8 \left(x^{1/3}+\frac{1}{4}x^{-1/3}\right)\,dx$

$= \left[\frac{3}{4}x^{4/3} + \frac{3}{8}x^{2/3}\right]_1^8 = \frac{3}{8}\left[2x^{4/3} + x^{2/3}\right]_1^8$

$= \frac{3}{8}\left[(2 \cdot 2^4 + 2^2) - (2+1)\right] = \frac{3}{8}(32+4-3) = \frac{99}{8}$

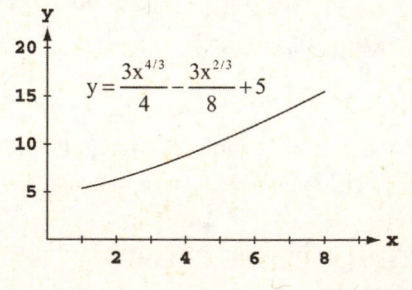

15. $\frac{dx}{dy} = \sqrt{\sec^4 y - 1} \Rightarrow \left(\frac{dx}{dy}\right)^2 = \sec^4 y - 1$

$\Rightarrow L = \int_{-\pi/4}^{\pi/4} \sqrt{1+(\sec^4 y - 1)}\,dy = \int_{-\pi/4}^{\pi/4} \sec^2 y\,dy$

$= [\tan y]_{-\pi/4}^{\pi/4} = 1 - (-1) = 2$

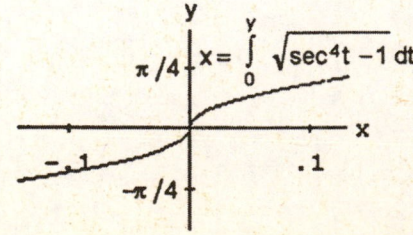

17. (a) $\frac{dy}{dx} = 2x \Rightarrow \left(\frac{dy}{dx}\right)^2 = 4x^2$

$\Rightarrow L = \int_{-1}^{2} \sqrt{1 + \left(\frac{dy}{dx}\right)^2}\, dx$

$= \int_{-1}^{2} \sqrt{1 + 4x^2}\, dx$

(c) $L \approx 6.13$

(b)
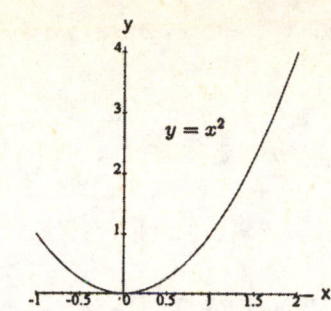

19. (a) $\frac{dx}{dy} = \cos y \Rightarrow \left(\frac{dx}{dy}\right)^2 = \cos^2 y$

$\Rightarrow L = \int_{0}^{\pi} \sqrt{1 + \cos^2 y}\, dy$

(c) $L \approx 3.82$

(b)
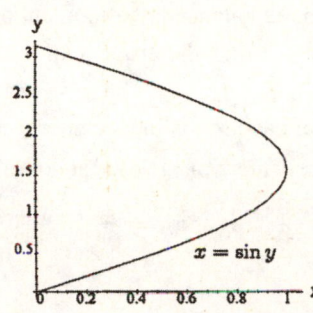

21. (a) $2y + 2 = 2\frac{dx}{dy} \Rightarrow \left(\frac{dx}{dy}\right)^2 = (y+1)^2$

$\Rightarrow L = \int_{-1}^{3} \sqrt{1 + (y+1)^2}\, dy$

(c) $L \approx 9.29$

(b)
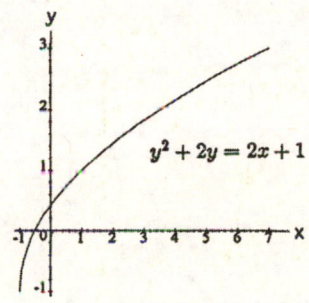

23. (a) $\frac{dy}{dx} = \tan x \Rightarrow \left(\frac{dy}{dx}\right)^2 = \tan^2 x$

$\Rightarrow L = \int_{0}^{\pi/6} \sqrt{1 + \tan^2 x}\, dx = \int_{0}^{\pi/6} \sqrt{\frac{\sin^2 x + \cos^2 x}{\cos^2 x}}\, dx$

$= \int_{0}^{\pi/6} \frac{dx}{\cos x} = \int_{0}^{\pi/6} \sec x\, dx$

(c) $L \approx 0.55$

(b)
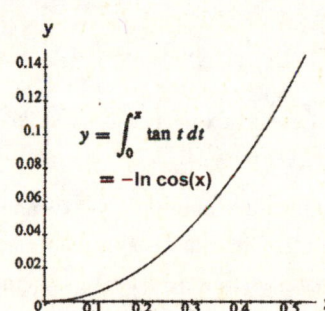

25. $\sqrt{2}\, x = \int_{0}^{x} \sqrt{1 + \left(\frac{dy}{dt}\right)^2}\, dt,\ x \geq 0 \Rightarrow \sqrt{2} = \sqrt{1 + \left(\frac{dy}{dx}\right)^2} \Rightarrow \frac{dy}{dx} = \pm 1 \Rightarrow y = f(x) = \pm x + C$ where C is any

real number.

27. (a) $\left(\frac{dy}{dx}\right)^2$ correspondes to $\frac{1}{4x}$ here, so take $\frac{dy}{dx}$ as $\frac{1}{2\sqrt{x}}$. Then $y = \sqrt{x} + C$ and since $(1, 1)$ lies on the curve, $C = 0$.

So $y = \sqrt{x}$ from $(1, 1)$ to $(4, 2)$.

(b) Only one. We know the derivative of the function and the value of the function at one value of x.

29. (a) $\frac{dx}{dt} = -2\sin 2t$ and $\frac{dy}{dt} = 2\cos 2t \Rightarrow \sqrt{\left(\frac{dx}{dt}\right)^2 + \left(\frac{dy}{dt}\right)^2} = \sqrt{(-2\sin 2t)^2 + (2\cos 2t)^2} = 2$

$\Rightarrow$ Length $= \int_0^{\pi/2} 2\,dt = [2t]_0^{\pi/2} = \pi$

(b) $\frac{dx}{dt} = \pi\cos\pi t$ and $\frac{dy}{dt} = -\pi\sin\pi t = \sqrt{\left(\frac{dx}{dt}\right)^2 + \left(\frac{dy}{dt}\right)^2} = \sqrt{(\pi\cos\pi t)^2 + (-\pi\sin\pi t)^2} = \pi$

$\Rightarrow$ Length $= \int_{-1/2}^{1/2} \pi\,dt = [\pi t]_{-1/2}^{1/2} = \pi$

6.4 MOMENTS AND CENTERS OF MASS

1. Because the children are balanced, the moment of the system about the origin must be equal to zero:
$5 \cdot 80 = x \cdot 100 \Rightarrow x = 4$ ft, the distance of the 100-lb child from the fulcrum.

3. The center of mass of each rod is in its center (see Example 1). The rod system is equivalent to two point masses located at the centers of the rods at coordinates $\left(\frac{L}{2}, 0\right)$ and $\left(0, \frac{L}{2}\right)$. Therefore $\bar{x} = \frac{m_y}{m}$

$= \frac{x_1 m_1 + x_2 m_2}{m_1 + m_2} = \frac{\frac{L}{2}\cdot m + 0}{m + m} = \frac{L}{4}$ and $\bar{y} = \frac{m_x}{m} = \frac{y_1 m_2 + y_2 m_2}{m_1 + m_2} = \frac{0 + \frac{L}{2}\cdot m}{m + m} = \frac{L}{4} \Rightarrow \left(\frac{L}{4}, \frac{L}{4}\right)$ is the center of mass location.

5. $M_0 = \int_0^2 x \cdot 4\,dx = \left[4\frac{x^2}{2}\right]_0^2 = 4 \cdot \frac{4}{2} = 8$; $M = \int_0^2 4\,dx = [4x]_0^2 = 4 \cdot 2 = 8 \Rightarrow \bar{x} = \frac{M_0}{M} = 1$

7. $M_0 = \int_0^3 x\left(1 + \frac{x}{3}\right)dx = \int_0^3 \left(x + \frac{x^2}{3}\right)dx = \left[\frac{x^2}{2} + \frac{x^3}{9}\right]_0^3 = \left(\frac{9}{2} + \frac{27}{9}\right) = \frac{15}{2}$; $M = \int_0^3 \left(1 + \frac{x}{3}\right)dx = \left[x + \frac{x^2}{6}\right]_0^3$

$= 3 + \frac{9}{6} = \frac{9}{2} \Rightarrow \bar{x} = \frac{M_0}{M} = \frac{\left(\frac{15}{2}\right)}{\left(\frac{9}{2}\right)} = \frac{15}{9} = \frac{5}{3}$

9. $M_0 = \int_1^4 x\left(1 + \frac{1}{\sqrt{x}}\right)dx = \int_1^4 (x + x^{1/2})\,dx = \left[\frac{x^2}{2} + \frac{2x^{3/2}}{3}\right]_1^4 = \left(8 + \frac{16}{3}\right) - \left(\frac{1}{2} + \frac{2}{3}\right) = \frac{15}{2} + \frac{14}{3} = \frac{45+28}{6} = \frac{73}{6}$;

$M = \int_1^4 (1 + x^{-1/2})\,dx = \left[x + 2x^{1/2}\right]_1^4 = (4 + 4) - (1 + 2) = 5 \Rightarrow \bar{x} = \frac{M_0}{M} = \frac{\left(\frac{73}{6}\right)}{5} = \frac{73}{30}$

11. $M_0 = \int_0^1 x(2 - x)\,dx + \int_1^2 x \cdot x\,dx = \int_0^1 (2x - x^2)\,dx + \int_1^2 x^2\,dx = \left[\frac{2x^2}{2} - \frac{x^3}{3}\right]_0^1 + \left[\frac{x^3}{3}\right]_1^2 = \left(1 - \frac{1}{3}\right) + \left(\frac{8}{3} - \frac{1}{3}\right)$

$= \frac{9}{3} = 3$; $M = \int_0^1 (2 - x)\,dx + \int_1^2 x\,dx = \left[2x - \frac{x^2}{2}\right]_0^1 + \left[\frac{x^2}{2}\right]_1^2 = \left(2 - \frac{1}{2}\right) + \left(\frac{4}{2} - \frac{1}{2}\right) = 3 \Rightarrow \bar{x} = \frac{M_0}{M} = 1$

13. Since the plate is symmetric about the y-axis and its density is constant, the distribution of mass is symmetric about the y-axis and the center of mass lies on the y-axis. This means that $\bar{x} = 0$. It remains to find $\bar{y} = \frac{M_x}{M}$. We model the distribution of mass with *vertical* strips. The typical strip has center of mass:

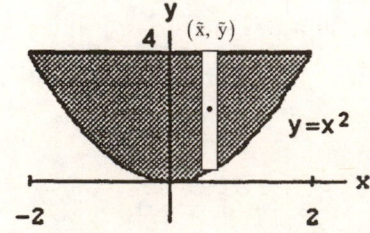

$(\tilde{x}, \tilde{y}) = \left(x, \frac{x^2 + 4}{2}\right)$, length: $4 - x^2$, width: dx, area:

$dA = (4 - x^2)\,dx$, mass: $dm = \delta\,dA = \delta(4 - x^2)\,dx$. The moment of the strip about the x-axis is

$\tilde{y}\,dm = \left(\frac{x^2+4}{2}\right)\delta(4 - x^2)\,dx = \frac{\delta}{2}(16 - x^4)\,dx$. The moment of the plate about the x-axis is $M_x = \int \tilde{y}\,dm$

$= \int_{-2}^2 \frac{\delta}{2}(16 - x^4)\,dx = \frac{\delta}{2}\left[16x - \frac{x^5}{5}\right]_{-2}^2 = \frac{\delta}{2}\left[\left(16 \cdot 2 - \frac{2^5}{5}\right) - \left(-16 \cdot 2 + \frac{2^5}{5}\right)\right] = \frac{\delta \cdot 2}{2}\left(32 - \frac{32}{5}\right) = \frac{128\delta}{5}$. The mass of the

plate is $M = \int \delta(4 - x^2)\,dx = \delta\left[4x - \frac{x^3}{3}\right]_{-2}^2 = 2\delta\left(8 - \frac{8}{3}\right) = \frac{32\delta}{3}$. Therefore $\bar{y} = \frac{M_x}{M} = \frac{\left(\frac{128\delta}{5}\right)}{\left(\frac{32\delta}{3}\right)} = \frac{12}{5}$. The plate's center of

mass is the point $(\bar{x}, \bar{y}) = \left(0, \frac{12}{5}\right)$.

15. Intersection points: $x - x^2 = -x \Rightarrow 2x - x^2 = 0$
    $\Rightarrow x(2 - x) = 0 \Rightarrow x = 0$ or $x = 2$. The typical *vertical*
    strip has center of mass: $(\tilde{x}, \tilde{y}) = \left(x, \frac{(x - x^2) + (-x)}{2}\right)$

    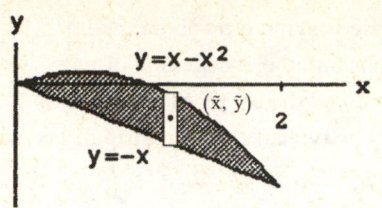

    $= \left(x, -\frac{x^2}{2}\right)$, length: $(x - x^2) - (-x) = 2x - x^2$, width: dx,
    area: $dA = (2x - x^2)\, dx$, mass: $dm = \delta\, dA = \delta(2x - x^2)\, dx$.
    The moment of the strip about the x-axis is

    $\tilde{y}\, dm = \left(-\frac{x^2}{2}\right)\delta(2x - x^2)\, dx$; about the y-axis it is $\tilde{x}\, dm = x \cdot \delta(2x - x^2)\, dx$. Thus, $M_x = \int \tilde{y}\, dm$

    $= -\int_0^2 \left(\frac{\delta}{2}x^2\right)(2x - x^2)\, dx = -\frac{\delta}{2}\int_0^2 (2x^3 - x^4)\, dx = -\frac{\delta}{2}\left[\frac{x^4}{2} - \frac{x^5}{5}\right]_0^2 = -\frac{\delta}{2}\left(2^3 - \frac{2^5}{5}\right) = -\frac{\delta}{2} \cdot 2^3\left(1 - \frac{4}{5}\right)$

    $= -\frac{4\delta}{5}$; $M_y = \int \tilde{x}\, dm = \int_0^2 x \cdot \delta(2x - x^2) = \delta\int_0^2 (2x^2 - x^3) = \delta\left[\frac{2}{3}x^3 - \frac{x^4}{4}\right]_0^2 = \delta\left(2 \cdot \frac{2^3}{3} - \frac{2^4}{4}\right) = \frac{\delta \cdot 2^4}{12} = \frac{4\delta}{3}$;

    $M = \int dm = \int_0^2 \delta(2x - x^2)\, dx = \delta\int_0^2 (2x - x^2)\, dx = \delta\left[x^2 - \frac{x^3}{3}\right]_0^2 = \delta\left(4 - \frac{8}{3}\right) = \frac{4\delta}{3}$. Therefore, $\bar{x} = \frac{M_y}{M}$

    $= \left(\frac{4\delta}{3}\right)\left(\frac{3}{4\delta}\right) = 1$ and $\bar{y} = \frac{M_x}{M} = \left(-\frac{4\delta}{5}\right)\left(\frac{3}{4\delta}\right) = -\frac{3}{5} \Rightarrow (\bar{x}, \bar{y}) = \left(1, -\frac{3}{5}\right)$ is the center of mass.

17. The typical *horizontal* strip has center of mass:

    $(\tilde{x}, \tilde{y}) = \left(\frac{y - y^3}{2}, y\right)$, length: $y - y^3$, width: dy,

    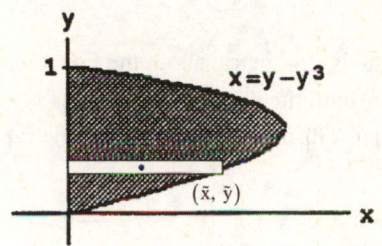

    area: $dA = (y - y^3)\, dy$, mass: $dm = \delta\, dA = \delta(y - y^3)\, dy$.
    The moment of the strip about the y-axis is

    $\tilde{x}\, dm = \delta\left(\frac{y - y^3}{2}\right)(y - y^3)\, dy = \frac{\delta}{2}(y - y^3)^2\, dy$

    $= \frac{\delta}{2}(y^2 - 2y^4 + y^6)\, dy$; the moment about the x-axis is

    $\tilde{y}\, dm = \delta y(y - y^3)\, dy = \delta(y^2 - y^4)\, dy$. Thus, $M_x = \int \tilde{y}\, dm = \delta\int_0^1 (y^2 - y^4)\, dy = \delta\left[\frac{y^3}{3} - \frac{y^5}{5}\right]_0^1 = \delta\left(\frac{1}{3} - \frac{1}{5}\right) = \frac{2\delta}{15}$;

    $M_y = \int \tilde{x}\, dm = \frac{\delta}{2}\int_0^1 (y^2 - 2y^4 + y^6)\, dy = \frac{\delta}{2}\left[\frac{y^3}{3} - \frac{2y^5}{5} + \frac{y^7}{7}\right]_0^1 = \frac{\delta}{2}\left(\frac{1}{3} - \frac{2}{5} + \frac{1}{7}\right) = \frac{\delta}{2}\left(\frac{35 - 42 + 15}{3 \cdot 5 \cdot 7}\right) = \frac{4\delta}{105}$; $M = \int dm$

    $= \delta\int_0^1 (y - y^3)\, dy = \delta\left[\frac{y^2}{2} - \frac{y^4}{4}\right]_0^1 = \delta\left(\frac{1}{2} - \frac{1}{4}\right) = \frac{\delta}{4}$. Therefore, $\bar{x} = \frac{M_y}{M} = \left(\frac{4\delta}{105}\right)\left(\frac{4}{\delta}\right) = \frac{16}{105}$ and $\bar{y} = \frac{M_x}{M} = \left(\frac{2\delta}{15}\right)\left(\frac{4}{\delta}\right)$

    $= \frac{8}{15} \Rightarrow (\bar{x}, \bar{y}) = \left(\frac{16}{105}, \frac{8}{15}\right)$ is the center of mass.

19. Applying the symmetry argument analogous to the one used
    in Exercise 13, we find $\bar{x} = 0$. The typical *vertical* strip has
    center of mass: $(\tilde{x}, \tilde{y}) = \left(x, \frac{\cos x}{2}\right)$, length: $\cos x$, width: dx,

    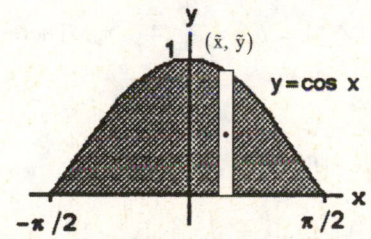

    area: $dA = \cos x\, dx$, mass: $dm = \delta\, dA = \delta \cos x\, dx$. The
    moment of the strip about the x-axis is $\tilde{y}\, dm = \delta \cdot \frac{\cos x}{2} \cdot \cos x\, dx$
    $= \frac{\delta}{2}\cos^2 x\, dx = \frac{\delta}{2}\left(\frac{1 + \cos 2x}{2}\right)dx = \frac{\delta}{4}(1 + \cos 2x)\, dx$; thus,

    $M_x = \int \tilde{y}\, dm = \int_{-\pi/2}^{\pi/2} \frac{\delta}{4}(1 + \cos 2x)\, dx = \frac{\delta}{4}\left[x + \frac{\sin 2x}{2}\right]_{-\pi/2}^{\pi/2} = \frac{\delta}{4}\left[\left(\frac{\pi}{2} + 0\right) - \left(-\frac{\pi}{2}\right)\right] = \frac{\delta\pi}{4}$; $M = \int dm = \delta\int_{-\pi/2}^{\pi/2}\cos x\, dx$

    $= \delta[\sin x]_{-\pi/2}^{\pi/2} = 2\delta$. Therefore, $\bar{y} = \frac{M_x}{M} = \frac{\delta\pi}{4 \cdot 2\delta} = \frac{\pi}{8} \Rightarrow (\bar{x}, \bar{y}) = \left(0, \frac{\pi}{8}\right)$ is the center of mass.

21. Since the plate is symmetric about the line x = 1 and its
density is constant, the distribution of mass is symmetric
about this line and the center of mass lies on it. This means
that $\bar{x} = 1$. The typical *vertical* strip has center of mass:

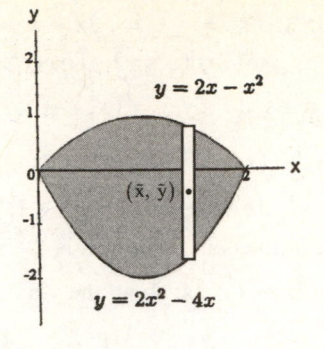

$\left(\tilde{x}, \tilde{y}\right) = \left(x, \frac{(2x - x^2) + (2x^2 - 4x)}{2}\right) = \left(x, \frac{x^2 - 2x}{2}\right)$,

length: $(2x - x^2) - (2x^2 - 4x) = -3x^2 + 6x = 3(2x - x^2)$,
width: dx, area: $dA = 3(2x - x^2)\,dx$, mass: $dm = \delta\,dA$
$= 3\delta(2x - x^2)\,dx$. The moment about the x-axis is

$\tilde{y}\,dm = \frac{3}{2}\delta(x^2 - 2x)(2x - x^2)\,dx = -\frac{3}{2}\delta(x^2 - 2x)^2\,dx$

$= -\frac{3}{2}\delta(x^4 - 4x^3 + 4x^2)\,dx$. Thus, $M_x = \int \tilde{y}\,dm = -\int_0^2 \frac{3}{2}\delta(x^4 - 4x^3 + 4x^2)\,dx = -\frac{3}{2}\delta\left[\frac{x^5}{5} - x^4 + \frac{4}{3}x^3\right]_0^2$

$= -\frac{3}{2}\delta\left(\frac{2^5}{5} - 2^4 + \frac{4}{3}\cdot 2^3\right) = -\frac{3}{2}\delta\cdot 2^4\left(\frac{2}{5} - 1 + \frac{2}{3}\right) = -\frac{3}{2}\delta\cdot 2^4\left(\frac{6 - 15 + 10}{15}\right) = -\frac{8\delta}{5}$; $M = \int dm$

$= \int_0^2 3\delta(2x - x^2)\,dx = 3\delta\left[x^2 - \frac{x^3}{3}\right]_0^2 = 3\delta\left(4 - \frac{8}{3}\right) = 4\delta$. Therefore, $\bar{y} = \frac{M_x}{M} = \left(-\frac{8\delta}{5}\right)\left(\frac{1}{4\delta}\right) = -\frac{2}{5}$

$\Rightarrow (\bar{x}, \bar{y}) = \left(1, -\frac{2}{5}\right)$ is the center of mass.

23. Since the plate is symmetric about the line x = y and its
density is constant, the distribution of mass is symmetric
about this line. This means that $\bar{x} = \bar{y}$. The typical *vertical*
strip has

center of mass: $\left(\tilde{x}, \tilde{y}\right) = \left(x, \frac{3 + \sqrt{9 - x^2}}{2}\right)$,

length: $3 - \sqrt{9 - x^2}$, width: dx,

area: $dA = \left(3 - \sqrt{9 - x^2}\right)dx$,

mass: $dm = \delta\,dA = \delta\left(3 - \sqrt{9 - x^2}\right)dx$.

The moment about the x-axis is

$\tilde{y}\,dm = \delta\frac{\left(3 + \sqrt{9 - x^2}\right)\left(3 - \sqrt{9 - x^2}\right)}{2}\,dx = \frac{\delta}{2}[9 - (9 - x^2)]\,dx = \frac{\delta x^2}{2}\,dx$. Thus, $M_x = \int_0^3 \frac{\delta x^2}{2}\,dx = \frac{\delta}{6}[x^3]_0^3 = \frac{9\delta}{2}$. The area

equals the area of a square with side length 3 minus one quarter the area of a disk with radius 3 $\Rightarrow A = 3^2 - \frac{\pi 9}{4}$

$= \frac{9}{4}(4 - \pi) \Rightarrow M = \delta A = \frac{9\delta}{4}(4 - \pi)$. Therefore, $\bar{y} = \frac{M_x}{M} = \left(\frac{9\delta}{2}\right)\left[\frac{4}{9\delta(4 - \pi)}\right] = \frac{2}{4 - \pi} \Rightarrow (\bar{x}, \bar{y}) = \left(\frac{2}{4 - \pi}, \frac{2}{4 - \pi}\right)$ is the

center of mass.

25. $M_x = \int \tilde{y}\,dm = \int_1^2 \frac{\left(\frac{2}{x^2}\right)}{2}\cdot\delta\cdot\left(\frac{2}{x^2}\right)dx$

$= \int_1^2 \left(\frac{1}{x^2}\right)(x^2)\left(\frac{2}{x^2}\right)dx = \int_1^2 \frac{2}{x^2}\,dx = 2\int_1^2 x^{-2}\,dx$

$= 2[-x^{-1}]_1^2 = 2\left[\left(-\frac{1}{2}\right) - (-1)\right] = 2\left(\frac{1}{2}\right) = 1$;

$M_y = \int \tilde{x}\,dm = \int_1^2 x\cdot\delta\cdot\left(\frac{2}{x^2}\right)dx$

$= \int_1^2 x(x^2)\left(\frac{2}{x^2}\right)dx = 2\int_1^2 x\,dx = 2\left[\frac{x^2}{2}\right]_1^2$

$= 2\left(2 - \frac{1}{2}\right) = 4 - 1 = 3$; $M = \int dm = \int_1^2 \delta\left(\frac{2}{x^2}\right)dx = \int_1^2 x^2\left(\frac{2}{x^2}\right)dx = 2\int_1^2 dx = 2[x]_1^2 = 2(2 - 1) = 2$. So

$\bar{x} = \frac{M_y}{M} = \frac{3}{2}$ and $\bar{y} = \frac{M_x}{M} = \frac{1}{2} \Rightarrow (\bar{x}, \bar{y}) = \left(\frac{3}{2}, \frac{1}{2}\right)$ is the center of mass.

27. (a) We use the shell method: $V = \int_a^b 2\pi\left(\begin{smallmatrix}\text{shell}\\\text{radius}\end{smallmatrix}\right)\left(\begin{smallmatrix}\text{shell}\\\text{height}\end{smallmatrix}\right)dx = \int_1^4 2\pi x\left[\frac{4}{\sqrt{x}} - \left(-\frac{4}{\sqrt{x}}\right)\right]dx$

$= 16\pi\int_1^4 \frac{x}{\sqrt{x}}\,dx = 16\pi\int_1^4 x^{1/2}\,dx = 16\pi\left[\frac{2}{3}x^{3/2}\right]_1^4 = 16\pi\left(\frac{2}{3}\cdot 8 - \frac{2}{3}\right) = \frac{32\pi}{3}(8 - 1) = \frac{224\pi}{3}$

(b) Since the plate is symmetric about the x-axis and its density $\delta(x) = \frac{1}{x}$ is a function of x alone, the distribution of its mass is symmetric about the x-axis. This means that $\bar{y} = 0$. We use the vertical strip approach to find $\bar{x}$: $M_y = \int \tilde{x}\ dm = \int_1^4 x \cdot \left[\frac{4}{\sqrt{x}} - \left(-\frac{4}{\sqrt{x}}\right)\right] \cdot \delta\ dx = \int_1^4 x \cdot \frac{8}{\sqrt{x}} \cdot \frac{1}{x}\ dx = 8\int_1^4 x^{-1/2}\ dx$

$= 8\left[2x^{1/2}\right]_1^4 = 8(2\cdot 2 - 2) = 16$; $M = \int dm = \int_1^4 \left[\frac{4}{\sqrt{x}} - \left(\frac{-4}{\sqrt{x}}\right)\right] \cdot \delta\ dx = 8\int_1^4 \left(\frac{1}{\sqrt{x}}\right)\left(\frac{1}{x}\right)\ dx = 8\int_1^4 x^{-3/2}\ dx$

$= 8\left[-2x^{-1/2}\right]_1^4 = 8[-1 - (-2)] = 8$. So $\bar{x} = \frac{M_y}{M} = \frac{16}{8} = 2 \Rightarrow (\bar{x}, \bar{y}) = (2, 0)$ is the center of mass.

(c)

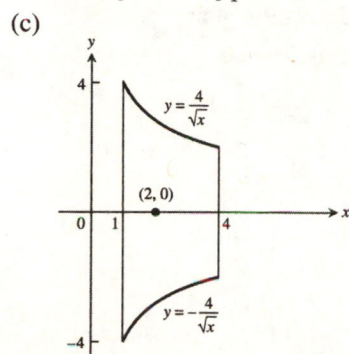

29. The mass of a horizontal strip is $dm = \delta\ dA = \delta L\ dy$, where L is the width of the triangle at a distance of y above its base on the x-axis as shown in the figure in the text. Also, by similar triangles we have $\frac{L}{b} = \frac{h-y}{h}$

$\Rightarrow L = \frac{b}{h}(h - y)$. Thus, $M_x = \int \tilde{y}\ dm = \int_0^h \delta y \left(\frac{b}{h}\right)(h - y)\ dy = \frac{\delta b}{h}\int_0^h (hy - y^2)\ dy = \frac{\delta b}{h}\left[\frac{hy^2}{2} - \frac{y^3}{3}\right]_0^h$

$= \frac{\delta b}{h}\left(\frac{h^3}{2} - \frac{h^3}{3}\right) = \delta bh^2\left(\frac{1}{2} - \frac{1}{3}\right) = \frac{\delta bh^2}{6}$; $M = \int dm = \int_0^h \delta\left(\frac{b}{h}\right)(h - y)\ dy = \frac{\delta b}{h}\int_0^h (h - y)\ dy = \frac{\delta b}{h}\left[hy - \frac{y^2}{2}\right]_0^h$

$= \frac{\delta b}{h}\left(h^2 - \frac{h^2}{2}\right) = \frac{\delta bh}{2}$. So $\bar{y} = \frac{M_x}{M} = \left(\frac{\delta bh^2}{6}\right)\left(\frac{2}{\delta bh}\right) = \frac{h}{3} \Rightarrow$ the center of mass lies above the base of the triangle one-third of the way toward the opposite vertex. Similarly the other two sides of the triangle can be placed on the x-axis and the same results will occur. Therefore the centroid does lie at the intersection of the medians, as claimed.

31. From the symmetry about the line $x = y$ it follows that $\bar{x} = \bar{y}$. It also follows that the line through the points $(0, 0)$ and $\left(\frac{1}{2}, \frac{1}{2}\right)$ is a median $\Rightarrow \bar{y} = \bar{x} = \frac{2}{3} \cdot \left(\frac{1}{2} - 0\right) = \frac{1}{3}$

$\Rightarrow (\bar{x}, \bar{y}) = \left(\frac{1}{3}, \frac{1}{3}\right)$.

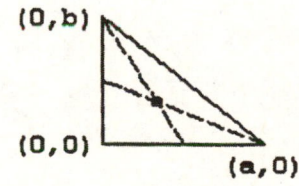

33. The point of intersection of the median from the vertex $(0, b)$ to the opposite side has coordinates $\left(0, \frac{a}{2}\right)$

$\Rightarrow \bar{y} = (b - 0) \cdot \frac{1}{3} = \frac{b}{3}$ and $\bar{x} = \left(\frac{a}{2} - 0\right) \cdot \frac{2}{3} = \frac{a}{3}$

$\Rightarrow (\bar{x}, \bar{y}) = \left(\frac{a}{3}, \frac{b}{3}\right)$.

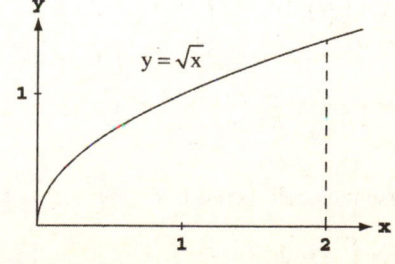

35. $y = x^{1/2} \Rightarrow dy = \frac{1}{2}x^{-1/2}\ dx$

$\Rightarrow ds = \sqrt{(dx)^2 + (dy)^2} = \sqrt{1 + \frac{1}{4x}}\ dx$;

$M_x = \delta\int_0^2 \sqrt{x}\sqrt{1 + \frac{1}{4x}}\ dx$

$= \delta\int_0^2 \sqrt{x + \frac{1}{4}}\ dx = \frac{2\delta}{3}\left[\left(x + \frac{1}{4}\right)^{3/2}\right]_0^2$

$= \frac{2\delta}{3}\left[\left(2 + \frac{1}{4}\right)^{3/2} - \left(\frac{1}{4}\right)^{3/2}\right]$

$= \frac{2\delta}{3}\left[\left(\frac{9}{4}\right)^{3/2} - \left(\frac{1}{4}\right)^{3/2}\right] = \frac{2\delta}{3}\left(\frac{27}{8} - \frac{1}{8}\right) = \frac{13\delta}{6}$

37. From Example 6 we have $M_x = \int_0^\pi a(a\sin\theta)(k\sin\theta)\,d\theta = a^2 k\int_0^\pi \sin^2\theta\,d\theta = \frac{a^2 k}{2}\int_0^\pi (1 - \cos 2\theta)\,d\theta$

$= \frac{a^2 k}{2}\left[\theta - \frac{\sin 2\theta}{2}\right]_0^\pi = \frac{a^2 k\pi}{2}$; $M_y = \int_0^\pi a(a\cos\theta)(k\sin\theta)\,d\theta = a^2 k\int_0^\pi \sin\theta\cos\theta\,d\theta = \frac{a^2 k}{2}\left[\sin^2\theta\right]_0^\pi = 0$;

$M = \int_0^\pi ak\sin\theta\,d\theta = ak[-\cos\theta]_0^\pi = 2ak$. Therefore, $\bar{x} = \frac{M_y}{M} = 0$ and $\bar{y} = \frac{M_x}{M} = \left(\frac{a^2 k\pi}{2}\right)\left(\frac{1}{2ak}\right) = \frac{a\pi}{4} \Rightarrow \left(0, \frac{a\pi}{4}\right)$

is the center of mass.

39. Consider the curve as an infinite number of line segments joined together. From the derivation of arc length we have that the length of a particular segment is $ds = \sqrt{(dx)^2 + (dy)^2}$. This implies that

$M_x = \int \delta y\,ds$, $M_y = \int \delta x\,ds$ and $M = \int \delta\,ds$. If δ is constant, then $\bar{x} = \frac{M_y}{M} = \frac{\int x\,ds}{\int ds} = \frac{\int x\,ds}{\text{length}}$ and

$\bar{y} = \frac{M_x}{M} = \frac{\int y\,ds}{\int ds} = \frac{\int y\,ds}{\text{length}}$.

41. Since the density is constant, its value will not affect our answers, so we can set $\delta = 1$.

A generalization of Example 6 yields $M_x = \int \tilde{y}\,dm = \int_{\pi/2-\alpha}^{\pi/2+\alpha} a^2 \sin\theta\,d\theta = a^2[-\cos\theta]_{\pi/2-\alpha}^{\pi/2+\alpha}$

$= a^2\left[-\cos\left(\frac{\pi}{2} + \alpha\right) + \cos\left(\frac{\pi}{2} - \alpha\right)\right] = a^2(\sin\alpha + \sin\alpha) = 2a^2\sin\alpha$; $M = \int dm = \int_{\pi/2-\alpha}^{\pi/2+\alpha} a\,d\theta = a[\theta]_{\pi/2-\alpha}^{\pi/2+\alpha}$

$= a\left[\left(\frac{\pi}{2} + \alpha\right) - \left(\frac{\pi}{2} - \alpha\right)\right] = 2a\alpha$. Thus, $\bar{y} = \frac{M_x}{M} = \frac{2a^2\sin\alpha}{2a\alpha} = \frac{a\sin\alpha}{\alpha}$. Now $s = a(2\alpha)$ and $a\sin\alpha = \frac{c}{2}$

$\Rightarrow c = 2a\sin\alpha$. Then $\bar{y} = \frac{a(2a\sin\alpha)}{2a\alpha} = \frac{ac}{s}$, as claimed.

6.5 AREAS OF SURFACES OF REVOLUTION AND THE THEOREMS OF PAPPUS

1. (a) $\frac{dy}{dx} = \sec^2 x \Rightarrow \left(\frac{dy}{dx}\right)^2 = \sec^4 x$

$\Rightarrow S = 2\pi\int_0^{\pi/4} (\tan x)\sqrt{1 + \sec^4 x}\,dx$

(c) $S \approx 3.84$

(b)

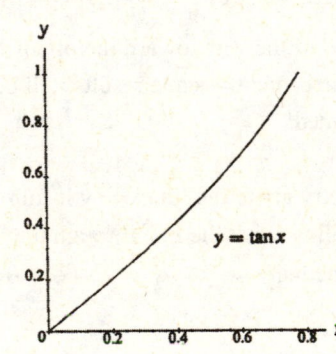

3. (a) $xy = 1 \Rightarrow x = \frac{1}{y} \Rightarrow \frac{dx}{dy} = -\frac{1}{y^2} \Rightarrow \left(\frac{dx}{dy}\right)^2 = \frac{1}{y^4}$

$\Rightarrow S = 2\pi\int_1^2 \frac{1}{y}\sqrt{1 + y^{-4}}\,dy$

(c) $S \approx 5.02$

(b)

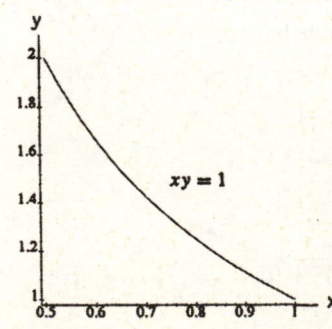

5. (a) $x^{1/2} + y^{1/2} = 3 \Rightarrow y = \left(3 - x^{1/2}\right)^2$ (b)

$\Rightarrow \frac{dy}{dx} = 2\left(3 - x^{1/2}\right)\left(-\frac{1}{2}x^{-1/2}\right)$

$\Rightarrow \left(\frac{dy}{dx}\right)^2 = \left(1 - 3x^{-1/2}\right)^2$

$\Rightarrow S = 2\pi \int_1^4 \left(3 - x^{1/2}\right)^2 \sqrt{1 + \left(1 - 3x^{-1/2}\right)^2}\, dx$

(c) $S \approx 63.37$

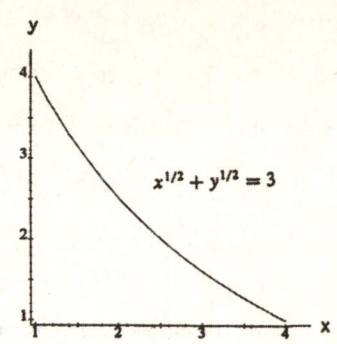

7. (a) $\frac{dx}{dy} = \tan y \Rightarrow \left(\frac{dx}{dy}\right)^2 = \tan^2 y$ (b)

$\Rightarrow S = 2\pi \int_0^{\pi/3} \left(\int_0^y \tan t\, dt\right)\sqrt{1 + \tan^2 y}\, dy$

$= 2\pi \int_0^{\pi/3} \left(\int_0^y \tan t\, dt\right)\sec y\, dy$

(c) $S \approx 2.08$

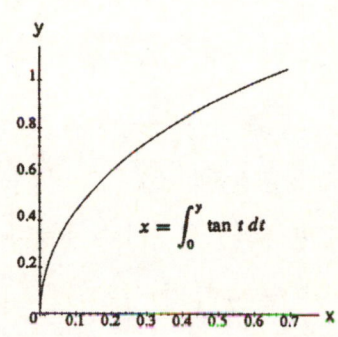

9. $y = \frac{x}{2} \Rightarrow \frac{dy}{dx} = \frac{1}{2}; S = \int_a^b 2\pi y \sqrt{1 + \left(\frac{dy}{dx}\right)^2}\, dx \Rightarrow S = \int_0^4 2\pi \left(\frac{x}{2}\right)\sqrt{1 + \frac{1}{4}}\, dx = \frac{\pi\sqrt{5}}{2}\int_0^4 x\, dx$

$= \frac{\pi\sqrt{5}}{2}\left[\frac{x^2}{2}\right]_0^4 = 4\pi\sqrt{5}$; Geometry formula: base circumference $= 2\pi(2)$, slant height $= \sqrt{4^2 + 2^2} = 2\sqrt{5}$

$\Rightarrow$ Lateral surface area $= \frac{1}{2}(4\pi)\left(2\sqrt{5}\right) = 4\pi\sqrt{5}$ in agreement with the integral value

11. $\frac{dy}{dx} = \frac{1}{2}; S = \int_a^b 2\pi y \sqrt{1 + \left(\frac{dy}{dx}\right)^2}\, dx = \int_1^3 2\pi \frac{(x+1)}{2}\sqrt{1 + \left(\frac{1}{2}\right)^2}\, dx = \frac{\pi\sqrt{5}}{2}\int_1^3 (x+1)\, dx = \frac{\pi\sqrt{5}}{2}\left[\frac{x^2}{2} + x\right]_1^3$

$= \frac{\pi\sqrt{5}}{2}\left[\left(\frac{9}{2} + 3\right) - \left(\frac{1}{2} + 1\right)\right] = \frac{\pi\sqrt{5}}{2}(4 + 2) = 3\pi\sqrt{5}$; Geometry formula: $r_1 = \frac{1}{2} + \frac{1}{2} = 1, r_2 = \frac{3}{2} + \frac{1}{2} = 2,$

slant height $= \sqrt{(2-1)^2 + (3-1)^2} = \sqrt{5} \Rightarrow$ Frustum surface area $= \pi(r_1 + r_2) \times$ slant height $= \pi(1+2)\sqrt{5}$

$= 3\pi\sqrt{5}$ in agreement with the integral value

13. $\frac{dy}{dx} = \frac{x^2}{3} \Rightarrow \left(\frac{dy}{dx}\right)^2 = \frac{x^4}{9} \Rightarrow S = \int_0^2 \frac{2\pi x^3}{9}\sqrt{1 + \frac{x^4}{9}}\, dx;$

$\left[u = 1 + \frac{x^4}{9} \Rightarrow du = \frac{4}{9}x^3\, dx \Rightarrow \frac{1}{4}du = \frac{x^3}{9}\, dx;\right.$

$x = 0 \Rightarrow u = 1, x = 2 \Rightarrow u = \frac{25}{9}\bigg]$

$\to S = 2\pi \int_1^{25/9} u^{1/2} \cdot \frac{1}{4}\, du = \frac{\pi}{2}\left[\frac{2}{3}u^{3/2}\right]_1^{25/9}$

$= \frac{\pi}{3}\left(\frac{125}{27} - 1\right) = \frac{\pi}{3}\left(\frac{125-27}{27}\right) = \frac{98\pi}{81}$

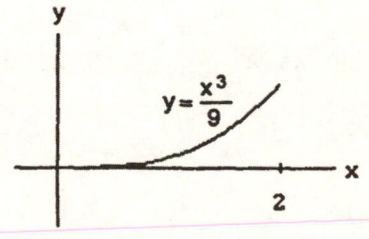

15. $\frac{dy}{dx} = \frac{1}{2}\frac{(2-2x)}{\sqrt{2x-x^2}} = \frac{1-x}{\sqrt{2x-x^2}} \Rightarrow \left(\frac{dy}{dx}\right)^2 = \frac{(1-x)^2}{2x-x^2}$

$\Rightarrow S = \int_{0.5}^{1.5} 2\pi\sqrt{2x-x^2}\sqrt{1 + \frac{(1-x)^2}{2x-x^2}}\, dx$

$= 2\pi \int_{0.5}^{1.5} \sqrt{2x-x^2}\, \frac{\sqrt{2x-x^2+1-2x+x^2}}{\sqrt{2x-x^2}}\, dx$

$= 2\pi \int_{0.5}^{1.5} dx = 2\pi[x]_{0.5}^{1.5} = 2\pi$

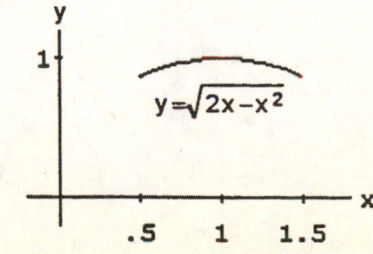

17. $\frac{dx}{dy} = y^2 \Rightarrow \left(\frac{dx}{dy}\right)^2 = y^4 \Rightarrow S = \int_0^1 \frac{2\pi y^3}{3} \sqrt{1+y^4}\, dy$;

$[u = 1+y^4 \Rightarrow du = 4y^3\, dy \Rightarrow \frac{1}{4}\, du = y^3\, dy;\ y = 0$

$\Rightarrow u = 1,\ y = 1 \Rightarrow u = 2] \rightarrow S = \int_1^2 2\pi \left(\frac{1}{3}\right) u^{1/2} \left(\frac{1}{4}\, du\right)$

$= \frac{\pi}{6} \int_1^2 u^{1/2}\, du = \frac{\pi}{6} \left[\frac{2}{3} u^{3/2}\right]_1^2 = \frac{\pi}{9} \left(\sqrt{8} - 1\right)$

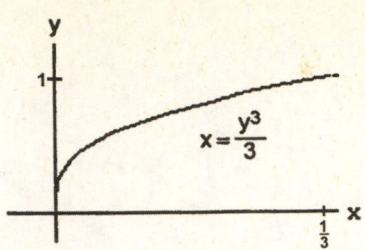

19. $\frac{dx}{dy} = \frac{-1}{\sqrt{4-y}} \Rightarrow \left(\frac{dx}{dy}\right)^2 = \frac{1}{4-y} \Rightarrow S = \int_0^{15/4} 2\pi \cdot 2\sqrt{4-y}\ \sqrt{1 + \frac{1}{4-y}}\, dy = 4\pi \int_0^{15/4} \sqrt{(4-y)+1}\, dy$

$= 4\pi \int_0^{15/4} \sqrt{5-y}\, dy = -4\pi \left[\frac{2}{3}(5-y)^{3/2}\right]_0^{15/4} = -\frac{8\pi}{3}\left[\left(5 - \frac{15}{4}\right)^{3/2} - 5^{3/2}\right] = -\frac{8\pi}{3}\left[\left(\frac{5}{4}\right)^{3/2} - 5^{3/2}\right]$

$= \frac{8\pi}{3}\left(5\sqrt{5} - \frac{5\sqrt{5}}{8}\right) = \frac{8\pi}{3}\left(\frac{40\sqrt{5} - 5\sqrt{5}}{8}\right) = \frac{35\pi\sqrt{5}}{3}$

21. $ds = \sqrt{dx^2 + dy^2} = \sqrt{\left(y^3 - \frac{1}{4y^3}\right)^2 + 1}\, dy = \sqrt{\left(y^6 - \frac{1}{2} + \frac{1}{16y^6}\right) + 1}\, dy = \sqrt{\left(y^6 + \frac{1}{2} + \frac{1}{16y^6}\right)}\, dy$

$= \sqrt{\left(y^3 + \frac{1}{4y^3}\right)^2}\, dy = \left(y^3 + \frac{1}{4y^3}\right)\, dy;\ S = \int_1^2 2\pi y\, ds = 2\pi \int_1^2 y\left(y^3 + \frac{1}{4y^3}\right)\, dy = 2\pi \int_1^2 \left(y^4 + \frac{1}{4}y^{-2}\right)\, dy$

$= 2\pi \left[\frac{y^5}{5} - \frac{1}{4}y^{-1}\right]_1^2 = 2\pi\left[\left(\frac{32}{5} - \frac{1}{8}\right) - \left(\frac{1}{5} - \frac{1}{4}\right)\right] = 2\pi\left(\frac{31}{5} + \frac{1}{8}\right) = \frac{2\pi}{40}(8 \cdot 31 + 5) = \frac{253\pi}{20}$

23. $y = \sqrt{a^2 - x^2} \Rightarrow \frac{dy}{dx} = \frac{1}{2}(a^2 - x^2)^{-1/2}(-2x) = \frac{-x}{\sqrt{a^2 - x^2}} \Rightarrow \left(\frac{dy}{dx}\right)^2 = \frac{x^2}{(a^2 - x^2)}$

$\Rightarrow S = 2\pi \int_{-a}^a \sqrt{a^2 - x^2}\ \sqrt{1 + \frac{x^2}{(a^2 - x^2)}}\, dx = 2\pi \int_{-a}^a \sqrt{(a^2 - x^2) + x^2}\, dx = 2\pi \int_{-a}^a a\, dx = 2\pi a[x]_{-a}^a$

$= 2\pi a[a - (-a)] = (2\pi a)(2a) = 4\pi a^2$

25. $y = \cos x \Rightarrow \frac{dy}{dx} = -\sin x \Rightarrow \left(\frac{dy}{dx}\right)^2 = \sin^2 x \Rightarrow S = 2\pi \int_{-\pi/2}^{\pi/2} (\cos x)\sqrt{1 + \sin^2 x}\, dx$

27. The area of the surface of one wok is $S = \int_c^d 2\pi x \sqrt{1 + \left(\frac{dx}{dy}\right)^2}\, dy$. Now, $x^2 + y^2 = 16^2 \Rightarrow x = \sqrt{16^2 - y^2}$

$\Rightarrow \frac{dx}{dy} = \frac{-y}{\sqrt{16^2 - y^2}} \Rightarrow \left(\frac{dx}{dy}\right)^2 = \frac{y^2}{16^2 - y^2};\ S = \int_{-16}^{-7} 2\pi \sqrt{16^2 - y^2}\ \sqrt{1 + \frac{y^2}{16^2 - y^2}}\, dy = 2\pi \int_{-16}^{-7} \sqrt{(16^2 - y^2) + y^2}\, dy$

$= 2\pi \int_{-16}^{-7} 16\, dy = 32\pi \cdot 9 = 288\pi \approx 904.78\ \text{cm}^2$. The enamel needed to cover one surface of one wok is

$V = S \cdot 0.5\ \text{mm} = S \cdot 0.05\ \text{cm} = (904.78)(0.05)\ \text{cm}^3 = 45.24\ \text{cm}^3$. For 5000 woks, we need

$5000 \cdot V = 5000 \cdot 45.24\ \text{cm}^3 = (5)(45.24)\text{L} = 226.2\text{L} \Rightarrow 226.2$ liters of each color are needed.

29. $y = \sqrt{R^2 - x^2} \Rightarrow \frac{dy}{dx} = -\frac{1}{2}\frac{2x}{\sqrt{R^2 - x^2}} = \frac{-x}{\sqrt{R^2 - x^2}} \Rightarrow \left(\frac{dx}{dy}\right)^2 = \frac{x^2}{R^2 - x^2};\ S = 2\pi \int_a^{a+h} \sqrt{R^2 - x^2}\ \sqrt{1 + \frac{x^2}{R^2 - x^2}}\, dx$

$= 2\pi \int_a^{a+h} \sqrt{(R^2 - x^2) + x^2}\, dx = 2\pi R \int_a^{a+h} dx = 2\pi R h$

31. $y = x \Rightarrow \left(\frac{dy}{dx}\right) = 1 \Rightarrow \left(\frac{dy}{dx}\right)^2 = 1 \Rightarrow S = 2\pi \int_{-1}^2 |x| \sqrt{1+1}\, dx = 2\pi \int_{-1}^0 (-x)\sqrt{2}\, dx + 2\pi \int_0^2 x\sqrt{2}\, dx$

$= -2\sqrt{2}\pi \left[\frac{x^2}{2}\right]_{-1}^0 + 2\sqrt{2}\pi \left[\frac{x^2}{2}\right]_0^2 = -2\sqrt{2}\pi\left(0 - \frac{1}{2}\right) + 2\sqrt{2}\pi(2 - 0) = 5\sqrt{2}\pi$

33. $\frac{dx}{dt} = -\sin t$ and $\frac{dy}{dt} = \cos t \Rightarrow \sqrt{\left(\frac{dx}{dt}\right)^2 + \left(\frac{dy}{dt}\right)^2} = \sqrt{(-\sin t)^2 + (\cos t)^2} = 1 \Rightarrow S = \int 2\pi y\, ds$

$= \int_0^{2\pi} 2\pi(2 + \sin t)(1)\, dt = 2\pi \left[2t - \cos t\right]_0^{2\pi} = 2\pi[(4\pi - 1) - (0 - 1)] = 8\pi^2$

35. $\frac{dx}{dt} = 1$ and $\frac{dy}{dt} = t + \sqrt{2}$ $\Rightarrow$ $\sqrt{\left(\frac{dx}{dt}\right)^2 + \left(\frac{dy}{dt}\right)^2} = \sqrt{1^2 + \left(t + \sqrt{2}\right)^2} = \sqrt{t^2 + 2\sqrt{2}t + 3}$ $\Rightarrow$ $S = \int 2\pi x \, ds$

$= \int_{-\sqrt{2}}^{\sqrt{2}} 2\pi \left(t + \sqrt{2}\right) \sqrt{t^2 + 2\sqrt{2}t + 3} \, dt; \left[u = t^2 + 2\sqrt{2}t + 3 \Rightarrow du = \left(2t + 2\sqrt{2}\right) dt; t = -\sqrt{2} \Rightarrow u = 1,\right.$

$t = \sqrt{2} \Rightarrow u = 9\Big] \to \int_1^9 \pi \sqrt{u} \, du = \left[\frac{2}{3}\pi u^{3/2}\right]_1^9 = \frac{2\pi}{3}(27 - 1) = \frac{52\pi}{3}$

37. $\frac{dx}{dt} = 2$ and $\frac{dy}{dt} = 1$ $\Rightarrow$ $\sqrt{\left(\frac{dx}{dt}\right)^2 + \left(\frac{dy}{dt}\right)^2} = \sqrt{2^2 + 1^2} = \sqrt{5}$ $\Rightarrow$ $S = \int 2\pi y \, ds = \int_0^1 2\pi(t+1)\sqrt{5} \, dt$

$= 2\pi\sqrt{5}\left[\frac{t^2}{2} + t\right]_0^1 = 3\pi\sqrt{5}$. Check: slant height is $\sqrt{5}$ $\Rightarrow$ Area is $\pi(1 + 2)\sqrt{5} = 3\pi\sqrt{5}$.

39. (a) An equation of the tangent line segment is
(see figure) $y = f(m_k) + f'(m_k)(x - m_k)$.
When $x = x_{k-1}$ we have
$r_1 = f(m_k) + f'(m_k)(x_{k-1} - m_k)$
$= f(m_k) + f'(m_k)\left(-\frac{\Delta x_k}{2}\right) = f(m_k) - f'(m_k)\frac{\Delta x_k}{2}$;
when $x = x_k$ we have
$r_2 = f(m_k) + f'(m_k)(x_k - m_k)$
$= f(m_k) + f'(m_k)\frac{\Delta x_k}{2}$;

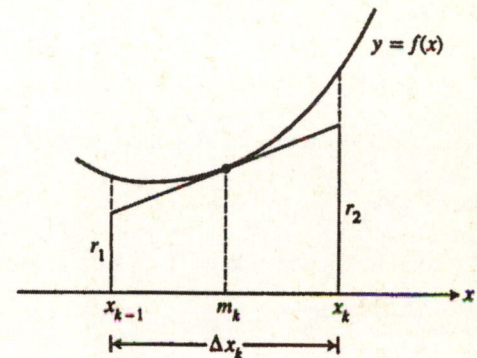

(b) $L_k^2 = (\Delta x_k)^2 + (r_2 - r_1)^2$
$= (\Delta x_k)^2 + \left[f'(m_k)\frac{\Delta x_k}{2} - \left(-f'(m_k)\frac{\Delta x_k}{2}\right)\right]^2$
$= (\Delta x_k)^2 + [f'(m_k)\Delta x_k]^2$ $\Rightarrow$ $L_k = \sqrt{(\Delta x_k)^2 + [f'(m_k)\Delta x_k]^2}$, as claimed

(c) From geometry it is a fact that the lateral surface area of the frustum obtained by revolving the tangent
line segment about the x-axis is given by $\Delta S_k = \pi(r_1 + r_2)L_k = \pi[2f(m_k)]\sqrt{(\Delta x_k)^2 + [f'(m_k)\Delta x_k]^2}$
using parts (a) and (b) above. Thus, $\Delta S_k = 2\pi f(m_k)\sqrt{1 + [f'(m_k)]^2}\, \Delta x_k$.

(d) $S = \lim_{n \to \infty} \sum_{k=1}^n \Delta S_k = \lim_{n \to \infty} \sum_{k=1}^n 2\pi f(m_k)\sqrt{1 + [f'(m_k)]^2}\, \Delta x_k = \int_a^b 2\pi f(x)\sqrt{1 + [f'(x)]^2}\, dx$

41. The centroid of the square is located at $(2, 2)$. The volume is $V = (2\pi)(\bar{y})(A) = (2\pi)(2)(8) = 32\pi$ and the
surface area is $S = (2\pi)(\bar{y})(L) = (2\pi)(2)\left(4\sqrt{8}\right) = 32\sqrt{2}\pi$ (where $\sqrt{8}$ is the length of a side).

43. The centroid is located at $(2, 0)$ $\Rightarrow$ $V = (2\pi)(\bar{x})(A) = (2\pi)(2)(\pi) = 4\pi^2$

45. $S = 2\pi\bar{y}L$ $\Rightarrow$ $4\pi a^2 = (2\pi\bar{y})(\pi a)$ $\Rightarrow$ $\bar{y} = \frac{2a}{\pi}$, and by symmetry $\bar{x} = 0$

47. $V = 2\pi\bar{y}A$ $\Rightarrow$ $\frac{4}{3}\pi ab^2 = (2\pi\bar{y})\left(\frac{\pi ab}{2}\right)$ $\Rightarrow$ $\bar{y} = \frac{4b}{3\pi}$ and by symmetry $\bar{x} = 0$

49. $V = 2\pi\rho A = (2\pi)(\text{area of the region}) \cdot (\text{distance from the centroid to the line } y = x - a)$. We must find the
distance from $\left(0, \frac{4a}{3\pi}\right)$ to $y = x - a$. The line containing the centroid and perpendicular to $y = x - a$ has slope
-1 and contains the point $\left(0, \frac{4a}{3\pi}\right)$. This line is $y = -x + \frac{4a}{3\pi}$. The intersection of $y = x - a$ and $y = -x + \frac{4a}{3\pi}$ is
the point $\left(\frac{4a + 3a\pi}{6\pi}, \frac{4a - 3a\pi}{6\pi}\right)$. Thus, the distance from the centroid to the line $y = x - a$ is
$\sqrt{\left(\frac{4a + 3a\pi}{6\pi}\right)^2 + \left(\frac{4a}{3\pi} - \frac{4a}{6\pi} + \frac{3a\pi}{6\pi}\right)^2} = \frac{\sqrt{2}(4a + 3a\pi)}{6\pi}$ $\Rightarrow$ $V = (2\pi)\left(\frac{\sqrt{2}(4a + 3a\pi)}{6\pi}\right)\left(\frac{\pi a^2}{2}\right) = \frac{\sqrt{2}\pi a^3(4 + 3\pi)}{6}$

51. From Example 4 and Pappus's Theorem for Volumes we have the moment about the x-axis is $M_x = \bar{y}M$
$= \left(\frac{4a}{3\pi}\right)\left(\frac{\pi a^2}{2}\right) = \frac{2a^3}{3}$.

6.6 WORK

1. The force required to stretch the spring from its natural length of 2 m to a length of 5 m is $F(x) = kx$. The

   work done by F is $W = \int_0^3 F(x)\, dx = k \int_0^3 x\, dx = \frac{k}{2} [x^2]_0^3 = \frac{9k}{2}$. This work is equal to 1800 J $\Rightarrow \frac{9}{2} k = 1800$

   $\Rightarrow k = 400$ N/m

3. We find the force constant from Hooke's law: $F = kx$. A force of 2 N stretches the spring to 0.02 m

   $\Rightarrow 2 = k \cdot (0.02) \Rightarrow k = 100 \frac{N}{m}$. The force of 4 N will stretch the rubber band y m, where $F = ky \Rightarrow y = \frac{F}{k}$

   $\Rightarrow y = \frac{4N}{100 \frac{N}{m}} \Rightarrow y = 0.04$ m = 4 cm. The work done to stretch the rubber band 0.04 m is $W = \int_0^{0.04} kx\, dx$

   $= 100 \int_0^{0.04} x\, dx = 100 \left[\frac{x^2}{2} \right]_0^{0.04} = \frac{(100)(0.04)^2}{2} = 0.08$ J

5. (a) We find the spring's constant from Hooke's law: $F = kx \Rightarrow k = \frac{F}{x} = \frac{21,714}{8-5} = \frac{21,714}{3} \Rightarrow k = 7238 \frac{lb}{in}$

   (b) The work done to compress the assembly the first half inch is $W = \int_0^{0.5} kx\, dx = 7238 \int_0^{0.5} x\, dx$

   $= 7238 \left[\frac{x^2}{2} \right]_0^{0.5} = (7238) \frac{(0.5)^2}{2} = \frac{(7238)(0.25)}{2} \approx 905$ in $\cdot$ lb. The work done to compress the assembly the

   second half inch is: $W = \int_{0.5}^{1.0} kx\, dx = 7238 \int_{0.5}^{1.0} x\, dx = 7238 \left[\frac{x^2}{2} \right]_{0.5}^{1.0} = \frac{7238}{2} [1 - (0.5)^2] = \frac{(7238)(0.75)}{2}$

   ≈ 2714 in $\cdot$ lb

7. The force required to haul up the rope is equal to the rope's weight, which varies steadily and is proportional to

   x, the length of the rope still hanging: $F(x) = 0.624x$. The work done is: $W = \int_0^{50} F(x)\, dx = \int_0^{50} 0.624x\, dx$

   $= 0.624 \left[\frac{x^2}{2} \right]_0^{50} = 780$ J

9. The force required to lift the cable is equal to the weight of the cable paid out: $F(x) = (4.5)(180 - x)$ where x

   is the position of the car off the first floor. The work done is: $W = \int_0^{180} F(x)\, dx = 4.5 \int_0^{180} (180 - x)\, dx$

   $= 4.5 \left[180x - \frac{x^2}{2} \right]_0^{180} = 4.5 \left(180^2 - \frac{180^2}{2} \right) = \frac{4.5 \cdot 180^2}{2} = 72,900$ ft $\cdot$ lb

11. The force against the piston is $F = pA$. If $V = Ax$, where x is the height of the cylinder, then $dV = A\, dx$

    $\Rightarrow$ Work $= \int F\, dx = \int pA\, dx = \int_{(p_1, V_1)}^{(p_2, V_2)} p\, dV$.

13. Let r = the constant rate of leakage. Since the bucket is leaking at a constant rate and the bucket is rising at a constant rate,

    the amount of water in the bucket is proportional to $(20 - x)$, the distance the bucket is being raised. The leakage rate of

    the water is 0.8 lb/ft raised and the weight of the water in the bucket is $F = 0.8(20 - x)$. So:

    $W = \int_0^{20} 0.8(20 - x)\, dx = 0.8 \left[20x - \frac{x^2}{2} \right]_0^{20} = 160$ ft $\cdot$ lb.

15. We will use the coordinate system given.

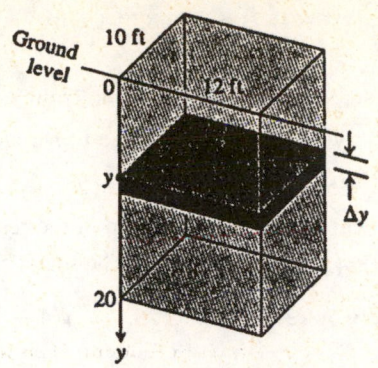

   (a) The typical slab between the planes at y and $y + \Delta y$ has a volume of $\Delta V = (10)(12)\,\Delta y = 120\,\Delta y$ ft^3. The force F required to lift the slab is equal to its weight: $F = 62.4\,\Delta V = 62.4 \cdot 120\,\Delta y$ lb. The distance through which F must act is about y ft, so the work done lifting the slab is about $\Delta W = $ force $\times$ distance $= 62.4 \cdot 120 \cdot y \cdot \Delta y$ ft $\cdot$ lb. The work it takes to lift all the water is approximately $W \approx \sum\limits_{0}^{20} \Delta W$

$= \sum\limits_{0}^{20} 62.4 \cdot 120y \cdot \Delta y$ ft $\cdot$ lb. This is a Riemann sum for

the function $62.4 \cdot 120y$ over the interval $0 \le y \le 20$. The work of pumping the tank empty is the limit of these sums:

$W = \int_{0}^{20} 62.4 \cdot 120y \, dy = (62.4)(120) \left[\frac{y^2}{2}\right]_{0}^{20} = (62.4)(120)\left(\frac{400}{2}\right) = (62.4)(120)(200) = 1{,}497{,}600$ ft $\cdot$ lb

   (b) The time t it takes to empty the full tank with $\left(\frac{5}{11}\right)$–hp motor is $t = \frac{W}{250 \frac{ft\cdot lb}{sec}} = \frac{1{,}497{,}600 \text{ ft·lb}}{250 \frac{ft\cdot lb}{sec}} = 5990.4$ sec

$= 1.664$ hr $\Rightarrow t \approx 1$ hr and 40 min

   (c) Following all the steps of part (a), we find that the work it takes to lower the water level 10 ft is

$W = \int_{0}^{10} 62.4 \cdot 120y \, dy = (62.4)(120)\left[\frac{y^2}{2}\right]_{0}^{10} = (62.4)(120)\left(\frac{100}{2}\right) = 374{,}400$ ft $\cdot$ lb and the time is $t = \frac{W}{250 \frac{ft\cdot lb}{sec}}$

$= 1497.6$ sec $= 0.416$ hr ≈ 25 min

   (d) In a location where water weighs $62.26\,\frac{lb}{ft^3}$:

   a) $W = (62.26)(24{,}000) = 1{,}494{,}240$ ft $\cdot$ lb.

   b) $t = \frac{1{,}494{,}240}{250} = 5976.96$ sec ≈ 1.660 hr $\Rightarrow t \approx 1$ hr and 40 min

   In a location where water weighs $62.59\,\frac{lb}{ft^3}$

   a) $W = (62.59)(24{,}000) = 1{,}502{,}160$ ft $\cdot$ lb

   b) $t = \frac{1{,}502{,}160}{250} = 6008.64$ sec ≈ 1.669 hr $\Rightarrow t \approx 1$ hr and 40.1 min

17. The slab is a disk of area $\pi x^2 = \pi\left(\frac{y}{2}\right)^2$, thickness $\triangle y$, and height below the top of the tank $(10 - y)$. So the work to pump the oil in this slab, $\triangle W$, is $57(10 - y)\pi\left(\frac{y}{2}\right)^2$. The work to pump all the oil to the top of the tank is

$W = \int_{0}^{10} \frac{57\pi}{4}(10y^2 - y^3)dy = \frac{57\pi}{4}\left[\frac{10y^3}{3} - \frac{y^4}{4}\right]_{0}^{10} = 11{,}875\pi$ ft $\cdot$ lb $\approx 37{,}306$ ft $\cdot$ lb.

19. The typical slab between the planes at y and and $y + \Delta y$ has a volume of $\Delta V = \pi(\text{radius})^2(\text{thickness}) = \pi\left(\frac{20}{2}\right)^2 \Delta y$
$= \pi \cdot 100\,\Delta y$ ft^3. The force F required to lift the slab is equal to its weight: $F = 51.2\,\Delta V = 51.2 \cdot 100\pi\,\Delta y$ lb
$\Rightarrow F = 5120\pi\,\Delta y$ lb. The distance through which F must act is about $(30 - y)$ ft. The work it takes to lift all the

kerosene is approximately $W \approx \sum\limits_{0}^{30} \Delta W = \sum\limits_{0}^{30} 5120\pi(30 - y)\,\Delta y$ ft $\cdot$ lb which is a Riemann sum. The work to pump the

tank dry is the limit of these sums: $W = \int_{0}^{30} 5120\pi(30 - y)\,dy = 5120\pi\left[30y - \frac{y^2}{2}\right]_{0}^{30} = 5120\pi\left(\frac{900}{2}\right) = (5120)(450\pi)$

$\approx 7{,}238{,}229.48$ ft $\cdot$ lb

21. (a) Follow all the steps of Example 5 but make the substitution of $64.5\,\frac{lb}{ft^3}$ for $57\,\frac{lb}{ft^3}$. Then,

$W = \int_{0}^{8} \frac{64.5\pi}{4}(10 - y)y^2 \, dy = \frac{64.5\pi}{4}\left[\frac{10y^3}{3} - \frac{y^4}{4}\right]_{0}^{8} = \frac{64.5\pi}{4}\left(\frac{10\cdot 8^3}{3} - \frac{8^4}{4}\right) = \left(\frac{64.5\pi}{4}\right)(8^3)\left(\frac{10}{3} - 2\right)$

$= \frac{64.5\pi \cdot 8^3}{3} = 21.5\pi \cdot 8^3 \approx 34{,}582.65$ ft $\cdot$ lb

   (b) Exactly as done in Example 5 but change the distance through which F acts to distance $\approx (13 - y)$ ft.

   Then $W = \int_{0}^{8} \frac{57\pi}{4}(13 - y)y^2 \, dy = \frac{57\pi}{4}\left[\frac{13y^3}{3} - \frac{y^4}{4}\right]_{0}^{8} = \frac{57\pi}{4}\left(\frac{13\cdot 8^3}{3} - \frac{8^4}{4}\right) = \left(\frac{57\pi}{4}\right)(8^3)\left(\frac{13}{3} - 2\right) = \frac{57\pi \cdot 8^3 \cdot 7}{3\cdot 4}$

$$= (19\pi)\,(8^2)\,(7)(2) \approx 53{,}482.5 \text{ ft} \cdot \text{lb}$$

23. The typical slab between the planes at y and y+Δy has a volume of about $\Delta V = \pi(\text{radius})^2(\text{thickness})$

$$= \pi \left(\sqrt{25 - y^2}\right)^2 \Delta y \text{ m}^3.$$ The force F(y) required to lift this slab is equal to its weight: $F(y) = 9800 \cdot \Delta V$

$$= 9800\pi \left(\sqrt{25 - y^2}\right)^2 \Delta y = 9800\pi\,(25 - y^2)\,\Delta y \text{ N}.$$ The distance through which F(y) must act to lift the

slab to the level of 4 m above the top of the reservoir is about $(4 - y)$ m, so the work done is approximately

$\Delta W \approx 9800\pi\,(25 - y^2)\,(4 - y)\,\Delta y \text{ N} \cdot \text{m}$. The work done lifting all the slabs from y = -5 m to y = 0 m is

approximately $W \approx \sum_{-5}^{0} 9800\pi\,(25 - y^2)\,(4 - y)\,\Delta y \text{ N} \cdot \text{m}$. Taking the limit of these Riemann sums, we get

$$W = \int_{-5}^{0} 9800\pi\,(25 - y^2)\,(4 - y)\,dy = 9800\pi \int_{-5}^{0}(100 - 25y - 4y^2 + y^3)\,dy = 9800\pi \left[100y - \tfrac{25}{2}y^2 - \tfrac{4}{3}y^3 + \tfrac{y^4}{4}\right]_{-5}^{0}$$

$$= -9800\pi \left(-500 - \tfrac{25 \cdot 25}{2} + \tfrac{4}{3} \cdot 125 + \tfrac{625}{4}\right) \approx 15{,}073{,}099.75 \text{ J}$$

25. $F = m\frac{dv}{dt} = mv\frac{dv}{dx}$ by the chain rule $\Rightarrow W = \int_{x_1}^{x_2} mv\frac{dv}{dx}\,dx = m\int_{x_1}^{x_2}\left(v\frac{dv}{dx}\right)dx = m\left[\tfrac{1}{2}v^2(x)\right]_{x_1}^{x_2}$

$$= \tfrac{1}{2}m\left[v^2(x_2) - v^2(x_1)\right] = \tfrac{1}{2}mv_2^2 - \tfrac{1}{2}mv_1^2, \text{ as claimed.}$$

27. 90 mph $= \frac{90 \text{ mi}}{1 \text{ hr}} \cdot \frac{1 \text{ hr}}{60 \text{ min}} \cdot \frac{1 \text{ min}}{60 \text{ sec}} \cdot \frac{5280 \text{ ft}}{1 \text{ mi}} = 132 \text{ ft/sec}$; $m = \frac{0.3125 \text{ lb}}{32 \text{ ft/sec}^2} = \frac{0.3125}{32} \text{ slugs}$;

$W = \left(\tfrac{1}{2}\right)\left(\frac{0.3125 \text{ lb}}{32 \text{ ft/sec}^2}\right)(132 \text{ ft/sec})^2 \approx 85.1 \text{ ft} \cdot \text{lb}$

29. weight $= 2 \text{ oz} = \tfrac{1}{8} \text{ lb} \Rightarrow m = \frac{\frac{1}{8}}{32} \text{ slugs} = \frac{1}{256} \text{ slugs}$; 124 mph $= \frac{(124)(5280)}{(60)(60)} \approx 181.87 \text{ ft/sec}$;

$W = \left(\tfrac{1}{2}\right)\left(\frac{1}{256} \text{ slugs}\right)(181.87 \text{ ft/sec})^2 \approx 64.6 \text{ ft} \cdot \text{lb}$

31. weight $= 6.5 \text{ oz} = \frac{6.5}{16} \text{ lb} \Rightarrow m = \frac{6.5}{(16)(32)} \text{ slugs}$; $W = \left(\tfrac{1}{2}\right)\left(\frac{6.5}{(16)(32)} \text{ slugs}\right)(132 \text{ ft/sec})^2 \approx 110.6 \text{ ft} \cdot \text{lb}$

33. (a) From the diagram,

$r(y) = 60 - x = 60 - \sqrt{50^2 - (y - 325)^2}$

for $325 \le y \le 375$ ft.

(b) The volume of a horizontal slice of the funnel

is $\Delta V \approx \pi\left[r(y)\right]^2 \Delta y$

$$= \pi\left[60 - \sqrt{50^2 - (y - 325)^2}\right]^2 \Delta y$$

(c) The work required to lift the single slice of

water is $\Delta W \approx 62.4 \Delta V(375 - y)$

$$= 62.4(375 - y)\pi\left[60 - \sqrt{50^2 - (y - 325)^2}\right]^2 \Delta y.$$

The total work to pump our the funnel is W

$$= \int_{325}^{375} 62.4(375 - y)\pi\left[60 - \sqrt{50^2 - (y - 325)^2}\right]^2 dy$$

$\approx 6.3358 \cdot 10^7 \text{ ft} \cdot \text{lb}.$

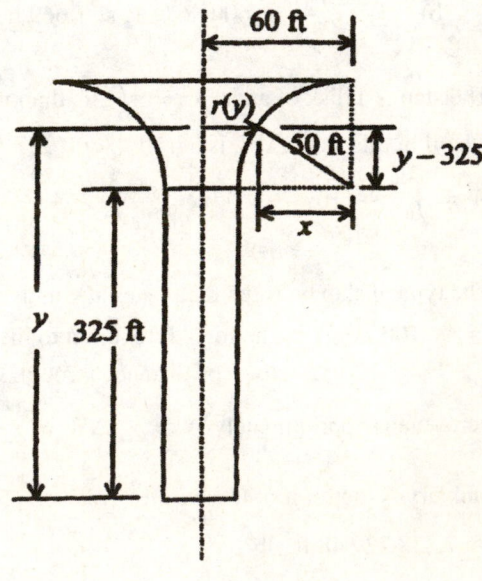

35. We imagine the milkshake divided into thin slabs by planes perpendicular to the y-axis at the points of a

partition of the interval $[0, 7]$. The typical slab between the planes at y and $y + \Delta y$ has a volume of about

$\Delta V = \pi(\text{radius})^2(\text{thickness}) = \pi\left(\frac{y+17.5}{14}\right)^2 \Delta y \text{ in}^3$. The force F(y) required to lift this slab is equal to its

weight: $F(y) = \tfrac{4}{9} \Delta V = \frac{4\pi}{9}\left(\frac{y+17.5}{14}\right)^2 \Delta y \text{ oz}$. The distance through which F(y) must act to lift this slab to

the level of 1 inch above the top is about $(8 - y)$ in. The work done lifting the slab is about

$\Delta W = \left(\frac{4\pi}{9}\right) \frac{(y+17.5)^2}{14^2} (8-y)\,\Delta y$ in · oz. The work done lifting all the slabs from $y=0$ to $y=7$ is

approximately $W = \sum_0^7 \frac{4\pi}{9 \cdot 14^2} (y+17.5)^2(8-y)\,\Delta y$ in · oz which is a Riemann sum. The work is the limit of

these sums as the norm of the partition goes to zero: $W = \int_0^7 \frac{4\pi}{9 \cdot 14^2} (y+17.5)^2(8-y)\,dy$

$= \frac{4\pi}{9 \cdot 14^2} \int_0^7 (2450 - 26.25y - 27y^2 - y^3)\,dy = \frac{4\pi}{9 \cdot 14^2} \left[-\frac{y^4}{4} - 9y^3 - \frac{26.25}{2} y^2 + 2450y\right]_0^7$

$= \frac{4\pi}{9 \cdot 14^2} \left[-\frac{7^4}{4} - 9 \cdot 7^3 - \frac{26.25}{2} \cdot 7^2 + 2450 \cdot 7\right] \approx 91.32$ in · oz

37. Work $= \int_{6,370,000}^{35,780,000} \frac{1000\,MG}{r^2}\,dr = 1000\,MG \int_{6,370,000}^{35,780,000} \frac{dr}{r^2} = 1000\,MG \left[-\frac{1}{r}\right]_{6,370,000}^{35,780,000}$

$= (1000)(5.975 \cdot 10^{24})(6.672 \cdot 10^{-11})\left(\frac{1}{6,370,000} - \frac{1}{35,780,000}\right) \approx 5.144 \times 10^{10}$ J

6.7 FLUID PRESSURES AND FORCES

1. To find the width of the plate at a typical depth y, we first find an equation for the line of the plate's
   right-hand edge: $y = x - 5$. If we let x denote the width of the right-hand half of the triangle at depth y, then
   $x = 5 + y$ and the total width is $L(y) = 2x = 2(5+y)$. The depth of the strip is $(-y)$. The force exerted by the
   water against one side of the plate is therefore $F = \int_{-5}^{-2} w(-y) \cdot L(y)\,dy = \int_{-5}^{-2} 62.4 \cdot (-y) \cdot 2(5+y)\,dy$

   $= 124.8 \int_{-5}^{-2} (-5y - y^2)\,dy = 124.8 \left[-\frac{5}{2} y^2 - \frac{1}{3} y^3\right]_{-5}^{-2} = 124.8 \left[\left(-\frac{5}{2} \cdot 4 + \frac{1}{3} \cdot 8\right) - \left(-\frac{5}{2} \cdot 25 + \frac{1}{3} \cdot 125\right)\right]$

   $= (124.8)\left(\frac{105}{2} - \frac{117}{3}\right) = (124.8)\left(\frac{315-234}{6}\right) = 1684.8$ lb

3. Using the coordinate system of Exercise 4, we find the equation for the line of the plate's right-hand edge is
   $y = x - 3 \Rightarrow x = y + 3$. Thus the total width is $L(y) = 2x = 2(y+3)$. The depth of the strip changes to $(4-y)$

   $\Rightarrow F = \int_{-3}^0 w(4-y)L(y)\,dy = \int_{-3}^0 62.4 \cdot (4-y) \cdot 2(y+3)\,dy = 124.8 \int_{-3}^0 (12 + y - y^2)\,dy$

   $= 124.8 \left[12y + \frac{y^2}{2} - \frac{y^3}{3}\right]_{-3}^0 = (-124.8)\left(-36 + \frac{9}{2} + 9\right) = (-124.8)\left(-\frac{45}{2}\right) = 2808$ lb

5. Using the coordinate system of Exercise 4, we find the equation for the line of the plate's right-hand edge to be
   $y = 2x - 4 \Rightarrow x = \frac{y+4}{2}$ and $L(y) = 2x = y + 4$. The depth of the strip is $(1-y)$.

   (a) $F = \int_{-4}^0 w(1-y)L(y)\,dy = \int_{-4}^0 62.4 \cdot (1-y)(y+4)\,dy = 62.4 \int_{-4}^0 (4 - 3y - y^2)\,dy = 62.4\left[4y - \frac{3y^2}{2} - \frac{y^3}{3}\right]_{-4}^0$

   $= (-62.4)\left[(-4)(4) - \frac{(3)(16)}{2} + \frac{64}{3}\right] = (-62.4)\left(-16 - 24 + \frac{64}{3}\right) = \frac{(-62.4)(-120+64)}{3} = 1164.8$ lb

   (b) $F = (-64.0)\left[(-4)(4) - \frac{(3)(16)}{2} + \frac{64}{3}\right] = \frac{(-64.0)(-120+64)}{3} \approx 1194.7$ lb

7. Using the coordinate system given in the accompanying
   figure, we see that the total width is $L(y) = 63$ and the depth
   of the strip is $(33.5 - y) \Rightarrow F = \int_0^{33} w(33.5 - y)L(y)\,dy$

   $= \int_0^{33} \frac{64}{12^3} \cdot (33.5 - y) \cdot 63\,dy = \left(\frac{64}{12^3}\right)(63)\int_0^{33} (33.5 - y)\,dy$

   $= \left(\frac{64}{12^3}\right)(63)\left[33.5y - \frac{y^2}{2}\right]_0^{33} = \left(\frac{64 \cdot 63}{12^3}\right)\left[(33.5)(33) - \frac{33^2}{2}\right]$

   $= \frac{(64)(63)(33)(67-33)}{(2)(12^3)} = 1309$ lb

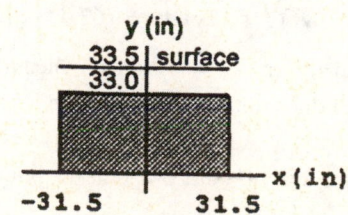

9. Using the coordinate system given in the accompanying figure, we see that the right-hand edge is $x = \sqrt{1 - y^2}$ so the total width is $L(y) = 2x = 2\sqrt{1 - y^2}$ and the depth of the strip is $(-y)$. The force exerted by the water is therefore $F = \int_{-1}^{0} w \cdot (-y) \cdot 2\sqrt{1 - y^2}\, dy$

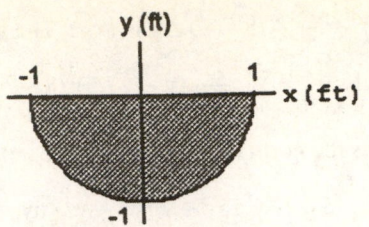

$$= 62.4\int_{-1}^{0} \sqrt{1 - y^2}\, d(1 - y^2) = 62.4 \left[\tfrac{2}{3}(1 - y^2)^{3/2}\right]_{-1}^{0} = (62.4)\left(\tfrac{2}{3}\right)(1 - 0) = 416 \text{ lb}$$

11. The coordinate system is given in the text. The right-hand edge is $x = \sqrt{y}$ and the total width is $L(y) = 2x = 2\sqrt{y}$.

(a) The depth of the strip is $(2 - y)$ so the force exerted by the liquid on the gate is $F = \int_{0}^{1} w(2 - y)L(y)\, dy$

$$= \int_{0}^{1} 50(2 - y) \cdot 2\sqrt{y}\, dy = 100 \int_{0}^{1} (2 - y)\sqrt{y}\, dy = 100\int_{0}^{1} (2y^{1/2} - y^{3/2})\, dy = 100\left[\tfrac{4}{3}y^{3/2} - \tfrac{2}{5}y^{5/2}\right]_{0}^{1}$$
$$= 100\left(\tfrac{4}{3} - \tfrac{2}{5}\right) = \left(\tfrac{100}{15}\right)(20 - 6) = 93.33 \text{ lb}$$

(b) We need to solve $160 = \int_{0}^{1} w(H - y) \cdot 2\sqrt{y}\, dy$ for h. $160 = 100\left(\tfrac{2H}{3} - \tfrac{2}{5}\right) \Rightarrow H = 3 \text{ ft}$.

13. Suppose that h is the maximum height. Using the coordinate system given in the text, we find an equation for the line of the end plate's right-hand edge is $y = \tfrac{5}{2}x \Rightarrow x = \tfrac{2}{5}y$. The total width is $L(y) = 2x = \tfrac{4}{5}y$ and the depth of the typical horizontal strip at level y is $(h - y)$. Then the force is $F = \int_{0}^{h} w(h - y)L(y)\, dy = F_{max}$,

where $F_{max} = 6667$ lb. Hence, $F_{max} = w\int_{0}^{h}(h - y) \cdot \tfrac{4}{5}y\, dy = (62.4)\left(\tfrac{4}{5}\right)\int_{0}^{h}(hy - y^2)\, dy$

$$= (62.4)\left(\tfrac{4}{5}\right)\left[\tfrac{hy^2}{2} - \tfrac{y^3}{3}\right]_{0}^{h} = (62.4)\left(\tfrac{4}{5}\right)\left(\tfrac{h^3}{2} - \tfrac{h^3}{3}\right) = (62.4)\left(\tfrac{4}{5}\right)\left(\tfrac{1}{6}\right)h^3 = (10.4)\left(\tfrac{4}{5}\right)h^3 \Rightarrow h = \sqrt[3]{\left(\tfrac{5}{4}\right)\left(\tfrac{F_{max}}{10.4}\right)}$$

$$= \sqrt[3]{\left(\tfrac{5}{4}\right)\left(\tfrac{6667}{10.4}\right)} \approx 9.288 \text{ ft.}$$ The volume of water which the tank can hold is $V = \tfrac{1}{2}(\text{Base})(\text{Height}) \cdot 30$, where Height $= h$ and $\tfrac{1}{2}(\text{Base}) = \tfrac{2}{5}h \Rightarrow V = \left(\tfrac{2}{5}h^2\right)(30) = 12h^2 \approx 12(9.288)^2 \approx 1035 \text{ ft}^3.$

15. The pressure at level y is $p(y) = w \cdot y \Rightarrow$ the average pressure is $\bar{p} = \tfrac{1}{b}\int_{0}^{b} p(y)\, dy = \tfrac{1}{b}\int_{0}^{b} w \cdot y\, dy = \tfrac{1}{b}w\left[\tfrac{y^2}{2}\right]_{0}^{b}$

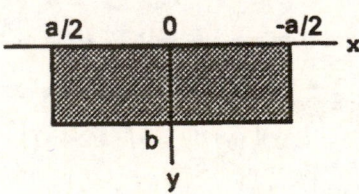

$= \left(\tfrac{w}{b}\right)\left(\tfrac{b^2}{2}\right) = \tfrac{wb}{2}$. This is the pressure at level $\tfrac{b}{2}$, which is the pressure at the middle of the plate.

17. When the water reaches the top of the tank the force on the movable side is $\int_{-2}^{0}(62.4)\left(2\sqrt{4 - y^2}\right)(-y)\, dy$

$$= (62.4)\int_{-2}^{0}(4 - y^2)^{1/2}(-2y)\, dy = (62.4)\left[\tfrac{2}{3}(4 - y^2)^{3/2}\right]_{-2}^{0} = (62.4)\left(\tfrac{2}{3}\right)\left(4^{3/2}\right) = 332.8 \text{ ft} \cdot \text{lb. The force}$$

compressing the spring is $F = 100x$, so when the tank is full we have $332.8 = 100x \Rightarrow x \approx 3.33 \text{ ft.}$ Therefore the movable end does not reach the required 5 ft to allow drainage $\Rightarrow$ the tank will overflow.

19. Use ac oordinate system with $y = 0$ at the bottom of the carton and with $L(y) = 3.75$ and the depth of a typical strip being $(7.75 - y)$. Then $F = \int_{0}^{7.75} w(7.75 - y)L(y)\, dy = \left(\tfrac{64.5}{12^3}\right)(3.75)\int_{0}^{7.75}(7.75 - y)\, dy = \left(\tfrac{64.5}{12^3}\right)(3.75)\left[7.75y - \tfrac{y^2}{2}\right]_{0}^{7.75}$

$$= \left(\tfrac{64.5}{12^3}\right)(3.75)\tfrac{(7.75)^2}{2} \approx 4.2 \text{ lb}$$

21. (a) An equation of the right-hand edge is $y = \tfrac{3}{2}x \Rightarrow x = \tfrac{2}{3}y$ and $L(y) = 2x = \tfrac{4y}{3}$. The depth of the strip is $(3 - y) \Rightarrow F = \int_{0}^{3} w(3 - y)L(y)\, dy = \int_{0}^{3}(62.4)(3 - y)\left(\tfrac{4}{3}y\right)\, dy = (62.4) \cdot \left(\tfrac{4}{3}\right)\int_{0}^{3}(3y - y^2)\, dy$

$= (62.4) \left(\frac{4}{3}\right) \left[\frac{3}{2} y^2 - \frac{y^3}{3}\right]_0^3 = (62.4) \left(\frac{4}{3}\right) \left[\frac{27}{2} - \frac{27}{3}\right] = (62.4) \left(\frac{4}{3}\right) \left(\frac{27}{6}\right) = 374.4$ lb

(b) We want to find a new water level Y such that $F_Y = \frac{1}{2}(374.4) = 187.2$ lb. The new depth of the strip is

(Y − y), and Y is the new upper limit of integration. Thus, $F_Y = \int_0^Y w(Y-y)L(y)\,dy$

$= 62.4 \int_0^Y (Y-y)\left(\frac{4}{3} y\right) dy = (62.4)\left(\frac{4}{3}\right)\int_0^Y (Yy - y^2)\,dy = (62.4)\left(\frac{4}{3}\right)\left[Y \cdot \frac{y^2}{2} - \frac{y^3}{3}\right]_0^Y = (62.4)\left(\frac{4}{3}\right)\left(\frac{Y^3}{2} - \frac{Y^3}{3}\right)$

$= (62.4)\left(\frac{2}{9}\right) Y^3$. Therefore $Y^3 = \frac{9 F_Y}{2 \cdot (62.4)} = \frac{(9)(187.2)}{124.8} \Rightarrow Y = \sqrt[3]{\frac{(9)(187.2)}{124.8}} = \sqrt[3]{13.5} \approx 2.3811$ ft. So,

$\Delta Y = 3 - Y \approx 3 - 2.3811 \approx 0.6189$ ft ≈ 7.5 in. to the nearest half inch.

(c) No, it does not matter how long the trough is. The fluid pressure and the resulting force depend only on depthe of the water.

CHAPTER 6 PRACTICE EXERCISES

1. $A(x) = \frac{\pi}{4} (\text{diameter})^2 = \frac{\pi}{4}\left(\sqrt{x} - x^2\right)^2$

$= \frac{\pi}{4}\left(x - 2\sqrt{x} \cdot x^2 + x^4\right)$; $a = 0, b = 1$

$\Rightarrow V = \int_a^b A(x)\,dx = \frac{\pi}{4}\int_0^1 \left(x - 2x^{5/2} + x^4\right) dx$

$= \frac{\pi}{4}\left[\frac{x^2}{2} - \frac{4}{7}x^{7/2} + \frac{x^5}{5}\right]_0^1 = \frac{\pi}{4}\left(\frac{1}{2} - \frac{4}{7} + \frac{1}{5}\right)$

$= \frac{\pi}{4 \cdot 70}(35 - 40 + 14) = \frac{9\pi}{280}$

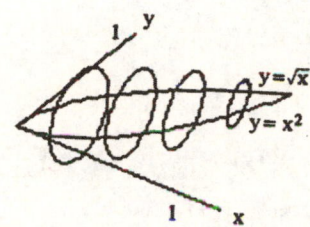

3. $A(x) = \frac{\pi}{4}(\text{diameter})^2 = \frac{\pi}{4}(2\sin x - 2\cos x)^2$

$= \frac{\pi}{4} \cdot 4\left(\sin^2 x - 2\sin x \cos x + \cos^2 x\right)$

$= \pi(1 - \sin 2x); a = \frac{\pi}{4}, b = \frac{5\pi}{4}$

$\Rightarrow V = \int_a^b A(x)\,dx = \pi\int_{\pi/4}^{5\pi/4} (1 - \sin 2x)\,dx$

$= \pi\left[x + \frac{\cos 2x}{2}\right]_{\pi/4}^{5\pi/4}$

$= \pi\left[\left(\frac{5\pi}{4} + \frac{\cos\frac{5\pi}{2}}{2}\right) - \left(\frac{\pi}{4} - \frac{\cos\frac{\pi}{2}}{2}\right)\right] = \pi^2$

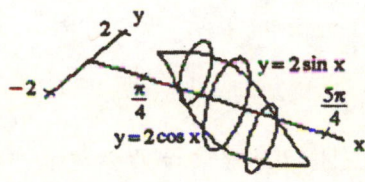

5. $A(x) = \frac{\pi}{4}(\text{diameter})^2 = \frac{\pi}{4}\left(2\sqrt{x} - \frac{x^2}{4}\right)^2 = \frac{\pi}{4}\left(4x - x^{5/2} + \frac{x^4}{16}\right)$; $a = 0, b = 4 \Rightarrow V = \int_a^b A(x)\,dx$

$= \frac{\pi}{4}\int_0^4 \left(4x - x^{5/2} + \frac{x^4}{16}\right) dx = \frac{\pi}{4}\left[2x^2 - \frac{2}{7}x^{7/2} + \frac{x^5}{5 \cdot 16}\right]_0^4 = \frac{\pi}{4}\left(32 - 32 \cdot \frac{8}{7} + \frac{2}{5} \cdot 32\right)$

$= \frac{32\pi}{4}\left(1 - \frac{8}{7} + \frac{2}{5}\right) = \frac{8\pi}{35}(35 - 40 + 14) = \frac{72\pi}{35}$

7. (a) *disk method*:

$V = \int_a^b \pi R^2(x)\,dx = \int_{-1}^1 \pi\left(3x^4\right)^2 dx = \pi\int_{-1}^1 9x^8\,dx$

$= \pi\left[x^9\right]_{-1}^1 = 2\pi$

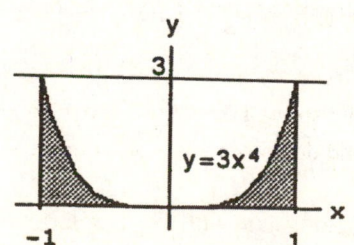

(b) *shell method*:

$V = \int_a^b 2\pi \left(\begin{array}{c}\text{shell}\\\text{radius}\end{array}\right)\left(\begin{array}{c}\text{shell}\\\text{height}\end{array}\right) dx = \int_0^1 2\pi x\left(3x^4\right) dx = 2\pi \cdot 3\int_0^1 x^5\,dx = 2\pi \cdot 3\left[\frac{x^6}{6}\right]_0^1 = \pi$

Note: The lower limit of integration is 0 rather than −1.

(c) *shell method*:

$V = \int_a^b 2\pi \left(\begin{array}{c}\text{shell}\\\text{radius}\end{array}\right)\left(\begin{array}{c}\text{shell}\\\text{height}\end{array}\right) dx = 2\pi\int_{-1}^1 (1 - x)\left(3x^4\right) dx = 2\pi\left[\frac{3x^5}{5} - \frac{x^6}{2}\right]_{-1}^1 = 2\pi\left[\left(\frac{3}{5} - \frac{1}{2}\right) - \left(-\frac{3}{5} - \frac{1}{2}\right)\right] = \frac{12\pi}{5}$

(d) *washer method*:

$$R(x) = 3, r(x) = 3 - 3x^4 = 3(1 - x^4) \implies V = \int_a^b \pi [R^2(x) - r^2(x)] \, dx = \int_{-1}^1 \pi \left[9 - 9(1 - x^4)^2\right] dx$$

$$= 9\pi \int_{-1}^1 [1 - (1 - 2x^4 + x^8)] \, dx = 9\pi \int_{-1}^1 (2x^4 - x^8) \, dx = 9\pi \left[\frac{2x^5}{5} - \frac{x^9}{9}\right]_{-1}^1 = 18\pi \left[\frac{2}{5} - \frac{1}{9}\right] = \frac{2\pi \cdot 13}{5} = \frac{26\pi}{5}$$

9. (a) *disk method*:

$$V = \pi \int_1^5 \left(\sqrt{x - 1}\right)^2 dx = \pi \int_1^5 (x - 1) \, dx = \pi \left[\frac{x^2}{2} - x\right]_1^5$$

$$= \pi \left[\left(\frac{25}{2} - 5\right) - \left(\frac{1}{2} - 1\right)\right] = \pi \left(\frac{24}{2} - 4\right) = 8\pi$$

(b) *washer method*:

$$R(y) = 5, r(y) = y^2 + 1 \implies V = \int_c^d \pi [R^2(y) - r^2(y)] \, dy = \pi \int_{-2}^2 \left[25 - (y^2 + 1)^2\right] dy$$

$$= \pi \int_{-2}^2 (25 - y^4 - 2y^2 - 1) \, dy = \pi \int_{-2}^2 (24 - y^4 - 2y^2) \, dy = \pi \left[24y - \frac{y^5}{5} - \frac{2}{3} y^3\right]_{-2}^2 = 2\pi \left(24 \cdot 2 - \frac{32}{5} - \frac{2}{3} \cdot 8\right)$$

$$= 32\pi \left(3 - \frac{2}{5} - \frac{1}{3}\right) = \frac{32\pi}{15}(45 - 6 - 5) = \frac{1088\pi}{15}$$

(c) *disk method*:

$$R(y) = 5 - (y^2 + 1) = 4 - y^2$$

$$\implies V = \int_c^d \pi R^2(y) \, dy = \int_{-2}^2 \pi (4 - y^2)^2 \, dy$$

$$= \pi \int_{-2}^2 (16 - 8y^2 + y^4) \, dy$$

$$= \pi \left[16y - \frac{8y^3}{3} + \frac{y^5}{5}\right]_{-2}^2 = 2\pi \left(32 - \frac{64}{3} + \frac{32}{5}\right)$$

$$= 64\pi \left(1 - \frac{2}{3} + \frac{1}{5}\right) = \frac{64\pi}{15}(15 - 10 + 3) = \frac{512\pi}{15}$$

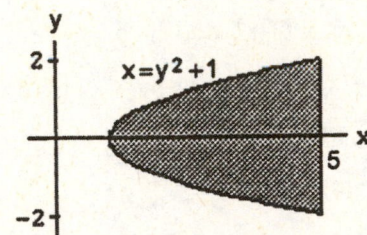

11. *disk method*:

$$R(x) = \tan x, a = 0, b = \frac{\pi}{3} \implies V = \pi \int_0^{\pi/3} \tan^2 x \, dx = \pi \int_0^{\pi/3} (\sec^2 x - 1) \, dx = \pi [\tan x - x]_0^{\pi/3} = \frac{\pi \left(3\sqrt{3} - \pi\right)}{3}$$

13. (a) *disk method*:

$$V = \pi \int_0^2 (x^2 - 2x)^2 \, dx = \pi \int_0^2 (x^4 - 4x^3 + 4x^2) \, dx = \pi \left[\frac{x^5}{5} - x^4 + \frac{4}{3} x^3\right]_0^2 = \pi \left(\frac{32}{5} - 16 + \frac{32}{3}\right)$$

$$= \frac{16\pi}{15}(6 - 15 + 10) = \frac{16\pi}{15}$$

(b) *washer method*:

$$V = \int_0^2 \pi \left[1^2 - (x^2 - 2x + 1)^2\right] dx = \int_0^2 \pi \, dx + \int_0^2 \pi (x - 1)^4 \, dx = 2\pi - \left[\pi \frac{(x-1)^5}{5}\right]_0^2 = 2\pi - \pi \cdot \frac{2}{5} = \frac{8\pi}{5}$$

(c) *shell method*:

$$V = \int_a^b 2\pi \left(\begin{smallmatrix}\text{shell}\\\text{radius}\end{smallmatrix}\right) \left(\begin{smallmatrix}\text{shell}\\\text{height}\end{smallmatrix}\right) dx = 2\pi \int_0^2 (2 - x) \left[-(x^2 - 2x)\right] dx = 2\pi \int_0^2 (2 - x)(2x - x^2) \, dx$$

$$= 2\pi \int_0^2 (4x - 2x^2 - 2x^2 + x^3) \, dx = 2\pi \int_0^2 (x^3 - 4x^2 + 4x) \, dx = 2\pi \left[\frac{x^4}{4} - \frac{4}{3} x^3 + 2x^2\right]_0^2 = 2\pi \left(4 - \frac{32}{3} + 8\right)$$

$$= \frac{2\pi}{3}(36 - 32) = \frac{8\pi}{3}$$

(d) *washer method*:

$$V = \pi \int_0^2 [2 - (x^2 - 2x)]^2 \, dx - \pi \int_0^2 2^2 \, dx = \pi \int_0^2 \left[4 - 4(x^2 - 2x) + (x^2 - 2x)^2\right] dx - 8\pi$$

$$= \pi \int_0^2 (4 - 4x^2 + 8x + x^4 - 4x^3 + 4x^2) \, dx - 8\pi = \pi \int_0^2 (x^4 - 4x^3 + 8x + 4) \, dx - 8\pi$$

$$= \pi \left[\frac{x^5}{5} - x^4 + 4x^2 + 4x\right]_0^2 - 8\pi = \pi \left(\frac{32}{5} - 16 + 16 + 8\right) - 8\pi = \frac{\pi}{5}(32 + 40) - 8\pi = \frac{72\pi}{5} - \frac{40\pi}{5} = \frac{32\pi}{5}$$

15. The material removed from the sphere consists of a cylinder
and two "caps." From the diagram, the height of the cylinder

is 2h, where $h^2 + \left(\sqrt{3}\right)^2 = 2^2$, i.e. $h = 1$. Thus

$V_{cyl} = (2h)\pi\left(\sqrt{3}\right)^2 = 6\pi$ ft^3. To get the volume of a cap,

use the disk method and $x^2 + y^2 = 2^2$: $V_{cap} = \int_1^2 \pi x^2 dy$

$= \int_1^2 \pi(4 - y^2)dy = \pi\left[4y - \frac{y^3}{3}\right]_1^2$

$= \pi\left[(8 - \frac{8}{3}) - (4 - \frac{1}{3})\right] = \frac{5\pi}{3}$ ft^3. Therefore,

$V_{removed} = V_{cyl} + 2V_{cap} = 6\pi + \frac{10\pi}{3} = \frac{28\pi}{3}$ ft^3.

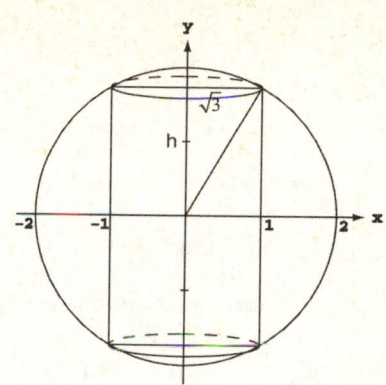

17. $y = x^{1/2} - \frac{x^{3/2}}{3} \Rightarrow \frac{dy}{dx} = \frac{1}{2}x^{-1/2} - \frac{1}{2}x^{1/2} \Rightarrow \left(\frac{dy}{dx}\right)^2 = \frac{1}{4}\left(\frac{1}{x} - 2 + x\right) \Rightarrow L = \int_1^4 \sqrt{1 + \frac{1}{4}\left(\frac{1}{x} - 2 + x\right)}\, dx$

$\Rightarrow L = \int_1^4 \sqrt{\frac{1}{4}\left(\frac{1}{x} + 2 + x\right)}\, dx = \int_1^4 \sqrt{\frac{1}{4}(x^{-1/2} + x^{1/2})^2}\, dx = \int_1^4 \frac{1}{2}(x^{-1/2} + x^{1/2})\, dx = \frac{1}{2}\left[2x^{1/2} + \frac{2}{3}x^{3/2}\right]_1^4$

$= \frac{1}{2}\left[(4 + \frac{2}{3}\cdot 8) - (2 + \frac{2}{3})\right] = \frac{1}{2}\left(2 + \frac{14}{3}\right) = \frac{10}{3}$

19. $y = \frac{5}{12}x^{6/5} - \frac{5}{8}x^{4/5} \Rightarrow \frac{dy}{dx} = \frac{1}{2}x^{1/5} - \frac{1}{2}x^{-1/5} \Rightarrow \left(\frac{dy}{dx}\right)^2 = \frac{1}{4}\left(x^{2/5} - 2 + x^{-2/5}\right)$

$\Rightarrow L = \int_1^{32}\sqrt{1 + \frac{1}{4}\left(x^{2/5} - 2 + x^{-2/5}\right)}\, dx \Rightarrow L = \int_1^{32}\sqrt{\frac{1}{4}\left(x^{2/5} + 2 + x^{-2/5}\right)}\, dx = \int_1^{32}\sqrt{\frac{1}{4}(x^{1/5} + x^{-1/5})^2}\, dx$

$= \int_1^{32}\frac{1}{2}\left(x^{1/5} + x^{-1/5}\right)dx = \frac{1}{2}\left[\frac{5}{6}x^{6/5} + \frac{5}{4}x^{4/5}\right]_1^{32} = \frac{1}{2}\left[\left(\frac{5}{6}\cdot 2^6 + \frac{5}{4}\cdot 2^4\right) - \left(\frac{5}{6} + \frac{5}{4}\right)\right] = \frac{1}{2}\left(\frac{315}{6} + \frac{75}{4}\right)$

$= \frac{1}{48}(1260 + 450) = \frac{1710}{48} = \frac{285}{8}$

21. $\frac{dx}{dt} = -5\sin t + 5\sin 5t$ and $\frac{dy}{dt} = 5\cos t - 5\cos 5t \Rightarrow \sqrt{\left(\frac{dx}{dt}\right)^2 + \left(\frac{dy}{dt}\right)^2}$

$= \sqrt{(-5\sin t + 5\sin 5t)^2 + (5\cos t - 5\cos 5t)^2}$

$= 5\sqrt{\sin^2 5t - 2\sin t\sin 5t + \sin^2 t + \cos^2 t - 2\cos t\cos 5t + \cos^2 5t} = 5\sqrt{2 - 2(\sin t\sin 5t + \cos t\cos 5t)}$

$= 5\sqrt{2(1 - \cos 4t)} = 5\sqrt{4\left(\frac{1}{2}\right)(1 - \cos 4t)} = 10\sqrt{\sin^2 2t} = 10|\sin 2t| = 10\sin 2t$ (since $0 \le t \le \frac{\pi}{2}$)

$\Rightarrow$ Length $= \int_0^{\pi/2} 10\sin 2t\, dt = [-5\cos 2t]_0^{\pi/2} = (-5)(-1) - (-5)(1) = 10$

23. $\frac{dx}{d\theta} = -3\sin\theta$ and $\frac{dy}{d\theta} = 3\cos\theta \Rightarrow \sqrt{\left(\frac{dx}{d\theta}\right)^2 + \left(\frac{dy}{d\theta}\right)^2} = \sqrt{(-3\sin\theta)^2 + (3\cos\theta)^2} = \sqrt{3(\sin^2\theta + \cos^2\theta)} = 3$

$\Rightarrow$ Length $= \int_0^{3\pi/2} 3\, d\theta = 3\int_0^{3\pi/2} d\theta = 3\left(\frac{3\pi}{2} - 0\right) = \frac{9\pi}{2}$

25. Intersection points: $3 - x^2 = 2x^2 \Rightarrow 3x^2 - 3 = 0$
$\Rightarrow 3(x - 1)(x + 1) = 0 \Rightarrow x = -1$ or $x = 1$. Symmetry
suggests that $\bar{x} = 0$. The typical *vertical* strip has
center of mass: $(\tilde{x}, \tilde{y}) = \left(x, \frac{2x^2 + (3 - x^2)}{2}\right) = \left(x, \frac{x^2 + 3}{2}\right)$,
length: $(3 - x^2) - 2x^2 = 3(1 - x^2)$, width: dx,
area: $dA = 3(1 - x^2)\, dx$, and mass: $dm = \delta \cdot dA$
$= 3\delta(1 - x^2)\, dx \Rightarrow$ the moment about the x-axis is

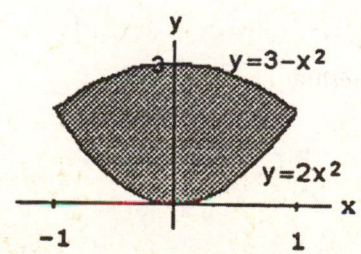

$\tilde{y}\, dm = \frac{3}{2}\delta(x^2 + 3)(1 - x^2)\, dx = \frac{3}{2}\delta(-x^4 - 2x^2 + 3)\, dx \Rightarrow M_x = \int \tilde{y}\, dm = \frac{3}{2}\delta\int_{-1}^1 (-x^4 - 2x^2 + 3)\, dx$

$= \frac{3}{2}\delta\left[-\frac{x^5}{5} - \frac{2x^3}{3} + 3x\right]_{-1}^1 = 3\delta\left(-\frac{1}{5} - \frac{2}{3} + 3\right) = \frac{3\delta}{15}(-3 - 10 + 45) = \frac{32\delta}{5}; M = \int dm = 3\delta\int_{-1}^1 (1 - x^2)\, dx$

$= 3\delta \left[x - \frac{x^3}{3}\right]_{-1}^{1} = 6\delta \left(1 - \frac{1}{3}\right) = 4\delta \Rightarrow \bar{y} = \frac{M_x}{M} = \frac{32\delta}{5 \cdot 4\delta} = \frac{8}{5}$. Therefore, the centroid is $(\bar{x}, \bar{y}) = \left(0, \frac{8}{5}\right)$.

27. The typical *vertical* strip has: center of mass: $(\tilde{x}, \tilde{y})$

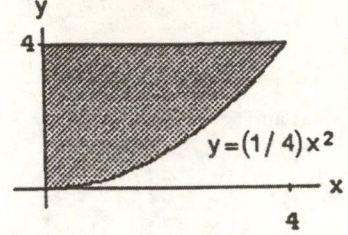

$= \left(x, \frac{4 + \frac{x^2}{4}}{2}\right)$, length: $4 - \frac{x^2}{4}$, width: dx,

area: $dA = \left(4 - \frac{x^2}{4}\right)dx$, mass: $dm = \delta \cdot dA$

$= \delta \left(4 - \frac{x^2}{4}\right) dx \Rightarrow$ the moment about the x-axis is

$\tilde{y}\, dm = \delta \cdot \frac{\left(4 + \frac{x^2}{4}\right)}{2}\left(4 - \frac{x^2}{4}\right) dx = \frac{\delta}{2}\left(16 - \frac{x^4}{16}\right) dx$; the

moment about the y-axis is $\tilde{x}\, dm = \delta \left(4 - \frac{x^2}{4}\right) \cdot x\, dx = \delta \left(4x - \frac{x^3}{4}\right) dx$. Thus, $M_x = \int \tilde{y}\, dm = \frac{\delta}{2}\int_0^4 \left(16 - \frac{x^4}{16}\right) dx$

$= \frac{\delta}{2}\left[16x - \frac{x^5}{5 \cdot 16}\right]_0^4 = \frac{\delta}{2}\left[64 - \frac{64}{5}\right] = \frac{128\delta}{5}$; $M_y = \int \tilde{x}\, dm = \delta \int_0^4 \left(4x - \frac{x^3}{4}\right) dx = \delta \left[2x^2 - \frac{x^4}{16}\right]_0^4$

$= \delta(32 - 16) = 16\delta$; $M = \int dm = \delta \int_0^4 \left(4 - \frac{x^2}{4}\right) dx = \delta \left[4x - \frac{x^3}{12}\right]_0^4 = \delta \left(16 - \frac{64}{12}\right) = \frac{32\delta}{3}$

$\Rightarrow \bar{x} = \frac{M_y}{M} = \frac{16 \cdot \delta \cdot 3}{32 \cdot \delta} = \frac{3}{2}$ and $\bar{y} = \frac{M_x}{M} = \frac{128 \cdot \delta \cdot 3}{5 \cdot 32 \cdot \delta} = \frac{12}{5}$. Therefore, the centroid is $(\bar{x}, \bar{y}) = \left(\frac{3}{2}, \frac{12}{5}\right)$.

29. A typical horizontal strip has: center of mass: $(\tilde{x}, \tilde{y})$

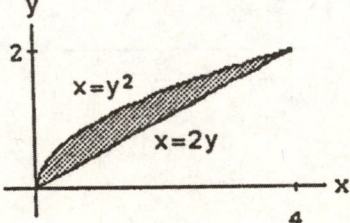

$= \left(\frac{y^2 + 2y}{2}, y\right)$, length: $2y - y^2$, width: dy,

area: $dA = (2y - y^2)\, dy$, mass: $dm = \delta \cdot dA$

$= (1 + y)(2y - y^2)\, dy \Rightarrow$ the moment about the

x-axis is $\tilde{y}\, dm = y(1 + y)(2y - y^2)\, dy$

$= (2y^2 + 2y^3 - y^3 - y^4)\, dy$

$= (2y^2 + y^3 - y^4)\, dy$; the moment about the y-axis is

$\tilde{x}\, dm = \left(\frac{y^2 + 2y}{2}\right)(1 + y)(2y - y^2)\, dy = \frac{1}{2}(4y^2 - y^4)(1 + y)\, dy = \frac{1}{2}(4y^2 + 4y^3 - y^4 - y^5)\, dy$

$\Rightarrow M_x = \int \tilde{y}\, dm = \int_0^2 (2y^2 + y^3 - y^4)\, dy = \left[\frac{2}{3}y^3 + \frac{y^4}{4} - \frac{y^5}{5}\right]_0^2 = \left(\frac{16}{3} + \frac{16}{4} - \frac{32}{5}\right) = 16\left(\frac{1}{3} + \frac{1}{4} - \frac{2}{5}\right)$

$= \frac{16}{60}(20 + 15 - 24) = \frac{4}{15}(11) = \frac{44}{15}$; $M_y = \int \tilde{x}\, dm = \int_0^2 \frac{1}{2}(4y^2 + 4y^3 - y^4 - y^5)\, dy = \frac{1}{2}\left[\frac{4}{3}y^3 + y^4 - \frac{y^5}{5} - \frac{y^6}{6}\right]_0^2$

$= \frac{1}{2}\left(\frac{4 \cdot 2^3}{3} + 2^4 - \frac{2^5}{5} - \frac{2^6}{6}\right) = 4\left(\frac{4}{3} + 2 - \frac{4}{5} - \frac{8}{6}\right) = 4\left(2 - \frac{4}{5}\right) = \frac{24}{5}$; $M = \int dm = \int_0^2 (1 + y)(2y - y^2)\, dy$

$= \int_0^2 (2y + y^2 - y^3)\, dy = \left[y^2 + \frac{y^3}{3} - \frac{y^4}{4}\right]_0^2 = \left(4 + \frac{8}{3} - \frac{16}{4}\right) = \frac{8}{3} \Rightarrow \bar{x} = \frac{M_y}{M} = \left(\frac{24}{5}\right)\left(\frac{3}{8}\right) = \frac{9}{5}$ and $\bar{y} = \frac{M_x}{M}$

$= \left(\frac{44}{15}\right)\left(\frac{3}{8}\right) = \frac{44}{40} = \frac{11}{10}$. Therefore, the center of mass is $(\bar{x}, \bar{y}) = \left(\frac{9}{5}, \frac{11}{10}\right)$.

31. $S = \int_a^b 2\pi y \sqrt{1 + \left(\frac{dy}{dx}\right)^2}\, dx$; $\frac{dy}{dx} = \frac{1}{\sqrt{2x+1}} \Rightarrow \left(\frac{dy}{dx}\right)^2 = \frac{1}{2x+1} \Rightarrow S = \int_0^3 2\pi \sqrt{2x+1}\sqrt{1 + \frac{1}{2x+1}}\, dx$

$= 2\pi \int_0^3 \sqrt{2x+1}\sqrt{\frac{2x+2}{2x+1}}\, dx = 2\sqrt{2}\pi \int_0^3 \sqrt{x+1}\, dx = 2\sqrt{2}\pi \left[\frac{2}{3}(x+1)^{3/2}\right]_0^3 = 2\sqrt{2}\pi \cdot \frac{2}{3}(8 - 1) = \frac{28\pi\sqrt{2}}{3}$

33. $S = \int_c^d 2\pi x \sqrt{1 + \left(\frac{dx}{dy}\right)^2}\, dy$; $\frac{dx}{dy} = \frac{\left(\frac{1}{2}\right)(4 - 2y)}{\sqrt{4y - y^2}} = \frac{2 - y}{\sqrt{4y - y^2}} \Rightarrow 1 + \left(\frac{dx}{dy}\right)^2 = \frac{4y - y^2 + 4 - 4y + y^2}{4y - y^2} = \frac{4}{4y - y^2}$

$\Rightarrow S = \int_1^2 2\pi \sqrt{4y - y^2}\sqrt{\frac{4}{4y - y^2}}\, dy = 4\pi \int_1^2 dx = 4\pi$

35. $x = \frac{t^2}{2}$ and $y = 2t$, $0 \le t \le \sqrt{5} \Rightarrow \frac{dx}{dt} = t$ and $\frac{dy}{dt} = 2 \Rightarrow$ Surface Area $= \int_0^{\sqrt{5}} 2\pi(2t)\sqrt{t^2 + 4}\, dt = \int_4^9 2\pi u^{1/2}\, du$

$= 2\pi \left[\frac{2}{3}u^{3/2}\right]_4^9 = \frac{76\pi}{3}$, where $u = t^2 + 4 \Rightarrow du = 2t\, dt$; $t = 0 \Rightarrow u = 4$, $t = \sqrt{5} \Rightarrow u = 9$

37. The equipment alone: the force required to lift the equipment is equal to its weight $\Rightarrow F_1(x) = 100$ N.

The work done is $W_1 = \int_a^b F_1(x)\,dx = \int_0^{40} 100\,dx = [100x]_0^{40} = 4000$ J; the rope alone: the force required

to lift the rope is equal to the weight of the rope paid out at elevation x $\Rightarrow F_2(x) = 0.8(40 - x)$. The work

done is $W_2 = \int_a^b F_2(x)\,dx = \int_0^{40} 0.8(40 - x)\,dx = 0.8\left[40x - \frac{x^2}{2}\right]_0^{40} = 0.8\left(40^2 - \frac{40^2}{2}\right) = \frac{(0.8)(1600)}{2} = 640$ J;

the total work is $W = W_1 + W_2 = 4000 + 640 = 4640$ J

39. Force constant: $F = kx \Rightarrow 20 = k \cdot 1 \Rightarrow k = 20$ lb/ft; the work to stretch the spring 1 ft is

$W = \int_0^1 kx\,dx = k\int_0^1 x\,dx = \left[20\frac{x^2}{2}\right]_0^1 = 10$ ft · lb; the work to stretch the spring an additional foot is

$W = \int_1^2 kx\,dx = k\int_1^2 x\,dx = 20\left[\frac{x^2}{2}\right]_1^2 = 20\left(\frac{4}{2} - \frac{1}{2}\right) = 20\left(\frac{3}{2}\right) = 30$ ft · lb

41. We imagine the water divided into thin slabs by planes
perpendicular to the y-axis at the points of a partition of the
interval $[0, 8]$. The typical slab between the planes at y and
$y + \Delta y$ has a volume of about $\Delta V = \pi(\text{radius})^2(\text{thickness})$
$= \pi\left(\frac{5}{4}y\right)^2\Delta y = \frac{25\pi}{16}y^2\,\Delta y$ ft³. The force $F(y)$ required to
lift this slab is equal to its weight: $F(y) = 62.4\,\Delta V$
$= \frac{(62.4)(25)}{16}\pi y^2\,\Delta y$ lb. The distance through which $F(y)$

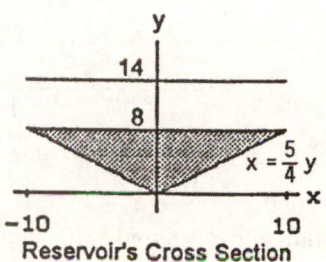

Reservoir's Cross Section

must act to lift this slab to the level 6 ft above the top is

about $(6 + 8 - y)$ ft, so the work done lifting the slab is about $\Delta W = \frac{(62.4)(25)}{16}\pi y^2(14 - y)\,\Delta y$ ft · lb. The work done

lifting all the slabs from $y = 0$ to $y = 8$ to the level 6 ft above the top is approximately

$W \approx \sum_0^8 \frac{(62.4)(25)}{16}\pi y^2(14 - y)\,\Delta y$ ft · lb so the work to pump the water is the limit of these Riemann sums as the norm of

the partition goes to zero: $W = \int_0^8 \frac{(62.4)(25)}{(16)}\pi y^2(14 - y)\,dy = \frac{(62.4)(25)\pi}{16}\int_0^8 (14y^2 - y^3)\,dy = (62.4)\left(\frac{25\pi}{16}\right)\left[\frac{14}{3}y^3 - \frac{y^4}{4}\right]_0^8$

$= (62.4)\left(\frac{25\pi}{16}\right)\left(\frac{14}{3}\cdot 8^3 - \frac{8^4}{4}\right) \approx 418{,}208.81$ ft · lb

43. The tank's cross section looks like the figure in Exercise 41 with right edge given by $x = \frac{5}{10}y = \frac{y}{2}$. A typical

horizontal slab has volume $\Delta V = \pi(\text{radius})^2(\text{thickness}) = \pi\left(\frac{y}{2}\right)^2\Delta y = \frac{\pi}{4}y^2\,\Delta y$. The force required to lift this

slab is its weight: $F(y) = 60 \cdot \frac{\pi}{4}y^2\,\Delta y$. The distance through which $F(y)$ must act is $(2 + 10 - y)$ ft, so the

work to pump the liquid is $W = 60\int_0^{10}\pi(12 - y)\left(\frac{y^2}{4}\right)dy = 15\pi\left[\frac{12y^3}{3} - \frac{y^4}{4}\right]_0^{10} = 22{,}500\pi$ ft · lb; the time needed

to empty the tank is $\frac{22{,}500\text{ ft·lb}}{275\text{ ft·lb/sec}} \approx 257$ sec

45. $F = \int_a^b W \cdot \left(\frac{\text{strip}}{\text{depth}}\right) \cdot L(y)\,dy \Rightarrow F = 2\int_0^2 (62.4)(2 - y)(2y)\,dy = 249.6\int_0^2 (2y - y^2)\,dy = 249.6\left[y^2 - \frac{y^3}{3}\right]_0^2$

$= (249.6)\left(4 - \frac{8}{3}\right) = (249.6)\left(\frac{4}{3}\right) = 332.8$ lb

47. $F = \int_a^b W \cdot \left(\frac{\text{strip}}{\text{depth}}\right) \cdot L(y)\,dy \Rightarrow F = 62.4\int_0^4 (9 - y)\left(2 \cdot \frac{\sqrt{y}}{2}\right)dy = 62.4\int_0^4 (9y^{1/2} - 3y^{3/2})\,dy$

$= 62.4\left[6y^{3/2} - \frac{2}{5}y^{5/2}\right]_0^4 = (62.4)\left(6 \cdot 8 - \frac{2}{5}\cdot 32\right) = \left(\frac{62.4}{5}\right)(48 \cdot 5 - 64) = \frac{(62.4)(176)}{5} = 2196.48$ lb

49. $F = w_1\int_0^6 (8 - y)(2)(6 - y)\,dy + w_2\int_{-6}^0 (8 - y)(2)(y + 6)\,dy = 2w_1\int_0^6 (48 - 14y + y^2)\,dy + 2w_2\int_{-6}^0 (48 + 2y - y^2)\,dy$

$= 2w_1\left[48y - 7y^2 + \frac{y^3}{3}\right]_0^6 + 2w_2\left[48y + y^2 - \frac{y^3}{3}\right]_{-6}^0 = 216w_1 + 360w_2$

CHAPTER 6 ADDITIONAL AND ADVANCED EXERCISES

1. $V = \pi \int_a^b [f(x)]^2 \, dx = b^2 - ab \Rightarrow \pi \int_a^x [f(t)]^2 \, dt = x^2 - ax$ for all $x > a \Rightarrow \pi [f(x)]^2 = 2x - a \Rightarrow f(x) = \pm \sqrt{\frac{2x-a}{\pi}}$

3. $s(x) = Cx \Rightarrow \int_0^x \sqrt{1 + [f'(t)]^2} \, dt = Cx \Rightarrow \sqrt{1 + [f'(x)]^2} = C \Rightarrow f'(x) = \sqrt{C^2 - 1}$ for $C \geq 1$

    $\Rightarrow f(x) = \int_0^x \sqrt{C^2 - 1} \, dt + k$. Then $f(0) = a \Rightarrow a = 0 + k \Rightarrow f(x) = \int_0^x \sqrt{C^2 - 1} \, dt + a \Rightarrow f(x) = x\sqrt{C^2 - 1} + a$,

    where $C \geq 1$.

5. From the symmetry of $y = 1 - x^n$, n even, about the y-axis for $-1 \leq x \leq 1$, we have $\bar{x} = 0$. To find $\bar{y} = \frac{M_x}{M}$, we
    use the vertical strips technique. The typical strip has center of mass: $(\tilde{x}, \tilde{y}) = \left(x, \frac{1-x^n}{2}\right)$, length: $1 - x^n$,
    width: dx, area: $dA = (1 - x^n) \, dx$, mass: $dm = 1 \cdot dA = (1 - x^n) \, dx$. The moment of the strip about the
    x-axis is $\tilde{y} \, dm = \frac{(1-x^n)^2}{2} \, dx \Rightarrow M_x = \int_{-1}^1 \frac{(1-x^n)^2}{2} \, dx = 2 \int_0^1 \frac{1}{2} (1 - 2x^n + x^{2n}) \, dx = \left[x - \frac{2x^{n+1}}{n+1} + \frac{x^{2n+1}}{2n+1}\right]_0^1$
    $= 1 - \frac{2}{n+1} + \frac{1}{2n+1} = \frac{(n+1)(2n+1) - 2(2n+1) + (n+1)}{(n+1)(2n+1)} = \frac{2n^2 + 3n + 1 - 4n - 2 + n + 1}{(n+1)(2n+1)} = \frac{2n^2}{(n+1)(2n+1)}$.
    Also, $M = \int_{-1}^1 dA = \int_{-1}^1 (1 - x^n) \, dx = 2\int_0^1 (1 - x^n) \, dx = 2\left[x - \frac{x^{n+1}}{n+1}\right]_0^1 = 2\left(1 - \frac{1}{n+1}\right) = \frac{2n}{n+1}$. Therefore,
    $\bar{y} = \frac{M_x}{M} = \frac{2n^2}{(n+1)(2n+1)} \cdot \frac{(n+1)}{2n} = \frac{n}{2n+1} \Rightarrow \left(0, \frac{n}{2n+1}\right)$ is the location of the centroid. As $n \to \infty$, $\bar{y} \to \frac{1}{2}$ so
    the limiting position of the centroid is $\left(0, \frac{1}{2}\right)$.

7. (a) Consider a single vertical strip with center of mass $(\tilde{x}, \tilde{y})$. If the plate lies to the right of the line, then
        the moment of this strip about the line $x = b$ is $(\tilde{x} - b) \, dm = (\tilde{x} - b)\delta \, dA \Rightarrow$ the plate's first moment
        about $x = b$ is the integral $\int (x - b)\delta \, dA = \int \delta x \, dA - \int \delta b \, dA = M_y - b\delta A$.

    (b) If the plate lies to the left of the line, the moment of a vertical strip about the line $x = b$ is
        $(b - \tilde{x}) \, dm = (b - \tilde{x})\delta \, dA \Rightarrow$ the plate's first moment about $x = b$ is $\int (b - x)\delta \, dA = \int b\delta \, dA - \int \delta x \, dA$
        $= b\delta A - M_y$.

9. (a) On $[0, a]$ a typical *vertical* strip has center of mass: $(\tilde{x}, \tilde{y}) = \left(x, \frac{\sqrt{b^2 - x^2} + \sqrt{a^2 - x^2}}{2}\right)$,
        length: $\sqrt{b^2 - x^2} - \sqrt{a^2 - x^2}$, width: dx, area: $dA = \left(\sqrt{b^2 - x^2} - \sqrt{a^2 - x^2}\right) dx$, mass: $dm = \delta \, dA$
        $= \delta \left(\sqrt{b^2 - x^2} - \sqrt{a^2 - x^2}\right) dx$. On $[a, b]$ a typical *vertical* strip has center of mass:
        $(\tilde{x}, \tilde{y}) = \left(x, \frac{\sqrt{b^2 - x^2}}{2}\right)$, length: $\sqrt{b^2 - x^2}$, width: dx, area: $dA = \sqrt{b^2 - x^2} \, dx$,
        mass: $dm = \delta \, dA = \delta \sqrt{b^2 - x^2} \, dx$. Thus, $M_x = \int \tilde{y} \, dm$
        $= \int_0^a \frac{1}{2} \left(\sqrt{b^2 - x^2} + \sqrt{a^2 - x^2}\right) \delta \left(\sqrt{b^2 - x^2} - \sqrt{a^2 - x^2}\right) dx + \int_a^b \frac{1}{2} \sqrt{b^2 - x^2} \, \delta \sqrt{b^2 - x^2} \, dx$
        $= \frac{\delta}{2} \int_0^a [(b^2 - x^2) - (a^2 - x^2)] \, dx + \frac{\delta}{2} \int_a^b (b^2 - x^2) \, dx = \frac{\delta}{2} \int_0^a (b^2 - a^2) \, dx + \frac{\delta}{2} \int_a^b (b^2 - x^2) \, dx$
        $= \frac{\delta}{2} [(b^2 - a^2) x]_0^a + \frac{\delta}{2} \left[b^2 x - \frac{x^3}{3}\right]_a^b = \frac{\delta}{2} [(b^2 - a^2) a] + \frac{\delta}{2} \left[\left(b^3 - \frac{b^3}{3}\right) - \left(b^2 a - \frac{a^3}{3}\right)\right]$
        $= \frac{\delta}{2} (ab^2 - a^3) + \frac{\delta}{2} \left(\frac{2}{3} b^3 - ab^2 + \frac{a^3}{3}\right) = \frac{\delta b^3}{3} - \frac{\delta a^3}{3} = \delta \left(\frac{b^3 - a^3}{3}\right)$; $M_y = \int \tilde{x} \, dm$
        $= \int_0^a x\delta \left(\sqrt{b^2 - x^2} - \sqrt{a^2 - x^2}\right) dx + \int_a^b x\delta \sqrt{b^2 - x^2} \, dx$
        $= \delta \int_0^a x (b^2 - x^2)^{1/2} \, dx - \delta \int_0^a x (a^2 - x^2)^{1/2} \, dx + \delta \int_a^b x (b^2 - x^2)^{1/2} \, dx$
        $= \frac{-\delta}{2} \left[\frac{2 (b^2 - x^2)^{3/2}}{3}\right]_0^a + \frac{\delta}{2} \left[\frac{2 (a^2 - x^2)^{3/2}}{3}\right]_0^a - \frac{\delta}{2} \left[\frac{2 (b^2 - x^2)^{3/2}}{3}\right]_a^b$
        $= -\frac{\delta}{3} \left[(b^2 - a^2)^{3/2} - (b^2)^{3/2}\right] + \frac{\delta}{3} \left[0 - (a^2)^{3/2}\right] - \frac{\delta}{3} \left[0 - (b^2 - a^2)^{3/2}\right] = \frac{\delta b^3}{3} - \frac{\delta a^3}{3} = \frac{\delta (b^3 - a^3)}{3} = M_x$;
        We calculate the mass geometrically: $M = \delta A = \delta \left(\frac{\pi b^2}{4}\right) - \delta \left(\frac{\pi a^2}{4}\right) = \frac{\delta \pi}{4} (b^2 - a^2)$. Thus, $\bar{x} = \frac{M_y}{M}$

$$= \frac{\delta(b^3 - a^3)}{3} \cdot \frac{4}{\delta\pi(b^2 - a^2)} = \frac{4}{3\pi}\left(\frac{b^3 - a^3}{b^2 - a^2}\right) = \frac{4}{3\pi}\frac{(b-a)(a^2 + ab + b^2)}{(b-a)(b+a)} = \frac{4(a^2 + ab + b^2)}{3\pi(a+b)} ; \text{ likewise}$$

$$\bar{y} = \frac{M_x}{M} = \frac{4(a^2 + ab + b^2)}{3\pi(a+b)} .$$

(b) $\lim\limits_{b \to a} \frac{4}{3\pi}\left(\frac{a^2 + ab + b^2}{a+b}\right) = \left(\frac{4}{3\pi}\right)\left(\frac{a^2 + a^2 + a^2}{a+a}\right) = \left(\frac{4}{3\pi}\right)\left(\frac{3a^2}{2a}\right) = \frac{2a}{\pi} \Rightarrow (\bar{x}, \bar{y}) = \left(\frac{2a}{\pi}, \frac{2a}{\pi}\right)$ is the limiting

position of the centroid as $b \to a$. This is the centroid of a circle of radius a (and we note the two circles coincide when $b = a$).

11. $y = 2\sqrt{x} \Rightarrow ds = \sqrt{\frac{1}{x} + 1} \, dx \Rightarrow A = \int_0^3 2\sqrt{x} \sqrt{\frac{1}{x} + 1} \, dx = \frac{4}{3}\left[(1+x)^{3/2}\right]_0^3 = \frac{28}{3}$

13. $F = ma = t^2 \Rightarrow \frac{d^2x}{dt^2} = a = \frac{t^2}{m} \Rightarrow v = \frac{dx}{dt} = \frac{t^3}{3m} + C; v = 0$ when $t = 0 \Rightarrow C = 0 \Rightarrow \frac{dx}{dt} = \frac{t^3}{3m} \Rightarrow x = \frac{t^4}{12m} + C_1;$

$x = 0$ when $t = 0 \Rightarrow C_1 = 0 \Rightarrow x = \frac{t^4}{12m}$. Then $x = h \Rightarrow t = (12mh)^{1/4}$. The work done is

$$W = \int F \, dx = \int_0^{(12mh)^{1/4}} F(t) \cdot \frac{dx}{dt} \, dt = \int_0^{(12mh)^{1/4}} t^2 \cdot \frac{t^3}{3m} \, dt = \frac{1}{3m}\left[\frac{t^6}{6}\right]_0^{(12mh)^{1/4}} = \left(\frac{1}{18m}\right)(12mh)^{6/4}$$

$$= \frac{(12mh)^{3/2}}{18m} = \frac{12mh\cdot\sqrt{12mh}}{18m} = \frac{2h}{3} \cdot 2\sqrt{3mh} = \frac{4h}{3}\sqrt{3mh}$$

15. The submerged triangular plate is depicted in the figure at the right. The hypotenuse of the triangle has slope -1
$\Rightarrow y - (-2) = -(x - 0) \Rightarrow x = -(y + 2)$ is an equation of the hypotenuse. Using a typical horizontal strip, the fluid
pressure is $F = \int (62.4) \cdot \left(\begin{smallmatrix}\text{strip}\\\text{depth}\end{smallmatrix}\right) \cdot \left(\begin{smallmatrix}\text{strip}\\\text{length}\end{smallmatrix}\right) dy$

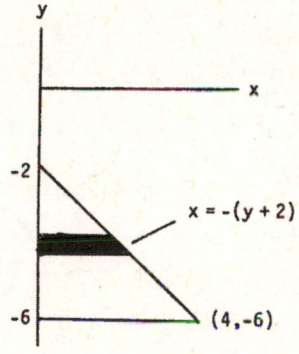

$$= \int_{-6}^{-2} (62.4)(-y)[-(y+2)] \, dy = 62.4 \int_{-6}^{-2} (y^2 + 2y) \, dy$$

$$= 62.4\left[\frac{y^3}{3} + y^2\right]_{-6}^{-2} = (62.4)\left[\left(-\frac{8}{3} + 4\right) - \left(-\frac{216}{3} + 36\right)\right]$$

$$= (62.4)\left(\frac{208}{3} - 32\right) = \frac{(62.4)(112)}{3} \approx 2329.6 \text{ lb}$$

17. (a) We establish a coordinate system as shown. A typical horizontal strip has: center of pressure: $(\tilde{x}, \tilde{y})$
$= \left(\frac{b}{2}, y\right)$, length: $L(y) = b$, width: dy, area: dA
$= b \, dy$, pressure: $dp = \omega |y| \, dA = \omega b |y| \, dy$
$\Rightarrow F_x = \int \tilde{y} \, dp = \int_{-h}^0 y \cdot \omega b |y| \, dy = -\omega b \int_{-h}^0 y^2 \, dy$
$= -\omega b \left[\frac{y^3}{3}\right]_{-h}^0 = -\omega b \left[0 - \left(\frac{-h^3}{3}\right)\right] = \frac{-\omega b h^3}{3};$
$F = \int dp = \int_{-h}^0 \omega |y| L(y) \, dy = -\omega b \int_{-h}^0 y \, dy$

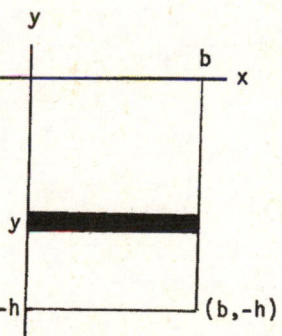

$$= -\omega b \left[\frac{y^2}{2}\right]_{-h}^0 = -\omega b \left[0 - \frac{h^2}{2}\right] = \frac{\omega b h^2}{2}. \text{ Thus, } \bar{y} = \frac{F_x}{F} = \frac{\left(\frac{-\omega b h^3}{3}\right)}{\left(\frac{\omega b h^2}{2}\right)} = \frac{-2h}{3} \Rightarrow \text{ the distance below the surface is } \frac{2}{3}h.$$

(b) A typical horizontal strip has length $L(y)$. By similar triangles from the figure at the right, $\frac{L(y)}{b} = \frac{-y-a}{h}$

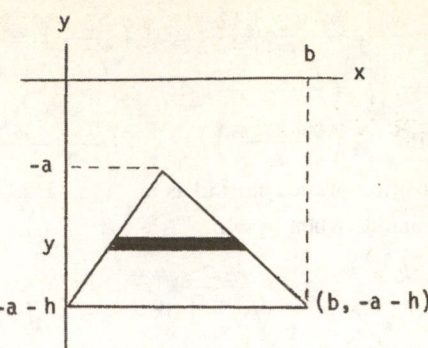

$\Rightarrow L(y) = -\frac{b}{h}(y + a)$. Thus, a typical strip has center of pressure: $(\tilde{x}, \tilde{y}) = (\tilde{x}, y)$, length: $L(y)$

$= -\frac{b}{h}(y + a)$, width: dy, area: $dA = -\frac{b}{h}(y + a)\, dy$,

pressure: $dp = \omega\, |y|\, dA = \omega(-y)\left(-\frac{b}{h}\right)(y + a)\, dy$

$= \frac{\omega b}{h}(y^2 + ay)\, dy \Rightarrow F_x = \int \tilde{y}\, dp$

$= \int_{-(a+h)}^{-a} y \cdot \frac{\omega b}{h}(y^2 + ay)\, dy = \frac{\omega b}{h}\int_{-(a+h)}^{-a}(y^3 + ay^2)\, dy$

$= \frac{\omega b}{h}\left[\frac{y^4}{4} + \frac{ay^3}{3}\right]_{-(a+h)}^{-a}$

$= \frac{\omega b}{h}\left[\left(\frac{a^4}{4} - \frac{a^4}{3}\right) - \left(\frac{(a+h)^4}{4} - \frac{a(a+h)^3}{3}\right)\right] = \frac{\omega b}{h}\left[\frac{a^4 - (a+h)^4}{4} - \frac{a^4 - a(a+h)^3}{3}\right]$

$= \frac{\omega b}{12h}[3(a^4 - (a^4 + 4a^3h + 6a^2h^2 + 4ah^3 + h^4)) - 4(a^4 - a(a^3 + 3a^2h + 3ah^2 + h^3))]$

$= \frac{\omega b}{12h}(12a^3h + 12a^2h^2 + 4ah^3 - 12a^3h - 18a^2h^2 - 12ah^3 - 3h^4) = \frac{\omega b}{12h}(-6a^2h^2 - 8ah^3 - 3h^4)$

$= \frac{-\omega bh}{12}(6a^2 + 8ah + 3h^2) ; F = \int dp = \int \omega\, |y|\, L(y)\, dy = \frac{\omega b}{h}\int_{-(a+h)}^{-a}(y^2 + ay)\, dy = \frac{\omega b}{h}\left[\frac{y^3}{3} + \frac{ay^2}{2}\right]_{-(a+h)}^{-a}$

$= \frac{\omega b}{h}\left[\left(\frac{-a^3}{3} + \frac{a^3}{2}\right) - \left(\frac{-(a+h)^3}{3} + \frac{a(a+h)^2}{2}\right)\right] = \frac{\omega b}{h}\left[\frac{(a+h)^3 - a^3}{3} + \frac{a^3 - a(a+h)^2}{2}\right]$

$= \frac{\omega b}{h}\left[\frac{a^3 + 3a^2h + 3ah^2 + h^3 - a^3}{3} + \frac{a^3 - (a^3 + 2a^2h + ah^2)}{2}\right] = \frac{\omega b}{6h}[2(3a^2h + 3ah^2 + h^3) - 3(2a^2h + ah^2)]$

$= \frac{\omega b}{6h}(6a^2h + 6ah^2 + 2h^3 - 6a^2h - 3ah^2) = \frac{\omega b}{6h}(3ah^2 + 2h^3) = \frac{\omega bh}{6}(3a + 2h)$. Thus, $\overline{y} = \frac{F_x}{F}$

$= \frac{\left(\frac{-\omega bh}{12}\right)(6a^2 + 8ah + 3h^2)}{\left(\frac{\omega bh}{6}\right)(3a + 2h)} = \left(\frac{-1}{2}\right)\left(\frac{6a^2 + 8ah + 3h^2}{3a + 2h}\right) \Rightarrow$ the distance below the surface is

$\frac{6a^2 + 8ah + 3h^2}{6a + 4h}$.

CHAPTER 7 TRANSCENDENTAL FUNCTIONS

7.1 INVERSE FUNCTIONS AND THEIR DERIVATIVES

1. Yes one-to-one, the graph passes the horizontal test.

3. Not one-to-one since (for example) the horizontal line $y = 2$ intersects the graph twice.

5. Yes one-to-one, the graph passes the horizontal test

7. Domain: $0 < x \leq 1$, Range: $0 \leq y$

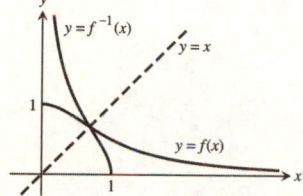

9. Domain: $-1 \leq x \leq 1$, Range: $-\frac{\pi}{2} \leq y \leq \frac{\pi}{2}$

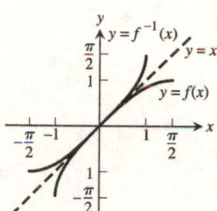

11. The graph is symmetric about $y = x$.

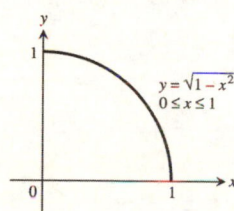

(b) $y = \sqrt{1 - x^2} \Rightarrow y^2 = 1 - x^2 \Rightarrow x^2 = 1 - y^2 \Rightarrow x = \sqrt{1 - y^2} \Rightarrow y = \sqrt{1 - x^2} = f^{-1}(x)$

13. Step 1: $y = x^2 + 1 \Rightarrow x^2 = y - 1 \Rightarrow x = \sqrt{y - 1}$
    Step 2: $y = \sqrt{x - 1} = f^{-1}(x)$

15. Step 1: $y = x^3 - 1 \Rightarrow x^3 = y + 1 \Rightarrow x = (y + 1)^{1/3}$
    Step 2: $y = \sqrt[3]{x + 1} = f^{-1}(x)$

17. Step 1: $y = (x + 1)^2 \Rightarrow \sqrt{y} = x + 1$, since $x \geq -1 \Rightarrow x = \sqrt{y} - 1$
    Step 2: $y = \sqrt{x} - 1 = f^{-1}(x)$

19. Step 1: $y = x^5 \Rightarrow x = y^{1/5}$
    Step 2: $y = \sqrt[5]{x} = f^{-1}(x)$;
    Domain and Range of f^{-1}: all reals;
    $f\left(f^{-1}(x)\right) = \left(x^{1/5}\right)^5 = x$ and $f^{-1}(f(x)) = \left(x^5\right)^{1/5} = x$

21. Step 1: $y = x^3 + 1 \Rightarrow x^3 = y - 1 \Rightarrow x = (y - 1)^{1/3}$
    Step 2: $y = \sqrt[3]{x - 1} = f^{-1}(x)$;
    Domain and Range of f^{-1}: all reals;

$$f\left(f^{-1}(x)\right) = \left((x-1)^{1/3}\right)^3 + 1 = (x-1) + 1 = x \text{ and } f^{-1}(f(x)) = \left((x^3+1)-1\right)^{1/3} = (x^3)^{1/3} = x$$

23. Step 1: $y = \frac{1}{x^2} \Rightarrow x^2 = \frac{1}{y} \Rightarrow x = \frac{1}{\sqrt{y}}$

    Step 2: $y = \frac{1}{\sqrt{x}} = f^{-1}(x)$

    Domain of f^{-1}: $x > 0$, Range of f^{-1}: $y > 0$;

    $f\left(f^{-1}(x)\right) = \frac{1}{\left(\frac{1}{\sqrt{x}}\right)^2} = \frac{1}{\left(\frac{1}{x}\right)} = x$ and $f^{-1}(f(x)) = \frac{1}{\sqrt{\frac{1}{x^2}}} = \frac{1}{\left(\frac{1}{x}\right)} = x$ since $x > 0$

25. (a) $y = 2x + 3 \Rightarrow 2x = y - 3$

    $\Rightarrow x = \frac{y}{2} - \frac{3}{2} \Rightarrow f^{-1}(x) = \frac{x}{2} - \frac{3}{2}$

    (c) $\left.\frac{df}{dx}\right|_{x=-1} = 2, \left.\frac{df^{-1}}{dx}\right|_{x=1} = \frac{1}{2}$

    (b)

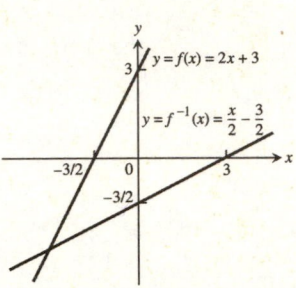

27. (a) $y = 5 - 4x \Rightarrow 4x = 5 - y$

    $\Rightarrow x = \frac{5}{4} - \frac{y}{4} \Rightarrow f^{-1}(x) = \frac{5}{4} - \frac{x}{4}$

    (c) $\left.\frac{df}{dx}\right|_{x=1/2} = -4, \left.\frac{df^{-1}}{dx}\right|_{x=3} = -\frac{1}{4}$

    (b)

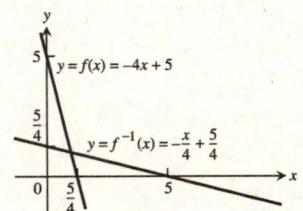

29. (a) $f(g(x)) = \left(\sqrt[3]{x}\right)^3 = x, g(f(x)) = \sqrt[3]{x^3} = x$

    (c) $f'(x) = 3x^2 \Rightarrow f'(1) = 3, f'(-1) = 3$;

    $g'(x) = \frac{1}{3}x^{-2/3} \Rightarrow g'(1) = \frac{1}{3}, g'(-1) = \frac{1}{3}$

    (d) The line $y = 0$ is tangent to $f(x) = x^3$ at $(0,0)$;

    the line $x = 0$ is tangent to $g(x) = \sqrt[3]{x}$ at $(0,0)$

    (b)

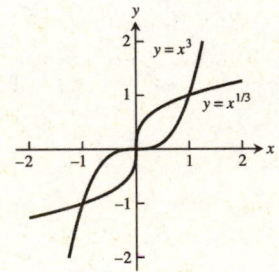

31. $\frac{df}{dx} = 3x^2 - 6x \Rightarrow \left.\frac{df^{-1}}{dx}\right|_{x=f(3)} = \left.\frac{1}{\frac{df}{dx}}\right|_{x=3} = \frac{1}{9}$

33. $\left.\frac{df^{-1}}{dx}\right|_{x=4} = \left.\frac{df^{-1}}{dx}\right|_{x=f(2)} = \left.\frac{1}{\frac{df}{dx}}\right|_{x=2} = \frac{1}{\left(\frac{1}{3}\right)} = 3$

35. (a) $y = mx \Rightarrow x = \frac{1}{m}y \Rightarrow f^{-1}(x) = \frac{1}{m}x$

    (b) The graph of $y = f^{-1}(x)$ is a line through the origin with slope $\frac{1}{m}$.

37. (a) $y = x + 1 \Rightarrow x = y - 1 \Rightarrow f^{-1}(x) = x - 1$

(b) $y = x + b \Rightarrow x = y - b \Rightarrow f^{-1}(x) = x - b$

(c) Their graphs will be parallel to one another and lie on opposite sides of the line $y = x$ equidistant from that line.

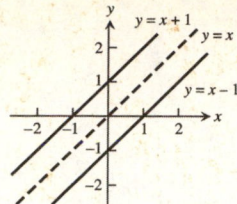

39. Let $x_1 \neq x_2$ be two numbers in the domain of an increasing function f. Then, either $x_1 < x_2$ or $x_1 > x_2$ which implies $f(x_1) < f(x_2)$ or $f(x_1) > f(x_2)$, since $f(x)$ is increasing. In either case, $f(x_1) \neq f(x_2)$ and f is one-to-one. Similar arguments hold if f is decreasing.

41. $f(x)$ is increasing since $x_2 > x_1 \Rightarrow 27x_2^3 > 27x_1^3$; $y = 27x^3 \Rightarrow x = \frac{1}{3}y^{1/3} \Rightarrow f^{-1}(x) = \frac{1}{3}x^{1/3}$;

$\frac{df}{dx} = 81x^2 \Rightarrow \frac{df^{-1}}{dx} = \frac{1}{81x^2}\big|_{\frac{1}{3}x^{1/3}} = \frac{1}{9x^{2/3}} = \frac{1}{9}x^{-2/3}$

43. $f(x)$ is decreasing since $x_2 > x_1 \Rightarrow (1 - x_2)^3 < (1 - x_1)^3$; $y = (1 - x)^3 \Rightarrow x = 1 - y^{1/3} \Rightarrow f^{-1}(x) = 1 - x^{1/3}$;

$\frac{df}{dx} = -3(1 - x)^2 \Rightarrow \frac{df^{-1}}{dx} = \frac{1}{-3(1-x)^2}\big|_{1-x^{1/3}} = \frac{-1}{3x^{2/3}} = -\frac{1}{3}x^{-2/3}$

45. The function $g(x)$ is also one-to-one. The reasoning: $f(x)$ is one-to-one means that if $x_1 \neq x_2$ then $f(x_1) \neq f(x_2)$, so $-f(x_1) \neq -f(x_2)$ and therefore $g(x_1) \neq g(x_2)$. Therefore $g(x)$ is one-to-one as well.

47. The composite is one-to-one also. The reasoning: If $x_1 \neq x_2$ then $g(x_1) \neq g(x_2)$ because g is one-to-one. Since $g(x_1) \neq g(x_2)$, we also have $f(g(x_1)) \neq f(g(x_2))$ because f is one-to-one. Thus, $f \circ g$ is one-to-one because $x_1 \neq x_2 \Rightarrow f(g(x_1)) \neq f(g(x_2))$.

49. The first integral is the area between $f(x)$ and the x-axis over $a \leq x \leq b$. The second integral is the area between $f(x)$ and the y-axis for $f(a) \leq y \leq f(b)$. The sum of the integrals is the area of the larger rectangle with corners at $(0, 0)$, $(b, 0)$, $(b, f(b))$ and $(0, f(b))$ minus the area of the smaller rectangle with vertices at $(0, 0)$, $(a, 0)$, $(a, f(a))$ and $(0, f(a))$. That is, the sum of the integrals is $bf(b) - af(a)$.

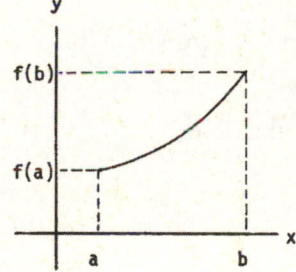

51. $(g \circ f)(x) = x \Rightarrow g(f(x)) = x \Rightarrow g'(f(x))f'(x) = 1$

7.2 NATURAL LOGARITHMS

1. (a) $\ln 0.75 = \ln \frac{3}{4} = \ln 3 - \ln 4 = \ln 3 - \ln 2^2 = \ln 3 - 2 \ln 2$

(b) $\ln \frac{4}{9} = \ln 4 - \ln 9 = \ln 2^2 - \ln 3^2 = 2 \ln 2 - 2 \ln 3$

(c) $\ln \frac{1}{2} = \ln 1 - \ln 2 = -\ln 2$

(d) $\ln \sqrt[3]{9} = \frac{1}{3}\ln 9 = \frac{1}{3}\ln 3^2 = \frac{2}{3}\ln 3$

(e) $\ln 3\sqrt{2} = \ln 3 + \ln 2^{1/2} = \ln 3 + \frac{1}{2}\ln 2$

(f) $\ln \sqrt{13.5} = \frac{1}{2}\ln 13.5 = \frac{1}{2}\ln \frac{27}{2} = \frac{1}{2}(\ln 3^3 - \ln 2) = \frac{1}{2}(3 \ln 3 - \ln 2)$

3. (a) $\ln \sin \theta - \ln \left(\frac{\sin \theta}{5}\right) = \ln \left(\frac{\sin \theta}{\frac{\sin \theta}{5}}\right) = \ln 5$

(b) $\ln (3x^2 - 9x) + \ln \left(\frac{1}{3x}\right) = \ln \left(\frac{3x^2 - 9x}{3x}\right) = \ln (x - 3)$

(c) $\frac{1}{2}\ln (4t^4) - \ln 2 = \ln \sqrt{4t^4} - \ln 2 = \ln 2t^2 - \ln 2 = \ln \left(\frac{2t^2}{2}\right) = \ln (t^2)$

5. $y = \ln 3x \Rightarrow y' = \left(\frac{1}{3x}\right)(3) = \frac{1}{x}$

7. $y = \ln(t^2) \Rightarrow \frac{dy}{dt} = \left(\frac{1}{t^2}\right)(2t) = \frac{2}{t}$

9. $y = \ln\frac{3}{x} = \ln 3x^{-1} \Rightarrow \frac{dy}{dx} = \left(\frac{1}{3x^{-1}}\right)(-3x^{-2}) = -\frac{1}{x}$

11. $y = \ln(\theta + 1) \Rightarrow \frac{dy}{d\theta} = \left(\frac{1}{\theta+1}\right)(1) = \frac{1}{\theta+1}$

13. $y = \ln x^3 \Rightarrow \frac{dy}{dx} = \left(\frac{1}{x^3}\right)(3x^2) = \frac{3}{x}$

15. $y = t(\ln t)^2 \Rightarrow \frac{dy}{dt} = (\ln t)^2 + 2t(\ln t) \cdot \frac{d}{dt}(\ln t) = (\ln t)^2 + \frac{2t \ln t}{t} = (\ln t)^2 + 2\ln t$

17. $y = \frac{x^4}{4}\ln x - \frac{x^4}{16} \Rightarrow \frac{dy}{dx} = x^3 \ln x + \frac{x^4}{4}\cdot\frac{1}{x} - \frac{4x^3}{16} = x^3 \ln x$

19. $y = \frac{\ln t}{t} \Rightarrow \frac{dy}{dt} = \frac{t\left(\frac{1}{t}\right) - (\ln t)(1)}{t^2} = \frac{1 - \ln t}{t^2}$

21. $y = \frac{\ln x}{1 + \ln x} \Rightarrow y' = \frac{(1 + \ln x)\left(\frac{1}{x}\right) - (\ln x)\left(\frac{1}{x}\right)}{(1 + \ln x)^2} = \frac{\frac{1}{x} + \frac{\ln x}{x} - \frac{\ln x}{x}}{(1 + \ln x)^2} = \frac{1}{x(1 + \ln x)^2}$

23. $y = \ln(\ln x) \Rightarrow y' = \left(\frac{1}{\ln x}\right)\left(\frac{1}{x}\right) = \frac{1}{x \ln x}$

25. $y = \theta[\sin(\ln \theta) + \cos(\ln \theta)] \Rightarrow \frac{dy}{d\theta} = [\sin(\ln \theta) + \cos(\ln \theta)] + \theta\left[\cos(\ln \theta)\cdot\frac{1}{\theta} - \sin(\ln \theta)\cdot\frac{1}{\theta}\right]$
$= \sin(\ln \theta) + \cos(\ln \theta) + \cos(\ln \theta) - \sin(\ln \theta) = 2\cos(\ln \theta)$

27. $y = \ln\frac{1}{x\sqrt{x+1}} = -\ln x - \frac{1}{2}\ln(x+1) \Rightarrow y' = -\frac{1}{x} - \frac{1}{2}\left(\frac{1}{x+1}\right) = -\frac{2(x+1)+x}{2x(x+1)} = -\frac{3x+2}{2x(x+1)}$

29. $y = \frac{1 + \ln t}{1 - \ln t} \Rightarrow \frac{dy}{dt} = \frac{(1 - \ln t)\left(\frac{1}{t}\right) - (1 + \ln t)\left(\frac{-1}{t}\right)}{(1 - \ln t)^2} = \frac{\frac{1}{t} - \frac{\ln t}{t} + \frac{1}{t} + \frac{\ln t}{t}}{(1 - \ln t)^2} = \frac{2}{t(1 - \ln t)^2}$

31. $y = \ln(\sec(\ln \theta)) \Rightarrow \frac{dy}{d\theta} = \frac{1}{\sec(\ln \theta)}\cdot\frac{d}{d\theta}(\sec(\ln \theta)) = \frac{\sec(\ln \theta)\tan(\ln \theta)}{\sec(\ln \theta)}\cdot\frac{d}{d\theta}(\ln \theta) = \frac{\tan(\ln \theta)}{\theta}$

33. $y = \ln\left(\frac{(x^2+1)^5}{\sqrt{1-x}}\right) = 5\ln(x^2+1) - \frac{1}{2}\ln(1-x) \Rightarrow y' = \frac{5\cdot 2x}{x^2+1} - \frac{1}{2}\left(\frac{1}{1-x}\right)(-1) = \frac{10x}{x^2+1} + \frac{1}{2(1-x)}$

35. $y = \int_{x^2/2}^{x^2} \ln\sqrt{t}\, dt \Rightarrow \frac{dy}{dx} = \left(\ln\sqrt{x^2}\right)\cdot\frac{d}{dx}(x^2) - \left(\ln\sqrt{\frac{x^2}{2}}\right)\cdot\frac{d}{dx}\left(\frac{x^2}{2}\right) = 2x\ln|x| - x\ln\frac{|x|}{\sqrt{2}}$

37. $\int_{-3}^{-2}\frac{1}{x}\, dx = [\ln|x|]_{-3}^{-2} = \ln 2 - \ln 3 = \ln\frac{2}{3}$

39. $\int\frac{2y}{y^2-25}\, dy = \ln|y^2 - 25| + C$

41. $\int_0^\pi \frac{\sin t}{2 - \cos t}\, dt = [\ln|2 - \cos t|]_0^\pi = \ln 3 - \ln 1 = \ln 3$; or let $u = 2 - \cos t \Rightarrow du = \sin t\, dt$ with $t = 0$
$\Rightarrow u = 1$ and $t = \pi \Rightarrow u = 3 \Rightarrow \int_0^\pi \frac{\sin t}{2 - \cos t}\, dt = \int_1^3 \frac{1}{u}\, du = [\ln|u|]_1^3 = \ln 3 - \ln 1 = \ln 3$

43. Let $u = \ln x \Rightarrow du = \frac{1}{x}\, dx$; $x = 1 \Rightarrow u = 0$ and $x = 2 \Rightarrow u = \ln 2$;
$\int_1^2 \frac{2\ln x}{x}\, dx = \int_0^{\ln 2} 2u\, du = [u^2]_0^{\ln 2} = (\ln 2)^2$

45. Let $u = \ln x \Rightarrow du = \frac{1}{x} dx; x = 2 \Rightarrow u = \ln 2$ and $x = 4 \Rightarrow u = \ln 4$;

$\int_2^4 \frac{dx}{x(\ln x)^2} = \int_{\ln 2}^{\ln 4} u^{-2} du = \left[-\frac{1}{u}\right]_{\ln 2}^{\ln 4} = -\frac{1}{\ln 4} + \frac{1}{\ln 2} = -\frac{1}{\ln 2^2} + \frac{1}{\ln 2} = -\frac{1}{2\ln 2} + \frac{1}{\ln 2} = \frac{1}{2\ln 2} = \frac{1}{\ln 4}$

47. Let $u = 6 + 3\tan t \Rightarrow du = 3\sec^2 t \, dt$;

$\int \frac{3\sec^2 t}{6 + 3\tan t} dt = \int \frac{du}{u} = \ln |u| + C = \ln |6 + 3\tan t| + C$

49. Let $u = \cos \frac{x}{2} \Rightarrow du = -\frac{1}{2}\sin \frac{x}{2} dx \Rightarrow -2 du = \sin \frac{x}{2} dx; x = 0 \Rightarrow u = 1$ and $x = \frac{\pi}{2} \Rightarrow u = \frac{1}{\sqrt{2}}$;

$\int_0^{\pi/2} \tan \frac{x}{2} dx = \int_0^{\pi/2} \frac{\sin \frac{x}{2}}{\cos \frac{x}{2}} dx = -2 \int_1^{1/\sqrt{2}} \frac{du}{u} = [-2\ln |u|]_1^{1/\sqrt{2}} = -2\ln \frac{1}{\sqrt{2}} = 2\ln \sqrt{2} = \ln 2$

51. Let $u = \sin \frac{\theta}{3} \Rightarrow du = \frac{1}{3}\cos \frac{\theta}{3} d\theta \Rightarrow 6 du = 2\cos \frac{\theta}{3} d\theta; \theta = \frac{\pi}{2} \Rightarrow u = \frac{1}{2}$ and $\theta = \pi \Rightarrow u = \frac{\sqrt{3}}{2}$;

$\int_{\pi/2}^{\pi} 2\cot \frac{\theta}{3} d\theta = \int_{\pi/2}^{\pi} \frac{2\cos \frac{\theta}{3}}{\sin \frac{\theta}{3}} d\theta = 6 \int_{1/2}^{\sqrt{3}/2} \frac{du}{u} = 6[\ln |u|]_{1/2}^{\sqrt{3}/2} = 6\left(\ln \frac{\sqrt{3}}{2} - \ln \frac{1}{2}\right) = 6\ln \sqrt{3} = \ln 27$

53. $\int \frac{dx}{2\sqrt{x} + 2x} = \int \frac{dx}{2\sqrt{x}(1 + \sqrt{x})}$; let $u = 1 + \sqrt{x} \Rightarrow du = \frac{1}{2\sqrt{x}} dx; \int \frac{dx}{2\sqrt{x}(1 + \sqrt{x})} = \int \frac{du}{u} = \ln |u| + C$

$= \ln |1 + \sqrt{x}| + C = \ln (1 + \sqrt{x}) + C$

55. $y = \sqrt{x(x+1)} = (x(x+1))^{1/2} \Rightarrow \ln y = \frac{1}{2}\ln (x(x+1)) \Rightarrow 2\ln y = \ln (x) + \ln (x+1) \Rightarrow \frac{2y'}{y} = \frac{1}{x} + \frac{1}{x+1}$

$\Rightarrow y' = \left(\frac{1}{2}\right)\sqrt{x(x+1)}\left(\frac{1}{x} + \frac{1}{x+1}\right) = \frac{\sqrt{x(x+1)}(2x+1)}{2x(x+1)} = \frac{2x+1}{2\sqrt{x(x+1)}}$

57. $y = \sqrt{\frac{t}{t+1}} = \left(\frac{t}{t+1}\right)^{1/2} \Rightarrow \ln y = \frac{1}{2}[\ln t - \ln (t+1)] \Rightarrow \frac{1}{y}\frac{dy}{dt} = \frac{1}{2}\left(\frac{1}{t} - \frac{1}{t+1}\right)$

$\Rightarrow \frac{dy}{dt} = \frac{1}{2}\sqrt{\frac{t}{t+1}}\left(\frac{1}{t} - \frac{1}{t+1}\right) = \frac{1}{2}\sqrt{\frac{t}{t+1}}\left[\frac{1}{t(t+1)}\right] = \frac{1}{2\sqrt{t}(t+1)^{3/2}}$

59. $y = \sqrt{\theta + 3}(\sin \theta) = (\theta + 3)^{1/2}\sin \theta \Rightarrow \ln y = \frac{1}{2}\ln (\theta + 3) + \ln (\sin \theta) \Rightarrow \frac{1}{y}\frac{dy}{d\theta} = \frac{1}{2(\theta + 3)} + \frac{\cos \theta}{\sin \theta}$

$\Rightarrow \frac{dy}{d\theta} = \sqrt{\theta + 3}(\sin \theta)\left[\frac{1}{2(\theta + 3)} + \cot \theta\right]$

61. $y = t(t+1)(t+2) \Rightarrow \ln y = \ln t + \ln (t+1) + \ln (t+2) \Rightarrow \frac{1}{y}\frac{dy}{dt} = \frac{1}{t} + \frac{1}{t+1} + \frac{1}{t+2}$

$\Rightarrow \frac{dy}{dt} = t(t+1)(t+2)\left(\frac{1}{t} + \frac{1}{t+1} + \frac{1}{t+2}\right) = t(t+1)(t+2)\left[\frac{(t+1)(t+2) + t(t+2) + t(t+1)}{t(t+1)(t+2)}\right] = 3t^2 + 6t + 2$

63. $y = \frac{\theta + 5}{\theta \cos \theta} \Rightarrow \ln y = \ln (\theta + 5) - \ln \theta - \ln (\cos \theta) \Rightarrow \frac{1}{y}\frac{dy}{d\theta} = \frac{1}{\theta + 5} - \frac{1}{\theta} + \frac{\sin \theta}{\cos \theta}$

$\Rightarrow \frac{dy}{d\theta} = \left(\frac{\theta + 5}{\theta \cos \theta}\right)\left(\frac{1}{\theta + 5} - \frac{1}{\theta} + \tan \theta\right)$

65. $y = \frac{x\sqrt{x^2 + 1}}{(x+1)^{2/3}} \Rightarrow \ln y = \ln x + \frac{1}{2}\ln (x^2 + 1) - \frac{2}{3}\ln (x+1) \Rightarrow \frac{y'}{y} = \frac{1}{x} + \frac{x}{x^2 + 1} - \frac{2}{3(x+1)}$

$\Rightarrow y' = \frac{x\sqrt{x^2 + 1}}{(x+1)^{2/3}}\left[\frac{1}{x} + \frac{x}{x^2 + 1} - \frac{2}{3(x+1)}\right]$

67. $y = \sqrt[3]{\frac{x(x-2)}{x^2 + 1}} \Rightarrow \ln y = \frac{1}{3}[\ln x + \ln (x-2) - \ln (x^2 + 1)] \Rightarrow \frac{y'}{y} = \frac{1}{3}\left(\frac{1}{x} + \frac{1}{x-2} - \frac{2x}{x^2 + 1}\right)$

$\Rightarrow y' = \frac{1}{3}\sqrt[3]{\frac{x(x-2)}{x^2 + 1}}\left(\frac{1}{x} + \frac{1}{x-2} - \frac{2x}{x^2 + 1}\right)$

69. (a) $f(x) = \ln (\cos x) \Rightarrow f'(x) = -\frac{\sin x}{\cos x} = -\tan x = 0 \Rightarrow x = 0; f'(x) > 0$ for $-\frac{\pi}{4} \le x < 0$ and $f'(x) < 0$ for

$0 < x \le \frac{\pi}{3} \Rightarrow$ there is a relative maximum at $x = 0$ with $f(0) = \ln (\cos 0) = \ln 1 = 0; f\left(-\frac{\pi}{4}\right) = \ln \left(\cos \left(-\frac{\pi}{4}\right)\right)$

$= \ln \left(\frac{1}{\sqrt{2}}\right) = -\frac{1}{2}\ln 2$ and $f\left(\frac{\pi}{3}\right) = \ln \left(\cos \left(\frac{\pi}{3}\right)\right) = \ln \frac{1}{2} = -\ln 2$. Therefore, the absolute minimum occurs at

$x = \frac{\pi}{3}$ with $f\left(\frac{\pi}{3}\right) = -\ln 2$ and the absolute maximum occurs at $x = 0$ with $f(0) = 0$.

(b) $f(x) = \cos(\ln x) \Rightarrow f'(x) = \frac{-\sin(\ln x)}{x} = 0 \Rightarrow x = 1; f'(x) > 0$ for $\frac{1}{2} \le x < 1$ and $f'(x) < 0$ for $1 < x \le 2$

$\Rightarrow$ there is a relative maximum at $x = 1$ with $f(1) = \cos(\ln 1) = \cos 0 = 1; f\left(\frac{1}{2}\right) = \cos\left(\ln\left(\frac{1}{2}\right)\right)$

$= \cos(-\ln 2) = \cos(\ln 2)$ and $f(2) = \cos(\ln 2)$. Therefore, the absolute minimum occurs at $x = \frac{1}{2}$ and

$x = 2$ with $f\left(\frac{1}{2}\right) = f(2) = \cos(\ln 2)$, and the absolute maximum occurs at $x = 1$ with $f(1) = 1$.

71. $\int_1^5 (\ln 2x - \ln x)\, dx = \int_1^5 (-\ln x + \ln 2 + \ln x)\, dx = (\ln 2)\int_1^5 dx = (\ln 2)(5 - 1) = \ln 2^4 = \ln 16$

73. $V = \pi \int_0^3 \left(\frac{2}{\sqrt{y+1}}\right)^2 dy = 4\pi \int_0^3 \frac{1}{y+1}\, dy = 4\pi \left[\ln|y + 1|\right]_0^3 = 4\pi(\ln 4 - \ln 1) = 4\pi \ln 4$

75. $V = 2\pi \int_{1/2}^2 x\left(\frac{1}{x^2}\right) dx = 2\pi \int_{1/2}^2 \frac{1}{x}\, dx = 2\pi \left[\ln|x|\right]_{1/2}^2 = 2\pi\left(\ln 2 - \ln\frac{1}{2}\right) = 2\pi(2\ln 2) = \pi \ln 2^4 = \pi \ln 16$

77. (a) $y = \frac{x^2}{8} - \ln x \Rightarrow 1 + (y')^2 = 1 + \left(\frac{x}{4} - \frac{1}{x}\right)^2 = 1 + \left(\frac{x^2 - 4}{4x}\right)^2 = \left(\frac{x^2 + 4}{4x}\right)^2 \Rightarrow L = \int_4^8 \sqrt{1 + (y')^2}\, dx$

$= \int_4^8 \frac{x^2 + 4}{4x}\, dx = \int_4^8 \left(\frac{x}{4} + \frac{1}{x}\right) dx = \left[\frac{x^2}{8} + \ln|x|\right]_4^8 = (8 + \ln 8) - (2 + \ln 4) = 6 + \ln 2$

(b) $x = \left(\frac{y}{4}\right)^2 - 2\ln\left(\frac{y}{4}\right) \Rightarrow \frac{dx}{dy} = \frac{y}{8} - \frac{2}{y} \Rightarrow 1 + \left(\frac{dx}{dy}\right)^2 = 1 + \left(\frac{y}{8} - \frac{2}{y}\right)^2 = 1 + \left(\frac{y^2 - 16}{8y}\right)^2 = \left(\frac{y^2 + 16}{8y}\right)^2$

$\Rightarrow L = \int_4^{12} \sqrt{1 + \left(\frac{dx}{dy}\right)^2}\, dy = \int_4^{12} \frac{y^2 + 16}{8y}\, dy = \int_4^{12} \left(\frac{y}{8} + \frac{2}{y}\right) dy = \left[\frac{y^2}{16} + 2\ln y\right]_4^{12} = (9 + 2\ln 12) - (1 + 2\ln 4)$

$= 8 + 2\ln 3 = 8 + \ln 9$

79. (a) $M_y = \int_1^2 x\left(\frac{1}{x}\right) dx = 1, M_x = \int_1^2 \left(\frac{1}{2x}\right)\left(\frac{1}{x}\right) dx = \frac{1}{2}\int_1^2 \frac{1}{x^2}\, dx = \left[-\frac{1}{2x}\right]_1^2 = \frac{1}{4}, M = \int_1^2 \frac{1}{x}\, dx = \left[\ln|x|\right]_1^2 = \ln 2$

$\Rightarrow \bar{x} = \frac{M_y}{M} = \frac{1}{\ln 2} \approx 1.44$ and $\bar{y} = \frac{M_x}{M} = \frac{\left(\frac{1}{4}\right)}{\ln 2} \approx 0.36$

(b)

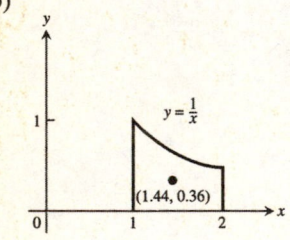

81. $\frac{dy}{dx} = 1 + \frac{1}{x}$ at $(1, 3) \Rightarrow y = x + \ln|x| + C; y = 3$ at $x = 1 \Rightarrow C = 2 \Rightarrow y = x + \ln|x| + 2$

83. (a) $L(x) = f(0) + f'(0) \cdot x$, and $f(x) = \ln(1 + x) \Rightarrow f'(x)|_{x=0} = \frac{1}{1+x}\big|_{x=0} = 1 \Rightarrow L(x) = \ln 1 + 1 \cdot x \Rightarrow L(x) = x$

(b) Let $f(x) = \ln(x + 1)$. Since $f''(x) = -\frac{1}{(x+1)^2} < 0$ on $[0, 0.1]$, the graph of f is concave down on this interval and the largest error in the linear approximation will occur when $x = 0.1$. This error is $0.1 - \ln(1.1) \approx 0.00469$ to five decimal places.

(c) The approximation $y = x$ for $\ln(1 + x)$ is best for smaller positive values of x; in particular for $0 \le x \le 0.1$ in the graph. As x increases, so does the error $x - \ln(1 + x)$. From the graph an upper bound for the error is $0.5 - \ln(1 + 0.5) \approx 0.095$; i.e., $|E(x)| \le 0.095$ for $0 \le x \le 0.5$. Note from the graph that $0.1 - \ln(1 + 0.1) \approx 0.00469$ estimates the error in replacing $\ln(1 + x)$ by x over $0 \le x \le 0.1$. This is consistent with the estimate given in part (b) above.

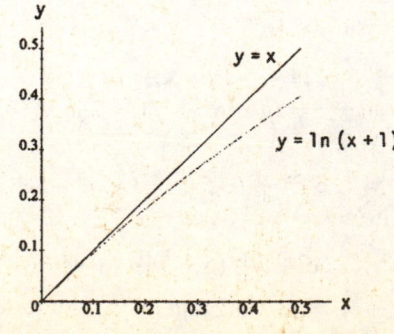

85. $y = \ln kx \Rightarrow y = \ln x + \ln k$; thus the graph of
     $y = \ln kx$ is the graph of $y = \ln x$ shifted vertically
     by $\ln k, k > 0$.

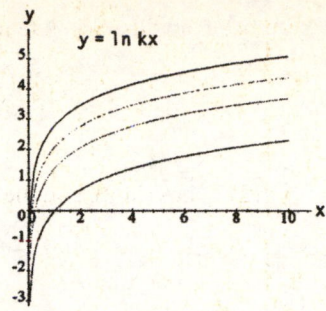

87. (a)

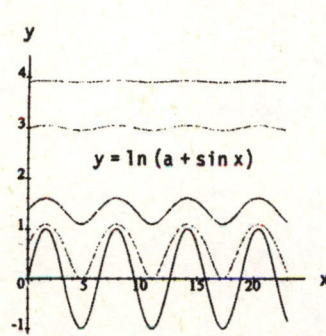

(b) $y' = \frac{\cos x}{a + \sin x}$. Since $|\sin x|$ and $|\cos x|$ are less than
     or equal to 1, we have for $a > 1$
     $\frac{-1}{a-1} \le y' \le \frac{1}{a-1}$ for all x.
     Thus, $\lim\limits_{a \to +\infty} y' = 0$ for all $x \Rightarrow$ the graph of y looks
     more and more horizontal as $a \to +\infty$.

7.3 THE EXPONENTIAL FUNCTION

1. (a) $e^{\ln 7.2} = 7.2$ (b) $e^{-\ln x^2} = \frac{1}{e^{\ln x^2}} = \frac{1}{x^2}$ (c) $e^{\ln x - \ln y} = e^{\ln(x/y)} = \frac{x}{y}$

3. (a) $2 \ln \sqrt{e} = 2 \ln e^{1/2} = (2)\left(\frac{1}{2}\right) \ln e = 1$ (b) $\ln(\ln e^e) = \ln(e \ln e) = \ln e = 1$
    (c) $\ln e^{(-x^2 - y^2)} = (-x^2 - y^2) \ln e = -x^2 - y^2$

5. $\ln y = 2t + 4 \Rightarrow e^{\ln y} = e^{2t+4} \Rightarrow y = e^{2t+4}$

7. $\ln(y - 40) = 5t \Rightarrow e^{\ln(y-40)} = e^{5t} \Rightarrow y - 40 = e^{5t} \Rightarrow y = e^{5t} + 40$

9. $\ln(y - 1) - \ln 2 = x + \ln x \Rightarrow \ln(y - 1) - \ln 2 - \ln x = x \Rightarrow \ln\left(\frac{y-1}{2x}\right) = x \Rightarrow e^{\ln\left(\frac{y-1}{2x}\right)} = e^x \Rightarrow \frac{y-1}{2x} = e^x$
    $\Rightarrow y - 1 = 2xe^x \Rightarrow y = 2xe^x + 1$

11. (a) $e^{2k} = 4 \Rightarrow \ln e^{2k} = \ln 4 \Rightarrow 2k \ln e = \ln 2^2 \Rightarrow 2k = 2 \ln 2 \Rightarrow k = \ln 2$
    (b) $100e^{10k} = 200 \Rightarrow e^{10k} = 2 \Rightarrow \ln e^{10k} = \ln 2 \Rightarrow 10k \ln e = \ln 2 \Rightarrow 10k = \ln 2 \Rightarrow k = \frac{\ln 2}{10}$
    (c) $e^{k/1000} = a \Rightarrow \ln e^{k/1000} = \ln a \Rightarrow \frac{k}{1000} \ln e = \ln a \Rightarrow \frac{k}{1000} = \ln a \Rightarrow k = 1000 \ln a$

13. (a) $e^{-0.3t} = 27 \Rightarrow \ln e^{-0.3t} = \ln 3^3 \Rightarrow (-0.3t) \ln e = 3 \ln 3 \Rightarrow -0.3t = 3 \ln 3 \Rightarrow t = -10 \ln 3$
    (b) $e^{kt} = \frac{1}{2} \Rightarrow \ln e^{kt} = \ln 2^{-1} = kt \ln e = -\ln 2 \Rightarrow t = -\frac{\ln 2}{k}$
    (c) $e^{(\ln 0.2)t} = 0.4 \Rightarrow \left(e^{\ln 0.2}\right)^t = 0.4 \Rightarrow 0.2^t = 0.4 \Rightarrow \ln 0.2^t = \ln 0.4 \Rightarrow t \ln 0.2 = \ln 0.4 \Rightarrow t = \frac{\ln 0.4}{\ln 0.2}$

15. $e^{\sqrt{t}} = x^2 \Rightarrow \ln e^{\sqrt{t}} = \ln x^2 \Rightarrow \sqrt{t} = 2 \ln x \Rightarrow t = 4(\ln x)^2$

17. $y = e^{-5x} \Rightarrow y' = e^{-5x} \frac{d}{dx}(-5x) \Rightarrow y' = -5e^{-5x}$

19. $y = e^{5-7x} \Rightarrow y' = e^{5-7x} \frac{d}{dx}(5 - 7x) \Rightarrow y' = -7e^{5-7x}$

21. $y = xe^x - e^x \Rightarrow y' = (e^x + xe^x) - e^x = xe^x$

23. $y = (x^2 - 2x + 2)\,e^x \Rightarrow y' = (2x - 2)e^x + (x^2 - 2x + 2)\,e^x = x^2 e^x$

25. $y = e^{\theta}(\sin\theta + \cos\theta) \Rightarrow y' = e^{\theta}(\sin\theta + \cos\theta) + e^{\theta}(\cos\theta - \sin\theta) = 2e^{\theta}\cos\theta$

27. $y = \cos\left(e^{-\theta^2}\right) \Rightarrow \frac{dy}{d\theta} = -\sin\left(e^{-\theta^2}\right) \frac{d}{d\theta}\left(e^{-\theta^2}\right) = \left(-\sin\left(e^{-\theta^2}\right)\right)\left(e^{-\theta^2}\right)\frac{d}{d\theta}\left(-\theta^2\right) = 2\theta e^{-\theta^2}\sin\left(e^{-\theta^2}\right)$

29. $y = \ln\left(3te^{-t}\right) = \ln 3 + \ln t + \ln e^{-t} = \ln 3 + \ln t - t \Rightarrow \frac{dy}{dt} = \frac{1}{t} - 1 = \frac{1-t}{t}$

31. $y = \ln\frac{e^{\theta}}{1+e^{\theta}} = \ln e^{\theta} - \ln\left(1 + e^{\theta}\right) = \theta - \ln\left(1+e^{\theta}\right) \Rightarrow \frac{dy}{d\theta} = 1 - \left(\frac{1}{1+e^{\theta}}\right)\frac{d}{d\theta}\left(1 + e^{\theta}\right) = 1 - \frac{e^{\theta}}{1+e^{\theta}} = \frac{1}{1+e^{\theta}}$

33. $y = e^{(\cos t + \ln t)} = e^{\cos t}\,e^{\ln t} = te^{\cos t} \Rightarrow \frac{dy}{dt} = e^{\cos t} + te^{\cos t}\frac{d}{dt}(\cos t) = (1 - t\sin t)\,e^{\cos t}$

35. $\int_0^{\ln x}\sin e^t\,dt \Rightarrow y' = \left(\sin e^{\ln x}\right)\cdot\frac{d}{dx}(\ln x) = \frac{\sin x}{x}$

37. $\ln y = e^y\sin x \Rightarrow \left(\frac{1}{y}\right)y' = (y'e^y)(\sin x) + e^y\cos x \Rightarrow y'\left(\frac{1}{y} - e^y\sin x\right) = e^y\cos x$

$\Rightarrow y'\left(\frac{1 - ye^y\sin x}{y}\right) = e^y\cos x \Rightarrow y' = \frac{ye^y\cos x}{1 - ye^y\sin x}$

39. $e^{2x} = \sin(x + 3y) \Rightarrow 2e^{2x} = (1 + 3y')\cos(x + 3y) \Rightarrow 1 + 3y' = \frac{2e^{2x}}{\cos(x+3y)} \Rightarrow 3y' = \frac{2e^{2x}}{\cos(x+3y)} - 1$

$\Rightarrow y' = \frac{2e^{2x} - \cos(x+3y)}{3\cos(x+3y)}$

41. $\int\left(e^{3x} + 5e^{-x}\right)dx = \frac{e^{3x}}{3} - 5e^{-x} + C$

43. $\int_{\ln 2}^{\ln 3}e^x\,dx = [e^x]_{\ln 2}^{\ln 3} = e^{\ln 3} - e^{\ln 2} = 3 - 2 = 1$

45. $\int 8e^{(x+1)}\,dx = 8e^{(x+1)} + C$

47. $\int_{\ln 4}^{\ln 9}e^{x/2}\,dx = \left[2e^{x/2}\right]_{\ln 4}^{\ln 9} = 2\left[e^{(\ln 9)/2} - e^{(\ln 4)/2}\right] = 2\left(e^{\ln 3} - e^{\ln 2}\right) = 2(3 - 2) = 2$

49. Let $u = r^{1/2} \Rightarrow du = \frac{1}{2}r^{-1/2}\,dr \Rightarrow 2\,du = r^{-1/2}\,dr;$

$\int\frac{e^{\sqrt{r}}}{\sqrt{r}}\,dr = \int e^{r^{1/2}}\cdot r^{-1/2}\,dr = 2\int e^u\,du = 2e^u + C = 2e^{r^{1/2}} + C = 2e^{\sqrt{r}} + C$

51. Let $u = -t^2 \Rightarrow du = -2t\,dt \Rightarrow -du = 2t\,dt;$

$\int 2te^{-t^2}\,dt = -\int e^u\,du = -e^u + C = -e^{-t^2} + C$

53. Let $u = \frac{1}{x} \Rightarrow du = -\frac{1}{x^2}\,dx \Rightarrow -du = \frac{1}{x^2}\,dx;$

$\int\frac{e^{1/x}}{x^2}\,dx = \int -e^u\,du = -e^u + C = -e^{1/x} + C$

55. Let $u = \tan\theta \Rightarrow du = \sec^2\theta\,d\theta; \theta = 0 \Rightarrow u = 0, \theta = \frac{\pi}{4} \Rightarrow u = 1;$

$\int_0^{\pi/4}\left(1 + e^{\tan\theta}\right)\sec^2\theta\,d\theta = \int_0^{\pi/4}\sec^2\theta\,d\theta + \int_0^1 e^u\,du = [\tan\theta]_0^{\pi/4} + [e^u]_0^1 = \left[\tan\left(\frac{\pi}{4}\right) - \tan(0)\right] + (e^1 - e^0)$

$= (1 - 0) + (e - 1) = e$

57. Let $u = \sec \pi t \Rightarrow du = \pi \sec \pi t \tan \pi t \, dt \Rightarrow \frac{du}{\pi} = \sec \pi t \tan \pi t \, dt$;

$\int e^{\sec(\pi t)} \sec(\pi t) \tan(\pi t) \, dt = \frac{1}{\pi} \int e^u \, du = \frac{e^u}{\pi} + C = \frac{e^{\sec(\pi t)}}{\pi} + C$

59. Let $u = e^v \Rightarrow du = e^v \, dv \Rightarrow 2 \, du = 2e^v \, dv; \; v = \ln \frac{\pi}{6} \Rightarrow u = \frac{\pi}{6}, \; v = \ln \frac{\pi}{2} \Rightarrow u = \frac{\pi}{2}$;

$\int_{\ln(\pi/6)}^{\ln(\pi/2)} 2e^v \cos e^v \, dv = 2 \int_{\pi/6}^{\pi/2} \cos u \, du = [2 \sin u]_{\pi/6}^{\pi/2} = 2 \left[\sin\left(\frac{\pi}{2}\right) - \sin\left(\frac{\pi}{6}\right) \right] = 2 \left(1 - \frac{1}{2} \right) = 1$

61. Let $u = 1 + e^r \Rightarrow du = e^r \, dr$;

$\int \frac{e^r}{1+e^r} \, dr = \int \frac{1}{u} \, du = \ln|u| + C = \ln(1 + e^r) + C$

63. $\frac{dy}{dt} = e^t \sin(e^t - 2) \Rightarrow y = \int e^t \sin(e^t - 2) \, dt$;

let $u = e^t - 2 \Rightarrow du = e^t \, dt \Rightarrow y = \int \sin u \, du = -\cos u + C = -\cos(e^t - 2) + C; \; y(\ln 2) = 0$

$\Rightarrow -\cos\left(e^{\ln 2} - 2\right) + C = 0 \Rightarrow -\cos(2 - 2) + C = 0 \Rightarrow C = \cos 0 = 1$; thus, $y = 1 - \cos(e^t - 2)$

65. $\frac{d^2y}{dx^2} = 2e^{-x} \Rightarrow \frac{dy}{dx} = -2e^{-x} + C; \; x = 0$ and $\frac{dy}{dx} = 0 \Rightarrow 0 = -2e^0 + C \Rightarrow C = 2$; thus $\frac{dy}{dx} = -2e^{-x} + 2$

$\Rightarrow y = 2e^{-x} + 2x + C_1; \; x = 0$ and $y = 1 \Rightarrow 1 = 2e^0 + C_1 \Rightarrow C_1 = -1 \Rightarrow y = 2e^{-x} + 2x - 1 = 2(e^{-x} + x) - 1$

67. $f(x) = e^x - 2x \Rightarrow f'(x) = e^x - 2; \; f'(x) = 0 \Rightarrow e^x = 2 \Rightarrow x = \ln 2; \; f(0) = 1$, the absolute maximum;

$f(\ln 2) = 2 - 2 \ln 2 \approx 0.613706$, the absolute minimum; $f(1) = e - 2 \approx 0.71828$, a relative or local maximum

since $f''(x) = e^x$ is always positive.

69. $f(x) = x^2 \ln \frac{1}{x} \Rightarrow f'(x) = 2x \ln \frac{1}{x} + x^2 \left(\frac{1}{\frac{1}{x}} \right) (-x^{-2}) = 2x \ln \frac{1}{x} - x = -x(2 \ln x + 1); \; f'(x) = 0 \Rightarrow x = 0$ or

$\ln x = -\frac{1}{2}$. Since $x = 0$ is not in the domain of f, $x = e^{-1/2} = \frac{1}{\sqrt{e}}$. Also, $f'(x) > 0$ for $0 < x < \frac{1}{\sqrt{e}}$ and

$f'(x) < 0$ for $x > \frac{1}{\sqrt{e}}$. Therefore, $f\left(\frac{1}{\sqrt{e}} \right) = \frac{1}{e} \ln \sqrt{e} = \frac{1}{e} \ln e^{1/2} = \frac{1}{2e} \ln e = \frac{1}{2e}$ is the absolute maximum value

of f assumed at $x = \frac{1}{\sqrt{e}}$.

71. $\int_0^{\ln 3} (e^{2x} - e^x) \, dx = \left[\frac{e^{2x}}{2} - e^x \right]_0^{\ln 3} = \left(\frac{e^{2 \ln 3}}{2} - e^{\ln 3} \right) - \left(\frac{e^0}{2} - e^0 \right) = \left(\frac{9}{2} - 3 \right) - \left(\frac{1}{2} - 1 \right) = \frac{8}{2} - 2 = 2$

73. $L = \int_0^1 \sqrt{1 + \frac{e^x}{4}} \, dx \Rightarrow \frac{dy}{dx} = \frac{e^{x/2}}{2} \Rightarrow y = e^{x/2} + C; \; y(0) = 0 \Rightarrow 0 = e^0 + C \Rightarrow C = -1 \Rightarrow y = e^{x/2} - 1$

75. (a) $\frac{d}{dx} (x \ln x - x + C) = x \cdot \frac{1}{x} + \ln x - 1 + 0 = \ln x$

(b) average value $= \frac{1}{e-1} \int_1^e \ln x \, dx = \frac{1}{e-1} [x \ln x - x]_1^e = \frac{1}{e-1} [(e \ln e - e) - (1 \ln 1 - 1)]$

$= \frac{1}{e-1} (e - e + 1) = \frac{1}{e-1}$

77. (a) $f(x) = e^x \Rightarrow f'(x) = e^x; \; L(x) = f(0) + f'(0)(x - 0) \Rightarrow L(x) = 1 + x$

(b) $f(0) = 1$ and $L(0) = 1 \Rightarrow$ error $= 0; \; f(0.2) = e^{0.2} \approx 1.22140$ and $L(0.2) = 1.2 \Rightarrow$ error ≈ 0.02140

(c) Since $y'' = e^x > 0$, the tangent line
approximation always lies below the curve $y = e^x$.
Thus $L(x) = x + 1$ never overestimates e^x.

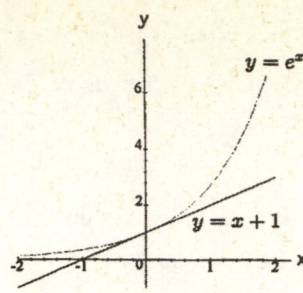

79. $f(x) = \ln(x) - 1 \Rightarrow f'(x) = \frac{1}{x} \Rightarrow x_{n+1} = x_n - \frac{\ln(x_n) - 1}{\left(\frac{1}{x_n}\right)} \Rightarrow x_{n+1} = x_n \left[2 - \ln(x_n)\right]$. Then $x_1 = 2$

$\Rightarrow x_2 = 2.61370564$, $x_3 = 2.71624393$ and $x_5 = 2.71828183$, where we have used Newton's method.

81. Note that $y = \ln x$ and $e^y = x$ are the same curve; $\int_1^a \ln x \, dx =$ area under the curve between 1 and a;

$\int_0^{\ln a} e^y \, dy =$ area to the left of the curve between 0 and ln a. The sum of these areas is equal to the area of the rectangle

$\Rightarrow \int_1^a \ln x \, dx + \int_0^{\ln a} e^y \, dy = a \ln a$.

7.4 a^x and $\log_a x$

1. (a) $5^{\log_5 7} = 7$ (b) $8^{\log_8 \sqrt{2}} = \sqrt{2}$ (c) $1.3^{\log_{1.3} 75} = 75$

   (d) $\log_4 16 = \log_4 4^2 = 2 \log_4 4 = 2 \cdot 1 = 2$ (e) $\log_3 \sqrt{3} = \log_3 3^{1/2} = \frac{1}{2} \log_3 3 = \frac{1}{2} \cdot 1 = \frac{1}{2} = 0.5$

   (f) $\log_4 \left(\frac{1}{4}\right) = \log_4 4^{-1} = -1 \log_4 4 = -1 \cdot 1 = -1$

3. (a) Let $z = \log_4 x \Rightarrow 4^z = x \Rightarrow 2^{2z} = x \Rightarrow (2^z)^2 = x \Rightarrow 2^z = \sqrt{x}$

   (b) Let $z = \log_3 x \Rightarrow 3^z = x \Rightarrow (3^z)^2 = x^2 \Rightarrow 3^{2z} = x^2 \Rightarrow 9^z = x^2$

   (c) $\log_2 \left(e^{(\ln 2) \sin x}\right) = \log_2 2^{\sin x} = \sin x$

5. (a) $\frac{\log_2 x}{\log_3 x} = \frac{\ln x}{\ln 2} \div \frac{\ln x}{\ln 3} = \frac{\ln x}{\ln 2} \cdot \frac{\ln 3}{\ln x} = \frac{\ln 3}{\ln 2}$ (b) $\frac{\log_2 x}{\log_8 x} = \frac{\ln x}{\ln 2} \div \frac{\ln x}{\ln 8} = \frac{\ln x}{\ln 2} \cdot \frac{\ln 8}{\ln x} = \frac{3 \ln 2}{\ln 2} = 3$

   (c) $\frac{\log_x a}{\log_{x^2} a} = \frac{\ln a}{\ln x} \div \frac{\ln a}{\ln x^2} = \frac{\ln a}{\ln x} \cdot \frac{\ln x^2}{\ln a} = \frac{2 \ln x}{\ln x} = 2$

7. $3^{\log_3 (7)} + 2^{\log_2 (5)} = 5^{\log_5 (x)} \Rightarrow 7 + 5 = x \Rightarrow x = 12$

9. $3^{\log_3 (x^2)} = 5e^{\ln x} - 3 \cdot 10^{\log_{10} (2)} \Rightarrow x^2 = 5x - 6 \Rightarrow x^2 - 5x + 6 = 0 \Rightarrow (x - 2)(x - 3) = 0 \Rightarrow x = 2$ or $x = 3$

11. $y = 2^x \Rightarrow y' = 2^x \ln 2$

13. $y = 5^{\sqrt{s}} \Rightarrow \frac{dy}{ds} = 5^{\sqrt{s}} (\ln 5) \left(\frac{1}{2} s^{-1/2}\right) = \left(\frac{\ln 5}{2\sqrt{s}}\right) 5^{\sqrt{s}}$

15. $y = x^\pi \Rightarrow y' = \pi x^{(\pi-1)}$

17. $y = (\cos \theta)^{\sqrt{2}} \Rightarrow \frac{dy}{d\theta} = -\sqrt{2} (\cos \theta)^{(\sqrt{2}-1)} (\sin \theta)$

19. $y = 7^{\sec \theta} \ln 7 \Rightarrow \frac{dy}{d\theta} = (7^{\sec \theta} \ln 7)(\ln 7)(\sec \theta \tan \theta) = 7^{\sec \theta} (\ln 7)^2 (\sec \theta \tan \theta)$

21. $y = 2^{\sin 3t} \Rightarrow \frac{dy}{dt} = (2^{\sin 3t} \ln 2)(\cos 3t)(3) = (3 \cos 3t) (2^{\sin 3t}) (\ln 2)$

23. $y = \log_2 5\theta = \frac{\ln 5\theta}{\ln 2} \Rightarrow \frac{dy}{d\theta} = \left(\frac{1}{\ln 2}\right)\left(\frac{1}{5\theta}\right)(5) = \frac{1}{\theta \ln 2}$

25. $y = \frac{\ln x}{\ln 4} + \frac{\ln x^2}{\ln 4} = \frac{\ln x}{\ln 4} + 2\frac{\ln x}{\ln 4} = 3\frac{\ln x}{\ln 4} \Rightarrow y' = \frac{3}{x \ln 4}$

27. $y = \log_2 r \cdot \log_4 r = \left(\frac{\ln r}{\ln 2}\right)\left(\frac{\ln r}{\ln 4}\right) = \frac{\ln^2 r}{(\ln 2)(\ln 4)} \Rightarrow \frac{dy}{dr} = \left[\frac{1}{(\ln 2)(\ln 4)}\right](2\ln r)\left(\frac{1}{r}\right) = \frac{2\ln r}{r(\ln 2)(\ln 4)}$

29. $y = \log_3\left(\left(\frac{x+1}{x-1}\right)^{\ln 3}\right) = \frac{\ln\left(\frac{x+1}{x-1}\right)^{\ln 3}}{\ln 3} = \frac{(\ln 3)\ln\left(\frac{x+1}{x-1}\right)}{\ln 3} = \ln\left(\frac{x+1}{x-1}\right) = \ln(x+1) - \ln(x-1)$

$\Rightarrow \frac{dy}{dx} = \frac{1}{x+1} - \frac{1}{x-1} = \frac{-2}{(x+1)(x-1)}$

31. $y = \theta \sin(\log_7 \theta) = \theta \sin\left(\frac{\ln \theta}{\ln 7}\right) \Rightarrow \frac{dy}{d\theta} = \sin\left(\frac{\ln \theta}{\ln 7}\right) + \theta\left[\cos\left(\frac{\ln \theta}{\ln 7}\right)\right]\left(\frac{1}{\theta \ln 7}\right) = \sin(\log_7 \theta) + \frac{1}{\ln 7}\cos(\log_7 \theta)$

33. $y = \log_5 e^x = \frac{\ln e^x}{\ln 5} = \frac{x}{\ln 5} \Rightarrow y' = \frac{1}{\ln 5}$

35. $y = 3^{\log_2 t} = 3^{(\ln t)/(\ln 2)} \Rightarrow \frac{dy}{dt} = [3^{(\ln t)/(\ln 2)}(\ln 3)]\left(\frac{1}{t \ln 2}\right) = \frac{1}{t}(\log_2 3)\, 3^{\log_2 t}$

37. $y = \log_2(8t^{\ln 2}) = \frac{\ln 8 + \ln(t^{\ln 2})}{\ln 2} = \frac{3\ln 2 + (\ln 2)(\ln t)}{\ln 2} = 3 + \ln t \Rightarrow \frac{dy}{dt} = \frac{1}{t}$

39. $y = (x+1)^x \Rightarrow \ln y = \ln(x+1)^x = x\ln(x+1) \Rightarrow \frac{y'}{y} = \ln(x+1) + x\cdot\frac{1}{(x+1)} \Rightarrow y' = (x+1)^x\left[\frac{x}{x+1} + \ln(x+1)\right]$

41. $y = \left(\sqrt{t}\right)^t = \left(t^{1/2}\right)^t = t^{t/2} \Rightarrow \ln y = \ln t^{t/2} = \left(\frac{t}{2}\right)\ln t \Rightarrow \frac{1}{y}\frac{dy}{dt} = \left(\frac{1}{2}\right)(\ln t) + \left(\frac{t}{2}\right)\left(\frac{1}{t}\right) = \frac{\ln t}{2} + \frac{1}{2}$

$\Rightarrow \frac{dy}{dt} = \left(\sqrt{t}\right)^t\left(\frac{\ln t}{2} + \frac{1}{2}\right)$

43. $y = (\sin x)^x \Rightarrow \ln y = \ln(\sin x)^x = x\ln(\sin x) \Rightarrow \frac{y'}{y} = \ln(\sin x) + x\left(\frac{\cos x}{\sin x}\right) \Rightarrow y' = (\sin x)^x[\ln(\sin x) + x\cot x]$

45. $y = x^{\ln x},\ x > 0 \Rightarrow \ln y = (\ln x)^2 \Rightarrow \frac{y'}{y} = 2(\ln x)\left(\frac{1}{x}\right) \Rightarrow y' = \left(x^{\ln x}\right)\left(\frac{\ln x^2}{x}\right)$

47. $\int 5^x\,dx = \frac{5^x}{\ln 5} + C$

49. $\int_0^1 2^{-\theta}\,d\theta = \int_0^1 \left(\frac{1}{2}\right)^\theta d\theta = \left[\frac{\left(\frac{1}{2}\right)^\theta}{\ln\left(\frac{1}{2}\right)}\right]_0^1 = \frac{\frac{1}{2}}{\ln\left(\frac{1}{2}\right)} - \frac{1}{\ln\left(\frac{1}{2}\right)} = -\frac{\frac{1}{2}}{\ln\left(\frac{1}{2}\right)} = \frac{-1}{2(\ln 1 - \ln 2)} = \frac{1}{2\ln 2}$

51. Let $u = x^2 \Rightarrow du = 2x\,dx \Rightarrow \frac{1}{2}du = x\,dx;\ x = 1 \Rightarrow u = 1,\ x = \sqrt{2} \Rightarrow u = 2;$

$\int_1^{\sqrt{2}} x2^{(x^2)}\,dx = \int_1^2 \left(\frac{1}{2}\right)2^u\,du = \frac{1}{2}\left[\frac{2^u}{\ln 2}\right]_1^2 = \left(\frac{1}{2\ln 2}\right)(2^2 - 2^1) = \frac{1}{\ln 2}$

53. Let $u = \cos t \Rightarrow du = -\sin t\,dt \Rightarrow -du = \sin t\,dt;\ t = 0 \Rightarrow u = 1,\ t = \frac{\pi}{2} \Rightarrow u = 0;$

$\int_0^{\pi/2} 7^{\cos t}\sin t\,dt = -\int_1^0 7^u\,du = \left[-\frac{7^u}{\ln 7}\right]_1^0 = \left(\frac{-1}{\ln 7}\right)(7^0 - 7) = \frac{6}{\ln 7}$

55. Let $u = x^{2x} \Rightarrow \ln u = 2x\ln x \Rightarrow \frac{1}{u}\frac{du}{dx} = 2\ln x + (2x)\left(\frac{1}{x}\right) \Rightarrow \frac{du}{dx} = 2u(\ln x + 1) \Rightarrow \frac{1}{2}du = x^{2x}(1 + \ln x)\,dx;$

$x = 2 \Rightarrow u = 2^4 = 16,\ x = 4 \Rightarrow u = 4^8 = 65{,}536;$

$\int_2^4 x^{2x}(1 + \ln x)\,dx = \frac{1}{2}\int_{16}^{65{,}536} du = \frac{1}{2}[u]_{16}^{65{,}536} = \frac{1}{2}(65{,}536 - 16) = \frac{65{,}520}{2} = 32{,}760$

57. $\int 3x^{\sqrt{3}}\, dx = \frac{3x^{(\sqrt{3}+1)}}{\sqrt{3}+1} + C$

59. $\int_0^3 \left(\sqrt{2}+1\right) x^{\sqrt{2}}\, dx = \left[x^{(\sqrt{2}+1)} \right]_0^3 = 3^{(\sqrt{2}+1)}$

61. $\int \frac{\log_{10} x}{x}\, dx = \int \left(\frac{\ln x}{\ln 10}\right)\left(\frac{1}{x}\right) dx; \left[u = \ln x \;\Rightarrow\; du = \frac{1}{x}\, dx \right]$
    $\rightarrow \int \left(\frac{\ln x}{\ln 10}\right)\left(\frac{1}{x}\right) dx = \frac{1}{\ln 10} \int u\, du = \left(\frac{1}{\ln 10}\right)\left(\frac{1}{2} u^2\right) + C = \frac{(\ln x)^2}{2 \ln 10} + C$

63. $\int_1^4 \frac{\ln 2 \log_2 x}{x}\, dx = \int_1^4 \left(\frac{\ln 2}{x}\right)\left(\frac{\ln x}{\ln 2}\right) dx = \int_1^4 \frac{\ln x}{x}\, dx = \left[\frac{1}{2} (\ln x)^2 \right]_1^4 = \frac{1}{2} \left[(\ln 4)^2 - (\ln 1)^2 \right] = \frac{1}{2} (\ln 4)^2$
    $= \frac{1}{2} (2 \ln 2)^2 = 2(\ln 2)^2$

65. $\int_0^2 \frac{\log_2 (x+2)}{x+2}\, dx = \frac{1}{\ln 2} \int_0^2 \left[\ln (x+2) \right] \left(\frac{1}{x+2}\right) dx = \left(\frac{1}{\ln 2}\right) \left[\frac{(\ln (x+2))^2}{2} \right]_0^2 = \left(\frac{1}{\ln 2}\right) \left[\frac{(\ln 4)^2}{2} - \frac{(\ln 2)^2}{2} \right]$
    $= \left(\frac{1}{\ln 2}\right) \left[\frac{4(\ln 2)^2}{2} - \frac{(\ln 2)^2}{2} \right] = \frac{3}{2} \ln 2$

67. $\int_0^9 \frac{2 \log_{10} (x+1)}{x+1}\, dx = \frac{2}{\ln 10} \int_0^9 \ln (x+1) \left(\frac{1}{x+1}\right) dx = \left(\frac{2}{\ln 10}\right) \left[\frac{(\ln (x+1))^2}{2} \right]_0^9 = \left(\frac{2}{\ln 10}\right) \left[\frac{(\ln 10)^2}{2} - \frac{(\ln 1)^2}{2} \right]$
    $= \ln 10$

69. $\int \frac{dx}{x \log_{10} x} = \int \left(\frac{\ln 10}{\ln x}\right)\left(\frac{1}{x}\right) dx = (\ln 10) \int \left(\frac{1}{\ln x}\right)\left(\frac{1}{x}\right) dx; \left[u = \ln x \;\Rightarrow\; du = \frac{1}{x}\, dx \right]$
    $\rightarrow (\ln 10) \int \left(\frac{1}{\ln x}\right)\left(\frac{1}{x}\right) dx = (\ln 10) \int \frac{1}{u}\, du = (\ln 10) \ln |u| + C = (\ln 10) \ln |\ln x| + C$

71. $\int_1^{\ln x} \frac{1}{t}\, dt = \left[\ln |t| \right]_1^{\ln x} = \ln |\ln x| - \ln 1 = \ln (\ln x), \; x > 1$

73. $\int_1^{1/x} \frac{1}{t}\, dt = \left[\ln |t| \right]_1^{1/x} = \ln \left| \frac{1}{x} \right| - \ln 1 = (\ln 1 - \ln |x|) - \ln 1 = -\ln x, \; x > 0$

75. $A = \int_{-2}^2 \frac{2x}{1+x^2}\, dx = 2\int_0^2 \frac{2x}{1+x^2}\, dx; \left[u = 1 + x^2 \;\Rightarrow\; du = 2x\, dx; x = 0 \;\Rightarrow\; u = 1, x = 2 \;\Rightarrow\; u = 5 \right]$
    $\rightarrow A = 2\int_1^5 \frac{1}{u}\, du = 2 \left[\ln |u| \right]_1^5 = 2(\ln 5 - \ln 1) = 2 \ln 5$

77. Let $[H_3O^+] = x$ and solve the equations $7.37 = -\log_{10} x$ and $7.44 = -\log_{10} x$. The solutions of these equations are $10^{-7.37}$ and $10^{-7.44}$. Consequently, the bounds for $[H_3O^+]$ are $[10^{-7.44}, 10^{-7.37}]$.

79. Let O = original sound level = $10 \log_{10} (I \times 10^{12})$ db from Equation (6) in the text. Solving
    $O + 10 = 10 \log_{10} (kI \times 10^{12})$ for $k \;\Rightarrow\; 10 \log_{10} (I \times 10^{12}) + 10 = 10 \log_{10} (kI \times 10^{12}) \;\Rightarrow\; \log_{10} (I \times 10^{12}) + 1$
    $= \log_{10} (kI \times 10^{12}) \;\Rightarrow\; \log_{10} (I \times 10^{12}) + 1 = \log_{10} k + \log_{10} (I \times 10^{12}) \;\Rightarrow\; 1 = \log_{10} k \;\Rightarrow\; 1 = \frac{\ln k}{\ln 10}$
    $\Rightarrow \ln k = \ln 10 \;\Rightarrow\; k = 10$

81. (a) If $x = [H_3O^+]$ and $S - x = [OH^-]$, then $x(S - x) = 10^{-14} \;\Rightarrow\; S = x + \frac{10^{-14}}{x} \;\Rightarrow\; \frac{dS}{dx} = 1 - \frac{10^{-14}}{x^2}$
    and $\frac{d^2S}{dx^2} = \frac{2 \cdot 10^{-14}}{x^3} > 0 \;\Rightarrow\;$ a minimum exists at $x = 10^{-7}$
    (b) $pH = -\log_{10} (10^{-7}) = 7$
    (c) $\frac{[OH^-]}{[H_3O^+]} = \frac{S-x}{x} = \frac{\left(x + \frac{10^{-14}}{x} \right) - x}{x} = \frac{10^{-14}}{x^2} \;\Rightarrow\;$ the ratio $\frac{[OH^-]}{[H_3O^+]}$ equals 1 at $x = 10^{-7}$

83. From zooming in on the graph at the right, we estimate
the third root to be x ≈ −0.76666

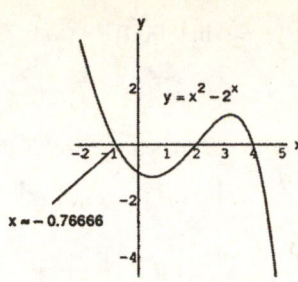

85. (a) $f(x) = 2^x \Rightarrow f'(x) = 2^x \ln 2; L(x) = (2^0 \ln 2) x + 2^0 = x \ln 2 + 1 \approx 0.69x + 1$
(b)

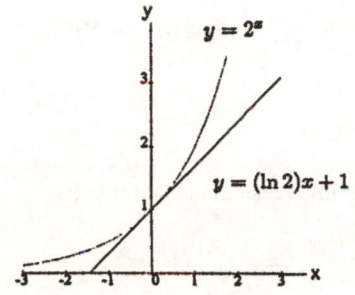

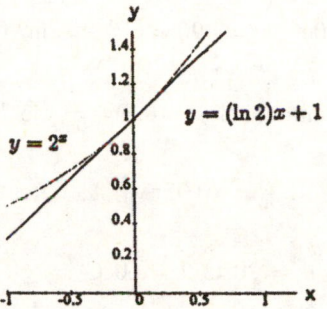

87. (a) $\log_3 8 = \frac{\ln 8}{\ln 3} \approx 1.89279$
(b) $\log_7 0.5 = \frac{\ln 0.5}{\ln 7} \approx -0.35621$
(c) $\log_{20} 17 = \frac{\ln 17}{\ln 20} \approx 0.94575$
(d) $\log_{0.5} 7 = \frac{\ln 7}{\ln 0.5} \approx -2.80735$
(e) $\ln x = (\log_{10} x)(\ln 10) = 2.3 \ln 10 \approx 5.29595$
(f) $\ln x = (\log_2 x)(\ln 2) = 1.4 \ln 2 \approx 0.97041$
(g) $\ln x = (\log_2 x)(\ln 2) = -1.5 \ln 2 \approx -1.03972$
(h) $\ln x = (\log_{10} x)(\ln 10) = -0.7 \ln 10 \approx -1.61181$

89. $\frac{d}{dx}\left(-\frac{1}{2}x^2 + k\right) = -x$ and $\frac{d}{dx}(\ln x + c) = \frac{1}{x}$.
Since $-x \cdot \frac{1}{x} = -1$ for any $x \neq 0$, these two curves will have perpendicular tangent lines.

91. Using Newton's Method: $f(x) = \ln(x) - 1 \Rightarrow f'(x) = \frac{1}{x} \Rightarrow x_{n+1} = x_n - \frac{\ln(x_n)-1}{\frac{1}{x_n}} \Rightarrow x_{n+1} = x_n\left[2 - \ln(x_n)\right]$.
Then, $x_1 = 2$, $x_2 = 2.61370564$, $x_3 = 2.71624393$, and $x_5 = 2.71828183$. Many other methods may be used. For example,
graph $y = \ln x - 1$ and determine the zero of y.

7.5 EXPONENTIAL GROWTH AND DECAY

1. (a) $y = y_0 e^{kt} \Rightarrow 0.99y_0 = y_0 e^{1000k} \Rightarrow k = \frac{\ln 0.99}{1000} \approx -0.00001$
(b) $0.9 = e^{(-0.00001)t} \Rightarrow (-0.00001)t = \ln(0.9) \Rightarrow t = \frac{\ln(0.9)}{-0.00001} \approx 10,536$ years
(c) $y = y_0 e^{(20,000)k} \approx y_0 e^{-0.2} = y_0(0.82) \Rightarrow 82\%$

3. $\frac{dy}{dt} = -0.6y \Rightarrow y = y_0 e^{-0.6t}; y_0 = 100 \Rightarrow y = 100e^{-0.6t} \Rightarrow y = 100e^{-0.6} \approx 54.88$ grams when $t = 1$ hr

5. $L(x) = L_0 e^{-kx} \Rightarrow \frac{L_0}{2} = L_0 e^{-18k} \Rightarrow \ln \frac{1}{2} = -18k \Rightarrow k = \frac{\ln 2}{18} \approx 0.0385 \Rightarrow L(x) = L_0 e^{-0.0385x}$; when the intensity
is one-tenth of the surface value, $\frac{L_0}{10} = L_0 e^{-0.0385x} \Rightarrow \ln 10 = 0.0385x \Rightarrow x \approx 59.8$ ft

7. $y = y_0 e^{kt}$ and $y_0 = 1 \Rightarrow y = e^{kt} \Rightarrow$ at $y = 2$ and $t = 0.5$ we have $2 = e^{0.5k} \Rightarrow \ln 2 = 0.5k \Rightarrow k = \frac{\ln 2}{0.5} = \ln 4$.
Therefore, $y = e^{(\ln 4)t} \Rightarrow y = e^{24 \ln 4} = 4^{24} = 2.81474978 \times 10^{14}$ at the end of 24 hrs

9. (a) $10,000e^{k(1)} = 7500 \Rightarrow e^k = 0.75 \Rightarrow k = \ln 0.75$ and $y = 10,000e^{(\ln 0.75)t}$. Now $1000 = 10,000e^{(\ln 0.75)t}$
$\Rightarrow \ln 0.1 = (\ln 0.75)t \Rightarrow t = \frac{\ln 0.1}{\ln 0.75} \approx 8.00$ years (to the nearest hundredth of a year)

(b) $1 = 10{,}000e^{(\ln 0.75)t} \Rightarrow \ln 0.0001 = (\ln 0.75)t \Rightarrow t = \frac{\ln 0.0001}{\ln 0.75} \approx 32.02$ years (to the nearest hundredth of a
   year)

11. $0.9P_0 = P_0 e^k \Rightarrow k = \ln 0.9$; when the well's output falls to one-fifth of its present value $P = 0.2P_0$
   $\Rightarrow 0.2P_0 = P_0 e^{(\ln 0.9)t} \Rightarrow 0.2 = e^{(\ln 0.9)t} \Rightarrow \ln(0.2) = (\ln 0.9)t \Rightarrow t = \frac{\ln 0.2}{\ln 0.9} \approx 15.28$ yr

13. (a) $A_0 e^{(0.04)5} = A_0 e^{0.2}$
   (b) $2A_0 = A_0 e^{(0.04)t} \Rightarrow \ln 2 = (0.04)t \Rightarrow t = \frac{\ln 2}{0.04} \approx 17.33$ years; $3A_0 = A_0 e^{(0.04)t} \Rightarrow \ln 3 = (0.04)t$
   $\Rightarrow t = \frac{\ln 3}{0.04} \approx 27.47$ years

15. $A(100) = 90{,}000 \Rightarrow 90{,}000 = 1000e^{r(100)} \Rightarrow 90 = e^{100r} \Rightarrow \ln 90 = 100r \Rightarrow r = \frac{\ln 90}{100} \approx 0.0450$ or 4.50%

17. $y = y_0 e^{-0.18t}$ represents the decay equation; solving $(0.9)y_0 = y_0 e^{-0.18t} \Rightarrow t = \frac{\ln(0.9)}{-0.18} \approx 0.585$ days

19. $y = y_0 e^{-kt} = y_0 e^{-(k)(3/k)} = y_0 e^{-3} = \frac{y_0}{e^3} < \frac{y_0}{20} = (0.05)(y_0) \Rightarrow$ after three mean lifetimes less than 5% remains

21. $T - T_s = (T_0 - T_s)e^{-kt}$, $T_0 = 90°C$, $T_s = 20°C$, $T = 60°C \Rightarrow 60 - 20 = 70e^{-10k} \Rightarrow \frac{4}{7} = e^{-10k}$
   $\Rightarrow k = \frac{\ln\left(\frac{7}{4}\right)}{10} \approx 0.05596$
   (a) $35 - 20 = 70e^{-0.05596t} \Rightarrow t \approx 27.5$ min is the total time $\Rightarrow$ it will take $27.5 - 10 = 17.5$ minutes longer to reach
       35°C
   (b) $T - T_s = (T_0 - T_s)e^{-kt}$, $T_0 = 90°C$, $T_s = -15°C \Rightarrow 35 + 15 = 105e^{-0.05596t} \Rightarrow t \approx 13.26$ min

23. $T - T_s = (T_0 - T_s)e^{-kt} \Rightarrow 39 - T_s = (46 - T_s)e^{-10k}$ and $33 - T_s = (46 - T_s)e^{-20k} \Rightarrow \frac{39 - T_s}{46 - T_s} = e^{-10k}$ and
   $\frac{33 - T_s}{46 - T_s} = e^{-20k} = (e^{-10k})^2 \Rightarrow \frac{33 - T_s}{46 - T_s} = \left(\frac{39 - T_s}{46 - T_s}\right)^2 \Rightarrow (33 - T_s)(46 - T_s) = (39 - T_s)^2 \Rightarrow 1518 - 79T_s + T_s^2$
   $= 1521 - 78T_s + T_s^2 \Rightarrow -T_s = 3 \Rightarrow T_s = -3°C$

25. From Example 5, the half-life of carbon-14 is 5700 yr $\Rightarrow \frac{1}{2}c_0 = c_0 e^{-k(5700)} \Rightarrow k = \frac{\ln 2}{5700} \approx 0.0001216$
   $\Rightarrow c = c_0 e^{-0.0001216t} \Rightarrow (0.445)c_0 = c_0 e^{-0.0001216t} \Rightarrow t = \frac{\ln(0.445)}{-0.0001216} \approx 6659$ years

27. From Exercise 25, $k \approx 0.0001216$ for carbon-14. Thus, $c = c_0 e^{-0.0001216t} \Rightarrow (0.995)c_0 = c_0 e^{-0.0001216t}$
   $\Rightarrow t = \frac{\ln(0.995)}{-0.0001216} \approx 41$ years old

7.6 RELATIVE RATES OF GROWTH

1. (a) slower, $\lim\limits_{x \to \infty} \frac{x+3}{e^x} = \lim\limits_{x \to \infty} \frac{1}{e^x} = 0$
   (b) slower, $\lim\limits_{x \to \infty} \frac{x^3 + \sin^2 x}{e^x} = \lim\limits_{x \to \infty} \frac{3x^2 + 2\sin x \cos x}{e^x} = \lim\limits_{x \to \infty} \frac{6x + 2\cos 2x}{e^x} = \lim\limits_{x \to \infty} \frac{6 - 4\sin 2x}{e^x} = 0$ by the
       Sandwich Theorem because $\frac{2}{e^x} \le \frac{6 - 4\sin 2x}{e^x} \le \frac{10}{e^x}$ for all reals and $\lim\limits_{x \to \infty} \frac{2}{e^x} = 0 = \lim\limits_{x \to \infty} \frac{10}{e^x}$
   (c) slower, $\lim\limits_{x \to \infty} \frac{\sqrt{x}}{e^x} = \lim\limits_{x \to \infty} \frac{x^{1/2}}{e^x} = \lim\limits_{x \to \infty} \frac{\left(\frac{1}{2}\right)x^{-1/2}}{e^x} = \lim\limits_{x \to \infty} \frac{1}{2\sqrt{x}\,e^x} = 0$
   (d) faster, $\lim\limits_{x \to \infty} \frac{4^x}{e^x} = \lim\limits_{x \to \infty} \left(\frac{4}{e}\right)^x = \infty$ since $\frac{4}{e} > 1$
   (e) slower, $\lim\limits_{x \to \infty} \frac{\left(\frac{3}{2}\right)^x}{e^x} = \lim\limits_{x \to \infty} \left(\frac{3}{2e}\right)^x = 0$ since $\frac{3}{2e} < 1$
   (f) slower, $\lim\limits_{x \to \infty} \frac{e^{x/2}}{e^x} = \lim\limits_{x \to \infty} \frac{1}{e^{x/2}} = 0$
   (g) same, $\lim\limits_{x \to \infty} \frac{\left(\frac{e^x}{2}\right)}{e^x} = \lim\limits_{x \to \infty} \frac{1}{2} = \frac{1}{2}$

(h) slower, $\displaystyle\lim_{x \to \infty} \frac{\log_{10} x}{e^x} = \lim_{x \to \infty} \frac{\ln x}{(\ln 10)\, e^x} = \lim_{x \to \infty} \frac{\frac{1}{x}}{(\ln 10)\, e^x} = \lim_{x \to \infty} \frac{1}{(\ln 10) x e^x} = 0$

3. (a) same, $\displaystyle\lim_{x \to \infty} \frac{x^2 + 4x}{x^2} = \lim_{x \to \infty} \frac{2x + 4}{2x} = \lim_{x \to \infty} \frac{2}{2} = 1$

(b) faster, $\displaystyle\lim_{x \to \infty} \frac{x^5 - x^2}{x^2} = \lim_{x \to \infty} (x^3 - 1) = \infty$

(c) same, $\displaystyle\lim_{x \to \infty} \frac{\sqrt{x^4 + x^3}}{x^2} = \sqrt{\lim_{x \to \infty} \frac{x^4 + x^3}{x^4}} = \sqrt{\lim_{x \to \infty} \left(1 + \frac{1}{x}\right)} = \sqrt{1} = 1$

(d) same, $\displaystyle\lim_{x \to \infty} \frac{(x+3)^2}{x^2} = \lim_{x \to \infty} \frac{2(x+3)}{2x} = \lim_{x \to \infty} \frac{2}{2} = 1$

(e) slower, $\displaystyle\lim_{x \to \infty} \frac{x \ln x}{x^2} = \lim_{x \to \infty} \frac{\ln x}{x} = \lim_{x \to \infty} \frac{\left(\frac{1}{x}\right)}{1} = 0$

(f) faster, $\displaystyle\lim_{x \to \infty} \frac{2^x}{x^2} = \lim_{x \to \infty} \frac{(\ln 2)\, 2^x}{2x} = \lim_{x \to \infty} \frac{(\ln 2)^2 \, 2^x}{2} = \infty$

(g) slower, $\displaystyle\lim_{x \to \infty} \frac{x^3 e^{-x}}{x^2} = \lim_{x \to \infty} \frac{x}{e^x} = \lim_{x \to \infty} \frac{1}{e^x} = 0$

(h) same, $\displaystyle\lim_{x \to \infty} \frac{8x^2}{x^2} = \lim_{x \to \infty} 8 = 8$

5. (a) same, $\displaystyle\lim_{x \to \infty} \frac{\log_3 x}{\ln x} = \lim_{x \to \infty} \frac{\left(\frac{\ln x}{\ln 3}\right)}{\ln x} = \lim_{x \to \infty} \frac{1}{\ln 3} = \frac{1}{\ln 3}$

(b) same, $\displaystyle\lim_{x \to \infty} \frac{\ln 2x}{\ln x} = \lim_{x \to \infty} \frac{\left(\frac{2}{2x}\right)}{\left(\frac{1}{x}\right)} = 1$

(c) same, $\displaystyle\lim_{x \to \infty} \frac{\ln \sqrt{x}}{\ln x} = \lim_{x \to \infty} \frac{\left(\frac{1}{2}\right)\ln x}{\ln x} = \lim_{x \to \infty} \frac{1}{2} = \frac{1}{2}$

(d) faster, $\displaystyle\lim_{x \to \infty} \frac{\sqrt{x}}{\ln x} = \lim_{x \to \infty} \frac{x^{1/2}}{\ln x} = \lim_{x \to \infty} \frac{\left(\frac{1}{2}\right)x^{-1/2}}{\left(\frac{1}{x}\right)} = \lim_{x \to \infty} \frac{x}{2\sqrt{x}} = \lim_{x \to \infty} \frac{\sqrt{x}}{2} = \infty$

(e) faster, $\displaystyle\lim_{x \to \infty} \frac{x}{\ln x} = \lim_{x \to \infty} \frac{1}{\left(\frac{1}{x}\right)} = \lim_{x \to \infty} x = \infty$

(f) same, $\displaystyle\lim_{x \to \infty} \frac{5 \ln x}{\ln x} = \lim_{x \to \infty} 5 = 5$

(g) slower, $\displaystyle\lim_{x \to \infty} \frac{\left(\frac{1}{x}\right)}{\ln x} = \lim_{x \to \infty} \frac{1}{x \ln x} = 0$

(h) faster, $\displaystyle\lim_{x \to \infty} \frac{e^x}{\ln x} = \lim_{x \to \infty} \frac{e^x}{\left(\frac{1}{x}\right)} = \lim_{x \to \infty} x e^x = \infty$

7. $\displaystyle\lim_{x \to \infty} \frac{e^x}{e^{x/2}} = \lim_{x \to \infty} e^{x/2} = \infty \Rightarrow e^x$ grows faster than $e^{x/2}$; since for $x > e^e$ we have $\ln x > e$ and $\displaystyle\lim_{x \to \infty} \frac{(\ln x)^x}{e^x}$
$= \displaystyle\lim_{x \to \infty} \left(\frac{\ln x}{e}\right)^x = \infty \Rightarrow (\ln x)^x$ grows faster than e^x; since $x > \ln x$ for all $x > 0$ and $\displaystyle\lim_{x \to \infty} \frac{x^x}{(\ln x)^x} = \lim_{x \to \infty} \left(\frac{x}{\ln x}\right)^x$
$= \infty \Rightarrow x^x$ grows faster than $(\ln x)^x$. Therefore, slowest to fastest are: $e^{x/2}, e^x, (\ln x)^x, x^x$ so the order is d, a, c, b

9. (a) false; $\displaystyle\lim_{x \to \infty} \frac{x}{x} = 1$

(b) false; $\displaystyle\lim_{x \to \infty} \frac{x}{x+5} = \frac{1}{1} = 1$

(c) true; $x < x + 5 \Rightarrow \frac{x}{x+5} < 1$ if $x > 1$ (or sufficiently large)

(d) true; $x < 2x \Rightarrow \frac{x}{2x} < 1$ if $x > 1$ (or sufficiently large)

(e) true; $\displaystyle\lim_{x \to \infty} \frac{e^x}{e^{2x}} = \lim_{x \to 0} \frac{1}{e^x} = 0$

(f) true; $\frac{x + \ln x}{x} = 1 + \frac{\ln x}{x} < 1 + \frac{\sqrt{x}}{x} = 1 + \frac{1}{\sqrt{x}} < 2$ if $x > 1$ (or sufficiently large)

(g) false; $\displaystyle\lim_{x \to \infty} \frac{\ln x}{\ln 2x} = \lim_{x \to \infty} \frac{\left(\frac{1}{x}\right)}{\left(\frac{2}{2x}\right)} = \lim_{x \to \infty} 1 = 1$

(h) true; $\frac{\sqrt{x^2 + 5}}{x} < \frac{\sqrt{(x+5)^2}}{x} < \frac{x+5}{x} = 1 + \frac{5}{x} < 6$ if $x > 1$ (or sufficiently large)

11. If $f(x)$ and $g(x)$ grow at the same rate, then $\displaystyle\lim_{x \to \infty} \frac{f(x)}{g(x)} = L \neq 0 \Rightarrow \lim_{x \to \infty} \frac{g(x)}{f(x)} = \frac{1}{L} \neq 0$. Then
$\left| \frac{f(x)}{g(x)} - L \right| < 1$ if x is sufficiently large $\Rightarrow L - 1 < \frac{f(x)}{g(x)} < L + 1 \Rightarrow \frac{f(x)}{g(x)} \leq |L| + 1$ if x is sufficiently large

$\Rightarrow f = O(g)$. Similarly, $\frac{g(x)}{f(x)} \le \left|\frac{1}{L}\right| + 1 \Rightarrow g = O(f)$.

13. When the degree of f is less than or equal to the degree of g since $\lim\limits_{x \to \infty} \frac{f(x)}{g(x)} = 0$ when the degree of f is smaller than the degree of g, and $\lim\limits_{x \to \infty} \frac{f(x)}{g(x)} = \frac{a}{b}$ (the ratio of the leading coefficients) when the degrees are the same.

15. $\lim\limits_{x \to \infty} \frac{\ln(x+1)}{\ln x} = \lim\limits_{x \to \infty} \frac{\left(\frac{1}{x+1}\right)}{\left(\frac{1}{x}\right)} = \lim\limits_{x \to \infty} \frac{x}{x+1} = \lim\limits_{x \to \infty} \frac{1}{1} = 1$ and $\lim\limits_{x \to \infty} \frac{\ln(x+999)}{\ln x} = \lim\limits_{x \to \infty} \frac{\left(\frac{1}{x+999}\right)}{\left(\frac{1}{x}\right)}$

$= \lim\limits_{x \to \infty} \frac{x}{x+999} = 1$

17. $\lim\limits_{x \to \infty} \frac{\sqrt{10x+1}}{\sqrt{x}} = \sqrt{\lim\limits_{x \to \infty} \frac{10x+1}{x}} = \sqrt{10}$ and $\lim\limits_{x \to \infty} \frac{\sqrt{x+1}}{\sqrt{x}} = \sqrt{\lim\limits_{x \to \infty} \frac{x+1}{x}} = \sqrt{1} = 1$. Since the growth rate is transitive, we conclude that $\sqrt{10x+1}$ and $\sqrt{x+1}$ have the same growth rate $\left(\text{that of } \sqrt{x}\right)$.

19. $\lim\limits_{x \to \infty} \frac{x^n}{e^x} = \lim\limits_{x \to \infty} \frac{nx^{n-1}}{e^x} = \ldots = \lim\limits_{x \to \infty} \frac{n!}{e^x} = 0 \Rightarrow x^n = o\left(e^x\right)$ for any non-negative integer n

21. (a) $\lim\limits_{x \to \infty} \frac{x^{1/n}}{\ln x} = \lim\limits_{x \to \infty} \frac{x^{(1-n)/n}}{n\left(\frac{1}{x}\right)} = \left(\frac{1}{n}\right) \lim\limits_{x \to \infty} x^{1/n} = \infty \Rightarrow \ln x = o\left(x^{1/n}\right)$ for any positive integer n

(b) $\ln\left(e^{17,000,000}\right) = 17,000,000 < \left(e^{17 \times 10^6}\right)^{1/10^6} = e^{17} \approx 24,154,952.75$

(c) $x \approx 3.430631121 \times 10^{15}$

(d) In the interval $[3.41 \times 10^{15}, 3.45 \times 10^{15}]$ we have $\ln x = 10 \ln(\ln x)$. The graphs cross at about 3.4306311×10^{15}.

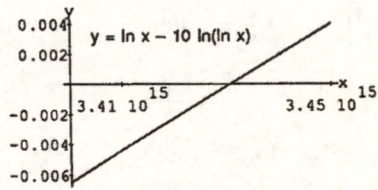

23. (a) $\lim\limits_{n \to \infty} \frac{n \log_2 n}{n(\log_2 n)^2} = \lim\limits_{n \to \infty} \frac{1}{\log_2 n} = 0 \Rightarrow n \log_2 n$ grows slower than $n(\log_2 n)^2$; $\lim\limits_{n \to \infty} \frac{n \log_2 n}{n^{3/2}} = \lim\limits_{n \to \infty} \frac{\left(\frac{\ln n}{\ln 2}\right)}{n^{1/2}}$

$= \frac{1}{\ln 2} \lim\limits_{n \to \infty} \frac{\left(\frac{1}{n}\right)}{\left(\frac{1}{2}\right)n^{-1/2}} = \frac{2}{\ln 2} \lim\limits_{n \to \infty} \frac{1}{n^{1/2}} = 0$

$\Rightarrow n \log_2 n$ grows slower than $n^{3/2}$. Therefore, $n \log_2 n$ grows at the slowest rate $\Rightarrow$ the algorithm that takes $O(n \log_2 n)$ steps is the most efficient in the long run.

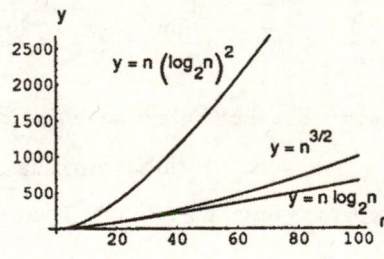

25. It could take one million steps for a sequential search, but at most 20 steps for a binary search because $2^{19} = 524,288 < 1,000,000 < 1,048,576 = 2^{20}$.

7.7 INVERSE TRIGONOMETRIC FUNCTIONS

1. (a) $\frac{\pi}{4}$ (b) $-\frac{\pi}{3}$ (c) $\frac{\pi}{6}$

3. (a) $-\frac{\pi}{6}$ (b) $\frac{\pi}{4}$ (c) $-\frac{\pi}{3}$

5. (a) $\frac{\pi}{3}$ (b) $\frac{3\pi}{4}$ (c) $\frac{\pi}{6}$

7. (a) $\frac{3\pi}{4}$ (b) $\frac{\pi}{6}$ (c) $\frac{2\pi}{3}$

9. (a) $\frac{\pi}{4}$ (b) $-\frac{\pi}{3}$ (c) $\frac{\pi}{6}$

11. (a) $\frac{3\pi}{4}$ (b) $\frac{\pi}{6}$ (c) $\frac{2\pi}{3}$

13. $\alpha = \sin^{-1}\left(\frac{5}{13}\right) \Rightarrow \cos\alpha = \frac{12}{13}, \tan\alpha = \frac{5}{12}, \sec\alpha = \frac{13}{12}, \csc\alpha = \frac{13}{5},$ and $\cot\alpha = \frac{12}{5}$

15. $\alpha = \sec^{-1}\left(-\sqrt{5}\right) \Rightarrow \sin\alpha = \frac{2}{\sqrt{5}}, \cos\alpha = -\frac{1}{\sqrt{5}}, \tan\alpha = -2, \csc\alpha = \frac{\sqrt{5}}{2},$ and $\cot\alpha = -\frac{1}{2}$

17. $\sin\left(\cos^{-1}\frac{\sqrt{2}}{2}\right) = \sin\left(\frac{\pi}{4}\right) = \frac{1}{\sqrt{2}}$

19. $\tan\left(\sin^{-1}\left(-\frac{1}{2}\right)\right) = \tan\left(-\frac{\pi}{6}\right) = -\frac{1}{\sqrt{3}}$

21. $\csc\left(\sec^{-1}2\right) + \cos\left(\tan^{-1}\left(-\sqrt{3}\right)\right) = \csc\left(\cos^{-1}\left(\frac{1}{2}\right)\right) + \cos\left(-\frac{\pi}{3}\right) = \csc\left(\frac{\pi}{3}\right) + \cos\left(-\frac{\pi}{3}\right) = \frac{2}{\sqrt{3}} + \frac{1}{2} = \frac{4+\sqrt{3}}{2\sqrt{3}}$

23. $\sin\left(\sin^{-1}\left(-\frac{1}{2}\right) + \cos^{-1}\left(-\frac{1}{2}\right)\right) = \sin\left(-\frac{\pi}{6} + \frac{2\pi}{3}\right) = \sin\left(\frac{\pi}{2}\right) = 1$

25. $\sec\left(\tan^{-1}1 + \csc^{-1}1\right) = \sec\left(\frac{\pi}{4} + \sin^{-1}\frac{1}{1}\right) = \sec\left(\frac{\pi}{4} + \frac{\pi}{2}\right) = \sec\left(\frac{3\pi}{4}\right) = -\sqrt{2}$

27. $\sec^{-1}\left(\sec\left(-\frac{\pi}{6}\right)\right) = \sec^{-1}\left(\frac{2}{\sqrt{3}}\right) = \cos^{-1}\left(\frac{\sqrt{3}}{2}\right) = \frac{\pi}{6}$

29. $\alpha = \tan^{-1}\frac{x}{2}$ indicates the diagram 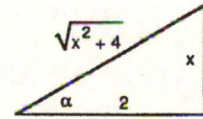 $\Rightarrow \sec\left(\tan^{-1}\frac{x}{2}\right) = \sec\alpha = \frac{\sqrt{x^2+4}}{2}$

31. $\alpha = \sec^{-1}3y$ indicates the diagram 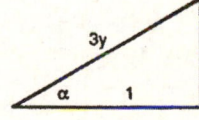 $\Rightarrow \tan\left(\sec^{-1}3y\right) = \tan\alpha = \sqrt{9y^2 - 1}$

33. $\alpha = \sin^{-1}x$ indicates the diagram 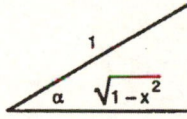 $\Rightarrow \cos\left(\sin^{-1}x\right) = \cos\alpha = \sqrt{1 - x^2}$

35. $\alpha = \tan^{-1}\sqrt{x^2 - 2x}$ indicates the diagram 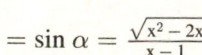 $\Rightarrow \sin\left(\tan^{-1}\sqrt{x^2 - 2x}\right)$

$= \sin\alpha = \frac{\sqrt{x^2 - 2x}}{x - 1}$

37. $\alpha = \sin^{-1}\frac{2y}{3}$ indicates the diagram 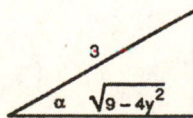 $\Rightarrow \cos\left(\sin^{-1}\frac{2y}{3}\right) = \cos\alpha = \frac{\sqrt{9 - 4y^2}}{3}$

39. $\alpha = \sec^{-1}\frac{x}{4}$ indicates the diagram

$\Rightarrow \sin\left(\sec^{-1}\frac{x}{4}\right) = \sin\alpha = \frac{\sqrt{x^2-16}}{x}$

41. $\lim_{x \to 1^-} \sin^{-1} x = \frac{\pi}{2}$

43. $\lim_{x \to \infty} \tan^{-1} x = \frac{\pi}{2}$

45. $\lim_{x \to \infty} \sec^{-1} x = \frac{\pi}{2}$

47. $\lim_{x \to \infty} \csc^{-1} x = \lim_{x \to \infty} \sin^{-1}\left(\frac{1}{x}\right) = 0$

49. $y = \cos^{-1}\left(x^2\right) \Rightarrow \frac{dy}{dx} = -\frac{2x}{\sqrt{1-(x^2)^2}} = \frac{-2x}{\sqrt{1-x^4}}$

51. $y = \sin^{-1}\sqrt{2t} \Rightarrow \frac{dy}{dt} = \frac{\sqrt{2}}{\sqrt{1-\left(\sqrt{2t}\right)^2}} = \frac{\sqrt{2}}{\sqrt{1-2t^2}}$

53. $y = \sec^{-1}(2s+1) \Rightarrow \frac{dy}{ds} = \frac{2}{|2s+1|\sqrt{(2s+1)^2-1}} = \frac{2}{|2s+1|\sqrt{4s^2+4s}} = \frac{1}{|2s+1|\sqrt{s^2+s}}$

55. $y = \csc^{-1}\left(x^2+1\right) \Rightarrow \frac{dy}{dx} = -\frac{2x}{|x^2+1|\sqrt{(x^2+1)^2-1}} = \frac{-2x}{(x^2+1)\sqrt{x^4+2x^2}}$

57. $y = \sec^{-1}\left(\frac{1}{t}\right) = \cos^{-1} t \Rightarrow \frac{dy}{dt} = \frac{-1}{\sqrt{1-t^2}}$

59. $y = \cot^{-1}\sqrt{t} = \cot^{-1} t^{1/2} \Rightarrow \frac{dy}{dt} = -\frac{\left(\frac{1}{2}\right)t^{-1/2}}{1+\left(t^{1/2}\right)^2} = \frac{-1}{2\sqrt{t}(1+t)}$

61. $y = \ln\left(\tan^{-1} x\right) \Rightarrow \frac{dy}{dx} = \frac{\left(\frac{1}{1+x^2}\right)}{\tan^{-1} x} = \frac{1}{(\tan^{-1} x)(1+x^2)}$

63. $y = \csc^{-1}\left(e^t\right) \Rightarrow \frac{dy}{dt} = -\frac{e^t}{|e^t|\sqrt{(e^t)^2-1}} = \frac{-1}{\sqrt{e^{2t}-1}}$

65. $y = s\sqrt{1-s^2} + \cos^{-1} s = s\left(1-s^2\right)^{1/2} + \cos^{-1} s \Rightarrow \frac{dy}{ds} = \left(1-s^2\right)^{1/2} + s\left(\frac{1}{2}\right)\left(1-s^2\right)^{-1/2}(-2s) - \frac{1}{\sqrt{1-s^2}}$

$= \sqrt{1-s^2} - \frac{s^2}{\sqrt{1-s^2}} - \frac{1}{\sqrt{1-s^2}} = \sqrt{1-s^2} - \frac{s^2+1}{\sqrt{1-s^2}} = \frac{1-s^2-s^2-1}{\sqrt{1-s^2}} = \frac{-2s^2}{\sqrt{1-s^2}}$

67. $y = \tan^{-1}\sqrt{x^2-1} + \csc^{-1} x = \tan^{-1}\left(x^2-1\right)^{1/2} + \csc^{-1} x \Rightarrow \frac{dy}{dx} = \frac{\left(\frac{1}{2}\right)\left(x^2-1\right)^{-1/2}(2x)}{1+\left[(x^2-1)^{1/2}\right]^2} - \frac{1}{|x|\sqrt{x^2-1}}$

$= \frac{1}{x\sqrt{x^2-1}} - \frac{1}{|x|\sqrt{x^2-1}} = 0$, for $x > 1$

69. $y = x\sin^{-1} x + \sqrt{1-x^2} = x\sin^{-1} x + \left(1-x^2\right)^{1/2} \Rightarrow \frac{dy}{dx} = \sin^{-1} x + x\left(\frac{1}{\sqrt{1-x^2}}\right) + \left(\frac{1}{2}\right)\left(1-x^2\right)^{-1/2}(-2x)$

$= \sin^{-1} x + \frac{x}{\sqrt{1-x^2}} - \frac{x}{\sqrt{1-x^2}} = \sin^{-1} x$

71. $\int \frac{1}{\sqrt{9-x^2}}\, dx = \sin^{-1}\left(\frac{x}{3}\right) + C$

73. $\int \frac{1}{17+x^2}\,dx = \int \frac{1}{\left(\sqrt{17}\right)^2 + x^2}\,dx = \frac{1}{\sqrt{17}}\tan^{-1}\frac{x}{\sqrt{17}} + C$

75. $\int \frac{dx}{x\sqrt{25x^2-2}} = \int \frac{du}{u\sqrt{u^2-2}}$, where $u = 5x$ and $du = 5\,dx$

$= \frac{1}{\sqrt{2}}\sec^{-1}\left|\frac{u}{\sqrt{2}}\right| + C = \frac{1}{\sqrt{2}}\sec^{-1}\left|\frac{5x}{\sqrt{2}}\right| + C$

77. $\int_0^1 \frac{4\,ds}{\sqrt{4-s^2}} = \left[4\sin^{-1}\frac{s}{2}\right]_0^1 = 4\left(\sin^{-1}\frac{1}{2} - \sin^{-1}0\right) = 4\left(\frac{\pi}{6} - 0\right) = \frac{2\pi}{3}$

79. $\int_0^2 \frac{dt}{8+2t^2} = \frac{1}{\sqrt{2}}\int_0^{2\sqrt{2}} \frac{du}{8+u^2}$, where $u = \sqrt{2}t$ and $du = \sqrt{2}\,dt$; $t = 0 \Rightarrow u = 0$, $t = 2 \Rightarrow u = 2\sqrt{2}$

$= \left[\frac{1}{\sqrt{2}} \cdot \frac{1}{\sqrt{8}}\tan^{-1}\frac{u}{\sqrt{8}}\right]_0^{2\sqrt{2}} = \frac{1}{4}\left(\tan^{-1}\frac{2\sqrt{2}}{\sqrt{8}} - \tan^{-1}0\right) = \frac{1}{4}\left(\tan^{-1}1 - \tan^{-1}0\right) = \frac{1}{4}\left(\frac{\pi}{4} - 0\right) = \frac{\pi}{16}$

81. $\int_{-1}^{-\sqrt{2}/2} \frac{dy}{y\sqrt{4y^2-1}} = \int_{-2}^{-\sqrt{2}} \frac{du}{u\sqrt{u^2-1}}$, where $u = 2y$ and $du = 2\,dy$; $y = -1 \Rightarrow u = -2$, $y = -\frac{\sqrt{2}}{2} \Rightarrow u = -\sqrt{2}$

$= \left[\sec^{-1}|u|\right]_{-2}^{-\sqrt{2}} = \sec^{-1}\left|-\sqrt{2}\right| - \sec^{-1}|-2| = \frac{\pi}{4} - \frac{\pi}{3} = -\frac{\pi}{12}$

83. $\int \frac{3\,dr}{\sqrt{1-4(r-1)^2}} = \frac{3}{2}\int \frac{du}{\sqrt{1-u^2}}$, where $u = 2(r-1)$ and $du = 2\,dr$

$= \frac{3}{2}\sin^{-1}u + C = \frac{3}{2}\sin^{-1}2(r-1) + C$

85. $\int \frac{dx}{2+(x-1)^2} = \int \frac{du}{2+u^2}$, where $u = x-1$ and $du = dx$

$= \frac{1}{\sqrt{2}}\tan^{-1}\frac{u}{\sqrt{2}} + C = \frac{1}{\sqrt{2}}\tan^{-1}\left(\frac{x-1}{\sqrt{2}}\right) + C$

87. $\int \frac{dx}{(2x-1)\sqrt{(2x-1)^2-4}} = \frac{1}{2}\int \frac{du}{u\sqrt{u^2-4}}$, where $u = 2x-1$ and $du = 2\,dx$

$= \frac{1}{2} \cdot \frac{1}{2}\sec^{-1}\left|\frac{u}{2}\right| + C = \frac{1}{4}\sec^{-1}\left|\frac{2x-1}{2}\right| + C$

89. $\int_{-\pi/2}^{\pi/2} \frac{2\cos\theta\,d\theta}{1+(\sin\theta)^2} = 2\int_{-1}^1 \frac{du}{1+u^2}$, where $u = \sin\theta$ and $du = \cos\theta\,d\theta$; $\theta = -\frac{\pi}{2} \Rightarrow u = -1$, $\theta = \frac{\pi}{2} \Rightarrow u = 1$

$= \left[2\tan^{-1}u\right]_{-1}^1 = 2\left(\tan^{-1}1 - \tan^{-1}(-1)\right) = 2\left[\frac{\pi}{4} - \left(-\frac{\pi}{4}\right)\right] = \pi$

91. $\int_0^{\ln\sqrt{3}} \frac{e^x\,dx}{1+e^{2x}} = \int_1^{\sqrt{3}} \frac{du}{1+u^2}$, where $u = e^x$ and $du = e^x\,dx$; $x = 0 \Rightarrow u = 1$, $x = \ln\sqrt{3} \Rightarrow u = \sqrt{3}$

$= \left[\tan^{-1}u\right]_1^{\sqrt{3}} = \tan^{-1}\sqrt{3} - \tan^{-1}1 = \frac{\pi}{3} - \frac{\pi}{4} = \frac{\pi}{12}$

93. $\int \frac{y\,dy}{\sqrt{1-y^4}} = \frac{1}{2}\int \frac{du}{\sqrt{1-u^2}}$, where $u = y^2$ and $du = 2y\,dy$

$= \frac{1}{2}\sin^{-1}u + C = \frac{1}{2}\sin^{-1}y^2 + C$

95. $\int \frac{dx}{\sqrt{-x^2+4x-3}} = \int \frac{dx}{\sqrt{1-(x^2-4x+4)}} = \int \frac{dx}{\sqrt{1-(x-2)^2}} = \sin^{-1}(x-2) + C$

97. $\int_{-1}^0 \frac{6\,dt}{\sqrt{3-2t-t^2}} = 6\int_{-1}^0 \frac{dt}{\sqrt{4-(t^2+2t+1)}} = 6\int_{-1}^0 \frac{dt}{\sqrt{2^2-(t+1)^2}} = 6\left[\sin^{-1}\left(\frac{t+1}{2}\right)\right]_{-1}^0$

$= 6\left[\sin^{-1}\left(\frac{1}{2}\right) - \sin^{-1}0\right] = 6\left(\frac{\pi}{6} - 0\right) = \pi$

99. $\int \frac{dy}{y^2-2y+5} = \int \frac{dy}{4+y^2-2y+1} = \int \frac{dy}{2^2+(y-1)^2} = \frac{1}{2}\tan^{-1}\left(\frac{y-1}{2}\right) + C$

101. $\int_1^2 \frac{8\,dx}{x^2-2x+2} = 8\int_1^2 \frac{dx}{1+(x^2-2x+1)} = 8\int_1^2 \frac{dx}{1+(x-1)^2} = 8\left[\tan^{-1}(x-1)\right]_1^2$

   $= 8\left(\tan^{-1}1 - \tan^{-1}0\right) = 8\left(\frac{\pi}{4} - 0\right) = 2\pi$

103. $\int \frac{dx}{(x+1)\sqrt{x^2+2x}} = \int \frac{dx}{(x+1)\sqrt{x^2+2x+1-1}} = \int \frac{dx}{(x+1)\sqrt{(x+1)^2-1}}$

   $= \int \frac{du}{u\sqrt{u^2-1}}$, where $u = x+1$ and $du = dx$

   $= \sec^{-1}|u| + C = \sec^{-1}|x+1| + C$

105. $\int \frac{e^{\sin^{-1}x}}{\sqrt{1-x^2}}\,dx = \int e^u\,du$, where $u = \sin^{-1}x$ and $du = \frac{dx}{\sqrt{1-x^2}}$

   $= e^u + C = e^{\sin^{-1}x} + C$

107. $\int \frac{(\sin^{-1}x)^2}{\sqrt{1-x^2}}\,dx = \int u^2\,du$, where $u = \sin^{-1}x$ and $du = \frac{dx}{\sqrt{1-x^2}}$

   $= \frac{u^3}{3} + C = \frac{(\sin^{-1}x)^3}{3} + C$

109. $\int \frac{1}{(\tan^{-1}y)(1+y^2)}\,dy = \int \frac{\left(\frac{1}{1+y^2}\right)}{\tan^{-1}y}\,dy = \int \frac{1}{u}\,du$, where $u = \tan^{-1}y$ and $du = \frac{dy}{1+y^2}$

   $= \ln|u| + C = \ln|\tan^{-1}y| + C$

111. $\int_{\sqrt{2}}^2 \frac{\sec^2(\sec^{-1}x)}{x\sqrt{x^2-1}}\,dx = \int_{\pi/4}^{\pi/3} \sec^2 u\,du$, where $u = \sec^{-1}x$ and $du = \frac{dx}{x\sqrt{x^2-1}}$; $x = \sqrt{2} \Rightarrow u = \frac{\pi}{4}$, $x = 2 \Rightarrow u = \frac{\pi}{3}$

   $= \left[\tan u\right]_{\pi/4}^{\pi/3} = \tan\frac{\pi}{3} - \tan\frac{\pi}{4} = \sqrt{3} - 1$

113. $\lim_{x \to 0} \frac{\sin^{-1}5x}{x} = \lim_{x \to 0} \frac{\left(\frac{5}{\sqrt{1-25x^2}}\right)}{1} = 5$

115. $\lim_{x \to \infty} x\tan^{-1}\left(\frac{2}{x}\right) = \lim_{x \to \infty} \frac{\tan^{-1}(2x^{-1})}{x^{-1}} = \lim_{x \to \infty} \frac{\left(\frac{-2x^{-2}}{1+4x^{-2}}\right)}{-x^{-2}} = \lim_{x \to \infty} \frac{2}{1+4x^{-2}} = 2$

117. If $y = \ln x - \frac{1}{2}\ln(1+x^2) - \frac{\tan^{-1}x}{x} + C$, then $dy = \left[\frac{1}{x} - \frac{x}{1+x^2} - \frac{\left(\frac{x}{1+x^2}\right) - \tan^{-1}x}{x^2}\right]dx$

   $= \left(\frac{1}{x} - \frac{x}{1+x^2} - \frac{1}{x(1+x^2)} + \frac{\tan^{-1}x}{x^2}\right)dx = \frac{x(1+x^2) - x^3 - x + (\tan^{-1}x)(1+x^2)}{x^2(1+x^2)}\,dx = \frac{\tan^{-1}x}{x^2}\,dx$,

   which verifies the formula

119. If $y = x(\sin^{-1}x)^2 - 2x + 2\sqrt{1-x^2}\sin^{-1}x + C$, then

   $dy = \left[(\sin^{-1}x)^2 + \frac{2x(\sin^{-1}x)}{\sqrt{1-x^2}} - 2 + \frac{-2x}{\sqrt{1-x^2}}\sin^{-1}x + 2\sqrt{1-x^2}\left(\frac{1}{\sqrt{1-x^2}}\right)\right]dx = (\sin^{-1}x)^2\,dx$, which verifies

   the formula

121. $\frac{dy}{dx} = \frac{1}{\sqrt{1-x^2}} \Rightarrow dy = \frac{dx}{\sqrt{1-x^2}} \Rightarrow y = \sin^{-1}x + C$; $x = 0$ and $y = 0 \Rightarrow 0 = \sin^{-1}0 + C \Rightarrow C = 0 \Rightarrow y = \sin^{-1}x$

123. $\frac{dy}{dx} = \frac{1}{x\sqrt{x^2-1}} \Rightarrow dy = \frac{dx}{x\sqrt{x^2-1}} \Rightarrow y = \sec^{-1}|x| + C$; $x = 2$ and $y = \pi \Rightarrow \pi = \sec^{-1}2 + C \Rightarrow C = \pi - \sec^{-1}2$

   $= \pi - \frac{\pi}{3} = \frac{2\pi}{3} \Rightarrow y = \sec^{-1}(x) + \frac{2\pi}{3}$, $x > 1$

125. The angle α is the large angle between the wall and the right end of the blackboard minus the small angle
   between the left end of the blackboard and the wall $\Rightarrow \alpha = \cot^{-1}\left(\frac{x}{15}\right) - \cot^{-1}\left(\frac{x}{3}\right)$.

127. $V = \left(\frac{1}{3}\right)\pi r^2 h = \left(\frac{1}{3}\right)\pi(3\sin\theta)^2(3\cos\theta) = 9\pi(\cos\theta - \cos^3\theta)$, where $0 \le \theta \le \frac{\pi}{2} \Rightarrow \frac{dV}{d\theta} = -9\pi(\sin\theta)(1 - 3\cos^2\theta)$

$= 0 \Rightarrow \sin\theta = 0$ or $\cos\theta = \pm\frac{1}{\sqrt{3}} \Rightarrow$ the critical points are: $0, \cos^{-1}\left(\frac{1}{\sqrt{3}}\right)$, and $\cos^{-1}\left(-\frac{1}{\sqrt{3}}\right)$; but

$\cos^{-1}\left(-\frac{1}{\sqrt{3}}\right)$ is not in the domain. When $\theta = 0$, we have a minimum and when $\theta = \cos^{-1}\left(\frac{1}{\sqrt{3}}\right) \approx 54.7°$, we

have a maximum volume.

129. Take each square as a unit square. From the diagram we have the following: the smallest angle α has a

tangent of $1 \Rightarrow \alpha = \tan^{-1}1$; the middle angle β has a tangent of $2 \Rightarrow \beta = \tan^{-1}2$; and the largest angle γ

has a tangent of $3 \Rightarrow \gamma = \tan^{-1}3$. The sum of these three angles is $\pi \Rightarrow \alpha + \beta + \gamma = \pi$

$\Rightarrow \tan^{-1}1 + \tan^{-1}2 + \tan^{-1}3 = \pi$.

131. $\sin^{-1}(1) + \cos^{-1}(1) = \frac{\pi}{2} + 0 = \frac{\pi}{2}$; $\sin^{-1}(0) + \cos^{-1}(0) = 0 + \frac{\pi}{2} = \frac{\pi}{2}$; and $\sin^{-1}(-1) + \cos^{-1}(-1) = -\frac{\pi}{2} + \pi = \frac{\pi}{2}$.

If $x \in (-1, 0)$ and $x = -a$, then $\sin^{-1}(x) + \cos^{-1}(x) = \sin^{-1}(-a) + \cos^{-1}(-a) = -\sin^{-1}a + (\pi - \cos^{-1}a)$

$= \pi - (\sin^{-1}a + \cos^{-1}a) = \pi - \frac{\pi}{2} = \frac{\pi}{2}$ from Equations (3) and (4) in the text.

133. (a) Defined; there is an angle whose tangent is 2.

(b) Not defined; there is no angle whose cosine is 2.

135. (a) Not defined; there is no angle whose secant is 0.

(b) Not defined; there is no angle whose sine is $\sqrt{2}$.

137. $\alpha(x) = \cot^{-1}\left(\frac{x}{15}\right) - \cot^{-1}\left(\frac{x}{3}\right), x > 0 \Rightarrow \alpha'(x) = \frac{-15}{225 + x^2} + \frac{3}{9 + x^2} = \frac{-15(9 + x^2) + 3(225 + x^2)}{(225 + x^2)(9 + x^2)}$; solving

$\alpha'(x) = 0 \Rightarrow -135 - 15x^2 + 675 + 3x^2 = 0 \Rightarrow x = 3\sqrt{5}$; $\alpha'(x) > 0$ when $0 < x < 3\sqrt{5}$ and $\alpha'(x) < 0$ for

$x > 3\sqrt{5} \Rightarrow$ there is a maximum at $3\sqrt{5}$ ft from the front of the room

139. Yes, $\sin^{-1}x$ and $-\cos^{-1}x$ differ by the constant $\frac{\pi}{2}$

141. $\csc^{-1}u = \frac{\pi}{2} - \sec^{-1}u \Rightarrow \frac{d}{dx}(\csc^{-1}u) = \frac{d}{dx}\left(\frac{\pi}{2} - \sec^{-1}u\right) = 0 - \frac{\frac{du}{dx}}{|u|\sqrt{u^2 - 1}} = -\frac{\frac{du}{dx}}{|u|\sqrt{u^2 - 1}}, |u| > 1$

143. $f(x) = \sec x \Rightarrow f'(x) = \sec x \tan x \Rightarrow \left.\frac{df^{-1}}{dx}\right|_{x=b} = \frac{1}{\left.\frac{df}{dx}\right|_{x=f^{-1}(b)}} = \frac{1}{\sec(\sec^{-1}b)\tan(\sec^{-1}b)} = \frac{1}{b\left(\pm\sqrt{b^2 - 1}\right)}$.

Since the slope of $\sec^{-1}x$ is always positive, we the right sign by writing $\frac{d}{dx}\sec^{-1}x = \frac{1}{|x|\sqrt{x^2 - 1}}$.

145. The functions f and g have the same derivative (for $x \ge 0$), namely $\frac{1}{\sqrt{x}(x + 1)}$. The functions therefore differ

by a constant. To identify the constant we can set x equal to 0 in the equation $f(x) = g(x) + C$, obtaining

$\sin^{-1}(-1) = 2\tan^{-1}(0) + C \Rightarrow -\frac{\pi}{2} = 0 + C \Rightarrow C = -\frac{\pi}{2}$. For $x \ge 0$, we have $\sin^{-1}\left(\frac{x-1}{x+1}\right) = 2\tan^{-1}\sqrt{x} - \frac{\pi}{2}$.

147. $V = \pi\int_{-\sqrt{3}/3}^{\sqrt{3}}\left(\frac{1}{\sqrt{1+x^2}}\right)^2 dx = \pi\int_{-\sqrt{3}/3}^{\sqrt{3}}\frac{1}{1+x^2}dx = \pi[\tan^{-1}x]_{-\sqrt{3}/3}^{\sqrt{3}} = \pi\left[\tan^{-1}\sqrt{3} - \tan^{-1}\left(-\frac{\sqrt{3}}{3}\right)\right]$

$= \pi\left[\frac{\pi}{3} - \left(-\frac{\pi}{6}\right)\right] = \frac{\pi^2}{2}$

149. (a) $A(x) = \frac{\pi}{4}(\text{diameter})^2 = \frac{\pi}{4}\left[\frac{1}{\sqrt{1+x^2}} - \left(-\frac{1}{\sqrt{1+x^2}}\right)\right]^2 = \frac{\pi}{1+x^2} \Rightarrow V = \int_a^b A(x)\,dx = \int_{-1}^1 \frac{\pi\,dx}{1+x^2}$

$= \pi[\tan^{-1}x]_{-1}^1 = (\pi)(2)\left(\frac{\pi}{4}\right) = \frac{\pi^2}{2}$

(b) $A(x) = (\text{edge})^2 = \left[\frac{1}{\sqrt{1+x^2}} - \left(-\frac{1}{\sqrt{1+x^2}}\right)\right]^2 = \frac{4}{1+x^2} \Rightarrow V = \int_a^b A(x)\,dx = \int_{-1}^1 \frac{4\,dx}{1+x^2}$

$= 4[\tan^{-1}x]_{-1}^1 = 4[\tan^{-1}(1) - \tan^{-1}(-1)] = 4\left[\frac{\pi}{4} - \left(-\frac{\pi}{4}\right)\right] = 2\pi$

151. (a) $\sec^{-1} 1.5 = \cos^{-1} \frac{1}{1.5} \approx 0.84107$ (b) $\csc^{-1}(-1.5) = \sin^{-1}\left(-\frac{1}{1.5}\right) \approx -0.72973$

    (c) $\cot^{-1} 2 = \frac{\pi}{2} - \tan^{-1} 2 \approx 0.46365$

153. (a) Domain: all real numbers except those having the form $\frac{\pi}{2} + k\pi$ where k is an integer.
Range: $-\frac{\pi}{2} < y < \frac{\pi}{2}$

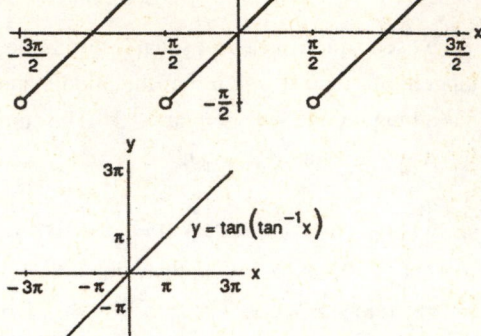

    (b) Domain: $-\infty < x < \infty$; Range: $-\infty < y < \infty$
The graph of $y = \tan^{-1}(\tan x)$ is periodic, the graph of $y = \tan(\tan^{-1} x) = x$ for $-\infty \le x < \infty$.

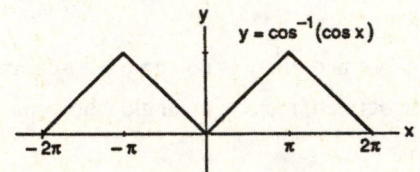

155. (a) Domain: $-\infty < x < \infty$; Range: $0 \le y \le \pi$

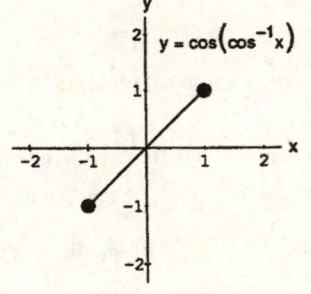

    (b) Domain: $-1 \le x \le 1$; Range: $-1 \le y \le 1$
The graph of $y = \cos^{-1}(\cos x)$ is periodic; the graph of $y = \cos(\cos^{-1} x) = x$ for $-1 \le x \le 1$.

157. The graphs are identical for $y = 2 \sin(2 \tan^{-1} x)$

$$= 4[\sin(\tan^{-1} x)][\cos(\tan^{-1} x)] = 4\left(\frac{x}{\sqrt{x^2+1}}\right)\left(\frac{1}{\sqrt{x^2+1}}\right)$$

$= \frac{4x}{x^2+1}$ from the triangle

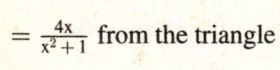
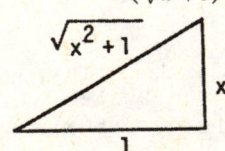
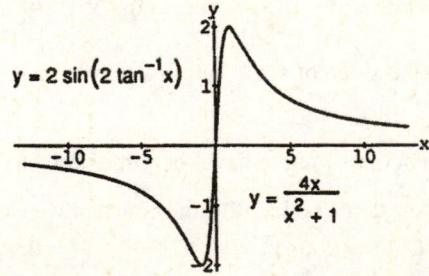

159. The values of f increase over the interval $[-1, 1]$ because $f' > 0$, and the graph of f steepens as the values of f' increase towards the ends of the interval. The graph of f is concave down to the left of the origin where $f'' < 0$, and concave up to the right of the origin where $f'' > 0$. There is an inflection point at $x = 0$ where $f'' = 0$ and f' has a local minimum value.

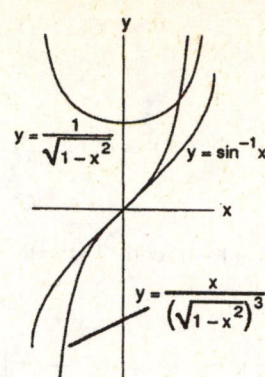

7.8 HYPERBOLIC FUNCTIONS

1. $\sinh x = -\frac{3}{4} \Rightarrow \cosh x = \sqrt{1 + \sinh^2 x} = \sqrt{1 + \left(-\frac{3}{4}\right)^2} = \sqrt{1 + \frac{9}{16}} = \sqrt{\frac{25}{16}} = \frac{5}{4}$, $\tanh x = \frac{\sinh x}{\cosh x} = \frac{\left(-\frac{3}{4}\right)}{\left(\frac{5}{4}\right)} = -\frac{3}{5}$,

   $\coth x = \frac{1}{\tanh x} = -\frac{5}{3}$, $\operatorname{sech} x = \frac{1}{\cosh x} = \frac{4}{5}$, and $\operatorname{csch} x = \frac{1}{\sin x} = -\frac{4}{3}$

3. $\cosh x = \frac{17}{15}, x > 0 \Rightarrow \sinh x = \sqrt{\cosh^2 x - 1} = \sqrt{\left(\frac{17}{15}\right)^2 - 1} = \sqrt{\frac{289}{225} - 1} = \sqrt{\frac{64}{225}} = \frac{8}{15}$, $\tanh x = \frac{\sinh x}{\cosh x} = \frac{\left(\frac{8}{15}\right)}{\left(\frac{17}{15}\right)}$

   $= \frac{8}{17}$, $\coth x = \frac{1}{\tanh x} = \frac{17}{8}$, $\operatorname{sech} x = \frac{1}{\cosh x} = \frac{15}{17}$, and $\operatorname{csch} x = \frac{1}{\sinh x} = \frac{15}{8}$

5. $2 \cosh(\ln x) = 2 \left(\frac{e^{\ln x} + e^{-\ln x}}{2}\right) = e^{\ln x} + \frac{1}{e^{\ln x}} = x + \frac{1}{x}$

7. $\cosh 5x + \sinh 5x = \frac{e^{5x} + e^{-5x}}{2} + \frac{e^{5x} - e^{-5x}}{2} = e^{5x}$

9. $(\sinh x + \cosh x)^4 = \left(\frac{e^x - e^{-x}}{2} + \frac{e^x + e^{-x}}{2}\right)^4 = (e^x)^4 = e^{4x}$

11. (a) $\sinh 2x = \sinh(x + x) = \sinh x \cosh x + \cosh x \sinh x = 2 \sinh x \cosh x$

    (b) $\cosh 2x = \cosh(x + x) = \cosh x \cosh x + \sinh x \sin x = \cosh^2 x + \sinh^2 x$

13. $y = 6 \sinh \frac{x}{3} \Rightarrow \frac{dy}{dx} = 6 \left(\cosh \frac{x}{3}\right) \left(\frac{1}{3}\right) = 2 \cosh \frac{x}{3}$

15. $y = 2\sqrt{t} \tanh \sqrt{t} = 2t^{1/2} \tanh t^{1/2} \Rightarrow \frac{dy}{dt} = \left[\operatorname{sech}^2\left(t^{1/2}\right)\right] \left(\frac{1}{2} t^{-1/2}\right) \left(2t^{1/2}\right) + \left(\tanh t^{1/2}\right) \left(t^{-1/2}\right)$

    $= \operatorname{sech}^2 \sqrt{t} + \frac{\tanh \sqrt{t}}{\sqrt{t}}$

17. $y = \ln(\sinh z) \Rightarrow \frac{dy}{dz} = \frac{\cosh z}{\sinh z} = \coth z$

19. $y = (\operatorname{sech} \theta)(1 - \ln \operatorname{sech} \theta) \Rightarrow \frac{dy}{d\theta} = \left(-\frac{-\operatorname{sech} \theta \tanh \theta}{\operatorname{sech} \theta}\right)(\operatorname{sech} \theta) + (-\operatorname{sech} \theta \tanh \theta)(1 - \ln \operatorname{sech} \theta)$

    $= \operatorname{sech} \theta \tanh \theta - (\operatorname{sech} \theta \tanh \theta)(1 - \ln \operatorname{sech} \theta) = (\operatorname{sech} \theta \tanh \theta)[1 - (1 - \ln \operatorname{sech} \theta)]$

    $= (\operatorname{sech} \theta \tanh \theta)(\ln \operatorname{sech} \theta)$

21. $y = \ln \cosh v - \frac{1}{2} \tanh^2 v \Rightarrow \frac{dy}{dv} = \frac{\sinh v}{\cosh v} - \left(\frac{1}{2}\right)(2 \tanh v)(\operatorname{sech}^2 v) = \tanh v - (\tanh v)(\operatorname{sech}^2 v)$

    $= (\tanh v)(1 - \operatorname{sech}^2 v) = (\tanh v)(\tanh^2 v) = \tanh^3 v$

23. $y = (x^2 + 1) \operatorname{sech}(\ln x) = (x^2 + 1) \left(\frac{2}{e^{\ln x} + e^{-\ln x}}\right) = (x^2 + 1) \left(\frac{2}{x + x^{-1}}\right) = (x^2 + 1) \left(\frac{2x}{x^2 + 1}\right) = 2x \Rightarrow \frac{dy}{dx} = 2$

25. $y = \sinh^{-1}\sqrt{x} = \sinh^{-1}\left(x^{1/2}\right) \Rightarrow \frac{dy}{dx} = \frac{\left(\frac{1}{2}\right)x^{-1/2}}{\sqrt{1+(x^{1/2})^2}} = \frac{1}{2\sqrt{x}\sqrt{1+x}} = \frac{1}{2\sqrt{x(1+x)}}$

27. $y = (1-\theta)\tanh^{-1}\theta \Rightarrow \frac{dy}{d\theta} = (1-\theta)\left(\frac{1}{1-\theta^2}\right) + (-1)\tanh^{-1}\theta = \frac{1}{1+\theta} - \tanh^{-1}\theta$

29. $y = (1-t)\coth^{-1}\sqrt{t} = (1-t)\coth^{-1}\left(t^{1/2}\right) \Rightarrow \frac{dy}{dt} = (1-t)\left[\frac{\left(\frac{1}{2}\right)t^{-1/2}}{1-(t^{1/2})^2}\right] + (-1)\coth^{-1}\left(t^{1/2}\right) = \frac{1}{2\sqrt{t}} - \coth^{-1}\sqrt{t}$

31. $y = \cos^{-1}x - x\,\text{sech}^{-1}x \Rightarrow \frac{dy}{dx} = \frac{-1}{\sqrt{1-x^2}} - \left[x\left(\frac{-1}{x\sqrt{1-x^2}}\right) + (1)\,\text{sech}^{-1}x\right] = \frac{-1}{\sqrt{1-x^2}} + \frac{1}{\sqrt{1-x^2}} - \text{sech}^{-1}x$

$= -\,\text{sech}^{-1}x$

33. $y = \text{csch}^{-1}\left(\frac{1}{2}\right)^{\theta} \Rightarrow \frac{dy}{d\theta} = -\frac{\left[\ln\left(\frac{1}{2}\right)\right]\left(\frac{1}{2}\right)^{\theta}}{\left(\frac{1}{2}\right)^{\theta}\sqrt{1+\left[\left(\frac{1}{2}\right)^{\theta}\right]^2}} = -\frac{\ln(1)-\ln(2)}{\sqrt{1+\left(\frac{1}{2}\right)^{2\theta}}} = \frac{\ln 2}{\sqrt{1+\left(\frac{1}{2}\right)^{2\theta}}}$

35. $y = \sinh^{-1}(\tan x) \Rightarrow \frac{dy}{dx} = \frac{\sec^2 x}{\sqrt{1+(\tan x)^2}} = \frac{\sec^2 x}{\sqrt{\sec^2 x}} = \frac{\sec^2 x}{|\sec x|} = \frac{|\sec x|\,|\sec x|}{|\sec x|} = |\sec x|$

37. (a) If $y = \tan^{-1}(\sinh x) + C$, then $\frac{dy}{dx} = \frac{\cosh x}{1+\sinh^2 x} = \frac{\cosh x}{\cosh^2 x} = \text{sech } x$, which verifies the formula

**(b)** If $y = \sin^{-1}(\tanh x) + C$, then $\frac{dy}{dx} = \frac{\text{sech}^2 x}{\sqrt{1-\tanh^2 x}} = \frac{\text{sech}^2 x}{\text{sech } x} = \text{sech } x$, which verifies the formula

39. If $y = \frac{x^2-1}{2}\coth^{-1}x + \frac{x}{2} + C$, then $\frac{dy}{dx} = x\coth^{-1}x + \left(\frac{x^2-1}{2}\right)\left(\frac{1}{1-x^2}\right) + \frac{1}{2} = x\coth^{-1}x$, which verifies

the formula

41. $\int \sinh 2x\, dx = \frac{1}{2}\int \sinh u\, du$, where $u = 2x$ and $du = 2\, dx$

$= \frac{\cosh u}{2} + C = \frac{\cosh 2x}{2} + C$

43. $\int 6\cosh\left(\frac{x}{2} - \ln 3\right) dx = 12\int \cosh u\, du$, where $u = \frac{x}{2} - \ln 3$ and $du = \frac{1}{2}\, dx$

$= 12\sinh u + C = 12\sinh\left(\frac{x}{2} - \ln 3\right) + C$

45. $\int \tanh\frac{x}{7}\, dx = 7\int \frac{\sinh u}{\cosh u}\, du$, where $u = \frac{x}{7}$ and $du = \frac{1}{7}\, dx$

$= 7\ln|\cosh u| + C_1 = 7\ln\left|\cosh\frac{x}{7}\right| + C_1 = 7\ln\left|\frac{e^{x/7}+e^{-x/7}}{2}\right| + C_1 = 7\ln\left|e^{x/7}+e^{-x/7}\right| - 7\ln 2 + C_1$

$= 7\ln\left|e^{x/7}+e^{-x/7}\right| + C$

47. $\int \text{sech}^2\left(x - \frac{1}{2}\right) dx = \int \text{sech}^2 u\, du$, where $u = \left(x - \frac{1}{2}\right)$ and $du = dx$

$= \tanh u + C = \tanh\left(x - \frac{1}{2}\right) + C$

49. $\int \frac{\text{sech }\sqrt{t}\,\tanh\sqrt{t}}{\sqrt{t}}\, dt = 2\int \text{sech } u\,\tanh u\, du$, where $u = \sqrt{t} = t^{1/2}$ and $du = \frac{dt}{2\sqrt{t}}$

$= 2(-\,\text{sech } u) + C = -2\,\text{sech }\sqrt{t} + C$

51. $\int_{\ln 2}^{\ln 4}\coth x\, dx = \int_{\ln 2}^{\ln 4}\frac{\cosh x}{\sinh x}\, dx = \int_{3/4}^{15/8}\frac{1}{u}\, du = \left[\ln|u|\right]_{3/4}^{15/8} = \ln\left|\frac{15}{8}\right| - \ln\left|\frac{3}{4}\right| = \ln\left|\frac{15}{8}\cdot\frac{4}{3}\right| = \ln\frac{5}{2}$,

where $u = \sinh x$, $du = \cosh x\, dx$, the lower limit is $\sinh(\ln 2) = \frac{e^{\ln 2}-e^{-\ln 2}}{2} = \frac{2-\left(\frac{1}{2}\right)}{2} = \frac{3}{4}$ and the upper

limit is $\sinh(\ln 4) = \frac{e^{\ln 4}-e^{-\ln 4}}{2} = \frac{4-\left(\frac{1}{4}\right)}{2} = \frac{15}{8}$

53. $\int_{-\ln 4}^{-\ln 2} 2e^\theta \cosh\theta \; d\theta = \int_{-\ln 4}^{-\ln 2} 2e^\theta \left(\frac{e^\theta + e^{-\theta}}{2}\right) d\theta = \int_{-\ln 4}^{-\ln 2} (e^{2\theta} + 1) \; d\theta = \left[\frac{e^{2\theta}}{2} + \theta\right]_{-\ln 4}^{-\ln 2}$

$= \left(\frac{e^{-2\ln 2}}{2} - \ln 2\right) - \left(\frac{e^{-2\ln 4}}{2} - \ln 4\right) = \left(\frac{1}{8} - \ln 2\right) - \left(\frac{1}{32} - \ln 4\right) = \frac{3}{32} - \ln 2 + 2\ln 2 = \frac{3}{32} + \ln 2$

55. $\int_{-\pi/4}^{\pi/4} \cosh(\tan\theta) \sec^2\theta \; d\theta = \int_{-1}^{1} \cosh u \; du = [\sinh u]_{-1}^{1} = \sinh(1) - \sinh(-1) = \left(\frac{e^1 - e^{-1}}{2}\right) - \left(\frac{e^{-1} - e^1}{2}\right)$

$= \frac{e - e^{-1} - e^{-1} + e}{2} = e - e^{-1}$, where $u = \tan\theta$, $du = \sec^2\theta \; d\theta$, the lower limit is $\tan\left(-\frac{\pi}{4}\right) = -1$ and the upper limit is $\tan\left(\frac{\pi}{4}\right) = 1$

57. $\int_{1}^{2} \frac{\cosh(\ln t)}{t} \; dt = \int_{0}^{\ln 2} \cosh u \; du = [\sinh u]_{0}^{\ln 2} = \sinh(\ln 2) - \sinh(0) = \frac{e^{\ln 2} - e^{-\ln 2}}{2} - 0 = \frac{2 - \frac{1}{2}}{2} = \frac{3}{4}$, where $u = \ln t$, $du = \frac{1}{t} \; dt$, the lower limit is $\ln 1 = 0$ and the upper limit is $\ln 2$

59. $\int_{-\ln 2}^{0} \cosh^2\left(\frac{x}{2}\right) dx = \int_{-\ln 2}^{0} \frac{\cosh x + 1}{2} \; dx = \frac{1}{2} \int_{-\ln 2}^{0} (\cosh x + 1) \; dx = \frac{1}{2} [\sinh x + x]_{-\ln 2}^{0}$

$= \frac{1}{2} [(\sinh 0 + 0) - (\sinh(-\ln 2) - \ln 2)] = \frac{1}{2}\left[(0+0) - \left(\frac{e^{-\ln 2} - e^{\ln 2}}{2} - \ln 2\right)\right] = \frac{1}{2}\left[-\frac{\left(\frac{1}{2}\right) - 2}{2} + \ln 2\right]$

$= \frac{1}{2}\left(1 - \frac{1}{4} + \ln 2\right) = \frac{3}{8} + \frac{1}{2}\ln 2 = \frac{3}{8} + \ln\sqrt{2}$

61. $\sinh^{-1}\left(\frac{-5}{12}\right) = \ln\left(-\frac{5}{12} + \sqrt{\frac{25}{144} + 1}\right) = \ln\left(\frac{2}{3}\right)$

63. $\tanh^{-1}\left(-\frac{1}{2}\right) = \frac{1}{2}\ln\left(\frac{1 - (1/2)}{1 + (1/2)}\right) = -\frac{\ln 3}{2}$

65. $\text{sech}^{-1}\left(\frac{3}{5}\right) = \ln\left(\frac{1 + \sqrt{1 - (9/25)}}{(3/5)}\right) = \ln 3$

67. (a) $\int_{0}^{2\sqrt{3}} \frac{dx}{\sqrt{4 + x^2}} = \left[\sinh^{-1}\frac{x}{2}\right]_{0}^{2\sqrt{3}} = \sinh^{-1}\sqrt{3} - \sinh^{-1}0 = \sinh^{-1}\sqrt{3}$

(b) $\sinh^{-1}\sqrt{3} = \ln\left(\sqrt{3} + \sqrt{3 + 1}\right) = \ln\left(\sqrt{3} + 2\right)$

69. (a) $\int_{5/4}^{2} \frac{1}{1 - x^2} \; dx = [\coth^{-1}x]_{5/4}^{2} = \coth^{-1}2 - \coth^{-1}\frac{5}{4}$

(b) $\coth^{-1}2 - \coth^{-1}\frac{5}{4} = \frac{1}{2}\left[\ln 3 - \ln\left(\frac{9/4}{1/4}\right)\right] = \frac{1}{2}\ln\frac{1}{3}$

71. (a) $\int_{1/5}^{3/13} \frac{dx}{x\sqrt{1 - 16x^2}} = \int_{4/5}^{12/13} \frac{du}{u\sqrt{a^2 - u^2}}$, where $u = 4x$, $du = 4 \; dx$, $a = 1$

$= [-\text{sech}^{-1}u]_{4/5}^{12/13} = -\text{sech}^{-1}\frac{12}{13} + \text{sech}^{-1}\frac{4}{5}$

(b) $-\text{sech}^{-1}\frac{12}{13} + \text{sech}^{-1}\frac{4}{5} = -\ln\left(\frac{1 + \sqrt{1 - (12/13)^2}}{(12/13)}\right) + \ln\left(\frac{1 + \sqrt{1 - (4/5)^2}}{(4/5)}\right)$

$= -\ln\left(\frac{13 + \sqrt{169 - 144}}{12}\right) + \ln\left(\frac{5 + \sqrt{25 - 16}}{4}\right) = \ln\left(\frac{5 + 3}{4}\right) - \ln\left(\frac{13 + 5}{12}\right) = \ln 2 - \ln\frac{3}{2}$

$= \ln\left(2 \cdot \frac{2}{3}\right) = \ln\frac{4}{3}$

73. (a) $\int_{0}^{\pi} \frac{\cos x}{\sqrt{1 + \sin^2 x}} \; dx = \int_{0}^{0} \frac{1}{\sqrt{1 + u^2}} \; du = [\sinh^{-1}u]_{0}^{0} = \sinh^{-1}0 - \sinh^{-1}0 = 0$, where $u = \sin x$, $du = \cos x \; dx$

(b) $\sinh^{-1}0 - \sinh^{-1}0 = \ln\left(0 + \sqrt{0 + 1}\right) - \ln\left(0 + \sqrt{0 + 1}\right) = 0$

75. (a) Let $E(x) = \frac{f(x) + f(-x)}{2}$ and $O(x) = \frac{f(x) - f(-x)}{2}$. Then $E(x) + O(x) = \frac{f(x) + f(-x)}{2} + \frac{f(x) - f(-x)}{2}$

$= \frac{2f(x)}{2} = f(x)$. Also, $E(-x) = \frac{f(-x) + f(-(-x))}{2} = \frac{f(x) + f(-x)}{2} = E(x) \Rightarrow E(x)$ is even, and

$O(-x) = \frac{f(-x) - f(-(-x))}{2} = -\frac{f(x) - f(-x)}{2} = -O(x) \Rightarrow O(x)$ is odd. Consequently, f(x) can be written as a sum of an even and an odd function.

(b) $f(x) = \frac{f(x) + f(-x)}{2}$ because $\frac{f(x) - f(-x)}{2} = 0$ if f is even and $f(x) = \frac{f(x) - f(-x)}{2}$ because $\frac{f(x) + f(-x)}{2} = 0$ if f is odd.

Thus, if f is even $f(x) = \frac{2f(x)}{2} + 0$ and if f is odd, $f(x) = 0 + \frac{2f(x)}{2}$

77. (a) $v = \sqrt{\frac{mg}{k}} \tanh\left(\sqrt{\frac{gk}{m}}\, t\right) \Rightarrow \frac{dv}{dt} = \sqrt{\frac{mg}{k}}\left[\text{sech}^2\left(\sqrt{\frac{gk}{m}}\, t\right)\right]\left(\sqrt{\frac{gk}{m}}\right) = g\,\text{sech}^2\left(\sqrt{\frac{gk}{m}}\, t\right).$

Thus $m\frac{dv}{dt} = mg\,\text{sech}^2\left(\sqrt{\frac{gk}{m}}\, t\right) = mg\left(1 - \tanh^2\left(\sqrt{\frac{gk}{m}}\, t\right)\right) = mg - kv^2$. Also, since tanh x = 0 when x = 0, v = 0 when t = 0.

(b) $\lim\limits_{t \to \infty} v = \lim\limits_{t \to \infty} \sqrt{\frac{mg}{k}} \tanh\left(\sqrt{\frac{kg}{m}}\, t\right) = \sqrt{\frac{mg}{k}} \lim\limits_{t \to \infty} \tanh\left(\sqrt{\frac{kg}{m}}\, t\right) = \sqrt{\frac{mg}{k}}\,(1) = \sqrt{\frac{mg}{k}}$

(c) $\sqrt{\frac{160}{0.005}} = \sqrt{\frac{160,000}{5}} = \frac{400}{\sqrt{5}} = 80\sqrt{5} \approx 178.89$ ft/sec

79. $\frac{dy}{dx} = \frac{-1}{x\sqrt{1-x^2}} + \frac{x}{\sqrt{1-x^2}} \Rightarrow y = \int \frac{-1}{x\sqrt{1-x^2}}\, dx + \int \frac{x}{\sqrt{1-x^2}}\, dx \Rightarrow y = \text{sech}^{-1}(x) - \sqrt{1-x^2} + C;\ x = 1$ and

$y = 0 \Rightarrow C = 0 \Rightarrow y = \text{sech}^{-1}(x) - \sqrt{1-x^2}$

81. $V = \pi \int_0^2 (\cosh^2 x - \sinh^2 x)\, dx = \pi \int_0^2 1\, dx = 2\pi$

83. (a) $y = \frac{1}{2}\cosh 2x \Rightarrow y' = \sinh 2x \Rightarrow L = \int_0^{\ln\sqrt{5}} \sqrt{1 + (\sinh 2x)^2}\, dx = \int_0^{\ln\sqrt{5}} \cosh 2x\, dx = \left[\frac{1}{2}\sinh 2x\right]_0^{\ln\sqrt{5}}$

$= \left[\frac{1}{2}\left(\frac{e^{2x} - e^{-2x}}{2}\right)\right]_0^{\ln\sqrt{5}} = \frac{1}{4}\left(5 - \frac{1}{5}\right) = \frac{6}{5}$

(b) $y = \frac{1}{a}\cosh ax \Rightarrow 1 + (y')^2 = 1 + \sinh^2 ax \Rightarrow \cosh^2 ax \Rightarrow L = \int_0^b \sqrt{\cosh^2 ax}\, dx = \int_0^b \cosh ax\, dx = \left[\frac{\sin ax}{a}\right]_0^b$

$= \frac{\sinh ab}{a}$

85. $y = 4\cosh\frac{x}{4} \Rightarrow 1 + \left(\frac{dy}{dx}\right)^2 = 1 + \sinh^2\left(\frac{x}{4}\right) = \cosh^2\left(\frac{x}{4}\right)$; the surface area is $S = \int_{-\ln 16}^{\ln 81} 2\pi y \sqrt{1 + \left(\frac{dy}{dx}\right)^2}\, dx$

$= 8\pi \int_{-\ln 16}^{\ln 81} \cosh^2\left(\frac{x}{4}\right) dx = 4\pi \int_{-\ln 16}^{\ln 81} \left(1 + \cosh\frac{x}{2}\right) dx = 4\pi\left[x + 2\sinh\frac{x}{2}\right]_{-\ln 16}^{\ln 81}$

$= 4\pi\left[\left(\ln 81 + 2\sinh\left(\frac{\ln 81}{2}\right)\right) - \left(-\ln 16 + 2\sinh\left(\frac{-\ln 16}{2}\right)\right)\right] = 4\pi\left[\ln(81 \cdot 16) + 2\sinh(\ln 9) + 2\sinh(\ln 4)\right]$

$= 4\pi\left[\ln(9 \cdot 4)^2 + (e^{\ln 9} - e^{-\ln 9}) + (e^{\ln 4} - e^{-\ln 4})\right] = 4\pi\left[2\ln 36 + \left(9 - \frac{1}{9}\right) + \left(4 - \frac{1}{4}\right)\right] = 4\pi\left(4\ln 6 + \frac{80}{9} + \frac{15}{4}\right)$

$= 4\pi\left(4\ln 6 + \frac{320 + 135}{36}\right) = 16\pi\ln 6 + \frac{455\pi}{9}$

87. (a) $y = \frac{H}{w}\cosh\left(\frac{w}{H}x\right) \Rightarrow \tan\phi = \frac{dy}{dx} = \left(\frac{H}{w}\right)\left[\frac{w}{H}\sinh\left(\frac{w}{H}x\right)\right] = \sinh\left(\frac{w}{H}x\right)$

(b) The tension at P is given by $T\cos\phi = H \Rightarrow T = H\sec\phi = H\sqrt{1 + \tan^2\phi} = H\sqrt{1 + \left(\sinh\frac{w}{H}x\right)^2}$

$= H\cosh\left(\frac{w}{H}x\right) = w\left(\frac{H}{w}\right)\cosh\left(\frac{w}{H}x\right) = wy$

89. (a) Since the cable is 32 ft long, s = 16 and x = 15. From Exercise 88, $x = \frac{1}{a}\sinh^{-1} as \Rightarrow 15a = \sinh^{-1} 16a$

$\Rightarrow \sinh 15a = 16a$.

(b) The intersection is near (0.042, 0.672).

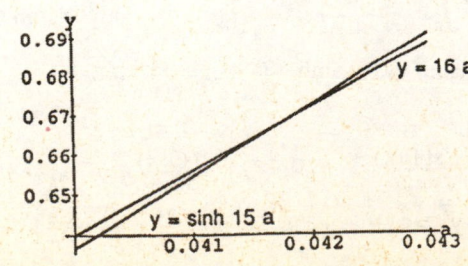

(c) Newton's method indicates that at a ≈ 0.0417525 the curves $y = 16a$ and $y = \sinh 15a$ intersect.

(d) $T = wy \approx (2\text{ lb}) \left(\frac{1}{0.0417525} \right) \approx 47.90 \text{ lb}$

(e) The sag is $\frac{1}{a}\cosh(15a) - \frac{1}{a} \approx 4.85 \text{ ft.}$

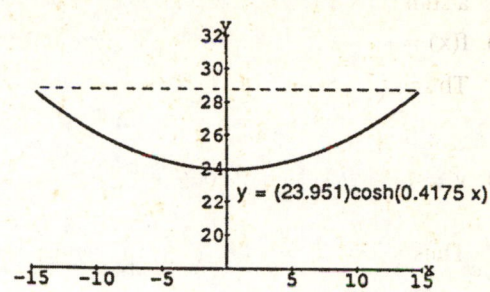

$y = (23.951)\cosh(0.4175\ x)$

CHAPTER 7 PRACTICE EXERCISES

1. $y = 10e^{-x/5} \ \Rightarrow \ \frac{dy}{dx} = (10)\left(-\frac{1}{5}\right) e^{-x/5} = -2e^{-x/5}$

3. $y = \frac{1}{4} xe^{4x} - \frac{1}{16} e^{4x} \ \Rightarrow \ \frac{dy}{dx} = \frac{1}{4}\left[x\left(4e^{4x}\right) + e^{4x}(1)\right] - \frac{1}{16}\left(4e^{4x}\right) = xe^{4x} + \frac{1}{4} e^{4x} - \frac{1}{4} e^{4x} = xe^{4x}$

5. $y = \ln\left(\sin^2 \theta\right) \ \Rightarrow \ \frac{dy}{d\theta} = \frac{2(\sin \theta)(\cos \theta)}{\sin^2 \theta} = \frac{2\cos \theta}{\sin \theta} = 2 \cot \theta$

7. $y = \log_2\left(\frac{x^2}{2}\right) = \frac{\ln\left(\frac{x^2}{2}\right)}{\ln 2} \ \Rightarrow \ \frac{dy}{dx} = \frac{1}{\ln 2}\left(\frac{x}{\left(\frac{x^2}{2}\right)}\right) = \frac{2}{(\ln 2)x}$

9. $y = 8^{-t} \ \Rightarrow \ \frac{dy}{dt} = 8^{-t}(\ln 8)(-1) = -8^{-t}(\ln 8)$

11. $y = 5x^{3.6} \ \Rightarrow \ \frac{dy}{dx} = 5(3.6)x^{2.6} = 18x^{2.6}$

13. $y = (x + 2)^{x+2} \ \Rightarrow \ \ln y = \ln(x + 2)^{x+2} = (x + 2)\ln(x + 2) \ \Rightarrow \ \frac{y'}{y} = (x + 2)\left(\frac{1}{x+2}\right) + (1)\ln(x + 2)$

   $\Rightarrow \ \frac{dy}{dx} = (x + 2)^{x+2}\left[\ln(x + 2) + 1\right]$

15. $y = \sin^{-1}\sqrt{1 - u^2} = \sin^{-1}\left(1 - u^2\right)^{1/2} \ \Rightarrow \ \frac{dy}{du} = \frac{\frac{1}{2}\left(1 - u^2\right)^{-1/2}(-2u)}{\sqrt{1 - \left[\left(1 - u^2\right)^{1/2}\right]^2}} = \frac{-u}{\sqrt{1 - u^2}\sqrt{1 - \left(1 - u^2\right)}} = \frac{-u}{|u|\sqrt{1 - u^2}}$

   $= \frac{-u}{u\sqrt{1 - u^2}} = \frac{-1}{\sqrt{1 - u^2}}, 0 < u < 1$

17. $y = \ln\left(\cos^{-1} x\right) \ \Rightarrow \ y' = \frac{\left(\frac{-1}{\sqrt{1 - x^2}}\right)}{\cos^{-1} x} = \frac{-1}{\sqrt{1 - x^2}\cos^{-1} x}$

19. $y = t\tan^{-1} t - \left(\frac{1}{2}\right)\ln t \ \Rightarrow \ \frac{dy}{dt} = \tan^{-1} t + t\left(\frac{1}{1 + t^2}\right) - \left(\frac{1}{2}\right)\left(\frac{1}{t}\right) = \tan^{-1} t + \frac{t}{1 + t^2} - \frac{1}{2t}$

21. $y = z\sec^{-1} z - \sqrt{z^2 - 1} = z\sec^{-1} z - \left(z^2 - 1\right)^{1/2} \ \Rightarrow \ \frac{dy}{dz} = z\left(\frac{1}{|z|\sqrt{z^2 - 1}}\right) + \left(\sec^{-1} z\right)(1) - \frac{1}{2}\left(z^2 - 1\right)^{-1/2}(2z)$

   $= \frac{z}{|z|\sqrt{z^2 - 1}} - \frac{z}{\sqrt{z^2 - 1}} + \sec^{-1} z = \frac{1 - z}{\sqrt{z^2 - 1}} + \sec^{-1} z, z > 1$

23. $y = \csc^{-1}(\sec \theta) \ \Rightarrow \ \frac{dy}{d\theta} = \frac{-\sec \theta \tan \theta}{|\sec \theta|\sqrt{\sec^2 \theta - 1}} = -\frac{\tan \theta}{|\tan \theta|} = -1, 0 < \theta < \frac{\pi}{2}$

25. $y = \frac{2\left(x^2 + 1\right)}{\sqrt{\cos 2x}} \ \Rightarrow \ \ln y = \ln\left(\frac{2\left(x^2 + 1\right)}{\sqrt{\cos 2x}}\right) = \ln(2) + \ln\left(x^2 + 1\right) - \frac{1}{2}\ln(\cos 2x) \ \Rightarrow \ \frac{y'}{y} = 0 + \frac{2x}{x^2 + 1} - \left(\frac{1}{2}\right)\frac{(-2\sin 2x)}{\cos 2x}$

   $\Rightarrow \ y' = \left(\frac{2x}{x^2 + 1} + \tan 2x\right)y = \frac{2\left(x^2 + 1\right)}{\sqrt{\cos 2x}}\left(\frac{2x}{x^2 + 1} + \tan 2x\right)$

27. $y = \left[\frac{(t+1)(t-1)}{(t-2)(t+3)}\right]^5 \Rightarrow \ln y = 5\left[\ln(t+1) + \ln(t-1) - \ln(t-2) - \ln(t+3)\right] \Rightarrow \left(\frac{1}{y}\right)\left(\frac{dy}{dt}\right)$

$= 5\left(\frac{1}{t+1} + \frac{1}{t-1} - \frac{1}{t-2} - \frac{1}{t+3}\right) \Rightarrow \frac{dy}{dt} = 5\left[\frac{(t+1)(t-1)}{(t-2)(t+3)}\right]^5 \left(\frac{1}{t+1} + \frac{1}{t-1} - \frac{1}{t-2} - \frac{1}{t+3}\right)$

29. $y = (\sin\theta)^{\sqrt{\theta}} \Rightarrow \ln y = \sqrt{\theta}\ln(\sin\theta) \Rightarrow \left(\frac{1}{y}\right)\left(\frac{dy}{d\theta}\right) = \sqrt{\theta}\left(\frac{\cos\theta}{\sin\theta}\right) + \frac{1}{2}\theta^{-1/2}\ln(\sin\theta)$

$\Rightarrow \frac{dy}{d\theta} = (\sin\theta)^{\sqrt{\theta}}\left(\sqrt{\theta}\cot\theta + \frac{\ln(\sin\theta)}{2\sqrt{\theta}}\right)$

31. $\int e^x \sin(e^x)\, dx = \int \sin u\, du$, where $u = e^x$ and $du = e^x\, dx$

$= -\cos u + C = -\cos(e^x) + C$

33. $\int e^x \sec^2(e^x - 7)\, dx = \int \sec^2 u\, du$, where $u = e^x - 7$ and $du = e^x\, dx$

$= \tan u + C = \tan(e^x - 7) + C$

35. $\int (\sec^2 x)\, e^{\tan x}\, dx = \int e^u\, du$, where $u = \tan x$ and $du = \sec^2 x\, dx$

$= e^u + C = e^{\tan x} + C$

37. $\int_{-1}^{1} \frac{1}{3x-4}\, dx = \frac{1}{3}\int_{-7}^{-1} \frac{1}{u}\, du$, where $u = 3x - 4$, $du = 3\, dx$; $x = -1 \Rightarrow u = -7$, $x = 1 \Rightarrow u = -1$

$= \frac{1}{3}\left[\ln|u|\right]_{-7}^{-1} = \frac{1}{3}\left[\ln|-1| - \ln|-7|\right] = \frac{1}{3}\left[0 - \ln 7\right] = -\frac{\ln 7}{3}$

39. $\int_0^{\pi} \tan\left(\frac{x}{3}\right)\, dx = \int_0^{\pi} \frac{\sin\left(\frac{x}{3}\right)}{\cos\left(\frac{x}{3}\right)}\, dx = -3\int_1^{1/2} \frac{1}{u}\, du$, where $u = \cos\left(\frac{x}{3}\right)$, $du = -\frac{1}{3}\sin\left(\frac{x}{3}\right)\, dx$; $x = 0 \Rightarrow u = 1$, $x = \pi$

$\Rightarrow u = \frac{1}{2}$

$= -3\left[\ln|u|\right]_1^{1/2} = -3\left[\ln\left|\frac{1}{2}\right| - \ln|1|\right] = -3\ln\frac{1}{2} = \ln 2^3 = \ln 8$

41. $\int_0^4 \frac{2t}{t^2 - 25}\, dt = \int_{-25}^{-9} \frac{1}{u}\, du$, where $u = t^2 - 25$, $du = 2t\, dt$; $t = 0 \Rightarrow u = -25$, $t = 4 \Rightarrow u = -9$

$= \left[\ln|u|\right]_{-25}^{-9} = \ln|-9| - \ln|-25| = \ln 9 - \ln 25 = \ln\frac{9}{25}$

43. $\int \frac{\tan(\ln v)}{v}\, dv = \int \tan u\, du = \int \frac{\sin u}{\cos u}\, du$, where $u = \ln v$ and $du = \frac{1}{v}\, dv$

$= -\ln|\cos u| + C = -\ln|\cos(\ln v)| + C$

45. $\int \frac{(\ln x)^{-3}}{x}\, dx = \int u^{-3}\, du$, where $u = \ln x$ and $du = \frac{1}{x}\, dx$

$= \frac{u^{-2}}{-2} + C = -\frac{1}{2}(\ln x)^{-2} + C$

47. $\int \frac{1}{r}\csc^2(1 + \ln r)\, dr = \int \csc^2 u\, du$, where $u = 1 + \ln r$ and $du = \frac{1}{r}\, dr$

$= -\cot u + C = -\cot(1 + \ln r) + C$

49. $\int x3^{x^2}\, dx = \frac{1}{2}\int 3^u\, du$, where $u = x^2$ and $du = 2x\, dx$

$= \frac{1}{2\ln 3}(3^u) + C = \frac{1}{2\ln 3}\left(3^{x^2}\right) + C$

51. $\int_1^7 \frac{3}{x}\, dx = 3\int_1^7 \frac{1}{x}\, dx = 3\left[\ln|x|\right]_1^7 = 3(\ln 7 - \ln 1) = 3\ln 7$

53. $\int_1^4 \left(\frac{x}{8} + \frac{1}{2x}\right)\, dx = \frac{1}{2}\int_1^4 \left(\frac{1}{4}x + \frac{1}{x}\right)\, dx = \frac{1}{2}\left[\frac{1}{8}x^2 + \ln|x|\right]_1^4 = \frac{1}{2}\left[\left(\frac{16}{8} + \ln 4\right) - \left(\frac{1}{8} + \ln 1\right)\right] = \frac{15}{16} + \frac{1}{2}\ln 4$

$= \frac{15}{16} + \ln\sqrt{4} = \frac{15}{16} + \ln 2$

55. $\int_{-2}^{-1} e^{-(x+1)} \, dx = -\int_{1}^{0} e^u \, du$, where $u = -(x+1)$, $du = -dx$; $x = -2 \Rightarrow u = 1$, $x = -1 \Rightarrow u = 0$

$\qquad = -\left[e^u\right]_1^0 = -\left(e^0 - e^1\right) = e - 1$

57. $\int_{1}^{\ln 5} e^r \left(3e^r + 1\right)^{-3/2} \, dr = \frac{1}{3} \int_{4}^{16} u^{-3/2} \, du$, where $u = 3e^r + 1$, $du = 3e^r dr$; $r = 0 \Rightarrow u = 4$, $r = \ln 5 \Rightarrow u = 16$

$\qquad = -\frac{2}{3} \left[u^{-1/2}\right]_4^{16} = -\frac{2}{3} \left(16^{-1/2} - 4^{-1/2}\right) = \left(-\frac{2}{3}\right)\left(\frac{1}{4} - \frac{1}{2}\right) = \left(-\frac{2}{3}\right)\left(-\frac{1}{4}\right) = \frac{1}{6}$

59. $\int_{1}^{e} \frac{1}{x} \left(1 + 7 \ln x\right)^{-1/3} \, dx = \frac{1}{7} \int_{1}^{8} u^{-1/3} \, du$, where $u = 1 + 7 \ln x$, $du = \frac{7}{x} \, dx$, $x = 1 \Rightarrow u = 1$, $x = e \Rightarrow u = 8$

$\qquad = \frac{3}{14} \left[u^{2/3}\right]_1^8 = \frac{3}{14} \left(8^{2/3} - 1^{2/3}\right) = \left(\frac{3}{14}\right)(4 - 1) = \frac{9}{14}$

61. $\int_{1}^{3} \frac{[\ln (v+1)]^2}{v+1} \, dv = \int_{1}^{3} \left[\ln(v+1)\right]^2 \frac{1}{v+1} \, dv = \int_{\ln 2}^{\ln 4} u^2 \, du$, where $u = \ln(v+1)$, $du = \frac{1}{v+1} \, dv$;

$\qquad\qquad\qquad\qquad\qquad\qquad\qquad\qquad\qquad\qquad\qquad v = 1 \Rightarrow u = \ln 2$, $v = 3 \Rightarrow u = \ln 4$;

$\qquad = \frac{1}{3} \left[u^3\right]_{\ln 2}^{\ln 4} = \frac{1}{3} \left[(\ln 4)^3 - (\ln 2)^3\right] = \frac{1}{3} \left[(2 \ln 2)^3 - (\ln 2)^3\right] = \frac{(\ln 2)^3}{3} (8 - 1) = \frac{7}{3} (\ln 2)^3$

63. $\int_{1}^{8} \frac{\log_4 \theta}{\theta} \, d\theta = \frac{1}{\ln 4} \int_{1}^{8} (\ln \theta) \left(\frac{1}{\theta}\right) d\theta = \frac{1}{\ln 4} \int_{0}^{\ln 8} u \, du$, where $u = \ln \theta$, $du = \frac{1}{\theta} \, d\theta$, $\theta = 1 \Rightarrow u = 0$, $\theta = 8 \Rightarrow u = \ln 8$

$\qquad = \frac{1}{2 \ln 4} \left[u^2\right]_0^{\ln 8} = \frac{1}{\ln 16} \left[(\ln 8)^2 - 0^2\right] = \frac{(3 \ln 2)^2}{4 \ln 2} = \frac{9 \ln 2}{4}$

65. $\int_{-3/4}^{3/4} \frac{6}{\sqrt{9 - 4x^2}} \, dx = 3 \int_{-3/4}^{3/4} \frac{2}{\sqrt{3^2 - (2x)^2}} \, dx = 3 \int_{-3/2}^{3/2} \frac{1}{\sqrt{3^2 - u^2}} \, du$, where $u = 2x$, $du = 2 \, dx$;

$\qquad\qquad\qquad\qquad\qquad\qquad\qquad\qquad\qquad\qquad\qquad\qquad x = -\frac{3}{4} \Rightarrow u = -\frac{3}{2}$, $x = \frac{3}{4} \Rightarrow u = \frac{3}{2}$

$\qquad = 3 \left[\sin^{-1}\left(\frac{u}{3}\right)\right]_{-3/2}^{3/2} = 3 \left[\sin^{-1}\left(\frac{1}{2}\right) - \sin^{-1}\left(-\frac{1}{2}\right)\right] = 3 \left[\frac{\pi}{6} - \left(-\frac{\pi}{6}\right)\right] = 3 \left(\frac{\pi}{3}\right) = \pi$

67. $\int_{-2}^{2} \frac{3}{4 + 3t^2} \, dt = \sqrt{3} \int_{-2}^{2} \frac{\sqrt{3}}{2^2 + \left(\sqrt{3}t\right)^2} \, dt = \sqrt{3} \int_{-2\sqrt{3}}^{2\sqrt{3}} \frac{1}{2^2 + u^2} \, du$, where $u = \sqrt{3}t$, $du = \sqrt{3} \, dt$;

$\qquad\qquad\qquad\qquad\qquad\qquad\qquad\qquad\qquad\qquad t = -2 \Rightarrow u = -2\sqrt{3}$, $t = 2 \Rightarrow u = 2\sqrt{3}$

$\qquad = \sqrt{3} \left[\frac{1}{2} \tan^{-1}\left(\frac{u}{2}\right)\right]_{-2\sqrt{3}}^{2\sqrt{3}} = \frac{\sqrt{3}}{2} \left[\tan^{-1}\left(\sqrt{3}\right) - \tan^{-1}\left(-\sqrt{3}\right)\right] = \frac{\sqrt{3}}{2} \left[\frac{\pi}{3} - \left(-\frac{\pi}{3}\right)\right] = \frac{\pi}{\sqrt{3}}$

69. $\int \frac{1}{y\sqrt{4y^2 - 1}} \, dy = \int \frac{2}{(2y)\sqrt{(2y)^2 - 1}} \, dy = \int \frac{1}{u\sqrt{u^2 - 1}} \, du$, where $u = 2y$ and $du = 2 \, dy$

$\qquad = \sec^{-1} |u| + C = \sec^{-1} |2y| + C$

71. $\int_{\sqrt{2}/3}^{2/3} \frac{1}{|y| \sqrt{9y^2 - 1}} \, dy = \int_{\sqrt{2}/3}^{2/3} \frac{3}{|3y| \sqrt{(3y)^2 - 1}} \, dy = \int_{\sqrt{2}}^{2} \frac{1}{|u| \sqrt{u^2 - 1}} \, du$, where $u = 3y$, $du = 3 \, dy$;

$\qquad\qquad\qquad\qquad\qquad\qquad\qquad\qquad\qquad\qquad\qquad y = \frac{\sqrt{2}}{3} \Rightarrow u = \sqrt{2}$, $y = \frac{2}{3} \Rightarrow u = 2$

$\qquad = \left[\sec^{-1} u\right]_{\sqrt{2}}^{2} = \left[\sec^{-1} 2 - \sec^{-1} \sqrt{2}\right] = \frac{\pi}{3} - \frac{\pi}{4} = \frac{\pi}{12}$

73. $\int \frac{1}{\sqrt{-2x - x^2}} \, dx = \int \frac{1}{\sqrt{1 - (x^2 + 2x + 1)}} \, dx = \int \frac{1}{\sqrt{1 - (x+1)^2}} \, dx = \int \frac{1}{\sqrt{1 - u^2}} \, du$, where $u = x + 1$ and

$\qquad\qquad\qquad\qquad\qquad\qquad\qquad\qquad\qquad\qquad\qquad\qquad\qquad\qquad du = dx$

$\qquad = \sin^{-1} u + C = \sin^{-1} (x+1) + C$

75. $\int_{-2}^{-1} \frac{2}{v^2 + 4v + 5} \, dv = 2 \int_{-2}^{-1} \frac{1}{1 + (v^2 + 4v + 4)} \, dv = 2 \int_{-2}^{-1} \frac{1}{1 + (v+2)^2} \, dv = 2 \int_{0}^{1} \frac{1}{1 + u^2} \, du$,

$\qquad\qquad\qquad\qquad\qquad\qquad\qquad\qquad\qquad\qquad\qquad\text{where } u = v + 2, \, du = dv; \, v = -2 \Rightarrow u = 0, \, v = -1 \Rightarrow u = 1$

$\qquad = 2 \left[\tan^{-1} u\right]_0^1 = 2 \left(\tan^{-1} 1 - \tan^{-1} 0\right) = 2 \left(\frac{\pi}{4} - 0\right) = \frac{\pi}{2}$

77. $\int \frac{1}{(t+1)\sqrt{t^2+2t-8}}\, dt = \int \frac{1}{(t+1)\sqrt{(t^2+2t+1)-9}}\, dt = \int \frac{1}{(t+1)\sqrt{(t+1)^2-3^2}}\, dt = \int \frac{1}{u\sqrt{u^2-3^2}}\, du$

where $u = t+1$ and $du = dt$

$\qquad = \frac{1}{3}\sec^{-1}\left|\frac{u}{3}\right| + C = \frac{1}{3}\sec^{-1}\left|\frac{t+1}{3}\right| + C$

79. $3^y = 2^{y+1} \Rightarrow \ln 3^y = \ln 2^{y+1} \Rightarrow y(\ln 3) = (y+1)\ln 2 \Rightarrow (\ln 3 - \ln 2)y = \ln 2 \Rightarrow \left(\ln \frac{3}{2}\right)y = \ln 2 \Rightarrow y = \frac{\ln 2}{\ln\left(\frac{3}{2}\right)}$

81. $9e^{2y} = x^2 \Rightarrow e^{2y} = \frac{x^2}{9} \Rightarrow \ln e^{2y} = \ln\left(\frac{x^2}{9}\right) \Rightarrow 2y(\ln e) = \ln\left(\frac{x^2}{9}\right) \Rightarrow y = \frac{1}{2}\ln\left(\frac{x^2}{9}\right) = \ln\sqrt{\frac{x^2}{9}} = \ln\left|\frac{x}{3}\right| = \ln|x| - \ln 3$

83. $\ln(y-1) = x + \ln y \Rightarrow e^{\ln(y-1)} = e^{(x+\ln y)} = e^x e^{\ln y} \Rightarrow y - 1 = ye^x \Rightarrow y - ye^x = 1 \Rightarrow y(1-e^x) = 1$

$\qquad \Rightarrow y = \frac{1}{1-e^x}$

85. The limit leads to the indeterminate form $\frac{0}{0}$: $\displaystyle\lim_{x\to 0}\frac{10^x - 1}{x} = \lim_{x\to 0}\frac{(\ln 10)10^x}{1} = \ln 10$

87. The limit leads to the indeterminate form $\frac{0}{0}$: $\displaystyle\lim_{x\to 0}\frac{2^{\sin x} - 1}{e^x - 1} = \lim_{x\to 0}\frac{2^{\sin x}(\ln 2)(\cos x)}{e^x} = \ln 2$

89. The limit leads to the indeterminate form $\frac{0}{0}$: $\displaystyle\lim_{x\to 0}\frac{5 - 5\cos x}{e^x - x - 1} = \lim_{x\to 0}\frac{5\sin x}{e^x - 1} = \lim_{x\to 0}\frac{5\cos x}{e^x} = 5$

91. The limit leads to the indeterminate form $\frac{0}{0}$: $\displaystyle\lim_{t\to 0^+}\frac{t - \ln(1+2t)}{t^2} = \lim_{t\to 0^+}\frac{\left(1 - \frac{2}{1+2t}\right)}{2t} = -\infty$

93. The limit leads to the indeterminate form $\frac{0}{0}$: $\displaystyle\lim_{t\to 0^+}\left(\frac{e^t}{t} - \frac{1}{t}\right) = \lim_{t\to 0^+}\left(\frac{e^t - 1}{t}\right) = \lim_{t\to 0^+}\frac{e^t}{1} = 1$

95. Let $f(x) = \left(1 + \frac{3}{x}\right)^x \Rightarrow \ln f(x) = \frac{\ln(1+3x^{-1})}{x^{-1}} \Rightarrow \displaystyle\lim_{x\to\infty}\ln f(x) = \lim_{x\to\infty}\frac{\ln(1+3x^{-1})}{x^{-1}}$; the limit leads to the

indeterminate form $\frac{0}{0}$: $\displaystyle\lim_{x\to\infty}\frac{\left(\frac{-3x^{-2}}{1+3x^{-1}}\right)}{-x^{-2}} = \lim_{x\to\infty}\frac{3}{1+\frac{3}{x}} = 3 \Rightarrow \lim_{x\to\infty}\left(1+\frac{3}{x}\right)^x = \lim_{x\to\infty}e^{\ln f(x)} = e^3$

97. (a) $\displaystyle\lim_{x\to\infty}\frac{\log_2 x}{\log_3 x} = \lim_{x\to\infty}\frac{\left(\frac{\ln x}{\ln 2}\right)}{\left(\frac{\ln x}{\ln 3}\right)} = \lim_{x\to\infty}\frac{\ln 3}{\ln 2} = \frac{\ln 3}{\ln 2} \Rightarrow$ same rate

(b) $\displaystyle\lim_{x\to\infty}\frac{x}{x+\left(\frac{1}{x}\right)} = \lim_{x\to\infty}\frac{x^2}{x^2+1} = \lim_{x\to\infty}\frac{2x}{2x} = \lim_{x\to\infty}1 = 1 \Rightarrow$ same rate

(c) $\displaystyle\lim_{x\to\infty}\frac{\left(\frac{x}{100}\right)}{xe^{-x}} = \lim_{x\to\infty}\frac{xe^x}{100x} = \lim_{x\to\infty}\frac{e^x}{100} = \infty \Rightarrow$ faster

(d) $\displaystyle\lim_{x\to\infty}\frac{x}{\tan^{-1}x} = \infty \Rightarrow$ faster

(e) $\displaystyle\lim_{x\to\infty}\frac{\csc^{-1}x}{\left(\frac{1}{x}\right)} = \lim_{x\to\infty}\frac{\sin^{-1}(x^{-1})}{x^{-1}} = \lim_{x\to\infty}\frac{\frac{(-x^{-2})}{\sqrt{1-(x^{-1})^2}}}{-x^{-2}} = \lim_{x\to\infty}\frac{1}{\sqrt{1-\left(\frac{1}{x^2}\right)}} = 1 \Rightarrow$ same rate

(f) $\displaystyle\lim_{x\to\infty}\frac{\sinh x}{e^x} = \lim_{x\to\infty}\frac{(e^x - e^{-x})}{2e^x} = \lim_{x\to\infty}\frac{1-e^{-2x}}{2} = \frac{1}{2} \Rightarrow$ same rate

99. (a) $\dfrac{\left(\frac{1}{x^2}+\frac{1}{x^4}\right)}{\left(\frac{1}{x^2}\right)} = 1 + \frac{1}{x^2} \le 2$ for x sufficiently large $\Rightarrow$ true

(b) $\dfrac{\left(\frac{1}{x^2}+\frac{1}{x^4}\right)}{\left(\frac{1}{x^4}\right)} = x^2 + 1 > M$ for any positive integer M whenever $x > \sqrt{M} \Rightarrow$ false

(c) $\displaystyle\lim_{x\to\infty}\frac{x}{x+\ln x} = \lim_{x\to\infty}\frac{1}{1+\frac{1}{x}} = 1 \Rightarrow$ the same growth rate $\Rightarrow$ false

(d) $\lim\limits_{x \to \infty} \frac{\ln(\ln x)}{\ln x} = \lim\limits_{x \to \infty} \frac{\left[\frac{\left(\frac{1}{x}\right)}{\ln x}\right]}{\left(\frac{1}{x}\right)} = \lim\limits_{x \to \infty} \frac{1}{\ln x} = 0 \Rightarrow$ grows slower $\Rightarrow$ true

(e) $\frac{\tan^{-1} x}{1} \leq \frac{\pi}{2}$ for all $x \Rightarrow$ true

(f) $\frac{\cosh x}{e^x} = \frac{1}{2}(1 + e^{-2x}) \leq \frac{1}{2}(1 + 1) = 1$ if $x > 0 \Rightarrow$ true

101. $\frac{df}{dx} = e^x + 1 \Rightarrow \left(\frac{df^{-1}}{dx}\right)_{x=f(\ln 2)} = \frac{1}{\left(\frac{df}{dx}\right)_{x=\ln 2}} \Rightarrow \left(\frac{df^{-1}}{dx}\right)_{x=f(\ln 2)} = \frac{1}{(e^x + 1)_{x=\ln 2}} = \frac{1}{2+1} = \frac{1}{3}$

103. $y = x \ln 2x - x \Rightarrow y' = x\left(\frac{2}{2x}\right) + \ln(2x) - 1 = \ln 2x$;

    solving $y' = 0 \Rightarrow x = \frac{1}{2}$; $y' > 0$ for $x > \frac{1}{2}$ and $y' < 0$ for

    $x < \frac{1}{2} \Rightarrow$ relative minimum of $-\frac{1}{2}$ at $x = \frac{1}{2}$; $f\left(\frac{1}{2e}\right) = -\frac{1}{e}$

    and $f\left(\frac{e}{2}\right) = 0 \Rightarrow$ absolute minimum is $-\frac{1}{2}$ at $x = \frac{1}{2}$ and the

    absolute maximum is 0 at $x = \frac{e}{2}$

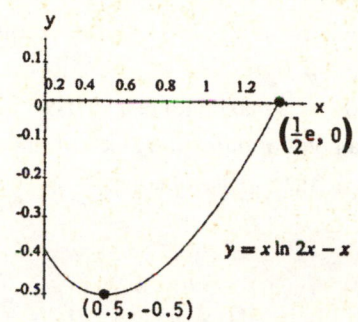

105. $A = \int_1^e \frac{2 \ln x}{x}\, dx = \int_0^1 2u\, du = [u^2]_0^1 = 1$, where

    $u = \ln x$ and $du = \frac{1}{x}\, dx$; $x = 1 \Rightarrow u = 0$, $x = e \Rightarrow u = 1$

107. $y = \ln x \Rightarrow \frac{dy}{dx} = \frac{1}{x}$; $\frac{dy}{dt} = \frac{dy}{dx} \frac{dx}{dt} \Rightarrow \frac{dy}{dt} = \left(\frac{1}{x}\right) \sqrt{x} = \frac{1}{\sqrt{x}} \Rightarrow \left.\frac{dy}{dt}\right|_{e^2} = \frac{1}{e}$ m/sec

109. $A = xy = xe^{-x^2} \Rightarrow \frac{dA}{dx} = e^{-x^2} + (x)(-2x)e^{-x^2} = e^{-x^2}(1 - 2x^2)$. Solving $\frac{dA}{dx} = 0 \Rightarrow 1 - 2x^2 = 0$

    $\Rightarrow x = \frac{1}{\sqrt{2}}$; $\frac{dA}{dx} < 0$ for $x > \frac{1}{\sqrt{2}}$ and $\frac{dA}{dx} > 0$ for $0 < x < \frac{1}{\sqrt{2}} \Rightarrow$ absolute maximum of $\frac{1}{\sqrt{2}} e^{-1/2} = \frac{1}{\sqrt{2e}}$ at

    $x = \frac{1}{\sqrt{2}}$ units long by $y = e^{-1/2} = \frac{1}{\sqrt{e}}$ units high.

111. $K = \ln(5x) - \ln(3x) = \ln 5 + \ln x - \ln 3 - \ln x = \ln 5 - \ln 3 = \ln \frac{5}{3}$

113. $\frac{\log_4 x}{\log_2 x} = \frac{\left(\frac{\ln x}{\ln 4}\right)}{\left(\frac{\ln x}{\ln 2}\right)} = \frac{\ln x}{\ln 4} \cdot \frac{\ln 2}{\ln x} = \frac{\ln 2}{\ln 4} = \frac{\ln 2}{2 \ln 2} = \frac{1}{2}$

115. (a) $y = \frac{\ln x}{\sqrt{x}} \Rightarrow y' = \frac{1}{x\sqrt{x}} - \frac{\ln x}{2x^{3/2}} = \frac{2 - \ln x}{2x\sqrt{x}}$

    $\Rightarrow y'' = -\frac{3}{4} x^{-5/2}(2 - \ln x) - \frac{1}{2} x^{-5/2} = x^{-5/2}\left(\frac{3}{4} \ln x - 2\right)$;

    solving $y' = 0 \Rightarrow \ln x = 2 \Rightarrow x = e^2$; $y' < 0$ for $x > e^2$ and

    and $y' > 0$ for $x < e^2 \Rightarrow$ a maximum of $\frac{2}{e}$; $y'' = 0$

    $\Rightarrow \ln x = \frac{8}{3} \Rightarrow x = e^{8/3}$; the curve is concave down on

    $\left(0, e^{8/3}\right)$ and concave up on $\left(e^{8/3}, \infty\right)$; so there is an

    inflection point at $\left(e^{8/3}, \frac{8}{3e^{4/3}}\right)$.

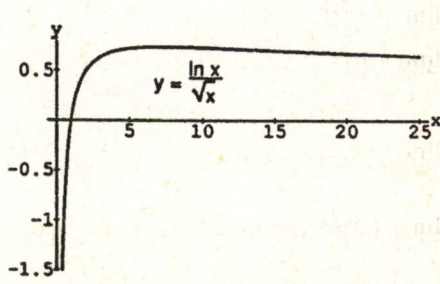

(b) $y = e^{-x^2} \Rightarrow y' = -2xe^{-x^2} \Rightarrow y'' = -2e^{-x^2} + 4x^2 e^{-x^2}$
$= (4x^2 - 2)e^{-x^2}$; solving $y' = 0 \Rightarrow x = 0$; $y' < 0$ for
$x > 0$ and $y' > 0$ for $x < 0 \Rightarrow$ a maximum at $x = 0$ of
$e^0 = 1$; there are points of inflection at $x = \pm \frac{1}{\sqrt{2}}$; the
curve is concave down for $-\frac{1}{\sqrt{2}} < x < \frac{1}{\sqrt{2}}$ and concave
up otherwise.

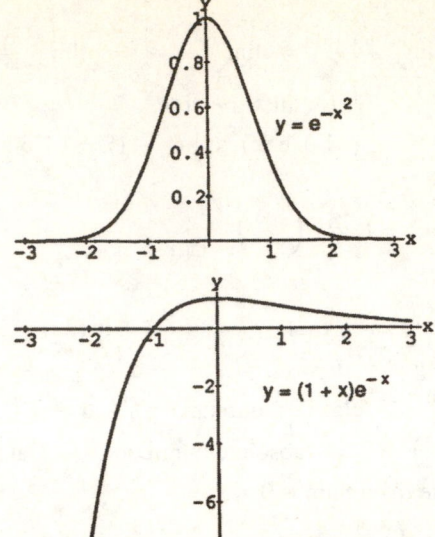

(c) $y = (1 + x)e^{-x} \Rightarrow y' = e^{-x} - (1 + x)e^{-x} = -xe^{-x}$
$\Rightarrow y'' = -e^{-x} + xe^{-x} = (x - 1)e^{-x}$; solving $y' = 0$
$\Rightarrow -xe^{-x} = 0 \Rightarrow x = 0$; $y' < 0$ for $x > 0$ and $y' > 0$
for $x < 0 \Rightarrow$ a maximum at $x = 0$ of $(1 + 0)e^0 = 1$;
there is a point of inflection at $x = 1$ and the curve is
concave up for $x > 1$ and concave down for $x < 1$.

117. Since the half life is 5700 years and $A(t) = A_0 e^{kt}$ we have $\frac{A_0}{2} = A_0 e^{5700k} \Rightarrow \frac{1}{2} = e^{5700k} \Rightarrow \ln(0.5) = 5700k$
$\Rightarrow k = \frac{\ln(0.5)}{5700}$. With 10% of the original carbon-14 remaining we have $0.1A_0 = A_0 e^{\frac{\ln(0.5)}{5700}t} \Rightarrow 0.1 = e^{\frac{\ln(0.5)}{5700}t}$
$\Rightarrow \ln(0.1) = \frac{\ln(0.5)}{5700}t \Rightarrow t = \frac{(5700)\ln(0.1)}{\ln(0.5)} \approx 18,935$ years (rounded to the nearest year).

119. $\theta = \pi - \cot^{-1}\left(\frac{x}{60}\right) - \cot^{-1}\left(\frac{5}{3} - \frac{x}{30}\right), 0 < x < 50 \Rightarrow \frac{d\theta}{dx} = \frac{\left(\frac{1}{60}\right)}{1 + \left(\frac{x}{60}\right)^2} + \frac{\left(-\frac{1}{30}\right)}{1 + \left(\frac{50-x}{30}\right)^2}$
$= 30\left[\frac{2}{60^2 + x^2} - \frac{1}{30^2 + (50-x)^2}\right]$; solving $\frac{d\theta}{dx} = 0 \Rightarrow x^2 - 200x + 3200 = 0 \Rightarrow x = 100 \pm 20\sqrt{17}$, but
$100 + 20\sqrt{17}$ is not in the domain; $\frac{d\theta}{dx} > 0$ for $x < 20\left(5 - \sqrt{17}\right)$ and $\frac{d\theta}{dx} < 0$ for $20\left(5 - \sqrt{17}\right) < x < 50$
$\Rightarrow x = 20\left(5 - \sqrt{17}\right) \approx 17.54$ m maximizes θ

CHAPTER 7 ADDITIONAL AND ADVANCED EXERCISES

1. $\lim\limits_{b \to 1^-} \int_0^b \frac{1}{\sqrt{1-x^2}}\, dx = \lim\limits_{b \to 1^-} \left[\sin^{-1}x\right]_0^b = \lim\limits_{b \to 1^-} \left(\sin^{-1}b - \sin^{-1}0\right) = \lim\limits_{b \to 1^-} \left(\sin^{-1}b - 0\right) = \lim\limits_{b \to 1^-} \sin^{-1}b = \frac{\pi}{2}$

3. $y = \left(\cos\sqrt{x}\right)^{1/x} \Rightarrow \ln y = \frac{1}{x}\ln\left(\cos\sqrt{x}\right)$ and $\lim\limits_{x \to 0^+} \frac{\ln\left(\cos\sqrt{x}\right)}{x} = \lim\limits_{x \to 0^+} \frac{-\sin\sqrt{x}}{2\sqrt{x}\cos\sqrt{x}} = \frac{-1}{2}\lim\limits_{x \to 0^+} \frac{\tan\sqrt{x}}{\sqrt{x}}$
$= -\frac{1}{2}\lim\limits_{x \to 0^+} \frac{\frac{1}{2}x^{-1/2}\sec^2\sqrt{x}}{\frac{1}{2}x^{-1/2}} = -\frac{1}{2} \Rightarrow \lim\limits_{x \to 0^+} \left(\cos\sqrt{x}\right)^{1/x} = e^{-1/2} = \frac{1}{\sqrt{e}}$

5. $\lim\limits_{x \to \infty} \left(\frac{1}{n+1} + \frac{1}{n+2} + \dots + \frac{1}{2n}\right) = \lim\limits_{x \to \infty} \left(\left(\frac{1}{n}\right)\left[\frac{1}{1+\left(\frac{1}{n}\right)}\right] + \left(\frac{1}{n}\right)\left[\frac{1}{1+2\left(\frac{1}{n}\right)}\right] + \dots + \left(\frac{1}{n}\right)\left[\frac{1}{1+n\left(\frac{1}{n}\right)}\right]\right)$
which can be interpreted as a Riemann sum with partitioning $\Delta x = \frac{1}{n} \Rightarrow \lim\limits_{x \to \infty} \left(\frac{1}{n+1} + \frac{1}{n+2} + \dots + \frac{1}{2n}\right)$
$= \int_0^1 \frac{1}{1+x}\, dx = \left[\ln(1+x)\right]_0^1 = \ln 2$

7. $A(t) = \int_0^t e^{-x}\, dx = \left[-e^{-x}\right]_0^t = 1 - e^{-t}$, $V(t) = \pi\int_0^t e^{-2x}\, dx = \left[-\frac{\pi}{2}e^{-2x}\right]_0^t = \frac{\pi}{2}\left(1 - e^{-2t}\right)$

   (a) $\lim\limits_{t \to \infty} A(t) = \lim\limits_{t \to \infty} \left(1 - e^{-t}\right) = 1$

   (b) $\lim\limits_{t \to \infty} \frac{V(t)}{A(t)} = \lim\limits_{t \to \infty} \frac{\frac{\pi}{2}\left(1 - e^{-2t}\right)}{1 - e^{-t}} = \frac{\pi}{2}$

   (c) $\lim\limits_{t \to 0^+} \frac{V(t)}{A(t)} = \lim\limits_{t \to 0^+} \frac{\frac{\pi}{2}\left(1 - e^{-2t}\right)}{1 - e^{-t}} = \lim\limits_{t \to 0^+} \frac{\frac{\pi}{2}\left(1 - e^{-t}\right)\left(1 + e^{-t}\right)}{\left(1 - e^{-t}\right)} = \lim\limits_{t \to 0^+} \frac{\pi}{2}\left(1 + e^{-t}\right) = \pi$

9. $A_1 = \int_1^e \frac{2\log_2 x}{x}\,dx = \frac{2}{\ln 2}\int_1^e \frac{\ln x}{x}\,dx = \left[\frac{(\ln x)^2}{\ln 2}\right]_1^e = \frac{1}{\ln 2}$; $A_2 = \int_1^e \frac{2\log_4 x}{4}\,dx = \frac{2}{\ln 4}\int_1^e \frac{\ln x}{x}\,dx$

$= \left[\frac{(\ln x)^2}{2\ln 2}\right]_1^e = \frac{1}{2\ln 2} \Rightarrow A_1 : A_2 = 2 : 1$

11. $\ln x^{(x^x)} = x^x \ln x$ and $\ln (x^x)^x = x \ln x^x = x^2 \ln x$; then, $x^x \ln x = x^2 \ln x \Rightarrow (x^x - x^2)\ln x = 0 \Rightarrow x^x = x^2$ or $\ln x = 0$.

$\ln x = 0 \Rightarrow x = 1$; $x^x = x^2 \Rightarrow x \ln x = 2\ln x \Rightarrow x = 2$. Therefore, $x^{(x^x)} = (x^x)^x$ when $x = 2$ or $x = 1$.

13. $f(x) = e^{g(x)} \Rightarrow f'(x) = e^{g(x)}\,g'(x)$, where $g'(x) = \frac{x}{1+x^4} \Rightarrow f'(2) = e^0\left(\frac{2}{1+16}\right) = \frac{2}{17}$

15. Triangle ABD is an isosceles right triangle with its right angle at B and an angle of measure $\frac{\pi}{4}$ at A. We therefore have $\frac{\pi}{4} = \angle DAB = \angle DAE + \angle CAB = \tan^{-1}\frac{1}{3} + \tan^{-1}\frac{1}{2}$.

17. The area of the shaded region is $\int_0^1 \sin^{-1} x\,dx = \int_0^1 \sin^{-1} y\,dy$, which is the same as the area of the region to the left of the curve $y = \sin x$ (and part of the rectangle formed by the coordinate axes and dashed lines $y = 1$, $x = \frac{\pi}{2}$). The area of the rectangle is $\frac{\pi}{2} = \int_0^1 \sin^{-1} y\,dy + \int_0^{\pi/2} \sin x\,dx$, so we have

$\frac{\pi}{2} = \int_0^1 \sin^{-1} x\,dx + \int_0^{\pi/2} \sin x\,dx \Rightarrow \int_0^{\pi/2} \sin x\,dx = \frac{\pi}{2} - \int_0^1 \sin^{-1} x\,dx.$

19. (a) $g(x) + h(x) = 0 \Rightarrow g(x) = -h(x)$; also $g(x) + h(x) = 0 \Rightarrow g(-x) + h(-x) = 0 \Rightarrow g(x) - h(x) = 0$

$\Rightarrow g(x) = h(x)$; therefore $-h(x) = h(x) \Rightarrow h(x) = 0 \Rightarrow g(x) = 0$

(b) $\frac{f(x) + f(-x)}{2} = \frac{[f_E(x) + f_O(x)] + [f_E(-x) + f_O(-x)]}{2} = \frac{f_E(x) + f_O(x) + f_E(x) - f_O(x)}{2} = f_E(x)$;

$\frac{f(x) - f(-x)}{2} = \frac{[f_E(x) + f_O(x)] - [f_E(-x) + f_O(-x)]}{2} = \frac{f_E(x) + f_O(x) - f_E(x) + f_O(x)}{2} = f_O(x)$

(c) Part b $\Rightarrow$ such a decomposition is unique.

21. $M = \int_0^1 \frac{2}{1+x^2}\,dx = 2\left[\tan^{-1} x\right]_0^1 = \frac{\pi}{2}$ and $M_y = \int_0^1 \frac{2x}{1+x^2}\,dx = \left[\ln(1+x^2)\right]_0^1 = \ln 2 \Rightarrow \bar{x} = \frac{M_y}{M}$

$= \frac{\ln 2}{\left(\frac{\pi}{2}\right)} = \frac{\ln 4}{\pi}$; $\bar{y} = 0$ by symmetry

23. $A(t) = A_0 e^{rt}$; $A(t) = 2A_0 \Rightarrow 2A_0 = A_0 e^{rt} \Rightarrow e^{rt} = 2 \Rightarrow rt = \ln 2 \Rightarrow t = \frac{\ln 2}{r} \Rightarrow t \approx \frac{.7}{r} = \frac{70}{100r} = \frac{70}{(r\%)}$

25. (a) $L = k\left(\frac{a - b\cot\theta}{R^4} + \frac{b\csc\theta}{r^4}\right) \Rightarrow \frac{dL}{d\theta} = k\left(\frac{b\csc^2\theta}{R^4} - \frac{b\csc\theta\cot\theta}{r^4}\right)$; solving $\frac{dL}{d\theta} = 0$

$\Rightarrow r^4 b\csc^2\theta - bR^4\csc\theta\cot\theta = 0 \Rightarrow (b\csc\theta)(r^4\csc\theta - R^4\cot\theta) = 0$; but $b\csc\theta \neq 0$ since

$\theta \neq \frac{\pi}{2} \Rightarrow r^4\csc\theta - R^4\cot\theta = 0 \Rightarrow \cos\theta = \frac{r^4}{R^4} \Rightarrow \theta = \cos^{-1}\left(\frac{r^4}{R^4}\right)$, the critical value of θ

(b) $\theta = \cos^{-1}\left(\frac{5}{6}\right)^4 \approx \cos^{-1}(0.48225) \approx 61°$

NOTES:

CHAPTER 8 TECHNIQUES OF INTEGRATION

8.1 BASIC INTEGRATION FORMULAS

1. $\int \frac{16x\,dx}{\sqrt{8x^2+1}}$; $\begin{bmatrix} u = 8x^2+1 \\ du = 16x\,dx \end{bmatrix} \to \int \frac{du}{\sqrt{u}} = 2\sqrt{u} + C = 2\sqrt{8x^2+1} + C$

3. $\int 3\sqrt{\sin v}\,\cos v\,dv$; $\begin{bmatrix} u = \sin v \\ du = \cos v\,dv \end{bmatrix} \to \int 3\sqrt{u}\,du = 3 \cdot \frac{2}{3}u^{3/2} + C = 2(\sin v)^{3/2} + C$

5. $\int_0^1 \frac{16x\,dx}{8x^2+2}$; $\begin{bmatrix} u = 8x^2+2 \\ du = 16x\,dx \\ x = 0 \Rightarrow u = 2, \ x = 1 \Rightarrow u = 10 \end{bmatrix} \to \int_2^{10} \frac{du}{u} = [\ln |u|]_2^{10} = \ln 10 - \ln 2 = \ln 5$

7. $\int \frac{dx}{\sqrt{x}\,(\sqrt{x}+1)}$; $\begin{bmatrix} u = \sqrt{x}+1 \\ du = \frac{1}{2\sqrt{x}}\,dx \\ 2\,du = \frac{dx}{\sqrt{x}} \end{bmatrix} \to \int \frac{2\,du}{u} = 2\ln|u| + C = 2\ln\left(\sqrt{x}+1\right) + C$

9. $\int \cot(3-7x)\,dx$; $\begin{bmatrix} u = 3-7x \\ du = -7\,dx \end{bmatrix} \to -\frac{1}{7}\int \cot u\,du = -\frac{1}{7}\ln|\sin u| + C = -\frac{1}{7}\ln|\sin(3-7x)| + C$

11. $\int e^\theta \csc(e^\theta+1)\,d\theta$; $\begin{bmatrix} u = e^\theta+1 \\ du = e^\theta\,d\theta \end{bmatrix} \to \int \csc u\,du = -\ln|\csc u + \cot u| + C = -\ln\left|\csc(e^\theta+1) + \cot(e^\theta+1)\right| + C$

13. $\int \sec \frac{t}{3}\,dt$; $\begin{bmatrix} u = \frac{t}{3} \\ du = \frac{dt}{3} \end{bmatrix} \to \int 3\sec u\,du = 3\ln|\sec u + \tan u| + C = 3\ln\left|\sec \frac{t}{3} + \tan \frac{t}{3}\right| + C$

15. $\int \csc(s-\pi)\,ds$; $\begin{bmatrix} u = s-\pi \\ du = ds \end{bmatrix} \to \int \csc u\,du = -\ln|\csc u + \cot u| + C = -\ln|\csc(s-\pi) + \cot(s-\pi)| + C$

17. $\int_0^{\sqrt{\ln 2}} 2xe^{x^2}\,dx$; $\begin{bmatrix} u = x^2 \\ du = 2x\,dx \\ x = 0 \Rightarrow u = 0, x = \sqrt{\ln 2} \Rightarrow u = \ln 2 \end{bmatrix} \to \int_0^{\ln 2} e^u\,du = [e^u]_0^{\ln 2} = e^{\ln 2} - e^0 = 2 - 1 = 1$

19. $\int e^{\tan v}\sec^2 v\,dv$; $\begin{bmatrix} u = \tan v \\ du = \sec^2 v\,dv \end{bmatrix} \to \int e^u\,du = e^u + C = e^{\tan v} + C$

21. $\int 3^{x+1}\,dx$; $\begin{bmatrix} u = x+1 \\ du = dx \end{bmatrix} \to \int 3^u\,du = \left(\frac{1}{\ln 3}\right)3^u + C = \frac{3^{(x+1)}}{\ln 3} + C$

23. $\int \frac{2^{\sqrt{w}}\,dw}{2\sqrt{w}}$; $\begin{bmatrix} u = \sqrt{w} \\ du = \frac{dw}{2\sqrt{w}} \end{bmatrix} \to \int 2^u\,du = \frac{2^u}{\ln 2} + C = \frac{2^{\sqrt{w}}}{\ln 2} + C$

25. $\int \frac{9\,du}{1+9u^2}$; $\begin{bmatrix} x = 3u \\ dx = 3\,du \end{bmatrix} \to \int \frac{3\,dx}{1+x^2} = 3\tan^{-1}x + C = 3\tan^{-1}3u + C$

27. $\int_0^{1/6} \frac{dx}{\sqrt{1-9x^2}}$; $\begin{bmatrix} u = 3x \\ du = 3\,dx \\ x = 0 \Rightarrow u = 0,\ x = \frac{1}{6} \Rightarrow u = \frac{1}{2} \end{bmatrix}$ $\rightarrow \int_0^{1/2} \frac{1}{3} \frac{du}{\sqrt{1-u^2}} = \left[\frac{1}{3}\sin^{-1} u\right]_0^{1/2} = \frac{1}{3}\left(\frac{\pi}{6} - 0\right) = \frac{\pi}{18}$

29. $\int \frac{2s\,ds}{\sqrt{1-s^4}}$; $\begin{bmatrix} u = s^2 \\ du = 2s\,ds \end{bmatrix}$ $\rightarrow \int \frac{du}{\sqrt{1-u^2}} = \sin^{-1} u + C = \sin^{-1} s^2 + C$

31. $\int \frac{6\,dx}{x\sqrt{25x^2-1}} = \int \frac{6\,dx}{5x\sqrt{x^2-\frac{1}{25}}} = \frac{6}{5}\cdot 5\sec^{-1}|5x| + C = 6\sec^{-1}|5x| + C$

33. $\int \frac{dx}{e^x + e^{-x}} = \int \frac{e^x\,dx}{e^{2x}+1}$; $\begin{bmatrix} u = e^x \\ du = e^x\,dx \end{bmatrix}$ $\rightarrow \int \frac{du}{u^2+1} = \tan^{-1} u + C = \tan^{-1} e^x + C$

35. $\int_1^{e^{\pi/3}} \frac{dx}{x\cos(\ln x)}$; $\begin{bmatrix} u = \ln x \\ du = \frac{dx}{x} \\ x = 1 \Rightarrow u = 0,\ x = e^{\pi/3} \Rightarrow u = \frac{\pi}{3} \end{bmatrix}$ $\rightarrow \int_0^{\pi/3} \frac{du}{\cos u} = \int_0^{\pi/3} \sec u\,du = [\ln|\sec u + \tan u|]_0^{\pi/3}$

$= \ln\left|\sec\frac{\pi}{3} + \tan\frac{\pi}{3}\right| - \ln|\sec 0 + \tan 0| = \ln\left(2 + \sqrt{3}\right) - \ln(1) = \ln\left(2 + \sqrt{3}\right)$

37. $\int_1^2 \frac{8\,dx}{x^2-2x+2} = 8\int_1^2 \frac{dx}{1+(x-1)^2}$; $\begin{bmatrix} u = x - 1 \\ du = dx \\ x = 1 \Rightarrow u = 0,\ x = 2 \Rightarrow u = 1 \end{bmatrix}$ $\rightarrow 8\int_0^1 \frac{du}{1+u^2} = 8\left[\tan^{-1} u\right]_0^1$

$= 8\left(\tan^{-1} 1 - \tan^{-1} 0\right) = 8\left(\frac{\pi}{4} - 0\right) = 2\pi$

39. $\int \frac{dt}{\sqrt{-t^2+4t-3}} = \int \frac{dt}{\sqrt{1-(t-2)^2}}$; $\begin{bmatrix} u = t - 2 \\ du = dt \end{bmatrix}$ $\rightarrow \int \frac{du}{\sqrt{1-u^2}} = \sin^{-1} u + C = \sin^{-1}(t-2) + C$

41. $\int \frac{dx}{(x+1)\sqrt{x^2+2x}} = \int \frac{dx}{(x+1)\sqrt{(x+1)^2-1}}$; $\begin{bmatrix} u = x+1 \\ du = dx \end{bmatrix}$ $\rightarrow \int \frac{du}{u\sqrt{u^2-1}} = \sec^{-1}|u| + C = \sec^{-1}|x+1| + C,$

$|u| = |x+1| > 1$

43. $\int (\sec x + \cot x)^2\,dx = \int (\sec^2 x + 2\sec x\cot x + \cot^2 x)\,dx = \int \sec^2 x\,dx + \int 2\csc x\,dx + \int (\csc^2 x - 1)\,dx$
$= \tan x - 2\ln|\csc x + \cot x| - \cot x - x + C$

45. $\int \csc x \sin 3x\,dx = \int (\csc x)(\sin 2x\cos x + \sin x\cos 2x)\,dx = \int (\csc x)(2\sin x\cos^2 x + \sin x\cos 2x)\,dx$
$= \int (2\cos^2 x + \cos 2x)\,dx = \int [(1 + \cos 2x) + \cos 2x]\,dx = \int (1 + 2\cos 2x)\,dx = x + \sin 2x + C$

47. $\int \frac{x}{x+1}\,dx = \int \left(1 - \frac{1}{x+1}\right)dx = x - \ln|x+1| + C$

49. $\int_{\sqrt{2}}^3 \frac{2x^3}{x^2-1}\,dx = \int_{\sqrt{2}}^3 \left(2x + \frac{2x}{x^2-1}\right)dx = \left[x^2 + \ln|x^2 - 1|\right]_{\sqrt{2}}^3 = (9 + \ln 8) - (2 + \ln 1) = 7 + \ln 8$

51. $\int \frac{4t^3 - t^2 + 16t}{t^2+4}\,dt = \int \left[(4t - 1) + \frac{4}{t^2+4}\right]dt = 2t^2 - t + 2\tan^{-1}\left(\frac{t}{2}\right) + C$

53. $\int \frac{1-x}{\sqrt{1-x^2}}\,dx = \int \frac{dx}{\sqrt{1-x^2}} - \int \frac{x\,dx}{\sqrt{1-x^2}} = \sin^{-1} x + \sqrt{1-x^2} + C$

55. $\int_0^{\pi/4} \frac{1+\sin x}{\cos^2 x}\,dx = \int_0^{\pi/4} (\sec^2 x + \sec x\tan x)\,dx = [\tan x + \sec x]_0^{\pi/4} = \left(1 + \sqrt{2}\right) - (0 + 1) = \sqrt{2}$

57. $\int \frac{dx}{1 + \sin x} = \int \frac{(1 - \sin x)}{(1 - \sin^2 x)}\,dx = \int \frac{(1 - \sin x)}{\cos^2 x}\,dx = \int (\sec^2 x - \sec x \tan x)\,dx = \tan x - \sec x + C$

59. $\int \frac{1}{\sec \theta + \tan \theta}\,d\theta = \int \frac{\cos \theta}{1 + \sin \theta}\,d\theta;\ \begin{bmatrix} u = 1 + \sin \theta \\ du = \cos \theta\,d\theta \end{bmatrix} \to \int \frac{du}{u} = \ln |u| + C = \ln |1 + \sin \theta| + C$

61. $\int \frac{1}{1 - \sec x}\,dx = \int \frac{\cos x}{\cos x - 1}\,dx = \int \left(1 + \frac{1}{\cos x - 1}\right)\,dx = \int \left(1 - \frac{1 + \cos x}{\sin^2 x}\right)\,dx = \int \left(1 - \csc^2 x - \frac{\cos x}{\sin^2 x}\right)\,dx$

$= \int (1 - \csc^2 x - \csc x \cot x)\,dx = x + \cot x + \csc x + C$

63. $\int_0^{2\pi} \sqrt{\frac{1 - \cos x}{2}}\,dx = \int_0^{2\pi} |\sin \tfrac{x}{2}|\,dx;\ \begin{bmatrix} \sin \tfrac{x}{2} \ge 0 \\ \text{for } 0 \le \tfrac{x}{2} \le 2\pi \end{bmatrix} \to \int_0^{2\pi} \sin \left(\tfrac{x}{2}\right)\,dx = \left[-2 \cos \tfrac{x}{2}\right]_0^{2\pi} = -2(\cos \pi - \cos 0)$

$= (-2)(-2) = 4$

65. $\int_{\pi/2}^{\pi} \sqrt{1 + \cos 2t}\,dt = \int_{\pi/2}^{\pi} \sqrt{2}\,|\cos t|\,dt;\ \begin{bmatrix} \cos t \le 0 \\ \text{for } \tfrac{\pi}{2} \le t \le \pi \end{bmatrix} \to \int_{\pi/2}^{\pi} -\sqrt{2} \cos t\,dt = \left[-\sqrt{2} \sin t\right]_{\pi/2}^{\pi}$

$= -\sqrt{2} \left(\sin \pi - \sin \tfrac{\pi}{2}\right) = \sqrt{2}$

67. $\int_{-\pi}^{0} \sqrt{1 - \cos^2 \theta}\,d\theta = \int_{-\pi}^{0} |\sin \theta|\,d\theta;\ \begin{bmatrix} \sin \theta \le 0 \\ \text{for } -\pi \le \theta \le 0 \end{bmatrix} \to \int_{-\pi}^{0} -\sin \theta\,d\theta = [\cos \theta]_{-\pi}^{0} = \cos 0 - \cos (-\pi)$

$= 1 - (-1) = 2$

69. $\int_{-\pi/4}^{\pi/4} \sqrt{\tan^2 y + 1}\,dy = \int_{-\pi/4}^{\pi/4} |\sec y|\,dy;\ \begin{bmatrix} \sec y \ge 0 \\ \text{for } -\tfrac{\pi}{4} \le y \le \tfrac{\pi}{4} \end{bmatrix} \to \int_{-\pi/4}^{\pi/4} \sec y\,dy = [\ln |\sec y + \tan y|]_{-\pi/4}^{\pi/4}$

$= \ln \left|\sqrt{2} + 1\right| - \ln \left|\sqrt{2} - 1\right|$

71. $\int_{\pi/4}^{3\pi/4} (\csc x - \cot x)^2\,dx = \int_{\pi/4}^{3\pi/4} (\csc^2 x - 2 \csc x \cot x + \cot^2 x)\,dx = \int_{\pi/4}^{3\pi/4} (2 \csc^2 x - 1 - 2 \csc x \cot x)\,dx$

$= [-2 \cot x - x + 2 \csc x]_{\pi/4}^{3\pi/4} = \left(-2 \cot \tfrac{3\pi}{4} - \tfrac{3\pi}{4} + 2 \csc \tfrac{3\pi}{4}\right) - \left(-2 \cot \tfrac{\pi}{4} - \tfrac{\pi}{4} + 2 \csc \tfrac{\pi}{4}\right)$

$= \left[-2(-1) - \tfrac{3\pi}{4} + 2\left(\sqrt{2}\right)\right] - \left[-2(1) - \tfrac{\pi}{4} + 2\left(\sqrt{2}\right)\right] = 4 - \tfrac{\pi}{2}$

73. $\int \cos \theta \csc (\sin \theta)\,d\theta;\ \begin{bmatrix} u = \sin \theta \\ du = \cos \theta\,d\theta \end{bmatrix} \to \int \csc u\,du = -\ln |\csc u + \cot u| + C$

$= -\ln |\csc (\sin \theta) + \cot (\sin \theta)| + C$

75. $\int (\csc x - \sec x)(\sin x + \cos x)\,dx = \int (1 + \cot x - \tan x - 1)\,dx = \int \cot x\,dx - \int \tan x\,dx$

$= \ln |\sin x| + \ln |\cos x| + C$

77. $\int \frac{6\,dy}{\sqrt{y}(1 + y)};\ \begin{bmatrix} u = \sqrt{y} \\ du = \frac{1}{2\sqrt{y}}\,dy \end{bmatrix} \to \int \frac{12\,du}{1 + u^2} = 12 \tan^{-1} u + C = 12 \tan^{-1} \sqrt{y} + C$

79. $\int \frac{7\,dx}{(x - 1)\sqrt{x^2 - 2x - 48}} = \int \frac{7\,dx}{(x - 1)\sqrt{(x - 1)^2 - 49}};\ \begin{bmatrix} u = x - 1 \\ du = dx \end{bmatrix} \to \int \frac{7\,du}{u\sqrt{u^2 - 49}} = 7 \cdot \tfrac{1}{7} \sec^{-1} \left|\tfrac{u}{7}\right| + C$

$= \sec^{-1} \left|\tfrac{x - 1}{7}\right| + C$

81. $\int \sec^2 t \tan (\tan t)\,dt;\ \begin{bmatrix} u = \tan t \\ du = \sec^2 t\,dt \end{bmatrix} \to \int \tan u\,du = -\ln |\cos u| + C = \ln |\sec u| + C = \ln |\sec (\tan t)| + C$

83. (a) $\int \cos^3 \theta \, d\theta = \int (\cos \theta)(1 - \sin^2 \theta) \, d\theta;$ $\begin{bmatrix} u = \sin \theta \\ du = \cos \theta \, d\theta \end{bmatrix}$ $\to \int (1 - u^2) \, du = u - \frac{u^3}{3} + C = \sin \theta - \frac{1}{3} \sin^3 \theta + C$

   (b) $\int \cos^5 \theta \, d\theta = \int (\cos \theta)(1 - \sin^2 \theta)^2 \, d\theta = \int (1 - u^2)^2 \, du = \int (1 - 2u^2 + u^4) \, du = u - \frac{2}{3} u^3 + \frac{u^5}{5} + C$

   $= \sin \theta - \frac{2}{3} \sin^3 \theta + \frac{1}{5} \sin^5 \theta + C$

   (c) $\int \cos^9 \theta \, d\theta = \int (\cos^8 \theta)(\cos \theta) \, d\theta = \int (1 - \sin^2 \theta)^4 (\cos \theta) \, d\theta$

85. (a) $\int \tan^3 \theta \, d\theta = \int (\sec^2 \theta - 1)(\tan \theta) \, d\theta = \int \sec^2 \theta \tan \theta \, d\theta - \int \tan \theta \, d\theta = \frac{1}{2} \tan^2 \theta - \int \tan \theta \, d\theta$

   $= \frac{1}{2} \tan^2 \theta + \ln |\cos \theta| + C$

   (b) $\int \tan^5 \theta \, d\theta = \int (\sec^2 \theta - 1)(\tan^3 \theta) \, d\theta = \int \tan^3 \theta \sec^2 \theta \, d\theta - \int \tan^3 \theta \, d\theta = \frac{1}{4} \tan^4 \theta - \int \tan^3 \theta \, d\theta$

   (c) $\int \tan^7 \theta \, d\theta = \int (\sec^2 \theta - 1)(\tan^5 \theta) \, d\theta = \int \tan^5 \theta \sec^2 \theta \, d\theta - \int \tan^5 \theta \, d\theta = \frac{1}{6} \tan^6 \theta - \int \tan^5 \theta \, d\theta$

   (d) $\int \tan^{2k+1} \theta \, d\theta = \int (\sec^2 \theta - 1)(\tan^{2k-1} \theta) \, d\theta = \int \tan^{2k-1} \theta \sec^2 \theta \, d\theta - \int \tan^{2k-1} \theta \, d\theta;$

   $\begin{bmatrix} u = \tan \theta \\ du = \sec^2 \theta \, d\theta \end{bmatrix}$ $\to \int u^{2k-1} \, du - \int \tan^{2k-1} \theta \, d\theta = \frac{1}{2k} u^{2k} - \int \tan^{2k-1} \theta \, d\theta = \frac{1}{2k} \tan^{2k} \theta - \int \tan^{2k-1} \theta \, d\theta$

87. $A = \int_{-\pi/4}^{\pi/4} (2 \cos x - \sec x) \, dx = [2 \sin x - \ln |\sec x + \tan x|]_{-\pi/4}^{\pi/4}$

   $= \left[\sqrt{2} - \ln \left(\sqrt{2} + 1\right)\right] - \left[-\sqrt{2} - \ln \left(\sqrt{2} - 1\right)\right]$

   $= 2\sqrt{2} - \ln \left(\frac{\sqrt{2}+1}{\sqrt{2}-1}\right) = 2\sqrt{2} - \ln \left(\frac{\left(\sqrt{2}+1\right)^2}{2-1}\right)$

   $= 2\sqrt{2} - \ln \left(3 + 2\sqrt{2}\right)$

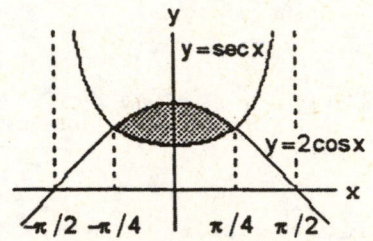

89. $V = \int_{-\pi/4}^{\pi/4} \pi (2 \cos x)^2 \, dx - \int_{-\pi/4}^{\pi/4} \pi \sec^2 x \, dx = 4\pi \int_{-\pi/4}^{\pi/4} \cos^2 x \, dx - \pi \int_{-\pi/4}^{\pi/4} \sec^2 x \, dx$

   $= 2\pi \int_{-\pi/4}^{\pi/4} (1 + \cos 2x) \, dx - \pi [\tan x]_{-\pi/4}^{\pi/4} = 2\pi \left[x + \frac{1}{2} \sin 2x\right]_{-\pi/4}^{\pi/4} - \pi [1 - (-1)]$

   $= 2\pi \left[\left(\frac{\pi}{4} + \frac{1}{2}\right) - \left(-\frac{\pi}{4} - \frac{1}{2}\right)\right] - 2\pi = 2\pi \left(\frac{\pi}{2} + 1\right) - 2\pi = \pi^2$

91. $y = \ln (\cos x) \Rightarrow \frac{dy}{dx} = -\frac{\sin x}{\cos x} \Rightarrow \left(\frac{dy}{dx}\right)^2 = \tan^2 x = \sec^2 x - 1; L = \int_a^b \sqrt{1 + \left(\frac{dy}{dx}\right)^2} \, dx$

   $= \int_0^{\pi/3} \sqrt{1 + (\sec^2 x - 1)} \, dx = \int_0^{\pi/3} \sec x \, dx = [\ln |\sec x + \tan x|]_0^{\pi/3} = \ln \left|2 + \sqrt{3}\right| - \ln |1 + 0| = \ln \left(2 + \sqrt{3}\right)$

93. $M_x = \int_{-\pi/4}^{\pi/4} \left(\frac{1}{2} \sec x\right)(\sec x) \, dx = \frac{1}{2} \int_{-\pi/4}^{\pi/4} \sec^2 x \, dx$

   $= \frac{1}{2} [\tan x]_{-\pi/4}^{\pi/4} = \frac{1}{2} [1 - (-1)] = 1;$

   $M = \int_{-\pi/4}^{\pi/4} \sec x \, dx = [\ln |\sec x + \tan x|]_{-\pi/4}^{\pi/4}$

   $= \ln \left|\sqrt{2} + 1\right| - \ln \left|\sqrt{2} - 1\right| = \ln \left(\frac{\sqrt{2}+1}{\sqrt{2}-1}\right)$

   $= \ln \left(\frac{\left(\sqrt{2}+1\right)^2}{2-1}\right) = \ln \left(3 + 2\sqrt{2}\right); \bar{x} = 0$ by

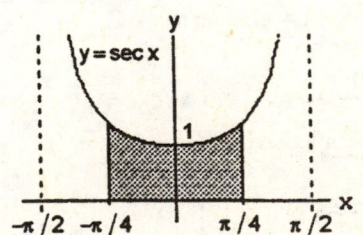

   symmetry of the region, and $\bar{y} = \frac{M_x}{M} = \frac{1}{\ln \left(3 + 2\sqrt{2}\right)}$

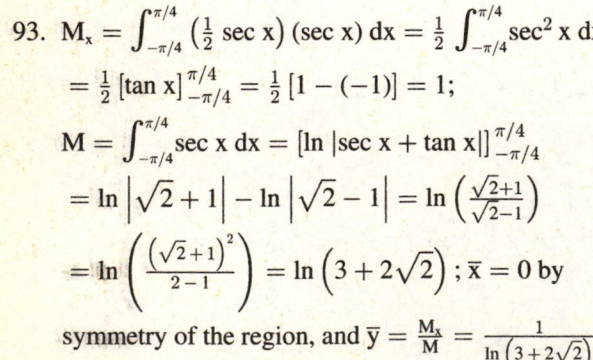

95. $\int \csc x \, dx = \int (\csc x)(1) \, dx = \int (\csc x) \left(\frac{\csc x + \cot x}{\csc x + \cot x}\right) \, dx = \int \frac{\csc^2 x + \csc x \cot x}{\csc x + \cot x} \, dx;$

   $\begin{bmatrix} u = \csc x + \cot x \\ du = (-\csc x \cot x - \csc^2 x) \, dx \end{bmatrix}$ $\to \int \frac{-du}{u} = -\ln |u| + C = -\ln |\csc x + \cot x| + C$

8.2 INTEGRATION BY PARTS

1. $u = x$, $du = dx$; $dv = \sin \frac{x}{2}\, dx$, $v = -2 \cos \frac{x}{2}$;

   $\int x \sin \frac{x}{2}\, dx = -2x \cos \frac{x}{2} - \int \left(-2 \cos \frac{x}{2}\right) dx = -2x \cos \left(\frac{x}{2}\right) + 4 \sin \left(\frac{x}{2}\right) + C$

3.

   $$
   \begin{array}{lll}
    & & \cos t \\
   t^2 & \xrightarrow{(+)} & \sin t \\
   2t & \xrightarrow{(-)} & -\cos t \\
   2 & \xrightarrow{(+)} & -\sin t \\
   0 & &
   \end{array}
   $$

   $\int t^2 \cos t\, dt = t^2 \sin t + 2t \cos t - 2 \sin t + C$

5. $u = \ln x$, $du = \frac{dx}{x}$; $dv = x\, dx$, $v = \frac{x^2}{2}$;

   $\int_1^2 x \ln x\, dx = \left[\frac{x^2}{2} \ln x\right]_1^2 - \int_1^2 \frac{x^2}{2} \frac{dx}{x} = 2 \ln 2 - \left[\frac{x^2}{4}\right]_1^2 = 2 \ln 2 - \frac{3}{4} = \ln 4 - \frac{3}{4}$

7. $u = \tan^{-1} y$, $du = \frac{dy}{1+y^2}$; $dv = dy$, $v = y$;

   $\int \tan^{-1} y\, dy = y \tan^{-1} y - \int \frac{y\, dy}{(1+y^2)} = y \tan^{-1} y - \frac{1}{2} \ln (1 + y^2) + C = y \tan^{-1} y - \ln \sqrt{1+y^2} + C$

9. $u = x$, $du = dx$; $dv = \sec^2 x\, dx$, $v = \tan x$;

   $\int x \sec^2 x\, dx = x \tan x - \int \tan x\, dx = x \tan x + \ln |\cos x| + C$

11.

   $$
   \begin{array}{lll}
    & & e^x \\
   x^3 & \xrightarrow{(+)} & e^x \\
   3x^2 & \xrightarrow{(-)} & e^x \\
   6x & \xrightarrow{(+)} & e^x \\
   6 & \xrightarrow{(-)} & e^x \\
   0 & &
   \end{array}
   $$

   $\int x^3 e^x\, dx = x^3 e^x - 3x^2 e^x + 6x e^x - 6 e^x + C = (x^3 - 3x^2 + 6x - 6) e^x + C$

13.

   $$
   \begin{array}{lll}
    & & e^x \\
   x^2 - 5x & \xrightarrow{(+)} & e^x \\
   2x - 5 & \xrightarrow{(-)} & e^x \\
   2 & \xrightarrow{(+)} & e^x \\
   0 & &
   \end{array}
   $$

   $\int (x^2 - 5x) e^x\, dx = (x^2 - 5x) e^x - (2x - 5)e^x + 2e^x + C = x^2 e^x - 7x e^x + 7 e^x + C$
   $= (x^2 - 7x + 7) e^x + C$

15.

$$x^5 \xrightarrow{(+)} e^x$$
$$5x^4 \xrightarrow{(-)} e^x$$
$$20x^3 \xrightarrow{(+)} e^x$$
$$60x^2 \xrightarrow{(-)} e^x$$
$$120x \xrightarrow{(+)} e^x$$
$$120 \xrightarrow{(-)} e^x$$
$$0$$

$$\int x^5 e^x \, dx = x^5 e^x - 5x^4 e^x + 20x^3 e^x - 60x^2 e^x + 120x e^x - 120 e^x + C$$
$$= (x^5 - 5x^4 + 20x^3 - 60x^2 + 120x - 120)\, e^x + C$$

17.

$$\theta^2 \xrightarrow{(+)} -\tfrac{1}{2}\cos 2\theta$$
$$2\theta \xrightarrow{(-)} -\tfrac{1}{4}\sin 2\theta$$
$$2 \xrightarrow{(+)} \tfrac{1}{8}\cos 2\theta$$
$$0$$

$$\int_0^{\pi/2} \theta^2 \sin 2\theta \, d\theta = \left[-\tfrac{\theta^2}{2}\cos 2\theta + \tfrac{\theta}{2}\sin 2\theta + \tfrac{1}{4}\cos 2\theta \right]_0^{\pi/2}$$
$$= \left[-\tfrac{\pi^2}{8}\cdot(-1) + \tfrac{\pi}{4}\cdot 0 + \tfrac{1}{4}\cdot(-1) \right] - \left[0 + 0 + \tfrac{1}{4}\cdot 1 \right] = \tfrac{\pi^2}{8} - \tfrac{1}{2} = \tfrac{\pi^2-4}{8}$$

19. $u = \sec^{-1} t$, $du = \dfrac{dt}{t\sqrt{t^2-1}}$; $dv = t\, dt$, $v = \tfrac{t^2}{2}$;

$$\int_{2/\sqrt{3}}^2 t \sec^{-1} t \, dt = \left[\tfrac{t^2}{2}\sec^{-1} t \right]_{2/\sqrt{3}}^2 - \int_{2/\sqrt{3}}^2 \left(\tfrac{t^2}{2} \right) \tfrac{dt}{t\sqrt{t^2-1}} = \left(2\cdot\tfrac{\pi}{3} - \tfrac{2}{3}\cdot\tfrac{\pi}{6} \right) - \int_{2/\sqrt{3}}^2 \tfrac{t\,dt}{2\sqrt{t^2-1}}$$
$$= \tfrac{5\pi}{9} - \left[\tfrac{1}{2}\sqrt{t^2-1} \right]_{2/\sqrt{3}}^2 = \tfrac{5\pi}{9} - \tfrac{1}{2}\left(\sqrt{3} - \sqrt{\tfrac{4}{3}-1} \right) = \tfrac{5\pi}{9} - \tfrac{1}{2}\left(\sqrt{3} - \tfrac{\sqrt{3}}{3} \right) = \tfrac{5\pi}{9} - \tfrac{\sqrt{3}}{3} = \tfrac{5\pi-3\sqrt{3}}{9}$$

21. $I = \int e^{\theta}\sin\theta\, d\theta$; $[u = \sin\theta,\ du = \cos\theta\, d\theta;\ dv = e^{\theta}\, d\theta,\ v = e^{\theta}] \Rightarrow I = e^{\theta}\sin\theta - \int e^{\theta}\cos\theta\, d\theta$;

$[u = \cos\theta,\ du = -\sin\theta\, d\theta;\ dv = e^{\theta}\, d\theta,\ v = e^{\theta}] \Rightarrow I = e^{\theta}\sin\theta - \left(e^{\theta}\cos\theta + \int e^{\theta}\sin\theta\, d\theta \right)$

$= e^{\theta}\sin\theta - e^{\theta}\cos\theta - I + C' \Rightarrow 2I = (e^{\theta}\sin\theta - e^{\theta}\cos\theta) + C' \Rightarrow I = \tfrac{1}{2}(e^{\theta}\sin\theta - e^{\theta}\cos\theta) + C$, where $C = \tfrac{C'}{2}$ is another arbitrary constant

23. $I = \int e^{2x}\cos 3x\, dx$; $\left[u = \cos 3x;\ du = -3\sin 3x\, dx,\ dv = e^{2x}\, dx;\ v = \tfrac{1}{2}e^{2x} \right]$

$\Rightarrow I = \tfrac{1}{2}e^{2x}\cos 3x + \tfrac{3}{2}\int e^{2x}\sin 3x\, dx$; $\left[u = \sin 3x,\ du = 3\cos 3x,\ dv = e^{2x}\, dx;\ v = \tfrac{1}{2}e^{2x} \right]$

$\Rightarrow I = \tfrac{1}{2}e^{2x}\cos 3x + \tfrac{3}{2}\left(\tfrac{1}{2}e^{2x}\sin 3x - \tfrac{3}{2}\int e^{2x}\cos 3x\, dx \right) = \tfrac{1}{2}e^{2x}\cos 3x + \tfrac{3}{4}e^{2x}\sin 3x - \tfrac{9}{4}I + C'$

$\Rightarrow \tfrac{13}{4}I = \tfrac{1}{2}e^{2x}\cos 3x + \tfrac{3}{4}e^{2x}\sin 3x + C' \Rightarrow \tfrac{e^{2x}}{13}(3\sin 3x + 2\cos 3x) + C$, where $C = \tfrac{4}{13}C'$

25. $\int e^{\sqrt{3s+9}}\, ds$; $\begin{bmatrix} 3s+9 = x^2 \\ ds = \tfrac{2}{3}x\, dx \end{bmatrix} \rightarrow \int e^x\cdot\tfrac{2}{3}x\, dx = \tfrac{2}{3}\int x e^x\, dx$; $[u = x,\ du = dx;\ dv = e^x\, dx,\ v = e^x]$;

$\tfrac{2}{3}\int x e^x\, dx = \tfrac{2}{3}\left(x e^x - \int e^x\, dx \right) = \tfrac{2}{3}(x e^x - e^x) + C = \tfrac{2}{3}\left(\sqrt{3s+9}\, e^{\sqrt{3s+9}} - e^{\sqrt{3s+9}} \right) + C$

27. $u = x$, $du = dx$; $dv = \tan^2 x\, dx$, $v = \int \tan^2 x\, dx = \int \tfrac{\sin^2 x}{\cos^2 x}\, dx = \int \tfrac{1-\cos^2 x}{\cos^2 x}\, dx = \int \tfrac{dx}{\cos^2 x} - \int dx$

$= \tan x - x$; $\int_0^{\pi/3} x\tan^2 x\, dx = [x(\tan x - x)]_0^{\pi/3} - \int_0^{\pi/3}(\tan x - x)\, dx = \tfrac{\pi}{3}\left(\sqrt{3} - \tfrac{\pi}{3} \right) + \left[\ln|\cos x| + \tfrac{x^2}{2} \right]_0^{\pi/3}$

$$= \frac{\pi}{3}\left(\sqrt{3} - \frac{\pi}{3}\right) + \ln\frac{1}{2} + \frac{\pi^2}{18} = \frac{\pi\sqrt{3}}{3} - \ln 2 - \frac{\pi^2}{18}$$

29. $\int \sin(\ln x)\,dx;$ $\begin{bmatrix} u = \ln x \\ du = \frac{1}{x}\,dx \\ dx = e^u\,du \end{bmatrix} \rightarrow \int (\sin u)\,e^u\,du.$ From Exercise 21, $\int (\sin u)\,e^u\,du = e^u\left(\frac{\sin u - \cos u}{2}\right) + C$

$$= \frac{1}{2}\left[-x\cos(\ln x) + x\sin(\ln x)\right] + C$$

31. (a) $u = x,\ du = dx;\ dv = \sin x\,dx,\ v = -\cos x;$

$$S_1 = \int_0^\pi x\sin x\,dx = [-x\cos x]_0^\pi + \int_0^\pi \cos x\,dx = \pi + [\sin x]_0^\pi = \pi$$

(b) $S_2 = -\int_\pi^{2\pi} x\sin x\,dx = -\left[[-x\cos x]_\pi^{2\pi} + \int_\pi^{2\pi}\cos x\,dx\right] = -[-3\pi + [\sin x]_\pi^{2\pi}] = 3\pi$

(c) $S_3 = \int_{2\pi}^{3\pi} x\sin x\,dx = [-x\cos x]_{2\pi}^{3\pi} + \int_{2\pi}^{3\pi}\cos x\,dx = 5\pi + [\sin x]_{2\pi}^{3\pi} = 5\pi$

(d) $S_{n+1} = (-1)^{n+1}\int_{n\pi}^{(n+1)\pi} x\sin x\,dx = (-1)^{n+1}\left[[-x\cos x]_{n\pi}^{(n+1)\pi} + [\sin x]_{n\pi}^{(n+1)\pi}\right]$

$$= (-1)^{n+1}\left[-(n+1)\pi(-1)^n + n\pi(-1)^{n+1}\right] + 0 = (2n+1)\pi$$

33. $V = \int_0^{\ln 2} 2\pi(\ln 2 - x)\,e^x\,dx = 2\pi\ln 2\int_0^{\ln 2} e^x\,dx - 2\pi\int_0^{\ln 2} xe^x\,dx$

$$= (2\pi\ln 2)\,[e^x]_0^{\ln 2} - 2\pi\left([xe^x]_0^{\ln 2} - \int_0^{\ln 2} e^x\,dx\right)$$

$$= 2\pi\ln 2 - 2\pi\left(2\ln 2 - [e^x]_0^{\ln 2}\right) = -2\pi\ln 2 + 2\pi = 2\pi(1 - \ln 2)$$

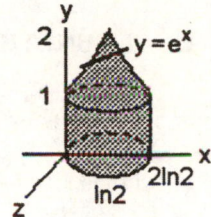

35. (a) $V = \int_0^{\pi/2} 2\pi x\cos x\,dx = 2\pi\left([x\sin x]_0^{\pi/2} - \int_0^{\pi/2}\sin x\,dx\right)$

$$= 2\pi\left(\frac{\pi}{2} + [\cos x]_0^{\pi/2}\right) = 2\pi\left(\frac{\pi}{2} + 0 - 1\right) = \pi(\pi - 2)$$

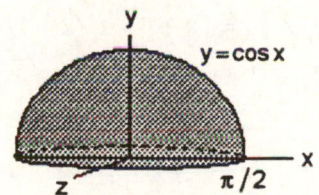

(b) $V = \int_0^{\pi/2} 2\pi\left(\frac{\pi}{2} - x\right)\cos x\,dx;\ u = \frac{\pi}{2} - x,\ du = -dx;\ dv = \cos x\,dx,\ v = \sin x;$

$$V = 2\pi\left[\left(\frac{\pi}{2} - x\right)\sin x\right]_0^{\pi/2} + 2\pi\int_0^{\pi/2}\sin x\,dx = 0 + 2\pi[-\cos x]_0^{\pi/2} = 2\pi(0+1) = 2\pi$$

37. (a) $av(y) = \frac{1}{2\pi}\int_0^{2\pi} 2e^{-t}\cos t\,dt$

$$= \frac{1}{\pi}\left[e^{-t}\left(\frac{\sin t - \cos t}{2}\right)\right]_0^{2\pi}$$

(see Exercise 22) $\Rightarrow av(y) = \frac{1}{2\pi}(1 - e^{-2\pi})$

(b)

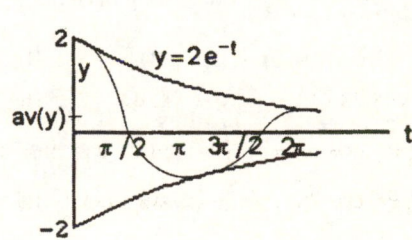

39. $I = \int x^n\cos x\,dx;\ [u = x^n,\ du = nx^{n-1}\,dx;\ dv = \cos x\,dx,\ v = \sin x]$

$$\Rightarrow I = x^n\sin x - \int nx^{n-1}\sin x\,dx$$

41. $I = \int x^n e^{ax}\,dx;\ \left[u = x^n,\ du = nx^{n-1}\,dx;\ dv = e^{ax}\,dx,\ v = \frac{1}{a}e^{ax}\right]$

$$\Rightarrow I = \frac{x^n e^{ax}}{a} - \frac{n}{a}\int x^{n-1} e^{ax}\,dx,\ a \neq 0$$

43. $\int \sin^{-1} x \, dx = x \sin^{-1} x - \int \sin y \, dy = x \sin^{-1} x + \cos y + C = x \sin^{-1} x + \cos(\sin^{-1} x) + C$

45. $\int \sec^{-1} x \, dx = x \sec^{-1} x - \int \sec y \, dy = x \sec^{-1} x - \ln |\sec y + \tan y| + C$

$= x \sec^{-1} x - \ln |\sec(\sec^{-1} x) + \tan(\sec^{-1} x)| + C = x \sec^{-1} x - \ln \left| x + \sqrt{x^2 - 1} \right| + C$

47. Yes, $\cos^{-1} x$ is the angle whose cosine is x which implies $\sin(\cos^{-1} x) = \sqrt{1 - x^2}$.

49. (a) $\int \sinh^{-1} x \, dx = x \sinh^{-1} x - \int \sinh y \, dy = x \sinh^{-1} x - \cosh y + C = x \sinh^{-1} x - \cosh(\sinh^{-1} x) + C$;

check: $d \left[x \sinh^{-1} x - \cosh(\sinh^{-1} x) + C \right] = \left[\sinh^{-1} x + \frac{x}{\sqrt{1+x^2}} - \sinh(\sinh^{-1} x) \frac{1}{\sqrt{1+x^2}} \right] dx$

$= \sinh^{-1} x \, dx$

(b) $\int \sinh^{-1} x \, dx = x \sinh^{-1} x - \int x \left(\frac{1}{\sqrt{1+x^2}} \right) dx = x \sinh^{-1} x - \frac{1}{2} \int (1 + x^2)^{-1/2} 2x \, dx$

$= x \sinh^{-1} x - (1 + x^2)^{1/2} + C$

check: $d \left[x \sinh^{-1} x - (1 + x^2)^{1/2} + C \right] = \left[\sinh^{-1} x + \frac{x}{\sqrt{1+x^2}} - \frac{x}{\sqrt{1+x^2}} \right] dx = \sinh^{-1} x \, dx$

8.3 INTEGRATION OF RATIONAL FUNCTIONS BY PARTIAL FRACTIONS

1. $\frac{5x - 13}{(x-3)(x-2)} = \frac{A}{x-3} + \frac{B}{x-2} \Rightarrow 5x - 13 = A(x-2) + B(x-3) = (A+B)x - (2A+3B)$

$\Rightarrow \left. \begin{array}{r} A + B = 5 \\ 2A + 3B = 13 \end{array} \right\} \Rightarrow -B = (10 - 13) \Rightarrow B = 3 \Rightarrow A = 2$; thus, $\frac{5x-13}{(x-3)(x-2)} = \frac{2}{x-3} + \frac{3}{x-2}$

3. $\frac{x+4}{(x+1)^2} = \frac{A}{x+1} + \frac{B}{(x+1)^2} \Rightarrow x + 4 = A(x+1) + B = Ax + (A+B) \Rightarrow \left. \begin{array}{r} A = 1 \\ A + B = 4 \end{array} \right\} \Rightarrow A = 1$ and $B = 3$;

thus, $\frac{x+4}{(x+1)^2} = \frac{1}{x+1} + \frac{3}{(x+1)^2}$

5. $\frac{z+1}{z^2(z-1)} = \frac{A}{z} + \frac{B}{z^2} + \frac{C}{z-1} \Rightarrow z + 1 = Az(z-1) + B(z-1) + Cz^2 \Rightarrow z + 1 = (A+C)z^2 + (-A+B)z - B$

$\Rightarrow \left. \begin{array}{r} A + C = 0 \\ -A + B = 1 \\ -B = 1 \end{array} \right\} \Rightarrow B = -1 \Rightarrow A = -2 \Rightarrow C = 2$; thus, $\frac{z+1}{z^2(z-1)} = \frac{-2}{z} + \frac{-1}{z^2} + \frac{2}{z-1}$

7. $\frac{t^2+8}{t^2-5t+6} = 1 + \frac{5t+2}{t^2-5t+6}$ (after long division); $\frac{5t+2}{t^2-5t+6} = \frac{5t+2}{(t-3)(t-2)} = \frac{A}{t-3} + \frac{B}{t-2}$

$\Rightarrow 5t + 2 = A(t-2) + B(t-3) = (A+B)t + (-2A-3B) \Rightarrow \left. \begin{array}{r} A + B = 5 \\ -2A - 3B = 2 \end{array} \right\} \Rightarrow -B = (10+2) = 12$

$\Rightarrow B = -12 \Rightarrow A = 17$; thus, $\frac{t^2+8}{t^2-5t+6} = 1 + \frac{17}{t-3} + \frac{-12}{t-2}$

9. $\frac{1}{1-x^2} = \frac{A}{1-x} + \frac{B}{1+x} \Rightarrow 1 = A(1+x) + B(1-x)$; $x = 1 \Rightarrow A = \frac{1}{2}$; $x = -1 \Rightarrow B = \frac{1}{2}$;

$\int \frac{dx}{1-x^2} = \frac{1}{2} \int \frac{dx}{1-x} + \frac{1}{2} \int \frac{dx}{1+x} = \frac{1}{2} \left[\ln |1 + x| - \ln |1 - x| \right] + C$

11. $\frac{x+4}{x^2+5x-6} = \frac{A}{x+6} + \frac{B}{x-1} \Rightarrow x + 4 = A(x-1) + B(x+6)$; $x = 1 \Rightarrow B = \frac{5}{7}$; $x = -6 \Rightarrow A = \frac{-2}{-7} = \frac{2}{7}$;

$\int \frac{x+4}{x^2+5x-6} \, dx = \frac{2}{7} \int \frac{dx}{x+6} + \frac{5}{7} \int \frac{dx}{x-1} = \frac{2}{7} \ln |x+6| + \frac{5}{7} \ln |x-1| + C = \frac{1}{7} \ln |(x+6)^2(x-1)^5| + C$

13. $\frac{y}{y^2-2y-3} = \frac{A}{y-3} + \frac{B}{y+1} \Rightarrow y = A(y+1) + B(y-3)$; $y = -1 \Rightarrow B = \frac{-1}{-4} = \frac{1}{4}$; $y = 3 \Rightarrow A = \frac{3}{4}$;

$\int_4^8 \frac{y \, dy}{y^2-2y-3} = \frac{3}{4} \int_4^8 \frac{dy}{y-3} + \frac{1}{4} \int_4^8 \frac{dy}{y+1} = \left[\frac{3}{4} \ln |y-3| + \frac{1}{4} \ln |y+1| \right]_4^8 = \left(\frac{3}{4} \ln 5 + \frac{1}{4} \ln 9 \right) - \left(\frac{3}{4} \ln 1 + \frac{1}{4} \ln 5 \right)$

$= \frac{1}{2} \ln 5 + \frac{1}{2} \ln 3 = \frac{\ln 15}{2}$

15. $\frac{1}{t^3+t^2-2t} = \frac{A}{t} + \frac{B}{t+2} + \frac{C}{t-1} \Rightarrow 1 = A(t+2)(t-1) + Bt(t-1) + Ct(t+2); t = 0 \Rightarrow A = -\frac{1}{2}; t = -2$

$\Rightarrow B = \frac{1}{6}; t = 1 \Rightarrow C = \frac{1}{3}; \int \frac{dt}{t^3+t^2-2t} = -\frac{1}{2}\int \frac{dt}{t} + \frac{1}{6}\int \frac{dt}{t+2} + \frac{1}{3}\int \frac{dt}{t-1}$

$= -\frac{1}{2}\ln|t| + \frac{1}{6}\ln|t+2| + \frac{1}{3}\ln|t-1| + C$

17. $\frac{x^3}{x^2+2x+1} = (x-2) + \frac{3x+2}{(x+1)^2}$ (after long division); $\frac{3x+2}{(x+1)^2} = \frac{A}{x+1} + \frac{B}{(x+1)^2} \Rightarrow 3x+2 = A(x+1) + B$

$= Ax + (A+B) \Rightarrow A = 3, A+B = 2 \Rightarrow A = 3, B = -1; \int_0^1 \frac{x^3\,dx}{x^2+2x+1}$

$= \int_0^1 (x-2)\,dx + 3\int_0^1 \frac{dx}{x+1} - \int_0^1 \frac{dx}{(x+1)^2} = \left[\frac{x^2}{2} - 2x + 3\ln|x+1| + \frac{1}{x+1}\right]_0^1$

$= \left(\frac{1}{2} - 2 + 3\ln 2 + \frac{1}{2}\right) - (1) = 3\ln 2 - 2$

19. $\frac{1}{(x^2-1)^2} = \frac{A}{x+1} + \frac{B}{x-1} + \frac{C}{(x+1)^2} + \frac{D}{(x-1)^2} \Rightarrow 1 = A(x+1)(x-1)^2 + B(x-1)(x+1)^2 + C(x-1)^2 + D(x+1)^2;$

$x = -1 \Rightarrow C = \frac{1}{4}; x = 1 \Rightarrow D = \frac{1}{4};$ coefficient of $x^3 = A + B \Rightarrow A + B = 0;$ constant $= A - B + C + D$

$\Rightarrow A - B + C + D = 1 \Rightarrow A - B = \frac{1}{2};$ thus, $A = \frac{1}{4} \Rightarrow B = -\frac{1}{4}; \int \frac{dx}{(x^2-1)^2}$

$= \frac{1}{4}\int \frac{dx}{x+1} - \frac{1}{4}\int \frac{dx}{x-1} + \frac{1}{4}\int \frac{dx}{(x+1)^2} + \frac{1}{4}\int \frac{dx}{(x-1)^2} = \frac{1}{4}\ln\left|\frac{x+1}{x-1}\right| - \frac{x}{2(x^2-1)} + C$

21. $\frac{1}{(x+1)(x^2+1)} = \frac{A}{x+1} + \frac{Bx+C}{x^2+1} \Rightarrow 1 = A(x^2+1) + (Bx+C)(x+1); x = -1 \Rightarrow A = \frac{1}{2};$ coefficient of x^2

$= A + B \Rightarrow A + B = 0 \Rightarrow B = -\frac{1}{2};$ constant $= A + C \Rightarrow A + C = 1 \Rightarrow C = \frac{1}{2}; \int_0^1 \frac{dx}{(x+1)(x^2+1)}$

$= \frac{1}{2}\int_0^1 \frac{dx}{x+1} + \frac{1}{2}\int_0^1 \frac{(-x+1)}{x^2+1}\,dx = \left[\frac{1}{2}\ln|x+1| - \frac{1}{4}\ln(x^2+1) + \frac{1}{2}\tan^{-1}x\right]_0^1$

$= \left(\frac{1}{2}\ln 2 - \frac{1}{4}\ln 2 + \frac{1}{2}\tan^{-1}1\right) - \left(\frac{1}{2}\ln 1 - \frac{1}{4}\ln 1 + \frac{1}{2}\tan^{-1}0\right) = \frac{1}{4}\ln 2 + \frac{1}{2}\left(\frac{\pi}{4}\right) = \frac{(\pi + 2\ln 2)}{8}$

23. $\frac{y^2+2y+1}{(y^2+1)^2} = \frac{Ay+B}{y^2+1} + \frac{Cy+D}{(y^2+1)^2} \Rightarrow y^2 + 2y + 1 = (Ay+B)(y^2+1) + Cy + D$

$= Ay^3 + By^2 + (A+C)y + (B+D) \Rightarrow A = 0, B = 1; A + C = 2 \Rightarrow C = 2; B + D = 1 \Rightarrow D = 0;$

$\int \frac{y^2+2y+1}{(y^2+1)^2}\,dy = \int \frac{1}{y^2+1}\,dy + 2\int \frac{y}{(y^2+1)^2}\,dy = \tan^{-1}y - \frac{1}{y^2+1} + C$

25. $\frac{2s+2}{(s^2+1)(s-1)^3} = \frac{As+B}{s^2+1} + \frac{C}{s-1} + \frac{D}{(s-1)^2} + \frac{E}{(s-1)^3} \Rightarrow 2s+2$

$= (As+B)(s-1)^3 + C(s^2+1)(s-1)^2 + D(s^2+1)(s-1) + E(s^2+1)$

$= [As^4 + (-3A+B)s^3 + (3A-3B)s^2 + (-A+3B)s - B] + C(s^4 - 2s^3 + 2s^2 - 2s + 1) + D(s^3 - s^2 + s - 1)$

$\quad + E(s^2+1)$

$= (A+C)s^4 + (-3A+B-2C+D)s^3 + (3A-3B+2C-D+E)s^2 + (-A+3B-2C+D)s + (-B+C-D+E)$

$\Rightarrow \left.\begin{array}{rcl} A \quad + C & = & 0 \\ -3A + B - 2C + D & = & 0 \\ 3A - 3B + 2C - D + E & = & 0 \\ -A + 3B - 2C + D & = & 2 \\ -B + C - D + E & = & 2 \end{array}\right\}$ summing all equations $\Rightarrow 2E = 4 \Rightarrow E = 2;$

summing eqs (2) and (3) $\Rightarrow -2B + 2 = 0 \Rightarrow B = 1;$ summing eqs (3) and (4) $\Rightarrow 2A + 2 = 2 \Rightarrow A = 0; C = 0$
from eq (1); then $-1 + 0 - D + 2 = 2$ from eq (5) $\Rightarrow D = -1;$

$\int \frac{2s+2}{(s^2+1)(s-1)^3}\,ds = \int \frac{ds}{s^2+1} - \int \frac{ds}{(s-1)^2} + 2\int \frac{ds}{(s-1)^3} = -(s-1)^{-2} + (s-1)^{-1} + \tan^{-1}s + C$

27. $\frac{2\theta^3+5\theta^2+8\theta+4}{(\theta^2+2\theta+2)^2} = \frac{A\theta+B}{\theta^2+2\theta+2} + \frac{C\theta+D}{(\theta^2+2\theta+2)^2} \Rightarrow 2\theta^3 + 5\theta^2 + 8\theta + 4 = (A\theta+B)(\theta^2+2\theta+2) + C\theta + D$

$= A\theta^3 + (2A+B)\theta^2 + (2A+2B+C)\theta + (2B+D) \Rightarrow A = 2; 2A+B = 5 \Rightarrow B = 1; 2A+2B+C = 8 \Rightarrow C = 2;$

$2B+D = 4 \Rightarrow D = 2; \int \frac{2\theta^3+5\theta^2+8\theta+4}{(\theta^2+2\theta+2)^2}\,d\theta = \int \frac{2\theta+1}{(\theta^2+2\theta+2)}\,d\theta + \int \frac{2\theta+2}{(\theta^2+2\theta+2)^2}\,d\theta$

$= \int \frac{2\theta+2}{\theta^2+2\theta+2}\,d\theta - \int \frac{d\theta}{\theta^2+2\theta+2} + \int \frac{d(\theta^2+2\theta+2)}{(\theta^2+2\theta+2)^2} = \int \frac{d(\theta^2+2\theta+2)}{\theta^2+2\theta+2} - \int \frac{d\theta}{(\theta+1)^2+1} - \frac{1}{\theta^2+2\theta+2}$

$$= \frac{-1}{\theta^2 + 2\theta + 2} + \ln(\theta^2 + 2\theta + 2) - \tan^{-1}(\theta + 1) + C$$

29. $\frac{2x^3 - 2x^2 + 1}{x^2 - x} = 2x + \frac{1}{x^2 - x} = 2x + \frac{1}{x(x-1)}$; $\frac{1}{x(x-1)} = \frac{A}{x} + \frac{B}{x-1} \Rightarrow 1 = A(x-1) + Bx$; $x = 0 \Rightarrow A = -1$;

$x = 1 \Rightarrow B = 1$; $\int \frac{2x^3 - 2x^2 + 1}{x^2 - x} = \int 2x\,dx - \int \frac{dx}{x} + \int \frac{dx}{x-1} = x^2 - \ln|x| + \ln|x-1| + C = x^2 + \ln\left|\frac{x-1}{x}\right| + C$

31. $\frac{9x^3 - 3x + 1}{x^3 - x^2} = 9 + \frac{9x^2 - 3x + 1}{x^2(x-1)}$ (after long division); $\frac{9x^2 - 3x + 1}{x^2(x-1)} = \frac{A}{x} + \frac{B}{x^2} + \frac{C}{x-1}$

$\Rightarrow 9x^2 - 3x + 1 = Ax(x-1) + B(x-1) + Cx^2$; $x = 1 \Rightarrow C = 7$; $x = 0 \Rightarrow B = -1$; $A + C = 9 \Rightarrow A = 2$;

$\int \frac{9x^3 - 3x + 1}{x^3 - x^2}\,dx = \int 9\,dx + 2\int \frac{dx}{x} - \int \frac{dx}{x^2} + 7\int \frac{dx}{x-1} = 9x + 2\ln|x| + \frac{1}{x} + 7\ln|x-1| + C$

33. $\frac{y^4 + y^2 - 1}{y^3 + y} = y - \frac{1}{y(y^2+1)}$; $\frac{1}{y(y^2+1)} = \frac{A}{y} + \frac{By+C}{y^2+1} \Rightarrow 1 = A(y^2+1) + (By+C)y = (A+B)y^2 + Cy + A$

$\Rightarrow A = 1$; $A + B = 0 \Rightarrow B = -1$; $C = 0$; $\int \frac{y^4 + y^2 - 1}{y^3 + y}\,dy = \int y\,dy - \int \frac{dy}{y} + \int \frac{y\,dy}{y^2+1}$

$= \frac{y^2}{2} - \ln|y| + \frac{1}{2}\ln(1 + y^2) + C$

35. $\int \frac{e^t\,dt}{e^{2t} + 3e^t + 2} = [e^t = y]\int \frac{dy}{y^2 + 3y + 2} = \int \frac{dy}{y+1} - \int \frac{dy}{y+2} = \ln\left|\frac{y+1}{y+2}\right| + C = \ln\left(\frac{e^t+1}{e^t+2}\right) + C$

37. $\int \frac{\cos y\,dy}{\sin^2 y + \sin y - 6}$; $[\sin y = t, \cos y\,dy = dt] \to \int \frac{dy}{t^2 + t - 6} = \frac{1}{5}\int \left(\frac{1}{t-2} - \frac{1}{t+3}\right)dt = \frac{1}{5}\ln\left|\frac{t-2}{t+3}\right| + C$

$= \frac{1}{5}\ln\left|\frac{\sin y - 2}{\sin y + 3}\right| + C$

39. $\int \frac{(x-2)^2 \tan^{-1}(2x) - 12x^3 - 3x}{(4x^2+1)(x-2)^2}\,dx = \int \frac{\tan^{-1}(2x)}{4x^2+1}\,dx - 3\int \frac{x}{(x-2)^2}\,dx$

$= \frac{1}{2}\int \tan^{-1}(2x)\,d(\tan^{-1}(2x)) - 3\int \frac{dx}{x-2} - 6\int \frac{dx}{(x-2)^2} = \frac{(\tan^{-1}2x)^2}{4} - 3\ln|x-2| + \frac{6}{x-2} + C$

41. $(t^2 - 3t + 2)\frac{dx}{dt} = 1$; $x = \int \frac{dt}{t^2 - 3t + 2} = \int \frac{dt}{t-2} - \int \frac{dt}{t-1} = \ln\left|\frac{t-2}{t-1}\right| + C$; $\frac{t-2}{t-1} = Ce^x$; $t = 3$ and $x = 0$

$\Rightarrow \frac{1}{2} = C \Rightarrow \frac{t-2}{t-1} = \frac{1}{2}e^x \Rightarrow x = \ln\left|2\left(\frac{t-2}{t-1}\right)\right| = \ln|t-2| - \ln|t-1| + \ln 2$

43. $(t^2 + 2t)\frac{dx}{dt} = 2x + 2$; $\frac{1}{2}\int \frac{dx}{x+1} = \int \frac{dt}{t^2 + 2t} \Rightarrow \frac{1}{2}\ln|x+1| = \frac{1}{2}\int \frac{dt}{t} - \frac{1}{2}\int \frac{dt}{t+2} \Rightarrow \ln|x+1| = \ln\left|\frac{t}{t+2}\right| + C$;

$t = 1$ and $x = 1 \Rightarrow \ln 2 = \ln\frac{1}{3} + C \Rightarrow C = \ln 2 + \ln 3 = \ln 6 \Rightarrow \ln|x+1| = \ln 6\left|\frac{t}{t+2}\right| \Rightarrow x + 1 = \frac{6t}{t+2}$

$\Rightarrow x = \frac{6t}{t+2} - 1, t > 0$

45. $V = \pi \int_{0.5}^{2.5} y^2\,dx = \pi \int_{0.5}^{2.5} \frac{9}{3x - x^2}\,dx = 3\pi \left(\int_{0.5}^{2.5}\left(-\frac{1}{x-3} + \frac{1}{x}\right)\right)dx = \left[3\pi \ln\left|\frac{x}{x-3}\right|\right]_{0.5}^{2.5} = 3\pi \ln 25$

47. $A = \int_0^{\sqrt{3}} \tan^{-1}x\,dx = [x\tan^{-1}x]_0^{\sqrt{3}} - \int_0^{\sqrt{3}} \frac{x}{1+x^2}\,dx$

$= \frac{\pi\sqrt{3}}{3} - \left[\frac{1}{2}\ln(x^2 + 1)\right]_0^{\sqrt{3}} = \frac{\pi\sqrt{3}}{3} - \ln 2$;

$\bar{x} = \frac{1}{A}\int_0^{\sqrt{3}} x\tan^{-1}x\,dx$

$= \frac{1}{A}\left(\left[\frac{1}{2}x^2 \tan^{-1}x\right]_0^{\sqrt{3}} - \frac{1}{2}\int_0^{\sqrt{3}} \frac{x^2}{1+x^2}\,dx\right)$

$= \frac{1}{A}\left[\frac{\pi}{2} - \left[\frac{1}{2}(x - \tan^{-1}x)\right]_0^{\sqrt{3}}\right]$

$= \frac{1}{A}\left(\frac{\pi}{2} - \frac{\sqrt{3}}{2} + \frac{\pi}{6}\right) = \frac{1}{A}\left(\frac{2\pi}{3} - \frac{\sqrt{3}}{2}\right) \cong 1.10$

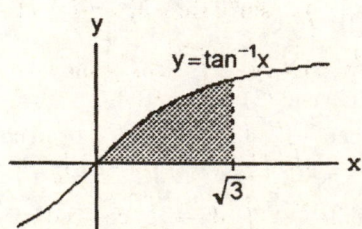

49. (a) $\frac{dx}{dt} = kx(N - x) \Rightarrow \int \frac{dx}{x(N-x)} = \int k\,dt \Rightarrow \frac{1}{N}\int \frac{dx}{x} + \frac{1}{N}\int \frac{dx}{N-x} = \int k\,dt \Rightarrow \frac{1}{N}\ln\left|\frac{x}{N-x}\right| = kt + C;$

$k = \frac{1}{250}, N = 1000, t = 0$ and $x = 2 \Rightarrow \frac{1}{1000}\ln\left|\frac{2}{998}\right| = C \Rightarrow \frac{1}{1000}\ln\left|\frac{x}{1000-x}\right| = \frac{t}{250} + \frac{1}{1000}\ln\left(\frac{1}{499}\right)$

$\Rightarrow \ln\left|\frac{499x}{1000-x}\right| = 4t \Rightarrow \frac{499x}{1000-x} = e^{4t} \Rightarrow 499x = e^{4t}(1000 - x) \Rightarrow (499 + e^{4t})x = 1000e^{4t} \Rightarrow x = \frac{1000e^{4t}}{499 + e^{4t}}$

(b) $x = \frac{1}{2}N = 500 \Rightarrow 500 = \frac{1000e^{4t}}{499 + e^{4t}} \Rightarrow 500 \cdot 499 + 500e^{4t} = 1000e^{4t} \Rightarrow e^{4t} = 499 \Rightarrow t = \frac{1}{4}\ln 499 \approx 1.55$ days

51. (a) $\int_0^1 \frac{x^4(x-1)^4}{x^2+1}\,dx = \int_0^1 \left(x^6 - 4x^5 + 5x^4 - 4x^2 + 4 - \frac{4}{x^2+1}\right)dx = \frac{22}{7} - \pi$

(b) $\frac{\frac{22}{7} - \pi}{\pi} \cdot 100\% \cong 0.04\%$

(c) The area is less than 0.003

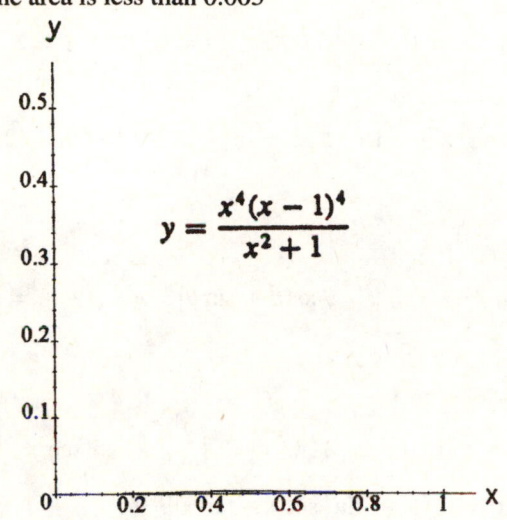
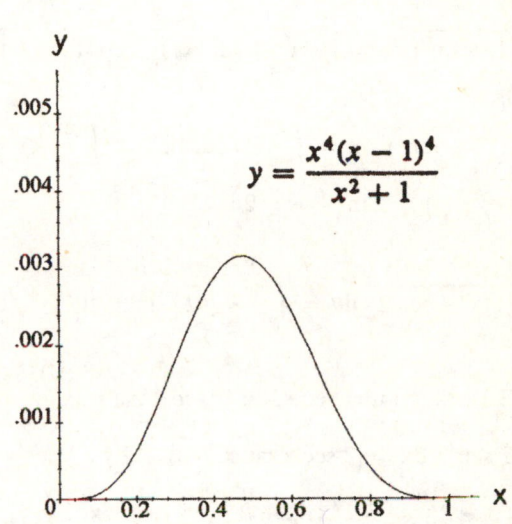

8.4 TRIGONOMETRIC INTEGRALS

1. $\int_0^{\pi/2} \sin^5 x\,dx = \int_0^{\pi/2}(\sin^2 x)^2 \sin x\,dx = \int_0^{\pi/2}(1 - \cos^2 x)^2 \sin x\,dx = \int_0^{\pi/2}(1 - 2\cos^2 x + \cos^4 x)\sin x\,dx$

$= \int_0^{\pi/2}\sin x\,dx - \int_0^{\pi/2} 2\cos^2 x \sin x\,dx + \int_0^{\pi/2}\cos^4 x \sin x\,dx = \left[-\cos x + 2\frac{\cos^3 x}{3} - \frac{\cos^5 x}{5}\right]_0^{\pi/2}$

$= (0) - \left(-1 + \frac{2}{3} - \frac{1}{5}\right) = \frac{8}{15}$

3. $\int_{-\pi/2}^{\pi/2}\cos^3 x\,dx = \int_{-\pi/2}^{\pi/2}(\cos^2 x)\cos x\,dx = \int_{-\pi/2}^{\pi/2}(1 - \sin^2 x)\cos x\,dx = \int_{-\pi/2}^{\pi/2}\cos x\,dx - \int_{-\pi/2}^{\pi/2}\sin^2 x \cos x\,dx$

$= \left[\sin x - \frac{\sin^3 x}{3}\right]_{-\pi/2}^{\pi/2} = \left(1 - \frac{1}{3}\right) - \left(-1 + \frac{1}{3}\right) = \frac{4}{3}$

5. $\int_0^{\pi/2}\sin^7 y\,dy = \int_0^{\pi/2}\sin^6 y \sin y\,dy = \int_0^{\pi/2}(1 - \cos^2 y)^3 \sin y\,dy = \int_0^{\pi/2}\sin y\,dy - 3\int_0^{\pi/2}\cos^2 y \sin y\,dy$

$+ 3\int_0^{\pi/2}\cos^4 y \sin y\,dy - \int_0^{\pi/2}\cos^6 y \sin y\,dy = \left[-\cos y + 3\frac{\cos^3 y}{3} - 3\frac{\cos^5 y}{5} + \frac{\cos^7 y}{7}\right]_0^{\pi/2} = (0) - \left(-1 + 1 - \frac{3}{5} + \frac{1}{7}\right) = \frac{16}{35}$

7. $\int_0^\pi 8\sin^4 x\,dx = 8\int_0^\pi \left(\frac{1 - \cos 2x}{2}\right)^2 dx = 2\int_0^\pi (1 - 2\cos 2x + \cos^2 2x)dx = 2\int_0^\pi dx - 2\int_0^\pi \cos 2x \cdot 2\,dx + 2\int_0^\pi \frac{1 + \cos 4x}{2}\,dx$

$= [2x - 2\sin 2x]_0^\pi + \int_0^\pi dx + \int_0^\pi \cos 4x\,dx = 2\pi + \left[x + \frac{1}{2}\sin 4x\right]_0^\pi = 2\pi + \pi = 3\pi$

9. $\int_{-\pi/4}^{\pi/4} 16\sin^2 x \cos^2 x\,dx = 16\int_{-\pi/4}^{\pi/4}\left(\frac{1 - \cos 2x}{2}\right)\left(\frac{1 + \cos 2x}{2}\right)dx = 4\int_{-\pi/4}^{\pi/4}(1 - \cos^2 2x)dx = 4\int_{-\pi/4}^{\pi/4}dx - 4\int_{-\pi/4}^{\pi/4}\left(\frac{1 + \cos 4x}{2}\right)dx$

$= [4x]_{-\pi/4}^{\pi/4} - 2\int_{-\pi/4}^{\pi/4}dx - 2\int_{-\pi/4}^{\pi/4}\cos 4x\,dx = \pi + \pi - \left[2x + \frac{\sin 4x}{2}\right]_{-\pi/4}^{\pi/4} = 2\pi - \left(\frac{\pi}{2} - \left(-\frac{\pi}{2}\right)\right) = \pi$

11. $\int_0^{\pi/2} 35 \sin^4 x \cos^3 x \, dx = \int_0^{\pi/2} 35 \sin^4 x (1 - \sin^2 x) \cos x \, dx = 35 \int_0^{\pi/2} \sin^4 x \cos x \, dx - 35 \int_0^{\pi/2} \sin^6 x \cos x \, dx$

$= \left[35 \frac{\sin^5 x}{5} - 35 \frac{\sin^7 x}{7} \right]_0^{\pi/2} = (7 - 5) - (0) = 2$

13. $\int_0^{\pi/4} 8\cos^3 2\theta \sin 2\theta \, d\theta = \left[8 \left(-\frac{1}{2} \right) \frac{\cos^4 2\theta}{4} \right]_0^{\pi/4} = \left[-\cos^4 2\theta \right]_0^{\pi/4} = (0) - (-1) = 1$

15. $\int_0^{2\pi} \sqrt{\frac{1 - \cos x}{2}} \, dx = \int_0^{2\pi} \left| \sin \frac{x}{2} \right| dx = \int_0^{2\pi} \sin \frac{x}{2} \, dx = \left[-2\cos \frac{x}{2} \right]_0^{2\pi} = 2 + 2 = 4$

17. $\int_0^{\pi} \sqrt{1 - \sin^2 t} \, dt = \int_0^{\pi} |\cos t| \, dt = \int_0^{\pi/2} \cos t \, dt - \int_{\pi/2}^{\pi} \cos t \, dt = [\sin t]_0^{\pi/2} - [\sin t]_{\pi/2}^{\pi} = 1 - 0 - 0 + 1 = 2$

19. $\int_{-\pi/4}^{\pi/4} \sqrt{1 + \tan^2 x} \, dx = \int_{-\pi/4}^{\pi/4} |\sec x| \, dx = \int_{-\pi/4}^{\pi/4} \sec x \, dx = [\ln|\sec x + \tan x|]_{-\pi/4}^{\pi/4} = \ln\left(\sqrt{2} + 1 \right) - \ln\left(\sqrt{2} - 1 \right)$

$= \ln\left(\frac{\sqrt{2} + 1}{\sqrt{2} - 1} \right) = 2\ln\left(1 + \sqrt{2} \right)$

21. $\int_0^{\pi/2} \theta \sqrt{1 - \cos 2\theta} \, d\theta = \int_0^{\pi/2} \theta \sqrt{2} |\sin \theta| \, d\theta = \sqrt{2} \int_0^{\pi/2} \theta \sin \theta \, d\theta = \sqrt{2} \left[-\theta \cos \theta + \sin \theta \right]_0^{\pi/2} = \sqrt{2}(1) = \sqrt{2}$

23. $\int_{-\pi/3}^{0} 2 \sec^3 x \, dx$; $u = \sec x$, $du = \sec x \tan x \, dx$, $dv = \sec^2 x \, dx$, $v = \tan x$;

$\int_{-\pi/3}^{0} 2 \sec^3 x \, dx = [2 \sec x \tan x]_{-\pi/3}^{0} - 2 \int_{-\pi/3}^{0} \sec x \tan^2 x \, dx = 2 \cdot 1 \cdot 0 - 2 \cdot 2 \cdot \sqrt{3} - 2 \int_{-\pi/3}^{0} \sec x (\sec^2 x - 1) dx$

$= 4\sqrt{3} - 2 \int_{-\pi/3}^{0} \sec^3 x \, dx + 2 \int_{-\pi/3}^{0} \sec x \, dx$; $2 \int_{-\pi/3}^{0} 2 \sec^3 x \, dx = 4\sqrt{3} + [2\ln|\sec x + \tan x|]_{-\pi/3}^{0}$

$2 \int_{-\pi/3}^{0} 2 \sec^3 x \, dx = 4\sqrt{3} + 2\ln|1 + 0| - 2\ln|2 - \sqrt{3}| = 4\sqrt{3} - 2\ln\left(2 - \sqrt{3} \right)$

$\int_{-\pi/3}^{0} 2 \sec^3 x \, dx = 2\sqrt{3} - \ln\left(2 - \sqrt{3} \right)$

25. $\int_0^{\pi/4} \sec^4 \theta \, d\theta = \int_0^{\pi/4} (1 + \tan^2 \theta) \sec^2 \theta \, d\theta = \int_0^{\pi/4} \sec^2 \theta \, d\theta + \int_0^{\pi/4} \tan^2 \theta \sec^2 \theta \, d\theta = \left[\tan \theta + \frac{\tan^3 \theta}{3} \right]_0^{\pi/4}$

$= \left(1 + \frac{1}{3} \right) - (0) = \frac{4}{3}$

27. $\int_{\pi/4}^{\pi/2} \csc^4 \theta \, d\theta = \int_{\pi/4}^{\pi/2} (1 + \cot^2 \theta) \csc^2 \theta \, d\theta = \int_{\pi/4}^{\pi/2} \csc^2 \theta \, d\theta + \int_{\pi/4}^{\pi/2} \cot^2 \theta \csc^2 \theta \, d\theta = \left[-\cot \theta - \frac{\cot^3 \theta}{3} \right]_{\pi/4}^{\pi/2}$

$= (0) - \left(-1 - \frac{1}{3} \right) = \frac{4}{3}$

29. $\int_0^{\pi/4} 4 \tan^3 x \, dx = 4 \int_0^{\pi/4} (\sec^2 x - 1) \tan x \, dx = 4 \int_0^{\pi/4} \sec^2 x \tan x \, dx - 4 \int_0^{\pi/4} \tan x \, dx = \left[4 \frac{\tan^2 x}{2} - 4 \ln|\sec x| \right]_0^{\pi/4}$

$= 2(1) - 4\ln\sqrt{2} - 2 \cdot 0 + 4\ln 1 = 2 - 2\ln 2$

31. $\int_{\pi/6}^{\pi/3} \cot^3 x \, dx = \int_{\pi/6}^{\pi/3} (\csc^2 x - 1) \cot x \, dx = \int_{\pi/6}^{\pi/3} \csc^2 x \cot x \, dx - \int_{\pi/6}^{\pi/3} \cot x \, dx = \left[-\frac{\cot^2 x}{2} + \ln|\csc x| \right]_{\pi/6}^{\pi/3}$

$= -\frac{1}{2} \left(\frac{1}{3} - 3 \right) + \left(\ln \frac{2}{\sqrt{3}} - \ln 2 \right) = \frac{4}{3} - \ln\sqrt{3}$

33. $\int_{-\pi}^{0} \sin 3x \cos 2x \, dx = \frac{1}{2} \int_{-\pi}^{0} (\sin x + \sin 5x) \, dx = \frac{1}{2} \left[-\cos x - \frac{1}{5}\cos 5x \right]_{-\pi}^{0} = \frac{1}{2} \left(-1 - \frac{1}{5} - 1 - \frac{1}{5} \right) = -\frac{6}{5}$

35. $\int_{-\pi}^{\pi} \sin 3x \sin 3x \, dx = \frac{1}{2} \int_{-\pi}^{\pi} (\cos 0 - \cos 6x) \, dx = \frac{1}{2} \int_{-\pi}^{\pi} dx - \frac{1}{2} \int_{-\pi}^{\pi} \cos 6x \, dx = \frac{1}{2} \left[x - \frac{1}{12}\sin 6x \right]_{-\pi}^{\pi} = \frac{\pi}{2} + \frac{\pi}{2} - 0 = \pi$

37. $\int_0^\pi \cos 3x \cos 4x \, dx = \frac{1}{2}\int_0^\pi (\cos(-x) + \cos 7x) \, dx = \frac{1}{2}\left[-\sin(-x) + \frac{1}{7}\sin 7x\right]_0^\pi = \frac{1}{2}(0) = 0$

39. $x = t^{2/3} \Rightarrow t^2 = x^3; y = \frac{t^2}{2} \Rightarrow y = \frac{x^3}{2}; 0 \le t \le 2 \Rightarrow 0 \le x \le 2^{2/3};$

$A = \int_0^{2^{2/3}} 2\pi \left(\frac{x^3}{2}\right)\sqrt{1 + \frac{9}{4}x^4} \, dx; \begin{bmatrix} u = \frac{9}{4}x^4 \\ du = 9x^3 dx \end{bmatrix} \rightarrow \frac{\pi}{9}\int_0^{9(2^{2/3})} \sqrt{1 + u} \, du = \left[\frac{\pi}{9} \cdot \frac{2}{3}(1 + u)^{3/2}\right]_0^{9(2^{2/3})}$

$= \frac{2\pi}{27}\left[\left(1 + 9(2^{2/3})\right)^{3/2} - 1\right]$

41. $y = \ln(\sec x); y' = \frac{\sec x \tan x}{\sec x} = \tan x; (y')^2 = \tan^2 x; \int_0^{\pi/4} \sqrt{1 + \tan^2 x} \, dx = \int_0^{\pi/4} |\sec x| \, dx = [\ln|\sec x + \tan x|]_0^{\pi/4}$

$= \ln\left(\sqrt{2} + 1\right) - \ln(0 + 1) = \ln\left(\sqrt{2} + 1\right)$

43. $V = \pi\int_0^\pi \sin^2 x \, dx = \pi\int_0^\pi \frac{1 - \cos 2x}{2} \, dx = \frac{\pi}{2}\int_0^\pi dx - \frac{\pi}{2}\int_0^\pi \cos 2x \, dx = \frac{\pi}{2}[x]_0^\pi - \frac{\pi}{4}[\sin 2x]_0^\pi = \frac{\pi}{2}(\pi - 0) - \frac{\pi}{4}(0 - 0) = \frac{\pi^2}{2}$

45. (a) $m^2 \ne n^2 \Rightarrow m + n \ne 0$ and $m - n \ne 0 \Rightarrow \int_k^{k+2\pi} \sin mx \sin nx \, dx = \frac{1}{2}\int_k^{k+2\pi} [\cos(m - n)x - \cos(m + n)x]dx$

$= \frac{1}{2}\left[\frac{1}{m-n}\sin(m - n)x - \frac{1}{m+n}\sin(m + n)x\right]_k^{k+2\pi}$

$= \frac{1}{2}\left(\frac{1}{m-n}\sin((m - n)(k + 2\pi)) - \frac{1}{m+n}\sin((m + n)(k + 2\pi))\right) - \frac{1}{2}\left(\frac{1}{m-n}\sin((m - n)k) - \frac{1}{m+n}\sin((m + n)k)\right)$

$= \frac{1}{2(m-n)}\sin((m - n)k) - \frac{1}{2(m+n)}\sin((m + n)k) - \frac{1}{2(m-n)}\sin((m - n)k) + \frac{1}{2(m+n)}\sin((m + n)k) = 0$

$\Rightarrow \sin mx$ and $\sin nx$ are orthogonal.

(b) Same as part since $\frac{1}{2}\int_k^{k+2\pi} \cos 0 \, dx = \pi$. $m^2 \ne n^2 \Rightarrow m + n \ne 0$ and $m - n \ne 0 \Rightarrow \int_k^{k+2\pi} \cos mx \cos nx \, dx$

$= \frac{1}{2}\int_k^{k+2\pi} [\cos(m - n)x + \cos(m + n)x]dx = \frac{1}{2}\left[\frac{1}{m-n}\sin(m - n)x + \frac{1}{m+n}\sin(m + n)x\right]_k^{k+2\pi}$

$= \frac{1}{2(m-n)}\sin((m - n)(k + 2\pi)) + \frac{1}{2(m+n)}\sin((m + n)(k + 2\pi)) - \frac{1}{2(m-n)}\sin((m - n)k) - \frac{1}{2(m+n)}\sin((m + n)k)$

$= \frac{1}{2(m-n)}\sin((m - n)k) + \frac{1}{2(m+n)}\sin((m + n)k) - \frac{1}{2(m-n)}\sin((m - n)k) - \frac{1}{2(m+n)}\sin((m + n)k) = 0$

$\Rightarrow \cos mx$ and $\cos nx$ are orthogonal.

(c) Let $m = n \Rightarrow \sin mx \cos nx = \frac{1}{2}(\sin 0 + \sin((m + n)x))$ and $\frac{1}{2}\int_k^{k+2\pi} \sin 0 \, dx = 0$ and $\frac{1}{2}\int_k^{k+2\pi} \sin((m + n)x) \, dx = 0$

$\Rightarrow \sin mx$ and $\cos nx$ are orthogonal if $m = n$.

Let $m \ne n$.

$\int_k^{k+2\pi} \sin mx \cos nx \, dx = \frac{1}{2}\int_k^{k+2\pi} [\sin(m - n)x + \sin(m + n)x]dx = \frac{1}{2}\left[-\frac{1}{m-n}\cos(m - n)x - \frac{1}{m+n}\cos(m + n)x\right]_k^{k+2\pi}$

$= -\frac{1}{2(m-n)}\cos((m - n)(k + 2\pi)) - \frac{1}{2(m+n)}\cos((m + n)(k + 2\pi)) + \frac{1}{2(m-n)}\cos((m - n)k) + \frac{1}{2(m+n)}\cos((m + n)k)$

$= -\frac{1}{2(m-n)}\cos((m - n)k) - \frac{1}{2(m+n)}\cos((m + n)k) + \frac{1}{2(m-n)}\cos((m - n)k) + \frac{1}{2(m+n)}\cos((m + n)k) = 0$

$\Rightarrow \sin mx$ and $\cos nx$ are orthogonal.

8.5 TRIGONOMETRIC SUBSTITUTIONS

1. $y = 3\tan\theta, -\frac{\pi}{2} < \theta < \frac{\pi}{2}, dy = \frac{3 \, d\theta}{\cos^2\theta}, 9 + y^2 = 9(1 + \tan^2\theta) = \frac{9}{\cos^2\theta} \Rightarrow \frac{1}{\sqrt{9+y^2}} = \frac{|\cos\theta|}{3} = \frac{\cos\theta}{3}$

(because $\cos\theta > 0$ when $-\frac{\pi}{2} < \theta < \frac{\pi}{2}$);

$\int \frac{dy}{\sqrt{9+y^2}} = 3\int \frac{\cos\theta \, d\theta}{3\cos^2\theta} = \int \frac{d\theta}{\cos\theta} = \ln|\sec\theta + \tan\theta| + C' = \ln\left|\frac{\sqrt{9+y^2}}{3} + \frac{y}{3}\right| + C' = \ln\left|\sqrt{9+y^2} + y\right| + C$

3. $\int_{-2}^2 \frac{dx}{4+x^2} = \left[\frac{1}{2}\tan^{-1}\frac{x}{2}\right]_{-2}^2 = \frac{1}{2}\tan^{-1}1 - \frac{1}{2}\tan^{-1}(-1) = \left(\frac{1}{2}\right)\left(\frac{\pi}{4}\right) - \left(\frac{1}{2}\right)\left(-\frac{\pi}{4}\right) = \frac{\pi}{4}$

5. $\int_0^{3/2} \frac{dx}{\sqrt{9-x^2}} = \left[\sin^{-1}\frac{x}{3}\right]_0^{3/2} = \sin^{-1}\frac{1}{2} - \sin^{-1}0 = \frac{\pi}{6} - 0 = \frac{\pi}{6}$

7. $t = 5 \sin \theta, -\frac{\pi}{2} < \theta < \frac{\pi}{2}, dt = 5 \cos \theta \, d\theta, \sqrt{25 - t^2} = 5 \cos \theta;$

$\int \sqrt{25 - t^2} \, dt = \int (5 \cos \theta)(5 \cos \theta) \, d\theta = 25 \int \cos^2 \theta \, d\theta = 25 \int \frac{1 + \cos 2\theta}{2} \, d\theta = 25 \left(\frac{\theta}{2} + \frac{\sin 2\theta}{4} \right) + C$

$= \frac{25}{2}(\theta + \sin \theta \cos \theta) + C = \frac{25}{2} \left[\sin^{-1} \left(\frac{t}{5} \right) + \left(\frac{t}{5} \right) \left(\frac{\sqrt{25 - t^2}}{5} \right) \right] + C = \frac{25}{2} \sin^{-1} \left(\frac{t}{5} \right) + \frac{t\sqrt{25 - t^2}}{2} + C$

9. $x = \frac{7}{2} \sec \theta, 0 < \theta < \frac{\pi}{2}, dx = \frac{7}{2} \sec \theta \tan \theta \, d\theta, \sqrt{4x^2 - 49} = \sqrt{49 \sec^2 \theta - 49} = 7 \tan \theta;$

$\int \frac{dx}{\sqrt{4x^2 - 49}} = \int \frac{\left(\frac{7}{2} \sec \theta \tan \theta \right) d\theta}{7 \tan \theta} = \frac{1}{2} \int \sec \theta \, d\theta = \frac{1}{2} \ln |\sec \theta + \tan \theta| + C = \frac{1}{2} \ln \left| \frac{2x}{7} + \frac{\sqrt{4x^2 - 49}}{7} \right| + C$

11. $y = 7 \sec \theta, 0 < \theta < \frac{\pi}{2}, dy = 7 \sec \theta \tan \theta \, d\theta, \sqrt{y^2 - 49} = 7 \tan \theta;$

$\int \frac{\sqrt{y^2 - 49}}{y} \, dy = \int \frac{(7 \tan \theta)(7 \sec \theta \tan \theta) \, d\theta}{7 \sec \theta} = 7 \int \tan^2 \theta \, d\theta = 7 \int (\sec^2 \theta - 1) \, d\theta = 7(\tan \theta - \theta) + C$

$= 7 \left[\frac{\sqrt{y^2 - 49}}{7} - \sec^{-1} \left(\frac{y}{7} \right) \right] + C$

13. $x = \sec \theta, 0 < \theta < \frac{\pi}{2}, dx = \sec \theta \tan \theta \, d\theta, \sqrt{x^2 - 1} = \tan \theta;$

$\int \frac{dx}{x^2 \sqrt{x^2 - 1}} = \int \frac{\sec \theta \tan \theta \, d\theta}{\sec^2 \theta \tan \theta} = \int \frac{d\theta}{\sec \theta} = \sin \theta + C = \frac{\sqrt{x^2 - 1}}{x} + C$

15. $x = 2 \tan \theta, -\frac{\pi}{2} < \theta < \frac{\pi}{2}, dx = \frac{2 \, d\theta}{\cos^2 \theta}, \sqrt{x^2 + 4} = \frac{2}{\cos \theta};$

$\int \frac{x^3 \, dx}{\sqrt{x^2 + 4}} = \int \frac{(8 \tan^3 \theta)(\cos \theta) \, d\theta}{\cos^2 \theta} = 8 \int \frac{\sin^3 \theta \, d\theta}{\cos^4 \theta} = 8 \int \frac{(\cos^2 \theta - 1)(-\sin \theta) \, d\theta}{\cos^4 \theta};$

$[t = \cos \theta] \rightarrow 8 \int \frac{t^2 - 1}{t^4} \, dt = 8 \int \left(\frac{1}{t^2} - \frac{1}{t^4} \right) dt = 8 \left(-\frac{1}{t} + \frac{1}{3t^3} \right) + C = 8 \left(-\sec \theta + \frac{\sec^3 \theta}{3} \right) + C$

$= 8 \left(-\frac{\sqrt{x^2 + 4}}{2} + \frac{(x^2 + 4)^{3/2}}{8 \cdot 3} \right) + C = \frac{1}{3} (x^2 + 4)^{3/2} - 4\sqrt{x^2 + 4} + C$

17. $w = 2 \sin \theta, -\frac{\pi}{2} < \theta < \frac{\pi}{2}, dw = 2 \cos \theta \, d\theta, \sqrt{4 - w^2} = 2 \cos \theta;$

$\int \frac{8 \, dw}{w^2 \sqrt{4 - w^2}} = \int \frac{8 \cdot 2 \cos \theta \, d\theta}{4 \sin^2 \theta \cdot 2 \cos \theta} = 2 \int \frac{d\theta}{\sin^2 \theta} = -2 \cot \theta + C = \frac{-2\sqrt{4 - w^2}}{w} + C$

19. $x = \sin \theta, 0 \leq \theta \leq \frac{\pi}{3}, dx = \cos \theta \, d\theta, (1 - x^2)^{3/2} = \cos^3 \theta;$

$\int_0^{\sqrt{3}/2} \frac{4x^2 \, dx}{(1 - x^2)^{3/2}} = \int_0^{\pi/3} \frac{4 \sin^2 \theta \cos \theta \, d\theta}{\cos^3 \theta} = 4 \int_0^{\pi/3} \left(\frac{1 - \cos^2 \theta}{\cos^2 \theta} \right) d\theta = 4 \int_0^{\pi/3} (\sec^2 \theta - 1) \, d\theta$

$= 4 \left[\tan \theta - \theta \right]_0^{\pi/3} = 4\sqrt{3} - \frac{4\pi}{3}$

21. $x = \sec \theta, 0 < \theta < \frac{\pi}{2}, dx = \sec \theta \tan \theta \, d\theta, (x^2 - 1)^{3/2} = \tan^3 \theta;$

$\int \frac{dx}{(x^2 - 1)^{3/2}} = \int \frac{\sec \theta \tan \theta \, d\theta}{\tan^3 \theta} = \int \frac{\cos \theta \, d\theta}{\sin^2 \theta} = -\frac{1}{\sin \theta} + C = -\frac{x}{\sqrt{x^2 - 1}} + C$

23. $x = \sin \theta, -\frac{\pi}{2} < \theta < \frac{\pi}{2}, dx = \cos \theta \, d\theta, (1 - x^2)^{3/2} = \cos^3 \theta;$

$\int \frac{(1 - x^2)^{3/2} \, dx}{x^6} = \int \frac{\cos^3 \theta \cdot \cos \theta \, d\theta}{\sin^6 \theta} = \int \cot^4 \theta \csc^2 \theta \, d\theta = -\frac{\cot^5 \theta}{5} + C = -\frac{1}{5} \left(\frac{\sqrt{1 - x^2}}{x} \right)^5 + C$

25. $x = \frac{1}{2} \tan \theta, -\frac{\pi}{2} < \theta < \frac{\pi}{2}, dx = \frac{1}{2} \sec^2 \theta \, d\theta, (4x^2 + 1)^2 = \sec^4 \theta;$

$\int \frac{8 \, dx}{(4x^2 + 1)^2} = \int \frac{8 \left(\frac{1}{2} \sec^2 \theta \right) d\theta}{\sec^4 \theta} = 4 \int \cos^2 \theta \, d\theta = 2(\theta + \sin \theta \cos \theta) + C = 2 \tan^{-1} 2x + \frac{4x}{(4x^2 + 1)} + C$

27. $v = \sin \theta, -\frac{\pi}{2} < \theta < \frac{\pi}{2}, dv = \cos \theta \, d\theta, (1 - v^2)^{5/2} = \cos^5 \theta;$

$\int \frac{v^2 \, dv}{(1 - v^2)^{5/2}} = \int \frac{\sin^2 \theta \cos \theta \, d\theta}{\cos^5 \theta} = \int \tan^2 \theta \sec^2 \theta \, d\theta = \frac{\tan^3 \theta}{3} + C = \frac{1}{3} \left(\frac{v}{\sqrt{1 - v^2}} \right)^3 + C$

29. Let $e^t = 3\tan\theta$, $t = \ln(3\tan\theta)$, $\tan^{-1}\left(\frac{1}{3}\right) \le \theta \le \tan^{-1}\left(\frac{4}{3}\right)$, $dt = \frac{\sec^2\theta}{\tan\theta}\,d\theta$, $\sqrt{e^{2t}+9} = \sqrt{9\tan^2\theta+9} = 3\sec\theta$;

$$\int_0^{\ln 4} \frac{e^t\,dt}{\sqrt{e^{2t}+9}} = \int_{\tan^{-1}(1/3)}^{\tan^{-1}(4/3)} \frac{3\tan\theta\cdot\sec^2\theta\,d\theta}{\tan\theta\cdot 3\sec\theta} = \int_{\tan^{-1}(1/3)}^{\tan^{-1}(4/3)} \sec\theta\,d\theta = [\ln|\sec\theta+\tan\theta|]_{\tan^{-1}(1/3)}^{\tan^{-1}(4/3)}$$

$$= \ln\left(\frac{5}{3}+\frac{4}{3}\right) - \ln\left(\frac{\sqrt{10}}{3}+\frac{1}{3}\right) = \ln 9 - \ln\left(1+\sqrt{10}\right)$$

31. $\int_{1/12}^{1/4} \frac{2\,dt}{\sqrt{t}+4t\sqrt{t}}$; $\left[u = 2\sqrt{t}, du = \frac{1}{\sqrt{t}}\,dt\right] \to \int_{1/\sqrt{3}}^1 \frac{2\,du}{1+u^2}$; $u = \tan\theta$, $\frac{\pi}{6} \le \theta \le \frac{\pi}{4}$, $du = \sec^2\theta\,d\theta$, $1+u^2 = \sec^2\theta$;

$$\int_{1/\sqrt{3}}^1 \frac{2\,du}{1+u^2} = \int_{\pi/6}^{\pi/4} \frac{2\sec^2\theta\,d\theta}{\sec^2\theta} = [2\theta]_{\pi/6}^{\pi/4} = 2\left(\frac{\pi}{4}-\frac{\pi}{6}\right) = \frac{\pi}{6}$$

33. $x = \sec\theta$, $0 < \theta < \frac{\pi}{2}$, $dx = \sec\theta\tan\theta\,d\theta$, $\sqrt{x^2-1} = \sqrt{\sec^2\theta-1} = \tan\theta$;

$$\int \frac{dx}{x\sqrt{x^2-1}} = \int \frac{\sec\theta\tan\theta\,d\theta}{\sec\theta\tan\theta} = \theta + C = \sec^{-1}x + C$$

35. $x = \sec\theta$, $dx = \sec\theta\tan\theta\,d\theta$, $\sqrt{x^2-1} = \sqrt{\sec^2\theta-1} = \tan\theta$;

$$\int \frac{x\,dx}{\sqrt{x^2-1}} = \int \frac{\sec\theta\cdot\sec\theta\tan\theta\,d\theta}{\tan\theta} = \int \sec^2\theta\,d\theta = \tan\theta + C = \sqrt{x^2-1} + C$$

37. $x\frac{dy}{dx} = \sqrt{x^2-4}$; $dy = \sqrt{x^2-4}\,\frac{dx}{x}$; $y = \int \frac{\sqrt{x^2-4}}{x}\,dx$; $\begin{bmatrix} x = 2\sec\theta, 0 < \theta < \frac{\pi}{2} \\ dx = 2\sec\theta\tan\theta\,d\theta \\ \sqrt{x^2-4} = 2\tan\theta \end{bmatrix}$

$$\to y = \int \frac{(2\tan\theta)(2\sec\theta\tan\theta)\,d\theta}{2\sec\theta} = 2\int \tan^2\theta\,d\theta = 2\int(\sec^2\theta-1)\,d\theta = 2(\tan\theta-\theta)+C$$

$$= 2\left[\frac{\sqrt{x^2-4}}{2} - \sec^{-1}\left(\frac{x}{2}\right)\right]+C; \; x = 2 \text{ and } y = 0 \Rightarrow 0 = 0+C \Rightarrow C = 0 \Rightarrow y = 2\left[\frac{\sqrt{x^2-4}}{2}-\sec^{-1}\frac{x}{2}\right]$$

39. $(x^2+4)\frac{dy}{dx} = 3$, $dy = \frac{3\,dx}{x^2+4}$; $y = 3\int \frac{dx}{x^2+4} = \frac{3}{2}\tan^{-1}\frac{x}{2}+C$; $x = 2$ and $y = 0 \Rightarrow 0 = \frac{3}{2}\tan^{-1}1 + C$

$$\Rightarrow C = -\frac{3\pi}{8} \Rightarrow y = \frac{3}{2}\tan^{-1}\left(\frac{x}{2}\right) - \frac{3\pi}{8}$$

41. $A = \int_0^3 \frac{\sqrt{9-x^2}}{3}\,dx$; $x = 3\sin\theta$, $0 \le \theta \le \frac{\pi}{2}$, $dx = 3\cos\theta\,d\theta$, $\sqrt{9-x^2} = \sqrt{9-9\sin^2\theta} = 3\cos\theta$;

$$A = \int_0^{\pi/2} \frac{3\cos\theta\cdot 3\cos\theta\,d\theta}{3} = 3\int_0^{\pi/2} \cos^2\theta\,d\theta = \frac{3}{2}[\theta+\sin\theta\cos\theta]_0^{\pi/2} = \frac{3\pi}{4}$$

43. $\int \frac{dx}{1-\sin x} = \int \frac{\left(\frac{2\,dz}{1+z^2}\right)}{1-\left(\frac{2z}{1+z^2}\right)} = \int \frac{2\,dz}{(1-z)^2} = \frac{2}{1-z}+C = \frac{2}{1-\tan\left(\frac{x}{2}\right)}+C$

45. $\int_0^{\pi/2} \frac{dx}{1+\sin x} = \int_0^1 \frac{\left(\frac{2\,dz}{1+z^2}\right)}{1+\left(\frac{2z}{1+z^2}\right)} = \int_0^1 \frac{2\,dz}{(1+z)^2} = -\left[\frac{2}{1+z}\right]_0^1 = -(1-2) = 1$

47. $\int_0^{\pi/2} \frac{d\theta}{2+\cos\theta} = \int_0^1 \frac{\left(\frac{2\,dz}{1+z^2}\right)}{2+\left(\frac{1-z^2}{1+z^2}\right)} = \int_0^1 \frac{2\,dz}{2+2z^2+1-z^2} = \int_0^1 \frac{2\,dz}{z^2+3} = \frac{2}{\sqrt{3}}\left[\tan^{-1}\frac{z}{\sqrt{3}}\right]_0^1 = \frac{2}{\sqrt{3}}\tan^{-1}\frac{1}{\sqrt{3}}$

$$= \frac{\pi}{3\sqrt{3}} = \frac{\sqrt{3}\pi}{9}$$

49. $\int \frac{dt}{\sin t - \cos t} = \int \frac{\left(\frac{2\,dz}{1+z^2}\right)}{\left(\frac{2z}{1+z^2} - \frac{1-z^2}{1+z^2}\right)} = \int \frac{2\,dz}{2z-1+z^2} = \int \frac{2\,dz}{(z+1)^2-2} = \frac{1}{\sqrt{2}}\ln\left|\frac{z+1-\sqrt{2}}{z+1+\sqrt{2}}\right|+C$

$$= \frac{1}{\sqrt{2}}\ln\left|\frac{\tan\left(\frac{t}{2}\right)+1-\sqrt{2}}{\tan\left(\frac{t}{2}\right)+1+\sqrt{2}}\right|+C$$

51. $\int \sec \theta \, d\theta = \int \frac{d\theta}{\cos \theta} = \int \frac{\left(\frac{2\,dz}{1+z^2}\right)}{\left(\frac{1-z^2}{1+z^2}\right)} = \int \frac{2\,dz}{1-z^2} = \int \frac{2\,dz}{(1+z)(1-z)} = \int \frac{dz}{1+z} + \int \frac{dz}{1-z}$

$= \ln|1+z| - \ln|1-z| + C = \ln \left| \frac{1 + \tan\left(\frac{\theta}{2}\right)}{1 - \tan\left(\frac{\theta}{2}\right)} \right| + C$

8.6 INTEGRAL TABLES AND COMPUTER ALGEBRA SYSTEMS

1. $\int \frac{dx}{x\sqrt{x-3}} = \frac{2}{\sqrt{3}} \tan^{-1} \sqrt{\frac{x-3}{3}} + C$

   (We used FORMULA 13(a) with a = 1, b = 3)

3. $\int \frac{x\,dx}{\sqrt{x-2}} = \int \frac{(x-2)\,dx}{\sqrt{x-2}} + 2\int \frac{dx}{\sqrt{x-2}} = \int \left(\sqrt{x-2}\right)^1 dx + 2\int \left(\sqrt{x-2}\right)^{-1} dx$

   $= \left(\frac{2}{1}\right) \frac{\left(\sqrt{x-2}\right)^3}{3} + 2\left(\frac{2}{1}\right) \frac{\left(\sqrt{x-2}\right)^1}{1} = \sqrt{x-2}\left[\frac{2(x-2)}{3} + 4\right] + C$

   (We used FORMULA 11 with a = 1, b = -2, n = 1 and a = 1, b = -2, n = -1)

5. $\int x\sqrt{2x-3}\,dx = \frac{1}{2}\int (2x-3)\sqrt{2x-3}\,dx + \frac{3}{2}\int \sqrt{2x-3}\,dx = \frac{1}{2}\int \left(\sqrt{2x-3}\right)^3 dx + \frac{3}{2}\int \left(\sqrt{2x-3}\right)^1 dx$

   $= \left(\frac{1}{2}\right)\left(\frac{2}{2}\right) \frac{\left(\sqrt{2x-3}\right)^5}{5} + \left(\frac{3}{2}\right)\left(\frac{2}{2}\right) \frac{\left(\sqrt{2x-3}\right)^3}{3} + C = \frac{(2x-3)^{3/2}}{2}\left[\frac{2x-3}{5} + 1\right] + C = \frac{(2x-3)^{3/2}(x+1)}{5} + C$

   (We used FORMULA 11 with a = 2, b = -3, n = 3 and a = 2, b = -3, n = 1)

7. $\int \frac{\sqrt{9-4x}}{x^2}\,dx = -\frac{\sqrt{9-4x}}{x} + \frac{(-4)}{2}\int \frac{dx}{x\sqrt{9-4x}} + C$

   (We used FORMULA 14 with a = -4, b = 9)

   $= -\frac{\sqrt{9-4x}}{x} - 2\left(\frac{1}{\sqrt{9}}\right) \ln \left|\frac{\sqrt{9-4x}-\sqrt{9}}{\sqrt{9-4x}+\sqrt{9}}\right| + C$

   (We used FORMULA 13(b) with a = -4, b = 9)

   $= \frac{-\sqrt{9-4x}}{x} - \frac{2}{3}\ln \left|\frac{\sqrt{9-4x}-3}{\sqrt{9-4x}+3}\right| + C$

9. $\int x\sqrt{4x-x^2}\,dx = \int x\sqrt{2\cdot 2x - x^2}\,dx = \frac{(x+2)(2x-3\cdot 2)\sqrt{2\cdot 2\cdot x - x^2}}{6} + \frac{2^3}{2}\sin^{-1}\left(\frac{x-2}{2}\right) + C$

   $= \frac{(x+2)(2x-6)\sqrt{4x-x^2}}{6} + 4\sin^{-1}\left(\frac{x-2}{2}\right) + C = \frac{(x+2)(x-3)\sqrt{4x-x^2}}{3} + 4\sin^{-1}\left(\frac{x-2}{2}\right) + C$

   (We used FORMULA 51 with a = 2)

11. $\int \frac{dx}{x\sqrt{7+x^2}} = \int \frac{dx}{x\sqrt{\left(\sqrt{7}\right)^2 + x^2}} = -\frac{1}{\sqrt{7}}\ln \left|\frac{\sqrt{7}+\sqrt{\left(\sqrt{7}\right)^2 + x^2}}{x}\right| + C = -\frac{1}{\sqrt{7}}\ln \left|\frac{\sqrt{7}+\sqrt{7+x^2}}{x}\right| + C$

   $\left(\text{We used FORMULA 26 with } a = \sqrt{7}\right)$

13. $\int \frac{\sqrt{4-x^2}}{x}\,dx = \int \frac{\sqrt{2^2 - x^2}}{x}\,dx = \sqrt{2^2 - x^2} - 2\ln \left|\frac{2 + \sqrt{2^2 - x^2}}{x}\right| + C = \sqrt{4-x^2} - 2\ln \left|\frac{2 + \sqrt{4-x^2}}{x}\right| + C$

   (We used FORMULA 31 with a = 2)

15. $\int \sqrt{25-p^2}\,dp = \int \sqrt{5^2 - p^2}\,dp = \frac{p}{2}\sqrt{5^2 - p^2} + \frac{5^2}{2}\sin^{-1}\frac{p}{5} + C = \frac{p}{2}\sqrt{25-p^2} + \frac{25}{2}\sin^{-1}\frac{p}{5} + C$

   (We used FORMULA 29 with a = 5)

17. $\int \frac{r^2}{\sqrt{4-r^2}}\,dr = \int \frac{r^2}{\sqrt{2^2 - r^2}}\,dr = \frac{2^2}{2}\sin^{-1}\left(\frac{r}{2}\right) - \frac{1}{2}r\sqrt{2^2 - r^2} + C = 2\sin^{-1}\left(\frac{r}{2}\right) - \frac{1}{2}r\sqrt{4-r^2} + C$

   (We used FORMULA 33 with a = 2)

19. $\int \frac{d\theta}{5+4\sin 2\theta} = \frac{-2}{2\sqrt{25-16}} \tan^{-1}\left[\sqrt{\frac{5-4}{5+4}} \tan\left(\frac{\pi}{4}-\frac{2\theta}{2}\right)\right] + C = -\frac{1}{3}\tan^{-1}\left[\frac{1}{3}\tan\left(\frac{\pi}{4}-\theta\right)\right] + C$

(We used FORMULA 70 with b = 5, c = 4, a = 2)

21. $\int e^{2t}\cos 3t\, dt = \frac{e^{2t}}{2^2+3^2}(2\cos 3t + 3\sin 3t) + C = \frac{e^{2t}}{13}(2\cos 3t + 3\sin 3t) + C$

(We used FORMULA 108 with a = 2, b = 3)

23. $\int x\cos^{-1}x\, dx = \int x^1 \cos^{-1}x\, dx = \frac{x^{1+1}}{1+1}\cos^{-1}x + \frac{1}{1+1}\int \frac{x^{1+1}\, dx}{\sqrt{1-x^2}} = \frac{x^2}{2}\cos^{-1}x + \frac{1}{2}\int \frac{x^2\, dx}{\sqrt{1-x^2}}$

(We used FORMULA 100 with a = 1, n = 1)

$= \frac{x^2}{2}\cos^{-1}x + \frac{1}{2}\left(\frac{1}{2}\sin^{-1}x\right) - \frac{1}{2}\left(\frac{1}{2}x\sqrt{1-x^2}\right) + C = \frac{x^2}{2}\cos^{-1}x + \frac{1}{4}\sin^{-1}x - \frac{1}{4}x\sqrt{1-x^2} + C$

(We used FORMULA 33 with a = 1)

25. $\int \frac{ds}{(9-s^2)^2} = \int \frac{ds}{(3^2-s^2)^2} = \frac{s}{2\cdot 3^2\cdot(3^2-s^2)} + \frac{1}{4\cdot 3^3}\ln\left|\frac{s+3}{s-3}\right| + C$

(We used FORMULA 19 with a = 3)

$= \frac{s}{18(9-s^2)} + \frac{1}{108}\ln\left|\frac{s+3}{s-3}\right| + C$

27. $\int \frac{\sqrt{4x+9}}{x^2}\, dx = -\frac{\sqrt{4x+9}}{x} + \frac{4}{2}\int \frac{dx}{x\sqrt{4x+9}}$

(We used FORMULA 14 with a = 4, b = 9)

$= -\frac{\sqrt{4x+9}}{x} + 2\left(\frac{1}{\sqrt{9}}\ln\left|\frac{\sqrt{4x+9}-\sqrt{9}}{\sqrt{4x+9}+\sqrt{9}}\right|\right) + C = -\frac{\sqrt{4x+9}}{x} + \frac{2}{3}\ln\left|\frac{\sqrt{4x+9}-3}{\sqrt{4x+9}+3}\right| + C$

(We used FORMULA 13(b) with a = 4, b = 9)

29. $\int \frac{\sqrt{3t-4}}{t}\, dt = 2\sqrt{3t-4} + (-4)\int \frac{dt}{t\sqrt{3t-4}}$

(We used FORMULA 12 with a = 3, b = −4)

$= 2\sqrt{3t-4} - 4\left(\frac{2}{\sqrt{4}}\tan^{-1}\sqrt{\frac{3t-4}{4}}\right) + C = 2\sqrt{3t-4} - 4\tan^{-1}\frac{\sqrt{3t-4}}{2} + C$

(We used FORMULA 13(a) with a = 3, b = 4)

31. $\int x^2\tan^{-1}x\, dx = \frac{x^{2+1}}{2+1}\tan^{-1}x - \frac{1}{2+1}\int \frac{x^{2+1}}{1+x^2}\, dx = \frac{x^3}{3}\tan^{-1}x - \frac{1}{3}\int \frac{x^3}{1+x^2}\, dx$

(We used FORMULA 101 with a = 1, n = 2);

$\int \frac{x^3}{1+x^2}\, dx = \int x\, dx - \int \frac{x\, dx}{1+x^2} = \frac{x^2}{2} - \frac{1}{2}\ln(1+x^2) + C \Rightarrow \int x^2\tan^{-1}x\, dx$

$= \frac{x^3}{3}\tan^{-1}x - \frac{x^2}{6} + \frac{1}{6}\ln(1+x^2) + C$

33. $\int \sin 3x\cos 2x\, dx = -\frac{\cos 5x}{10} - \frac{\cos x}{2} + C$

(We used FORMULA 62(a) with a = 3, b = 2)

35. $\int 8\sin 4t\sin\frac{t}{2}\, dx = \frac{8}{7}\sin\left(\frac{7t}{2}\right) - \frac{8}{9}\sin\left(\frac{9t}{2}\right) + C = 8\left[\frac{\sin\left(\frac{7t}{2}\right)}{7} - \frac{\sin\left(\frac{9t}{2}\right)}{9}\right] + C$

(We used FORMULA 62(b) with a = 4, b = $\frac{1}{2}$)

37. $\int \cos\frac{\theta}{3}\cos\frac{\theta}{4}\, d\theta = 6\sin\left(\frac{\theta}{12}\right) + \frac{6}{7}\sin\left(\frac{7\theta}{12}\right) + C$

(We used FORMULA 62(c) with a = $\frac{1}{3}$, b = $\frac{1}{4}$)

39. $\int \frac{x^3+x+1}{(x^2+1)^2}\, dx = \int \frac{x\, dx}{x^2+1} + \int \frac{dx}{(x^2+1)^2} = \frac{1}{2}\int \frac{d(x^2+1)}{x^2+1} + \int \frac{dx}{(x^2+1)^2}$

$= \frac{1}{2} \ln(x^2 + 1) + \frac{x}{2(1 + x^2)} + \frac{1}{2} \tan^{-1} x + C$

(For the second integral we used FORMULA 17 with $a = 1$)

41. $\int \sin^{-1} \sqrt{x} \, dx; \quad \begin{bmatrix} u = \sqrt{x} \\ x = u^2 \\ dx = 2u \, du \end{bmatrix} \rightarrow 2 \int u^1 \sin^{-1} u \, du = 2 \left(\frac{u^{1+1}}{1+1} \sin^{-1} u - \frac{1}{1+1} \int \frac{u^{1+1}}{\sqrt{1 - u^2}} \, du \right)$

$= u^2 \sin^{-1} u - \int \frac{u^2 \, du}{\sqrt{1 - u^2}}$

(We used FORMULA 99 with $a = 1$, $n = 1$)

$= u^2 \sin^{-1} u - \left(\frac{1}{2} \sin^{-1} u - \frac{1}{2} u \sqrt{1 - u^2} \right) + C = \left(u^2 - \frac{1}{2} \right) \sin^{-1} u + \frac{1}{2} u \sqrt{1 - u^2} + C$

(We used FORMULA 33 with $a = 1$)

$= \left(x - \frac{1}{2} \right) \sin^{-1} \sqrt{x} + \frac{1}{2} \sqrt{x - x^2} + C$

43. $\int \frac{\sqrt{x}}{\sqrt{1 - x}} \, dx; \quad \begin{bmatrix} u = \sqrt{x} \\ x = u^2 \\ dx = 2u \, du \end{bmatrix} \rightarrow \int \frac{u \cdot 2u}{\sqrt{1 - u^2}} \, du = 2 \int \frac{u^2}{\sqrt{1 - u^2}} \, du = 2 \left(\frac{1}{2} \sin^{-1} u - \frac{1}{2} u \sqrt{1 - u^2} \right) + C$

$= \sin^{-1} u - u \sqrt{1 - u^2} + C$

(We used FORMULA 33 with $a = 1$)

$= \sin^{-1} \sqrt{x} - \sqrt{x} \sqrt{1 - x} + C = \sin^{-1} \sqrt{x} - \sqrt{x - x^2} + C$

45. $\int (\cot t) \sqrt{1 - \sin^2 t} \, dt = \int \frac{\sqrt{1 - \sin^2 t} \, (\cos t) \, dt}{\sin t}; \quad \begin{bmatrix} u = \sin t \\ du = \cos t \, dt \end{bmatrix} \rightarrow \int \frac{\sqrt{1 - u^2} \, du}{u}$

$= \sqrt{1 - u^2} - \ln \left| \frac{1 + \sqrt{1 - u^2}}{u} \right| + C$

(We used FORMULA 31 with $a = 1$)

$= \sqrt{1 - \sin^2 t} - \ln \left| \frac{1 + \sqrt{1 - \sin^2 t}}{\sin t} \right| + C$

47. $\int \frac{dy}{y \sqrt{3 + (\ln y)^2}}; \quad \begin{bmatrix} u = \ln y \\ y = e^u \\ dy = e^u \, du \end{bmatrix} \rightarrow \int \frac{e^u \, du}{e^u \sqrt{3 + u^2}} = \int \frac{du}{\sqrt{3 + u^2}} = \ln \left| u + \sqrt{3 + u^2} \right| + C$

$= \ln \left| \ln y + \sqrt{3 + (\ln y)^2} \right| + C$

$\left(\text{We used FORMULA 20 with } a = \sqrt{3} \right)$

49. $\int \frac{3 \, dr}{\sqrt{9r^2 - 1}}; \quad \begin{bmatrix} u = 3r \\ du = 3 \, dr \end{bmatrix} \rightarrow \int \frac{du}{\sqrt{u^2 - 1}} = \ln \left| u + \sqrt{u^2 - 1} \right| + C = \ln \left| 3r + \sqrt{9r^2 - 1} \right| + C$

(We used FORMULA 36 with $a = 1$)

51. $\int \cos^{-1} \sqrt{x} \, dx; \quad \begin{bmatrix} t = \sqrt{x} \\ x = t^2 \\ dx = 2t \, dt \end{bmatrix} \rightarrow 2 \int t \cos^{-1} t \, dt = 2 \left(\frac{t^2}{2} \cos^{-1} t + \frac{1}{2} \int \frac{t^2}{\sqrt{1 - t^2}} \, dt \right) = t^2 \cos^{-1} t + \int \frac{t^2}{\sqrt{1 - t^2}} \, dt$

(We used FORMULA 100 with $a = 1$, $n = 1$)

$= t^2 \cos^{-1} t + \frac{1}{2} \sin^{-1} t - \frac{1}{2} t \sqrt{1 - t^2} + C$

(We used FORMULA 33 with $a = 1$)

$= x \cos^{-1} \sqrt{x} + \frac{1}{2} \sin^{-1} \sqrt{x} - \frac{1}{2} \sqrt{x} \sqrt{1 - x} + C = x \cos^{-1} \sqrt{x} + \frac{1}{2} \sin^{-1} \sqrt{x} - \frac{1}{2} \sqrt{x - x^2} + C$

53. $\int \sin^5 2x \, dx = - \frac{\sin^4 2x \cos 2x}{5 \cdot 2} + \frac{5 - 1}{5} \int \sin^3 2x \, dx = - \frac{\sin^4 2x \cos 2x}{10} + \frac{4}{5} \left[- \frac{\sin^2 2x \cos 2x}{3 \cdot 2} + \frac{3 - 1}{3} \int \sin 2x \, dx \right]$

(We used FORMULA 60 with $a = 2$, $n = 5$ and $a = 2$, $n = 3$)

$$= -\frac{\sin^4 2x \cos 2x}{10} - \frac{2}{15} \sin^2 2x \cos 2x + \frac{8}{15}\left(-\frac{1}{2}\right)\cos 2x + C = -\frac{\sin^4 2x \cos 2x}{10} - \frac{2\sin^2 2x \cos 2x}{15} - \frac{4\cos 2x}{15} + C$$

55. $\int 8\cos^4 2\pi t \, dt = 8\left(\frac{\cos^3 2\pi t \sin 2\pi t}{4\cdot 2\pi} + \frac{4-1}{4}\int \cos^2 2\pi t \, dt\right)$

(We used FORMULA 61 with a = 2π, n = 4)

$$= \frac{\cos^3 2\pi t \sin 2\pi t}{\pi} + 6\left[\frac{t}{2} + \frac{\sin(2\cdot 2\pi \cdot t)}{4\cdot 2\pi}\right] + C$$

(We used FORMULA 59 with a = 2π)

$$= \frac{\cos^3 2\pi t \sin 2\pi t}{\pi} + 3t + \frac{3\sin 4\pi t}{4\pi} + C = \frac{\cos^3 2\pi t \sin 2\pi t}{\pi} + \frac{3\cos 2\pi t \sin 2\pi t}{2\pi} + 3t + C$$

57. $\int \sin^2 2\theta \cos^3 2\theta \, d\theta = \frac{\sin^3 2\theta \cos^2 2\theta}{2(2+3)} + \frac{3-1}{3+2}\int \sin^2 2\theta \cos 2\theta \, d\theta$

(We used FORMULA 69 with a = 2, m = 3, n = 2)

$$= \frac{\sin^3 2\theta \cos^2 2\theta}{10} + \frac{2}{5}\int \sin^2 2\theta \cos 2\theta \, d\theta = \frac{\sin^3 2\theta \cos^2 2\theta}{10} + \frac{2}{5}\left[\frac{1}{2}\int \sin^2 2\theta \, d(\sin 2\theta)\right] = \frac{\sin^3 2\theta \cos^2 2\theta}{10} + \frac{\sin^3 2\theta}{15} + C$$

59. $\int 2\sin^2 t \sec^4 t \, dt = \int 2\sin^2 t \cos^{-4} t \, dt = 2\left(-\frac{\sin t \cos^{-3} t}{2-4} + \frac{2-1}{2-4}\int \cos^{-4} t \, dt\right)$

(We used FORMULA 68 with a = 1, n = 2, m = -4)

$$= \sin t \cos^{-3} t - \int \cos^{-4} t \, dt = \sin t \cos^{-3} t - \int \sec^4 t \, dt = \sin t \cos^{-3} t - \left(\frac{\sec^2 t \tan t}{4-1} + \frac{4-2}{4-1}\int \sec^2 t \, dt\right)$$

(We used FORMULA 92 with a = 1, n = 4)

$$= \sin t \cos^{-3} t - \left(\frac{\sec^2 t \tan t}{3}\right) - \frac{2}{3}\tan t + C = \frac{2}{3}\sec^2 t \tan t - \frac{2}{3}\tan t + C = \frac{2}{3}\tan t(\sec^2 t - 1) + C$$

$$= \frac{2}{3}\tan^3 t + C$$

An easy way to find the integral using substitution:

$$\int 2\sin^2 t \cos^{-4} t \, dt = \int 2\tan^2 t \sec^2 t \, dt = \int 2\tan^2 t \, d(\tan t) = \frac{2}{3}\tan^3 t + C$$

61. $\int 4\tan^3 2x \, dx = 4\left(\frac{\tan^2 2x}{2\cdot 2} - \int \tan 2x \, dx\right) = \tan^2 2x - 4\int \tan 2x \, dx$

(We used FORMULA 86 with n = 3, a = 2)

$$= \tan^2 2x - \frac{4}{2}\ln|\sec 2x| + C = \tan^2 2x - 2\ln|\sec 2x| + C$$

63. $\int 8\cot^4 t \, dt = 8\left(-\frac{\cot^3 t}{3} - \int \cot^2 t \, dt\right)$

(We used FORMULA 87 with a = 1, n = 4)

$$= 8\left(-\frac{1}{3}\cot^3 t + \cot t + t\right) + C$$

(We used FORMULA 85 with a = 1)

65. $\int 2\sec^3 \pi x \, dx = 2\left[\frac{\sec \pi x \tan \pi x}{\pi(3-1)} + \frac{3-2}{3-1}\int \sec \pi x \, dx\right]$

(We used FORMULA 92 with n = 3, a = π)

$$= \frac{1}{\pi}\sec \pi x \tan \pi x + \frac{1}{\pi}\ln|\sec \pi x + \tan \pi x| + C$$

(We used FORMULA 88 with a = π)

67. $\int 3\sec^4 3x \, dx = 3\left[\frac{\sec^2 3x \tan 3x}{3(4-1)} + \frac{4-2}{4-1}\int \sec^2 3x \, dx\right]$

(We used FORMULA 92 with n = 4, a = 3)

$$= \frac{\sec^2 3x \tan 3x}{3} + \frac{2}{3}\tan 3x + C$$

(We used FORMULA 90 with a = 3)

69. $\int \csc^5 x \, dx = -\frac{\csc^3 x \cot x}{5-1} + \frac{5-2}{5-1}\int \csc^3 x \, dx = -\frac{\csc^3 x \cot x}{4} + \frac{3}{4}\left(-\frac{\csc x \cot x}{3-1} + \frac{3-2}{3-1}\int \csc x \, dx\right)$

(We used FORMULA 93 with n = 5, a = 1 and n = 3, a = 1)

$$= -\tfrac{1}{4} \csc^3 x \cot x - \tfrac{3}{8} \csc x \cot x - \tfrac{3}{8} \ln |\csc x + \cot x| + C$$

(We used FORMULA 89 with $a = 1$)

71. $\int 16x^3 (\ln x)^2 \, dx = 16 \left[\frac{x^4 (\ln x)^2}{4} - \frac{2}{4} \int x^3 \ln x \, dx \right] = 16 \left[\frac{x^4 (\ln x)^2}{4} - \frac{1}{2} \left[\frac{x^4 (\ln x)}{4} - \frac{1}{4} \int x^3 \, dx \right] \right]$

(We used FORMULA 110 with $a = 1$, $n = 3$, $m = 2$ and $a = 1$, $n = 3$, $m = 1$)

$$= 16 \left(\frac{x^4 (\ln x)^2}{4} - \frac{x^4 (\ln x)}{8} + \frac{x^4}{32} \right) + C = 4x^4 (\ln x)^2 - 2x^4 \ln x + \frac{x^4}{2} + C$$

73. $\int xe^{3x} \, dx = \frac{e^{3x}}{3^2} (3x - 1) + C = \frac{e^{3x}}{9} (3x - 1) + C$

(We used FORMULA 104 with $a = 3$)

75. $\int x^3 e^{x/2} \, dx = 2x^3 e^{x/2} - 3 \cdot 2 \int x^2 e^{x/2} \, dx = 2x^3 e^{x/2} - 6 \left(2x^2 e^{x/2} - 2 \cdot 2 \int xe^{x/2} \, dx \right)$

$$= 2x^3 e^{x/2} - 12x^2 e^{x/2} + 24 \cdot 4e^{x/2} \left(\tfrac{x}{2} - 1 \right) + C = 2x^3 e^{x/2} - 12x^2 e^{x/2} + 96 e^{x/2} \left(\tfrac{x}{2} - 1 \right) + C$$

(We used FORMULA 105 with $a = \frac{1}{2}$ twice and FORMULA 104 with $a = \frac{1}{2}$)

77. $\int x^2 2^x \, dx = \frac{x^2 2^x}{\ln 2} - \frac{2}{\ln 2} \int x 2^x \, dx = \frac{x^2 2^x}{\ln 2} - \frac{2}{\ln 2} \left(\frac{x 2^x}{\ln 2} - \frac{1}{\ln 2} \int 2^x \, dx \right) = \frac{x^2 2^x}{\ln 2} - \frac{2}{\ln 2} \left[\frac{x 2^x}{\ln 2} - \frac{2^x}{(\ln 2)^2} \right] + C$

(We used FORMULA 106 with $a = 1$, $b = 2$, $n = 2$, $n = 1$)

79. $\int x \pi^x \, dx = \frac{x \pi^x}{\ln \pi} - \frac{1}{\ln \pi} \int \pi^x \, dx = \frac{x \pi^x}{\ln \pi} - \frac{1}{\ln \pi} \left(\frac{\pi^x}{\ln \pi} \right) + C = \frac{x \pi^x}{\ln \pi} - \frac{\pi^x}{(\ln \pi)^2} + C$

(We used FORMULA 106 with $n = 1$, $b = \pi$, $a = 1$)

81. $\int e^t \sec^3 (e^t - 1) \, dt$; $\begin{bmatrix} x = e^t - 1 \\ dx = e^t \, dt \end{bmatrix} \rightarrow \int \sec^3 x \, dx = \frac{\sec x \tan x}{3 - 1} + \frac{3 - 2}{3 - 1} \int \sec x \, dx$

(We used FORMULA 92 with $a = 1$, $n = 3$)

$$= \frac{\sec x \tan x}{2} + \tfrac{1}{2} \ln |\sec x + \tan x| + C = \tfrac{1}{2} \left[\sec (e^t - 1) \tan (e^t - 1) + \ln |\sec (e^t - 1) + \tan (e^t - 1)| \right] + C$$

83. $\int_0^1 2\sqrt{x^2 + 1} \, dx$; $[x = \tan t] \rightarrow 2 \int_0^{\pi/4} \sec t \cdot \sec^2 t \, dt = 2 \int_0^{\pi/4} \sec^3 t \, dt = 2 \left[\left[\frac{\sec t \cdot \tan t}{3 - 1} \right]_0^{\pi/4} + \frac{3 - 2}{3 - 1} \int_0^{\pi/4} \sec t \, dt \right]$

(We used FORMULA 92 with $n = 3$, $a = 1$)

$$= [\sec t \cdot \tan t + \ln |\sec t + \tan t|]_0^{\pi/4} = \sqrt{2} + \ln \left(\sqrt{2} + 1 \right)$$

85. $\int_1^2 \frac{(r^2 - 1)^{3/2}}{r} \, dr$; $[r = \sec \theta] \rightarrow \int_0^{\pi/3} \frac{\tan^3 \theta}{\sec \theta} (\sec \theta \tan \theta) \, d\theta = \int_0^{\pi/3} \tan^4 \theta \, d\theta = \left[\frac{\tan^3 \theta}{4 - 1} \right]_0^{\pi/3} - \int_0^{\pi/3} \tan^2 \theta \, d\theta$

$$= \left[\frac{\tan^3 \theta}{3} - \tan \theta + \theta \right]_0^{\pi/3} = \frac{3\sqrt{3}}{3} - \sqrt{3} + \frac{\pi}{3} = \frac{\pi}{3}$$

(We used FORMULA 86 with $a = 1$, $n = 4$ and FORMULA 84 with $a = 1$)

87. $\int \tfrac{1}{8} \sinh^5 3x \, dx = \tfrac{1}{8} \left(\frac{\sinh^4 3x \cosh 3x}{5 \cdot 3} - \frac{5 - 1}{5} \int \sinh^3 3x \, dx \right)$

$$= \frac{\sinh^4 3x \cosh 3x}{120} - \frac{1}{10} \left(\frac{\sinh 3x \cosh 3x}{3 \cdot 3} - \frac{3 - 1}{3} \int \sinh 3x \, dx \right)$$

(We used FORMULA 117 with $a = 3$, $n = 5$ and $a = 3$, $n = 3$)

$$= \frac{\sinh^4 3x \cosh 3x}{120} - \frac{\sinh 3x \cosh 3x}{90} + \frac{2}{30} \left(\tfrac{1}{3} \cosh 3x \right) + C$$

$$= \tfrac{1}{120} \sinh^4 3x \cosh 3x - \tfrac{1}{90} \sinh 3x \cosh 3x + \tfrac{1}{45} \cosh 3x + C$$

89. $\int x^2 \cosh 3x \, dx = \frac{x^2}{3} \sinh 3x - \frac{2}{3} \int x \sinh 3x \, dx = \frac{x^2}{3} \sinh 3x - \frac{2}{3} \left(\frac{x}{3} \cosh 3x - \tfrac{1}{3} \int \cosh 3x \, dx \right)$

(We used FORMULA 122 with $a = 3$, $n = 2$ and FORMULA 121 with $a = 3$, $n = 1$)

$$= \tfrac{x^2}{3} \sinh 3x - \tfrac{2x}{9} \cosh 3x + \tfrac{2}{27} \sinh 3x + C$$

91. $\displaystyle\int \operatorname{sech}^7 x \tanh x \, dx = -\frac{\operatorname{sech}^7 x}{7} + C$

(We used FORMULA 135 with $a = 1$, $n = 7$)

93. $u = ax + b \Rightarrow x = \frac{u-b}{a} \Rightarrow dx = \frac{du}{a}$;

$\displaystyle\int \frac{x \, dx}{(ax+b)^2} = \int \frac{(u-b)}{au^2} \frac{du}{a} = \frac{1}{a^2} \int \left(\frac{1}{u} - \frac{b}{u^2} \right) du = \frac{1}{a^2} \left[\ln |u| + \frac{b}{u} \right] + C = \frac{1}{a^2} \left[\ln |ax + b| + \frac{b}{ax+b} \right] + C$

95. $x = a \sin \theta \Rightarrow a^2 - x^2 = a^2 \cos^2 \theta \Rightarrow -2x \, dx = -2a^2 \cos \theta \sin \theta \, d\theta \Rightarrow dx = a \cos \theta \, d\theta$;

$\displaystyle\int \sqrt{a^2 - x^2} \, dx = \int a \cos \theta (a \cos \theta) \, d\theta = a^2 \int \cos^2 \theta \, d\theta = \frac{a^2}{2} \int (1 + \cos 2\theta) \, d\theta = \frac{a^2}{2} \left(\theta + \frac{\sin 2\theta}{2} \right) + C$

$= \frac{a^2}{2} (\theta + \cos \theta \sin \theta) + C = \frac{a^2}{2} \left(\theta + \sqrt{1 - \sin^2 \theta} \cdot \sin \theta \right) + C = \frac{a^2}{2} \left(\sin^{-1} \frac{x}{a} + \frac{\sqrt{a^2 - x^2}}{a} \cdot \frac{x}{a} \right) + C$

$= \frac{a^2}{2} \sin^{-1} \frac{x}{a} + \frac{x}{2} \sqrt{a^2 - x^2} + C$

97. $\displaystyle\int x^n \sin ax \, dx = -\int x^n \left(\frac{1}{a} \right) d(\cos ax) = (\cos ax) x^n \left(-\frac{1}{a} \right) + \frac{1}{a} \int \cos ax \cdot nx^{n-1} \, dx$

$= -\frac{x^n}{a} \cos ax + \frac{n}{a} \int x^{n-1} \cos ax \, dx$

$\left(\text{We used integration by parts } \int u \, dv = uv - \int v \, du \text{ with } u = x^n, v = -\frac{1}{a} \cos ax\right)$

99. $\displaystyle\int x^n \sin^{-1} ax \, dx = \int \sin^{-1} ax \, d\left(\frac{x^{n+1}}{n+1} \right) = \frac{x^{n+1}}{n+1} \sin^{-1} ax - \int \left(\frac{x^{n+1}}{n+1} \right) \frac{a}{\sqrt{1 - (ax)^2}} \, dx$

$= \frac{x^{n+1}}{n+1} \sin^{-1} ax - \frac{a}{n+1} \int \frac{x^{n+1} \, dx}{\sqrt{1 - a^2 x^2}}, n \neq -1$

$\left(\text{We used integration by parts } \int u \, dv = uv - \int v \, du \text{ with } u = \sin^{-1} ax, v = \frac{x^{n+1}}{n+1}\right)$

101. $S = \displaystyle\int_0^{\sqrt{2}} 2\pi y \sqrt{1 + (y')^2} \, dx$

$= 2\pi \displaystyle\int_0^{\sqrt{2}} \sqrt{x^2 + 2} \sqrt{1 + \frac{x^2}{x^2 + 2}} \, dx$

$= 2\sqrt{2}\pi \displaystyle\int_0^{\sqrt{2}} \sqrt{x^2 + 1} \, dx$

$= 2\sqrt{2}\pi \left[\frac{x\sqrt{x^2+1}}{2} + \frac{1}{2} \ln \left| x + \sqrt{x^2 + 1} \right| \right]_0^{\sqrt{2}}$

(We used FORMULA 21 with $a = 1$)

$= \sqrt{2}\pi \left[\sqrt{6} + \ln \left(\sqrt{2} + \sqrt{3} \right) \right] = 2\pi\sqrt{3} + \pi\sqrt{2} \ln \left(\sqrt{2} + \sqrt{3} \right)$

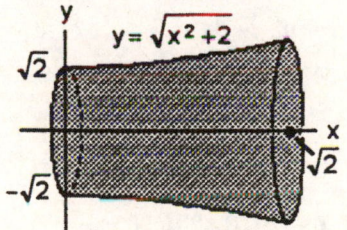

$y = \sqrt{x^2 + 2}$

103. $A = \displaystyle\int_0^3 \frac{dx}{\sqrt{x+1}} = \left[2\sqrt{x+1} \right]_0^3 = 2; \bar{x} = \frac{1}{A} \int_0^3 \frac{x \, dx}{\sqrt{x+1}}$

$= \frac{1}{A} \displaystyle\int_0^3 \sqrt{x + 1} \, dx - \frac{1}{A} \int_0^3 \frac{dx}{\sqrt{x+1}}$

$= \frac{1}{2} \cdot \frac{2}{3} \left[(x+1)^{3/2} \right]_0^3 - 1 = \frac{4}{3}$;

(We used FORMULA 11 with $a = 1$, $b = 1$, $n = 1$ and
$a = 1$, $b = 1$, $n = -1$)

$\bar{y} = \frac{1}{2A} \displaystyle\int_0^3 \frac{dx}{x+1} = \frac{1}{4} \left[\ln (x+1) \right]_0^3 = \frac{1}{4} \ln 4 = \frac{1}{2} \ln 2 = \ln \sqrt{2}$

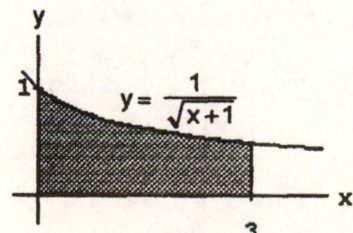

$y = \dfrac{1}{\sqrt{x+1}}$

105. $S = 2\pi \int_{-1}^{1} x^2 \sqrt{1 + 4x^2}\, dx;$

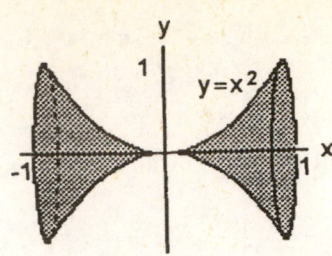

$\begin{bmatrix} u = 2x \\ du = 2\, dx \end{bmatrix} \rightarrow \frac{\pi}{4} \int_{-2}^{2} u^2 \sqrt{1 + u^2}\, du$

$= \frac{\pi}{4} \left[\frac{u}{8} (1 + 2u^2) \sqrt{1 + u^2} - \frac{1}{8} \ln \left(u + \sqrt{1 + u^2} \right) \right]_{-2}^{2}$

(We used FORMULA 22 with $a = 1$)

$= \frac{\pi}{4} \left[\frac{2}{8} (1 + 2 \cdot 4) \sqrt{1 + 4} - \frac{1}{8} \ln \left(2 + \sqrt{1 + 4} \right) \right.$

$\left. + \frac{2}{8} (1 + 2 \cdot 4) \sqrt{1 + 4} + \frac{1}{8} \ln \left(-2 + \sqrt{1 + 4} \right) \right]$

$= \frac{\pi}{4} \left[\frac{9}{2} \sqrt{5} - \frac{1}{8} \ln \left(\frac{2 + \sqrt{5}}{-2 + \sqrt{5}} \right) \right] \approx 7.62$

107. The integrand $f(x) = \sqrt{x - x^2}$ is nonnegative, so the integral is maximized by integrating over the function's entire domain, which runs from $x = 0$ to $x = 1$

$\Rightarrow \int_{0}^{1} \sqrt{x - x^2}\, dx = \int_{0}^{1} \sqrt{2 \cdot \frac{1}{2} x - x^2}\, dx = \left[\frac{(x - \frac{1}{2})}{2} \sqrt{2 \cdot \frac{1}{2} x - x^2} + \frac{(\frac{1}{2})^2}{2} \sin^{-1} \left(\frac{x - \frac{1}{2}}{\frac{1}{2}} \right) \right]_{0}^{1}$

(We used FORMULA 48 with $a = \frac{1}{2}$)

$= \left[\frac{(x - \frac{1}{2})}{2} \sqrt{x - x^2} + \frac{1}{8} \sin^{-1} (2x - 1) \right]_{0}^{1} = \frac{1}{8} \cdot \frac{\pi}{2} - \frac{1}{8} \left(-\frac{\pi}{2} \right) = \frac{\pi}{8}$

109. (e) $\int x^n \ln x\, dx = \frac{x^{n+1} \ln x}{n+1} - \frac{1}{n+1} \int x^n\, dx, \ n \neq -1$

(We used FORMULA 110 with $a = 1, m = 1$)

$= \frac{x^{n+1} \ln x}{n+1} - \frac{x^{n+1}}{(n+1)^2} + C = \frac{x^{n+1}}{n+1} \left(\ln x - \frac{1}{n+1} \right) + C$

111. (a) Neither MAPLE nor MATHEMATICA can find this integral for arbitrary n.

(b) MAPLE and MATHEMATICA get stuck at about $n = 5$.

(c) Let $x = \frac{\pi}{2} - u \Rightarrow dx = -du; \ x = 0 \Rightarrow u = \frac{\pi}{2}, \ x = \frac{\pi}{2} \Rightarrow u = 0;$

$I = \int_{0}^{\pi/2} \frac{\sin^n x\, dx}{\sin^n x + \cos^n x} = \int_{\pi/2}^{0} \frac{-\sin^n \left(\frac{\pi}{2} - u \right) du}{\sin^n \left(\frac{\pi}{2} - u \right) + \cos^n \left(\frac{\pi}{2} - u \right)} = \int_{0}^{\pi/2} \frac{\cos^n u\, du}{\cos^n u + \sin^n u} = \int_{0}^{\pi/2} \frac{\cos^n x\, dx}{\cos^n x + \sin^n x}$

$\Rightarrow I + I = \int_{0}^{\pi/2} \left(\frac{\sin^n x + \cos^n x}{\sin^n x + \cos^n x} \right) dx = \int_{0}^{\pi/2} dx = \frac{\pi}{2} \Rightarrow I = \frac{\pi}{4}$

8.7 NUMERICAL INTEGRATION

1. $\int_{1}^{2} x\, dx$

I. (a) For $n = 4$, $\Delta x = \frac{b - a}{n} = \frac{2 - 1}{4} = \frac{1}{4} \Rightarrow \frac{\Delta x}{2} = \frac{1}{8}$;

$\sum mf(x_i) = 12 \Rightarrow T = \frac{1}{8} (12) = \frac{3}{2}$;

$f(x) = x \Rightarrow f'(x) = 1 \Rightarrow f'' = 0 \Rightarrow M = 0$

$\Rightarrow |E_T| = 0$

	x_i	$f(x_i)$	m	$mf(x_i)$
x_0	1	1	1	1
x_1	5/4	5/4	2	5/2
x_2	3/2	3/2	2	3
x_3	7/4	7/4	2	7/2
x_4	2	2	1	2

(b) $\int_{1}^{2} x\, dx = \left[\frac{x^2}{2} \right]_{1}^{2} = 2 - \frac{1}{2} = \frac{3}{2} \Rightarrow |E_T| = \int_{1}^{2} x\, dx - T = 0$

(c) $\frac{|E_T|}{\text{True Value}} \times 100 = 0\%$

II. (a) For $n = 4$, $\Delta x = \frac{b - a}{n} = \frac{2 - 1}{4} = \frac{1}{4} \Rightarrow \frac{\Delta x}{3} = \frac{1}{12}$;

$\sum mf(x_i) = 18 \Rightarrow S = \frac{1}{12} (18) = \frac{3}{2}$;

$f^{(4)}(x) = 0 \Rightarrow M = 0 \Rightarrow |E_S| = 0$

	x_i	$f(x_i)$	m	$mf(x_i)$
x_0	1	1	1	1
x_1	5/4	5/4	4	5
x_2	3/2	3/2	2	3
x_3	7/4	7/4	4	7
x_4	2	2	1	2

(b) $\int_{1}^{2} x\, dx = \frac{3}{2} \Rightarrow |E_S| = \int_{1}^{2} x\, dx - S = \frac{3}{2} - \frac{3}{2} = 0$

(c) $\frac{|E_S|}{\text{True Value}} \times 100 = 0\%$

3. $\int_{-1}^{1} (x^2 + 1)\, dx$

I. (a) For $n = 4$, $\Delta x = \frac{b-a}{n} = \frac{1-(-1)}{4} = \frac{2}{4} = \frac{1}{2} \Rightarrow \frac{\Delta x}{2} = \frac{1}{4}$;

$\sum mf(x_i) = 11 \Rightarrow T = \frac{1}{4}(11) = 2.75$;

$f(x) = x^2 + 1 \Rightarrow f'(x) = 2x \Rightarrow f''(x) = 2 \Rightarrow M = 2$

$\Rightarrow |E_T| \le \frac{1-(-1)}{12}\left(\frac{1}{2}\right)^2(2) = \frac{1}{12}$ or 0.08333

	x_i	$f(x_i)$	m	$mf(x_i)$
x_0	-1	2	1	2
x_1	$-1/2$	5/4	2	5/2
x_2	0	1	2	2
x_3	1/2	5/4	2	5/2
x_4	1	2	1	2

(b) $\int_{-1}^{1} (x^2 + 1)\, dx = \left[\frac{x^3}{3} + x\right]_{-1}^{1} = \left(\frac{1}{3} + 1\right) - \left(-\frac{1}{3} - 1\right) = \frac{8}{3} \Rightarrow E_T = \int_{-1}^{1}(x^2 + 1)\, dx - T = \frac{8}{3} - \frac{11}{4} = -\frac{1}{12}$

$\Rightarrow |E_T| = \left|-\frac{1}{12}\right| \approx 0.08333$

(c) $\frac{|E_T|}{\text{True Value}} \times 100 = \left(\frac{\frac{1}{12}}{\frac{8}{3}}\right) \times 100 \approx 3\%$

II. (a) For $n = 4$, $\Delta x = \frac{b-a}{n} = \frac{1-(-1)}{4} = \frac{2}{4} = \frac{1}{2} \Rightarrow \frac{\Delta x}{3} = \frac{1}{6}$;

$\sum mf(x_i) = 16 \Rightarrow S = \frac{1}{6}(16) = \frac{8}{3} = 2.66667$;

$f^{(3)}(x) = 0 \Rightarrow f^{(4)}(x) = 0 \Rightarrow M = 0 \Rightarrow |E_S| = 0$

	x_i	$f(x_i)$	m	$mf(x_i)$
x_0	-1	2	1	2
x_1	$-1/2$	5/4	4	5
x_2	0	1	2	2
x_3	1/2	5/4	4	5
x_4	1	2	1	2

(b) $\int_{-1}^{1} (x^2 + 1)\, dx = \left[\frac{x^3}{3} + x\right]_{-1}^{1} = \frac{8}{3}$

$\Rightarrow |E_S| = \int_{-1}^{1}(x^2 + 1)\, dx - S = \frac{8}{3} - \frac{8}{3} = 0$

(c) $\frac{|E_S|}{\text{True Value}} \times 100 = 0\%$

5. $\int_{0}^{2} (t^3 + t)\, dt$

I. (a) For $n = 4$, $\Delta x = \frac{b-a}{n} = \frac{2-0}{4} = \frac{2}{4} = \frac{1}{2}$

$\Rightarrow \frac{\Delta x}{2} = \frac{1}{4}$; $\sum mf(t_i) = 25 \Rightarrow T = \frac{1}{4}(25) = \frac{25}{4}$;

$f(t) = t^3 + t \Rightarrow f'(t) = 3t^2 + 1 \Rightarrow f''(t) = 6t$

$\Rightarrow M = 12 = f''(2) \Rightarrow |E_T| \le \frac{2-0}{12}\left(\frac{1}{2}\right)^2(12) = \frac{1}{2}$

	t_i	$f(t_i)$	m	$mf(t_i)$
t_0	0	0	1	0
t_1	1/2	5/8	2	5/4
t_2	1	2	2	4
t_3	3/2	39/8	2	39/4
t_4	2	10	1	10

(b) $\int_{0}^{2}(t^3 + t)\, dt = \left[\frac{t^4}{4} + \frac{t^2}{2}\right]_{0}^{2} = \left(\frac{2^4}{4} + \frac{2^2}{2}\right) - 0 = 6 \Rightarrow |E_T| = \int_{0}^{2}(t^3 + t)\, dt - T = 6 - \frac{25}{4} = -\frac{1}{4} \Rightarrow |E_T| = \frac{1}{4}$

(c) $\frac{|E_T|}{\text{True Value}} \times 100 = \frac{\left|-\frac{1}{4}\right|}{6} \times 100 \approx 4\%$

II. (a) For $n = 4$, $\Delta x = \frac{b-a}{n} = \frac{2-0}{4} = \frac{2}{4} = \frac{1}{2} \Rightarrow \frac{\Delta x}{3} = \frac{1}{6}$;

$\sum mf(t_i) = 36 \Rightarrow S = \frac{1}{6}(36) = 6$;

$f^{(3)}(t) = 6 \Rightarrow f^{(4)}(t) = 0 \Rightarrow M = 0 \Rightarrow |E_S| = 0$

	t_i	$f(t_i)$	m	$mf(t_i)$
t_0	0	0	1	0
t_1	1/2	5/8	4	5/2
t_2	1	2	2	4
t_3	3/2	39/8	4	39/2
t_4	2	10	1	10

(b) $\int_{0}^{2}(t^3 + t)\, dt = 6 \Rightarrow |E_S| = \int_{0}^{2}(t^3 + t)\, dt - S$

$= 6 - 6 = 0$

(c) $\frac{|E_S|}{\text{True Value}} \times 100 = 0\%$

7. $\int_{1}^{2} \frac{1}{s^2}\, ds$

I. (a) For $n = 4$, $\Delta x = \frac{b-a}{n} = \frac{2-1}{4} = \frac{1}{4} \Rightarrow \frac{\Delta x}{2} = \frac{1}{8}$;

$\sum mf(s_i) = \frac{179,573}{44,100} \Rightarrow T = \frac{1}{8}\left(\frac{179,573}{44,100}\right) = \frac{179,573}{352,800}$

≈ 0.50899; $f(s) = \frac{1}{s^2} \Rightarrow f'(s) = -\frac{2}{s^3}$

$\Rightarrow f''(s) = \frac{6}{s^4} \Rightarrow M = 6 = f''(1)$

$\Rightarrow |E_T| \le \frac{2-1}{12}\left(\frac{1}{4}\right)^2(6) = \frac{1}{32} = 0.03125$

	s_i	$f(s_i)$	m	$mf(s_i)$
s_0	1	1	1	1
s_1	5/4	16/25	2	32/25
s_2	3/2	4/9	2	8/9
s_3	7/4	16/49	2	32/49
s_4	2	1/4	1	1/4

(b) $\int_{1}^{2} \frac{1}{s^2}\, ds = \int_{1}^{2} s^{-2}\, ds = \left[-\frac{1}{s}\right]_{1}^{2} = -\frac{1}{2} - \left(-\frac{1}{1}\right) = \frac{1}{2} \Rightarrow E_T = \int_{1}^{2} \frac{1}{s^2}\, ds - T = \frac{1}{2} - 0.50899 = -0.00899$

$\Rightarrow |E_T| = 0.00899$

(c) $\frac{|E_T|}{\text{True Value}} \times 100 = \frac{0.00899}{0.5} \times 100 \approx 2\%$

II. (a) For $n = 4$, $\Delta x = \frac{b-a}{n} = \frac{2-1}{4} = \frac{1}{4} \Rightarrow \frac{\Delta x}{3} = \frac{1}{12}$;

	s_i	$f(s_i)$	m	$mf(s_i)$
s_0	1	1	1	1
s_1	5/4	16/25	4	64/25
s_2	3/2	4/9	2	8/9
s_3	7/4	16/49	4	64/49
s_4	2	1/4	1	1/4

$\sum mf(s_i) = \frac{264{,}821}{44{,}100} \Rightarrow S = \frac{1}{12}\left(\frac{264{,}821}{44{,}100}\right) = \frac{264{,}821}{529{,}200}$

≈ 0.50042; $f^{(3)}(s) = -\frac{24}{s^5} \Rightarrow f^{(4)}(s) = \frac{120}{s^6}$

$\Rightarrow M = 120 \Rightarrow |E_s| \le \left|\frac{2-1}{180}\right|\left(\frac{1}{4}\right)^4(120)$

$= \frac{1}{384} \approx 0.00260$

(b) $\int_1^2 \frac{1}{s^2}\,ds = \frac{1}{2} \Rightarrow E_s = \int_1^2 \frac{1}{s^2}\,ds - S = \frac{1}{2} - 0.50042 = -0.00042 \Rightarrow |E_s| = 0.00042$

(c) $\frac{|E_s|}{\text{True Value}} \times 100 = \frac{0.0004}{0.5} \times 100 \approx 0.08\%$

9. $\int_0^\pi \sin t\,dt$

I. (a) For $n = 4$, $\Delta x = \frac{b-a}{n} = \frac{\pi-0}{4} = \frac{\pi}{4} \Rightarrow \frac{\Delta x}{2} = \frac{\pi}{8}$;

	t_i	$f(t_i)$	m	$mf(t_i)$
t_0	0	0	1	0
t_1	$\pi/4$	$\sqrt{2}/2$	2	$\sqrt{2}$
t_2	$\pi/2$	1	2	2
t_3	$3\pi/4$	$\sqrt{2}/2$	2	$\sqrt{2}$
t_4	π	0	1	0

$\sum mf(t_i) = 2 + 2\sqrt{2} \approx 4.8284$

$\Rightarrow T = \frac{\pi}{8}\left(2 + 2\sqrt{2}\right) \approx 1.89612$;

$f(t) = \sin t \Rightarrow f'(t) = \cos t \Rightarrow f''(t) = -\sin t$

$\Rightarrow M = 1 \Rightarrow |E_T| \le \frac{\pi-0}{12}\left(\frac{\pi}{4}\right)^2(1) = \frac{\pi^3}{192}$

≈ 0.16149

(b) $\int_0^\pi \sin t\,dt = [-\cos t]_0^\pi = (-\cos \pi) - (-\cos 0) = 2 \Rightarrow |E_T| = \int_0^\pi \sin t\,dt - T \approx 2 - 1.89612 = 0.10388$

(c) $\frac{|E_T|}{\text{True Value}} \times 100 = \frac{0.10388}{2} \times 100 \approx 5\%$

II. (a) For $n = 4$, $\Delta x = \frac{b-a}{n} = \frac{\pi-0}{4} = \frac{\pi}{4} \Rightarrow \frac{\Delta x}{3} = \frac{\pi}{12}$;

	t_i	$f(t_i)$	m	$mf(t_i)$
t_0	0	0	1	0
t_1	$\pi/4$	$\sqrt{2}/2$	4	$2\sqrt{2}$
t_2	$\pi/2$	1	2	2
t_3	$3\pi/4$	$\sqrt{2}/2$	4	$2\sqrt{2}$
t_4	π	0	1	0

$\sum mf(t_i) = 2 + 4\sqrt{2} \approx 7.6569$

$\Rightarrow S = \frac{\pi}{12}\left(2 + 4\sqrt{2}\right) \approx 2.00456$;

$f^{(3)}(t) = -\cos t \Rightarrow f^{(4)}(t) = \sin t$

$\Rightarrow M = 1 \Rightarrow |E_s| \le \frac{\pi-0}{180}\left(\frac{\pi}{4}\right)^4(1) \approx 0.00664$

(b) $\int_0^\pi \sin t\,dt \Rightarrow E_s = \int_0^\pi \sin t\,dt - S \approx 2 - 2.00456 = -0.00456 \Rightarrow |E_s| \approx 0.00456$

(c) $\frac{|E_s|}{\text{True Value}} \times 100 = \frac{0.00456}{2} \times 100 \approx 0\%$

11. (a) $n = 8 \Rightarrow \Delta x = \frac{1}{8} \Rightarrow \frac{\Delta x}{2} = \frac{1}{16}$;

$\sum mf(x_i) = 1(0.0) + 2(0.12402) + 2(0.24206) + 2(0.34763) + 2(0.43301) + 2(0.48789) + 2(0.49608)$
$+ 2(0.42361) + 1(0) = 5.1086 \Rightarrow T = \frac{1}{16}(5.1086) = 0.31929$

(b) $n = 8 \Rightarrow \Delta x = \frac{1}{8} \Rightarrow \frac{\Delta x}{3} = \frac{1}{24}$;

$\sum mf(x_i) = 1(0.0) + 4(0.12402) + 2(0.24206) + 4(0.34763) + 2(0.43301) + 4(0.48789) + 2(0.49608)$
$+ 4(0.42361) + 1(0) = 7.8749 \Rightarrow S = \frac{1}{24}(7.8749) = 0.32812$

(c) Let $u = 1 - x^2 \Rightarrow du = -2x\,dx \Rightarrow -\frac{1}{2}du = x\,dx$; $x = 0 \Rightarrow u = 1, x = 1 \Rightarrow u = 0$

$\int_0^1 x\sqrt{1-x^2}\,dx = \int_1^0 \sqrt{u}\left(-\frac{1}{2}du\right) = \frac{1}{2}\int_0^1 u^{1/2}\,du = \left[\frac{1}{2}\left(\frac{u^{3/2}}{\frac{3}{2}}\right)\right]_0^1 = \left[\frac{1}{3}u^{3/2}\right]_0^1 = \frac{1}{3}\left(\sqrt{1}\right)^3 - \frac{1}{3}\left(\sqrt{0}\right)^3 = \frac{1}{3}$;

$E_T = \int_0^1 x\sqrt{1-x^2}\,dx - T \approx \frac{1}{3} - 0.31929 = 0.01404$; $E_s = \int_0^1 x\sqrt{1-x^2}\,dx - S \approx \frac{1}{3} - 0.32812 = 0.00521$

13. (a) $n = 8 \Rightarrow \Delta x = \frac{\pi}{8} \Rightarrow \frac{\Delta x}{2} = \frac{\pi}{16}$;

$\sum mf(t_i) = 1(0.0) + 2(0.99138) + 2(1.26906) + 2(1.05961) + 2(0.75) + 2(0.48821) + 2(0.28946) + 2(0.13429)$
$+ 1(0) = 9.96402 \Rightarrow T = \frac{\pi}{16}(9.96402) \approx 1.95643$

(b) $n = 8 \Rightarrow \Delta x = \frac{\pi}{8} \Rightarrow \frac{\Delta x}{3} = \frac{\pi}{24}$;

$\sum mf(t_i) = 1(0.0) + 4(0.99138) + 2(1.26906) + 4(1.05961) + 2(0.75) + 4(0.48821) + 2(0.28946) + 4(0.13429)$
$+ 1(0) = 15.311 \Rightarrow S \approx \frac{\pi}{24}(15.311) \approx 2.00421$

(c) Let $u = 2 + \sin t \Rightarrow du = \cos t \, dt$; $t = -\frac{\pi}{2} \Rightarrow u = 2 + \sin\left(-\frac{\pi}{2}\right) = 1$, $t = \frac{\pi}{2} \Rightarrow u = 2 + \sin\frac{\pi}{2} = 3$

$$\int_{-\pi/2}^{\pi/2} \frac{3\cos t}{(2+\sin t)^2}\, dt = \int_1^3 \frac{3}{u^2}\, du = 3\int_1^3 u^{-2}\, du = \left[3\left(\frac{u^{-1}}{-1}\right)\right]_1^3 = 3\left(-\frac{1}{3}\right) - 3\left(-\frac{1}{1}\right) = 2;$$

$$E_T = \int_{-\pi/2}^{\pi/2} \frac{3\cos t}{(2+\sin t)^2}\, dt - T \approx 2 - 1.95643 = 0.04357; \quad E_S = \int_{-\pi/2}^{\pi/2} \frac{3\cos t}{(2+\sin t)^2}\, dt - S$$

$$\approx 2 - 2.00421 = -0.00421$$

15. (a) $M = 0$ (see Exercise 1): Then $n = 1 \Rightarrow \Delta x = 1 \Rightarrow |E_T| = \frac{1}{12}(1)^2(0) = 0 < 10^{-4}$

    (b) $M = 0$ (see Exercise 1): Then $n = 2$ (n must be even) $\Rightarrow \Delta x = \frac{1}{2} \Rightarrow |E_S| = \frac{1}{180}\left(\frac{1}{2}\right)^4(0) = 0 < 10^{-4}$

17. (a) $M = 2$ (see Exercise 3): Then $\Delta x = \frac{2}{n} \Rightarrow |E_T| \le \frac{2}{12}\left(\frac{2}{n}\right)^2(2) = \frac{4}{3n^2} < 10^{-4} \Rightarrow n^2 > \frac{4}{3}(10^4) \Rightarrow n > \sqrt{\frac{4}{3}(10^4)}$

    $\Rightarrow n > 115.4$, so let $n = 116$

    (b) $M = 0$ (see Exercise 3): Then $n = 2$ (n must be even) $\Rightarrow \Delta x = 1 \Rightarrow |E_S| = \frac{2}{180}(1)^4(0) = 0 < 10^{-4}$

19. (a) $M = 12$ (see Exercise 5): Then $\Delta x = \frac{2}{n} \Rightarrow |E_T| \le \frac{2}{12}\left(\frac{2}{n}\right)^2(12) = \frac{8}{n^2} < 10^{-4} \Rightarrow n^2 > 8(10^4) \Rightarrow n > \sqrt{8(10^4)}$

    $\Rightarrow n > 282.8$, so let $n = 283$

    (b) $M = 0$ (see Exercise 5): Then $n = 2$ (n must be even) $\Rightarrow \Delta x = 1 \Rightarrow |E_S| = \frac{2}{180}(1)^4(0) = 0 < 10^{-4}$

21. (a) $M = 6$ (see Exercise 7): Then $\Delta x = \frac{1}{n} \Rightarrow |E_T| \le \frac{1}{12}\left(\frac{1}{n}\right)^2(6) = \frac{1}{2n^2} < 10^{-4} \Rightarrow n^2 > \frac{1}{2}(10^4) \Rightarrow n > \sqrt{\frac{1}{2}(10^4)}$

    $\Rightarrow n > 70.7$, so let $n = 71$

    (b) $M = 120$ (see Exercise 7): Then $\Delta x = \frac{1}{n} \Rightarrow |E_S| = \frac{1}{180}\left(\frac{1}{n}\right)^4(120) = \frac{2}{3n^4} < 10^{-4} \Rightarrow n^4 > \frac{2}{3}(10^4)$

    $\Rightarrow n > \sqrt[4]{\frac{2}{3}(10^4)} \Rightarrow n > 9.04$, so let $n = 10$ (n must be even)

23. (a) $f(x) = \sqrt{x+1} \Rightarrow f'(x) = \frac{1}{2}(x+1)^{-1/2} \Rightarrow f''(x) = -\frac{1}{4}(x+1)^{-3/2} = -\frac{1}{4\left(\sqrt{x+1}\right)^3} \Rightarrow M = \frac{1}{4\left(\sqrt{1}\right)^3} = \frac{1}{4}.$

    Then $\Delta x = \frac{3}{n} \Rightarrow |E_T| \le \frac{3}{12}\left(\frac{3}{n}\right)^2\left(\frac{1}{4}\right) = \frac{9}{16n^2} < 10^{-4} \Rightarrow n^2 > \frac{9}{16}(10^4) \Rightarrow n > \sqrt{\frac{9}{16}(10^4)} \Rightarrow n > 75$,

    so let $n = 76$

    (b) $f^{(3)}(x) = \frac{3}{8}(x+1)^{-5/2} \Rightarrow f^{(4)}(x) = -\frac{15}{16}(x+1)^{-7/2} = -\frac{15}{16\left(\sqrt{x+1}\right)^7} \Rightarrow M = \frac{15}{16\left(\sqrt{1}\right)^7} = \frac{15}{16}$. Then $\Delta x = \frac{3}{n}$

    $\Rightarrow |E_S| \le \frac{3}{180}\left(\frac{3}{n}\right)^4\left(\frac{15}{16}\right) = \frac{3^5(15)}{16(180)n^4} < 10^{-4} \Rightarrow n^4 > \frac{3^5(15)(10^4)}{16(180)} \Rightarrow n > \sqrt[4]{\frac{3^5(15)(10^4)}{16(180)}} \Rightarrow n > 10.6$, so let

    $n = 12$ (n must be even)

25. (a) $f(x) = \sin(x+1) \Rightarrow f'(x) = \cos(x+1) \Rightarrow f''(x) = -\sin(x+1) \Rightarrow M = 1$. Then $\Delta x = \frac{2}{n} \Rightarrow |E_T| \le \frac{2}{12}\left(\frac{2}{n}\right)^2(1)$

    $= \frac{8}{12n^2} < 10^{-4} \Rightarrow n^2 > \frac{8(10^4)}{12} \Rightarrow n > \sqrt{\frac{8(10^4)}{12}} \Rightarrow n > 81.6$, so let $n = 82$

    (b) $f^{(3)}(x) = -\cos(x+1) \Rightarrow f^{(4)}(x) = \sin(x+1) \Rightarrow M = 1$. Then $\Delta x = \frac{2}{n} \Rightarrow |E_S| \le \frac{2}{180}\left(\frac{2}{n}\right)^4(1) = \frac{32}{180n^4} < 10^{-4}$

    $\Rightarrow n^4 > \frac{32(10^4)}{180} \Rightarrow n > \sqrt[4]{\frac{32(10^4)}{180}} \Rightarrow n > 6.49$, so let $n = 8$ (n must be even)

27. $\frac{5}{2}(6.0 + 2(8.2) + 2(9.1)\ldots + 2(12.7) + 13.0)(30) = 15{,}990 \text{ ft}^3.$

29. Use the conversion 30 mph = 44 fps (ft per sec) since time is measured in seconds. The distance traveled as the car accelerates from, say, 40 mph = 58.67 fps to 50 mph = 73.33 fps in $(4.5 - 3.2) = 1.3$ sec is the area of the trapezoid (see figure) associated with that time interval: $\frac{1}{2}(58.67 + 73.33)(1.3) = 85.8$ ft. The total distance traveled by the Ford Mustang Cobra is the sum of all these eleven trapezoids (using $\frac{\Delta t}{2}$ and the table below):

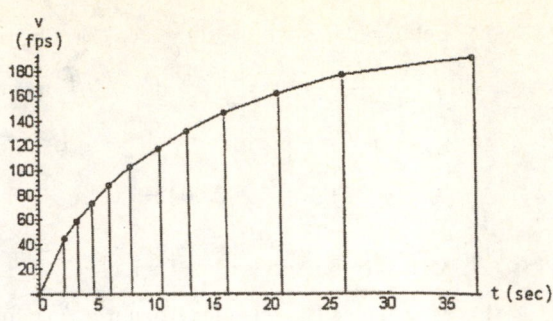

$s = (44)(1.1) + (102.67)(0.5) + (132)(0.65) + (161.33)(0.7) + (190.67)(0.95) + (220)(1.2) + (249.33)(1.25)$
$+ (278.67)(1.65) + (308)(2.3) + (337.33)(2.8) + (366.67)(5.45) = 5166.346$ ft ≈ 0.9785 mi

v (mph)	0	30	40	50	60	70	80	90	100	110	120	130
v (fps)	0	44	58.67	73.33	88	102.67	117.33	132	146.67	161.33	176	190.67
t (sec)	0	2.2	3.2	4.5	5.9	7.8	10.2	12.7	16	20.6	26.2	37.1
$\Delta t/2$	0	1.1	0.5	0.65	0.7	0.95	1.2	1.25	1.65	2.3	2.8	5.45

31. Using Simpson's Rule, $\Delta x = 1 \Rightarrow \frac{\Delta x}{3} = \frac{1}{3}$;
$\sum my_i = 33.6 \Rightarrow$ Cross Section Area $\approx \frac{1}{3}(33.6)$
$= 11.2$ ft^2. Let x be the length of the tank. Then the
Volume V = (Cross Sectional Area) x = 11.2x.
Now 5000 lb of gasoline at 42 lb/ft^3
$\Rightarrow V = \frac{5000}{42} = 119.05$ ft^3
$\Rightarrow 119.05 = 11.2x \Rightarrow x \approx 10.63$ ft

	x_i	y_i	m	my_i
x_0	0	1.5	1	1.5
x_1	1	1.6	4	6.4
x_2	2	1.8	2	3.6
x_3	3	1.9	4	7.6
x_4	4	2.0	2	4.0
x_5	5	2.1	4	8.4
x_6	6	2.1	1	2.1

33. (a) $|E_s| \le \frac{b-a}{180}(\Delta x^4)M$; $n = 4 \Rightarrow \Delta x = \frac{\frac{\pi}{2}-0}{4} = \frac{\pi}{8}$; $|f^{(4)}| \le 1 \Rightarrow M = 1 \Rightarrow |E_s| \le \frac{(\frac{\pi}{2}-0)}{180}\left(\frac{\pi}{8}\right)^4(1) \approx 0.00021$

   (b) $\Delta x = \frac{\pi}{8} \Rightarrow \frac{\Delta x}{3} = \frac{\pi}{24}$;
   $\sum mf(x_i) = 10.47208705$
   $\Rightarrow S = \frac{\pi}{24}(10.47208705) \approx 1.37079$

	x_i	$f(x_i)$	m	$mf(x_{1i})$
x_0	0	1	1	1
x_1	$\pi/8$	0.974495358	4	3.897981432
x_2	$\pi/4$	0.900316316	2	1.800632632
x_3	$3\pi/8$	0.784213303	4	3.136853212
x_4	$\pi/2$	0.636619772	1	0.636619772

   (c) $\approx \left(\frac{0.00021}{1.37079}\right) \times 100 \approx 0.015\%$

35. (a) $n = 10 \Rightarrow \Delta x = \frac{\pi-0}{10} = \frac{\pi}{10} \Rightarrow \frac{\Delta x}{2} = \frac{\pi}{20}$;
   $\sum mf(x_i) = 1(0) + 2(0.09708) + 2(0.36932) + 2(0.76248) + 2(1.19513) + 2(1.57080) + 2(1.79270)$
   $+ 2(1.77912) + 2(1.47727) + 2(0.87372) + 1(0) = 19.83524 \Rightarrow T = \frac{\pi}{20}(19.83524) = 3.11571$
   (b) $\pi - 3.11571 \approx 0.02588$
   (c) With M = 3.11, we get $|E_T| \le \frac{\pi}{12}\left(\frac{\pi}{10}\right)^2(3.11) = \frac{\pi^3}{1200}(3.11) < 0.08036$

37. $T = \frac{\Delta x}{2}(y_0 + 2y_1 + 2y_2 + 2y_3 + \ldots + 2y_{n-1} + y_n)$ where $\Delta x = \frac{b-a}{n}$ and f is continuous on [a, b]. So
   $T = \frac{b-a}{n}\frac{(y_0 + y_1 + y_1 + y_2 + y_2 + \ldots + y_{n-1} + y_{n-1} + y_n)}{2} = \frac{b-a}{n}\left(\frac{f(x_0)+f(x_1)}{2} + \frac{f(x_1)+f(x_2)}{2} + \ldots + \frac{f(x_{n-1})+f(x_n)}{2}\right)$.

   Since f is continuous on each interval $[x_{k-1}, x_k]$, and $\frac{f(x_{k-1})+f(x_k)}{2}$ is always between $f(x_{k-1})$ and $f(x_k)$, there is a point c_k in $[x_{k-1}, x_k]$ with $f(c_k) = \frac{f(x_{k-1})+f(x_k)}{2}$; this is a consequence of the Intermediate Value Theorem. Thus our sum is $\sum_{k=1}^{n}\left(\frac{b-a}{n}\right)f(c_k)$ which has the form $\sum_{k=1}^{n}\Delta x_k f(c_k)$ with $\Delta x_k = \frac{b-a}{n}$ for all k. This is a Riemann Sum for f on [a, b].

Exercises 39-42 were done using a graphing calculator with $n = 50$

39. 1.08943

41. 0.82812

43. (a) $T_{10} \approx 1.983523538$

$T_{100} \approx 1.999835504$

$T_{1000} \approx 1.999998355$

(b)

n	$|E_T| = 2 - T_n$
10	$0.016476462 = 1.6476462 \times 10^{-2}$
100	1.64496×10^{-4}
1000	1.646×10^{-6}

(c) $|E_{T_{10n}}| \approx 10^{-2}|E_{T_n}|$

(d) $b - a = \pi$, $(\Delta x)^2 = \frac{\pi^2}{n^2}$, $M = 1$

$|E_{T_n}| \leq \frac{\pi}{12}\left(\frac{\pi^2}{n^2}\right) = \frac{\pi^3}{12n^2}$

$|E_{T_{10n}}| \leq \frac{\pi^3}{12(10n)^2} \leq 10^{-2}|E_{T_n}|$

45. (a) $f'(x) = 2x\cos(x^2)$, $f''(x) = 2x \cdot (-2x)\sin(x^2) + 2\cos(x^2) = -4x^2\sin(x^2) + 2\cos(x^2)$

(b)

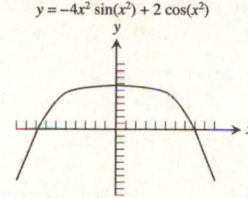

$y = -4x^2 \sin(x^2) + 2\cos(x^2)$

(c) The graph shows that $3 \leq f''(x) \leq 2$ so $|f''(x)| \leq 3$ for $-1 \leq x \leq 1$.

(d) $|E_T| \leq \frac{1-(-1)}{12}(\Delta x)^2(3) = \frac{(\Delta x)^2}{2}$

(e) For $0 < \Delta x < 0.1$, $|E_T| \leq \frac{(\Delta x)^2}{2} \leq \frac{0.1^2}{2} = 0.005 < 0.01$

(f) $n \geq \frac{1-(-1)}{\Delta x} \geq \frac{2}{0.1} = 20$

47. (a) Using $d = \frac{C}{\pi}$, and $A = \pi\left(\frac{d}{2}\right)^2 = \frac{C^2}{4\pi}$ yields the following areas (in square inches, rounded to the nearest tenth):

2.3, 1.6, 1.5, 2.1, 3.2, 4.8, 7.0, 9.3, 10.7, 10.7, 9.3, 6.4, 3.2

(b) If $C(y)$ is the circumference as a function of y, then the area of a cross section is

$A(y) = \pi\left(\frac{C(y)/\pi}{2}\right)^2 = \frac{C^2(y)}{4\pi}$, and the volume is $\frac{1}{4\pi}\int_0^6 C^2(y)\, dy$.

(c) $\int_0^6 A(y)\, dy = \frac{1}{4\pi}\int_0^6 C^2(y)\, dy$

$\approx \frac{1}{4\pi}\left(\frac{6-0}{24}\right)[5.4^2 + 2(4.5^2 + 4.4^2 + 5.1^2 + 6.3^2 + 7.8^2 + 9.4^2 + 10.8^2 + 11.6^2 + 11.6^2 + 10.8^2 + 9.0^2) + 6.3^2]$

≈ 34.7 in^3

(d) $V = \frac{1}{4\pi}\int_0^6 C^2(y)\, dy \approx \frac{1}{4\pi}\left(\frac{6-0}{36}\right)\Big[5.4^2 + 4(4.5^2) + 2(4.4^2) + 4(5.1^2) + 2(6.3^2) + 4(7.8^2) + 2(9.4^2) + 4(10.8^2)$

$+ 2(11.6^2) + 4(11.6^2) + 2(10.8^2) + 4(9.0^2) + 6.3^2\Big] = 34.792$ in^3

by Simpson's Rule. The Simpson's Rule estimate should be more accurate than the trapezoid estimate. The error in the Simpson's estimate is proportional to $(\Delta y)^4 = 0.0625$ whereas the error in the trapezoid estimate is proportional to $(\Delta y)^2 = 0.25$, a larger number when $\Delta y = 0.5$ in.

49. (a) $a = 1, e = \frac{1}{2} \Rightarrow$ Length $= 4 \int_0^{\pi/2} \sqrt{1 - \frac{1}{4} \cos^2 t}\, dt$

$= 2 \int_0^{\pi/2} \sqrt{4 - \cos^2 t}\, dt = \int_0^{\pi/2} f(t)\, dt$; use the

Trapezoid Rule with $n = 10 \Rightarrow \Delta t = \frac{b-a}{n} = \frac{\left(\frac{\pi}{2}\right) - 0}{10}$

$= \frac{\pi}{20}$. $\int_0^{\pi/2} \sqrt{4 - \cos^2 t}\, dt \approx \sum_{n=0}^{10} mf(x_n) = 37.3686183$

$\Rightarrow T = \frac{\Delta t}{2} (37.3686183) = \frac{\pi}{40} (37.3686183)$

$= 2.934924419 \Rightarrow$ Length $= 2(2.934924419)$

≈ 5.870

(b) $|f''(t)| < 1 \Rightarrow M = 1$

$\Rightarrow |E_T| \le \frac{b-a}{12} (\Delta t^2 M) \le \frac{\left(\frac{\pi}{2}\right) - 0}{12} \left(\frac{\pi}{20}\right)^2 1 \le 0.0032$

	x_i	$f(x_i)$	m	$mf(x_i)$
x_0	0	1.732050808	1	1.732050808
x_1	$\pi/20$	1.739100843	2	3.478201686
x_2	$\pi/10$	1.759400893	2	3.518801786
x_3	$3\pi/20$	1.790560631	2	3.581121262
x_4	$\pi/5$	1.82906848	1	3.658136959
x_5	$\pi/4$	1.870828693	1	3.741657387
x_6	$3\pi/10$	1.911676881	2	3.823353762
x_7	$7\pi/20$	1.947791731	2	3.895583461
x_8	$2\pi/5$	1.975982919	2	3.951965839
x_9	$9\pi/20$	1.993872679	2	3.987745357
x_{10}	$\pi/2$	2	1	2

51. The length of the curve $y = \sin\left(\frac{3\pi}{20} x\right)$ from 0 to 20 is: $L = \int_0^{20} \sqrt{1 + \left(\frac{dy}{dx}\right)^2}\, dx$; $\frac{dy}{dx} = \frac{3\pi}{20} \cos\left(\frac{3\pi}{20} x\right) \Rightarrow \left(\frac{dy}{dx}\right)^2$

$= \frac{9\pi^2}{400} \cos^2\left(\frac{3\pi}{20} x\right) \Rightarrow L = \int_0^{20} \sqrt{1 + \frac{9\pi^2}{400} \cos^2\left(\frac{3\pi}{20} x\right)}\, dx$. Using numerical integration we find $L \approx 21.07$ in

53. $y = \sin x \Rightarrow \frac{dy}{dx} = \cos x \Rightarrow \left(\frac{dy}{dx}\right)^2 = \cos^2 x \Rightarrow S = \int_0^\pi 2\pi(\sin x) \sqrt{1 + \cos^2 x}\, dx$; a numerical integration gives

$S \approx 14.4$

55. $y = x + \sin 2x \Rightarrow \frac{dy}{dx} = 1 + 2\cos 2x \Rightarrow \left(\frac{dy}{dx}\right)^2 = (1 + 2\cos 2x)^2$; by symmetry of the graph we have that

$S = 2 \int_0^{2\pi/3} 2\pi(x + \sin 2x) \sqrt{1 + (1 + 2\cos 2x)^2}\, dx$; a numerical integration gives $S \approx 54.9$

57. A calculator or computer numerical integrator yields $\sin^{-1} 0.6 \approx 0.643501109$.

8.8 IMPROPER INTEGRALS

1. $\int_0^\infty \frac{dx}{x^2 + 1} = \lim_{b \to \infty} \int_0^b \frac{dx}{x^2 + 1} = \lim_{b \to \infty} \left[\tan^{-1} x\right]_0^b = \lim_{b \to \infty} (\tan^{-1} b - \tan^{-1} 0) = \frac{\pi}{2} - 0 = \frac{\pi}{2}$

3. $\int_0^1 \frac{dx}{\sqrt{x}} = \lim_{b \to 0^+} \int_b^1 x^{-1/2}\, dx = \lim_{b \to 0^+} \left[2x^{1/2}\right]_b^1 = \lim_{b \to 0^+} \left(2 - 2\sqrt{b}\right) = 2 - 0 = 2$

5. $\int_{-1}^1 \frac{dx}{x^{2/3}} = \int_{-1}^0 \frac{dx}{x^{2/3}} + \int_0^1 \frac{dx}{x^{2/3}} = \lim_{b \to 0^-} \left[3x^{1/3}\right]_{-1}^b + \lim_{c \to 0^+} \left[3x^{1/3}\right]_c^1$

$= \lim_{b \to 0^-} \left[3b^{1/3} - 3(-1)^{1/3}\right] + \lim_{c \to 0^+} \left[3(1)^{1/3} - 3c^{1/3}\right] = (0 + 3) + (3 - 0) = 6$

7. $\int_0^1 \frac{dx}{\sqrt{1 - x^2}} = \lim_{b \to 1^-} \left[\sin^{-1} x\right]_0^b = \lim_{b \to 1^-} (\sin^{-1} b - \sin^{-1} 0) = \frac{\pi}{2} - 0 = \frac{\pi}{2}$

9. $\int_{-\infty}^{-2} \frac{2\, dx}{x^2 - 1} = \int_{-\infty}^{-2} \frac{dx}{x - 1} - \int_{-\infty}^{-2} \frac{dx}{x + 1} = \lim_{b \to -\infty} \left[\ln|x - 1|\right]_b^{-2} - \lim_{b \to -\infty} \left[\ln|x + 1|\right]_b^{-2} = \lim_{b \to -\infty} \left[\ln\left|\frac{x-1}{x+1}\right|\right]_b^{-2}$

$= \lim_{b \to -\infty} \left(\ln\left|\frac{-3}{-1}\right| - \ln\left|\frac{b-1}{b+1}\right|\right) = \ln 3 - \ln\left(\lim_{b \to -\infty} \frac{b-1}{b+1}\right) = \ln 3 - \ln 1 = \ln 3$

11. $\int_2^\infty \frac{2\, dv}{v^2 - v} = \lim_{b \to \infty} \left[2\ln\left|\frac{v-1}{v}\right|\right]_2^b = \lim_{b \to \infty} \left(2\ln\left|\frac{b-1}{b}\right| - 2\ln\left|\frac{2-1}{2}\right|\right) = 2\ln(1) - 2\ln\left(\frac{1}{2}\right) = 0 + 2\ln 2 = \ln 4$

13. $\int_{-\infty}^{\infty} \frac{2x\,dx}{(x^2+1)^2} = \int_{-\infty}^{0} \frac{2x\,dx}{(x^2+1)^2} + \int_{0}^{\infty} \frac{2x\,dx}{(x^2+1)^2}$; $\begin{bmatrix} u = x^2+1 \\ du = 2x\,dx \end{bmatrix} \rightarrow \int_{\infty}^{1} \frac{du}{u^2} + \int_{1}^{\infty} \frac{du}{u^2} = \lim_{b\to\infty} \left[-\frac{1}{u} \right]_{b}^{1} + \lim_{c\to\infty} \left[-\frac{1}{u} \right]_{1}^{c}$

$= \lim_{b\to\infty} \left(-1 + \frac{1}{b} \right) + \lim_{c\to\infty} \left[-\frac{1}{c} - (-1) \right] = (-1+0) + (0+1) = 0$

15. $\int_{0}^{1} \frac{\theta+1}{\sqrt{\theta^2+2\theta}}\,d\theta$; $\begin{bmatrix} u = \theta^2+2\theta \\ du = 2(\theta+1)\,d\theta \end{bmatrix} \rightarrow \int_{0}^{3} \frac{du}{2\sqrt{u}} = \lim_{b\to 0^+} \int_{b}^{3} \frac{du}{2\sqrt{u}} = \lim_{b\to 0^+} \left[\sqrt{u} \right]_{b}^{3} = \lim_{b\to 0^+} \left(\sqrt{3} - \sqrt{b} \right)$

$= \sqrt{3} - 0 = \sqrt{3}$

17. $\int_{0}^{\infty} \frac{dx}{(1+x)\sqrt{x}}$; $\begin{bmatrix} u = \sqrt{x} \\ du = \frac{dx}{2\sqrt{x}} \end{bmatrix} \rightarrow \int_{0}^{\infty} \frac{2\,du}{u^2+1} = \lim_{b\to\infty} \int_{0}^{b} \frac{2\,du}{u^2+1} = \lim_{b\to\infty} \left[2\tan^{-1} u \right]_{0}^{b}$

$= \lim_{b\to\infty} \left(2\tan^{-1} b - 2\tan^{-1} 0 \right) = 2\left(\frac{\pi}{2} \right) - 2(0) = \pi$

19. $\int_{0}^{\infty} \frac{dv}{(1+v^2)(1+\tan^{-1} v)} = \lim_{b\to\infty} \left[\ln|1+\tan^{-1} v| \right]_{0}^{b} = \lim_{b\to\infty} \left[\ln|1+\tan^{-1} b| \right] - \ln|1+\tan^{-1} 0|$

$= \ln\left(1 + \frac{\pi}{2} \right) - \ln(1+0) = \ln\left(1 + \frac{\pi}{2} \right)$

21. $\int_{-\infty}^{0} \theta e^{\theta}\,d\theta = \lim_{b\to -\infty} \left[\theta e^{\theta} - e^{\theta} \right]_{b}^{0} = (0 \cdot e^0 - e^0) - \lim_{b\to -\infty} \left[be^{b} - e^{b} \right] = -1 - \lim_{b\to -\infty} \left(\frac{b-1}{e^{-b}} \right)$

$= -1 - \lim_{b\to -\infty} \left(\frac{1}{-e^{-b}} \right)$ (l'Hôpital's rule for $\frac{\infty}{\infty}$ form)

$= -1 - 0 = -1$

23. $\int_{-\infty}^{0} e^{-|x|}\,dx = \int_{-\infty}^{0} e^{x}\,dx = \lim_{b\to -\infty} \left[e^{x} \right]_{b}^{0} = \lim_{b\to -\infty} (1 - e^{b}) = (1-0) = 1$

25. $\int_{0}^{1} x\ln x\,dx = \lim_{b\to 0^+} \left[\frac{x^2}{2}\ln x - \frac{x^2}{4} \right]_{b}^{1} = \left(\frac{1}{2}\ln 1 - \frac{1}{4} \right) - \lim_{b\to 0^+} \left(\frac{b^2}{2}\ln b - \frac{b^2}{4} \right) = -\frac{1}{4} - \lim_{b\to 0^+} \frac{\ln b}{\left(\frac{2}{b^2} \right)} + 0$

$= -\frac{1}{4} - \lim_{b\to 0^+} \frac{\left(\frac{1}{b} \right)}{\left(-\frac{4}{b^3} \right)} = -\frac{1}{4} + \lim_{b\to 0^+} \left(\frac{b^2}{4} \right) = -\frac{1}{4} + 0 = -\frac{1}{4}$

27. $\int_{0}^{2} \frac{ds}{\sqrt{4-s^2}} = \lim_{b\to 2^-} \left[\sin^{-1} \frac{s}{2} \right]_{0}^{b} = \lim_{b\to 2^-} \left(\sin^{-1} \frac{b}{2} \right) - \sin^{-1} 0 = \frac{\pi}{2} - 0 = \frac{\pi}{2}$

29. $\int_{1}^{2} \frac{ds}{s\sqrt{s^2-1}} = \lim_{b\to 1^+} \left[\sec^{-1} s \right]_{b}^{2} = \sec^{-1} 2 - \lim_{b\to 1^+} \sec^{-1} b = \frac{\pi}{3} - 0 = \frac{\pi}{3}$

31. $\int_{-1}^{4} \frac{dx}{\sqrt{|x|}} = \lim_{b\to 0^-} \int_{-1}^{b} \frac{dx}{\sqrt{-x}} + \lim_{c\to 0^+} \int_{c}^{4} \frac{dx}{\sqrt{x}} = \lim_{b\to 0^-} \left[-2\sqrt{-x} \right]_{-1}^{b} + \lim_{c\to 0^+} \left[2\sqrt{x} \right]_{c}^{4}$

$= \lim_{b\to 0^-} \left(-2\sqrt{-b} \right) - \left(-2\sqrt{-(-1)} \right) + 2\sqrt{4} - \lim_{c\to 0^+} 2\sqrt{c} = 0 + 2 + 2\cdot 2 - 0 = 6$

33. $\int_{-1}^{\infty} \frac{d\theta}{\theta^2+5\theta+6} = \lim_{b\to\infty} \left[\ln \left| \frac{\theta+2}{\theta+3} \right| \right]_{-1}^{b} = \lim_{b\to\infty} \left[\ln \left| \frac{b+2}{b+3} \right| \right] - \ln \left| \frac{-1+2}{-1+3} \right| = 0 - \ln \left(\frac{1}{2} \right) = \ln 2$

35. $\int_{0}^{\pi/2} \tan\theta\,d\theta = \lim_{b\to \frac{\pi}{2}^-} \left[-\ln|\cos\theta| \right]_{0}^{b} = \lim_{b\to \frac{\pi}{2}^-} \left[-\ln|\cos b| \right] + \ln 1 = \lim_{b\to \frac{\pi}{2}^-} \left[-\ln|\cos b| \right] = +\infty$,

the integral diverges

37. $\int_{0}^{\pi} \frac{\sin\theta\,d\theta}{\sqrt{\pi-\theta}}$; $[\pi-\theta = x] \rightarrow -\int_{\pi}^{0} \frac{\sin x\,dx}{\sqrt{x}} = \int_{0}^{\pi} \frac{\sin x\,dx}{\sqrt{x}}$. Since $0 \le \frac{\sin x}{\sqrt{x}} \le \frac{1}{\sqrt{x}}$ for all $0 \le x \le \pi$ and $\int_{0}^{\pi} \frac{dx}{\sqrt{x}}$

converges, then $\int_{0}^{\pi} \frac{\sin x}{\sqrt{x}}\,dx$ converges by the Direct Comparison Test.

39. $\int_0^{\ln 2} x^{-2} e^{-1/x} \, dx$; $\left[\frac{1}{x} = y\right] \rightarrow \int_\infty^{1/\ln 2} \frac{y^2 e^{-y} \, dy}{-y^2} = \int_{1/\ln 2}^\infty e^{-y} \, dy = \lim_{b \to \infty} \left[-e^{-y}\right]_{1/\ln 2}^b = \lim_{b \to \infty} \left[-e^{-b}\right] - \left[-e^{-1/\ln 2}\right]$
$= 0 + e^{-1/\ln 2} = e^{-1/\ln 2}$, so the integral converges.

41. $\int_0^\pi \frac{dt}{\sqrt{t + \sin t}}$. Since for $0 \le t \le \pi$, $0 \le \frac{1}{\sqrt{t + \sin t}} \le \frac{1}{\sqrt{t}}$ and $\int_0^\pi \frac{dt}{\sqrt{t}}$ converges, then the original integral converges as well by the Direct Comparison Test.

43. $\int_0^2 \frac{dx}{1 - x^2} = \int_0^1 \frac{dx}{1 - x^2} + \int_1^2 \frac{dx}{1 - x^2}$ and $\int_0^1 \frac{dx}{1 - x^2} = \lim_{b \to 1^-} \left[\frac{1}{2} \ln \left|\frac{1+x}{1-x}\right|\right]_0^b = \lim_{b \to 1^-} \left[\frac{1}{2} \ln \left|\frac{1+b}{1-b}\right|\right] - 0 = \infty$, which
diverges $\Rightarrow \int_0^2 \frac{dx}{1 - x^2}$ diverges as well.

45. $\int_{-1}^1 \ln |x| \, dx = \int_{-1}^0 \ln(-x) \, dx + \int_0^1 \ln x \, dx$; $\int_0^1 \ln x \, dx = \lim_{b \to 0^+} [x \ln x - x]_b^1 = [1 \cdot 0 - 1] - \lim_{b \to 0^+} [b \ln b - b]$
$= -1 - 0 = -1$; $\int_{-1}^0 \ln(-x) \, dx = -1 \Rightarrow \int_{-1}^1 \ln |x| \, dx = -2$ converges.

47. $\int_1^\infty \frac{dx}{1 + x^3}$; $0 \le \frac{1}{x^3 + 1} \le \frac{1}{x^3}$ for $1 \le x < \infty$ and $\int_1^\infty \frac{dx}{x^3}$ converges $\Rightarrow \int_1^\infty \frac{dx}{1 + x^3}$ converges by the Direct
Comparison Test.

49. $\int_2^\infty \frac{dv}{\sqrt{v - 1}}$; $\lim_{v \to \infty} \frac{\left(\frac{1}{\sqrt{v-1}}\right)}{\left(\frac{1}{\sqrt{v}}\right)} = \lim_{v \to \infty} \frac{\sqrt{v}}{\sqrt{v - 1}} = \lim_{v \to \infty} \frac{1}{\sqrt{1 - \frac{1}{v}}} = \frac{1}{\sqrt{1 - 0}} = 1$ and $\int_2^\infty \frac{dv}{\sqrt{v}} = \lim_{b \to \infty} \left[2\sqrt{v}\right]_2^b = \infty$,
which diverges $\Rightarrow \int_2^\infty \frac{dv}{\sqrt{v - 1}}$ diverges by the Limit Comparison Test.

51. $\int_0^\infty \frac{dx}{\sqrt{x^6 + 1}} = \int_0^1 \frac{dx}{\sqrt{x^6 + 1}} + \int_1^\infty \frac{dx}{\sqrt{x^6 + 1}} < \int_0^1 \frac{dx}{\sqrt{x^6 + 1}} + \int_1^\infty \frac{dx}{x^3}$ and $\int_1^\infty \frac{dx}{x^3} = \lim_{b \to \infty} \left[-\frac{1}{2x^2}\right]_1^b$
$= \lim_{b \to \infty} \left(-\frac{1}{2b^2} + \frac{1}{2}\right) = \frac{1}{2} \Rightarrow \int_0^\infty \frac{dx}{\sqrt{x^6 + 1}}$ converges by the Direct Comparison Test.

53. $\int_1^\infty \frac{\sqrt{x + 1}}{x^2} \, dx$; $\lim_{x \to \infty} \frac{\left(\frac{\sqrt{x}}{x^2}\right)}{\left(\frac{\sqrt{x+1}}{x^2}\right)} = \lim_{x \to \infty} \frac{\sqrt{x}}{\sqrt{x + 1}} = \lim_{x \to \infty} \frac{1}{\sqrt{1 + \frac{1}{x}}} = 1$; $\int_1^\infty \frac{\sqrt{x}}{x^2} \, dx = \int_1^\infty \frac{dx}{x^{3/2}}$
$= \lim_{b \to \infty} \left[-2x^{-1/2}\right]_1^b = \lim_{b \to \infty} \left(\frac{-2}{\sqrt{b}} + 2\right) = 2 \Rightarrow \int_1^\infty \frac{\sqrt{x + 1}}{x^2} \, dx$ converges by the Limit Comparison Test.

55. $\int_\pi^\infty \frac{2 + \cos x}{x} \, dx$; $0 < \frac{1}{x} \le \frac{2 + \cos x}{x}$ for $x \ge \pi$ and $\int_\pi^\infty \frac{dx}{x} = \lim_{b \to \infty} [\ln x]_\pi^b = \infty$, which diverges
$\Rightarrow \int_\pi^\infty \frac{2 + \cos x}{x} \, dx$ diverges by the Direct Comparison Test.

57. $\int_4^\infty \frac{2 \, dt}{t^{3/2} - 1}$; $\lim_{t \to \infty} \frac{t^{3/2}}{t^{3/2} - 1} = 1$ and $\int_4^\infty \frac{2 \, dt}{t^{3/2}} = \lim_{b \to \infty} \left[-4t^{-1/2}\right]_4^b = \lim_{b \to \infty} \left(\frac{-4}{\sqrt{b}} + 2\right) = 2 \Rightarrow \int_4^\infty \frac{2 \, dt}{t^{3/2}}$ converges
$\Rightarrow \int_4^\infty \frac{2 \, dt}{t^{3/2} + 1}$ converges by the Limit Comparison Test.

59. $\int_1^\infty \frac{e^x}{x} \, dx$; $0 < \frac{1}{x} < \frac{e^x}{x}$ for $x > 1$ and $\int_1^\infty \frac{dx}{x}$ diverges $\Rightarrow \int_1^\infty \frac{e^x \, dx}{x}$ diverges by the Direct Comparison Test.

61. $\int_1^\infty \frac{dx}{\sqrt{e^x - x}}$; $\lim_{x \to \infty} \frac{\left(\frac{1}{\sqrt{e^x - x}}\right)}{\left(\frac{1}{\sqrt{e^x}}\right)} = \lim_{x \to \infty} \frac{\sqrt{e^x}}{\sqrt{e^x - x}} = \lim_{x \to \infty} \frac{1}{\sqrt{1 - \frac{x}{e^x}}} = \frac{1}{\sqrt{1 - 0}} = 1$; $\int_1^\infty \frac{dx}{\sqrt{e^x}} = \int_1^\infty e^{-x/2} \, dx$
$= \lim_{b \to \infty} \left[-2e^{-x/2}\right]_1^b = \lim_{b \to \infty} \left(-2e^{-b/2} + 2e^{-1/2}\right) = \frac{2}{\sqrt{e}} \Rightarrow \int_1^\infty e^{-x/2} \, dx$ converges $\Rightarrow \int_1^\infty \frac{dx}{\sqrt{e^x - x}}$ converges
by the Limit Comparison Test.

63. $\int_{-\infty}^{\infty} \frac{dx}{\sqrt{x^4+1}} = 2\int_0^{\infty} \frac{dx}{\sqrt{x^4+1}}$; $\int_0^{\infty} \frac{dx}{\sqrt{x^4+1}} = \int_0^1 \frac{dx}{\sqrt{x^4+1}} + \int_1^{\infty} \frac{dx}{\sqrt{x^4+1}} < \int_0^1 \frac{dx}{\sqrt{x^4+1}} + \int_1^{\infty} \frac{dx}{x^2}$ and

$\int_1^{\infty} \frac{dx}{x^2} = \lim_{b\to\infty} \left[-\frac{1}{x}\right]_1^b = \lim_{b\to\infty} \left(-\frac{1}{b}+1\right) = 1 \Rightarrow \int_{-\infty}^{\infty} \frac{dx}{\sqrt{x^4+1}}$ converges by the Direct Comparison Test.

65. (a) $\int_1^2 \frac{dx}{x(\ln x)^p}$; $[t = \ln x] \to \int_0^{\ln 2} \frac{dt}{t^p} = \lim_{b\to 0^+} \left[\frac{1}{-p+1}t^{1-p}\right]_b^{\ln 2} = \lim_{b\to 0^+} \frac{b^{1-p}}{p-1} + \frac{1}{1-p}(\ln 2)^{1-p}$

$\Rightarrow$ the integral converges for $p < 1$ and diverges for $p \geq 1$

(b) $\int_2^{\infty} \frac{dx}{x(\ln x)^p}$; $[t = \ln x] \to \int_{\ln 2}^{\infty} \frac{dt}{t^p}$ and this integral is essentially the same as in Exercise 65(a): it converges for $p > 1$ and diverges for $p \leq 1$

67. $A = \int_0^{\infty} e^{-x}\, dx = \lim_{b\to\infty} \left[-e^{-x}\right]_0^b = \lim_{b\to\infty} \left(-e^{-b}\right) - \left(-e^{-0}\right)$

$= 0 + 1 = 1$

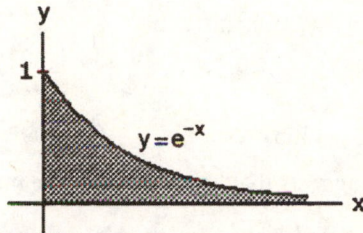

69. $V = \int_0^{\infty} 2\pi x e^{-x}\, dx = 2\pi \int_0^{\infty} x e^{-x}\, dx = 2\pi \lim_{b\to\infty} \left[-x e^{-x} - e^{-x}\right]_0^b = 2\pi \left[\lim_{b\to\infty} \left(-b e^{-b} - e^{-b}\right) - 1\right] = 2\pi$

71. $A = \int_0^{\pi/2} (\sec x - \tan x)\, dx = \lim_{b\to\frac{\pi}{2}^-} \left[\ln|\sec x + \tan x| - \ln|\sec x|\right]_0^b = \lim_{b\to\frac{\pi}{2}^-} \left(\ln\left|1 + \frac{\tan b}{\sec b}\right| - \ln|1 + 0|\right)$

$= \lim_{b\to\frac{\pi}{2}^-} \ln|1 + \sin b| = \ln 2$

73. (a) $\int_3^{\infty} e^{-3x}\, dx = \lim_{b\to\infty} \left[-\frac{1}{3}e^{-3x}\right]_3^b = \lim_{b\to\infty} \left(-\frac{1}{3}e^{-3b}\right) - \left(-\frac{1}{3}e^{-3\cdot 3}\right) = 0 + \frac{1}{3}\cdot e^{-9} = \frac{1}{3}e^{-9}$

$\approx 0.0000411 < 0.000042$. Since $e^{-x^2} \leq e^{-3x}$ for $x > 3$, then $\int_3^{\infty} e^{-x^2}\, dx < 0.000042$ and therefore

$\int_0^{\infty} e^{-x^2}\, dx$ can be replaced by $\int_0^3 e^{-x^2}\, dx$ without introducing an error greater than 0.000042.

(b) $\int_0^3 e^{-x^2}\, dx \cong 0.88621$

75. (a)

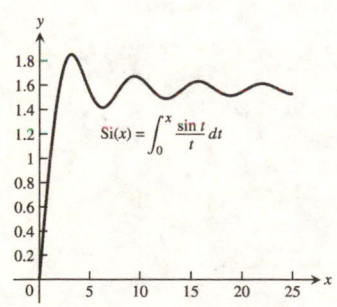

(b) > int((sin(t))/t, t=0..infinity); (answer is $\frac{\pi}{2}$)

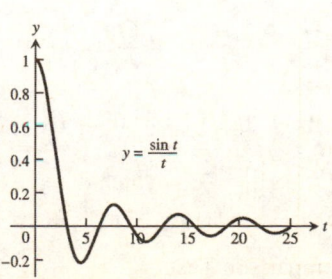

77. (a) $f(x) = \frac{1}{\sqrt{2\pi}} e^{-x^2/2}$

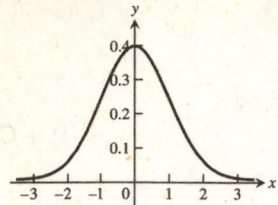

f is increasing on $(-\infty, 0]$. f is decreasing on $[0, \infty)$. f has a local maximum at $(0, f(0)) = \left(0, \frac{1}{\sqrt{2\pi}}\right)$

(b) Maple commands:

>f: = exp($-x^2/2$)(sqrt(2*pi);

>int(f, x = $-1..1$); ≈ 0.683

>int(f, x = $-2..2$); ≈ 0.954

>int(f, x = $-3..3$); ≈ 0.997

(c) Part (b) suggests that as n increases, the integral approaches 1. We can take $\int_{-n}^{n} f(x)\,dx$ as close to 1 as we want by choosing n > 1 large enough. Also, we can make $\int_{n}^{\infty} f(x)\,dx$ and $\int_{-\infty}^{-n} f(x)\,dx$ as small as we want by choosing n large enough. This is because $0 < f(x) < e^{-x/2}$ for x > 1. (Likewise, $0 < f(x) < e^{x/2}$ for x < -1.)

Thus, $\int_{n}^{\infty} f(x)\,dx < \int_{n}^{\infty} e^{-x/2}\,dx$.

$\int_{n}^{\infty} e^{-x/2}\,dx = \lim_{c \to \infty} \int_{n}^{c} e^{-x/2}\,dx = \lim_{c \to \infty} \left[-2e^{-x/2}\right]_{n}^{c} = \lim_{c \to \infty}\left[-2e^{-c/2} + 2e^{-n/2}\right] = 2e^{-n/2}$

As $n \to \infty$, $2e^{-n/2} \to 0$, for large enough n, $\int_{n}^{\infty} f(x)\,dx$ is as small as we want. Likewise for large enough n, $\int_{-\infty}^{-n} f(x)\,dx$ is as small as we want.

79. (a) The statement is true since $\int_{-\infty}^{b} f(x)\,dx = \int_{-\infty}^{a} f(x)\,dx + \int_{a}^{b} f(x)\,dx$, $\int_{b}^{\infty} f(x)\,dx = \int_{a}^{\infty} f(x)\,dx - \int_{a}^{b} f(x)\,dx$

and $\int_{a}^{b} f(x)\,dx$ exists since f(x) is integrable on every interval [a, b].

(b) $\int_{-\infty}^{a} f(x)\,dx + \int_{a}^{\infty} f(x)\,dx = \int_{-\infty}^{a} f(x)\,dx + \int_{a}^{b} f(x)\,dx - \int_{a}^{b} f(x)\,dx + \int_{a}^{\infty} f(x)\,dx$

$= \int_{-\infty}^{b} f(x)\,dx + \int_{b}^{a} f(x)\,dx + \int_{a}^{\infty} f(x)\,dx = \int_{-\infty}^{b} f(x)\,dx + \int_{b}^{\infty} f(x)\,dx$

81. $\int_{-\infty}^{\infty} \frac{dx}{\sqrt{x^2+1}} = \int_{-\infty}^{1} \frac{dx}{\sqrt{x^2+1}} + \int_{1}^{\infty} \frac{dx}{\sqrt{x^2+1}}$; $\int_{1}^{\infty} \frac{dx}{\sqrt{x^2+1}}$ diverges because $\lim_{x \to \infty} \frac{\left(\frac{1}{x}\right)}{\left(\frac{1}{\sqrt{x^2+1}}\right)}$

$= \lim_{x \to \infty} \frac{\sqrt{x^2+1}}{x} = \lim_{x \to \infty} \sqrt{1 + \frac{1}{x^2}} = 1$ and $\int_{1}^{\infty} \frac{dx}{x}$ diverges; therefore, $\int_{-\infty}^{\infty} \frac{dx}{\sqrt{x^2+1}}$ diverges

83. $\int_{-\infty}^{\infty} \frac{dx}{e^x + e^{-x}} = \int_{-\infty}^{\infty} \frac{e^x\,dx}{e^{2x}+1}$; $\frac{e^x}{e^{2x}+1} = \frac{1}{e^x + e^{-x}} < \frac{1}{e^x}$ and $\int_{0}^{\infty} \frac{dx}{e^x} = \lim_{c \to \infty} \left[-e^{-x}\right]_{0}^{c} = \lim_{c \to \infty} \left(-e^{-c} + 1\right) = 1$

$\Rightarrow \int_{-\infty}^{\infty} \frac{e^x\,dx}{e^{2x}+1} = 2\int_{0}^{\infty} \frac{dx}{e^x + e^{-x}}$ converges

85. $\int_{-\infty}^{\infty} e^{-|x|}\,dx = 2\int_{0}^{\infty} e^{-x}\,dx = 2\lim_{b \to \infty} \int_{0}^{b} e^{-x}\,dx = -2\lim_{b \to \infty} \left[e^{-x}\right]_{0}^{b} = 2$, so the integral converges.

87. $\int_{-\infty}^{\infty} \frac{|\sin x| + |\cos x|}{|x| + 1}\,dx = 2\int_{0}^{\infty} \frac{|\sin x| + |\cos x|}{x + 1}\,dx \geq 2\int_{0}^{\infty} \frac{\sin^2 x + \cos^2 x}{x + 1}\,dx = 2\lim_{b \to \infty} \int_{0}^{b} \frac{dx}{x + 1}\,dx$

$= 2\lim_{b \to \infty} \left[\ln|x + 1|\right]_{0}^{b} = \infty$, which diverges $\Rightarrow \int_{-\infty}^{\infty} \frac{|\sin x| + |\cos x|}{|x| + 1}\,dx$ diverges

CHAPTER 8 PRACTICE EXERCISES

1. $\int x\sqrt{4x^2 - 9}\,dx;\ \begin{bmatrix} u = 4x^2 - 9 \\ du = 8x\,dx \end{bmatrix} \rightarrow \frac{1}{8}\int \sqrt{u}\,du = \frac{1}{8} \cdot \frac{2}{3} u^{3/2} + C = \frac{1}{12}\left(4x^2 - 9\right)^{3/2} + C$

3. $\int x(2x+1)^{1/2}\,dx;\ \begin{bmatrix} u = 2x+1 \\ du = 2\,dx \end{bmatrix} \rightarrow \frac{1}{2}\int \left(\frac{u-1}{2}\right)\sqrt{u}\,du = \frac{1}{4}\left(\int u^{3/2}\,du - \int u^{1/2}\,du\right) = \frac{1}{4}\left(\frac{2}{5}u^{5/2} - \frac{2}{3}u^{3/2}\right) + C$

   $= \frac{(2x+1)^{5/2}}{10} - \frac{(2x+1)^{3/2}}{6} + C$

5. $\int \frac{x\,dx}{\sqrt{8x^2+1}};\ \begin{bmatrix} u = 8x^2 + 1 \\ du = 16x\,dx \end{bmatrix} \rightarrow \frac{1}{16}\int \frac{du}{\sqrt{u}} = \frac{1}{16}\cdot 2u^{1/2} + C = \frac{\sqrt{8x^2+1}}{8} + C$

7. $\int \frac{y\,dy}{25+y^2};\ \begin{bmatrix} u = 25 + y^2 \\ du = 2y\,dy \end{bmatrix} \rightarrow \frac{1}{2}\int \frac{du}{u} = \frac{1}{2}\ln|u| + C = \frac{1}{2}\ln\left(25+y^2\right) + C$

9. $\int \frac{t^3\,dt}{\sqrt{9-4t^4}};\ \begin{bmatrix} u = 9 - 4t^4 \\ du = -16t^3\,dt \end{bmatrix} \rightarrow -\frac{1}{16}\int \frac{du}{\sqrt{u}} = -\frac{1}{16}\cdot 2u^{1/2} + C = -\frac{\sqrt{9-4t^4}}{8} + C$

11. $\int z^{2/3}\left(z^{5/3}+1\right)^{2/3}\,dz;\ \begin{bmatrix} u = z^{5/3} + 1 \\ du = \frac{5}{3}z^{2/3}\,dz \end{bmatrix} \rightarrow \frac{3}{5}\int u^{2/3}\,du = \frac{3}{5}\cdot\frac{3}{5}u^{5/3} + C = \frac{9}{25}\left(z^{5/3}+1\right)^{5/3} + C$

13. $\int \frac{\sin 2\theta\,d\theta}{(1-\cos 2\theta)^2};\ \begin{bmatrix} u = 1 - \cos 2\theta \\ du = 2\sin 2\theta\,d\theta \end{bmatrix} \rightarrow \frac{1}{2}\int \frac{du}{u^2} = -\frac{1}{2u} + C = -\frac{1}{2(1-\cos 2\theta)} + C$

15. $\int \frac{\sin t\,dt}{3+4\cos t};\ \begin{bmatrix} u = 3 + 4\cos t \\ du = -4\sin t\,dt \end{bmatrix} \rightarrow -\frac{1}{4}\int \frac{du}{u} = -\frac{1}{4}\ln|u| + C = -\frac{1}{4}\ln|3+4\cos t| + C$

17. $\int (\sin 2x)\,e^{\cos 2x}\,dx;\ \begin{bmatrix} u = \cos 2x \\ du = -2\sin 2x\,dx \end{bmatrix} \rightarrow -\frac{1}{2}\int e^u\,du = -\frac{1}{2}e^u + C = -\frac{1}{2}e^{\cos 2x} + C$

19. $\int e^{\theta}\sin\left(e^{\theta}\right)\cos^2\left(e^{\theta}\right)\,d\theta;\ \begin{bmatrix} u = \cos\left(e^{\theta}\right) \\ du = -\sin\left(e^{\theta}\right)\cdot e^{\theta}\,d\theta \end{bmatrix} \rightarrow \int -u^2\,du = -\frac{1}{3}u^3 + C = -\frac{1}{3}\cos^3\left(e^{\theta}\right) + C$

21. $\int 2^{x-1}\,dx = \frac{2^{x-1}}{\ln 2} + C$

23. $\int \frac{dv}{v\ln v};\ \begin{bmatrix} u = \ln v \\ du = \frac{1}{v}\,dv \end{bmatrix} \rightarrow \int \frac{du}{u} = \ln|u| + C = \ln|\ln v| + C$

25. $\int \frac{dx}{(x^2+1)\left(2+\tan^{-1}x\right)};\ \begin{bmatrix} u = 2 + \tan^{-1}x \\ du = \frac{dx}{x^2+1} \end{bmatrix} \rightarrow \int \frac{du}{u} = \ln|u| + C = \ln\left|2+\tan^{-1}x\right| + C$

27. $\int \frac{2\,dx}{\sqrt{1-4x^2}};\ \begin{bmatrix} u = 2x \\ du = 2\,dx \end{bmatrix} \rightarrow \int \frac{du}{\sqrt{1-u^2}} = \sin^{-1}u + C = \sin^{-1}(2x) + C$

29. $\int \frac{dt}{\sqrt{16-9t^2}} = \frac{1}{4}\int \frac{dt}{\sqrt{1-\left(\frac{3t}{4}\right)^2}};\ \begin{bmatrix} u = \frac{3}{4}t \\ du = \frac{3}{4}\,dt \end{bmatrix} \rightarrow \frac{1}{3}\int \frac{du}{\sqrt{1-u^2}} = \frac{1}{3}\sin^{-1}u + C = \frac{1}{3}\sin^{-1}\left(\frac{3t}{4}\right) + C$

31. $\int \frac{dt}{9+t^2} = \frac{1}{9}\int \frac{dt}{1+\left(\frac{t}{3}\right)^2}$; $\left[\begin{array}{l} u = \frac{1}{3}t \\ du = \frac{1}{3}dt \end{array}\right] \rightarrow \frac{1}{3}\int \frac{du}{1+u^2} = \frac{1}{3}\tan^{-1}u + C = \frac{1}{3}\tan^{-1}\left(\frac{t}{3}\right) + C$

33. $\int \frac{4\,dx}{5x\sqrt{25x^2-16}} = \frac{4}{25}\int \frac{dx}{x\sqrt{x^2 - \frac{16}{25}}} = \frac{1}{5}\sec^{-1}\left|\frac{5x}{4}\right| + C$

35. $\int \frac{dx}{\sqrt{4x-x^2}} = \int \frac{d(x-2)}{\sqrt{4-(x-2)^2}} = \sin^{-1}\left(\frac{x-2}{2}\right) + C$

37. $\int \frac{dy}{y^2-4y+8} = \int \frac{d(y-2)}{(y-2)^2+4} = \frac{1}{2}\tan^{-1}\left(\frac{y-2}{2}\right) + C$

39. $\int \frac{dx}{(x-1)\sqrt{x^2-2x}} = \int \frac{d(x-1)}{(x-1)\sqrt{(x-1)^2-1}} = \sec^{-1}|x-1| + C$

41. $\int \sin^2 x\,dx = \int \frac{1-\cos 2x}{2}\,dx = \frac{x}{2} - \frac{\sin 2x}{4} + C$

43. $\int \sin^3 \frac{\theta}{2}\,d\theta = \int \left(1-\cos^2 \frac{\theta}{2}\right)\left(\sin \frac{\theta}{2}\right)d\theta$; $\left[\begin{array}{l} u = \cos \frac{\theta}{2} \\ du = -\frac{1}{2}\sin \frac{\theta}{2}\,d\theta \end{array}\right] \rightarrow -2\int \left(1-u^2\right)du = \frac{2u^3}{3} - 2u + C$

$= \frac{2}{3}\cos^3 \frac{\theta}{2} - 2\cos \frac{\theta}{2} + C$

45. $\int \tan^3 2t\,dt = \int (\tan 2t)(\sec^2 2t - 1)\,dt = \int \tan 2t \sec^2 2t\,dt - \int \tan 2t\,dt$; $\left[\begin{array}{l} u = 2t \\ du = 2\,dt \end{array}\right]$

$\rightarrow \frac{1}{2}\int \tan u \sec^2 u\,du - \frac{1}{2}\int \tan u\,du = \frac{1}{4}\tan^2 u + \frac{1}{2}\ln|\cos u| + C = \frac{1}{4}\tan^2 2t + \frac{1}{2}\ln|\cos 2t| + C$

$= \frac{1}{4}\tan^2 2t - \frac{1}{2}\ln|\sec 2t| + C$

47. $\int \frac{dx}{2\sin x \cos x} = \int \frac{dx}{\sin 2x} = \int \csc 2x\,dx = -\frac{1}{2}\ln|\csc 2x + \cot 2x| + C$

49. $\int_{\pi/4}^{\pi/2} \sqrt{\csc^2 y - 1}\,dy = \int_{\pi/4}^{\pi/2} \cot y\,dy = [\ln|\sin y|]_{\pi/4}^{\pi/2} = \ln 1 - \ln \frac{1}{\sqrt{2}} = \ln \sqrt{2}$

51. $\int_0^{\pi} \sqrt{1-\cos^2 2x}\,dx = \int_0^{\pi} |\sin 2x|\,dx = \int_0^{\pi/2} \sin 2x\,dx - \int_{\pi/2}^{\pi} \sin 2x\,dx = -\left[\frac{\cos 2x}{2}\right]_0^{\pi/2} + \left[\frac{\cos 2x}{2}\right]_{\pi/2}^{\pi}$

$= -\left(-\frac{1}{2} - \frac{1}{2}\right) + \left[\frac{1}{2} - \left(-\frac{1}{2}\right)\right] = 2$

53. $\int_{-\pi/2}^{\pi/2} \sqrt{1-\cos 2t}\,dt = \sqrt{2}\int_{-\pi/2}^{\pi/2} |\sin t|\,dt = 2\sqrt{2}\int_0^{\pi/2} \sin t\,dt = \left[-2\sqrt{2}\cos t\right]_0^{\pi/2} = 2\sqrt{2}\,[0-(-1)] = 2\sqrt{2}$

55. $\int \frac{x^2\,dx}{x^2+4} = x - \int \frac{4\,dx}{x^2+4} = x - 2\tan^{-1}\left(\frac{x}{2}\right) + C$

57. $\int \frac{4x^2+3}{2x-1}\,dx = \int \left[(2x+1) + \frac{4}{2x-1}\right]dx = x + x^2 + 2\ln|2x-1| + C$

59. $\int \frac{2y-1}{y^2+4}\,dy = \int \frac{2y\,dy}{y^2+4} - \int \frac{dy}{y^2+4} = \ln(y^2+4) - \frac{1}{2}\tan^{-1}\left(\frac{y}{2}\right) + C$

61. $\int \frac{t+2}{\sqrt{4-t^2}}\,dt = \int \frac{t\,dt}{\sqrt{4-t^2}} + 2\int \frac{dt}{\sqrt{4-t^2}} = -\sqrt{4-t^2} + 2\sin^{-1}\left(\frac{t}{2}\right) + C$

63. $\int \frac{\tan x\, dx}{\tan x + \sec x} = \int \frac{\sin x\, dx}{\sin x + 1} = \int \frac{(\sin x)(1 - \sin x)}{1 - \sin^2 x}\, dx = \int \frac{\sin x - 1 + \cos^2 x}{\cos^2 x}\, dx$

$= -\int \frac{d(\cos x)}{\cos^2 x} - \int \frac{dx}{\cos^2 x} + \int dx = \frac{1}{\cos x} - \tan x + x + C = x - \tan x + \sec x + C$

65. $\int \sec(5 - 3x)\, dx; \begin{bmatrix} y = 5 - 3x \\ dy = -3\, dx \end{bmatrix} \rightarrow \int \sec y \cdot \left(-\frac{dy}{3}\right) = -\frac{1}{3}\int \sec y\, dy = -\frac{1}{3}\ln|\sec y + \tan y| + C$

$= -\frac{1}{3}\ln|\sec(5 - 3x) + \tan(5 - 3x)| + C$

67. $\int \cot\left(\frac{x}{4}\right) dx = 4\int \cot\left(\frac{x}{4}\right) d\left(\frac{x}{4}\right) = 4\ln\left|\sin\left(\frac{x}{4}\right)\right| + C$

69. $\int x\sqrt{1 - x}\, dx; \begin{bmatrix} u = 1 - x \\ du = -dx \end{bmatrix} \rightarrow -\int (1 - u)\sqrt{u}\, du = \int \left(u^{3/2} - u^{1/2}\right) du = \frac{2}{5}u^{5/2} - \frac{2}{3}u^{3/2} + C$

$= \frac{2}{5}(1 - x)^{5/2} - \frac{2}{3}(1 - x)^{3/2} + C = -2\left[\frac{\left(\sqrt{1-x}\right)^3}{3} - \frac{\left(\sqrt{1-x}\right)^5}{5}\right] + C$

71. $\int \sqrt{z^2 + 1}\, dz; \begin{bmatrix} z = \tan\theta \\ dz = \sec^2\theta\, d\theta \end{bmatrix} \rightarrow \int \sqrt{\tan^2\theta + 1} \cdot \sec^2\theta\, d\theta = \int \sec^3\theta\, d\theta$

$= \frac{\sec\theta \tan\theta}{3 - 1} + \frac{3 - 2}{3 - 1}\int \sec\theta\, d\theta$ (FORMULA 92)

$= \frac{\sin\theta}{2\cos^2\theta} + \frac{1}{2}\ln|\sec\theta + \tan\theta| + C = \frac{z\sqrt{z^2 + 1}}{2} + \frac{1}{2}\ln\left|z + \sqrt{1 + z^2}\right| + C$

73. $\int \frac{dy}{\sqrt{25 + y^2}} = \frac{1}{5}\int \frac{dy}{\sqrt{1 + \left(\frac{y}{5}\right)^2}} = \int \frac{du}{\sqrt{1 + u^2}}, \left[u = \frac{y}{5}\right]; \begin{bmatrix} u = \tan\theta \\ du = \sec^2\theta\, d\theta \end{bmatrix} \rightarrow \int \frac{\sec^2\theta\, d\theta}{\sqrt{1 + \tan^2\theta}} = \int \sec\theta\, d\theta$

$= \ln|\sec\theta + \tan\theta| + C_1 = \ln\left|\sqrt{1 + u^2} + u\right| + C_1 = \ln\left|\sqrt{1 + \left(\frac{y}{5}\right)^2} + \frac{y}{5}\right| + C_1 = \ln\left|\frac{\sqrt{25 + y^2} + y}{5}\right| + C_1$

$= \ln\left|y + \sqrt{25 + y^2}\right| + C$

75. $\int \frac{dx}{x^2\sqrt{1 - x^2}}; \begin{bmatrix} x = \sin\theta \\ dx = \cos\theta\, d\theta \end{bmatrix} \rightarrow \int \frac{\cos\theta\, d\theta}{\sin^2\theta \cos\theta} = \int \csc^2\theta\, d\theta = -\cot\theta + C = \frac{-\sqrt{1 - x^2}}{x} + C$

77. $\int \frac{x^2\, dx}{\sqrt{1 - x^2}}; \begin{bmatrix} x = \sin\theta \\ dx = \cos\theta\, d\theta \end{bmatrix} \rightarrow \int \frac{\sin^2\theta \cos\theta\, d\theta}{\cos\theta} = \int \sin^2\theta\, d\theta = \int \frac{1 - \cos 2\theta}{2}\, d\theta = \frac{1}{2}\theta - \frac{1}{4}\sin 2\theta + C$

$= \frac{1}{2}\theta - \frac{1}{2}\sin\theta \cos\theta = \frac{\sin^{-1} x}{2} - \frac{x\sqrt{1 - x^2}}{2} + C$

79. $\int \frac{dx}{\sqrt{x^2 - 9}}; \begin{bmatrix} x = 3\sec\theta \\ dx = 3\sec\theta \tan\theta\, d\theta \end{bmatrix} \rightarrow \int \frac{3\sec\theta \tan\theta\, d\theta}{\sqrt{9\sec^2\theta - 9}} = \int \frac{3\sec\theta \tan\theta\, d\theta}{3\tan\theta} = \int \sec\theta\, d\theta$

$= \ln|\sec\theta + \tan\theta| + C_1 = \ln\left|\frac{x}{3} + \sqrt{\left(\frac{x}{3}\right)^2 - 1}\right| + C_1 = \ln\left|\frac{x + \sqrt{x^2 - 9}}{3}\right| + C_1 = \ln\left|x + \sqrt{x^2 - 9}\right| + C$

81. $\int \frac{\sqrt{w^2 - 1}}{w}\, dw; \begin{bmatrix} w = \sec\theta \\ dw = \sec\theta \tan\theta\, d\theta \end{bmatrix} \rightarrow \int \left(\frac{\tan\theta}{\sec\theta}\right) \cdot \sec\theta \tan\theta\, d\theta = \int \tan^2\theta\, d\theta = \int (\sec^2\theta - 1)\, d\theta$

$= \tan\theta - \theta + C = \sqrt{w^2 - 1} - \sec^{-1} w + C$

83. $u = \ln(x + 1), du = \frac{dx}{x + 1}; dv = dx, v = x;$

$\int \ln(x + 1)\, dx = x\ln(x + 1) - \int \frac{x}{x + 1}\, dx = x\ln(x + 1) - \int dx + \int \frac{dx}{x + 1} = x\ln(x + 1) - x + \ln(x + 1) + C_1$

$= (x + 1)\ln(x + 1) - x + C_1 = (x + 1)\ln(x + 1) - (x + 1) + C, \text{ where } C = C_1 + 1$

85. $u = \tan^{-1} 3x,\ du = \frac{3\,dx}{1+9x^2}$; $dv = dx,\ v = x$;

$$\int \tan^{-1} 3x\,dx = x \tan^{-1} 3x - \int \frac{3x\,dx}{1+9x^2}\ ;\ \begin{bmatrix} y = 1+9x^2 \\ dy = 18x\,dx \end{bmatrix} \rightarrow x \tan^{-1} 3x - \frac{1}{6}\int \frac{dy}{y}$$

$$= x \tan^{-1}(3x) - \frac{1}{6}\ln(1+9x^2) + C$$

87.

$$e^x$$

$(x+1)^2 \xrightarrow{(+)} e^x$

$2(x+1) \xrightarrow{(-)} e^x$

$2 \xrightarrow{(+)} e^x$

$0 \qquad\qquad \Rightarrow \int (x+1)^2 e^x\,dx = [(x+1)^2 - 2(x+1) + 2]\,e^x + C$

89. $u = \cos 2x,\ du = -2\sin 2x\,dx;\ dv = e^x\,dx,\ v = e^x$;

$I = \int e^x \cos 2x\,dx = e^x \cos 2x + 2\int e^x \sin 2x\,dx$;

$u = \sin 2x,\ du = 2\cos 2x\,dx;\ dv = e^x\,dx,\ v = e^x$;

$I = e^x \cos 2x + 2\left[e^x \sin 2x - 2\int e^x \cos 2x\,dx \right] = e^x \cos 2x + 2e^x \sin 2x - 4I \Rightarrow I = \frac{e^x \cos 2x}{5} + \frac{2e^x \sin 2x}{5} + C$

91. $\int \frac{x\,dx}{x^2 - 3x + 2} = \int \frac{2\,dx}{x-2} - \int \frac{dx}{x-1} = 2\ln|x-2| - \ln|x-1| + C$

93. $\int \frac{dx}{x(x+1)^2} = \int \left(\frac{1}{x} - \frac{1}{x+1} + \frac{-1}{(x+1)^2} \right) dx = \ln|x| - \ln|x+1| + \frac{1}{x+1} + C$

95. $\int \frac{\sin\theta\,d\theta}{\cos^2\theta + \cos\theta - 2}$; $[\cos\theta = y] \rightarrow -\int \frac{dy}{y^2 + y - 2} = -\frac{1}{3}\int \frac{dy}{y-1} + \frac{1}{3}\int \frac{dy}{y+2} = \frac{1}{3}\ln\left|\frac{y+2}{y-1}\right| + C$

$= \frac{1}{3}\ln\left|\frac{\cos\theta+2}{\cos\theta-1}\right| + C = -\frac{1}{3}\ln\left|\frac{\cos\theta-1}{\cos\theta+2}\right| + C$

97. $\int \frac{3x^2 + 4x + 4}{x^3 + x}\,dx = \int \frac{4}{x}\,dx - \int \frac{x-4}{x^2+1}\,dx = 4\ln|x| - \frac{1}{2}\ln(x^2+1) + 4\tan^{-1} x + C$

99. $\int \frac{(v+3)\,dv}{2v^3 - 8v} = \frac{1}{2}\int \left(-\frac{3}{4v} + \frac{5}{8(v-2)} + \frac{1}{8(v+2)} \right) dv = -\frac{3}{8}\ln|v| + \frac{5}{16}\ln|v-2| + \frac{1}{16}\ln|v+2| + C$

$= \frac{1}{16}\ln\left| \frac{(v-2)^5 (v+2)}{v^6} \right| + C$

101. $\int \frac{dt}{t^4 + 4t^2 + 3} = \frac{1}{2}\int \frac{dt}{t^2+1} - \frac{1}{2}\int \frac{dt}{t^2+3} = \frac{1}{2}\tan^{-1} t - \frac{1}{2\sqrt{3}}\tan^{-1}\left(\frac{t}{\sqrt{3}} \right) + C = \frac{1}{2}\tan^{-1} t - \frac{\sqrt{3}}{6}\tan^{-1}\frac{t}{\sqrt{3}} + C$

103. $\int \frac{x^3 + x^2}{x^2 + x - 2}\,dx = \int \left(x + \frac{2x}{x^2 + x - 2} \right) dx = \int x\,dx + \frac{2}{3}\int \frac{dx}{x-1} + \frac{4}{3}\int \frac{dx}{x+2}$

$= \frac{x^2}{2} + \frac{4}{3}\ln|x+2| + \frac{2}{3}\ln|x-1| + C$

105. $\int \frac{x^3 + 4x^2}{x^2 + 4x + 3}\,dx = \int \left(x - \frac{3x}{x^2 + 4x + 3} \right) dx = \int x\,dx + \frac{3}{2}\int \frac{dx}{x+1} - \frac{9}{2}\int \frac{dx}{x+3}$

$= \frac{x^2}{2} - \frac{9}{2}\ln|x+3| + \frac{3}{2}\ln|x+1| + C$

107. $\int \frac{dx}{x(3\sqrt{x+1})}$; $\begin{bmatrix} u = \sqrt{x+1} \\ du = \frac{dx}{2\sqrt{x+1}} \\ dx = 2u\,du \end{bmatrix} \rightarrow \frac{2}{3}\int \frac{u\,du}{(u^2-1)u} = \frac{1}{3}\int \frac{du}{u-1} - \frac{1}{3}\int \frac{du}{u+1} = \frac{1}{3}\ln|u-1| - \frac{1}{3}\ln|u+1| + C$

$= \frac{1}{3}\ln\left| \frac{\sqrt{x+1}-1}{\sqrt{x+1}+1} \right| + C$

109. $\int \frac{ds}{e^s - 1}$; $\begin{bmatrix} u = e^s - 1 \\ du = e^s\, ds \\ ds = \frac{du}{u + 1} \end{bmatrix}$ $\rightarrow$ $\int \frac{du}{u(u + 1)} = -\int \frac{du}{u+1} + \int \frac{du}{u} = \ln \left| \frac{u}{u + 1} \right| + C = \ln \left| \frac{e^s - 1}{e^s} \right| + C = \ln \left| 1 - e^{-s} \right| + C$

111. (a) $\int \frac{y\, dy}{\sqrt{16 - y^2}} = -\frac{1}{2} \int \frac{d(16 - y^2)}{\sqrt{16 - y^2}} = -\sqrt{16 - y^2} + C$

(b) $\int \frac{y\, dy}{\sqrt{16 - y^2}}$; $[y = 4 \sin x]$ $\rightarrow$ $4 \int \frac{\sin x \cos x\, dx}{\cos x} = -4 \cos x + C = -\frac{4\sqrt{16 - y^2}}{4} + C = -\sqrt{16 - y^2} + C$

113. (a) $\int \frac{x\, dx}{4 - x^2} = -\frac{1}{2} \int \frac{d(4 - x^2)}{4 - x^2} = -\frac{1}{2} \ln |4 - x^2| + C$

(b) $\int \frac{x\, dx}{4 - x^2}$; $[x = 2 \sin \theta]$ $\rightarrow$ $\int \frac{2 \sin \theta \cdot 2 \cos \theta\, d\theta}{4 \cos^2 \theta} = \int \tan \theta\, d\theta = -\ln |\cos \theta| + C = -\ln \left(\frac{\sqrt{4 - x^2}}{2} \right) + C$

$= -\frac{1}{2} \ln |4 - x^2| + C$

115. $\int \frac{x\, dx}{9 - x^2}$; $\begin{bmatrix} u = 9 - x^2 \\ du = -2x\, dx \end{bmatrix}$ $\rightarrow$ $-\frac{1}{2} \int \frac{du}{u} = -\frac{1}{2} \ln |u| + C = \ln \frac{1}{\sqrt{u}} + C = \ln \frac{1}{\sqrt{9 - x^2}} + C$

117. $\int \frac{dx}{9 - x^2} = \frac{1}{6} \int \frac{dx}{3 - x} + \frac{1}{6} \int \frac{dx}{3 + x} = -\frac{1}{6} \ln |3 - x| + \frac{1}{6} \ln |3 + x| + C = \frac{1}{6} \ln \left| \frac{x + 3}{x - 3} \right| + C$

119. $\int \sin^3 x \cos^4 x\, dx = \int \cos^4 x (1 - \cos^2 x) \sin x\, dx = \int \cos^4 x \sin x\, dx - \int \cos^6 x \sin x\, dx = -\frac{\cos^5 x}{5} + \frac{\cos^7 x}{7} + C$

121. $\int \tan^4 x \sec^2 x\, dx = \frac{\tan^5 x}{5} + C$

123. $\int \sin 5\theta \cos 6\theta\, d\theta = \frac{1}{2} \int (\sin(-\theta) + \sin(11\theta))\, d\theta = \frac{1}{2} \int \sin(-\theta)\, d\theta + \frac{1}{2} \int \sin(11\theta)\, d\theta = \frac{1}{2} \cos(-\theta) - \frac{1}{22} \cos 11\theta + C$

$= \frac{1}{2} \cos \theta - \frac{1}{22} \cos 11\theta + C$

125. $\int \sqrt{1 + \cos\left(\frac{t}{2}\right)}\, dt = \int \sqrt{2} \left| \cos \frac{t}{4} \right|\, dt = 4\sqrt{2} \left| \sin \frac{t}{4} \right| + C$

127. $|E_s| \le \frac{3 - 1}{180} (\triangle x)^4 M$ where $\triangle x = \frac{3 - 1}{n} = \frac{2}{n}$; $f(x) = \frac{1}{x} = x^{-1}$ $\Rightarrow$ $f'(x) = -x^{-2}$ $\Rightarrow$ $f''(x) = 2x^{-3}$ $\Rightarrow$ $f'''(x) = -6x^{-4}$

$\Rightarrow$ $f^{(4)}(x) = 24x^{-5}$ which is decreasing on $[1, 3]$ $\Rightarrow$ maximum of $f^{(4)}(x)$ on $[1, 3]$ is $f^{(4)}(1) = 24$ $\Rightarrow$ $M = 24$. Then

$|E_s| \le 0.0001$ $\Rightarrow$ $\left(\frac{3 - 1}{180} \right) \left(\frac{2}{n} \right)^4 (24) \le 0.0001$ $\Rightarrow$ $\left(\frac{768}{180} \right) \left(\frac{1}{n^4} \right) \le 0.0001$ $\Rightarrow$ $\frac{1}{n^4} \le (0.0001) \left(\frac{180}{768} \right)$ $\Rightarrow$ $n^4 \ge 10{,}000 \left(\frac{768}{180} \right)$

$\Rightarrow$ $n \ge 14.37$ $\Rightarrow$ $n \ge 16$ (n must be even)

129. $\triangle x = \frac{b - a}{n} = \frac{\pi - 0}{6} = \frac{\pi}{6}$ $\Rightarrow$ $\frac{\triangle x}{2} = \frac{\pi}{12}$;

$\sum_{i=0}^{6} mf(x_i) = 12$ $\Rightarrow$ $T = \left(\frac{\pi}{12} \right)(12) = \pi$;

	x_i	$f(x_i)$	m	$mf(x_i)$
x_0	0	0	1	0
x_1	$\pi/6$	1/2	2	1
x_2	$\pi/3$	3/2	2	3
x_3	$\pi/2$	2	2	4
x_4	$2\pi/3$	3/2	2	3
x_5	$5\pi/6$	1/2	2	1
x_6	π	0	1	0

$\sum_{i=0}^{6} mf(x_i) = 18$ and $\frac{\triangle x}{3} = \frac{\pi}{18}$ $\Rightarrow$

$S = \left(\frac{\pi}{18} \right)(18) = \pi$.

	x_i	$f(x_i)$	m	$mf(x_i)$
x_0	0	0	1	0
x_1	$\pi/6$	1/2	4	2
x_2	$\pi/3$	3/2	2	3
x_3	$\pi/2$	2	4	8
x_4	$2\pi/3$	3/2	2	3
x_5	$5\pi/6$	1/2	4	2
x_6	π	0	1	0

131. $y_{av} = \frac{1}{365-0} \int_0^{365} \left[37 \sin\left(\frac{2\pi}{365}(x-101)\right) + 25\right] dx = \frac{1}{365}\left[-37\left(\frac{365}{2\pi}\right)\cos\left(\frac{2\pi}{365}(x-101)\right) + 25x\right]_0^{365}$

$= \frac{1}{365}\left[\left(-37\left(\frac{365}{2\pi}\right)\cos\left[\frac{2\pi}{365}(365-101)\right] + 25(365)\right) - \left(-37\left(\frac{365}{2\pi}\right)\cos\left[\frac{2\pi}{365}(0-101)\right] + 25(0)\right)\right]$

$= -\frac{37}{2\pi}\cos\left(\frac{2\pi}{365}(264)\right) + 25 + \frac{37}{2\pi}\cos\left(\frac{2\pi}{365}(-101)\right) = -\frac{37}{2\pi}\left(\cos\left(\frac{2\pi}{365}(264)\right) - \cos\left(\frac{2\pi}{365}(-101)\right)\right) + 25$

$\approx -\frac{37}{2\pi}(0.16705 - 0.16705) + 25 = 25°\,F$

133. (a) Each interval is 5 min $= \frac{1}{12}$ hour.

$\frac{1}{24}[2.5 + 2(2.4) + 2(2.3) + \ldots + 2(2.4) + 2.3] = \frac{29}{12} \approx 2.42$ gal

(b) $(60 \text{ mph})\left(\frac{12}{29} \text{ hours/gal}\right) \approx 24.83$ mi/gal

135. $\int_0^3 \frac{dx}{\sqrt{9-x^2}} = \lim_{b\to 3^-} \int_0^b \frac{dx}{\sqrt{9-x^2}} = \lim_{b\to 3^-}\left[\sin^{-1}\left(\frac{x}{3}\right)\right]_0^b = \lim_{b\to 3^-} \sin^{-1}\left(\frac{b}{3}\right) - \sin^{-1}\left(\frac{0}{3}\right) = \frac{\pi}{2} - 0 = \frac{\pi}{2}$

137. $\int_{-1}^1 \frac{dy}{y^{2/3}} = \int_{-1}^0 \frac{dy}{y^{2/3}} + \int_0^1 \frac{dy}{y^{2/3}} = 2\int_0^1 \frac{dy}{y^{2/3}} = 2\cdot 3 \lim_{b\to 0^+} \left[y^{1/3}\right]_b^1 = 6\left(1 - \lim_{b\to 0^+} b^{1/3}\right) = 6$

139. $\int_3^\infty \frac{2\,du}{u^2 - 2u} = \int_3^\infty \frac{du}{u-2} - \int_3^\infty \frac{du}{u} = \lim_{b\to\infty}\left[\ln\left|\frac{u-2}{u}\right|\right]_3^b = \lim_{b\to\infty}\left[\ln\left|\frac{b-2}{b}\right|\right] - \ln\left|\frac{3-2}{3}\right| = 0 - \ln\left(\frac{1}{3}\right) = \ln 3$

141. $\int_0^\infty x^2 e^{-x}\,dx = \lim_{b\to\infty}\left[-x^2 e^{-x} - 2xe^{-x} - 2e^{-x}\right]_0^b = \lim_{b\to\infty}(-b^2 e^{-b} - 2be^{-b} - 2e^{-b}) - (-2) = 0 + 2 = 2$

143. $\int_{-\infty}^\infty \frac{dx}{4x^2+9} = 2\int_0^\infty \frac{dx}{4x^2+9} = \frac{1}{2}\int_0^\infty \frac{dx}{x^2+\frac{9}{4}} = \frac{1}{2}\lim_{b\to\infty}\left[\frac{2}{3}\tan^{-1}\left(\frac{2x}{3}\right)\right]_0^b = \frac{1}{2}\lim_{b\to\infty}\left[\frac{2}{3}\tan^{-1}\left(\frac{2b}{3}\right)\right] - \frac{1}{3}\tan^{-1}(0)$

$= \frac{1}{2}\left(\frac{2}{3}\cdot\frac{\pi}{2}\right) - 0 = \frac{\pi}{6}$

145. $\lim_{\theta\to\infty} \frac{\theta}{\sqrt{\theta^2+1}} = 1$ and $\int_6^\infty \frac{d\theta}{\theta}$ diverges $\Rightarrow \int_6^\infty \frac{d\theta}{\sqrt{\theta^2+1}}$ diverges

147. $\int_1^\infty \frac{\ln z}{z}\,dz = \int_1^e \frac{\ln z}{z}\,dz + \int_e^\infty \frac{\ln z}{z}\,dz = \left[\frac{(\ln z)^2}{2}\right]_1^e + \lim_{b\to\infty}\left[\frac{(\ln z)^2}{2}\right]_e^b = \left(\frac{1^2}{2} - 0\right) + \lim_{b\to\infty}\left[\frac{(\ln b)^2}{2} - \frac{1}{2}\right]$

$= \infty \Rightarrow$ diverges

149. $\int_{-\infty}^\infty \frac{2\,dx}{e^x + e^{-x}} = 2\int_0^\infty \frac{2\,dx}{e^x + e^{-x}} < \int_0^\infty \frac{4\,dx}{e^x}$ converges $\Rightarrow \int_{-\infty}^\infty \frac{2\,dx}{e^x + e^{-x}}$ converges

151. $\int \frac{x\,dx}{1+\sqrt{x}}; \begin{bmatrix} u = \sqrt{x} \\ du = \frac{dx}{2\sqrt{x}} \end{bmatrix} \to \int \frac{u^2 \cdot 2u\,du}{1+u} = \int \left(2u^2 - 2u + 2 - \frac{2}{1+u}\right) du = \frac{2}{3}u^3 - u^2 + 2u - 2\ln|1+u| + C$

$= \frac{2x^{3/2}}{3} - x + 2\sqrt{x} - 2\ln(1+\sqrt{x}) + C$

153. $\int \frac{dx}{x(x^2+1)^2}; \begin{bmatrix} x = \tan\theta \\ dx = \sec^2\theta\,d\theta \end{bmatrix} \to \int \frac{\sec^2\theta\,d\theta}{\tan\theta\sec^4\theta} = \int \frac{\cos^3\theta\,d\theta}{\sin\theta} = \int\left(\frac{1-\sin^2\theta}{\sin\theta}\right) d(\sin\theta)$

$= \ln|\sin\theta| - \frac{1}{2}\sin^2\theta + C = \ln\left|\frac{x}{\sqrt{x^2+1}}\right| - \frac{1}{2}\left(\frac{x}{\sqrt{x^2+1}}\right)^2 + C$

155. $\int \frac{dx}{\sqrt{-2x-x^2}} = \int \frac{d(x+1)}{\sqrt{1-(x+1)^2}} = \sin^{-1}(x+1) + C$

157. $\int \frac{du}{\sqrt{1+u^2}}; [u = \tan\theta] \to \int \frac{\sec^2\theta\,d\theta}{\sec\theta} = \ln|\sec\theta + \tan\theta| + C = \ln\left|\sqrt{1+u^2} + u\right| + C$

159. $\int \frac{2 - \cos x + \sin x}{\sin^2 x}\,dx = \int 2\csc^2 x\,dx - \int \frac{\cos x\,dx}{\sin^2 x} + \int \csc x\,dx = -2\cot x + \frac{1}{\sin x} - \ln|\csc x + \cot x| + C$

$= -2\cot x + \csc x - \ln|\csc x + \cot x| + C$

161. $\int \frac{9\,dv}{81-v^4} = \frac{1}{2}\int \frac{dv}{v^2+9} + \frac{1}{12}\int \frac{dv}{3-v} + \frac{1}{12}\int \frac{dv}{3+v} = \frac{1}{12}\ln\left|\frac{3+v}{3-v}\right| + \frac{1}{6}\tan^{-1}\frac{v}{3} + C$

163.

$\cos(2\theta+1)$

$\theta \xrightarrow{\quad(+)\quad} \frac{1}{2}\sin(2\theta+1)$

$1 \xrightarrow{\quad(-)\quad} -\frac{1}{4}\cos(2\theta+1)$

$0 \qquad\qquad\qquad\qquad \Rightarrow \int \theta\cos(2\theta+1)\,d\theta = \frac{\theta}{2}\sin(2\theta+1) + \frac{1}{4}\cos(2\theta+1) + C$

165. $\int \frac{x^3\,dx}{x^2-2x+1} = \int\left(x+2+\frac{3x-2}{x^2-2x+1}\right)dx = \int(x+2)\,dx + 3\int\frac{dx}{x-1} + \int\frac{dx}{(x-1)^2}$

$\qquad = \frac{x^2}{2} + 2x + 3\ln|x-1| - \frac{1}{x-1} + C$

167. $\int \frac{2\sin\sqrt{x}\,dx}{\sqrt{x}\sec\sqrt{x}};\ \begin{bmatrix} y=\sqrt{x} \\ dy=\frac{dx}{2\sqrt{x}} \end{bmatrix} \rightarrow \int \frac{2\sin y\cdot 2y\,dy}{y\sec y} = \int 2\sin 2y\,dy = -\cos(2y) + C = -\cos\left(2\sqrt{x}\right) + C$

169. $\int \frac{dy}{\sin y\cos y} = \int \frac{2\,dy}{\sin 2y} = \int 2\csc(2y)\,dy = -\ln|\csc(2y) + \cot(2y)| + C$

171. $\int \frac{\tan x}{\cos^2 x}\,dx = \int \tan x\sec^2 x\,dx = \int \tan x\cdot d(\tan x) = \frac{1}{2}\tan^2 x + C$

173. $\int \frac{(r+2)\,dr}{\sqrt{-r^2-4r}} = \int \frac{(r+2)\,dr}{\sqrt{4-(r+2)^2}};\ \begin{bmatrix} u=4-(r+2)^2 \\ du=-2(r+2)\,dr \end{bmatrix} \rightarrow -\int\frac{du}{2\sqrt{u}} = -\sqrt{u} + C = -\sqrt{4-(r+2)^2} + C$

175. $\int \frac{\sin 2\theta\,d\theta}{(1+\cos 2\theta)^2} = -\frac{1}{2}\int\frac{d(1+\cos 2\theta)}{(1+\cos 2\theta)^2} = \frac{1}{2(1+\cos 2\theta)} + C = \frac{1}{4}\sec^2\theta + C$

177. $\int_{\pi/4}^{\pi/2}\sqrt{1+\cos 4x}\,dx = -\sqrt{2}\int_{\pi/4}^{\pi/2}\cos 2x\,dx = \left[-\frac{\sqrt{2}}{2}\sin 2x\right]_{\pi/4}^{\pi/2} = \frac{\sqrt{2}}{2}$

179. $\int \frac{x\,dx}{\sqrt{2-x}};\ \begin{bmatrix} y=2-x \\ dy=-dx \end{bmatrix} \rightarrow -\int\frac{(2-y)\,dy}{\sqrt{y}} = \frac{2}{3}y^{3/2} - 4y^{1/2} + C = \frac{2}{3}(2-x)^{3/2} - 4(2-x)^{1/2} + C$

$\qquad = 2\left[\frac{\left(\sqrt{2-x}\right)^3}{3} - 2\sqrt{2-x}\right] + C$

181. $\int \frac{dy}{y^2-2y+2} = \int \frac{d(y-1)}{(y-1)^2+1} = \tan^{-1}(y-1) + C$

183. $\int \theta^2\tan(\theta^3)\,d\theta = \frac{1}{3}\int\tan(\theta^3)\,d(\theta^3) = \frac{1}{3}\ln|\sec\theta^3| + C$

185. $\int \frac{z+1}{z^2(z^2+4)}\,dz = \frac{1}{4}\int\left(\frac{1}{z} + \frac{1}{z^2} - \frac{z+1}{z^2+4}\right)dz = \frac{1}{4}\ln|z| - \frac{1}{4z} - \frac{1}{8}\ln(z^2+4) - \frac{1}{8}\tan^{-1}\frac{z}{2} + C$

187. $\int \frac{t\,dt}{\sqrt{9-4t^2}} = -\frac{1}{8}\int\frac{d(9-4t^2)}{\sqrt{9-4t^2}} = -\frac{1}{4}\sqrt{9-4t^2} + C$

189. $\int \frac{\cot\theta\,d\theta}{1+\sin^2\theta} = \int \frac{\cos\theta\,d\theta}{(\sin\theta)(1+\sin^2\theta)};\ \begin{bmatrix} x=\sin\theta \\ dx=\cos\theta\,d\theta \end{bmatrix} \rightarrow \int\frac{dx}{x(1+x^2)} = \int\frac{dx}{x} - \int\frac{x\,dx}{x^2+1}$

$\qquad = \ln|\sin\theta| - \frac{1}{2}\ln(1+\sin^2\theta) + C$

191. $\int \frac{\tan\sqrt{y}\,dy}{2\sqrt{y}};\ \left[\sqrt{y}=x\right] \rightarrow \int \frac{\tan x\cdot 2x\,dx}{2x} = \ln|\sec x| + C = \ln\left|\sec\sqrt{y}\right| + C$

193. $\int \frac{\theta^2 \, d\theta}{4 - \theta^2} = \int \left(-1 + \frac{4}{4 - \theta^2}\right) d\theta = -\int d\theta - \int \frac{d\theta}{\theta - 2} + \int \frac{d\theta}{\theta + 2} = -\theta - \ln|\theta - 2| + \ln|\theta + 2| + C$

  $= -\theta + \ln\left|\frac{\theta + 2}{\theta - 2}\right| + C$

195. $\int \frac{\cos(\sin^{-1} x)\, dx}{\sqrt{1 - x^2}} ; \quad \begin{bmatrix} u = \sin^{-1} x \\ du = \frac{dx}{\sqrt{1 - x^2}} \end{bmatrix} \rightarrow \int \cos u \, du = \sin u + C = \sin(\sin^{-1} x) + C = x + C$

197. $\int \sin \frac{x}{2} \cos \frac{x}{2} \, dx = \int \frac{1}{2} \sin\left(\frac{x}{2} + \frac{x}{2}\right) dx = \frac{1}{2} \int \sin x \, dx = -\frac{1}{2} \cos x + C$

199. $\int \frac{e^t \, dt}{1 + e^t} = \ln(1 + e^t) + C$

201. $\int_1^\infty \frac{\ln y \, dy}{y^3} ; \quad \begin{bmatrix} x = \ln y \\ dx = \frac{dy}{y} \\ dy = e^x \, dx \end{bmatrix} \rightarrow \int_0^\infty \frac{x \cdot e^x}{e^{3x}} \, dx = \int_0^\infty x e^{-2x} \, dx = \lim_{b \to \infty} \left[-\frac{x}{2} e^{-2x} - \frac{1}{4} e^{-2x}\right]_0^b$

  $= \lim_{b \to \infty} \left(\frac{-b}{2e^{2b}} - \frac{1}{4e^{2b}}\right) - \left(0 - \frac{1}{4}\right) = \frac{1}{4}$

203. $\int \frac{\cot v \, dv}{\ln(\sin v)} = \int \frac{\cos v \, dv}{(\sin v) \ln(\sin v)} ; \quad \begin{bmatrix} u = \ln(\sin v) \\ du = \frac{\cos v \, dv}{\sin v} \end{bmatrix} \rightarrow \int \frac{du}{u} = \ln|u| + C = \ln|\ln(\sin v)| + C$

205. $\int e^{\ln \sqrt{x}} \, dx = \int \sqrt{x} \, dx = \frac{2}{3} x^{3/2} + C$

207. $\int \frac{\sin 5t \, dt}{1 + (\cos 5t)^2} ; \quad \begin{bmatrix} u = \cos 5t \\ du = -5 \sin 5t \, dt \end{bmatrix} \rightarrow -\frac{1}{5} \int \frac{du}{1 + u^2} = -\frac{1}{5} \tan^{-1} u + C = -\frac{1}{5} \tan^{-1}(\cos 5t) + C$

209. $\int (27)^{3\theta + 1} \, d\theta = \frac{1}{3} \int (27)^{3\theta + 1} \, d(3\theta + 1) = \frac{1}{3 \ln 27} (27)^{3\theta + 1} + C = \frac{1}{3} \left(\frac{27^{3\theta + 1}}{\ln 27}\right) + C$

211. $\int \frac{dr}{1 + \sqrt{r}} ; \quad \begin{bmatrix} u = \sqrt{r} \\ du = \frac{dr}{2\sqrt{r}} \end{bmatrix} \rightarrow \int \frac{2u \, du}{1 + u} = \int \left(2 - \frac{2}{1 + u}\right) du = 2u - 2 \ln|1 + u| + C = 2\sqrt{r} - 2 \ln(1 + \sqrt{r}) + C$

213. $\int \frac{8 \, dy}{y^3(y + 2)} = \int \frac{dy}{y} - \int \frac{2 \, dy}{y^2} + \int \frac{4 \, dy}{y^3} - \int \frac{dy}{(y + 2)} = \ln\left|\frac{y}{y + 2}\right| + \frac{2}{y} - \frac{2}{y^2} + C$

215. $\int \frac{8 \, dm}{m\sqrt{49m^2 - 4}} = \frac{8}{7} \int \frac{dm}{m\sqrt{m^2 - \left(\frac{2}{7}\right)^2}} = 4 \sec^{-1}\left|\frac{7m}{2}\right| + C$

217. If $u = \int_0^x \sqrt{1 + (t - 1)^4} \, dt$ and $dv = 3(x - 1)^2 \, dx$, then $du = \sqrt{1 + (x - 1)^4} \, dx$, and $v = (x - 1)^3$ so integration

  by parts $\Rightarrow \int_0^1 3(x - 1)^2 \left[\int_0^x \sqrt{1 + (t - 1)^4} \, dt\right] dx = \left[(x - 1)^3 \int_0^x \sqrt{1 + (t - 1)^4} \, dt\right]_0^1$

  $- \int_0^1 (x - 1)^3 \sqrt{1 + (x - 1)^4} \, dx = \left[-\frac{1}{6} (1 + (x - 1)^4)^{3/2}\right]_0^1 = \frac{\sqrt{8} - 1}{6}$

219. $u = f(x), \, du = f'(x) \, dx; \, dv = dx, \, v = x;$

  $\int_{\pi/2}^{3\pi/2} f(x) \, dx = \left[x \, f(x)\right]_{\pi/2}^{3\pi/2} - \int_{\pi/2}^{3\pi/2} x f'(x) \, dx = \left[\frac{3\pi}{2} f\left(\frac{3\pi}{2}\right) - \frac{\pi}{2} f\left(\frac{\pi}{2}\right)\right] - \int_{\pi/2}^{3\pi/2} \cos x \, dx$

  $= \left(\frac{3\pi b}{2} - \frac{\pi a}{2}\right) - \left[\sin x\right]_{\pi/2}^{3\pi/2} = \frac{\pi}{2}(3b - a) - \left[(-1) - 1\right] = \frac{\pi}{2}(3b - a) + 2$

CHAPTER 8 ADDITIONAL AND ADVANCED EXERCISES

1. $u = \left(\sin^{-1} x\right)^2$, $du = \frac{2 \sin^{-1} x \, dx}{\sqrt{1-x^2}}$; $dv = dx$, $v = x$;

$$\int \left(\sin^{-1} x\right)^2 dx = x \left(\sin^{-1} x\right)^2 - \int \frac{2x \sin^{-1} x \, dx}{\sqrt{1-x^2}} ;$$

$u = \sin^{-1} x$, $du = \frac{dx}{\sqrt{1-x^2}}$; $dv = -\frac{2x \, dx}{\sqrt{1-x^2}}$, $v = 2\sqrt{1-x^2}$;

$$-\int \frac{2x \sin^{-1} x \, dx}{\sqrt{1-x^2}} = 2 \left(\sin^{-1} x\right) \sqrt{1-x^2} - \int 2 \, dx = 2 \left(\sin^{-1} x\right) \sqrt{1-x^2} - 2x + C; \text{ therefore}$$

$$\int \left(\sin^{-1} x\right)^2 dx = x \left(\sin^{-1} x\right)^2 + 2 \left(\sin^{-1} x\right) \sqrt{1-x^2} - 2x + C$$

3. $u = \sin^{-1} x$, $du = \frac{dx}{\sqrt{1-x^2}}$; $dv = x \, dx$, $v = \frac{x^2}{2}$;

$$\int x \sin^{-1} x \, dx = \frac{x^2}{2} \sin^{-1} x - \int \frac{x^2 \, dx}{2\sqrt{1-x^2}} ; \begin{bmatrix} x = \sin \theta \\ dx = \cos \theta \, d\theta \end{bmatrix} \rightarrow \int x \sin^{-1} x \, dx = \frac{x^2}{2} \sin^{-1} x - \int \frac{\sin^2 \theta \cos \theta \, d\theta}{2 \cos \theta}$$

$$= \frac{x^2}{2} \sin^{-1} x - \frac{1}{2} \int \sin^2 \theta \, d\theta = \frac{x^2}{2} \sin^{-1} x - \frac{1}{2} \left(\frac{\theta}{2} - \frac{\sin 2\theta}{4}\right) + C = \frac{x^2}{2} \sin^{-1} x + \frac{\sin \theta \cos \theta - \theta}{4} + C$$

$$= \frac{x^2}{2} \sin^{-1} x + \frac{x\sqrt{1-x^2} - \sin^{-1} x}{4} + C$$

5. $\int \frac{d\theta}{1 - \tan^2 \theta} = \int \frac{\cos^2 \theta}{\cos^2 \theta - \sin^2 \theta} \, d\theta = \int \frac{1 + \cos 2\theta}{2 \cos 2\theta} \, d\theta = \frac{1}{2} \int (\sec 2\theta + 1) \, d\theta = \frac{\ln |\sec 2\theta + \tan 2\theta| + 2\theta}{4} + C$

7. $\int \frac{dt}{t - \sqrt{1-t^2}}$; $\begin{bmatrix} t = \sin \theta \\ dt = \cos \theta \, d\theta \end{bmatrix} \rightarrow \int \frac{\cos \theta \, d\theta}{\sin \theta - \cos \theta} = \int \frac{d\theta}{\tan \theta - 1}$; $\begin{bmatrix} u = \tan \theta \\ du = \sec^2 \theta \, d\theta \\ d\theta = \frac{du}{u^2 + 1} \end{bmatrix} \rightarrow \int \frac{du}{(u-1)(u^2+1)}$

$$= \frac{1}{2} \int \frac{du}{u-1} - \frac{1}{2} \int \frac{du}{u^2+1} - \frac{1}{2} \int \frac{u \, du}{u^2+1} = \frac{1}{2} \ln \left| \frac{u-1}{\sqrt{u^2+1}} \right| - \frac{1}{2} \tan^{-1} u + C = \frac{1}{2} \ln \left| \frac{\tan \theta - 1}{\sec \theta} \right| - \frac{1}{2} \theta + C$$

$$= \frac{1}{2} \ln \left(t - \sqrt{1-t^2} \right) - \frac{1}{2} \sin^{-1} t + C$$

9. $\int \frac{1}{x^4 + 4} \, dx = \int \frac{1}{(x^2+2)^2 - 4x^2} \, dx = \int \frac{1}{(x^2 + 2x + 2)(x^2 - 2x + 2)} \, dx$

$$= \frac{1}{16} \int \left[\frac{2x+2}{x^2+2x+2} + \frac{2}{(x+1)^2+1} - \frac{2x-2}{x^2-2x+2} + \frac{2}{(x-1)^2+1} \right] dx$$

$$= \frac{1}{16} \ln \left| \frac{x^2+2x+2}{x^2-2x+2} \right| + \frac{1}{8} \left[\tan^{-1} (x+1) + \tan^{-1} (x-1) \right] + C$$

11. $\lim\limits_{x \to \infty} \int_{-x}^{x} \sin t \, dt = \lim\limits_{x \to \infty} \left[-\cos t \right]_{-x}^{x} = \lim\limits_{x \to \infty} \left[-\cos x + \cos (-x) \right] = \lim\limits_{x \to \infty} \left(-\cos x + \cos x \right) = \lim\limits_{x \to \infty} 0 = 0$

13. $\lim\limits_{n \to \infty} \sum\limits_{k=1}^{n} \ln \sqrt[n]{1 + \frac{k}{n}} = \lim\limits_{n \to \infty} \sum\limits_{k=1}^{n} \ln \left(1 + k \left(\frac{1}{n}\right)\right) \left(\frac{1}{n}\right) = \int_0^1 \ln (1 + x) \, dx$; $\begin{bmatrix} u = 1 + x, \, du = dx \\ x = 0 \Rightarrow u = 1, x = 1 \Rightarrow u = 2 \end{bmatrix}$

$$\rightarrow \int_1^2 \ln u \, du = \left[u \ln u - u \right]_1^2 = (2 \ln 2 - 2) - (\ln 1 - 1) = 2 \ln 2 - 1 = \ln 4 - 1$$

15. $\frac{dy}{dx} = \sqrt{\cos 2x} \Rightarrow 1 + \left(\frac{dy}{dx}\right)^2 = 1 + \cos 2x = 2 \cos^2 x$; $L = \int_0^{\pi/4} \sqrt{1 + \left(\sqrt{\cos 2t}\right)^2} \, dt = \sqrt{2} \int_0^{\pi/4} \sqrt{\cos^2 t} \, dt$

$$= \sqrt{2} \left[\sin t \right]_0^{\pi/4} = 1$$

17. $V = \int_a^b 2\pi \left(\begin{smallmatrix} \text{shell} \\ \text{radius} \end{smallmatrix}\right) \left(\begin{smallmatrix} \text{shell} \\ \text{height} \end{smallmatrix}\right) dx = \int_0^1 2\pi xy \, dx$

$= 6\pi \int_0^1 x^2 \sqrt{1-x} \, dx; \quad \begin{bmatrix} u = 1 - x \\ du = -dx \\ x^2 = (1-u)^2 \end{bmatrix}$

$\to -6\pi \int_1^0 (1-u)^2 \sqrt{u} \, du$

$= -6\pi \int_1^0 \left(u^{1/2} - 2u^{3/2} + u^{5/2}\right) du$

$= -6\pi \left[\frac{2}{3} u^{3/2} - \frac{4}{5} u^{5/2} + \frac{2}{7} u^{7/2}\right]_1^0 = 6\pi \left(\frac{2}{3} - \frac{4}{5} + \frac{2}{7}\right)$

$= 6\pi \left(\frac{70 - 84 + 30}{105}\right) = 6\pi \left(\frac{16}{105}\right) = \frac{32\pi}{35}$

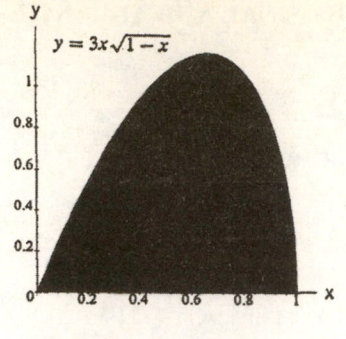

19. $V = \int_a^b 2\pi \left(\begin{smallmatrix} \text{shell} \\ \text{radius} \end{smallmatrix}\right) \left(\begin{smallmatrix} \text{shell} \\ \text{height} \end{smallmatrix}\right) dx = \int_0^1 2\pi xe^x \, dx$

$= 2\pi \left[xe^x - e^x\right]_0^1 = 2\pi$

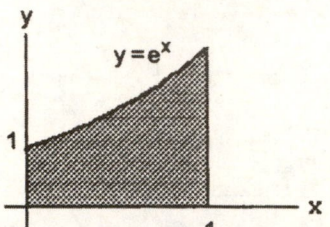

21. (a) $V = \int_1^e \pi \left[1 - (\ln x)^2\right] dx$

$= \pi \left[x - x(\ln x)^2\right]_1^e + 2\pi \int_1^e \ln x \, dx$

(FORMULA 110)

$= \pi \left[x - x(\ln x)^2 + 2(x \ln x - x)\right]_1^e$

$= \pi \left[-x - x(\ln x)^2 + 2x \ln x\right]_1^e$

$= \pi \left[-e - e + 2e - (-1)\right] = \pi$

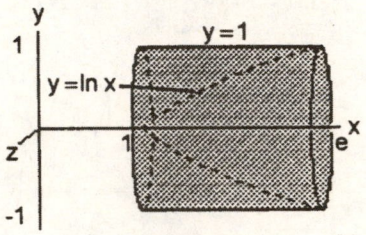

(b) $V = \int_1^e \pi (1 - \ln x)^2 \, dx = \pi \int_1^e \left[1 - 2\ln x + (\ln x)^2\right] dx$

$= \pi \left[x - 2(x \ln x - x) + x(\ln x)^2\right]_1^e - 2\pi \int_1^e \ln x \, dx$

$= \pi \left[x - 2(x \ln x - x) + x(\ln x)^2 - 2(x \ln x - x)\right]_1^e$

$= \pi \left[5x - 4x \ln x + x(\ln x)^2\right]_1^e$

$= \pi \left[(5e - 4e + e) - (5)\right] = \pi(2e - 5)$

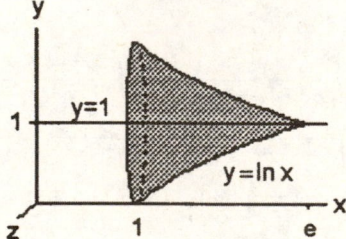

23. (a) $\lim_{x \to 0^+} x \ln x = 0 \Rightarrow \lim_{x \to 0^+} f(x) = 0 = f(0) \Rightarrow f$ is continuous

(b) $V = \int_0^2 \pi x^2 (\ln x)^2 \, dx; \quad \begin{bmatrix} u = (\ln x)^2 \\ du = (2 \ln x) \frac{dx}{x} \\ dv = x^2 dx \\ v = \frac{x^3}{3} \end{bmatrix} \to \pi \left(\lim_{b \to 0^+} \left[\frac{x^3}{3} (\ln x)^2\right]_b^2 - \int_0^2 \left(\frac{x^3}{3}\right) (2 \ln x) \frac{dx}{x}\right)$

$= \pi \left[\left(\frac{8}{3}\right) (\ln 2)^2 - \left(\frac{2}{3}\right) \lim_{b \to 0^+} \left[\frac{x^3}{3} \ln x - \frac{x^3}{9}\right]_b^2\right] = \pi \left[\frac{8(\ln 2)^2}{3} - \frac{16(\ln 2)}{9} + \frac{16}{27}\right]$

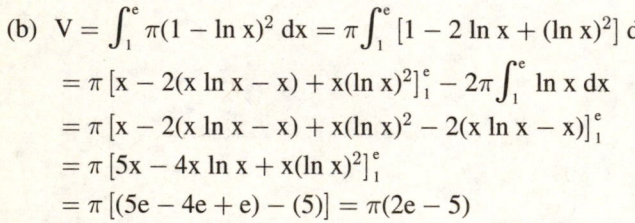

25. $M = \int_1^e \ln x \, dx = [x \ln x - x]_1^e = (e - e) - (0 - 1) = 1;$

$M_x = \int_1^e (\ln x) \left(\frac{\ln x}{2}\right) dx = \frac{1}{2} \int_1^e (\ln x)^2 \, dx$

$= \frac{1}{2} \left([x(\ln x)^2]_1^e - 2 \int_1^e \ln x \, dx\right) = \frac{1}{2} (e - 2);$

$M_y = \int_1^e x \ln x \, dx = \left[\frac{x^2 \ln x}{2}\right]_1^e - \frac{1}{2} \int_1^e x \, dx$

$= \frac{1}{2} \left[x^2 \ln x - \frac{x^2}{2}\right]_1^e = \frac{1}{2} \left[\left(e^2 - \frac{e^2}{2}\right) + \frac{1}{2}\right] = \frac{1}{4} (e^2 + 1);$

therefore, $\bar{x} = \frac{M_y}{M} = \frac{e^2 + 1}{4}$ and $\bar{y} = \frac{M_x}{M} = \frac{e - 2}{2}$

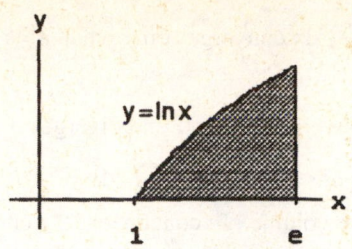

$y = \ln x$

27. $L = \int_1^e \sqrt{1 + \frac{1}{x^2}} \, dx = \int_1^e \frac{\sqrt{x^2 + 1}}{x} \, dx; \begin{bmatrix} x = \tan \theta \\ dx = \sec^2 \theta \, d\theta \end{bmatrix} \rightarrow L = \int_{\pi/4}^{\tan^{-1} e} \frac{\sec \theta \cdot \sec^2 \theta \, d\theta}{\tan \theta}$

$= \int_{\pi/4}^{\tan^{-1} e} \frac{(\sec \theta)(\tan^2 \theta + 1)}{\tan \theta} \, d\theta = \int_{\pi/4}^{\tan^{-1} e} (\tan \theta \sec \theta + \csc \theta) \, d\theta = [\sec \theta - \ln |\csc \theta + \cot \theta|]_{\pi/4}^{\tan^{-1} e}$

$= \left(\sqrt{1 + e^2} - \ln \left|\frac{\sqrt{1 + e^2}}{e} + \frac{1}{e}\right|\right) - \left[\sqrt{2} - \ln \left(1 + \sqrt{2}\right)\right] = \sqrt{1 + e^2} - \ln \left(\frac{\sqrt{1 + e^2}}{e} + \frac{1}{e}\right) - \sqrt{2} + \ln \left(1 + \sqrt{2}\right)$

29. $L = 4 \int_0^1 \sqrt{1 + \left(\frac{dy}{dx}\right)^2} \, dx; x^{2/3} + y^{2/3} = 1 \Rightarrow y = \left(1 - x^{2/3}\right)^{3/2} \Rightarrow \frac{dy}{dx} = -\frac{3}{2} \left(1 - x^{2/3}\right)^{1/2} \left(x^{-1/3}\right) \left(\frac{2}{3}\right)$

$\Rightarrow \left(\frac{dy}{dx}\right)^2 = \frac{1 - x^{2/3}}{x^{2/3}} \Rightarrow L = 4 \int_0^1 \sqrt{1 + \left(\frac{1 - x^{2/3}}{x^{2/3}}\right)} \, dx = 4 \int_0^1 \frac{dx}{x^{1/3}} = 6 \left[x^{2/3}\right]_0^1 = 6$

31. $\left(\frac{dy}{dx}\right)^2 = \frac{1}{4x} \Rightarrow \frac{dy}{dx} = \frac{\pm 1}{2\sqrt{x}} \Rightarrow y = \sqrt{x} \text{ or } y = -\sqrt{x}, 0 \leq x \leq 4$

33. (b) $\int_{-\infty}^{\infty} e^{(x - e^x)} \, dx = \int_{-\infty}^{\infty} e^{(-e^x)} e^x \, dx$

$= \lim_{a \to -\infty} \int_a^0 e^{(-e^x)} e^x \, dx + \lim_{b \to +\infty} \int_0^b e^{(-e^x)} e^x \, dx;$

$\begin{bmatrix} u = e^x \\ du = e^x \, dx \end{bmatrix} \rightarrow$

$\lim_{a \to -\infty} \int_{e^a}^1 e^{-u} \, du + \lim_{b \to +\infty} \int_1^{e^b} e^{-u} \, du$

$= \lim_{a \to -\infty} [-e^{-u}]_{e^a}^1 + \lim_{b \to -\infty} [-e^{-u}]_1^{e^b}$

$= \lim_{a \to -\infty} \left[-\frac{1}{e} + e^{-(e^a)}\right] + \lim_{b \to +\infty} \left[-e^{-(e^b)} + \frac{1}{e}\right]$

$= \left(-\frac{1}{e} + e^0\right) + \left(0 + \frac{1}{e}\right) = 1$

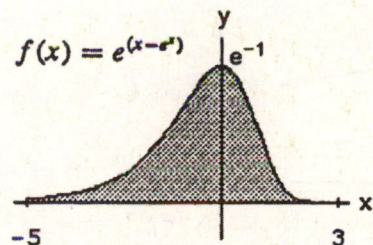

$f(x) = e^{(x - e^x)}$

35. $u = x^2 - a^2 \Rightarrow du = 2x \, dx;$

$\int x \left(\sqrt{x^2 - a^2}\right)^n \, dx = \frac{1}{2} \int \left(\sqrt{u}\right)^n \, du = \frac{1}{2} \int u^{n/2} \, du = \frac{1}{2} \left(\frac{u^{n/2 + 1}}{\frac{n}{2} + 1}\right) + C, n \neq -2$

$= \frac{u^{(n+2)/2}}{n + 2} + C = \frac{\left(\sqrt{u}\right)^{n+2}}{n + 2} + C = \frac{\left(\sqrt{x^2 - a^2}\right)^{n+2}}{n + 2} + C$

37. $\int_1^{\infty} \left(\frac{ax}{x^2 + 1} - \frac{1}{2x}\right) dx = \lim_{b \to \infty} \int_1^b \left(\frac{ax}{x^2 + 1} - \frac{1}{2x}\right) dx = \lim_{b \to \infty} \left[\frac{a}{2} \ln (x^2 + 1) - \frac{1}{2} \ln x\right]_1^b = \lim_{b \to \infty} \left[\frac{1}{2} \ln \frac{(x^2 + 1)^a}{x}\right]_1^b$

$= \lim_{b \to \infty} \frac{1}{2} \left[\ln \frac{(b^2 + 1)^a}{b} - \ln 2^a\right]; \lim_{b \to \infty} \frac{(b^2 + 1)^a}{b} > \lim_{b \to \infty} \frac{b^{2a}}{b} = \lim_{b \to \infty} b^{2(a - \frac{1}{2})} = \infty \text{ if } a > \frac{1}{2} \Rightarrow \text{ the improper}$

integral diverges if $a > \frac{1}{2}$; for $a = \frac{1}{2}$: $\lim_{b \to \infty} \frac{\sqrt{b^2 + 1}}{b} = \lim_{b \to \infty} \sqrt{1 + \frac{1}{b^2}} = 1 \Rightarrow \lim_{b \to \infty} \frac{1}{2} \left[\ln \frac{(b^2 + 1)^{1/2}}{b} - \ln 2^{1/2}\right]$

$= \frac{1}{2} \left(\ln 1 - \frac{1}{2} \ln 2\right) = -\frac{\ln 2}{4}; \text{ if } a < \frac{1}{2}: 0 \leq \lim_{b \to \infty} \frac{(b^2 + 1)^a}{b} < \lim_{b \to \infty} \frac{(b + 1)^{2a}}{b + 1} = \lim_{b \to \infty} (b + 1)^{2a - 1} = 0$

$\Rightarrow \lim_{b \to \infty} \ln \frac{(b^2 + 1)^a}{b} = -\infty \Rightarrow \text{ the improper integral diverges if } a < \frac{1}{2}; \text{ in summary, the improper integral}$

$\int_1^\infty \left(\frac{ax}{x^2+1} - \frac{1}{2x}\right)\, dx$ converges only when $a = \frac{1}{2}$ and has the value $-\frac{\ln 2}{4}$

39. $A = \int_1^\infty \frac{dx}{x^p}$ converges if $p > 1$ and diverges if $p \le 1$. Thus, $p \le 1$ for infinite area. The volume of the solid of revolution about the x-axis is $V = \int_1^\infty \pi \left(\frac{1}{x^p}\right)^2 dx = \pi \int_1^\infty \frac{dx}{x^{2p}}$ which converges if $2p > 1$ and diverges if $2p \le 1$. Thus we want $p > \frac{1}{2}$ for finite volume. In conclusion, the curve $y = x^{-p}$ gives infinite area and finite volume for values of p satisfying $\frac{1}{2} < p \le 1$.

41. e^{2x} (+) $\cos 3x$

    $2e^{2x}$ (−) ⟶ $\frac{1}{3}\sin 3x$

    $4e^{2x}$ (+) ⟶ $-\frac{1}{9}\cos 3x$

    $I = \frac{e^{2x}}{3}\sin 3x + \frac{2e^{2x}}{9}\cos 3x - \frac{4}{9}I \Rightarrow \frac{13}{9}I = \frac{e^{2x}}{9}(3\sin 3x + 2\cos 3x) \Rightarrow I = \frac{e^{2x}}{13}(3\sin 3x + 2\cos 3x) + C$

43. $\sin 3x$ (+) $\sin x$

    $3\cos 3x$ (−) ⟶ $-\cos x$

    $-9\sin 3x$ (+) ⟶ $-\sin x$

    $I = -\sin 3x \cos x + 3\cos 3x \sin x + 9I \Rightarrow -8I = -\sin 3x \cos x + 3\cos 3x \sin x$

    $\Rightarrow I = \frac{\sin 3x \cos x - 3\cos 3x \sin x}{8} + C$

45. e^{ax} (+) $\sin bx$

    ae^{ax} (−) ⟶ $-\frac{1}{b}\cos bx$

    $a^2 e^{ax}$ (+) ⟶ $-\frac{1}{b^2}\sin bx$

    $I = -\frac{e^{ax}}{b}\cos bx + \frac{ae^{ax}}{b^2}\sin bx - \frac{a^2}{b^2}I \Rightarrow \left(\frac{a^2+b^2}{b^2}\right)I = \frac{e^{ax}}{b^2}(a\sin bx - b\cos bx)$

    $\Rightarrow I = \frac{e^{ax}}{a^2+b^2}(a\sin bx - b\cos bx) + C$

47. $\ln(ax)$ (+)   1

    $\frac{1}{x}$ (+) ⟶ x

    $I = x\ln(ax) - \int \left(\frac{1}{x}\right) x\, dx = x\ln(ax) - x + C$

49. (a) $\Gamma(1) = \int_0^\infty e^{-t}\, dt = \lim_{b \to \infty} \int_0^b e^{-t}\, dt = \lim_{b \to \infty} \left[-e^{-t}\right]_0^b = \lim_{b \to \infty} \left[-\frac{1}{e^b} - (-1)\right] = 0 + 1 = 1$

    (b) $u = t^x$, $du = xt^{x-1}\, dt$; $dv = e^{-t}\, dt$, $v = -e^{-t}$; $x = $ fixed positive real

    $\Rightarrow \Gamma(x+1) = \int_0^\infty t^x e^{-t}\, dt = \lim_{b \to \infty} \left[-t^x e^{-t}\right]_0^b + x \int_0^\infty t^{x-1} e^{-t}\, dt = \lim_{b \to \infty} \left(-\frac{b^x}{e^b} + 0^x e^0\right) + x\Gamma(x) = x\Gamma(x)$

    (c) $\Gamma(n+1) = n\Gamma(n) = n!$:

    $n = 0$: $\Gamma(0+1) = \Gamma(1) = 0!$;

    $n = k$: Assume $\Gamma(k+1) = k!$ for some $k > 0$;

    $n = k+1$: $\Gamma(k+1+1) = (k+1)\Gamma(k+1)$ from part (b)

                      $= (k+1)k!$ induction hypothesis

                      $= (k+1)!$ definition of factorial

    Thus, $\Gamma(n+1) = n\Gamma(n) = n!$ for every positive integer n.

CHAPTER 9 FURTHER APPLICATIONS OF INTEGRATION

9.1 SLOPE FIELDS AND SEPARABLE DIFFERENTIAL EQUATIONS

1. (a) $y = e^{-x} \Rightarrow y' = -e^{-x} \Rightarrow 2y' + 3y = 2(-e^{-x}) + 3e^{-x} = e^{-x}$

   (b) $y = e^{-x} + e^{-3x/2} \Rightarrow y' = -e^{-x} - \frac{3}{2}e^{-3x/2} \Rightarrow 2y' + 3y = 2\left(-e^{-x} - \frac{3}{2}e^{-3x/2}\right) + 3\left(e^{-x} + e^{-3x/2}\right) = e^{-x}$

   (c) $y = e^{-x} + Ce^{-3x/2} \Rightarrow y' = -e^{-x} - \frac{3}{2}Ce^{-3x/2} \Rightarrow 2y' + 3y = 2\left(-e^{-x} - \frac{3}{2}Ce^{-3x/2}\right) + 3\left(e^{-x} + Ce^{-3x/2}\right) = e^{-x}$

3. $y = \frac{1}{x}\int_1^x \frac{e^t}{t}\,dt \Rightarrow y' = -\frac{1}{x^2}\int_1^x \frac{e^t}{t}\,dt + \left(\frac{1}{x}\right)\left(\frac{e^x}{x}\right) \Rightarrow x^2 y' = -\int_1^x \frac{e^t}{t}\,dt + e^x = -x\left(\frac{1}{x}\int_1^x \frac{e^t}{t}\,dt\right) + e^x = -xy + e^x$

   $\Rightarrow x^2 y' + xy = e^x$

5. $y = e^{-x}\tan^{-1}(2e^x) \Rightarrow y' = -e^{-x}\tan^{-1}(2e^x) + e^{-x}\left[\frac{1}{1+(2e^x)^2}\right](2e^x) = -e^{-x}\tan^{-1}(2e^x) + \frac{2}{1+4e^{2x}}$

   $\Rightarrow y' = -y + \frac{2}{1+4e^{2x}} \Rightarrow y' + y = \frac{2}{1+4e^{2x}}$; $y(-\ln 2) = e^{-(-\ln 2)}\tan^{-1}(2e^{-\ln 2}) = 2\tan^{-1} 1 = 2\left(\frac{\pi}{4}\right) = \frac{\pi}{2}$

7. $y = \frac{\cos x}{x} \Rightarrow y' = \frac{-x\sin x - \cos x}{x^2} \Rightarrow y' = -\frac{\sin x}{x} - \frac{1}{x}\left(\frac{\cos x}{x}\right) \Rightarrow y' = -\frac{\sin x}{x} - \frac{y}{x} \Rightarrow xy' = -\sin x - y$

   $\Rightarrow xy' + y = -\sin x$; $y\left(\frac{\pi}{2}\right) = \frac{\cos(\pi/2)}{(\pi/2)} = 0$

9. $2\sqrt{xy}\,\frac{dy}{dx} = 1 \Rightarrow 2x^{1/2}y^{1/2}\,dy = dx \Rightarrow 2y^{1/2}\,dy = x^{-1/2}\,dx \Rightarrow \int 2y^{1/2}\,dy = \int x^{-1/2}\,dx \Rightarrow 2\left(\frac{2}{3}y^{3/2}\right)$

   $= 2x^{1/2} + C_1 \Rightarrow \frac{2}{3}y^{3/2} - x^{1/2} = C$, where $C = \frac{1}{2}C_1$

11. $\frac{dy}{dx} = e^{x-y} \Rightarrow dy = e^x e^{-y}\,dx \Rightarrow e^y\,dy = e^x\,dx \Rightarrow \int e^y\,dy = \int e^x\,dx \Rightarrow e^y = e^x + C \Rightarrow e^y - e^x = C$

13. $\frac{dy}{dx} = \sqrt{y}\cos^2\sqrt{y} \Rightarrow dy = \left(\sqrt{y}\cos^2\sqrt{y}\right)dx \Rightarrow \frac{\sec^2\sqrt{y}}{\sqrt{y}}\,dy = dx \Rightarrow \int \frac{\sec^2\sqrt{y}}{\sqrt{y}}\,dy = \int dx$. In the integral on the left-hand

   side, substitute $u = \sqrt{y} \Rightarrow du = \frac{1}{2\sqrt{y}}\,dy \Rightarrow 2\,du = \frac{1}{\sqrt{y}}\,dy$, and we have $\int \sec^2 u\,du = \int dx \Rightarrow 2\tan u = x + C$

   $\Rightarrow -x + 2\tan\sqrt{y} = C$

15. $\sqrt{x}\,\frac{dy}{dx} = e^{y+\sqrt{x}} \Rightarrow \frac{dy}{dx} = \frac{e^y e^{\sqrt{x}}}{\sqrt{x}} \Rightarrow dy = \frac{e^y e^{\sqrt{x}}}{\sqrt{x}}\,dx \Rightarrow e^{-y}\,dy = \frac{e^{\sqrt{x}}}{\sqrt{x}}\,dx \Rightarrow \int e^{-y}\,dy = \int \frac{e^{\sqrt{x}}}{\sqrt{x}}\,dx$. In the integral on the

   right-hand side, substitute $u = \sqrt{x} \Rightarrow du = \frac{1}{2\sqrt{x}}\,dx \Rightarrow 2\,du = \frac{1}{\sqrt{x}}\,dx$, and we have $\int e^{-y}\,dy = 2\int e^u\,du$

   $\Rightarrow -e^{-y} = 2e^u + C_1 \Rightarrow -e^{-y} = 2e^{\sqrt{x}} + C$, where $C = -C_1$

17. $\frac{dy}{dx} = 2x\sqrt{1-y^2} \Rightarrow dy = 2x\sqrt{1-y^2}\,dx \Rightarrow \frac{dy}{\sqrt{1-y^2}} = 2x\,dx \Rightarrow \int \frac{dy}{\sqrt{1-y^2}} = \int 2x\,dx \Rightarrow \sin^{-1}y = x^2 + C$ since $|y| < 1$

   $\Rightarrow y = \sin(x^2 + C)$

19. $y' = x + y \Rightarrow$ slope of 0 for the line $y = -x$.

For $x, y > 0$, $y' = x + y \Rightarrow$ slope > 0 in Quadrant I.

For $x, y < 0$, $y' = x + y \Rightarrow$ slope < 0 in Quadrant III.

For $|y| > |x|$, $y > 0$, $x < 0$, $y' = x + y \Rightarrow$ slope > 0 in Quadrant II above $y = -x$.

For $|y| < |x|$, $y > 0$, $x < 0$, $y' = x + y \Rightarrow$ slope < 0 in Quadrant II below $y = -x$.

For $|y| < |x|$, $x > 0$, $y < 0$, $y' = x + y \Rightarrow$ slope > 0 in Quadrant IV above $y = -x$.

For $|y| > |x|$, $x > 0$, $y < 0$, $y' = x + y \Rightarrow$ slope < 0 in Quadrant IV below $y = -x$.

All of the conditions are seen in slope field (d).

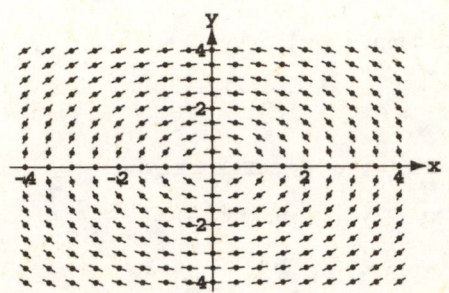

21. $y' = -\frac{x}{y} \Rightarrow$ slope $= 1$ on $y = -x$ and -1 on $y = x$.

$y' = -\frac{x}{y} \Rightarrow$ slope $= 0$ on the y-axis, excluding $(0, 0)$, and is undefined on the x-axis. Slopes are positive for $x > 0$, $y < 0$ and $x < 0$, $y > 0$ (Quadrants II and IV), otherwise negative. Field (a) is consistent with these conditions.

23.

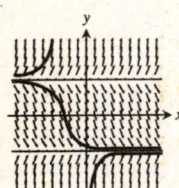

25.

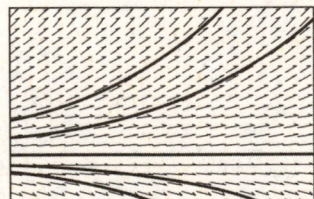

27.

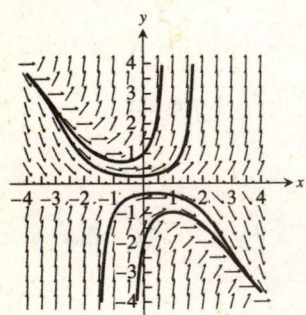

29.

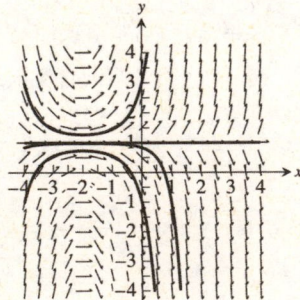

9.2 FIRST-ORDER LINEAR DIFFERENTIAL EQUATIONS

1. $x\frac{dy}{dx} + y = e^x \Rightarrow \frac{dy}{dx} + \left(\frac{1}{x}\right)y = \frac{e^x}{x}$, $P(x) = \frac{1}{x}$, $Q(x) = \frac{e^x}{x}$

$\int P(x)\,dx = \int \frac{1}{x}\,dx = \ln|x| = \ln x$, $x > 0 \Rightarrow v(x) = e^{\int P(x)\,dx} = e^{\ln x} = x$

$y = \frac{1}{v(x)}\int v(x)\,Q(x)\,dx = \frac{1}{x}\int x\left(\frac{e^x}{x}\right)dx = \frac{1}{x}\left(e^x + C\right) = \frac{e^x + C}{x}$, $x > 0$

3. $xy' + 3y = \frac{\sin x}{x^2}$, $x > 0 \Rightarrow \frac{dy}{dx} + \left(\frac{3}{x}\right)y = \frac{\sin x}{x^3}$, $P(x) = \frac{3}{x}$, $Q(x) = \frac{\sin x}{x^3}$

$\int \frac{3}{x}\, dx = 3\ln|x| = \ln x^3$, $x > 0 \Rightarrow v(x) = e^{\ln x^3} = x^3$

$y = \frac{1}{x^3}\int x^3\left(\frac{\sin x}{x^3}\right)dx = \frac{1}{x^3}\int \sin x\, dx = \frac{1}{x^3}(-\cos x + C) = \frac{C - \cos x}{x^3}$, $x > 0$

5. $x\frac{dy}{dx} + 2y = 1 - \frac{1}{x}$, $x > 0 \Rightarrow \frac{dy}{dx} + \left(\frac{2}{x}\right)y = \frac{1}{x} - \frac{1}{x^2}$, $P(x) = \frac{2}{x}$, $Q(x) = \frac{1}{x} - \frac{1}{x^2}$

$\int \frac{2}{x}\, dx = 2\ln|x| = \ln x^2$, $x > 0 \Rightarrow v(x) = e^{\ln x^2} = x^2$

$y = \frac{1}{x^2}\int x^2\left(\frac{1}{x} - \frac{1}{x^2}\right)dx = \frac{1}{x^2}\int(x - 1)\, dx = \frac{1}{x^2}\left(\frac{x^2}{2} - x + C\right) = \frac{1}{2} - \frac{1}{x} + \frac{C}{x^2}$, $x > 0$

7. $\frac{dy}{dx} - \frac{1}{2}y = \frac{1}{2}e^{x/2} \Rightarrow P(x) = -\frac{1}{2}$, $Q(x) = \frac{1}{2}e^{x/2} \Rightarrow \int P(x)\, dx = -\frac{1}{2}x \Rightarrow v(x) = e^{-x/2}$

$\Rightarrow y = \frac{1}{e^{-x/2}}\int e^{-x/2}\left(\frac{1}{2}e^{x/2}\right)dx = e^{x/2}\int \frac{1}{2}\, dx = e^{x/2}\left(\frac{1}{2}x + C\right) = \frac{1}{2}xe^{x/2} + Ce^{x/2}$

9. $\frac{dy}{dx} - \left(\frac{1}{x}\right)y = 2\ln x \Rightarrow P(x) = -\frac{1}{x}$, $Q(x) = 2\ln x \Rightarrow \int P(x)\, dx = -\int \frac{1}{x}\, dx = -\ln x$, $x > 0$

$\Rightarrow v(x) = e^{-\ln x} = \frac{1}{x} \Rightarrow y = x\int \left(\frac{1}{x}\right)(2\ln x)\, dx = x\left[(\ln x)^2 + C\right] = x(\ln x)^2 + Cx$

11. $\frac{ds}{dt} + \left(\frac{4}{t-1}\right)s = \frac{t+1}{(t-1)^3} \Rightarrow P(t) = \frac{4}{t-1}$, $Q(t) = \frac{t+1}{(t-1)^3} \Rightarrow \int P(t)\, dt = \int \frac{4}{t-1}\, dt = 4\ln|t-1| = \ln(t-1)^4$

$\Rightarrow v(t) = e^{\ln(t-1)^4} = (t-1)^4 \Rightarrow s = \frac{1}{(t-1)^4}\int(t-1)^4\left[\frac{t+1}{(t-1)^3}\right]dt = \frac{1}{(t-1)^4}\int(t^2 - 1)\, dt$

$= \frac{1}{(t-1)^4}\left(\frac{t^3}{3} - t + C\right) = \frac{t^3}{3(t-1)^4} - \frac{t}{(t-1)^4} + \frac{C}{(t-1)^4}$

13. $\frac{dr}{d\theta} + (\cot\theta)r = \sec\theta \Rightarrow P(\theta) = \cot\theta$, $Q(\theta) = \sec\theta \Rightarrow \int P(\theta)\, d\theta = \int \cot\theta\, d\theta = \ln|\sin\theta| \Rightarrow v(\theta) = e^{\ln|\sin\theta|}$

$= \sin\theta$ because $0 < \theta < \frac{\pi}{2} \Rightarrow r = \frac{1}{\sin\theta}\int(\sin\theta)(\sec\theta)\, d\theta = \frac{1}{\sin\theta}\int \tan\theta\, d\theta = \frac{1}{\sin\theta}(\ln|\sec\theta| + C)$

$= (\csc\theta)(\ln|\sec\theta| + C)$

15. $\frac{dy}{dt} + 2y = 3 \Rightarrow P(t) = 2$, $Q(t) = 3 \Rightarrow \int P(t)\, dt = \int 2\, dt = 2t \Rightarrow v(t) = e^{2t} \Rightarrow y = \frac{1}{e^{2t}}\int 3e^{2t}\, dt$

$= \frac{1}{e^{2t}}\left(\frac{3}{2}e^{2t} + C\right)$; $y(0) = 1 \Rightarrow \frac{3}{2} + C = 1 \Rightarrow C = -\frac{1}{2} \Rightarrow y = \frac{3}{2} - \frac{1}{2}e^{-2t}$

17. $\frac{dy}{d\theta} + \left(\frac{1}{\theta}\right)y = \frac{\sin\theta}{\theta} \Rightarrow P(\theta) = \frac{1}{\theta}$, $Q(\theta) = \frac{\sin\theta}{\theta} \Rightarrow \int P(\theta)\, d\theta = \ln|\theta| \Rightarrow v(\theta) = e^{\ln|\theta|} = |\theta|$

$\Rightarrow y = \frac{1}{|\theta|}\int |\theta|\left(\frac{\sin\theta}{\theta}\right)d\theta = \frac{1}{\theta}\int \theta\left(\frac{\sin\theta}{\theta}\right)d\theta$ for $\theta \neq 0 \Rightarrow y = \frac{1}{\theta}\int \sin\theta\, d\theta = \frac{1}{\theta}(-\cos\theta + C)$

$= -\frac{1}{\theta}\cos\theta + \frac{C}{\theta}$; $y\left(\frac{\pi}{2}\right) = 1 \Rightarrow C = \frac{\pi}{2} \Rightarrow y = -\frac{1}{\theta}\cos\theta + \frac{\pi}{2\theta}$

19. $(x+1)\frac{dy}{dx} - 2(x^2 + x)y = \frac{e^{x^2}}{x+1} \Rightarrow \frac{dy}{dx} - 2\left[\frac{x(x+1)}{x+1}\right]y = \frac{e^{x^2}}{(x+1)^2} \Rightarrow \frac{dy}{dx} - 2xy = \frac{e^{x^2}}{(x+1)^2} \Rightarrow P(x) = -2x$,

$Q(x) = \frac{e^{x^2}}{(x+1)^2} \Rightarrow \int P(x)\, dx = \int -2x\, dx = -x^2 \Rightarrow v(x) = e^{-x^2} \Rightarrow y = \frac{1}{e^{-x^2}}\int e^{-x^2}\left[\frac{e^{x^2}}{(x+1)^2}\right]dx$

$= e^{x^2}\int \frac{1}{(x+1)^2}\, dx = e^{x^2}\left[\frac{(x+1)^{-1}}{-1} + C\right] = -\frac{e^{x^2}}{x+1} + Ce^{x^2}$; $y(0) = 5 \Rightarrow -\frac{1}{0+1} + C = 5 \Rightarrow -1 + C = 5$

$\Rightarrow C = 6 \Rightarrow y = 6e^{x^2} - \frac{e^{x^2}}{x+1}$

21. $\frac{dy}{dt} - ky = 0 \Rightarrow P(t) = -k$, $Q(t) = 0 \Rightarrow \int P(t)\, dt = \int -k\, dt = -kt \Rightarrow v(t) = e^{-kt}$

$\Rightarrow y = \frac{1}{e^{-kt}}\int (e^{-kt})(0)\, dt = e^{kt}(0 + C) = Ce^{kt}$; $y(0) = y_0 \Rightarrow C = y_0 \Rightarrow y = y_0 e^{kt}$

23. $x\int \frac{1}{x}\, dx = x(\ln|x| + C) = x\ln|x| + Cx \Rightarrow$ (b) is correct

25. Let $y(t) =$ the amount of salt in the container and $V(t) =$ the total volume of liquid in the tank at time t. Then, the departure rate is $\frac{y(t)}{V(t)}$ (the outflow rate).

(a) Rate entering $= \frac{2 \text{ lb}}{\text{gal}} \cdot \frac{5 \text{ gal}}{\text{min}} = 10$ lb/min

(b) Volume $= V(t) = 100 \text{ gal} + (5t \text{ gal} - 4t \text{ gal}) = (100 + t)$ gal

(c) The volume at time t is $(100 + t)$ gal. The amount of salt in the tank at time t is y lbs. So the concentration at any time t is $\frac{y}{100+t}$ lbs/gal. Then, the rate leaving $= \frac{y}{100+t}$ (lbs/gal) $\cdot 4$ (gal/min)

$= \frac{4y}{100+t}$ lbs/min

(d) $\frac{dy}{dt} = 10 - \frac{4y}{100+t} \Rightarrow \frac{dy}{dt} + \left(\frac{4}{100+t}\right)y = 10 \Rightarrow P(t) = \frac{4}{100+t}, Q(t) = 10 \Rightarrow \int P(t)\,dt = \int \frac{4}{100+t}\,dt$

$= 4\ln(100+t) \Rightarrow v(t) = e^{4\ln(100+t)} = (100+t)^4 \Rightarrow y = \frac{1}{(100+t)^4}\int (100+t)^4 (10\,dt)$

$= \frac{10}{(100+t)^4}\left(\frac{(100+t)^5}{5} + C\right) = 2(100+t) + \frac{C}{(100+t)^4}; y(0) = 50 \Rightarrow 2(100+0) + \frac{C}{(100+0)^4} = 50$

$\Rightarrow C = -(150)(100)^4 \Rightarrow y = 2(100+t) - \frac{(150)(100)^4}{(100+t)^4} \Rightarrow y = 2(100+t) - \frac{150}{\left(1+\frac{t}{100}\right)^4}$

(e) $y(25) = 2(100+25) - \frac{(150)(100)^4}{(100+25)^4} \approx 188.56$ lbs $\Rightarrow$ concentration $= \frac{y(25)}{\text{volume}} \approx \frac{188.6}{125} \approx 1.5$ lb/gal

27. Let y be the amount of fertilizer in the tank at time t. Then rate entering $= 1\,\frac{\text{lb}}{\text{gal}} \cdot 1\,\frac{\text{gal}}{\text{min}} = 1\,\frac{\text{lb}}{\text{min}}$ and the volume in the tank at time t is $V(t) = 100\,(\text{gal}) + [1\,(\text{gal/min}) - 3\,(\text{gal/min})]t\,\text{min} = (100 - 2t)$ gal. Hence rate out $= \left(\frac{y}{100-2t}\right)3 = \frac{3y}{100-2t}$ lbs/min $\Rightarrow \frac{dy}{dt} = \left(1 - \frac{3y}{100-2t}\right)$ lbs/min $\Rightarrow \frac{dy}{dt} + \left(\frac{3}{100-2t}\right)y = 1$

$\Rightarrow P(t) = \frac{3}{100-2t}, Q(t) = 1 \Rightarrow \int P(t)\,dt = \int \frac{3}{100-2t}\,dt = \frac{3\ln(100-2t)}{-2} \Rightarrow v(t) = e^{(-3\ln(100-2t))2}$

$= (100-2t)^{-3/2} \Rightarrow y = \frac{1}{(100-2t)^{-3/2}}\int (100-2t)^{-3/2}\,dt = (100-2t)^{-3/2}\left[\frac{-2(100-2t)^{-1/2}}{-2} + C\right]$

$= (100-2t) + C(100-2t)^{3/2}; y(0) = 0 \Rightarrow [100-2(0)] + C[100-2(0)]^{3/2} \Rightarrow C(100)^{3/2} = -100$

$\Rightarrow C = -(100)^{-1/2} = -\frac{1}{10} \Rightarrow y = (100-2t) - \frac{(100-2t)^{3/2}}{10}$. Let $\frac{dy}{dt} = 0 \Rightarrow \frac{dy}{dt} = -2 - \frac{\left(\frac{3}{2}\right)(100-2t)^{1/2}(-2)}{10}$

$= -2 + \frac{3\sqrt{100-2t}}{10} = 0 \Rightarrow 20 = 3\sqrt{100-2t} \Rightarrow 400 = 9(100-2t) \Rightarrow 400 = 900 - 18t \Rightarrow -500 = -18t$

$\Rightarrow t \approx 27.8$ min, the time to reach the maximum. The maximum amount is then

$y(27.8) = [100 - 2(27.8)] - \frac{[100-2(27.8)]^{3/2}}{10} \approx 14.8$ lb

29. Steady State $= \frac{V}{R}$ and we want $i = \frac{1}{2}\left(\frac{V}{R}\right) \Rightarrow \frac{1}{2}\left(\frac{V}{R}\right) = \frac{V}{R}\left(1 - e^{-Rt/L}\right) \Rightarrow \frac{1}{2} = 1 - e^{-Rt/L} \Rightarrow -\frac{1}{2} = -e^{-Rt/L}$

$\Rightarrow \ln\frac{1}{2} = -\frac{Rt}{L} \Rightarrow -\frac{L}{R}\ln\frac{1}{2} = t \Rightarrow t = \frac{L}{R}\ln 2$ sec

31. (a) $t = \frac{3L}{R} \Rightarrow i = \frac{V}{R}\left(1 - e^{(-R/L)(3L/R)}\right) = \frac{V}{R}\left(1 - e^{-3}\right) \approx 0.9502\,\frac{V}{R}$ amp, or about 95% of the steady state value

(b) $t = \frac{2L}{R} \Rightarrow i = \frac{V}{R}\left(1 - e^{(-R/L)(2L/R)}\right) = \frac{V}{R}\left(1 - e^{-2}\right) \approx 0.8647\,\frac{V}{R}$ amp, or about 86% of the steady state value

33. $y' - y = -y^2$; we have $n = 2$, so let $u = y^{1-2} = y^{-1}$. Then $y = u^{-1}$ and $\frac{du}{dx} = -1y^{-2}\frac{dy}{dx} \Rightarrow \frac{dy}{dx} = -y^2\frac{du}{dx}$

$\Rightarrow -u^{-2}\frac{du}{dx} - u^{-1} = -u^{-2} \Rightarrow \frac{du}{dx} + u = 1$. With $e^{\int dx} = e^x$ as the integrating factor, we have

$e^x\left(\frac{du}{dx} + u\right) = \frac{d}{dx}(e^x u) = e^x$. Integrating, we get $e^x u = e^x + C \Rightarrow u = 1 + \frac{C}{e^x} = \frac{1}{y} \Rightarrow y = \frac{1}{1+\frac{C}{e^x}} = \frac{e^x}{e^x + C}$

35. $xy' + y = y^{-2} \Rightarrow y' + \left(\frac{1}{x}\right)y = \left(\frac{1}{x}\right)y^{-2}$. Let $u = y^{1-(-2)} = y^3 \Rightarrow y = u^{1/3}$ and $y^{-2} = u^{-2/3}$.

$\frac{du}{dx} = 3y^2\frac{dy}{dx} \Rightarrow y' = \frac{dy}{dx} = \left(\frac{1}{3}\right)\left(\frac{du}{dx}\right)(y^{-2}) = \left(\frac{1}{3}\right)\left(\frac{du}{dx}\right)(u^{-2/3})$. Thus we have

$\left(\frac{1}{3}\right)\left(\frac{du}{dx}\right)(u^{-2/3}) + \left(\frac{1}{x}\right)u^{1/3} = \left(\frac{1}{x}\right)u^{-2/3} \Rightarrow \frac{du}{dx} + \left(\frac{3}{x}\right)u = \left(\frac{3}{x}\right)1$. The integrating factor, v(x), is

$e^{\int \frac{3}{x}dx} = e^{3\ln x} = e^{\ln x^3} = x^3$. Thus $\frac{d}{dx}(x^3 u) = \left(\frac{3}{x}\right)x^3 = 3x^2 \Rightarrow x^3 u = x^3 + C \Rightarrow u = 1 + \frac{C}{x^3} = y^3$

$\Rightarrow y = \left(1 + \frac{C}{x^3}\right)^{1/3}$

9.3 EULER'S METHOD

1. $y_1 = y_0 + \left(1 - \frac{y_0}{x_0}\right) dx = -1 + \left(1 - \frac{-1}{2}\right)(.5) = -0.25,$

   $y_2 = y_1 + \left(1 - \frac{y_1}{x_1}\right) dx = -0.25 + \left(1 - \frac{-0.25}{2.5}\right)(.5) = 0.3,$

   $y_3 = y_2 + \left(1 - \frac{y_2}{x_2}\right) dx = 0.3 + \left(1 - \frac{0.3}{3}\right)(.5) = 0.75;$

   $\frac{dy}{dx} + \left(\frac{1}{x}\right) y = 1 \Rightarrow P(x) = \frac{1}{x}, Q(x) = 1 \Rightarrow \int P(x)\, dx = \int \frac{1}{x}\, dx = \ln|x| = \ln x, x > 0 \Rightarrow v(x) = e^{\ln x} = x$

   $\Rightarrow y = \frac{1}{x} \int x \cdot 1\, dx = \frac{1}{x}\left(\frac{x^2}{2} + C\right); x = 2, y = -1 \Rightarrow -1 = 1 + \frac{C}{2} \Rightarrow C = -4 \Rightarrow y = \frac{x}{2} - \frac{4}{x}$

   $\Rightarrow y(3.5) = \frac{3.5}{2} - \frac{4}{3.5} = \frac{4.25}{7} \approx 0.6071$

3. $y_1 = y_0 + (2x_0 y_0 + 2y_0)\, dx = 3 + [2(0)(3) + 2(3)](.2) = 4.2,$

   $y_2 = y_1 + (2x_1 y_1 + 2y_1)\, dx = 4.2 + [2(.2)(4.2) + 2(4.2)](.2) = 6.216,$

   $y_3 = y_2 + (2x_2 y_2 + 2y_2)\, dx = 6.216 + [2(.4)(6.216) + 2(6.216)](.2) = 9.6969;$

   $\frac{dy}{dx} = 2y(x + 1) \Rightarrow \frac{dy}{y} = 2(x + 1)\, dx \Rightarrow \ln|y| = (x + 1)^2 + C; x = 0, y = 3 \Rightarrow \ln 3 = 1 + C \Rightarrow C = \ln 3 - 1$

   $\Rightarrow \ln y = (x + 1)^2 + \ln 3 - 1 \Rightarrow y = e^{(x+1)^2 + \ln 3 - 1} = e^{\ln 3} e^{x^2 + 2x} = 3e^{x(x+2)} \Rightarrow y(.6) \approx 14.2765$

5. $y_1 = y_0 + 2x_0 e^{x_0^2}\, dx = 2 + 2(0)(.1) = 2,$

   $y_2 = y_1 + 2x_1 e^{x_1^2}\, dx = 2 + 2(.1)\, e^{.1^2}(.1) = 2.0202,$

   $y_3 = y_2 + 2x_2 e^{x_2^2}\, dx = 2.0202 + 2(.2)\, e^{.2^2}(.1) = 2.0618,$

   $dy = 2xe^{x^2}\, dx \Rightarrow y = e^{x^2} + C; y(0) = 2 \Rightarrow 2 = 1 + C \Rightarrow C = 1 \Rightarrow y = e^{x^2} + 1 \Rightarrow y(.3) = e^{.3^2} + 1 \approx 2.0942$

7. $y_1 = 1 + 1(.2) = 1.2,$

   $y_2 = 1.2 + (1.2)(.2) = 1.44,$

   $y_3 = 1.44 + (1.44)(.2) = 1.728,$

   $y_4 = 1.728 + (1.728)(.2) = 2.0736,$

   $y_5 = 2.0736 + (2.0736)(.2) = 2.48832;$

   $\frac{dy}{y} = dx \Rightarrow \ln y = x + C_1 \Rightarrow y = Ce^x; y(0) = 1 \Rightarrow 1 = Ce^0 \Rightarrow C = 1 \Rightarrow y = e^x \Rightarrow y(1) = e \approx 2.7183$

9. $y_1 = -1 + \left[\frac{(-1)^2}{\sqrt{1}}\right](.5) = -.5,$

   $y_2 = -.5 + \left[\frac{(-.5)^2}{\sqrt{1.5}}\right](.5) = -.39794,$

   $y_3 = -.39794 + \left[\frac{(-.39794)^2}{\sqrt{2}}\right](.5) = -.34195,$

   $y_4 = -.34195 + \left[\frac{(-.34195)^2}{\sqrt{2.5}}\right](.5) = -.30497,$

   $y_5 = -.27812, y_6 = -.25745, y_7 = -.24088, y_8 = -.2272;$

   $\frac{dy}{y^2} = \frac{dx}{\sqrt{x}} \Rightarrow -\frac{1}{y} = 2\sqrt{x} + C; y(1) = -1 \Rightarrow 1 = 2 + C \Rightarrow C = -1 \Rightarrow y = \frac{1}{1 - 2\sqrt{x}} \Rightarrow y(5) = \frac{1}{1 - 2\sqrt{5}} \approx -.2880$

11. Let $z_n = y_{n-1} + 2y_{n-1}(x_{n-1} + 1)dx$ and $y_n = y_{n-1} + (y_{n-1}(x_{n-1} + 1) + z_n(x_n + 1))dx$ with $x_0 = 0$, $y_0 = 3$, and $dx = 0.2$.
The exact solution is $y = 3e^{x(x+2)}$. Using a programmable calculator or a spreadsheet (I used a spreadsheet) gives the
values in the following table.

x	z	y-approx	y-exact	Error
0	---	3	3	0
0.2	4.2	4.608	4.658122	0.050122
0.4	6.81984	7.623475	7.835089	0.211614
0.6	11.89262	13.56369	14.27646	0.712777

13. $\frac{dy}{dx} = 2xe^{x^2}, y(0) = 2 \Rightarrow y_{n+1} = y_n + 2x_n e^{x_n^2} dx = y_n + 2x_n e^{x_n^2}(0.1) = y_n + 0.2x_n e^{x_n^2}$

On a TI-92 Plus calculator home screen, type the following commands:

2 STO > y: 0 STO > x: y (enter)

y + 0.2*x*e^(x^2) STO > y: x + 0.1 STO > x: y (enter, 10 times)

The last value displayed gives $y_{Euler}(1) \approx 3.45835$

The exact solution: $dy = 2xe^{x^2} dx \Rightarrow y = e^{x^2} + C; y(0) = 2 = e^0 + C \Rightarrow C = 1 \Rightarrow y = 1 + e^{x^2}$

$\Rightarrow y_{exact}(1) = 1 + e \approx 3.71828$

15. $\frac{dy}{dx} = \frac{\sqrt{x}}{y}, y > 0, y(0) = 1 \Rightarrow y_{n+1} = y_n + \frac{\sqrt{x_n}}{y_n} dx = y_n + \frac{\sqrt{x_n}}{y_n}(0.1) = y_n + 0.1\frac{\sqrt{x_n}}{y_n}$

On a TI-92 Plus calculator home screen, type the following commands:

1 STO > y: 0 STO > x: y (enter)

y + 0.1*($\sqrt{x}$ /y) STO > y: x + 0.1 STO > x: y (enter, 10 times)

The last value displayed gives $y_{Euler}(1) \approx 1.5000$

The exact solution: $dy = \frac{\sqrt{x}}{y} dx \Rightarrow y\,dy = \sqrt{x}\,dx \Rightarrow \frac{y^2}{2} = \frac{2}{3}x^{3/2} + C; \frac{(y(0))^2}{2} = \frac{1^2}{2} = \frac{1}{2} = \frac{2}{3}(0)^{3/2} + C \Rightarrow C = \frac{1}{2}$

$\Rightarrow \frac{y^2}{2} = \frac{2}{3}x^{3/2} + \frac{1}{2} \Rightarrow y = \sqrt{\frac{4}{3}x^{3/2} + 1} \Rightarrow y_{exact}(1) = \sqrt{\frac{4}{3}(1)^{3/2} + 1} \approx 1.5275$

17. (a) $\frac{dy}{dx} = 2y^2(x - 1) \Rightarrow \frac{dy}{y^2} = 2(x - 1)dx \Rightarrow \int y^{-2}dy = \int (2x - 2)dx \Rightarrow -y^{-1} = x^2 - 2x + C$

Initial value: $y(2) = -\frac{1}{2} \Rightarrow 2 = 2^2 - 2(2) + C \Rightarrow C = 2$

Solution: $-y^{-1} = x^2 - 2x + 2$ or $y = -\frac{1}{x^2 - 2x + 2}$

$y(3) = -\frac{1}{3^2 - 2(3) + 2} = -\frac{1}{5} = -0.2$

(b) To find the approximation, set $y_1 = 2y^2(x - 1)$ and use EULERT with initial values $x = 2$ and $y = -\frac{1}{2}$ and step size 0.2 for 5 Points. This gives $y(3) \approx -0.1851$; error ≈ 0.0149.

(c) Use step size 0.1 for 10 points. This gives $y(3) \approx -0.1929$; error ≈ 0.0071.

(d) Use step size 0.05 for 20 points. This gives $y(3) \approx -0.1965$; error ≈ 0.0035.

19. The exact solution is $y = \frac{-1}{x^2 - 2x + 2}$, so $y(3) = -0.2$. To find the approximation, let $z_n = y_{n-1} + 2y_{n-1}^2(x_{n-1} - 1)dx$ and $y_n = y_{n-1} + (y_{n-1}^2(x_{n-1} - 1) + z_n^2(x_n^2 - 1))dx$ with initial values $x_0 = 2$ and $y_0 = -\frac{1}{2}$. Use a spreadsheet, graphing calculator, or CAS as indicated in parts (a) through (d).

(a) Use $dx = 0.2$ with 5 steps to obtain $y(3) \approx -0.2024 \Rightarrow$ error ≈ 0.0024.

(b) Use $dx = 0.1$ with 10 steps to obtain $y(3) \approx -0.2005 \Rightarrow$ error ≈ 0.0005.

(c) Use $dx = 0.05$ with 20 steps to obtain $y(3) \approx -0.2001 \Rightarrow$ error ≈ 0.0001.

(d) Each time the step size is cut in half, the error is reduced to approximately one-fourth of what it was for the larger step size.

9.4 GRAPHICAL SOUTIONS OF AUTONOMOUS DIFFERENTIAL EQUATIONS

1. $y' = (y + 2)(y - 3)$

(a) $y = -2$ is a stable equilibrium value and $y = 3$ is an unstable equilibrium.

(b) $y'' = (2y - 1)y' = 2(y + 2)\left(y - \frac{1}{2}\right)(y - 3)$

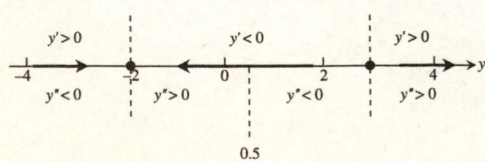

(c)

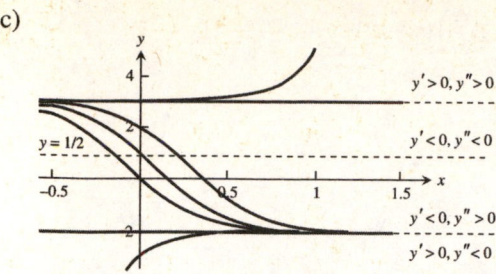

3. $y' = y^3 - y = (y + 1)y(y - 1)$

(a) $y = -1$ and $y = 1$ is an unstable equilibrium and $y = 0$ is a stable equilibrium value.

(b) $y'' = (3y^2 - 1)y' = 3(y + 1)\left(y + \frac{1}{\sqrt{3}}\right)y\left(y - \frac{1}{\sqrt{3}}\right)(y - 1)$

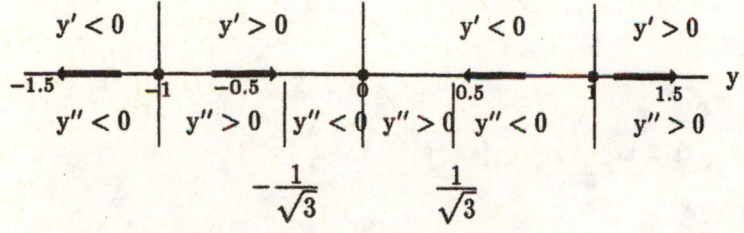

(c)

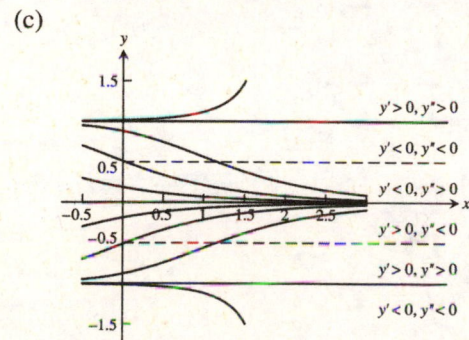

5. $y' = \sqrt{y}, y > 0$

(a) There are no equilibrium values.

(b) $y'' = \frac{1}{2\sqrt{y}} y' = \frac{1}{2\sqrt{y}} \sqrt{y} = \frac{1}{2}$

(c)

7. $y' = (y - 1)(y - 2)(y - 3)$

(a) $y = 1$ and $y = 3$ is an unstable equilibrium and $y = 2$ is a stable equilibrium value.

(b) $y'' = (3y^2 - 12y + 11)(y-1)(y-2)(y-3) = 3(y-1)\left(y - \frac{6-\sqrt{3}}{3}\right)(y-2)\left(y - \frac{6+\sqrt{3}}{3}\right)(y-3)$

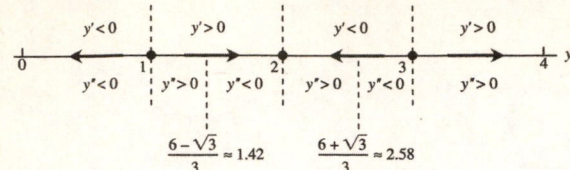

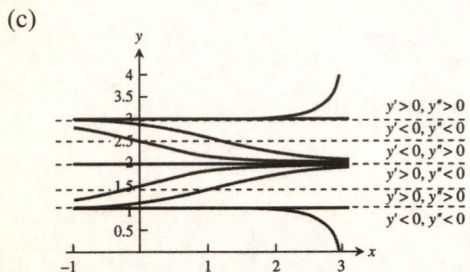

(c)

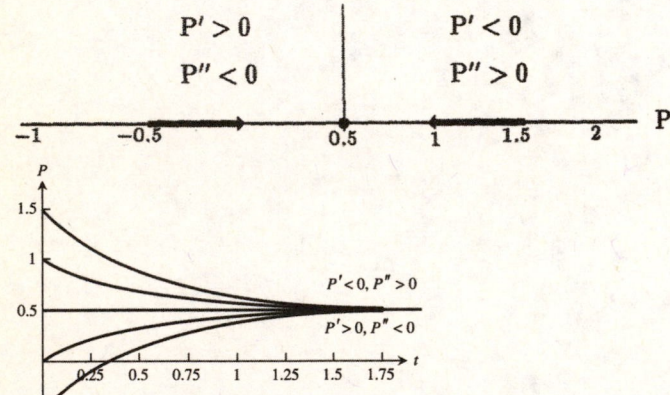

9. $\frac{dP}{dt} = 1 - 2P$ has a stable equilibrium at $P = \frac{1}{2}$. $\frac{d^2P}{dt^2} = -2\frac{dP}{dt} = -2(1 - 2P)$

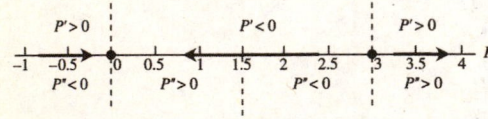

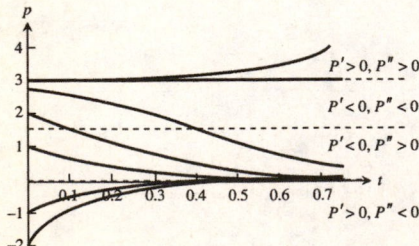

11. $\frac{dP}{dt} = 2P(P-3)$ has a stable equilibrium at $P = 0$ and an unstable equilibrium at $P = 3$.

$\frac{d^2P}{dt^2} = 2(2P-3)\frac{dP}{dt} = 4P(2P-3)(P-3)$

13.

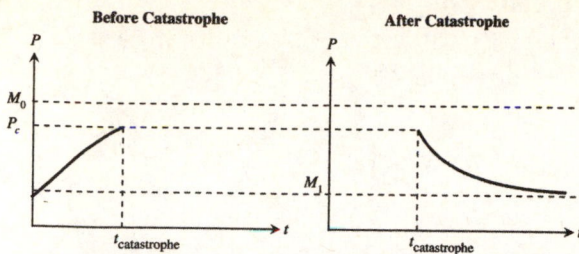

Before Catastrophe   **After Catastrophe**

Before the catastrophe, the population exhibits logistic growth and $P(t) \to M_0$, the stable equilibrium. After the catastrophe, the population declines logistically and $P(t) \to M_1$, the new stable equilibrium.

15. $\frac{dv}{dt} = g - \frac{k}{m}v^2$, g, k, m > 0 and $v(t) \geq 0$

Equilibrium: $\frac{dv}{dt} = g - \frac{k}{m}v^2 = 0 \Rightarrow v = \sqrt{\frac{mg}{k}}$

Concavity: $\frac{d^2v}{dt^2} = -2\left(\frac{k}{m}v\right)\frac{dv}{dt} = -2\left(\frac{k}{m}v\right)\left(g - \frac{k}{m}v^2\right)$

(a)

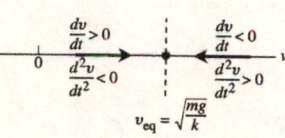

(b)

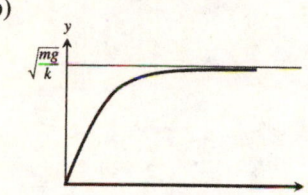

(c) $v_{\text{terminal}} = \sqrt{\frac{160}{0.005}} = 178.9 \frac{\text{ft}}{\text{s}} = 122$ mph

17. $F = F_p - F_r$

$ma = 50 - 5|v|$

$\frac{dv}{dt} = \frac{1}{m}(50 - 5|v|)$

The maximum velocity occurs when $\frac{dv}{dt} = 0$ or $v = 10 \frac{\text{ft}}{\text{sec}}$.

19. $L\frac{di}{dt} + Ri = V \Rightarrow \frac{di}{dt} = \frac{V}{L} - \frac{R}{L}i = \frac{R}{L}\left(\frac{V}{R} - i\right)$, V, L, R > 0

Equilibrium: $\frac{di}{dt} = \frac{R}{L}\left(\frac{V}{R} - i\right) = 0 \Rightarrow i = \frac{V}{R}$

Concavity: $\frac{d^2i}{dt^2} = -\left(\frac{R}{L}\right)\frac{di}{dt} = -\left(\frac{R}{L}\right)^2\left(\frac{V}{R} - i\right)$

Phase Line:

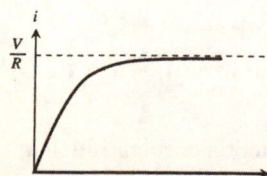

If the switch is closed at $t = 0$, then $i(0) = 0$, and the graph of the solution looks like this:

As $t \to \infty$, it $\to i_{\text{steady state}} = \frac{V}{R}$. (In the steady state condition, the self-inductance acts like a simple wire connector and, as a result, the current throught the resistor can be calculated using the familiar version of Ohm's Law.)

9.5 APPLICATIONS OF FIRST ORDER DIFFERENTIAL EQUATIONS

1. Note that the total mass is $66 + 7 = 73$ kg, therefore, $v = v_0 e^{-(k/m)t} \Rightarrow v = 9e^{-3.9t/73}$

    (a) $s(t) = \int 9e^{-3.9t/73} dt = -\frac{2190}{13} e^{-3.9t/73} + C$

    Since $s(0) = 0$ we have $C = \frac{2190}{13}$ and $\lim_{t \to \infty} s(t) = \lim_{t \to \infty} \frac{2190}{13}\left(1 - e^{-3.9t/73}\right) = \frac{2190}{13} \approx 168.5$

    The cyclist will coast about 168.5 meters.

    (b) $1 = 9e^{-3.9t/73} \Rightarrow \frac{3.9t}{73} = \ln 9 \Rightarrow t = \frac{73 \ln 9}{3.9} \approx 41.13$ sec

    It will take about 41.13 seconds.

3. The total distance traveled $= \frac{v_0 m}{k} \Rightarrow \frac{(2.75)(39.92)}{k} = 4.91 \Rightarrow k = 22.36$. Therefore, the distance traveled is given by the

    function $s(t) = 4.91\left(1 - e^{-(22.36/39.92)t}\right)$. The graph shows $s(t)$ and the data points.

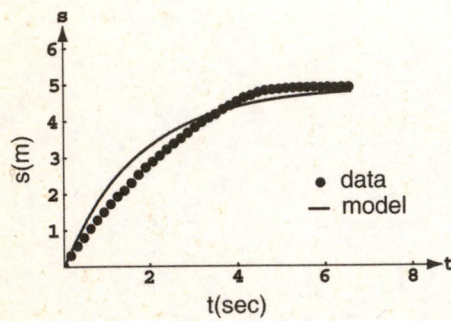

5. (a) $\frac{dP}{dt} = 0.0015P(150 - P) = \frac{0.255}{150}P(150 - P) = \frac{k}{M}P(M - P)$

    Thus, $k = 0.255$ and $M = 150$, and $P = \frac{M}{1 + Ae^{-kt}} = \frac{150}{1 + Ae^{-0.255t}}$

    Initial condition: $P(0) = 6 \Rightarrow 6 = \frac{150}{1 + Ae^0} \Rightarrow 1 + A = 25 \Rightarrow A = 24$

    Formula: $P = \frac{150}{1 + 24e^{-0.255t}}$

    (b) $100 = \frac{150}{1 + 24e^{-0.255t}} \Rightarrow 1 + 24e^{-0.255t} = \frac{3}{2} \Rightarrow 24e^{-0.255t} = \frac{1}{2} \Rightarrow e^{-0.255t} = \frac{1}{48} \Rightarrow -0.255t = -\ln 48$

    $\Rightarrow t = \frac{\ln 48}{0.255} \approx 17.21$ weeks

    $125 = \frac{150}{1 + 24e^{-0.255t}} \Rightarrow 1 + 24e^{-0.255t} = \frac{6}{5} \Rightarrow 24e^{-0.255t} = \frac{1}{5} \Rightarrow e^{-0.255t} = \frac{1}{120} \Rightarrow -0.255t = -\ln 120$

    $\Rightarrow t = \frac{\ln 120}{0.255} \approx 21.28$

    It will take about 17.21 weeks to reach 100 guppies, and about 21.28 weeks to reach 125 guppies.

7. (a) Using the general solution form Example 2, part (c),

    $\frac{dy}{dt} = (0.08875 \times 10^{-7})(8 \times 10^7 - y)y \Rightarrow y(t) = \frac{M}{1 + Ae^{-rMt}} = \frac{8 \times 10^7}{1 + Ae^{-(0.08875)(8)t}} = \frac{8 \times 10^7}{1 + Ae^{-0.71t}}$

    Apply the initial condition:

    $y(0) = 1.6 \times 10^7 = \frac{8 \times 10^7}{1 + A} \Rightarrow \frac{8}{1.6} - 1 = 4 \Rightarrow y(1) = \frac{8 \times 10^7}{1 + 4e^{-0.71}} \approx 2.69671 \times 10^7$ kg.

    (b) $y(t) = 4 \times 10^7 = \frac{8 \times 10^7}{1 + 4e^{-0.71t}} \Rightarrow 4e^{-0.71t} = 1 \Rightarrow t = -\frac{\ln\left(\frac{1}{4}\right)}{0.71} \approx 1.95253$ years.

9. (a) $\frac{dy}{dt} = 1 + y \Rightarrow dy = (1 + y)dt \Rightarrow \frac{dy}{1 + y} = dt \Rightarrow \ln|1 + y| = t + C_1 \Rightarrow e^{\ln|1+y|} = e^{t+C_1} \Rightarrow |1 + y| = e^t e^{C_1}$

    $1 + y = \pm C_2 e^t \Rightarrow y = Ce^t - 1$, where $C_2 = e^{C_1}$ and $C = \pm C_2$. Apply the initial condition: $y(0) = 1 = Ce^0 - 1$

    $\Rightarrow C = 2 \Rightarrow y = 2e^t - 1$.

    (b) $\frac{dy}{dt} = 0.5(400 - y)y \Rightarrow dy = 0.5(400 - y)y\, dt \Rightarrow \frac{dy}{(400 - y)y} = 0.5\, dt$. Using the partial fraction decomposition in

    Example 2, part (c), we obtain $\frac{1}{400}\left(\frac{1}{y} + \frac{1}{400 - y}\right)dy = 0.5\, dt \Rightarrow \left(\frac{1}{y} + \frac{1}{400 - y}\right)dy = 200\, dt$

    $\Rightarrow \int\left(\frac{1}{y} + \frac{1}{400 - y}\right)dy = \int 200\, dt \Rightarrow \ln|y| - \ln|y - 400| = 200t + C_1 \Rightarrow \ln\left|\frac{y}{y - 400}\right| = 200t + C_1$

$$\Rightarrow e^{\ln\left|\frac{y}{y-400}\right|} = e^{200t+C_1} = e^{200t}e^{C_1} \Rightarrow \left|\frac{y}{y-400}\right| = C_2 e^{200t} \text{ (where } C_2 = e^{C_1}) \Rightarrow \frac{y}{y-400} = \pm C_2 e^{200t}$$

$$\Rightarrow \frac{y}{y-400} = Ce^{200t} \text{ (where } C = \pm C_2) \Rightarrow y = Ce^{200t}y - 400\,Ce^{200t} \Rightarrow (1 - Ce^{200t})y = -400\,Ce^{200t}$$

$$\Rightarrow y = \frac{400\,Ce^{200t}}{Ce^{200t}-1} \Rightarrow y = \frac{400}{1-\frac{1}{C}e^{-200t}} = \frac{400}{1+Ae^{-200t}}, \text{ where } A = -\frac{1}{C}. \text{ Apply the initial condition:}$$

$$y(0) = 2 = \frac{400}{1+Ae^0} \Rightarrow A = 199 \Rightarrow y(t) = \frac{400}{1+199e^{-200t}}$$

11. (a) $\frac{dP}{dt} = kP^2 \Rightarrow \int P^{-2}dP = \int k\,dt \Rightarrow -P^{-1} = kt + C \Rightarrow P = \frac{-1}{kt+C}$

Initial condition: $P(0) = P_0 \Rightarrow P_0 = -\frac{1}{C} \Rightarrow C = \frac{-1}{P_0}$

Solution: $P = -\frac{1}{kt-(1/P_0)} = \frac{P_0}{1-kP_0t}$

(b) There is a vertical asymptote at $t = \frac{1}{kP_O}$

13. $y = mx \Rightarrow \frac{y}{x} = m \Rightarrow \frac{xy'-y}{x^2} = 0 \Rightarrow y' = \frac{y}{x}$. So for

orthogonals: $\frac{dy}{dx} = -\frac{x}{y} \Rightarrow y\,dy = -x\,dx \Rightarrow \frac{y^2}{2} + \frac{x^2}{2} = C$

$\Rightarrow x^2 + y^2 = C_1$

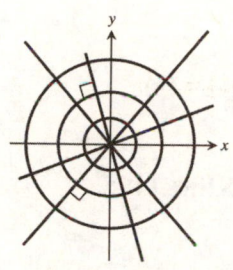

15. $kx^2 + y^2 = 1 \Rightarrow 1 - y^2 = kx^2 \Rightarrow \frac{1-y^2}{x^2} = k$

$\Rightarrow \frac{x^2(2y)y' - (1-y^2)2x}{x^4} = 0 \Rightarrow -2yx^2y' = (1-y^2)(2x)$

$\Rightarrow y' = \frac{(1-y^2)(2x)}{-2xy^2} = \frac{(1-y^2)}{-xy}$. So for the orthogonals:

$\frac{dy}{dx} = \frac{xy}{1-y^2} \Rightarrow \frac{(1-y^2)}{y}dy = x\,dx \Rightarrow \ln y - \frac{y^2}{2} = \frac{x^2}{2} + C$

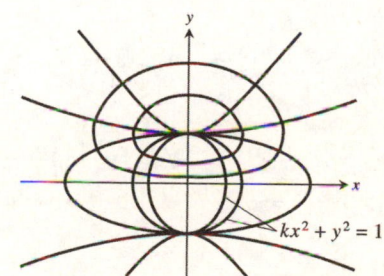

17. $y = ce^{-x} \Rightarrow \frac{y}{e^{-x}} = c \Rightarrow \frac{e^{-x}y' - y(e^{-x})(-1)}{(e^{-x})^2} = 0$

$\Rightarrow e^{-x}y' = -ye^{-x} \Rightarrow y' = -y$. So for the orthogonals:

$\frac{dy}{dx} = \frac{1}{y} \Rightarrow y\,dy = dx \Rightarrow \frac{y^2}{2} = x + C$

$\Rightarrow y^2 = 2x + C_1 \Rightarrow y = \pm\sqrt{2x + C_1}$

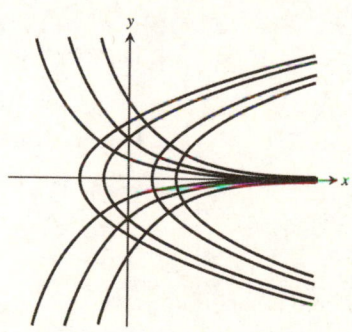

19. $2x^2 + 3y^2 = 5$ and $y^2 = x^3$ intersect at $(1, 1)$. Also, $2x^2 + 3y^2 = 5 \Rightarrow 4x + 6y\,y' = 0 \Rightarrow y' = -\frac{4x}{6y} \Rightarrow y'(1, 1) = -\frac{2}{3}$

$y_1^2 = x^3 \Rightarrow 2y_1y_1' = 3x^2 \Rightarrow y_1' = \frac{3x^2}{2y_1} \Rightarrow y_1'(1, 1) = \frac{3}{2}$. Since $y' \cdot y_1' = \left(-\frac{2}{3}\right)\left(\frac{3}{2}\right) = -1$, the curves are orthogonal.

21. $y^2 = 4a^2 - 4ax$ and $y^2 = 4b^2 + 4bx \Rightarrow$ (at intersection) $4a^2 - 4ax = 4b^2 + 4bx \Rightarrow a^2 - b^2 = x(a + b)$

$\Rightarrow (a + b)(a - b) = (a + b)x \Rightarrow x = a - b$. Now, $y^2 = 4a^2 - 4a(a - b) = 4a^2 - 4a^2 + 4ab = 4ab \Rightarrow y = \pm 2\sqrt{ab}$.

Thus the intersections are at $\left(a - b, \pm 2\sqrt{ab}\right)$. So, $y^2 = 4a^2 - 4ax \Rightarrow y_1' = -\frac{4a}{2y}$ which are equal to $-\frac{4a}{2\left(2\sqrt{ab}\right)}$ and

$-\frac{4a}{2\left(-2\sqrt{ab}\right)} = -\sqrt{\frac{a}{b}}$ and $\sqrt{\frac{a}{b}}$ at the intersections. Also, $y^2 = 4b^2 + 4bx \Rightarrow y_2' = \frac{4b}{2y}$ which are equal to $\frac{4b}{2\left(2\sqrt{ab}\right)}$ and

$\frac{4b}{2(-2\sqrt{ab})} = -\sqrt{\frac{b}{a}}$ and $\sqrt{\frac{b}{a}}$ at the intersections. $(y_1') \cdot (y_2') = -1$. Thus the curves are orthogonal.

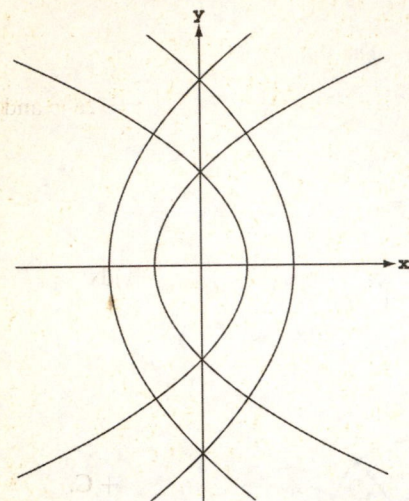

CHAPTER 9 PRACTICE EXERCISES

1. $\frac{dy}{dx} = \sqrt{y}\cos^2\sqrt{y} \Rightarrow \frac{dy}{\sqrt{y}\cos^2\sqrt{y}} = dx \Rightarrow 2\tan\sqrt{y} = x + C \Rightarrow y = \left(\tan^{-1}\left(\frac{x+C}{2}\right)\right)^2$

3. $yy' = \sec(y^2)\sec^2 x \Rightarrow \frac{y\,dy}{\sec(y^2)} = \sec^2 x\,dx \Rightarrow \frac{\sin(y^2)}{2} = \tan x + C \Rightarrow \sin(y^2) = 2\tan x + C_1$

5. $y' = xe^y\sqrt{x-2} \Rightarrow e^{-y}dy = x\sqrt{x-2}\,dx \Rightarrow -e^{-y} = \frac{2(x-2)^{3/2}(3x+4)}{15} + C \Rightarrow e^{-y} = \frac{-2(x-2)^{3/2}(3x+4)}{15} - C$
    $\Rightarrow -y = \ln\left[\frac{-2(x-2)^{3/2}(3x+4)}{15} - C\right] \Rightarrow y = -\ln\left[\frac{-2(x-2)^{3/2}(3x+4)}{15} - C\right]$

7. $\sec x\,dy + x\cos^2 y\,dx = 0 \Rightarrow \frac{dy}{\cos^2 y} = -\frac{x\,dx}{\sec x} \Rightarrow \tan y = -\cos x - x\sin x + C$

9. $y' = \frac{e^y}{xy} \Rightarrow ye^{-y}dy = \frac{dx}{x} \Rightarrow (y+1)e^{-y} = -\ln|x| + C$

11. $x(x-1)dy - y\,dx = 0 \Rightarrow x(x-1)dy = y\,dx \Rightarrow \frac{dy}{y} = \frac{dx}{x(x-1)} \Rightarrow \ln y = \ln(x-1) - \ln(x) + C$
    $\Rightarrow \ln y = \ln(x-1) - \ln(x) + \ln C_1 \Rightarrow \ln y = \ln\left(\frac{C_1(x-1)}{x}\right) \Rightarrow y = \frac{C_1(x-1)}{x}$

13. $2y' - y = xe^{x/2} \Rightarrow y' - \frac{1}{2}y = \frac{x}{2}e^{x/2}$.
    $p(x) = -\frac{1}{2},\ v(x) = e^{\int\left(-\frac{1}{2}\right)dx} = e^{-x/2}$.
    $e^{-x/2}y' - \frac{1}{2}e^{-x/2}y = \left(e^{-x/2}\right)\left(\frac{x}{2}\right)\left(e^{x/2}\right) = \frac{x}{2} \Rightarrow \frac{d}{dx}\left(e^{-x/2}y\right) = \frac{x}{2} \Rightarrow e^{-x/2}y = \frac{x^2}{4} + C \Rightarrow y = e^{x/2}\left(\frac{x^2}{4} + C\right)$

15. $xy' + 2y = 1 - x^{-1} \Rightarrow y' + \left(\frac{2}{x}\right)y = \frac{1}{x} - \frac{1}{x^2}$.
    $v(x) = e^{2\int\frac{dx}{x}} = e^{2\ln x} = e^{\ln x^2} = x^2$.
    $x^2y' + 2xy = x - 1 \Rightarrow \frac{d}{dx}(x^2y) = x - 1 \Rightarrow x^2y = \frac{x^2}{2} - x + C \Rightarrow y = \frac{1}{2} - \frac{1}{x} + \frac{C}{x^2}$

17. $(1+e^x)dy + (ye^x + e^{-x})dx = 0 \Rightarrow (1+e^x)y' + e^xy = -e^{-x} \Rightarrow y' = \frac{e^x}{1+e^x}y = \frac{-e^{-x}}{(1+e^x)}$.
    $v(x) = e^{\int\frac{e^x dx}{(1+e^x)}} = e^{\ln(e^x+1)} = e^x + 1$.
    $(e^x+1)y' + (e^x+1)\left(\frac{e^x}{1+e^x}\right)y = \frac{-e^{-x}}{(1+e^x)}(e^x+1) \Rightarrow \frac{d}{dx}[(e^x+1)y] = -e^{-x} \Rightarrow (e^x+1)y = e^{-x} + C$

$\Rightarrow y = \frac{e^{-x}+C}{e^x+1} = \frac{e^{-x}+C}{1+e^x}$

19. $(x+3y^2)\,dy + y\,dx = 0 \Rightarrow x\,dy + y\,dx = -3y^2dy \Rightarrow \frac{d}{dx}(xy) = -3y^2dy \Rightarrow xy = -y^3 + C$

21. $\frac{dy}{dx} = e^{-x-y-2} \Rightarrow e^y dy = e^{-(x+2)}dx \Rightarrow e^y = -e^{-(x+2)} + C.$ We have $y(0) = -2$, so $e^{-2} = -e^{-2} + C \Rightarrow C = 2e^{-2}$ and
$e^y = -e^{-(x+2)} + 2e^{-2} \Rightarrow y = \ln(-e^{-(x+2)} + 2e^{-2})$

23. $(x+1)\frac{dy}{dx} + 2y = x \Rightarrow y' + \left(\frac{2}{x+1}\right)y = \frac{x}{x+1}.$ Let $v(x) = e^{\int \frac{2}{x+1}dx} = e^{2\ln(x+1)} = e^{\ln(x+1)^2} = (x+1)^2.$
So $y'(x+1)^2 + \frac{2}{(x+1)}(x+1)^2y = \frac{x}{(x+1)}(x+1)^2 \Rightarrow \frac{d}{dx}\left[y(x+1)^2\right] = x(x+1) \Rightarrow y(x+1)^2 = \int x(x+1)dx$
$\Rightarrow y(x+1)^2 = \frac{x^3}{3} + \frac{x^2}{2} + C \Rightarrow y = (x+1)^{-2}\left(\frac{x^3}{3} + \frac{x^2}{2} + C\right).$ We have $y(0) = 1 \Rightarrow 1 = C.$ So
$y = (x+1)^{-2}\left(\frac{x^3}{3} + \frac{x^2}{2} + 1\right)$

25. $\frac{dy}{dx} + 3x^2y = x^2.$ Let $v(x) = e^{\int 3x^2dx} = e^{x^3}.$ So $e^{x^3}y' + 3x^2e^{x^3}y = x^2e^{x^3} \Rightarrow \frac{d}{dx}\left(e^{x^3}y\right) = x^2e^{x^3} \Rightarrow e^{x^3}y = \frac{1}{3}e^{x^3} + C.$
We have $y(0) = -1 \Rightarrow e^{0^3}(-1) = \frac{1}{3}e^{0^3} + C \Rightarrow -1 = \frac{1}{3} + C \Rightarrow C = -\frac{4}{3}$ and $e^{x^3}y = \frac{1}{3}e^{x^3} - \frac{4}{3} \Rightarrow y = \frac{1}{3} - \frac{4}{3}e^{-x^3}$

27. $x\,dy - (y+\sqrt{y})dx = 0 \Rightarrow \frac{dy}{(y+\sqrt{y})} = \frac{dx}{x} \Rightarrow 2\ln(\sqrt{y}+1) = \ln x + C.$ We have $y(1) = 1 \Rightarrow 2\ln(\sqrt{1}+1) = \ln 1 + C$
$\Rightarrow 2\ln 2 = C = \ln 2^2 = \ln 4.$ So $2\ln(\sqrt{y}+1) = \ln x + \ln 4 = \ln(4x) \Rightarrow \ln(\sqrt{y}+1) = \frac{1}{2}\ln(4x) = \ln(4x)^{1/2}$
$\Rightarrow e^{\ln(\sqrt{y}+1)} = e^{\ln(4x)^{1/2}} \Rightarrow \sqrt{y}+1 = 2\sqrt{x} \Rightarrow y = (2\sqrt{x}-1)^2$

29. $xy' + (x-2)y = 3x^3e^{-x} \Rightarrow y' + \left(\frac{x-2}{x}\right)y = 3x^2e^{-x}.$ Let $v(x) = e^{\int \left(\frac{x-2}{x}\right)dx} = e^{x-2\ln x} = \frac{e^x}{x^2}.$ So
$\frac{e^x}{x^2}y' + \frac{e^x}{x^2}\left(\frac{x-2}{x}\right)y = 3 \Rightarrow \frac{d}{dx}\left(y \cdot \frac{e^x}{x^2}\right) = 3 \Rightarrow y \cdot \frac{e^x}{x^2} = 3x + C.$ We have $y(1) = 0 \Rightarrow 0 = 3(1) + C \Rightarrow C = -3$
$\Rightarrow y \cdot \frac{e^x}{x^2} = 3x - 3 \Rightarrow y = x^2e^{-x}(3x-3)$

31. To find the approximate values let $y_n = y_{n-1} + (y_{n-1} + \cos x_{n-1})(0.1)$ with $x_0 = 0$, $y_0 = 0$, and 20 steps. Use a
spreadsheet, graphing calculator, or CAS to obtain the values in the following table.

x	y		x	y
0	0		1.1	1.6241
0.1	0.1000		1.2	1.8319
0.2	0.2095		1.3	2.0513
0.3	0.3285		1.4	2.2832
0.4	0.4568		1.5	2.5285
0.5	0.5946		1.6	2.7884
0.6	0.7418		1.7	3.0643
0.7	0.8986		1.8	3.3579
0.8	1.0649		1.9	3.6709
0.9	1.2411		2.0	4.0057
1.0	1.4273			

33. To estimate $y(3)$, let $z_n = y_{n-1} + \left(\frac{x_{n-1}-2y_{n-1}}{x_{n-1}+1}\right)(0.05)$ and $y_n = y_{n-1} + \frac{1}{2}\left(\frac{x_{n-1}-2y_{n-1}}{x_{n-1}+1} + \frac{x_n-2z_n}{x_n+1}\right)(0.05)$ with initial values
$x_0 = 0$, $y_0 = 1$, and 60 steps. Use a spreadsheet, graphing calculator, or CAS to obtain $y(3) \approx 0.9063.$

35. Let $y_n = y_{n-1} + \left(\frac{1}{e^{x_{n-1}+y_{n-1}+2}}\right)(dx)$ with starting values $x_0 = 0$ and $y_0 = 2$, and steps of 0.1 and -0.1. Use a spreadsheet,
programmable calculator, or CAS to generate the following graphs.

(a)

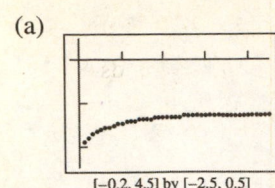

[−0.2, 4.5] by [−2.5, 0.5]

(b) Note that we choose a small interval of x-values because the y-values decrease very rapidly and our calculator cannot handle the calculations for $x \leq -1$. (This occurs because the analytic solution is $y = -2 + \ln(2 - e^{-x})$, which has an asymptote at $x = -\ln 2 \approx 0.69$. Obviously, the Euler approximations are misleading for $x \leq -0.7$.)

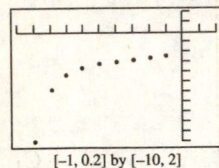

[−1, 0.2] by [−10, 2]

37.

x	1	1.2	1.4	1.6	1.8	2.0
y	−1	−0.8	−0.56	−0.28	0.04	0.4

$\frac{dy}{dx} = x \Rightarrow dy = x\,dx \Rightarrow y = \frac{x^2}{2} + C;\ x = 1$ and $y = -1$

$\Rightarrow -1 = \frac{1}{2} + C \Rightarrow C = -\frac{3}{2} \Rightarrow y(\text{exact}) = \frac{x^2}{2} - \frac{3}{2}$

$\Rightarrow y(2) = \frac{2^2}{2} - \frac{3}{2} = \frac{1}{2}$ is the exact value.

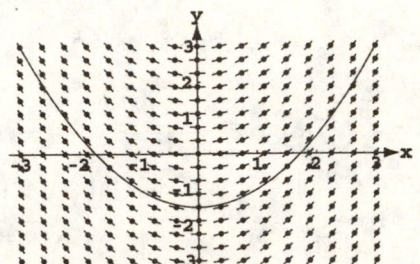

39.

x	1	1.2	1.4	1.6	1.8	2.0
y	−1	−1.2	−0.488	−1.9046	−2.5141	−3.4192

$\frac{dy}{dx} = xy \Rightarrow \frac{dy}{y} = x\,dx \Rightarrow \ln|y| = \frac{x^2}{2} + C$

$\Rightarrow y = e^{\frac{x^2}{2}+C} = e^{\frac{x^2}{2}} \cdot e^C = C_1 e^{\frac{x^2}{2}};\ x = 1$ and $y = -1$

$\Rightarrow -1 = C_1 e^{1/2} \Rightarrow C_1 = -e^{1/2} y(\text{exact}) = -e^{1/2} \cdot e^{\frac{x^2}{2}}$

$= -e^{(x^2-1)/2} \Rightarrow y(2) = -e^{3/2} \approx -4.4817$ is the

exact value.

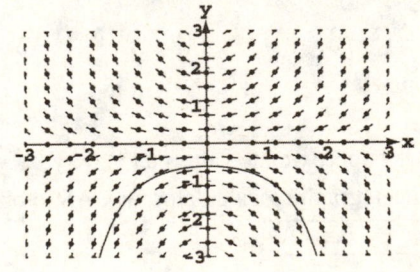

41. $\frac{dy}{dx} = y^2 - 1 \Rightarrow y' = (y+1)(y-1)$. We have $y' = 0 \Rightarrow (y+1) = 0, (y-1) = 0 \Rightarrow y = -1, 1$.

(a) Equilibrium points are -1 (stable) and 1 (unstable)

(b) $y' = y^2 - 1 \Rightarrow y'' = 2yy' \Rightarrow y'' = 2y(y^2 - 1) = 2y(y+1)(y-1)$. So $y'' = 0 \Rightarrow y = 0, y = -1, y = 1$.

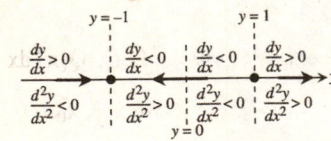

(c)

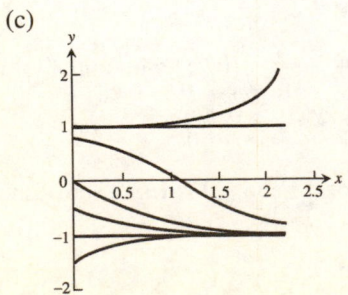

43. (a) Force $=$ Mass times Acceleration (Newton's Second Law) or $F = ma$. Let $a = \frac{dv}{dt} = \frac{dv}{ds} \cdot \frac{ds}{dt} = v\frac{dv}{ds}$. Then

$ma = -mgR^2s^{-2} \Rightarrow a = -gR^2s^{-2} \Rightarrow v\frac{dv}{ds} = -gR^2s^{-2} \Rightarrow v\,dv = -gR^2s^{-2}ds \Rightarrow \int v\,dv = \int -gR^2s^{-2}ds$

$\Rightarrow \frac{v^2}{2} = \frac{gR^2}{s} + C_1 \Rightarrow v^2 = \frac{2gR^2}{s} + 2C_1 = \frac{2gR^2}{s} + C$. When $t = 0$, $v = v_0$ and $s = R \Rightarrow v_0^2 = \frac{2gR^2}{R} + C$

$\Rightarrow C = v_0^2 - 2gR \Rightarrow v^2 = \frac{2gR^2}{s} + v_0^2 - 2gR$

(b) If $v_0 = \sqrt{2gR}$, then $v^2 = \frac{2gR^2}{s} \Rightarrow v = \sqrt{\frac{2gR^2}{s}}$, since $v \geq 0$ if $v_0 \geq \sqrt{2gR}$. Then $\frac{ds}{dt} = \frac{\sqrt{2gR^2}}{\sqrt{s}} \Rightarrow \sqrt{s}\,ds = \sqrt{2gR^2}\,dt$

$\Rightarrow \int s^{1/2}ds = \int \sqrt{2gR^2}\,dt \Rightarrow \frac{2}{3}s^{3/2} = \sqrt{2gR^2}\,t + C_1 \Rightarrow s^{3/2} = \left(\frac{3}{2}\sqrt{2gR^2}\right)t + C; t = 0$ and $s = R$

$\Rightarrow R^{3/2} = \left(\frac{3}{2}\sqrt{2gR^2}\right)(0) + C \Rightarrow C = R^{3/2} \Rightarrow s^{3/2} = \left(\frac{3}{2}\sqrt{2gR^2}\right)t + R^{3/2} = \left(\frac{3}{2}R\sqrt{2g}\right)t + R^{3/2}$

$= R^{3/2}\left[\left(\frac{3}{2}R^{-1/2}\sqrt{2g}\right)t + 1\right] = R^{3/2}\left[\left(\frac{3\sqrt{2gR}}{2R}\right)t + 1\right] = R^{3/2}\left[\left(\frac{3v_0}{2R}\right)t + 1\right] \Rightarrow s = R\left[1 + \left(\frac{3v_0}{2R}\right)t\right]^{2/3}$

CHAPTER 9 ADDITIONAL AND ADVANCED EXERCISES

1. (a) $\frac{dy}{dt} = k\frac{A}{V}(c - y) \Rightarrow dy = -k\frac{A}{V}(y - c)dt \Rightarrow \frac{dy}{y-c} = -k\frac{A}{V}dt \Rightarrow \int \frac{dy}{y-c} = -\int k\frac{A}{V}dt \Rightarrow \ln|y - c| = -k\frac{A}{V}t + C_1$

$\Rightarrow y - c = \pm e^{C_1}e^{-k\frac{A}{V}t}$. Apply the initial condition, $y(0) = y_0 \Rightarrow y_0 = c + C \Rightarrow C = y_0 - c$

$\Rightarrow y = c + (y_0 - c)e^{-k\frac{A}{V}t}$.

(b) Steady state solution: $y_\infty = \lim_{t\to\infty} y(t) = \lim_{t\to\infty}\left[c + (y_0 - c)e^{-k\frac{A}{V}t}\right] = c + (y_0 - c)(0) = c$

3. The amount of CO_2 in the room at time t is $A(t)$. The rate of change in the amount of CO_2, $\frac{dA}{dt}$ is the rate of internal production (R_1) plus the inflow rate (R_2) minus the outflow rate (R_3).

$R_1 = \left(20 \frac{\text{breaths/min}}{\text{student}}\right)(30 \text{ students})\left(\frac{100}{1728} \text{ ft}^3\right)\left(0.04 \frac{\text{ft}^3 \, CO_2}{\text{ft}^3}\right) \approx 1.39 \frac{\text{ft}^3 \, CO_2}{\text{min}}$

$R_2 = \left(1000 \frac{\text{ft}^3}{\text{min}}\right)\left(0.0004 \frac{\text{ft}^3 \, CO_2}{\text{min}}\right) = 0.4 \frac{\text{ft}^3 \, CO_2}{\text{min}}$

$R_3 = \left(\frac{A}{10,000}\right)1000 = 0.1A \frac{\text{ft}^3 \, CO_2}{\text{min}}$

$\frac{dA}{dt} = 1.39 + 0.4 - 0.1A = 1.79 - 0.1A \Rightarrow A' + 0.1A = 1.79$. Let $v(t) = e^{\int 0.1dt}$. We have $\frac{d}{dt}\left(Ae^{\int 0.1dt}\right) = 1.79e^{\int 0.1dt}$

$\Rightarrow Ae^{0.1t} = \int 1.79e^{0.1t}dt = 17.9e^{0.1t} + C$. At $t = 0$, $A = (10,000)(0.0004) = 4 \text{ ft}^3 \, CO_2 \Rightarrow C = -13.9$

$\Rightarrow A = 17.9 - 13.9e^{-0.1t}$. So $A(60) = 17.9 - 13.9e^{-0.1(60)} \approx 17.87 \text{ ft}^3$ of CO_2 in the 10,000 ft^3 room. The percent of CO_2 is $\frac{17.87}{10,000} \times 100 = 0.18\%$

5. (a) Let y be any function such that $v(x)y = \int v(x)Q(x)\,dx + C$, $v(x) = e^{\int P(x)\,dx}$. Then

$\frac{d}{dx}(v(x) \cdot y) = v(x) \cdot y' + y \cdot v'(x) = v(x)Q(x)$. We have $v(x) = e^{\int P(x)\,dx} \Rightarrow v'(x) = e^{\int P(x)\,dx}P(x) = v(x)P(x)$.

Thus $v(x) \cdot y' + y \cdot v(x)P(x) = v(x)Q(x) \Rightarrow y' + yP(x) = Q(x) \Rightarrow$ the given y is a solution.

(b) If v and Q are continuous on $[a, b]$ and $x \in (a, b)$, then $\frac{d}{dx}\left[\int_{x_0}^x v(t)Q(t)\,dt\right] = v(x)Q(x)$

$\Rightarrow \int_{x_0}^x v(t)Q(t)\,dt = \int v(x)Q(x)\,dx$. So $C = y_0v(x_0) - \int v(x)Q(x)\,dx$. From part (a), $v(x)y = \int v(x)Q(x)\,dx + C$.

Substituting for C: $v(x)y = \int v(x)Q(x)\,dx + y_0v(x_0) - \int v(x)Q(x)\,dx \Rightarrow v(x)y = y_0v(x_0)$ when $x = x_0$.

NOTES:

CHAPTER 10   CONIC SECTIONS AND POLAR COORDINATES

10.1 CONIC SECTIONS AND QUADRATIC EQUATIONS

1. $x = \frac{y^2}{8} \Rightarrow 4p = 8 \Rightarrow p = 2$; focus is $(2, 0)$, directrix is $x = -2$

3. $y = -\frac{x^2}{6} \Rightarrow 4p = 6 \Rightarrow p = \frac{3}{2}$; focus is $\left(0, -\frac{3}{2}\right)$, directrix is $y = \frac{3}{2}$

5. $\frac{x^2}{4} - \frac{y^2}{9} = 1 \Rightarrow c = \sqrt{4+9} = \sqrt{13} \Rightarrow$ foci are $\left(\pm\sqrt{13}, 0\right)$; vertices are $(\pm 2, 0)$; asymptotes are $y = \pm\frac{3}{2}x$

7. $\frac{x^2}{2} + y^2 = 1 \Rightarrow c = \sqrt{2-1} = 1 \Rightarrow$ foci are $(\pm 1, 0)$; vertices are $\left(\pm\sqrt{2}, 0\right)$

9. $y^2 = 12x \Rightarrow x = \frac{y^2}{12} \Rightarrow 4p = 12 \Rightarrow p = 3$; focus is $(3, 0)$, directrix is $x = -3$

11. $x^2 = -8y \Rightarrow y = \frac{x^2}{-8} \Rightarrow 4p = 8 \Rightarrow p = 2$; focus is $(0, -2)$, directrix is $y = 2$

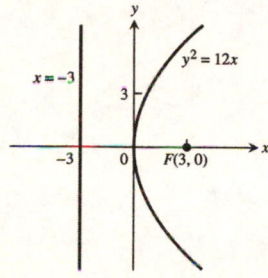

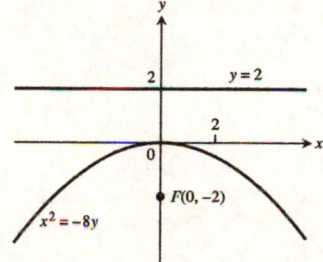

13. $y = 4x^2 \Rightarrow y = \frac{x^2}{\left(\frac{1}{4}\right)} \Rightarrow 4p = \frac{1}{4} \Rightarrow p = \frac{1}{16}$; focus is $\left(0, \frac{1}{16}\right)$, directrix is $y = -\frac{1}{16}$

15. $x = -3y^2 \Rightarrow x = -\frac{y^2}{\left(\frac{1}{3}\right)} \Rightarrow 4p = \frac{1}{3} \Rightarrow p = \frac{1}{12}$; focus is $\left(-\frac{1}{12}, 0\right)$, directrix is $x = \frac{1}{12}$

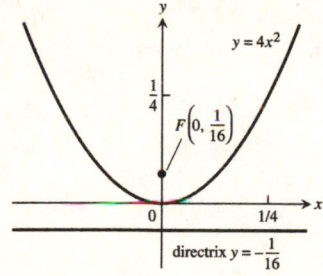

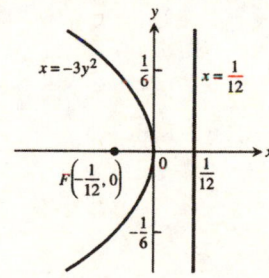

17. $16x^2 + 25y^2 = 400 \Rightarrow \frac{x^2}{25} + \frac{y^2}{16} = 1$
$\Rightarrow c = \sqrt{a^2 - b^2} = \sqrt{25 - 16} = 3$

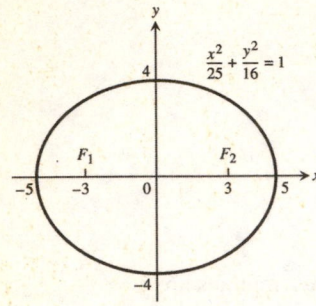

19. $2x^2 + y^2 = 2 \Rightarrow x^2 + \frac{y^2}{2} = 1$
$\Rightarrow c = \sqrt{a^2 - b^2} = \sqrt{2 - 1} = 1$

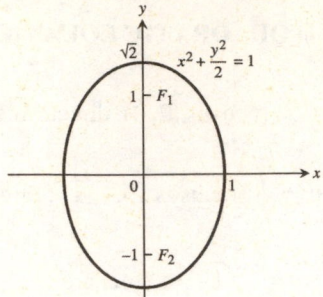

21. $3x^2 + 2y^2 = 6 \Rightarrow \frac{x^2}{2} + \frac{y^2}{3} = 1$
$\Rightarrow c = \sqrt{a^2 - b^2} = \sqrt{3 - 2} = 1$

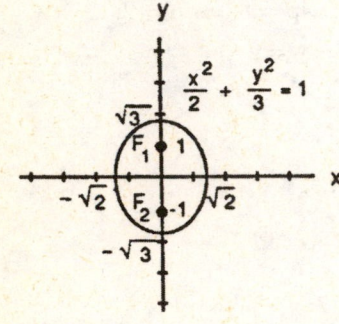

23. $6x^2 + 9y^2 = 54 \Rightarrow \frac{x^2}{9} + \frac{y^2}{6} = 1$
$\Rightarrow c = \sqrt{a^2 - b^2} = \sqrt{9 - 6} = \sqrt{3}$

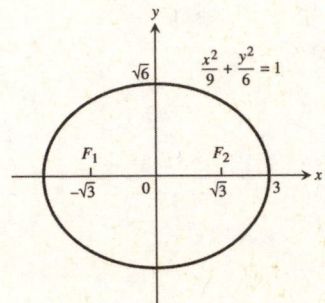

25. Foci: $\left(\pm\sqrt{2}, 0\right)$, Vertices: $(\pm 2, 0) \Rightarrow a = 2, c = \sqrt{2} \Rightarrow b^2 = a^2 - c^2 = 4 - \left(\sqrt{2}\right)^2 = 2 \Rightarrow \frac{x^2}{4} + \frac{y^2}{2} = 1$

27. $x^2 - y^2 = 1 \Rightarrow c = \sqrt{a^2 + b^2} = \sqrt{1 + 1} = \sqrt{2}$;
asymptotes are $y = \pm x$

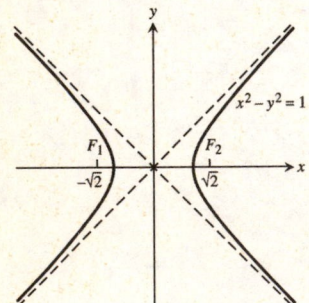

29. $y^2 - x^2 = 8 \Rightarrow \frac{y^2}{8} - \frac{x^2}{8} = 1 \Rightarrow c = \sqrt{a^2 + b^2}$
$= \sqrt{8 + 8} = 4$; asymptotes are $y = \pm x$

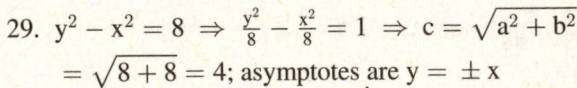
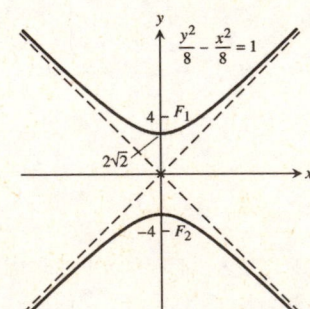

31. $8x^2 - 2y^2 = 16 \Rightarrow \frac{x^2}{2} - \frac{y^2}{8} = 1 \Rightarrow c = \sqrt{a^2 + b^2}$
$= \sqrt{2 + 8} = \sqrt{10}$; asymptotes are $y = \pm 2x$

33. $8y^2 - 2x^2 = 16 \Rightarrow \frac{y^2}{2} - \frac{x^2}{8} = 1 \Rightarrow c = \sqrt{a^2 + b^2}$
$= \sqrt{2 + 8} = \sqrt{10}$; asymptotes are $y = \pm \frac{x}{2}$

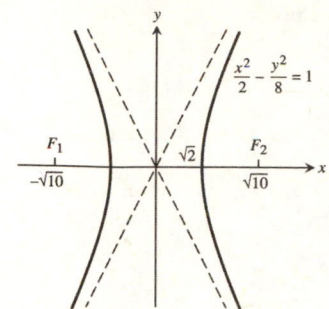

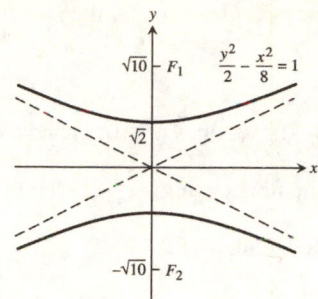

35. Foci: $\left(0, \pm \sqrt{2}\right)$, Asymptotes: $y = \pm x \Rightarrow c = \sqrt{2}$ and $\frac{a}{b} = 1 \Rightarrow a = b \Rightarrow c^2 = a^2 + b^2 = 2a^2 \Rightarrow 2 = 2a^2$
$\Rightarrow a = 1 \Rightarrow b = 1 \Rightarrow y^2 - x^2 = 1$

37. Vertices: $(\pm 3, 0)$, Asymptotes: $y = \pm \frac{4}{3}x \Rightarrow a = 3$ and $\frac{b}{a} = \frac{4}{3} \Rightarrow b = \frac{4}{3}(3) = 4 \Rightarrow \frac{x^2}{9} - \frac{y^2}{16} = 1$

39. (a) $y^2 = 8x \Rightarrow 4p = 8 \Rightarrow p = 2 \Rightarrow$ directrix is $x = -2$,
    focus is $(2, 0)$, and vertex is $(0, 0)$; therefore the new
    directrix is $x = -1$, the new focus is $(3, -2)$, and the
    new vertex is $(1, -2)$

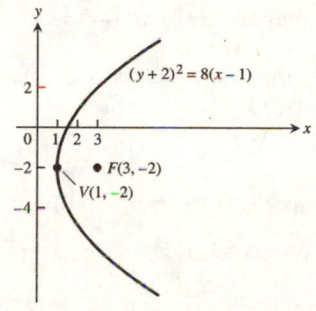

41. (a) $\frac{x^2}{16} + \frac{y^2}{9} = 1 \Rightarrow$ center is $(0, 0)$, vertices are $(-4, 0)$
    and $(4, 0)$; $c = \sqrt{a^2 - b^2} = \sqrt{7} \Rightarrow$ foci are $\left(\sqrt{7}, 0\right)$
    and $\left(-\sqrt{7}, 0\right)$; therefore the new center is $(4, 3)$, the
    new vertices are $(0, 3)$ and $(8, 3)$, and the new foci are
    $\left(4 \pm \sqrt{7}, 3\right)$

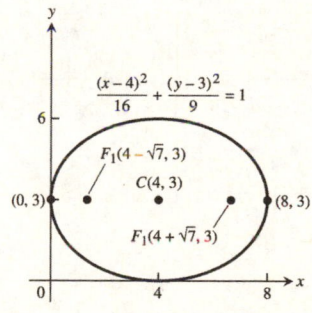

43. (a) $\frac{x^2}{16} - \frac{y^2}{9} = 1 \Rightarrow$ center is $(0, 0)$, vertices are $(-4, 0)$
    and $(4, 0)$, and the asymptotes are $\frac{x}{4} = \pm \frac{y}{3}$ or
    $y = \pm \frac{3x}{4}$; $c = \sqrt{a^2 + b^2} = \sqrt{25} = 5 \Rightarrow$ foci are
    $(-5, 0)$ and $(5, 0)$; therefore the new center is $(2, 0)$, the
    new vertices are $(-2, 0)$ and $(6, 0)$, the new foci
    are $(-3, 0)$ and $(7, 0)$, and the new asymptotes are
    $y = \pm \frac{3(x - 2)}{4}$

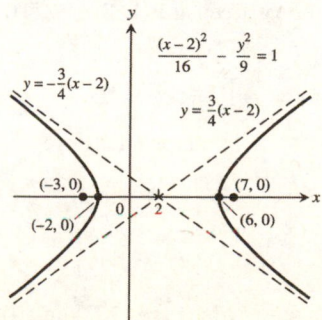

45. $y^2 = 4x \Rightarrow 4p = 4 \Rightarrow p = 1 \Rightarrow$ focus is $(1, 0)$, directrix is $x = -1$, and vertex is $(0, 0)$; therefore the new
    vertex is $(-2, -3)$, the new focus is $(-1, -3)$, and the new directrix is $x = -3$; the new equation is

$(y+3)^2 = 4(x+2)$

47. $x^2 = 8y \Rightarrow 4p = 8 \Rightarrow p = 2 \Rightarrow$ focus is $(0,2)$, directrix is $y = -2$, and vertex is $(0,0)$; therefore the new vertex is $(1,-7)$, the new focus is $(1,-5)$, and the new directrix is $y = -9$; the new equation is $(x-1)^2 = 8(y+7)$

49. $\frac{x^2}{6} + \frac{y^2}{9} = 1 \Rightarrow$ center is $(0,0)$, vertices are $(0,3)$ and $(0,-3)$; $c = \sqrt{a^2-b^2} = \sqrt{9-6} = \sqrt{3} \Rightarrow$ foci are $\left(0, \sqrt{3}\right)$ and $\left(0, -\sqrt{3}\right)$; therefore the new center is $(-2,-1)$, the new vertices are $(-2,2)$ and $(-2,-4)$, and the new foci are $\left(-2, -1 \pm \sqrt{3}\right)$; the new equation is $\frac{(x+2)^2}{6} + \frac{(y+1)^2}{9} = 1$

51. $\frac{x^2}{3} + \frac{y^2}{2} = 1 \Rightarrow$ center is $(0,0)$, vertices are $\left(\sqrt{3},0\right)$ and $\left(-\sqrt{3},0\right)$; $c = \sqrt{a^2-b^2} = \sqrt{3-2} = 1 \Rightarrow$ foci are $(-1,0)$ and $(1,0)$; therefore the new center is $(2,3)$, the new vertices are $\left(2 \pm \sqrt{3}, 3\right)$, and the new foci are $(1,3)$ and $(3,3)$; the new equation is $\frac{(x-2)^2}{3} + \frac{(y-3)^2}{2} = 1$

53. $\frac{x^2}{4} - \frac{y^2}{5} = 1 \Rightarrow$ center is $(0,0)$, vertices are $(2,0)$ and $(-2,0)$; $c = \sqrt{a^2+b^2} = \sqrt{4+5} = 3 \Rightarrow$ foci are $(3,0)$ and $(-3,0)$; the asymptotes are $\pm \frac{x}{2} = \frac{y}{\sqrt{5}} \Rightarrow y = \pm \frac{\sqrt{5}x}{2}$; therefore the new center is $(2,2)$, the new vertices are $(4,2)$ and $(0,2)$, and the new foci are $(5,2)$ and $(-1,2)$; the new asymptotes are $y - 2 = \pm \frac{\sqrt{5}(x-2)}{2}$; the new equation is $\frac{(x-2)^2}{4} - \frac{(y-2)^2}{5} = 1$

55. $y^2 - x^2 = 1 \Rightarrow$ center is $(0,0)$, vertices are $(0,1)$ and $(0,-1)$; $c = \sqrt{a^2+b^2} = \sqrt{1+1} = \sqrt{2} \Rightarrow$ foci are $\left(0, \pm \sqrt{2}\right)$; the asymptotes are $y = \pm x$; therefore the new center is $(-1,-1)$, the new vertices are $(-1,0)$ and $(-1,-2)$, and the new foci are $\left(-1, -1 \pm \sqrt{2}\right)$; the new asymptotes are $y + 1 = \pm(x+1)$; the new equation is $(y+1)^2 - (x+1)^2 = 1$

57. $x^2 + 4x + y^2 = 12 \Rightarrow x^2 + 4x + 4 + y^2 = 12 + 4 \Rightarrow (x+2)^2 + y^2 = 16$; this is a circle: center at $C(-2,0)$, $a = 4$

59. $x^2 + 2x + 4y - 3 = 0 \Rightarrow x^2 + 2x + 1 = -4y + 3 + 1 \Rightarrow (x+1)^2 = -4(y-1)$; this is a parabola: $V(-1,1)$, $F(-1,0)$

61. $x^2 + 5y^2 + 4x = 1 \Rightarrow x^2 + 4x + 4 + 5y^2 = 5 \Rightarrow (x+2)^2 + 5y^2 = 5 \Rightarrow \frac{(x+2)^2}{5} + y^2 = 1$; this is an ellipse: the center is $(-2,0)$, the vertices are $\left(-2 \pm \sqrt{5}, 0\right)$; $c = \sqrt{a^2-b^2} = \sqrt{5-1} = 2 \Rightarrow$ the foci are $(-4,0)$ and $(0,0)$

63. $x^2 + 2y^2 - 2x - 4y = -1 \Rightarrow x^2 - 2x + 1 + 2(y^2 - 2y + 1) = 2 \Rightarrow (x-1)^2 + 2(y-1)^2 = 2$ $\Rightarrow \frac{(x-1)^2}{2} + (y-1)^2 = 1$; this is an ellipse: the center is $(1,1)$, the vertices are $\left(1 \pm \sqrt{2}, 1\right)$; $c = \sqrt{a^2-b^2} = \sqrt{2-1} = 1 \Rightarrow$ the foci are $(2,1)$ and $(0,1)$

65. $x^2 - y^2 - 2x + 4y = 4 \Rightarrow x^2 - 2x + 1 - (y^2 - 4y + 4) = 1 \Rightarrow (x-1)^2 - (y-2)^2 = 1$; this is a hyperbola: the center is $(1,2)$, the vertices are $(2,2)$ and $(0,2)$; $c = \sqrt{a^2+b^2} = \sqrt{1+1} = \sqrt{2} \Rightarrow$ the foci are $\left(1 \pm \sqrt{2}, 2\right)$; the asymptotes are $y - 2 = \pm(x-1)$

67. $2x^2 - y^2 + 6y = 3 \Rightarrow 2x^2 - (y^2 - 6y + 9) = -6 \Rightarrow \frac{(y-3)^2}{6} - \frac{x^2}{3} = 1$; this is a hyperbola: the center is $(0, 3)$, the vertices are $\left(0, 3 \pm \sqrt{6}\right)$; $c = \sqrt{a^2 + b^2} = \sqrt{6 + 3} = 3 \Rightarrow$ the foci are $(0, 6)$ and $(0, 0)$; the asymptotes are $\frac{y-3}{\sqrt{6}} = \pm \frac{x}{\sqrt{3}} \Rightarrow y = \pm \sqrt{2}x + 3$

69.

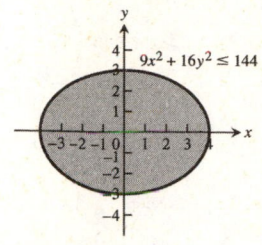

71.

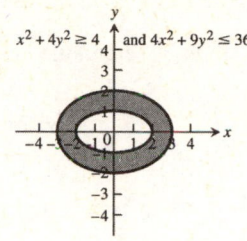

73.

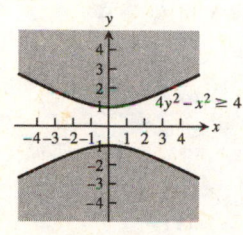

75. Volume of the Parabolic Solid: $V_1 = \int_0^{b/2} 2\pi x \left(h - \frac{4h}{b^2} x^2\right) dx = 2\pi h \int_0^{b/2} \left(x - \frac{4x^3}{b^2}\right) dx = 2\pi h \left[\frac{x^2}{2} - \frac{x^4}{b^2}\right]_0^{b/2}$

$= \frac{\pi h b^2}{8}$; Volume of the Cone: $V_2 = \frac{1}{3} \pi \left(\frac{b}{2}\right)^2 h = \frac{1}{3} \pi \left(\frac{b^2}{4}\right) h = \frac{\pi h b^2}{12}$; therefore $V_1 = \frac{3}{2} V_2$

77. A general equation of the circle is $x^2 + y^2 + ax + by + c = 0$, so we will substitute the three given points into

this equation and solve the resulting system: $\left. \begin{array}{r} a \quad\quad + c = -1 \\ b + c = -1 \\ 2a + 2b + c = -8 \end{array} \right\} \Rightarrow c = \frac{4}{3}$ and $a = b = -\frac{7}{3}$; therefore

$3x^2 + 3y^2 - 7x - 7y + 4 = 0$ represents the circle

79. $r^2 = (-2 - 1)^2 + (1 - 3)^2 = 13 \Rightarrow (x + 2)^2 + (y - 1)^2 = 13$ is an equation of the circle; the distance from the center to $(1.1, 2.8)$ is $\sqrt{(-2 - 1.1)^2 + (1 - 2.8)^2} = \sqrt{12.85} < \sqrt{13}$, the radius $\Rightarrow$ the point is inside the circle

81. (a) $y^2 = kx \Rightarrow x = \frac{y^2}{k}$; the volume of the solid formed by

revolving R_1 about the y-axis is $V_1 = \int_0^{\sqrt{kx}} \pi \left(\frac{y^2}{k}\right)^2 dy$

$= \frac{\pi}{k^2} \int_0^{\sqrt{kx}} y^4 \, dy = \frac{\pi x^2 \sqrt{kx}}{5}$; the volume of the right

circular cylinder formed by revolving PQ about the

y-axis is $V_2 = \pi x^2 \sqrt{kx} \Rightarrow$ the volume of the solid

formed by revolving R_2 about the y-axis is

$V_3 = V_2 - V_1 = \frac{4\pi x^2 \sqrt{kx}}{5}$. Therefore we can see the

ratio of V_3 to V_1 is 4:1.

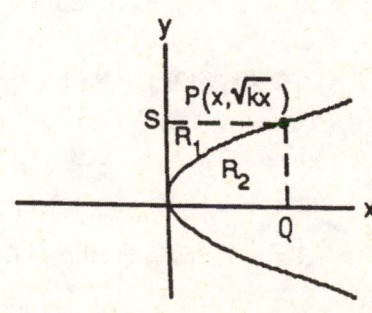

(b) The volume of the solid formed by revolving R_2 about the x-axis is $V_1 = \int_0^x \pi \left(\sqrt{kt}\right)^2 dt = \pi k \int_0^x t \, dt$

$= \frac{\pi k x^2}{2}$. The volume of the right circular cylinder formed by revolving PS about the x-axis is

$V_2 = \pi \left(\sqrt{kx}\right)^2 x = \pi k x^2 \Rightarrow$ the volume of the solid formed by revolving R_1 about the x-axis is

$V_3 = V_2 - V_1 = \pi k x^2 - \frac{\pi k x^2}{2} = \frac{\pi k x^2}{2}$. Therefore the ratio of V_3 to V_1 is 1:1.

83. Let $y = \sqrt{1 - \frac{x^2}{4}}$ on the interval $0 \le x \le 2$. The area of the inscribed rectangle is given by

$A(x) = 2x\left(2\sqrt{1 - \frac{x^2}{4}}\right) = 4x\sqrt{1 - \frac{x^2}{4}}$ (since the length is 2x and the height is 2y)

$\Rightarrow A'(x) = 4\sqrt{1 - \frac{x^2}{4}} - \frac{x^2}{\sqrt{1 - \frac{x^2}{4}}}$. Thus $A'(x) = 0 \Rightarrow 4\sqrt{1 - \frac{x^2}{4}} - \frac{x^2}{\sqrt{1 - \frac{x^2}{4}}} = 0 \Rightarrow 4\left(1 - \frac{x^2}{4}\right) - x^2 = 0 \Rightarrow x^2 = 2$

$\Rightarrow x = \sqrt{2}$ (only the positive square root lies in the interval). Since $A(0) = A(2) = 0$ we have that $A\left(\sqrt{2}\right) = 4$

is the maximum area when the length is $2\sqrt{2}$ and the height is $\sqrt{2}$.

85. $9x^2 - 4y^2 = 36 \Rightarrow y^2 = \frac{9x^2 - 36}{4} \Rightarrow y = \pm\frac{3}{2}\sqrt{x^2 - 4}$ on the interval $2 \le x \le 4 \Rightarrow V = \int_2^4 \pi\left(\frac{3}{2}\sqrt{x^2 - 4}\right)^2 dx$

$= \frac{9\pi}{4}\int_2^4 (x^2 - 4)\, dx = \frac{9\pi}{4}\left[\frac{x^3}{3} - 4x\right]_2^4 = \frac{9\pi}{4}\left[\left(\frac{64}{3} - 16\right) - \left(\frac{8}{3} - 8\right)\right] = \frac{9\pi}{4}\left(\frac{56}{3} - 8\right) = \frac{3\pi}{4}(56 - 24) = 24\pi$

87. Let $y = \sqrt{16 - \frac{16}{9}x^2}$ on the interval $-3 \le x \le 3$. Since the plate is symmetric about the y-axis, $\bar{x} = 0$. For a

vertical strip: $(\tilde{x}, \tilde{y}) = \left(x, \frac{\sqrt{16 - \frac{16}{9}x^2}}{2}\right)$, length $= \sqrt{16 - \frac{16}{9}x^2}$, width $= dx \Rightarrow$ area $= dA = \sqrt{16 - \frac{16}{9}x^2}\, dx$

$\Rightarrow$ mass $= dm = \delta\, dA = \delta\sqrt{16 - \frac{16}{9}x^2}\, dx$. Moment of the strip about the x-axis:

$\tilde{y}\, dm = \frac{\sqrt{16 - \frac{16}{9}x^2}}{2}\left(\delta\sqrt{16 - \frac{16}{9}x^2}\right) dx = \delta\left(8 - \frac{8}{9}x^2\right) dx$ so the moment of the plate about the x-axis is

$M_x = \int \tilde{y}\, dm = \int_{-3}^3 \delta\left(8 - \frac{8}{9}x^2\right) dx = \delta\left[8x - \frac{8}{27}x^3\right]_{-3}^3 = 32\delta$; also the mass of the plate is

$M = \int_{-3}^3 \delta\sqrt{16 - \frac{16}{9}x^2}\, dx = \int_{-3}^3 4\delta\sqrt{1 - \left(\frac{1}{3}x\right)^2}\, dx = 4\delta\int_{-1}^1 3\sqrt{1 - u^2}\, du$ where $u = \frac{x}{3} \Rightarrow 3\, du = dx; x = -3$

$\Rightarrow u = -1$ and $x = 3 \Rightarrow u = 1$. Hence, $4\delta\int_{-1}^1 3\sqrt{1 - u^2}\, du = 12\delta\int_{-1}^1 \sqrt{1 - u^2}\, du$

$= 12\delta\left[\frac{1}{2}\left(u\sqrt{1 - u^2} + \sin^{-1} u\right)\right]_{-1}^1 = 6\pi\delta \Rightarrow \bar{y} = \frac{M_x}{M} = \frac{32\delta}{6\pi\delta} = \frac{16}{3\pi}$. Therefore the center of mass is $\left(0, \frac{16}{3\pi}\right)$.

89. $\frac{dr_A}{dt} = \frac{dr_B}{dt} \Rightarrow \frac{d}{dt}(r_A - r_B) = 0 \Rightarrow r_A - r_B = C$, a constant $\Rightarrow$ the points P(t) lie on a hyperbola with foci at A
and B

91. PF will always equal PB because the string has constant length $AB = FP + PA = AP + PB$.

93. $x^2 = 4py$ and $y = p \Rightarrow x^2 = 4p^2 \Rightarrow x = \pm 2p$. Therefore the line $y = p$ cuts the parabola at points $(-2p, p)$ and
$(2p, p)$, and these points are $\sqrt{[2p - (-2p)]^2 + (p - p)^2} = 4p$ units apart.

10.2 CLASSIFYING CONIC SECTIONS BY ECCENTRICITY

1. $16x^2 + 25y^2 = 400 \Rightarrow \frac{x^2}{25} + \frac{y^2}{16} = 1 \Rightarrow c = \sqrt{a^2 - b^2}$

$= \sqrt{25 - 16} = 3 \Rightarrow e = \frac{c}{a} = \frac{3}{5}; F(\pm 3, 0);$

directrices are $x = 0 \pm \frac{a}{e} = \pm\frac{5}{\left(\frac{3}{5}\right)} = \pm\frac{25}{3}$

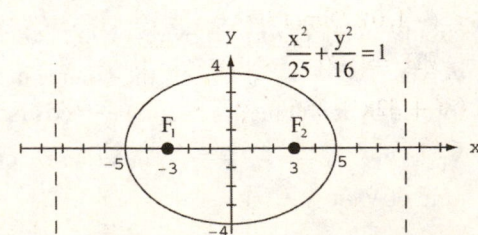

3. $2x^2 + y^2 = 2 \Rightarrow x^2 + \frac{y^2}{2} = 1 \Rightarrow c = \sqrt{a^2 - b^2}$

$= \sqrt{2 - 1} = 1 \Rightarrow e = \frac{c}{a} = \frac{1}{\sqrt{2}} ; F(0, \pm 1) ;$

directrices are $y = 0 \pm \frac{a}{e} = \pm \frac{\sqrt{2}}{\left(\frac{1}{\sqrt{2}}\right)} = \pm 2$

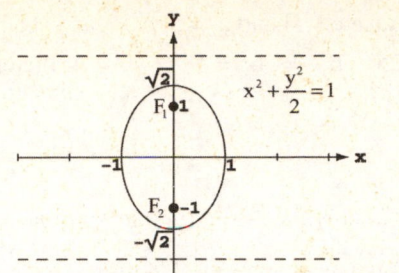

5. $3x^2 + 2y^2 = 6 \Rightarrow \frac{x^2}{2} + \frac{y^2}{3} = 1 \Rightarrow c = \sqrt{a^2 - b^2}$

$= \sqrt{3 - 2} = 1 \Rightarrow e = \frac{c}{a} = \frac{1}{\sqrt{3}} ; F(0, \pm 1) ;$

directrices are $y = 0 \pm \frac{a}{e} = \pm \frac{\sqrt{3}}{\left(\frac{1}{\sqrt{3}}\right)} = \pm 3$

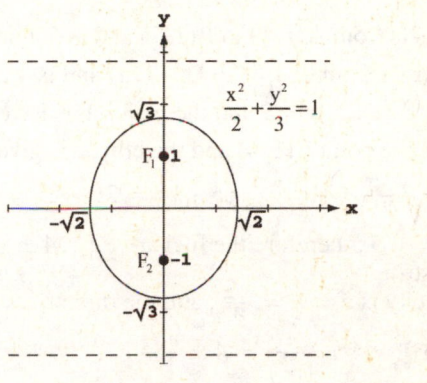

7. $6x^2 + 9y^2 = 54 \Rightarrow \frac{x^2}{9} + \frac{y^2}{6} = 1 \Rightarrow c = \sqrt{a^2 - b^2}$

$= \sqrt{9 - 6} = \sqrt{3} \Rightarrow e = \frac{c}{a} = \frac{\sqrt{3}}{3} ; F\left(\pm\sqrt{3}, 0\right) ;$

directrices are $x = 0 \pm \frac{a}{e} = \pm \frac{3}{\left(\frac{\sqrt{3}}{3}\right)} = \pm 3\sqrt{3}$

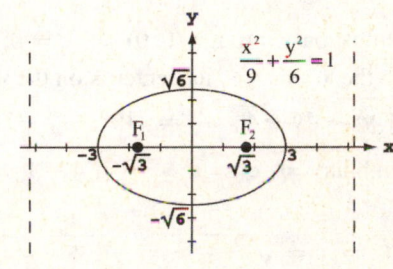

9. Foci: $(0, \pm 3), e = 0.5 \Rightarrow c = 3$ and $a = \frac{c}{e} = \frac{3}{0.5} = 6 \Rightarrow b^2 = 36 - 9 = 27 \Rightarrow \frac{x^2}{27} + \frac{y^2}{36} = 1$

11. Vertices: $(0, \pm 70), e = 0.1 \Rightarrow a = 70$ and $c = ae = 70(0.1) = 7 \Rightarrow b^2 = 4900 - 49 = 4851 \Rightarrow \frac{x^2}{4851} + \frac{y^2}{4900} = 1$

13. Focus: $\left(\sqrt{5}, 0\right)$, Directrix: $x = \frac{9}{\sqrt{5}} \Rightarrow c = ae = \sqrt{5}$ and $\frac{a}{e} = \frac{9}{\sqrt{5}} \Rightarrow \frac{ae}{e^2} = \frac{9}{\sqrt{5}} \Rightarrow \frac{\sqrt{5}}{e^2} = \frac{9}{\sqrt{5}} \Rightarrow e^2 = \frac{5}{9}$

$\Rightarrow e = \frac{\sqrt{5}}{3}$. Then $PF = \frac{\sqrt{5}}{3} PD \Rightarrow \sqrt{\left(x - \sqrt{5}\right)^2 + (y - 0)^2} = \frac{\sqrt{5}}{3}\left|x - \frac{9}{\sqrt{5}}\right| \Rightarrow \left(x - \sqrt{5}\right)^2 + y^2 = \frac{5}{9}\left(x - \frac{9}{\sqrt{5}}\right)^2$

$\Rightarrow x^2 - 2\sqrt{5}x + 5 + y^2 = \frac{5}{9}\left(x^2 - \frac{18}{\sqrt{5}}x + \frac{81}{5}\right) \Rightarrow \frac{4}{9}x^2 + y^2 = 4 \Rightarrow \frac{x^2}{9} + \frac{y^2}{4} = 1$

15. Focus: $(-4, 0)$, Directrix: $x = -16 \Rightarrow c = ae = 4$ and $\frac{a}{e} = 16 \Rightarrow \frac{ae}{e^2} = 16 \Rightarrow \frac{4}{e^2} = 16 \Rightarrow e^2 = \frac{1}{4} \Rightarrow e = \frac{1}{2}$. Then

$PF = \frac{1}{2} PD \Rightarrow \sqrt{(x + 4)^2 + (y - 0)^2} = \frac{1}{2}|x + 16| \Rightarrow (x + 4)^2 + y^2 = \frac{1}{4}(x + 16)^2 \Rightarrow x^2 + 8x + 16 + y^2$

$= \frac{1}{4}(x^2 + 32x + 256) \Rightarrow \frac{3}{4}x^2 + y^2 = 48 \Rightarrow \frac{x^2}{64} + \frac{y^2}{48} = 1$

17. $e = \frac{4}{5} \Rightarrow$ take $c = 4$ and $a = 5$; $c^2 = a^2 - b^2$

$\Rightarrow 16 = 25 - b^2 \Rightarrow b^2 = 9 \Rightarrow b = 3$; therefore

$\frac{x^2}{25} + \frac{y^2}{9} = 1$

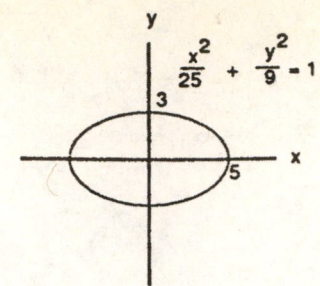

19. One axis is from $A(1, 1)$ to $B(1, 7)$ and is 6 units long; the other axis is from $C(3, 4)$ to $D(-1, 4)$ and is 4 units long. Therefore $a = 3$, $b = 2$ and the major axis is vertical. The center is the point $C(1, 4)$ and the ellipse is given by $\frac{(x-1)^2}{4} + \frac{(y-4)^2}{9} = 1$; $c^2 = a^2 - b^2 = 3^2 - 2^2 = 5$

$\Rightarrow c = \sqrt{5}$; therefore the foci are $F\left(1, 4 \pm \sqrt{5}\right)$, the eccentricity is $e = \frac{c}{a} = \frac{\sqrt{5}}{3}$, and the directrices are

$y = 4 \pm \frac{a}{e} = 4 \pm \frac{3}{\left(\frac{\sqrt{5}}{3}\right)} = 4 \pm \frac{9\sqrt{5}}{5}$.

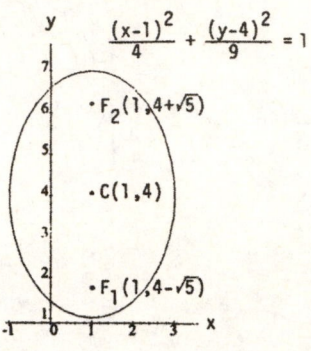

21. The ellipse must pass through $(0, 0) \Rightarrow c = 0$; the point $(-1, 2)$ lies on the ellipse $\Rightarrow -a + 2b = -8$. The ellipse is tangent to the x-axis $\Rightarrow$ its center is on the y-axis, so $a = 0$ and $b = -4 \Rightarrow$ the equation is $4x^2 + y^2 - 4y = 0$. Next, $4x^2 + y^2 - 4y + 4 = 4 \Rightarrow 4x^2 + (y - 24)^2 = 4 \Rightarrow x^2 + \frac{(y-2)^2}{4} = 1 \Rightarrow a = 2$ and $b = 1$ (now using the standard symbols) $\Rightarrow c^2 = a^2 - b^2 = 4 - 1 = 3 \Rightarrow c = \sqrt{3} \Rightarrow e = \frac{c}{a} = \frac{\sqrt{3}}{2}$.

23. $x^2 - y^2 = 1 \Rightarrow c = \sqrt{a^2 + b^2} = \sqrt{1 + 1} = \sqrt{2} \Rightarrow e = \frac{c}{a}$

$= \frac{\sqrt{2}}{1} = \sqrt{2}$; asymptotes are $y = \pm x$; $F\left(\pm \sqrt{2}, 0\right)$;

directrices are $x = 0 \pm \frac{a}{e} = \pm \frac{1}{\sqrt{2}}$

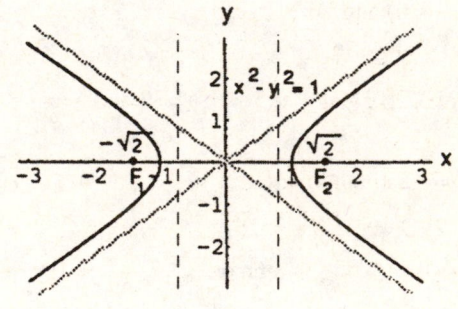

25. $y^2 - x^2 = 8 \Rightarrow \frac{y^2}{8} - \frac{x^2}{8} = 1 \Rightarrow c = \sqrt{a^2 + b^2}$

$= \sqrt{8 + 8} = 4 \Rightarrow e = \frac{c}{a} = \frac{4}{\sqrt{8}} = \sqrt{2}$; asymptotes are

$y = \pm x$; $F(0, \pm 4)$; directrices are $y = 0 \pm \frac{a}{e}$

$= \pm \frac{\sqrt{8}}{\sqrt{2}} = \pm 2$

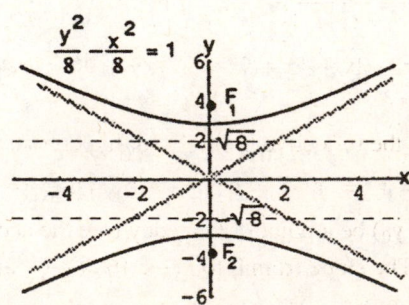

27. $8x^2 - 2y^2 = 16 \Rightarrow \frac{x^2}{2} - \frac{y^2}{8} = 1 \Rightarrow c = \sqrt{a^2 + b^2}$

   $= \sqrt{2 + 8} = \sqrt{10} \Rightarrow e = \frac{c}{a} = \frac{\sqrt{10}}{\sqrt{2}} = \sqrt{5}$; asymptotes

   are $y = \pm 2x$; $F\left(\pm \sqrt{10}, 0\right)$; directrices are $x = 0 \pm \frac{a}{e}$

   $= \pm \frac{\sqrt{2}}{\sqrt{5}} = \pm \frac{2}{\sqrt{10}}$

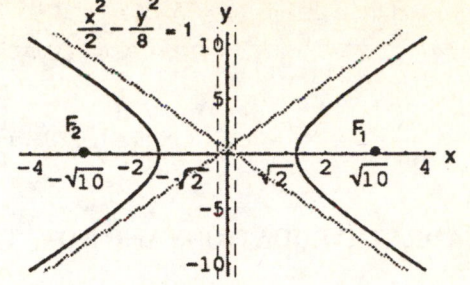

29. $8y^2 - 2x^2 = 16 \Rightarrow \frac{y^2}{2} - \frac{x^2}{8} = 1 \Rightarrow c = \sqrt{a^2 + b^2}$

   $= \sqrt{2 + 8} = \sqrt{10} \Rightarrow e = \frac{c}{a} = \frac{\sqrt{10}}{\sqrt{2}} = \sqrt{5}$; asymptotes

   are $y = \pm \frac{x}{2}$; $F\left(0, \pm \sqrt{10}\right)$; directrices are $y = 0 \pm \frac{a}{e}$

   $= \pm \frac{\sqrt{2}}{\sqrt{5}} = \pm \frac{2}{\sqrt{10}}$

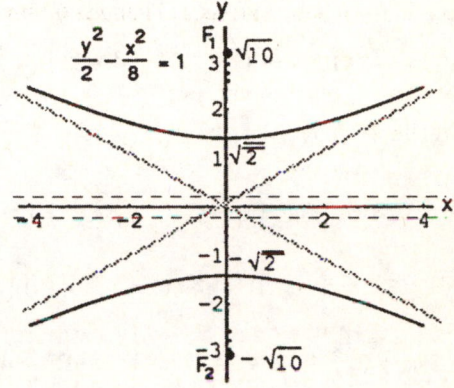

31. Vertices $(0, \pm 1)$ and $e = 3 \Rightarrow a = 1$ and $e = \frac{c}{a} = 3 \Rightarrow c = 3a = 3 \Rightarrow b^2 = c^2 - a^2 = 9 - 1 = 8 \Rightarrow y^2 - \frac{x^2}{8} = 1$

33. Foci $(\pm 3, 0)$ and $e = 3 \Rightarrow c = 3$ and $e = \frac{c}{a} = 3 \Rightarrow c = 3a \Rightarrow a = 1 \Rightarrow b^2 = c^2 - a^2 = 9 - 1 = 8 \Rightarrow x^2 - \frac{y^2}{8} = 1$

35. Focus $(4, 0)$ and Directrix $x = 2 \Rightarrow c = ae = 4$ and $\frac{a}{e} = 2 \Rightarrow \frac{ae}{e^2} = 2 \Rightarrow \frac{4}{e^2} = 2 \Rightarrow e^2 = 2 \Rightarrow e = \sqrt{2}$. Then

   $PF = \sqrt{2}\, PD \Rightarrow \sqrt{(x - 4)^2 + (y - 0)^2} = \sqrt{2}\, |x - 2| \Rightarrow (x - 4)^2 + y^2 = 2(x - 2)^2 \Rightarrow x^2 - 8x + 16 + y^2$

   $= 2\left(x^2 - 4x + 4\right) \Rightarrow -x^2 + y^2 = -8 \Rightarrow \frac{x^2}{8} - \frac{y^2}{8} = 1$

37. Focus $(-2, 0)$ and Directrix $x = -\frac{1}{2} \Rightarrow c = ae = 2$ and $\frac{a}{e} = \frac{1}{2} \Rightarrow \frac{ae}{e^2} = \frac{1}{2} \Rightarrow \frac{2}{e^2} = \frac{1}{2} \Rightarrow e^2 = 4 \Rightarrow e = 2$. Then

   $PF = 2PD \Rightarrow \sqrt{(x + 2)^2 + (y - 0)^2} = 2\left|x + \frac{1}{2}\right| \Rightarrow (x + 2)^2 + y^2 = 4\left(x + \frac{1}{2}\right)^2 \Rightarrow x^2 + 4x + 4 + y^2$

   $= 4\left(x^2 + x + \frac{1}{4}\right) \Rightarrow -3x^2 + y^2 = -3 \Rightarrow x^2 - \frac{y^2}{3} = 1$

39. $\sqrt{(x - 1)^2 + (y + 3)^2} = \frac{3}{2}|y - 2| \Rightarrow x^2 - 2x + 1 + y^2 + 6y + 9 = \frac{9}{4}\left(y^2 - 4y + 4\right) \Rightarrow 4x^2 - 5y^2 - 8x + 60y + 4 = 0$

   $\Rightarrow 4\left(x^2 - 2x + 1\right) - 5\left(y^2 - 12y + 36\right) = -4 + 4 - 180 \Rightarrow \frac{(y-6)^2}{36} - \frac{(x-1)^2}{45} = 1$

41. To prove the reflective property for hyperbolas:

   $\frac{x^2}{a^2} - \frac{y^2}{b^2} = 1 \Rightarrow a^2 y^2 = b^2 x^2 - a^2 b^2$ and $\frac{dy}{dx} = \frac{xb^2}{ya^2}$.

   Let $P(x_0, y_0)$ be a point of tangency (see the accompanying
   figure). The slope from P to $F(-c, 0)$ is $\frac{y_0}{x_0 + c}$ and from
   P to $F_2(c, 0)$ it is $\frac{y_0}{x_0 - c}$. Let the tangent through P meet
   the x-axis in point A, and define the angles $\angle F_1 PA = \alpha$
   and $\angle F_2 PA = \beta$. We will show that $\tan \alpha = \tan \beta$. From
   the preliminary result in Exercise 22,

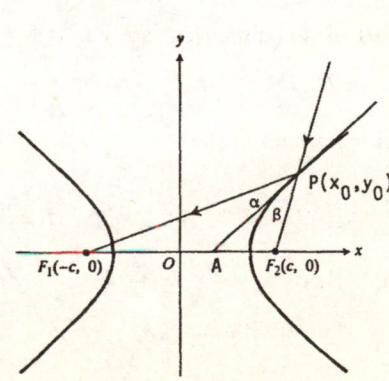

$\tan \alpha = \dfrac{\left(\dfrac{x_0 b^2}{y_0 a^2} - \dfrac{y_0}{x_0 + c}\right)}{1 + \left(\dfrac{x_0 b^2}{y_0 a^2}\right)\left(\dfrac{y_0}{x_0 + c}\right)} = \dfrac{x_0^2 b^2 + x_0 b^2 c - y_0^2 a^2}{x_0 y_0 a^2 + y_0 a^2 c + x_0 y_0 b^2} = \dfrac{a^2 b^2 + x_0 b^2 c}{x_0 y_0 c^2 + y_0 a^2 c} = \dfrac{b^2}{y_0 c}$. In a similar manner,

$\tan \beta = \dfrac{\left(\dfrac{y_0}{x_0 - c} - \dfrac{x_0 b^2}{y_0 a^2}\right)}{1 + \left(\dfrac{y_0}{x_0 - c}\right)\left(\dfrac{x_0 b^2}{y_0 a^2}\right)} = \dfrac{b^2}{y_0 c}$. Since $\tan \alpha = \tan \beta$, and α and β are acute angles, we have $\alpha = \beta$.

10.3 QUADRATIC EQUATIONS AND ROTATIONS

1. $x^2 - 3xy + y^2 - x = 0 \Rightarrow B^2 - 4AC = (-3)^2 - 4(1)(1) = 5 > 0 \Rightarrow$ Hyperbola

3. $3x^2 - 7xy + \sqrt{17}y^2 = 1 \Rightarrow B^2 - 4AC = (-7)^2 - 4(3)\sqrt{17} \approx -0.477 < 0 \Rightarrow$ Ellipse

5. $x^2 + 2xy + y^2 + 2x - y + 2 = 0 \Rightarrow B^2 - 4AC = 2^2 - 4(1)(1) = 0 \Rightarrow$ Parabola

7. $x^2 + 4xy + 4y^2 - 3x = 6 \Rightarrow B^2 - 4AC = 4^2 - 4(1)(4) = 0 \Rightarrow$ Parabola

9. $xy + y^2 - 3x = 5 \Rightarrow B^2 - 4AC = 1^2 - 4(0)(1) = 1 > 0 \Rightarrow$ Hyperbola

11. $3x^2 - 5xy + 2y^2 - 7x - 14y = -1 \Rightarrow B^2 - 4AC = (-5)^2 - 4(3)(2) = 1 > 0 \Rightarrow$ Hyperbola

13. $x^2 - 3xy + 3y^2 + 6y = 7 \Rightarrow B^2 - 4AC = (-3)^2 - 4(1)(3) = -3 < 0 \Rightarrow$ Ellipse

15. $6x^2 + 3xy + 2y^2 + 17y + 2 = 0 \Rightarrow B^2 - 4AC = 3^2 - 4(6)(2) = -39 < 0 \Rightarrow$ Ellipse

17. $\cot 2\alpha = \frac{A-C}{B} = \frac{0}{1} = 0 \Rightarrow 2\alpha = \frac{\pi}{2} \Rightarrow \alpha = \frac{\pi}{4}$; therefore $x = x' \cos \alpha - y' \sin \alpha$,

   $y = x' \sin \alpha + y' \cos \alpha \Rightarrow x = x' \frac{\sqrt{2}}{2} - y' \frac{\sqrt{2}}{2} , y = x' \frac{\sqrt{2}}{2} + y' \frac{\sqrt{2}}{2}$

   $\Rightarrow \left(\frac{\sqrt{2}}{2} x' - \frac{\sqrt{2}}{2} y'\right)\left(\frac{\sqrt{2}}{2} x' + \frac{\sqrt{2}}{2} y'\right) = 2 \Rightarrow \frac{1}{2} x'^2 - \frac{1}{2} y'^2 = 2 \Rightarrow x'^2 - y'^2 = 4 \Rightarrow$ Hyperbola

19. $\cot 2\alpha = \frac{A-C}{B} = \frac{3-1}{2\sqrt{3}} = \frac{1}{\sqrt{3}} \Rightarrow 2\alpha = \frac{\pi}{3} \Rightarrow \alpha = \frac{\pi}{6}$; therefore $x = x' \cos \alpha - y' \sin \alpha$,

   $y = x' \sin \alpha + y' \cos \alpha \Rightarrow x = \frac{\sqrt{3}}{2} x' - \frac{1}{2} y', y = \frac{1}{2} x' + \frac{\sqrt{3}}{2} y'$

   $\Rightarrow 3\left(\frac{\sqrt{3}}{2} x' - \frac{1}{2} y'\right)^2 + 2\sqrt{3}\left(\frac{\sqrt{3}}{2} x' + \frac{1}{2} y'\right)\left(\frac{1}{2} x' + \frac{\sqrt{3}}{2} y'\right) + \left(\frac{1}{2} x' + \frac{\sqrt{3}}{2} y'\right)^2 - 8\left(\frac{\sqrt{3}}{2} x' - \frac{1}{2} y'\right)$

   $+ 8\sqrt{3}\left(\frac{1}{2} x' + \frac{\sqrt{3}}{2} y'\right) = 0 \Rightarrow 4x'^2 + 16y' = 0 \Rightarrow$ Parabola

21. $\cot 2\alpha = \frac{A-C}{B} = \frac{1-1}{-2} = 0 \Rightarrow 2\alpha = \frac{\pi}{2} \Rightarrow \alpha = \frac{\pi}{4}$; therefore $x = x' \cos \alpha - y' \sin \alpha$,

   $y = x' \sin \alpha + y' \cos \alpha \Rightarrow x = \frac{\sqrt{2}}{2} x' - \frac{\sqrt{2}}{2} y', y = \frac{\sqrt{2}}{2} x' + \frac{\sqrt{2}}{2} y'$

   $\Rightarrow \left(\frac{\sqrt{2}}{2} x' - \frac{\sqrt{2}}{2} y'\right)^2 - 2\left(\frac{\sqrt{2}}{2} x' - \frac{\sqrt{2}}{2} y'\right)\left(\frac{\sqrt{2}}{2} x' + \frac{\sqrt{2}}{2} y'\right) + \left(\frac{\sqrt{2}}{2} x' + \frac{\sqrt{2}}{2} y'\right)^2 = 2 \Rightarrow y'^2 = 1$

   $\Rightarrow$ Parallel horizontal lines

23. $\cot 2\alpha = \frac{A-C}{B} = \frac{\sqrt{2}-\sqrt{2}}{2\sqrt{2}} = 0 \Rightarrow 2\alpha = \frac{\pi}{2} \Rightarrow \alpha = \frac{\pi}{4}$; therefore $x = x' \cos \alpha - y' \sin \alpha$,

   $y = x' \sin \alpha + y' \cos \alpha \Rightarrow x = \frac{\sqrt{2}}{2} x' - \frac{\sqrt{2}}{2} y', y = \frac{\sqrt{2}}{2} x' + \frac{\sqrt{2}}{2} y'$

   $\Rightarrow \sqrt{2}\left(\frac{\sqrt{2}}{2} x' - \frac{\sqrt{2}}{2} y'\right)^2 + 2\sqrt{2}\left(\frac{\sqrt{2}}{2} x' - \frac{\sqrt{2}}{2} y'\right)\left(\frac{\sqrt{2}}{2} x' + \frac{\sqrt{2}}{2} y'\right) + \sqrt{2}\left(\frac{\sqrt{2}}{2} x' + \frac{\sqrt{2}}{2} y'\right)^2$

   $- 8\left(\frac{\sqrt{2}}{2} x' - \frac{\sqrt{2}}{2} y'\right) + 8\left(\frac{\sqrt{2}}{2} x' + \frac{\sqrt{2}}{2} y'\right) = 0 \Rightarrow 2\sqrt{2}x'^2 + 8\sqrt{2}y' = 0 \Rightarrow$ Parabola

25. $\cot 2\alpha = \frac{A-C}{B} = \frac{3-3}{2} = 0 \Rightarrow 2\alpha = \frac{\pi}{2} \Rightarrow \alpha = \frac{\pi}{4}$; therefore $x = x'\cos\alpha - y'\sin\alpha$,

$y = x'\sin\alpha + y'\cos\alpha \Rightarrow x = \frac{\sqrt{2}}{2}x' - \frac{\sqrt{2}}{2}y', y = \frac{\sqrt{2}}{2}x' + \frac{\sqrt{2}}{2}y'$

$\Rightarrow 3\left(\frac{\sqrt{2}}{2}x' - \frac{\sqrt{2}}{2}y'\right)^2 + 2\left(\frac{\sqrt{2}}{2}x' - \frac{\sqrt{2}}{2}y'\right)\left(\frac{\sqrt{2}}{2}x' + \frac{\sqrt{2}}{2}y'\right) + 3\left(\frac{\sqrt{2}}{2}x' + \frac{\sqrt{2}}{2}y'\right)^2 = 19 \Rightarrow 4x'^2 + 2y'^2 = 19$

$\Rightarrow$ Ellipse

27. $\cot 2\alpha = \frac{14-2}{16} = \frac{3}{4} \Rightarrow \cos 2\alpha = \frac{3}{5}$ (if we choose 2α in Quadrant I); thus $\sin\alpha = \sqrt{\frac{1-\cos 2\alpha}{2}} = \sqrt{\frac{1-\left(\frac{3}{5}\right)}{2}} = \frac{1}{\sqrt{5}}$

and $\cos\alpha = \sqrt{\frac{1+\cos 2\alpha}{2}} = \sqrt{\frac{1+\left(\frac{3}{5}\right)}{2}} = \frac{2}{\sqrt{5}}$ (or $\sin\alpha = \frac{2}{\sqrt{5}}$ and $\cos\alpha = \frac{-1}{\sqrt{5}}$)

29. $\tan 2\alpha = \frac{-1}{1-3} = \frac{1}{2} \Rightarrow 2\alpha \approx 26.57° \Rightarrow \alpha \approx 13.28° \Rightarrow \sin\alpha \approx 0.23, \cos\alpha \approx 0.97$; then $A' \approx 0.9, B' \approx 0.0$,

$C' \approx 3.1, D' \approx 0.7, E' \approx -1.2$, and $F' = -3 \Rightarrow 0.9x'^2 + 3.1y'^2 + 0.7x' - 1.2y' - 3 = 0$, an ellipse

31. $\tan 2\alpha = \frac{-4}{1-4} = \frac{4}{3} \Rightarrow 2\alpha \approx 53.13° \Rightarrow \alpha \approx 26.57° \Rightarrow \sin\alpha \approx 0.45, \cos\alpha \approx 0.89$; then $A' \approx 0.0, B' \approx 0.0$,

$C' \approx 5.0, D' \approx 0, E' \approx 0$, and $F' = -5 \Rightarrow 5.0y'^2 - 5 = 0$ or $y' = \pm 1.0$, parallel lines

33. $\tan 2\alpha = \frac{5}{3-2} = 5 \Rightarrow 2\alpha \approx 78.69° \Rightarrow \alpha \approx 39.35° \Rightarrow \sin\alpha \approx 0.63, \cos\alpha \approx 0.77$; then $A' \approx 5.0, B' \approx 0.0$,

$C' \approx -0.05, D' \approx -5.0, E' \approx -6.2$, and $F' = -1 \Rightarrow 5.0x'^2 - 0.05y'^2 - 5.0x' - 6.2y' - 1 = 0$, a hyperbola

35. $\alpha = 90° \Rightarrow x = x'\cos 90° - y'\sin 90° = -y'$ and $y = x'\sin 90° + y'\cos 90° = x'$

(a) $\frac{x'^2}{b^2} + \frac{y'^2}{a^2} = 1$ (b) $\frac{y'^2}{a^2} - \frac{x'^2}{b^2} = 1$ (c) $x'^2 + y'^2 = a^2$

(d) $y = mx \Rightarrow y - mx = 0 \Rightarrow D = -m$ and $E = 1; \alpha = 90° \Rightarrow D' = 1$ and $E' = m \Rightarrow my' + x' = 0 \Rightarrow y' = -\frac{1}{m}x'$

(e) $y = mx + b \Rightarrow y - mx - b = 0 \Rightarrow D = -m$ and $E = 1; \alpha = 90° \Rightarrow D' = 1, E' = m$ and $F' = -b$

$\Rightarrow my' + x' - b = 0 \Rightarrow y' = -\frac{1}{m}x' + \frac{b}{m}$

37. (a) $A' = \cos 45°\sin 45° = \left(\frac{\sqrt{2}}{2}\right)\left(\frac{\sqrt{2}}{2}\right) = \frac{1}{2}, B' = 0, C' = -\cos 45°\sin 45° = -\frac{1}{2}, F' = -1$

$\Rightarrow \frac{1}{2}x'^2 - \frac{1}{2}y'^2 = 1 \Rightarrow x'^2 - y'^2 = 2$

(b) $A' = \frac{1}{2}, C' = -\frac{1}{2}$ (see part (a) above), $D' = E' = B' = 0, F' = -a \Rightarrow \frac{1}{2}x'^2 - \frac{1}{2}y'^2 = a \Rightarrow x'^2 - y'^2 = 2a$

39. Yes, the graph is a hyperbola: with $AC < 0$ we have $-4AC > 0$ and $B^2 - 4AC > 0$.

41. Let α be any angle. Then $A' = \cos^2\alpha + \sin^2\alpha = 1, B' = 0, C' = \sin^2\alpha + \cos^2\alpha = 1, D' = E' = 0$ and $F' = -a^2$

$\Rightarrow x'^2 + y'^2 = a^2$.

43. (a) $B^2 - 4AC = 4^2 - 4(1)(4) = 0$, so the discriminant indicates this conic is a parabola

(b) The left-hand side of $x^2 + 4xy + 4y^2 + 6x + 12y + 9 = 0$ factors as a perfect square: $(x + 2y + 3)^2 = 0$

$\Rightarrow x + 2y + 3 = 0 \Rightarrow 2y = -x - 3$; thus the curve is a degenerate parabola (i.e., a straight line).

45. (a) $B^2 - 4AC = 1 - 4(0)(0) = 1 \Rightarrow$ hyperbola

(b) $xy + 2x - y = 0 \Rightarrow y(x - 1) = -2x \Rightarrow y = \frac{-2x}{x-1}$

(c) $y = \frac{-2x}{x-1} \Rightarrow \frac{dy}{dx} = \frac{2}{(x-1)^2}$ and we want $\frac{-1}{\left(\frac{dy}{dx}\right)} = -2$,

the slope of $y = -2x \Rightarrow -2 = -\frac{(x-1)^2}{2}$

$\Rightarrow (x-1)^2 = 4 \Rightarrow x = 3$ or $x = -1; x = 3$

$\Rightarrow y = -3 \Rightarrow (3, -3)$ is a point on the hyperbola

where the line with slope $m = -2$ is normal

$\Rightarrow$ the line is $y + 3 = -2(x - 3)$ or $y = -2x + 3$;

$x = -1 \Rightarrow y = -1 \Rightarrow (-1, -1)$ is a point on the

hyperbola where the line with slope $m = -2$ is

normal $\Rightarrow$ the line is $y + 1 = -2(x + 1)$ or

$y = -2x - 3$

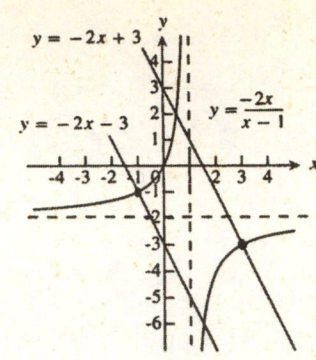

47. Assume the ellipse has been rotated to eliminate the xy-term $\Rightarrow$ the new equation is $A'x'^2 + C'y'^2 = 1 \Rightarrow$ the semi-axes are $\sqrt{\frac{1}{A'}}$ and $\sqrt{\frac{1}{C'}} \Rightarrow$ the area is $\pi \left(\sqrt{\frac{1}{A'}}\right)\left(\sqrt{\frac{1}{C'}}\right) = \frac{\pi}{\sqrt{A'C'}} = \frac{2\pi}{\sqrt{4A'C'}}$. Since $B^2 - 4AC$

$= B'^2 - 4A'C' = -4A'C'$ (because $B' = 0$) we find that the area is $\frac{2\pi}{\sqrt{4AC - B^2}}$ as claimed.

49. $B'^2 - 4A'C'$

$= (B \cos 2\alpha + (C - A) \sin 2\alpha)^2 - 4 (A \cos^2 \alpha + B \cos \alpha \sin \alpha + C \sin^2 \alpha)(A \sin^2 \alpha - B \cos \alpha \sin \alpha + C \cos^2 \alpha)$

$= B^2 \cos^2 2\alpha + 2B(C - A) \sin 2\alpha \cos 2\alpha + (C - A)^2 \sin^2 2\alpha - 4A^2 \cos^2 \alpha \sin^2 \alpha + 4AB \cos^3 \alpha \sin \alpha$

$\quad - 4AC \cos^4 \alpha - 4AB \cos \alpha \sin^3 \alpha + 4B^2 \cos^2 \alpha \sin^2 \alpha - 4BC \cos^3 \alpha \sin \alpha - 4AC \sin^4 \alpha + 4BC \cos \alpha \sin^3 \alpha$

$\quad - 4C^2 \cos^2 \alpha \sin^2 \alpha$

$= B^2 \cos^2 2\alpha + 2BC \sin 2\alpha \cos 2\alpha - 2AB \sin 2\alpha \cos 2\alpha + C^2 \sin^2 2\alpha - 2AC \sin^2 2\alpha + A^2 \sin^2 2\alpha$

$\quad - 4A^2 \cos^2 \alpha \sin^2 \alpha + 4AB \cos^3 \alpha \sin \alpha - 4AC \cos^4 \alpha - 4AB \cos \alpha \sin^3 \alpha + B^2 \sin^2 2\alpha - 4BC \cos^3 \alpha \sin \alpha$

$\quad - 4AC \sin^4 \alpha + 4BC \cos \alpha \sin^3 \alpha - 4C^2 \cos^2 \alpha \sin^2 \alpha$

$= B^2 + 2BC(2 \sin \alpha \cos \alpha)(\cos^2 \alpha - \sin^2 \alpha) - 2AB(2 \sin \alpha \cos \alpha)(\cos^2 \alpha - \sin^2 \alpha) + C^2 (4 \sin^2 \alpha \cos^2 \alpha)$

$\quad - 2AC (4 \sin^2 \alpha \cos^2 \alpha) + A^2 (4 \sin^2 \alpha \cos^2 \alpha) - 4A^2 \cos^2 \alpha \sin^2 \alpha + 4AB \cos^3 \alpha \sin \alpha - 4AC \cos^4 \alpha$

$\quad - 4AB \cos \alpha \sin^3 \alpha - 4BC \cos^3 \alpha \sin \alpha - 4AC \sin^4 \alpha + 4BC \cos \alpha \sin^3 \alpha - 4C^2 \cos^2 \alpha \sin^2 \alpha$

$= B^2 - 8AC \sin^2 \alpha \cos^2 \alpha - 4AC \cos^4 \alpha - 4AC \sin^4 \alpha$

$= B^2 - 4AC (\cos^4 \alpha + 2 \sin^2 \alpha \cos^2 \alpha + \sin^4 \alpha)$

$= B^2 - 4AC (\cos^2 \alpha + \sin^2 \alpha)^2$

$= B^2 - 4AC$

10.4 CONICS AND PARAMETRIC EQUATIONS; THE CYCLOID

1. $x = \cos t, y = \sin t, 0 \le t \le \pi$

$\Rightarrow \cos^2 t + \sin^2 t = 1 \Rightarrow x^2 + y^2 = 1$

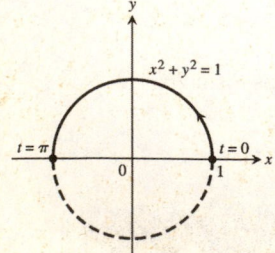

3. $x = 4 \cos t, y = 5 \sin t, 0 \le t \le \pi$

$\Rightarrow \frac{16 \cos^2 t}{16} + \frac{25 \sin^2 t}{25} = 1 \Rightarrow \frac{x^2}{16} + \frac{y^2}{25} = 1$

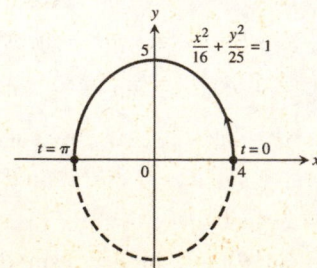

5. $x = t, y = \sqrt{t}, t \geq 0 \Rightarrow y = \sqrt{x}$

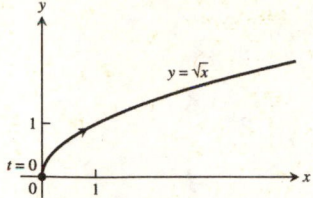

7. $x = -\sec t, y = \tan t, -\frac{\pi}{2} < t < \frac{\pi}{2}$

$\Rightarrow \sec^2 t - \tan^2 t = 1 \Rightarrow x^2 - y^2 = 1$

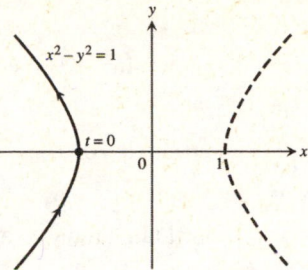

9. $x = t, y = \sqrt{4 - t^2}, 0 \leq t \leq 2$

$\Rightarrow y = \sqrt{4 - x^2}$

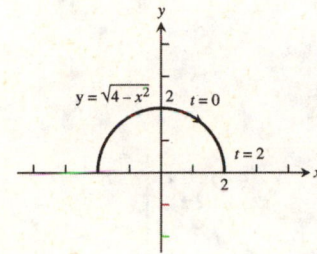

11. $x = -\cosh t, y = \sinh t, -\infty < 1 < \infty$

$\Rightarrow \cosh^2 t - \sinh^2 t = 1 \Rightarrow x^2 - y^2 = 1$

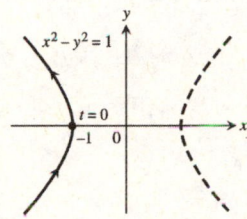

13. Arc PF = Arc AF since each is the distance rolled and
$\frac{\text{Arc PF}}{b} = \angle FCP \Rightarrow \text{Arc PF} = b(\angle FCP); \frac{\text{Arc AF}}{a} = \theta$
$\Rightarrow \text{Arc AF} = a\theta \Rightarrow a\theta = b(\angle FCP) \Rightarrow \angle FCP = \frac{a}{b}\theta;$
$\angle OCG = \frac{\pi}{2} - \theta; \angle OCG = \angle OCP + \angle PCE$
$= \angle OCP + \left(\frac{\pi}{2} - \alpha\right).$ Now $\angle OCP = \pi - \angle FCP$
$= \pi - \frac{a}{b}\theta.$ Thus $\angle OCG = \pi - \frac{a}{b}\theta + \frac{\pi}{2} - \alpha \Rightarrow \frac{\pi}{2} - \theta$
$= \pi - \frac{a}{b}\theta + \frac{\pi}{2} - \alpha \Rightarrow \alpha = \pi - \frac{a}{b}\theta + \theta = \pi - \left(\frac{a-b}{b}\theta\right).$

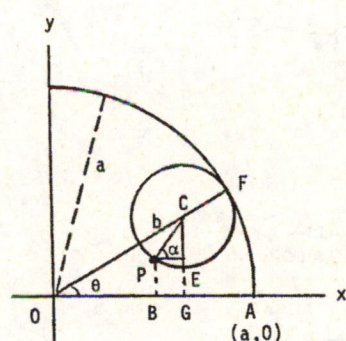

Then $x = OG - BG = OG - PE = (a - b)\cos\theta - b\cos\alpha = (a - b)\cos\theta - b\cos\left(\pi - \frac{a-b}{b}\theta\right)$
$= (a - b)\cos\theta + b\cos\left(\frac{a-b}{b}\theta\right).$ Also $y = EG = CG - CE = (a - b)\sin\theta - b\sin\alpha$
$= (a - b)\sin\theta - b\sin\left(\pi - \frac{a-b}{b}\theta\right) = (a - b)\sin\theta - b\sin\left(\frac{a-b}{b}\theta\right).$ Therefore
$x = (a - b)\cos\theta + b\cos\left(\frac{a-b}{b}\theta\right)$ and $y = (a - b)\sin\theta - b\sin\left(\frac{a-b}{b}\theta\right).$

If $b = \frac{a}{4}$, then $x = \left(a - \frac{a}{4}\right)\cos\theta + \frac{a}{4}\cos\left(\frac{a-\left(\frac{a}{4}\right)}{\left(\frac{a}{4}\right)}\theta\right)$
$= \frac{3a}{4}\cos\theta + \frac{a}{4}\cos 3\theta = \frac{3a}{4}\cos\theta + \frac{a}{4}(\cos\theta\cos 2\theta - \sin\theta\sin 2\theta)$
$= \frac{3a}{4}\cos\theta + \frac{a}{4}((\cos\theta)(\cos^2\theta - \sin^2\theta) - (\sin\theta)(2\sin\theta\cos\theta))$
$= \frac{3a}{4}\cos\theta + \frac{a}{4}\cos^3\theta - \frac{a}{4}\cos\theta\sin^2\theta - \frac{2a}{4}\sin^2\theta\cos\theta$
$= \frac{3a}{4}\cos\theta + \frac{a}{4}\cos^3\theta - \frac{3a}{4}(\cos\theta)(1 - \cos^2\theta) = a\cos^3\theta;$
$y = \left(a - \frac{a}{4}\right)\sin\theta - \frac{a}{4}\sin\left(\frac{a-\left(\frac{a}{4}\right)}{\left(\frac{a}{4}\right)}\theta\right) = \frac{3a}{4}\sin\theta - \frac{a}{4}\sin 3\theta = \frac{3a}{4}\sin\theta - \frac{a}{4}(\sin\theta\cos 2\theta + \cos\theta\sin 2\theta)$
$= \frac{3a}{4}\sin\theta - \frac{a}{4}((\sin\theta)(\cos^2\theta - \sin^2\theta) + (\cos\theta)(2\sin\theta\cos\theta))$
$= \frac{3a}{4}\sin\theta - \frac{a}{4}\sin\theta\cos^2\theta + \frac{a}{4}\sin^3\theta - \frac{2a}{4}\cos^2\theta\sin\theta$
$= \frac{3a}{4}\sin\theta - \frac{3a}{4}\sin\theta\cos^2\theta + \frac{a}{4}\sin^3\theta$
$= \frac{3a}{4}\sin\theta - \frac{3a}{4}(\sin\theta)(1 - \sin^2\theta) + \frac{a}{4}\sin^3\theta = a\sin^3\theta.$

15. Draw line AM in the figure and note that $\angle AMO$ is a right angle since it is an inscribed angle which spans the diameter of a circle. Then $AN^2 = MN^2 + AM^2$. Now, $OA = a$, $\frac{AN}{a} = \tan t$, and $\frac{AM}{a} = \sin t$. Next $MN = OP$

$\Rightarrow OP^2 = AN^2 - AM^2 = a^2 \tan^2 t - a^2 \sin^2 t$

$\Rightarrow OP = \sqrt{a^2 \tan^2 t - a^2 \sin^2 t}$

$= (a \sin t)\sqrt{\sec^2 t - 1} = \frac{a \sin^2 t}{\cos t}$. In triangle BPO,

$x = OP \sin t = \frac{a \sin^3 t}{\cos t} = a \sin^2 t \tan t$ and

$y = OP \cos t = a \sin^2 t \Rightarrow x = a \sin^2 t \tan t$ and $y = a \sin^2 t$.

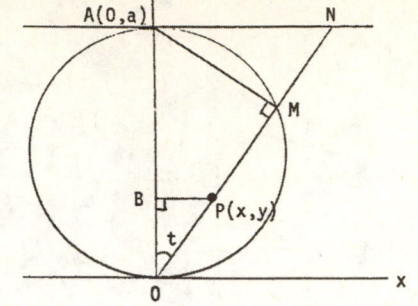

17. $D = \sqrt{(x-2)^2 + \left(y - \frac{1}{2}\right)^2} \Rightarrow D^2 = (x-2)^2 + \left(y - \frac{1}{2}\right)^2 = (t-2)^2 + \left(t^2 - \frac{1}{2}\right)^2 \Rightarrow D^2 = t^4 - 4t + \frac{17}{4}$

$\Rightarrow \frac{d(D^2)}{dt} = 4t^3 - 4 = 0 \Rightarrow t = 1$. The second derivative is always positive for $t \neq 0 \Rightarrow t = 1$ gives a local minimum for D^2 (and hence D) which is an absolute minimum since it is the only extremum $\Rightarrow$ the closest point on the parabola is $(1, 1)$.

19. (a)

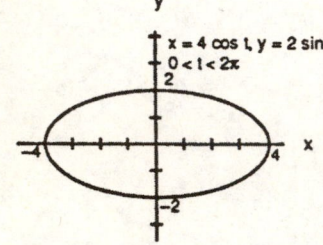

(b)

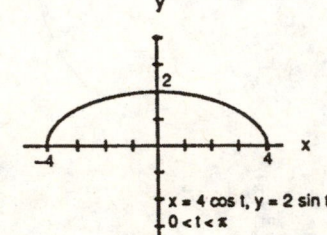

(c)

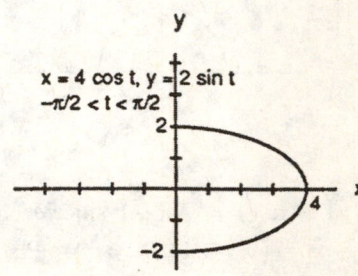

21.

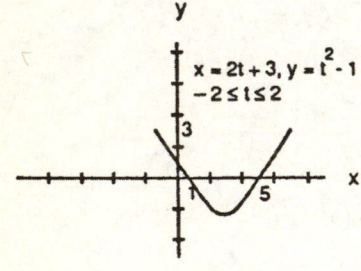

23. (a)

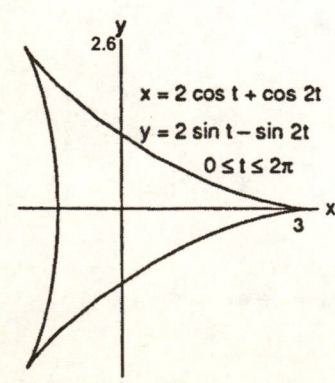

(b)

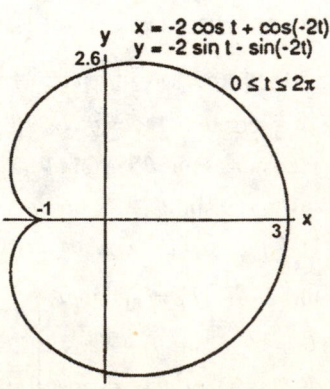

25. (a)

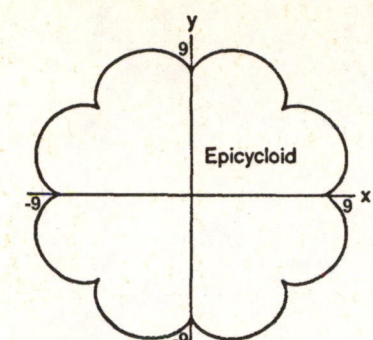

(b)

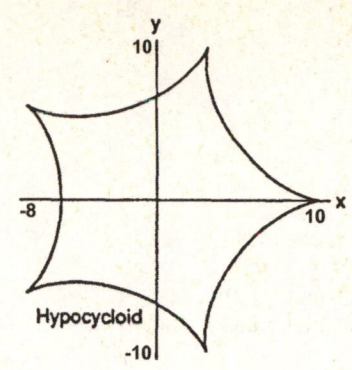

(c)

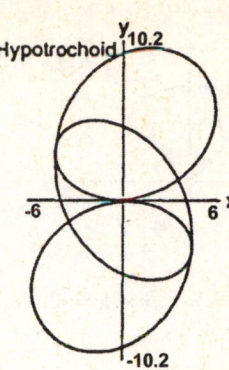

10.5 POLAR COORDINATES

1. a, e; b, g; c, h; d, f

3. (a) $\left(2, \frac{\pi}{2} + 2n\pi\right)$ and $\left(-2, \frac{\pi}{2} + (2n+1)\pi\right)$, n an integer

   (b) $(2, 2n\pi)$ and $(-2, (2n+1)\pi)$, n an integer

   (c) $\left(2, \frac{3\pi}{2} + 2n\pi\right)$ and $\left(-2, \frac{3\pi}{2} + (2n+1)\pi\right)$, n an integer

   (d) $(2, (2n+1)\pi)$ and $(-2, 2n\pi)$, n an integer

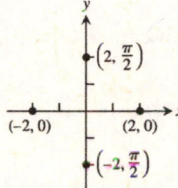

5. (a) $x = r \cos \theta = 3 \cos 0 = 3$, $y = r \sin \theta = 3 \sin 0 = 0 \Rightarrow$ Cartesian coordinates are $(3, 0)$

   (b) $x = r \cos \theta = -3 \cos 0 = -3$, $y = r \sin \theta = -3 \sin 0 = 0 \Rightarrow$ Cartesian coordinates are $(-3, 0)$

   (c) $x = r \cos \theta = 2 \cos \frac{2\pi}{3} = -1$, $y = r \sin \theta = 2 \sin \frac{2\pi}{3} = \sqrt{3} \Rightarrow$ Cartesian coordinates are $\left(-1, \sqrt{3}\right)$

   (d) $x = r \cos \theta = 2 \cos \frac{7\pi}{3} = 1$, $y = r \sin \theta = 2 \sin \frac{7\pi}{3} = \sqrt{3} \Rightarrow$ Cartesian coordinates are $\left(1, \sqrt{3}\right)$

   (e) $x = r \cos \theta = -3 \cos \pi = 3$, $y = r \sin \theta = -3 \sin \pi = 0 \Rightarrow$ Cartesian coordinates are $(3, 0)$

   (f) $x = r \cos \theta = 2 \cos \frac{\pi}{3} = 1$, $y = r \sin \theta = 2 \sin \frac{\pi}{3} = \sqrt{3} \Rightarrow$ Cartesian coordinates are $\left(1, \sqrt{3}\right)$

   (g) $x = r \cos \theta = -3 \cos 2\pi = -3$, $y = r \sin \theta = -3 \sin 2\pi = 0 \Rightarrow$ Cartesian coordinates are $(-3, 0)$

   (h) $x = r \cos \theta = -2 \cos \left(-\frac{\pi}{3}\right) = -1$, $y = r \sin \theta = -2 \sin \left(-\frac{\pi}{3}\right) = \sqrt{3} \Rightarrow$ Cartesian coordinates are $\left(-1, \sqrt{3}\right)$

7.

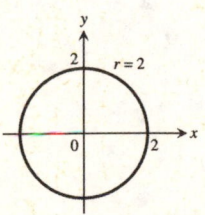

9.

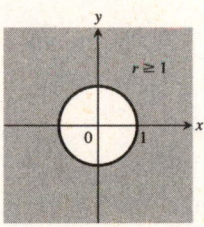

11.

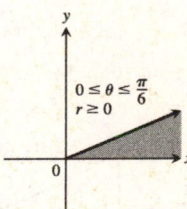

13.

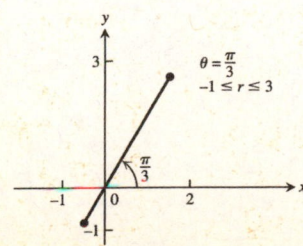

15.

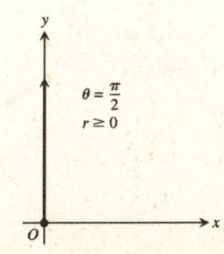

17.

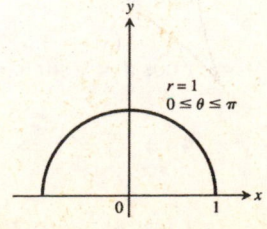

19.

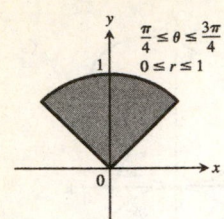

21.

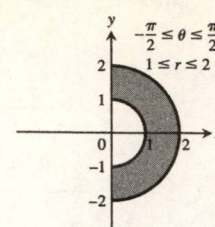

23. $r \cos \theta = 2 \Rightarrow x = 2$, vertical line through $(2, 0)$

25. $r \sin \theta = 0 \Rightarrow y = 0$, the x-axis

27. $r = 4 \csc \theta \Rightarrow r = \frac{4}{\sin \theta} \Rightarrow r \sin \theta = 4 \Rightarrow y = 4$, a horizontal line through $(0, 4)$

29. $r \cos \theta + r \sin \theta = 1 \Rightarrow x + y = 1$, line with slope $m = -1$ and intercept $b = 1$

31. $r^2 = 1 \Rightarrow x^2 + y^2 = 1$, circle with center $C = (0, 0)$ and radius 1

33. $r = \frac{5}{\sin \theta - 2 \cos \theta} \Rightarrow r \sin \theta - 2r \cos \theta = 5 \Rightarrow y - 2x = 5$, line with slope $m = 2$ and intercept $b = 5$

35. $r = \cot \theta \csc \theta = \left(\frac{\cos \theta}{\sin \theta}\right)\left(\frac{1}{\sin \theta}\right) \Rightarrow r \sin^2 \theta = \cos \theta \Rightarrow r^2 \sin^2 \theta = r \cos \theta \Rightarrow y^2 = x$, parabola with vertex $(0, 0)$
which opens to the right

37. $r = (\csc \theta) e^{r \cos \theta} \Rightarrow r \sin \theta = e^{r \cos \theta} \Rightarrow y = e^x$, graph of the natural exponential function

39. $r^2 + 2r^2 \cos \theta \sin \theta = 1 \Rightarrow x^2 + y^2 + 2xy = 1 \Rightarrow x^2 + 2xy + y^2 = 1 \Rightarrow (x + y)^2 = 1 \Rightarrow x + y = \pm 1$, two parallel
straight lines of slope -1 and y-intercepts $b = \pm 1$

41. $r^2 = -4r \cos \theta \Rightarrow x^2 + y^2 = -4x \Rightarrow x^2 + 4x + y^2 = 0 \Rightarrow x^2 + 4x + 4 + y^2 = 4 \Rightarrow (x + 2)^2 + y^2 = 4$, a circle with
center $C(-2, 0)$ and radius 2

43. $r = 8 \sin \theta \Rightarrow r^2 = 8r \sin \theta \Rightarrow x^2 + y^2 = 8y \Rightarrow x^2 + y^2 - 8y = 0 \Rightarrow x^2 + y^2 - 8y + 16 = 16$
$\Rightarrow x^2 + (y - 4)^2 = 16$, a circle with center $C(0, 4)$ and radius 4

45. $r = 2 \cos \theta + 2 \sin \theta \Rightarrow r^2 = 2r \cos \theta + 2r \sin \theta \Rightarrow x^2 + y^2 = 2x + 2y \Rightarrow x^2 - 2x + y^2 - 2y = 0$
$\Rightarrow (x - 1)^2 + (y - 1)^2 = 2$, a circle with center $C(1, 1)$ and radius $\sqrt{2}$

47. $r \sin\left(\theta + \frac{\pi}{6}\right) = 2 \Rightarrow r\left(\sin \theta \cos \frac{\pi}{6} + \cos \theta \sin \frac{\pi}{6}\right) = 2 \Rightarrow \frac{\sqrt{3}}{2} r \sin \theta + \frac{1}{2} r \cos \theta = 2 \Rightarrow \frac{\sqrt{3}}{2} y + \frac{1}{2} x = 2$
$\Rightarrow \sqrt{3} y + x = 4$, line with slope $m = -\frac{1}{\sqrt{3}}$ and intercept $b = \frac{4}{\sqrt{3}}$

49. $x = 7 \Rightarrow r \cos \theta = 7$

51. $x = y \Rightarrow r \cos \theta = r \sin \theta \Rightarrow \theta = \frac{\pi}{4}$

53. $x^2 + y^2 = 4 \Rightarrow r^2 = 4 \Rightarrow r = 2$ or $r = -2$

55. $\frac{x^2}{9} + \frac{y^2}{4} = 1 \Rightarrow 4x^2 + 9y^2 = 36 \Rightarrow 4r^2 \cos^2 \theta + 9r^2 \sin^2 \theta = 36$

57. $y^2 = 4x \Rightarrow r^2 \sin^2 \theta = 4r \cos \theta \Rightarrow r \sin^2 \theta = 4 \cos \theta$

59. $x^2 + (y - 2)^2 = 4 \Rightarrow x^2 + y^2 - 4y + 4 = 4 \Rightarrow x^2 + y^2 = 4y \Rightarrow r^2 = 4r \sin \theta \Rightarrow r = 4 \sin \theta$

61. $(x - 3)^2 + (y + 1)^2 = 4 \Rightarrow x^2 - 6x + 9 + y^2 + 2y + 1 = 4 \Rightarrow x^2 + y^2 = 6x - 2y - 6 \Rightarrow r^2 = 6r \cos \theta - 2r \sin \theta - 6$

63. $(0, \theta)$ where θ is any angle

10.6 GRAPHING IN POLAR COORDINATES

1. $1 + \cos(-\theta) = 1 + \cos \theta = r \Rightarrow$ symmetric about the
    x-axis; $1 + \cos(-\theta) \neq -r$ and $1 + \cos(\pi - \theta)$
    $= 1 - \cos \theta \neq r \Rightarrow$ not symmetric about the y-axis;
    therefore not symmetric about the origin

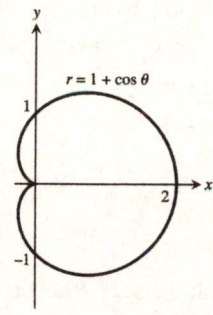

3. $1 - \sin(-\theta) = 1 + \sin \theta \neq r$ and $1 - \sin(\pi - \theta)$
    $= 1 - \sin \theta \neq -r \Rightarrow$ not symmetric about the x-axis;
    $1 - \sin(\pi - \theta) = 1 - \sin \theta = r \Rightarrow$ symmetric about
    the y-axis; therefore not symmetric about the origin

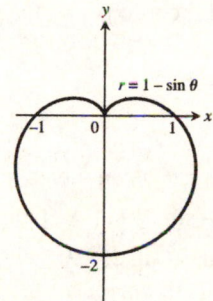

5. $2 + \sin(-\theta) = 2 - \sin \theta \neq r$ and $2 + \sin(\pi - \theta)$
    $= 2 + \sin \theta \neq -r \Rightarrow$ not symmetric about the x-axis;
    $2 + \sin(\pi - \theta) = 2 + \sin \theta = r \Rightarrow$ symmetric about the
    y-axis; therefore not symmetric about the origin

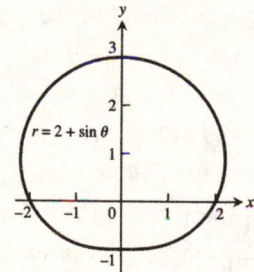

7. $\sin\left(-\frac{\theta}{2}\right) = -\sin\left(\frac{\theta}{2}\right) = -r \Rightarrow$ symmetric about the y-axis;
    $\sin\left(\frac{2\pi - \theta}{2}\right) = \sin\left(\frac{\theta}{2}\right)$, so the graph is symmetric about the
    x-axis, and hence the origin.

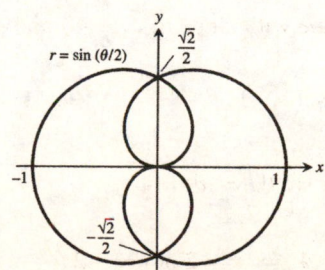

9. $\cos(-\theta) = \cos\theta = r^2 \Rightarrow (r, -\theta)$ and $(-r, -\theta)$ are on the graph when (r, θ) is on the graph $\Rightarrow$ symmetric about the x-axis and the y-axis; therefore symmetric about the origin

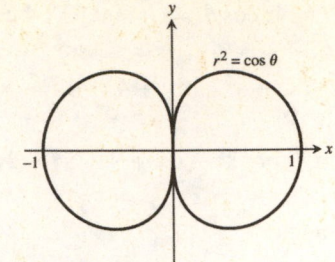

11. $-\sin(\pi - \theta) = -\sin\theta = r^2 \Rightarrow (r, \pi - \theta)$ and $(-r, \pi - \theta)$ are on the graph when (r, θ) is on the graph $\Rightarrow$ symmetric about the y-axis and the x-axis; therefore symmetric about the origin

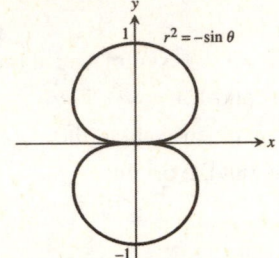

13. Since $(\pm r, -\theta)$ are on the graph when (r, θ) is on the graph $\left((\pm r)^2 = 4\cos 2(-\theta) \Rightarrow r^2 = 4\cos 2\theta\right)$, the graph is symmetric about the x-axis and the y-axis $\Rightarrow$ the graph is symmetric about the origin

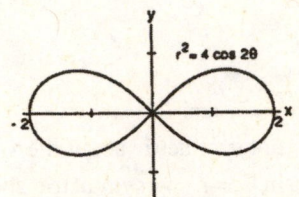

15. Since (r, θ) on the graph $\Rightarrow (-r, \theta)$ is on the graph $\left((\pm r)^2 = -\sin 2\theta \Rightarrow r^2 = -\sin 2\theta\right)$, the graph is symmetric about the origin. But $-\sin 2(-\theta) = -(-\sin 2\theta)$ $\sin 2\theta \neq r^2$ and $-\sin 2(\pi - \theta) = -\sin(2\pi - 2\theta)$ $= -\sin(-2\theta) = -(-\sin 2\theta) = \sin 2\theta \neq r^2 \Rightarrow$ the graph is not symmetric about the x-axis; therefore the graph is not symmetric about the y-axis

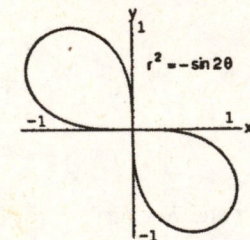

17. $\theta = \frac{\pi}{2} \Rightarrow r = -1 \Rightarrow \left(-1, \frac{\pi}{2}\right)$, and $\theta = -\frac{\pi}{2} \Rightarrow r = -1$ $\Rightarrow \left(-1, -\frac{\pi}{2}\right)$; $r' = \frac{dr}{d\theta} = -\sin\theta$; Slope $= \frac{r'\sin\theta + r\cos\theta}{r'\cos\theta - r\sin\theta}$ $= \frac{-\sin^2\theta + r\cos\theta}{-\sin\theta\cos\theta - r\sin\theta} \Rightarrow$ Slope at $\left(-1, \frac{\pi}{2}\right)$ is $\frac{-\sin^2\left(\frac{\pi}{2}\right) + (-1)\cos\frac{\pi}{2}}{-\sin\frac{\pi}{2}\cos\frac{\pi}{2} - (-1)\sin\frac{\pi}{2}} = -1$; Slope at $\left(-1, -\frac{\pi}{2}\right)$ is $\frac{-\sin^2\left(-\frac{\pi}{2}\right) + (-1)\cos\left(-\frac{\pi}{2}\right)}{-\sin\left(-\frac{\pi}{2}\right)\cos\left(-\frac{\pi}{2}\right) - (-1)\sin\left(-\frac{\pi}{2}\right)} = 1$

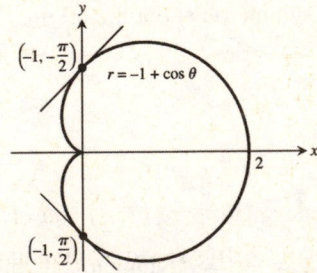

19. $\theta = \frac{\pi}{4} \Rightarrow r = 1 \Rightarrow \left(1, \frac{\pi}{4}\right)$; $\theta = -\frac{\pi}{4} \Rightarrow r = -1$ $\Rightarrow \left(-1, -\frac{\pi}{4}\right)$; $\theta = \frac{3\pi}{4} \Rightarrow r = -1 \Rightarrow \left(-1, \frac{3\pi}{4}\right)$; $\theta = -\frac{3\pi}{4} \Rightarrow r = 1 \Rightarrow \left(1, -\frac{3\pi}{4}\right)$; $r' = \frac{dr}{d\theta} = 2\cos 2\theta$; Slope $= \frac{r'\sin\theta + r\cos\theta}{r'\cos\theta - r\sin\theta} = \frac{2\cos 2\theta\sin\theta + r\cos\theta}{2\cos 2\theta\cos\theta - r\sin\theta}$ $\Rightarrow$ Slope at $\left(1, \frac{\pi}{4}\right)$ is $\frac{2\cos\left(\frac{\pi}{2}\right)\sin\left(\frac{\pi}{4}\right) + (1)\cos\left(\frac{\pi}{4}\right)}{2\cos\left(\frac{\pi}{2}\right)\cos\left(\frac{\pi}{4}\right) - (1)\sin\left(\frac{\pi}{4}\right)} = -1$;

Slope at $\left(-1, -\frac{\pi}{4}\right)$ is $\frac{2\cos\left(-\frac{\pi}{2}\right)\sin\left(-\frac{\pi}{4}\right) + (-1)\cos\left(-\frac{\pi}{4}\right)}{2\cos\left(-\frac{\pi}{2}\right)\cos\left(-\frac{\pi}{4}\right) - (-1)\sin\left(-\frac{\pi}{4}\right)} = 1$;

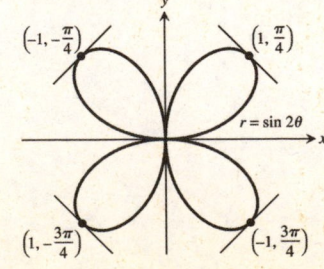

Slope at $\left(-1, \frac{3\pi}{4}\right)$ is $\dfrac{2\cos\left(\frac{3\pi}{2}\right)\sin\left(\frac{3\pi}{4}\right)+(-1)\cos\left(\frac{3\pi}{4}\right)}{2\cos\left(\frac{3\pi}{2}\right)\cos\left(\frac{3\pi}{4}\right)-(-1)\sin\left(\frac{3\pi}{4}\right)} = 1;$

Slope at $\left(1, -\frac{3\pi}{4}\right)$ is $\dfrac{2\cos\left(-\frac{3\pi}{2}\right)\sin\left(-\frac{3\pi}{4}\right)+(1)\cos\left(-\frac{3\pi}{4}\right)}{2\cos\left(-\frac{3\pi}{2}\right)\cos\left(-\frac{3\pi}{4}\right)-(1)\sin\left(-\frac{3\pi}{4}\right)} = -1$

21. (a)

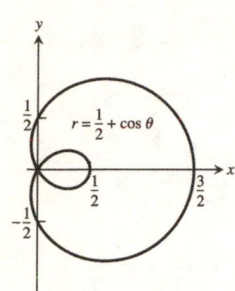

(b)

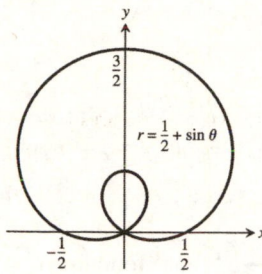

23. (a)

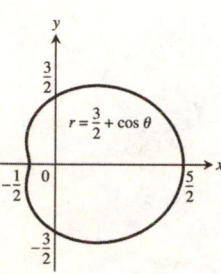

(b)

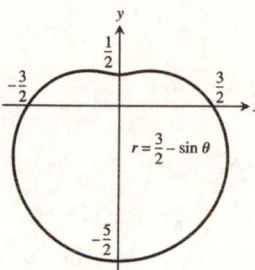

25.

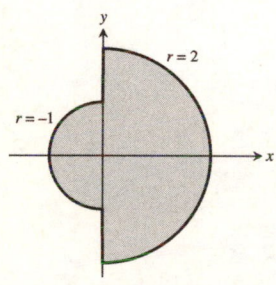

27.

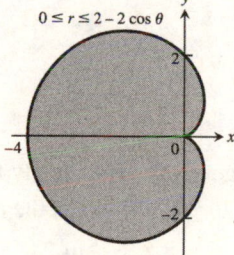

29. $\left(2, \frac{3\pi}{4}\right)$ is the same point as $\left(-2, -\frac{\pi}{4}\right)$; $r = 2\sin 2\left(-\frac{\pi}{4}\right) = 2\sin\left(-\frac{\pi}{2}\right) = -2 \Rightarrow \left(-2, -\frac{\pi}{4}\right)$ is on the graph
$\Rightarrow \left(2, \frac{3\pi}{4}\right)$ is on the graph

31. $1 + \cos\theta = 1 - \cos\theta \Rightarrow \cos\theta = 0 \Rightarrow \theta = \frac{\pi}{2}, \frac{3\pi}{2}$
$\Rightarrow r = 1$; points of intersection are $\left(1, \frac{\pi}{2}\right)$ and $\left(1, \frac{3\pi}{2}\right)$.
The point of intersection $(0, 0)$ is found by graphing.

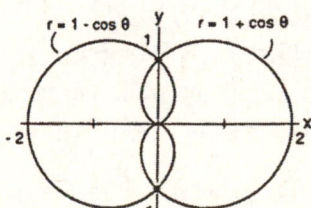

33. $2 \sin \theta = 2 \sin 2\theta \Rightarrow \sin \theta = \sin 2\theta \Rightarrow \sin \theta$
$= 2 \sin \theta \cos \theta \Rightarrow \sin \theta - 2 \sin \theta \cos \theta = 0$
$\Rightarrow (\sin \theta)(1 - 2 \cos \theta) = 0 \Rightarrow \sin \theta = 0$ or $\cos \theta = \frac{1}{2}$
$\Rightarrow \theta = 0, \pi, \frac{\pi}{3}$, or $-\frac{\pi}{3}$; $\theta = 0$ or $\pi \Rightarrow r = 0$,
$\theta = \frac{\pi}{3} \Rightarrow r = \sqrt{3}$, and $\theta = -\frac{\pi}{3} \Rightarrow r = -\sqrt{3}$; points of
intersection are $(0,0)$, $\left(\sqrt{3}, \frac{\pi}{3}\right)$, and $\left(-\sqrt{3}, -\frac{\pi}{3}\right)$

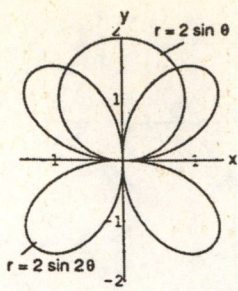

35. $\left(\sqrt{2}\right)^2 = 4 \sin \theta \Rightarrow \frac{1}{2} = \sin \theta \Rightarrow \theta = \frac{\pi}{6}, \frac{5\pi}{6}$; points
of intersection are $\left(\sqrt{2}, \frac{\pi}{6}\right)$ and $\left(\sqrt{2}, \frac{5\pi}{6}\right)$. The
points $\left(\sqrt{2}, -\frac{\pi}{6}\right)$ and $\left(\sqrt{2}, -\frac{5\pi}{6}\right)$ are found by
graphing.

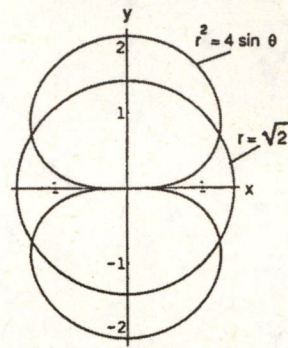

37. $1 = 2 \sin 2\theta \Rightarrow \sin 2\theta = \frac{1}{2} \Rightarrow 2\theta = \frac{\pi}{6}, \frac{5\pi}{6}, \frac{13\pi}{6}, \frac{17\pi}{6}$
$\Rightarrow \theta = \frac{\pi}{12}, \frac{5\pi}{12}, \frac{13\pi}{12}, \frac{17\pi}{12}$; points of intersection are
$\left(1, \frac{\pi}{12}\right), \left(1, \frac{5\pi}{12}\right), \left(1, \frac{13\pi}{12}\right)$, and $\left(1, \frac{17\pi}{12}\right)$. No other
points are found by graphing.

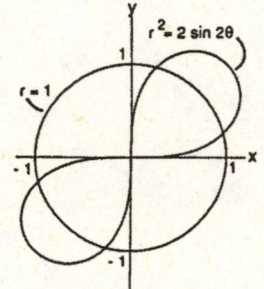

39. $r^2 = \sin 2\theta$ and $r^2 = \cos 2\theta$ are generated completely for
$0 \le \theta \le \frac{\pi}{2}$. Then $\sin 2\theta = \cos 2\theta \Rightarrow 2\theta = \frac{\pi}{4}$ is the only
solution on that interval $\Rightarrow \theta = \frac{\pi}{8} \Rightarrow r^2 = \sin 2\left(\frac{\pi}{8}\right) = \frac{1}{\sqrt{2}}$
$\Rightarrow r = \pm \frac{1}{\sqrt[4]{2}}$; points of intersection are $\left(\pm \frac{1}{\sqrt[4]{2}}, \frac{\pi}{8}\right)$.
The point of intersection $(0,0)$ is found by graphing.

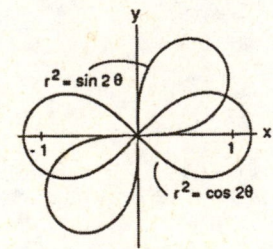

41. $1 = 2 \sin 2\theta \Rightarrow \sin 2\theta = \frac{1}{2} \Rightarrow 2\theta = \frac{\pi}{6}, \frac{5\pi}{6}, \frac{13\pi}{6}, \frac{17\pi}{6}$
$\Rightarrow \theta = \frac{\pi}{12}, \frac{5\pi}{12}, \frac{13\pi}{12}, \frac{17\pi}{12}$; points of intersection are
$\left(1, \frac{\pi}{12}\right), \left(1, \frac{5\pi}{12}\right), \left(1, \frac{13\pi}{12}\right)$, and $\left(1, \frac{17\pi}{12}\right)$. The points
of intersection $\left(1, \frac{7\pi}{12}\right), \left(1, \frac{11\pi}{12}\right), \left(1, \frac{19\pi}{12}\right)$ and
$\left(1, \frac{23\pi}{12}\right)$ are found by graphing and symmetry.

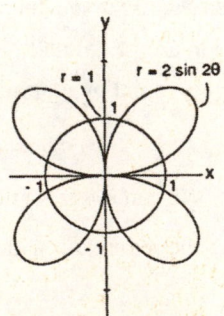

43. Note that (r, θ) and $(-r, \theta + \pi)$ describe the same point in the plane. Then $r = 1 - \cos \theta \Leftrightarrow -1 - \cos(\theta + \pi)$
$= -1 - (\cos \theta \cos \pi - \sin \theta \sin \pi) = -1 + \cos \theta = -(1 - \cos \theta) = -r$; therefore (r, θ) is on the graph of
$r = 1 - \cos \theta \Leftrightarrow (-r, \theta + \pi)$ is on the graph of $r = -1 - \cos \theta \Rightarrow$ the answer is (a).

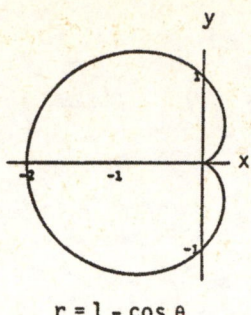

$r = 1 - \cos \theta$

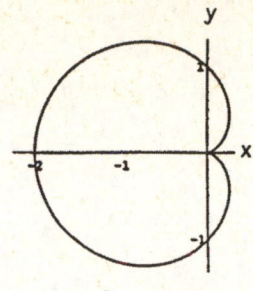

$r = -1 - \cos \theta$

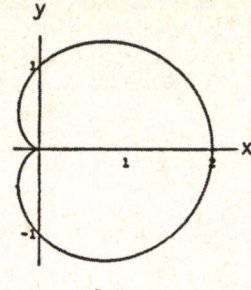

$r = 1 + \cos \theta$

45.

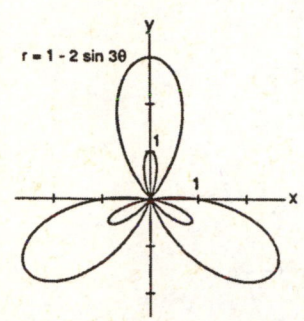

$r = 1 - 2 \sin 3\theta$

47. (a)

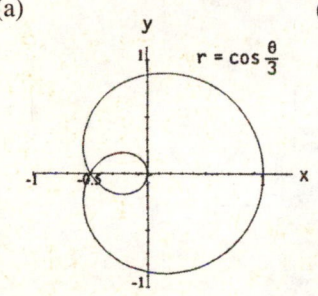

$r = \cos \frac{\theta}{3}$

(b)

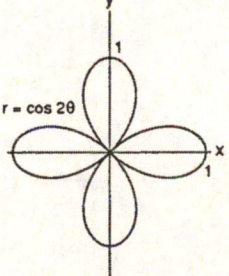

$r = \cos 2\theta$

(c)

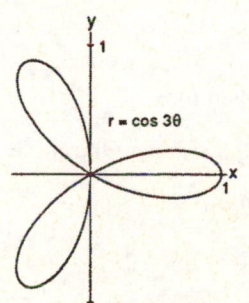

$r = \cos 3\theta$

(d)

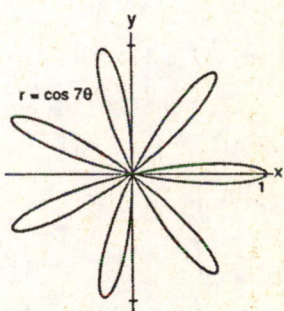

$r = \cos 7\theta$

49. (a) $r^2 = -4 \cos \theta \Rightarrow \cos \theta = -\frac{r^2}{4}$; $r = 1 - \cos \theta \Rightarrow r = 1 - \left(-\frac{r^2}{4} \right) \Rightarrow 0 = r^2 - 4r + 4 \Rightarrow (r-2)^2 = 0$

$\Rightarrow r = 2$; therefore $\cos \theta = -\frac{2^2}{4} = -1 \Rightarrow \theta = \pi \Rightarrow (2, \pi)$ is a point of intersection

(b) $r = 0 \Rightarrow 0^2 = 4 \cos \theta \Rightarrow \cos \theta = 0 \Rightarrow \theta = \frac{\pi}{2}, \frac{3\pi}{2} \Rightarrow \left(0, \frac{\pi}{2} \right)$ or $\left(0, \frac{3\pi}{2} \right)$ is on the graph; $r = 0 \Rightarrow 0 = 1 - \cos \theta$

$\Rightarrow \cos \theta = 1 \Rightarrow \theta = 0 \Rightarrow (0, 0)$ is on the graph. Since $(0, 0) = \left(0, \frac{\pi}{2} \right)$ for polar coordinates, the graphs intersect at the origin.

51. The maximum width of the petal of the rose which lies along the x-axis is twice the largest y value of the curve on the interval $0 \le \theta \le \frac{\pi}{4}$. So we wish to maximize $2y = 2r \sin \theta = 2 \cos 2\theta \sin \theta$ on $0 \le \theta \le \frac{\pi}{4}$. Let $f(\theta) = 2 \cos 2\theta \sin \theta = 2 \left(1 - 2 \sin^2 \theta \right) (\sin \theta) = 2 \sin \theta - 4 \sin^3 \theta \Rightarrow f'(\theta) = 2 \cos \theta - 12 \sin^2 \theta \cos \theta$. Then $f'(\theta) = 0 \Rightarrow 2 \cos \theta - 12 \sin^2 \theta \cos \theta = 0 \Rightarrow (\cos \theta) \left(1 - 6 \sin^2 \theta \right) = 0 \Rightarrow \cos \theta = 0$ or $1 - 6 \sin^2 \theta = 0 \Rightarrow \theta = \frac{\pi}{2}$ or $\sin \theta = \frac{\pm 1}{\sqrt{6}}$. Since we want $0 \le \theta \le \frac{\pi}{4}$, we choose $\theta = \sin^{-1} \left(\frac{1}{\sqrt{6}} \right) \Rightarrow f(\theta) = 2 \sin \theta - 4 \sin^3 \theta$

$= 2 \left(\frac{1}{\sqrt{6}} \right) - 4 \cdot \frac{1}{6\sqrt{6}} = \frac{2\sqrt{6}}{9}$. We can see from the graph of $r = \cos 2\theta$ that a maximum does occur in the interval $0 \le \theta \le \frac{\pi}{4}$. Therefore the maximum width occurs at $\theta = \sin^{-1} \left(\frac{1}{\sqrt{6}} \right)$, and the maximum width is $\frac{2\sqrt{6}}{9}$.

10.7 AREA AND LENGTHS IN POLAR COORDINATES

1. $A = \int_0^{2\pi} \frac{1}{2} (4 + 2 \cos \theta)^2 \, d\theta = \int_0^{2\pi} \frac{1}{2} (16 + 16 \cos \theta + 4 \cos^2 \theta) \, d\theta = \int_0^{2\pi} \left[8 + 8 \cos \theta + 2 \left(\frac{1 + \cos 2\theta}{2} \right) \right] d\theta$

    $= \int_0^{2\pi} (9 + 8 \cos \theta + \cos 2\theta) \, d\theta = \left[9\theta + 8 \sin \theta + \frac{1}{2} \sin 2\theta \right]_0^{2\pi} = 18\pi$

3. $A = 2 \int_0^{\pi/4} \frac{1}{2} \cos^2 2\theta \, d\theta = \int_0^{\pi/4} \frac{1 + \cos 4\theta}{2} \, d\theta = \frac{1}{2} \left[\theta + \frac{\sin 4\theta}{4} \right]_0^{\pi/4} = \frac{\pi}{8}$

5. $A = \int_0^{\pi/2} \frac{1}{2} (4 \sin 2\theta) \, d\theta = \int_0^{\pi/2} 2 \sin 2\theta \, d\theta = \left[-\cos 2\theta \right]_0^{\pi/2} = 2$

7. $r = 2 \cos \theta$ and $r = 2 \sin \theta \Rightarrow 2 \cos \theta = 2 \sin \theta$
    $\Rightarrow \cos \theta = \sin \theta \Rightarrow \theta = \frac{\pi}{4}$; therefore
    $A = 2 \int_0^{\pi/4} \frac{1}{2} (2 \sin \theta)^2 \, d\theta = \int_0^{\pi/4} 4 \sin^2 \theta \, d\theta$
    $= \int_0^{\pi/4} 4 \left(\frac{1 - \cos 2\theta}{2} \right) d\theta = \int_0^{\pi/4} (2 - 2 \cos 2\theta) \, d\theta$
    $= \left[2\theta - \sin 2\theta \right]_0^{\pi/4} = \frac{\pi}{2} - 1$

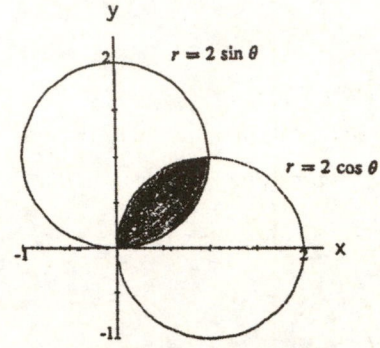

9. $r = 2$ and $r = 2(1 - \cos \theta) \Rightarrow 2 = 2(1 - \cos \theta)$
    $\Rightarrow \cos \theta = 0 \Rightarrow \theta = \pm \frac{\pi}{2}$; therefore
    $A = 2 \int_0^{\pi/2} \frac{1}{2} [2(1 - \cos \theta)]^2 \, d\theta + \frac{1}{2} \text{area of the circle}$
    $= \int_0^{\pi/2} 4 (1 - 2 \cos \theta + \cos^2 \theta) \, d\theta + \left(\frac{1}{2} \pi \right) (2)^2$
    $= \int_0^{\pi/2} 4 \left(1 - 2 \cos \theta + \frac{1 + \cos 2\theta}{2} \right) d\theta + 2\pi$
    $= \int_0^{\pi/2} (4 - 8 \cos \theta + 2 + 2 \cos 2\theta) \, d\theta + 2\pi$
    $= \left[6\theta - 8 \sin \theta + \sin 2\theta \right]_0^{\pi/2} + 2\pi = 5\pi - 8$

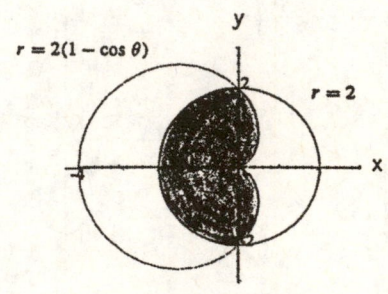

11. $r = \sqrt{3}$ and $r^2 = 6 \cos 2\theta \Rightarrow 3 = 6 \cos 2\theta \Rightarrow \cos 2\theta = \frac{1}{2}$
    $\Rightarrow \theta = \frac{\pi}{6}$ (in the 1st quadrant); we use symmetry of the
    graph to find the area, so
    $A = 4 \int_0^{\pi/6} \left[\frac{1}{2} (6 \cos 2\theta) - \frac{1}{2} \left(\sqrt{3} \right)^2 \right] d\theta$
    $= 2 \int_0^{\pi/6} (6 \cos 2\theta - 3) \, d\theta = 2 \left[3 \sin 2\theta - 3\theta \right]_0^{\pi/6}$
    $= 3\sqrt{3} - \pi$

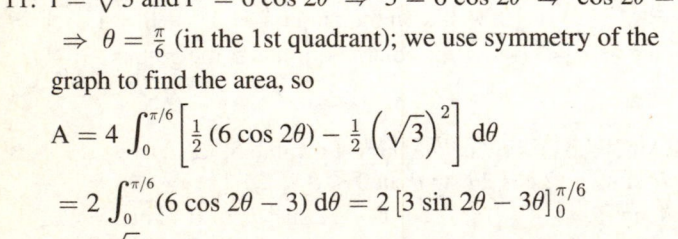

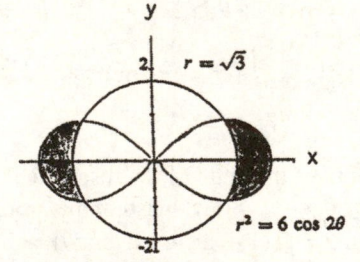

13. $r = 1$ and $r = -2 \cos \theta \Rightarrow 1 = -2 \cos \theta \Rightarrow \cos \theta = -\frac{1}{2}$
    $\Rightarrow \theta = \frac{2\pi}{3}$ in quadrant II; therefore
    $A = 2 \int_{2\pi/3}^{\pi} \frac{1}{2} [(-2 \cos \theta)^2 - 1^2] \, d\theta = \int_{2\pi/3}^{\pi} (4 \cos^2 \theta - 1) \, d\theta$
    $= \int_{2\pi/3}^{\pi} [2(1 + \cos 2\theta) - 1] \, d\theta = \int_{2\pi/3}^{\pi} (1 + 2 \cos 2\theta) \, d\theta$
    $= \left[\theta + \sin 2\theta \right]_{2\pi/3}^{\pi} = \frac{\pi}{3} + \frac{\sqrt{3}}{2}$

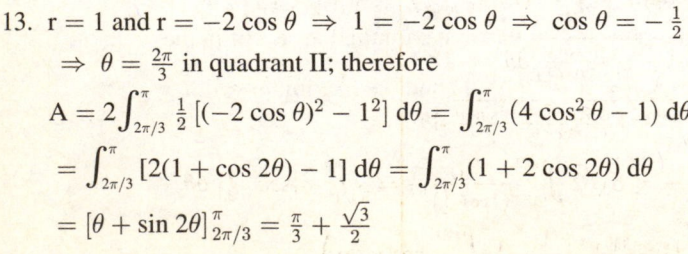

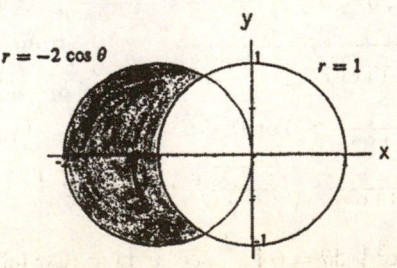

15. $r = 6$ and $r = 3 \csc \theta \Rightarrow 6 \sin \theta = 3 \Rightarrow \sin \theta = \frac{1}{2}$

$\Rightarrow \theta = \frac{\pi}{6}$ or $\frac{5\pi}{6}$; therefore $A = \int_{\pi/6}^{5\pi/6} \frac{1}{2} \left(6^2 - 9 \csc^2 \theta\right) d\theta$

$= \int_{\pi/6}^{5\pi/6} \left(18 - \frac{9}{2} \csc^2 \theta\right) d\theta = \left[18\theta + \frac{9}{2} \cot \theta\right]_{\pi/6}^{5\pi/6}$

$= \left(15\pi - \frac{9}{2} \sqrt{3}\right) - \left(3\pi + \frac{9}{2} \sqrt{3}\right) = 12\pi - 9\sqrt{3}$

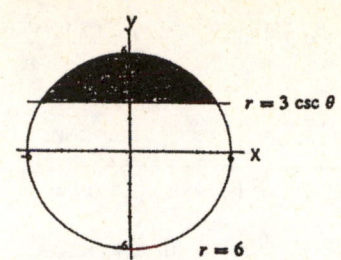

17. (a) $r = \tan \theta$ and $r = \left(\frac{\sqrt{2}}{2}\right) \csc \theta \Rightarrow \tan \theta = \left(\frac{\sqrt{2}}{2}\right) \csc \theta$

$\Rightarrow \sin^2 \theta = \left(\frac{\sqrt{2}}{2}\right) \cos \theta \Rightarrow 1 - \cos^2 \theta = \left(\frac{\sqrt{2}}{2}\right) \cos \theta$

$\Rightarrow \cos^2 \theta + \left(\frac{\sqrt{2}}{2}\right) \cos \theta - 1 = 0 \Rightarrow \cos \theta = -\sqrt{2}$ or

$\frac{\sqrt{2}}{2}$ (use the quadratic formula) $\Rightarrow \theta = \frac{\pi}{4}$ (the solution

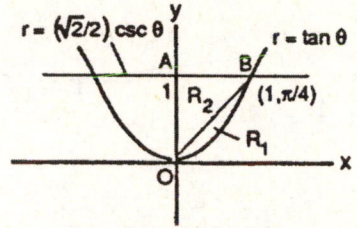

in the first quadrant); therefore the area of R_1 is

$A_1 = \int_0^{\pi/4} \frac{1}{2} \tan^2 \theta \, d\theta = \frac{1}{2} \int_0^{\pi/4} \left(\sec^2 \theta - 1\right) d\theta = \frac{1}{2} \left[\tan \theta - \theta\right]_0^{\pi/4} = \frac{1}{2} \left(\tan \frac{\pi}{4} - \frac{\pi}{4}\right) = \frac{1}{2} - \frac{\pi}{8}$;

$AO = \left(\frac{\sqrt{2}}{2}\right) \csc \frac{\pi}{2} = \frac{\sqrt{2}}{2}$ and $OB = \left(\frac{\sqrt{2}}{2}\right) \csc \frac{\pi}{4} = 1 \Rightarrow AB = \sqrt{1^2 - \left(\frac{\sqrt{2}}{2}\right)^2} = \frac{\sqrt{2}}{2}$

$\Rightarrow$ the area of R_2 is $A_2 = \frac{1}{2} \left(\frac{\sqrt{2}}{2}\right) \left(\frac{\sqrt{2}}{2}\right) = \frac{1}{4}$; therefore the area of the region shaded in the text is

$2 \left(\frac{1}{2} - \frac{\pi}{8} + \frac{1}{4}\right) = \frac{3}{2} - \frac{\pi}{4}$. Note: The area must be found this way since no common interval generates the region. For

example, the interval $0 \leq \theta \leq \frac{\pi}{4}$ generates the arc OB of $r = \tan \theta$ but does not generate the segment AB of the line

$r = \frac{\sqrt{2}}{2} \csc \theta$. Instead the interval generates the half-line from B to $+\infty$ on the line $r = \frac{\sqrt{2}}{2} \csc \theta$.

(b) $\lim_{\theta \to \pi/2^-} \tan \theta = \infty$ and the line $x = 1$ is $r = \sec \theta$ in polar coordinates; then $\lim_{\theta \to \pi/2^-} (\tan \theta - \sec \theta)$

$= \lim_{\theta \to \pi/2^-} \left(\frac{\sin \theta}{\cos \theta} - \frac{1}{\cos \theta}\right) = \lim_{\theta \to \pi/2^-} \left(\frac{\sin \theta - 1}{\cos \theta}\right) = \lim_{\theta \to \pi/2^-} \left(\frac{\cos \theta}{-\sin \theta}\right) = 0 \Rightarrow r = \tan \theta$ approaches

$r = \sec \theta$ as $\theta \to \frac{\pi^-}{2} \Rightarrow r = \sec \theta$ (or $x = 1$) is a vertical asymptote of $r = \tan \theta$. Similarly, $r = -\sec \theta$

(or $x = -1$) is a vertical asymptote of $r = \tan \theta$.

19. $r = \theta^2, 0 \leq \theta \leq \sqrt{5} \Rightarrow \frac{dr}{d\theta} = 2\theta$; therefore Length $= \int_0^{\sqrt{5}} \sqrt{\left(\theta^2\right)^2 + (2\theta)^2} \, d\theta = \int_0^{\sqrt{5}} \sqrt{\theta^4 + 4\theta^2} \, d\theta$

$= \int_0^{\sqrt{5}} |\theta| \sqrt{\theta^2 + 4} \, d\theta = \text{(since } \theta \geq 0\text{)} \int_0^{\sqrt{5}} \theta \sqrt{\theta^2 + 4} \, d\theta; \left[u = \theta^2 + 4 \Rightarrow \frac{1}{2} \, du = \theta \, d\theta; \theta = 0 \Rightarrow u = 4,\right.$

$\left.\theta = \sqrt{5} \Rightarrow u = 9\right] \to \int_4^9 \frac{1}{2} \sqrt{u} \, du = \frac{1}{2} \left[\frac{2}{3} u^{3/2}\right]_4^9 = \frac{19}{3}$

21. $r = 1 + \cos \theta \Rightarrow \frac{dr}{d\theta} = -\sin \theta$; therefore Length $= \int_0^{2\pi} \sqrt{(1 + \cos \theta)^2 + (-\sin \theta)^2} \, d\theta$

$= 2 \int_0^\pi \sqrt{2 + 2 \cos \theta} \, d\theta = 2 \int_0^\pi \sqrt{\frac{4(1 + \cos \theta)}{2}} \, d\theta = 4 \int_0^\pi \sqrt{\frac{1 + \cos \theta}{2}} \, d\theta = 4 \int_0^\pi \cos \left(\frac{\theta}{2}\right) d\theta = 4 \left[2 \sin \frac{\theta}{2}\right]_0^\pi = 8$

23. $r = \frac{6}{1 + \cos \theta}, 0 \leq \theta \leq \frac{\pi}{2} \Rightarrow \frac{dr}{d\theta} = \frac{6 \sin \theta}{(1 + \cos \theta)^2}$; therefore Length $= \int_0^{\pi/2} \sqrt{\left(\frac{6}{1 + \cos \theta}\right)^2 + \left(\frac{6 \sin \theta}{(1 + \cos \theta)^2}\right)^2} \, d\theta$

$= \int_0^{\pi/2} \sqrt{\frac{36}{(1 + \cos \theta)^2} + \frac{36 \sin^2 \theta}{(1 + \cos \theta)^4}} \, d\theta = 6 \int_0^{\pi/2} \left|\frac{1}{1 + \cos \theta}\right| \sqrt{1 + \frac{\sin^2 \theta}{(1 + \cos \theta)^2}} \, d\theta$

$= \text{(since } \frac{1}{1 + \cos \theta} > 0 \text{ on } 0 \leq \theta \leq \frac{\pi}{2}\text{)} \; 6 \int_0^{\pi/2} \left(\frac{1}{1 + \cos \theta}\right) \sqrt{\frac{1 + 2 \cos \theta + \cos^2 \theta + \sin^2 \theta}{(1 + \cos \theta)^2}} \, d\theta$

$= 6 \int_0^{\pi/2} \left(\frac{1}{1 + \cos \theta}\right) \sqrt{\frac{2 + 2 \cos \theta}{(1 + \cos \theta)^2}} \, d\theta = 6\sqrt{2} \int_0^{\pi/2} \frac{d\theta}{(1 + \cos \theta)^{3/2}} = 6\sqrt{2} \int_0^{\pi/2} \frac{d\theta}{\left(2 \cos^2 \frac{\theta}{2}\right)^{3/2}} = 3 \int_0^{\pi/2} \left|\sec^3 \frac{\theta}{2}\right| d\theta$

$= 3 \int_0^{\pi/2} \sec^3 \frac{\theta}{2} \, d\theta = 6 \int_0^{\pi/4} \sec^3 u \, du = \text{(use tables) } 6 \left(\left[\frac{\sec u \tan u}{2}\right]_0^{\pi/4} + \frac{1}{2} \int_0^{\pi/4} \sec u \, du\right)$

$$= 6\left(\frac{1}{\sqrt{2}} + \left[\frac{1}{2}\ln|\sec u + \tan u|\right]_0^{\pi/4}\right) = 3\left[\sqrt{2} + \ln\left(1 + \sqrt{2}\right)\right]$$

25. $r = \cos^3\frac{\theta}{3} \Rightarrow \frac{dr}{d\theta} = -\sin\frac{\theta}{3}\cos^2\frac{\theta}{3}$; therefore Length $= \int_0^{\pi/4}\sqrt{\left(\cos^3\frac{\theta}{3}\right)^2 + \left(-\sin\frac{\theta}{3}\cos^2\frac{\theta}{3}\right)^2}\, d\theta$

$$= \int_0^{\pi/4}\sqrt{\cos^6\left(\frac{\theta}{3}\right) + \sin^2\left(\frac{\theta}{3}\right)\cos^4\left(\frac{\theta}{3}\right)}\, d\theta = \int_0^{\pi/4}\left(\cos^2\frac{\theta}{3}\right)\sqrt{\cos^2\left(\frac{\theta}{3}\right) + \sin^2\left(\frac{\theta}{3}\right)}\, d\theta = \int_0^{\pi/4}\cos^2\left(\frac{\theta}{3}\right)d\theta$$

$$= \int_0^{\pi/4}\frac{1 + \cos\left(\frac{2\theta}{3}\right)}{2}\, d\theta = \frac{1}{2}\left[\theta + \frac{3}{2}\sin\frac{2\theta}{3}\right]_0^{\pi/4} = \frac{\pi}{8} + \frac{3}{8}$$

27. $r = \sqrt{1 + \cos 2\theta} \Rightarrow \frac{dr}{d\theta} = \frac{1}{2}(1 + \cos 2\theta)^{-1/2}(-2\sin 2\theta)$; therefore Length $= \int_0^{\pi\sqrt{2}}\sqrt{(1 + \cos 2\theta) + \frac{\sin^2 2\theta}{(1 + \cos 2\theta)}}\, d\theta$

$$= \int_0^{\pi\sqrt{2}}\sqrt{\frac{1 + 2\cos 2\theta + \cos^2 2\theta + \sin^2 2\theta}{1 + \cos 2\theta}}\, d\theta = \int_0^{\pi\sqrt{2}}\sqrt{\frac{2 + 2\cos 2\theta}{1 + \cos 2\theta}}\, d\theta = \int_0^{\pi\sqrt{2}}\sqrt{2}\, d\theta = \left[\sqrt{2}\,\theta\right]_0^{\pi\sqrt{2}} = 2\pi$$

29. $r = \sqrt{\cos 2\theta}, 0 \leq \theta \leq \frac{\pi}{4} \Rightarrow \frac{dr}{d\theta} = \frac{1}{2}(\cos 2\theta)^{-1/2}(-\sin 2\theta)(2) = \frac{-\sin 2\theta}{\sqrt{\cos 2\theta}}$; therefore Surface Area

$$= \int_0^{\pi/4}(2\pi r\cos\theta)\sqrt{\left(\sqrt{\cos 2\theta}\right)^2 + \left(\frac{-\sin 2\theta}{\sqrt{\cos 2\theta}}\right)^2}\, d\theta = \int_0^{\pi/4}\left(2\pi\sqrt{\cos 2\theta}\right)(\cos\theta)\sqrt{\cos 2\theta + \frac{\sin^2 2\theta}{\cos 2\theta}}\, d\theta$$

$$= \int_0^{\pi/4}\left(2\pi\sqrt{\cos 2\theta}\right)(\cos\theta)\sqrt{\frac{1}{\cos 2\theta}}\, d\theta = \int_0^{\pi/4}2\pi\cos\theta\, d\theta = [2\pi\sin\theta]_0^{\pi/4} = \pi\sqrt{2}$$

31. $r^2 = \cos 2\theta \Rightarrow r = \pm\sqrt{\cos 2\theta}$; use $r = \sqrt{\cos 2\theta}$ on $\left[0, \frac{\pi}{4}\right] \Rightarrow \frac{dr}{d\theta} = \frac{1}{2}(\cos 2\theta)^{-1/2}(-\sin 2\theta)(2) = \frac{-\sin 2\theta}{\sqrt{\cos 2\theta}}$;

therefore Surface Area $= 2\int_0^{\pi/4}\left(2\pi\sqrt{\cos 2\theta}\right)(\sin\theta)\sqrt{\cos 2\theta + \frac{\sin^2 2\theta}{\cos 2\theta}}\, d\theta = 4\pi\int_0^{\pi/4}\sqrt{\cos 2\theta}\,(\sin\theta)\sqrt{\frac{1}{\cos 2\theta}}\, d\theta$

$$= 4\pi\int_0^{\pi/4}\sin\theta\, d\theta = 4\pi[-\cos\theta]_0^{\pi/4} = 4\pi\left[-\frac{\sqrt{2}}{2} - (-1)\right] = 2\pi\left(2 - \sqrt{2}\right)$$

33. Let $r = f(\theta)$. Then $x = f(\theta)\cos\theta \Rightarrow \frac{dx}{d\theta} = f'(\theta)\cos\theta - f(\theta)\sin\theta \Rightarrow \left(\frac{dx}{d\theta}\right)^2 = [f'(\theta)\cos\theta - f(\theta)\sin\theta]^2$

$$= [f'(\theta)]^2\cos^2\theta - 2f'(\theta)f(\theta)\sin\theta\cos\theta + [f(\theta)]^2\sin^2\theta; y = f(\theta)\sin\theta \Rightarrow \frac{dy}{d\theta} = f'(\theta)\sin\theta + f(\theta)\cos\theta$$

$$\Rightarrow \left(\frac{dy}{d\theta}\right)^2 = [f'(\theta)\sin\theta + f(\theta)\cos\theta]^2 = [f'(\theta)]^2\sin^2\theta + 2f'(\theta)f(\theta)\sin\theta\cos\theta + [f(\theta)]^2\cos^2\theta. \text{ Therefore}$$

$$\left(\frac{dx}{d\theta}\right)^2 + \left(\frac{dy}{d\theta}\right)^2 = [f'(\theta)]^2(\cos^2\theta + \sin^2\theta) + [f(\theta)]^2(\cos^2\theta + \sin^2\theta) = [f'(\theta)]^2 + [f(\theta)]^2 = r^2 + \left(\frac{dr}{d\theta}\right)^2.$$

Thus, $L = \int_\alpha^\beta\sqrt{\left(\frac{dx}{d\theta}\right)^2 + \left(\frac{dy}{d\theta}\right)^2}\, d\theta = \int_\alpha^\beta\sqrt{r^2 + \left(\frac{dr}{d\theta}\right)^2}\, d\theta.$

35. $r = 2f(\theta), \alpha \leq \theta \leq \beta \Rightarrow \frac{dr}{d\theta} = 2f'(\theta) \Rightarrow r^2 + \left(\frac{dr}{d\theta}\right)^2 = [2f(\theta)]^2 + [2f'(\theta)]^2 \Rightarrow \text{Length} = \int_\alpha^\beta\sqrt{4[f(\theta)]^2 + 4[f'(\theta)]^2}\, d\theta$

$$= 2\int_\alpha^\beta\sqrt{[f(\theta)]^2 + [f'(\theta)]^2}\, d\theta \text{ which is twice the length of the curve } r = f(\theta) \text{ for } \alpha \leq \theta \leq \beta.$$

37. $\bar{x} = \dfrac{\frac{2}{3}\int_0^{2\pi}r^3\cos\theta\, d\theta}{\int_0^{2\pi}r^2\, d\theta} = \dfrac{\frac{2}{3}\int_0^{2\pi}[a(1 + \cos\theta)]^3(\cos\theta)\, d\theta}{\int_0^{2\pi}[a(1 + \cos\theta)]^2\, d\theta} = \dfrac{\frac{2}{3}a^3\int_0^{2\pi}(1 + 3\cos\theta + 3\cos^2\theta + \cos^3\theta)(\cos\theta)\, d\theta}{a^2\int_0^{2\pi}(1 + 2\cos\theta + \cos^2\theta)\, d\theta}$

$$= \dfrac{\frac{2}{3}a\int_0^{2\pi}\left[\cos\theta + 3\left(\frac{1 + \cos 2\theta}{2}\right) + 3(1 - \sin^2\theta)(\cos\theta) + \left(\frac{1 + \cos 2\theta}{2}\right)^2\right]d\theta}{\int_0^{2\pi}\left[1 + 2\cos\theta + \left(\frac{1 + \cos 2\theta}{2}\right)\right]d\theta} = \text{(After considerable algebra using}$$

the identity $\cos^2 A = \frac{1 + \cos 2A}{2}$) $\dfrac{a\int_0^{2\pi}\left(\frac{15}{12} + \frac{8}{3}\cos\theta + \frac{4}{3}\cos 2\theta - 2\cos\theta\sin^2\theta + \frac{1}{12}\cos 4\theta\right)d\theta}{\int_0^{2\pi}\left(\frac{3}{2} + 2\cos\theta + \frac{1}{2}\cos 2\theta\right)d\theta}$

$$= \dfrac{a\left[\frac{15}{12}\theta + \frac{8}{3}\sin\theta + \frac{2}{3}\sin 2\theta - \frac{2}{3}\sin^3\theta + \frac{1}{48}\sin 4\theta\right]_0^{2\pi}}{\left[\frac{3}{2}\theta + 2\sin\theta + \frac{1}{4}\sin 2\theta\right]_0^{2\pi}} = \dfrac{a\left(\frac{15}{6}\pi\right)}{3\pi} = \frac{5}{6}a;$$

$\bar{y} = \dfrac{\frac{2}{3}\int_0^{2\pi}r^3\sin\theta\, d\theta}{\int_0^{2\pi}r^2\, d\theta} = \dfrac{\frac{2}{3}\int_0^{2\pi}[a(1 + \cos\theta)]^3(\sin\theta)\, d\theta}{3\pi}$; $\left[u = a(1 + \cos\theta) \Rightarrow -\frac{1}{a}du = \sin\theta\, d\theta; \theta = 0 \Rightarrow u = 2a;\right.$

$\theta = 2\pi \Rightarrow u = 2a] \rightarrow \dfrac{\frac{2}{3}\int_{2a}^{2a}-\frac{1}{a}u^3\, du}{3\pi} = \dfrac{0}{3\pi} = 0.$ Therefore the centroid is $(\bar{x}, \bar{y}) = \left(\frac{5}{6}a, 0\right)$

10.8 CONIC SECTIONS IN POLAR COORDINATES

1. $r \cos \left(\theta - \frac{\pi}{6}\right) = 5 \Rightarrow r\left(\cos\theta \cos\frac{\pi}{6} + \sin\theta \sin\frac{\pi}{6}\right) = 5 \Rightarrow \frac{\sqrt{3}}{2} r \cos\theta + \frac{1}{2} r \sin\theta = 5 \Rightarrow \frac{\sqrt{3}}{2} x + \frac{1}{2} y = 5 \Rightarrow \sqrt{3}x + y$
   $= 10 \Rightarrow y = -\sqrt{3}x + 10$

3. $r \cos \left(\theta - \frac{4\pi}{3}\right) = 3 \Rightarrow r\left(\cos\theta \cos\frac{4\pi}{3} + \sin\theta \sin\frac{4\pi}{3}\right) = 3 \Rightarrow -\frac{1}{2} r \cos\theta - \frac{\sqrt{3}}{2} r \sin\theta = 3$
   $\Rightarrow -\frac{1}{2} x - \frac{\sqrt{3}}{2} y = 3 \Rightarrow x + \sqrt{3}y = -6 \Rightarrow y = -\frac{\sqrt{3}}{3} x - 2\sqrt{3}$

5. $r \cos \left(\theta - \frac{\pi}{4}\right) = \sqrt{2} \Rightarrow r\left(\cos\theta \cos\frac{\pi}{4} + \sin\theta \sin\frac{\pi}{4}\right)$
   $= \sqrt{2} \Rightarrow \frac{1}{\sqrt{2}} r \cos\theta + \frac{1}{\sqrt{2}} r \sin\theta = \sqrt{2} \Rightarrow \frac{1}{\sqrt{2}} x + \frac{1}{\sqrt{2}} y$
   $= \sqrt{2} \Rightarrow x + y = 2 \Rightarrow y = 2 - x$

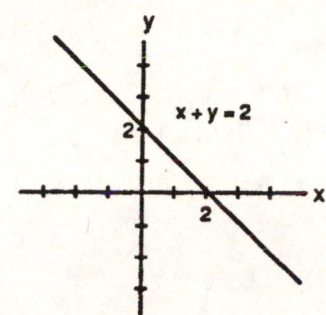

7. $r \cos \left(\theta - \frac{2\pi}{3}\right) = 3 \Rightarrow r\left(\cos\theta \cos\frac{2\pi}{3} + \sin\theta \sin\frac{2\pi}{3}\right) = 3$
   $\Rightarrow -\frac{1}{2} r \cos\theta + \frac{\sqrt{3}}{2} r \sin\theta = 3 \Rightarrow -\frac{1}{2} x + \frac{\sqrt{3}}{2} y = 3$
   $\Rightarrow -x + \sqrt{3}y = 6 \Rightarrow y = \frac{\sqrt{3}}{3} x + 2\sqrt{3}$

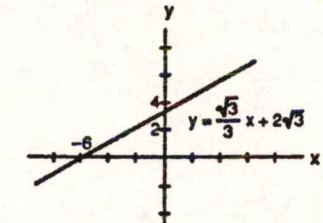

9. $\sqrt{2}x + \sqrt{2}y = 6 \Rightarrow \sqrt{2}r \cos\theta + \sqrt{2}r \sin\theta = 6 \Rightarrow r\left(\frac{\sqrt{2}}{2} \cos\theta + \frac{\sqrt{2}}{2} \sin\theta\right) = 3 \Rightarrow r\left(\cos\frac{\pi}{4} \cos\theta + \sin\frac{\pi}{4} \sin\theta\right)$
   $= 3 \Rightarrow r \cos \left(\theta - \frac{\pi}{4}\right) = 3$

11. $y = -5 \Rightarrow r \sin\theta = -5 \Rightarrow -r \sin\theta = 5 \Rightarrow r \sin(-\theta) = 5 \Rightarrow r \cos\left(\frac{\pi}{2} - (-\theta)\right) = 5 \Rightarrow r \cos\left(\theta + \frac{\pi}{2}\right) = 5$

13. $r = 2(4) \cos\theta = 8 \cos\theta$

15. $r = 2\sqrt{2} \sin\theta$

17.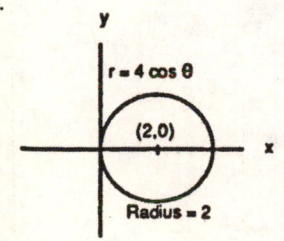

19.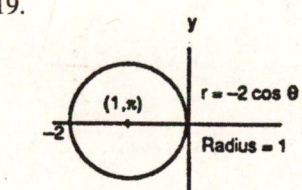

21. $(x - 6)^2 + y^2 = 36 \Rightarrow C = (6, 0), a = 6$
    $\Rightarrow r = 12 \cos \theta$ is the polar equation

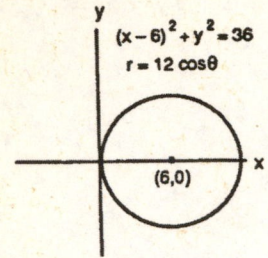

23. $x^2 + (y - 5)^2 = 25 \Rightarrow C = (0, 5), a = 5$
    $\Rightarrow r = 10 \sin \theta$ is the polar equation

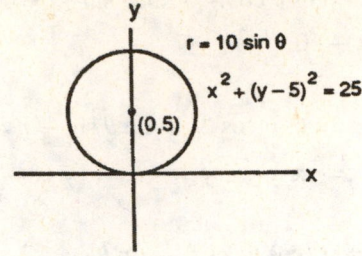

25. $x^2 + 2x + y^2 = 0 \Rightarrow (x + 1)^2 + y^2 = 1$
    $\Rightarrow C = (-1, 0), a = 1 \Rightarrow r = -2 \cos \theta$ is
    the polar equation

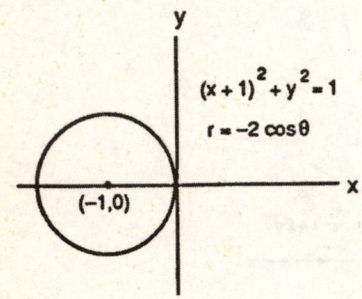

27. $x^2 + y^2 + y = 0 \Rightarrow x^2 + \left(y + \frac{1}{2}\right)^2 = \frac{1}{4}$
    $\Rightarrow C = \left(0, -\frac{1}{2}\right), a = \frac{1}{2} \Rightarrow r = -\sin \theta$ is the
    polar equation

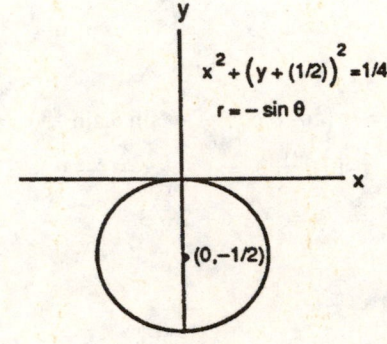

29. $e = 1, x = 2 \Rightarrow k = 2 \Rightarrow r = \frac{2(1)}{1 + (1) \cos \theta} = \frac{2}{1 + \cos \theta}$

31. $e = 5, y = -6 \Rightarrow k = 6 \Rightarrow r = \frac{6(5)}{1 - 5 \sin \theta} = \frac{30}{1 - 5 \sin \theta}$

33. $e = \frac{1}{2}, x = 1 \Rightarrow k = 1 \Rightarrow r = \frac{\left(\frac{1}{2}\right)(1)}{1 + \left(\frac{1}{2}\right) \cos \theta} = \frac{1}{2 + \cos \theta}$

35. $e = \frac{1}{5}, x = -10 \Rightarrow k = 10 \Rightarrow r = \frac{\left(\frac{1}{5}\right)(10)}{1 - \left(\frac{1}{5}\right) \sin \theta} = \frac{10}{5 - \sin \theta}$

37. $r = \frac{1}{1 + \cos \theta} \Rightarrow e = 1, k = 1 \Rightarrow x = 1$

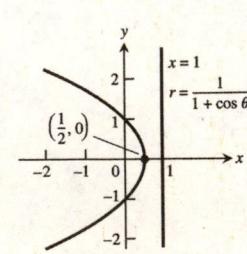

39. $r = \frac{25}{10 - 5 \cos \theta} \Rightarrow r = \frac{\left(\frac{25}{10}\right)}{1 - \left(\frac{5}{10}\right) \cos \theta} = \frac{\left(\frac{5}{2}\right)}{1 - \left(\frac{1}{2}\right) \cos \theta}$
    $\Rightarrow e = \frac{1}{2}, k = 5 \Rightarrow x = -5; a(1 - e^2) = ke$
    $\Rightarrow a\left[1 - \left(\frac{1}{2}\right)^2\right] = \frac{5}{2} \Rightarrow \frac{3}{4} a = \frac{5}{2} \Rightarrow a = \frac{10}{3} \Rightarrow ea = \frac{5}{3}$

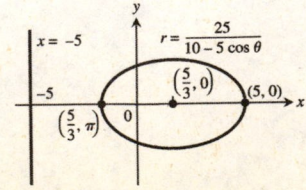

41. $r = \frac{400}{16 + 8\sin\theta} \Rightarrow r = \frac{\left(\frac{400}{16}\right)}{1 + \left(\frac{8}{16}\right)\sin\theta} \Rightarrow r = \frac{25}{1 + \left(\frac{1}{2}\right)\sin\theta}$

$e = \frac{1}{2}, k = 50 \Rightarrow y = 50; a(1 - e^2) = ke$

$\Rightarrow a\left[1 - \left(\frac{1}{2}\right)^2\right] = 25 \Rightarrow \frac{3}{4}a = 25 \Rightarrow a = \frac{100}{3}$

$\Rightarrow ea = \frac{50}{3}$

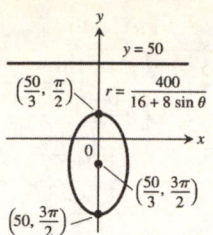

43. $r = \frac{8}{2 - 2\sin\theta} \Rightarrow r = \frac{4}{1 - \sin\theta} \Rightarrow e = 1,$
$k = 4 \Rightarrow y = -4$

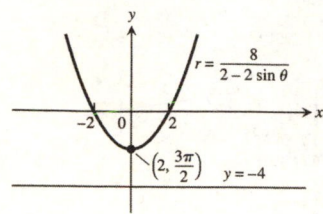

45.

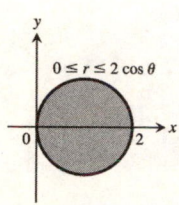

47.

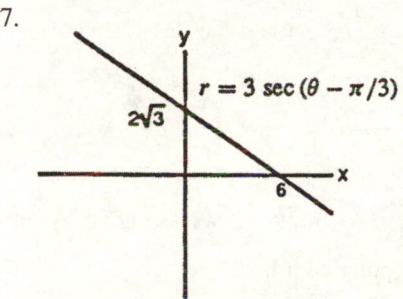

49.

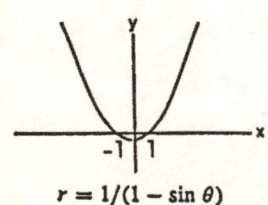

51.

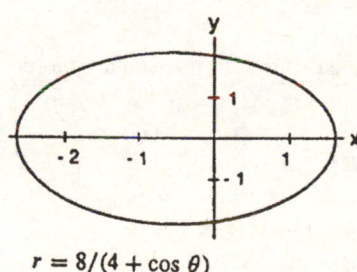

53.

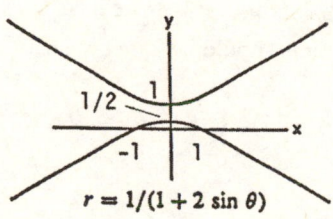

55.

57. (a) Perihelion $= a - ae = a(1 - e)$, Aphelion $= ea + a = a(1 + e)$

(b)

Planet	Perihelion	Aphelion
Mercury	0.3075 AU	0.4667 AU
Venus	0.7184 AU	0.7282 AU
Earth	0.9833 AU	1.0167 AU
Mars	1.3817 AU	1.6663 AU
Jupiter	4.9512 AU	5.4548 AU
Saturn	9.0210 AU	10.0570 AU
Uranus	18.2977 AU	20.0623 AU
Neptune	29.8135 AU	30.3065 AU
Pluto	29.6549 AU	49.2251 AU

59. (a) $r = 4 \sin\theta \Rightarrow r^2 = 4r \sin\theta \Rightarrow x^2 + y^2 = 4y$;

$r = \sqrt{3} \sec\theta \Rightarrow r = \frac{\sqrt{3}}{\cos\theta} \Rightarrow r\cos\theta = \sqrt{3}$

$\Rightarrow x = \sqrt{3}$; $x = \sqrt{3} \Rightarrow \left(\sqrt{3}\right)^2 + y^2 = 4y$

$\Rightarrow y^2 - 4y + 3 = 0 \Rightarrow (y-3)(y-1) = 0 \Rightarrow y = 3$

or $y = 1$. Therefore in Cartesian coordinates, the points

of intersection are $\left(\sqrt{3}, 3\right)$ and $\left(\sqrt{3}, 1\right)$. In polar

coordinates, $4\sin\theta = \sqrt{3}\sec\theta \Rightarrow 4\sin\theta\cos\theta = \sqrt{3}$

$\Rightarrow 2\sin\theta\cos\theta = \frac{\sqrt{3}}{2} \Rightarrow \sin 2\theta = \frac{\sqrt{3}}{2} \Rightarrow 2\theta = \frac{\pi}{3}$ or

$\frac{2\pi}{3} \Rightarrow \theta = \frac{\pi}{6}$ or $\frac{\pi}{3}$; $\theta = \frac{\pi}{6} \Rightarrow r = 2$, and $\theta = \frac{\pi}{3}$

$\Rightarrow r = 2\sqrt{3} \Rightarrow \left(2, \frac{\pi}{6}\right)$ and $\left(2\sqrt{3}, \frac{\pi}{3}\right)$ are the points

of intersection in polar coordinates.

(b)

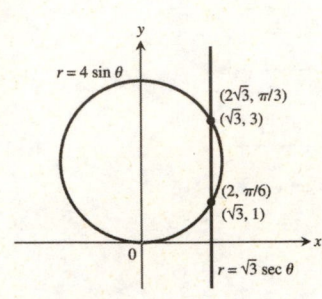

61. $r\cos\theta = 4 \Rightarrow x = 4 \Rightarrow k = 4$: parabola $\Rightarrow e = 1 \Rightarrow r = \frac{4}{1 + \cos\theta}$

63. (a) Let the ellipse be the orbit, with the Sun at one focus.

Then $r_{max} = a + c$ and $r_{min} = a - c \Rightarrow \frac{r_{max} - r_{min}}{r_{max} + r_{min}}$

$= \frac{(a+c) - (a-c)}{(a+c) + (a-c)} = \frac{2c}{2a} = \frac{c}{a} = e$

(b) Let F_1, F_2 be the foci. Then $PF_1 + PF_2 = 10$ where

P is any point on the ellipse. If P is a vertex, then

$PF_1 = a + c$ and $PF_2 = a - c$

$\Rightarrow (a + c) + (a - c) = 10$

$\Rightarrow 2a = 10 \Rightarrow a = 5$. Since $e = \frac{c}{a}$ we have $0.2 = \frac{c}{5}$

$\Rightarrow c = 1.0 \Rightarrow$ the pins should be 2 inches apart.

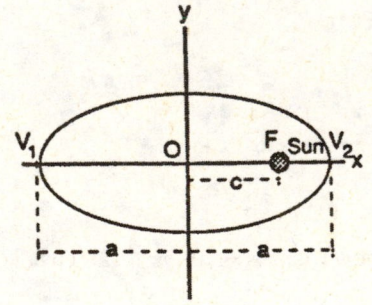

65. $x^2 + y^2 - 2ay = 0 \Rightarrow (r\cos\theta)^2 + (r\sin\theta)^2 - 2ar\sin\theta = 0$

$\Rightarrow r^2\cos^2\theta + r^2\sin^2\theta - 2ar\sin\theta = 0 \Rightarrow r^2 = 2ar\sin\theta$

$\Rightarrow r = 2a\sin\theta$

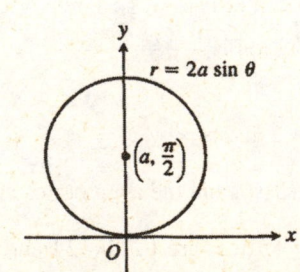

67. $x \cos \alpha + y \sin \alpha = p \Rightarrow r \cos \theta \cos \alpha + r \sin \theta \sin \alpha = p$
$\Rightarrow r(\cos \theta \cos \alpha + \sin \theta \sin \alpha) = p \Rightarrow r \cos(\theta - \alpha) = p$

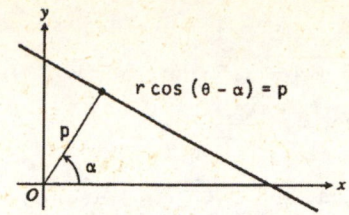

CHAPTER 10 PRACTICE EXERCISES

1. $x^2 = -4y \Rightarrow y = -\frac{x^2}{4} \Rightarrow 4p = 4 \Rightarrow p = 1$;
therefore Focus is $(0, -1)$, Directrix is $y = 1$

3. $y^2 = 3x \Rightarrow x = \frac{y^2}{3} \Rightarrow 4p = 3 \Rightarrow p = \frac{3}{4}$;
therefore Focus is $\left(\frac{3}{4}, 0\right)$, Directrix is $x = -\frac{3}{4}$

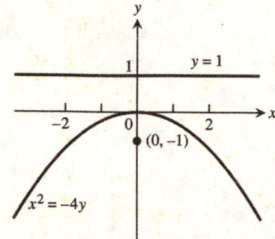

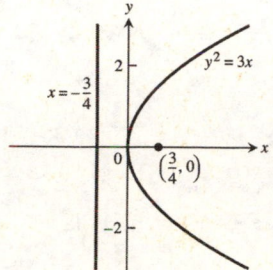

5. $16x^2 + 7y^2 = 112 \Rightarrow \frac{x^2}{7} + \frac{y^2}{16} = 1$
$\Rightarrow c^2 = 16 - 7 = 9 \Rightarrow c = 3; e = \frac{c}{a} = \frac{3}{4}$

7. $3x^2 - y^2 = 3 \Rightarrow x^2 - \frac{y^2}{3} = 1 \Rightarrow c^2 = 1 + 3 = 4$
$\Rightarrow c = 2; e = \frac{c}{a} = \frac{2}{1} = 2$; the asymptotes are
$y = \pm \sqrt{3}\, x$

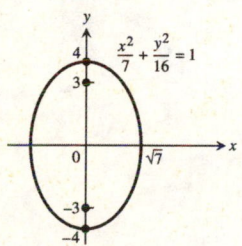

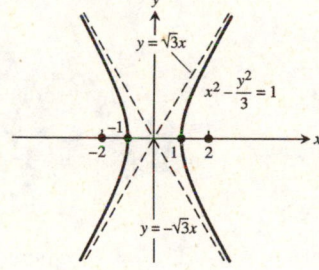

9. $x^2 = -12y \Rightarrow -\frac{x^2}{12} = y \Rightarrow 4p = 12 \Rightarrow p = 3 \Rightarrow$ focus is $(0, -3)$, directrix is $y = 3$, vertex is $(0, 0)$; therefore new vertex is $(2, 3)$, new focus is $(2, 0)$, new directrix is $y = 6$, and the new equation is $(x - 2)^2 = -12(y - 3)$

11. $\frac{x^2}{9} + \frac{y^2}{25} = 1 \Rightarrow a = 5$ and $b = 3 \Rightarrow c = \sqrt{25 - 9} = 4 \Rightarrow$ foci are $(0, \pm 4)$, vertices are $(0, \pm 5)$, center is $(0, 0)$; therefore the new center is $(-3, -5)$, new foci are $(-3, -1)$ and $(-3, -9)$, new vertices are $(-3, -10)$ and $(-3, 0)$, and the new equation is $\frac{(x+3)^2}{9} + \frac{(y+5)^2}{25} = 1$

13. $\frac{y^2}{8} - \frac{x^2}{2} = 1 \Rightarrow a = 2\sqrt{2}$ and $b = \sqrt{2} \Rightarrow c = \sqrt{8 + 2} = \sqrt{10} \Rightarrow$ foci are $\left(0, \pm \sqrt{10}\right)$, vertices are $\left(0, \pm 2\sqrt{2}\right)$, center is $(0, 0)$, and the asymptotes are $y = \pm 2x$; therefore the new center is $\left(2, 2\sqrt{2}\right)$, new foci are $\left(2, 2\sqrt{2} \pm \sqrt{10}\right)$, new vertices are $\left(2, 4\sqrt{2}\right)$ and $(2, 0)$, the new asymptotes are $y = 2x - 4 + 2\sqrt{2}$ and $y = -2x + 4 + 2\sqrt{2}$; the new equation is $\frac{\left(y - 2\sqrt{2}\right)^2}{8} - \frac{(x-2)^2}{2} = 1$

15. $x^2 - 4x - 4y^2 = 0 \Rightarrow x^2 - 4x + 4 - 4y^2 = 4 \Rightarrow (x-2)^2 - 4y^2 = 4 \Rightarrow \frac{(x-2)^2}{4} - y^2 = 1$, a hyperbola; $a = 2$ and

$b = 1 \Rightarrow c = \sqrt{1+4} = \sqrt{5}$; the center is $(2, 0)$, the vertices are $(0, 0)$ and $(4, 0)$; the foci are $\left(2 \pm \sqrt{5}, 0\right)$ and

the asymptotes are $y = \pm \frac{x-2}{2}$

17. $y^2 - 2y + 16x = -49 \Rightarrow y^2 - 2y + 1 = -16x - 48 \Rightarrow (y-1)^2 = -16(x+3)$, a parabola; the vertex is $(-3, 1)$;

$4p = 16 \Rightarrow p = 4 \Rightarrow$ the focus is $(-7, 1)$ and the directrix is $x = 1$

19. $9x^2 + 16y^2 + 54x - 64y = -1 \Rightarrow 9(x^2 + 6x) + 16(y^2 - 4y) = -1 \Rightarrow 9(x^2 + 6x + 9) + 16(y^2 - 4y + 4) = 144$

$\Rightarrow 9(x+3)^2 + 16(y-2)^2 = 144 \Rightarrow \frac{(x+3)^2}{16} + \frac{(y-2)^2}{9} = 1$, an ellipse; the center is $(-3, 2)$; $a = 4$ and $b = 3$

$\Rightarrow c = \sqrt{16-9} = \sqrt{7}$; the foci are $\left(-3 \pm \sqrt{7}, 2\right)$; the vertices are $(1, 2)$ and $(-7, 2)$

21. $x^2 + y^2 - 2x - 2y = 0 \Rightarrow x^2 - 2x + 1 + y^2 - 2y + 1 = 2 \Rightarrow (x-1)^2 + (y-1)^2 = 2$, a circle with center $(1, 1)$ and

radius $= \sqrt{2}$

23. $B^2 - 4AC = 1 - 4(1)(1) = -3 < 0 \Rightarrow$ ellipse

25. $B^2 - 4AC = 3^2 - 4(1)(2) = 1 > 0 \Rightarrow$ hyperbola

27. $x^2 - 2xy + y^2 = 0 \Rightarrow (x-y)^2 = 0 \Rightarrow x - y = 0$ or $y = x$, a straight line

29. $B^2 - 4AC = 1^2 - 4(2)(2) = -15 < 0 \Rightarrow$ ellipse; $\cot 2\alpha = \frac{A-C}{B} = 0 \Rightarrow 2\alpha = \frac{\pi}{2} \Rightarrow \alpha = \frac{\pi}{4}$; $x = \frac{\sqrt{2}}{2}x' - \frac{\sqrt{2}}{2}y'$ and

$y = \frac{\sqrt{2}}{2}x' + \frac{\sqrt{2}}{2}y' \Rightarrow 2\left(\frac{\sqrt{2}}{2}x' - \frac{\sqrt{2}}{2}y'\right)^2 + \left(\frac{\sqrt{2}}{2}x' - \frac{\sqrt{2}}{2}y'\right)\left(\frac{\sqrt{2}}{2}x' + \frac{\sqrt{2}}{2}y'\right) + 2\left(\frac{\sqrt{2}}{2}x' + \frac{\sqrt{2}}{2}y'\right)^2 - 15 = 0$

$\Rightarrow 5x'^2 + 3y'^2 = 30$

31. $B^2 - 4AC = \left(2\sqrt{3}\right)^2 - 4(1)(-1) = 16 \Rightarrow$ hyperbola; $\cot 2\alpha = \frac{A-C}{B} = \frac{1}{\sqrt{3}} \Rightarrow 2\alpha = \frac{\pi}{3} \Rightarrow \alpha = \frac{\pi}{6}$; $x = \frac{\sqrt{3}}{2}x' - \frac{1}{2}y'$

and $y = \frac{1}{2}x' + \frac{\sqrt{3}}{2}y' \Rightarrow \left(\frac{\sqrt{3}}{2}x' - \frac{1}{2}y'\right)^2 + 2\sqrt{3}\left(\frac{\sqrt{3}}{2}x' - \frac{1}{2}y'\right)\left(\frac{1}{2}x' + \frac{\sqrt{3}}{2}y'\right) - \left(\frac{1}{2}x' + \frac{\sqrt{3}}{2}y'\right)^2 = 4$

$\Rightarrow 2x'^2 - 2y'^2 = -4 \Rightarrow y'^2 - x'^2 = 2$

33. $x = \frac{1}{2}\tan t$ and $y = \frac{1}{2}\sec t \Rightarrow x^2 = \frac{1}{4}\tan^2 t$ 35. $x = -\cos t$ and $y = \cos^2 t \Rightarrow y = (-x)^2 = x^2$

and $y^2 = \frac{1}{4}\sec^2 t \Rightarrow 4x^2 = \tan^2 t$ and

$4y^2 = \sec^2 t \Rightarrow 4x^2 + 1 = 4y^2 \Rightarrow 4y^2 - 4x^2 = 1$

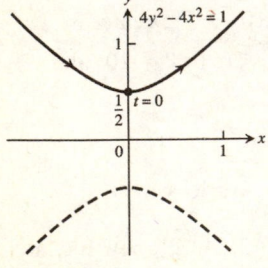

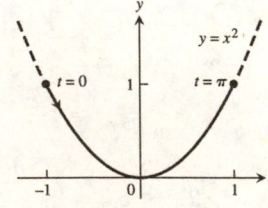

37.

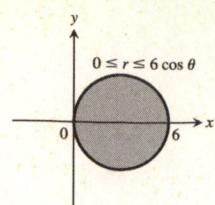

39. d

41. 1

43. k

45. i

47. $r = \sin \theta$ and $r = 1 + \sin \theta \Rightarrow \sin \theta = 1 + \sin \theta \Rightarrow 0 = 1$
so no solutions exist. There are no points of intersection
found by solving the system. The point of intersection
$(0, 0)$ is found by graphing.

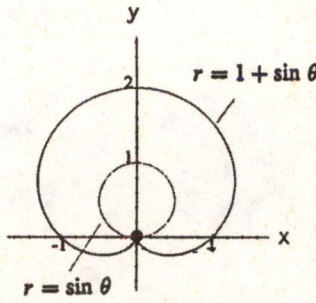

49. $r = 1 + \cos \theta$ and $r = 1 - \cos \theta \Rightarrow 1 + \cos \theta = 1 - \cos \theta$
$\Rightarrow 2 \cos \theta = 0 \Rightarrow \cos \theta = 0 \Rightarrow \theta = \frac{\pi}{2}, \frac{3\pi}{2}; \theta = \frac{\pi}{2}$ or $\frac{3\pi}{2}$
$\Rightarrow r = 1$. The points of intersection are $\left(1, \frac{\pi}{2}\right)$ and $\left(1, \frac{3\pi}{2}\right)$.
The point of intersection $(0, 0)$ is found by graphing.

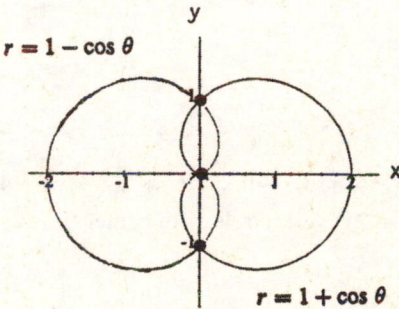

51. $r = 1 + \sin \theta$ and $r = -1 + \sin \theta$ intersect at all points of
$r = 1 + \sin \theta$ because the graphs coincide. This can be
seen by graphing them.

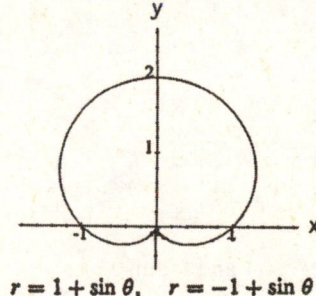

53. $r = \sec \theta$ and $r = 2 \sin \theta \Rightarrow \sec \theta = 2 \sin \theta$
$\Rightarrow 1 = 2 \sin \theta \cos \theta \Rightarrow 1 = \sin 2\theta \Rightarrow 2\theta = \frac{\pi}{2} \Rightarrow \theta = \frac{\pi}{4}$
$\Rightarrow r = 2 \sin \frac{\pi}{4} = \sqrt{2} \Rightarrow$ the point of intersection is
$\left(\sqrt{2}, \frac{\pi}{4}\right)$. No other points of intersection exist.

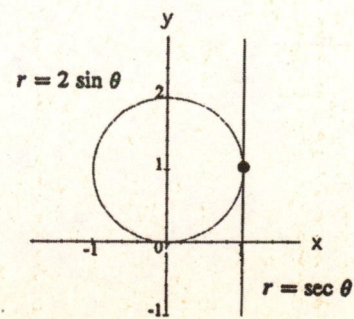

55. $r \cos\left(\theta + \frac{\pi}{3}\right) = 2\sqrt{3} \Rightarrow r\left(\cos\theta\cos\frac{\pi}{3} - \sin\theta\sin\frac{\pi}{3}\right)$

$= 2\sqrt{3} \Rightarrow \frac{1}{2}r\cos\theta - \frac{\sqrt{3}}{2}r\sin\theta = 2\sqrt{3}$

$\Rightarrow r\cos\theta - \sqrt{3}r\sin\theta = 4\sqrt{3} \Rightarrow x - \sqrt{3}y = 4\sqrt{3}$

$\Rightarrow y = \frac{\sqrt{3}}{3}x - 4$

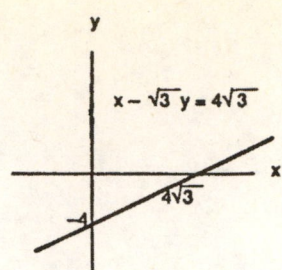

57. $r = 2\sec\theta \Rightarrow r = \frac{2}{\cos\theta} \Rightarrow r\cos\theta = 2 \Rightarrow x = 2$

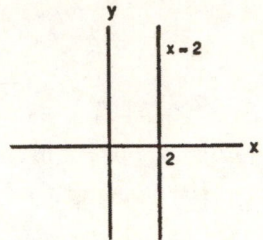

59. $r = -\frac{3}{2}\csc\theta \Rightarrow r\sin\theta = -\frac{3}{2} \Rightarrow y = -\frac{3}{2}$

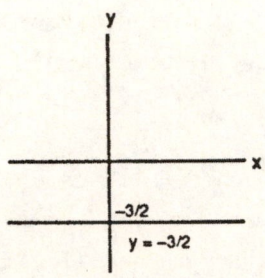

61. $r = -4\sin\theta \Rightarrow r^2 = -4r\sin\theta \Rightarrow x^2 + y^2 + 4y = 0$

$\Rightarrow x^2 + (y+2)^2 = 4$; circle with center $(0, -2)$ and radius 2.

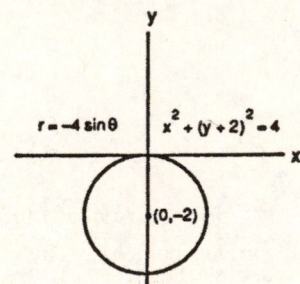

63. $r = 2\sqrt{2}\cos\theta \Rightarrow r^2 = 2\sqrt{2}r\cos\theta$

$\Rightarrow x^2 + y^2 - 2\sqrt{2}x = 0 \Rightarrow \left(x - \sqrt{2}\right)^2 + y^2 = 2$;

circle with center $\left(\sqrt{2}, 0\right)$ and radius $\sqrt{2}$

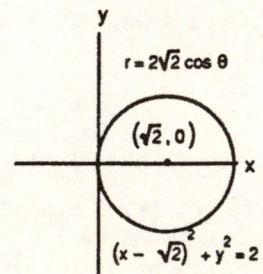

65. $x^2 + y^2 + 5y = 0 \Rightarrow x^2 + \left(y + \frac{5}{2}\right)^2 = \frac{25}{4} \Rightarrow C = \left(0, -\frac{5}{2}\right)$

and $a = \frac{5}{2}$; $r^2 + 5r \sin \theta = 0 \Rightarrow r = -5 \sin \theta$

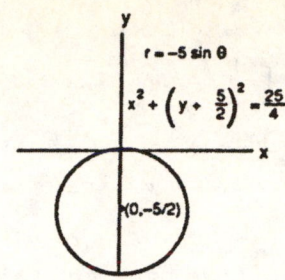

67. $x^2 + y^2 - 3x = 0 \Rightarrow \left(x - \frac{3}{2}\right)^2 + y^2 = \frac{9}{4} \Rightarrow C = \left(\frac{3}{2}, 0\right)$

and $a = \frac{3}{2}$; $r^2 - 3r \cos \theta = 0 \Rightarrow r = 3 \cos \theta$

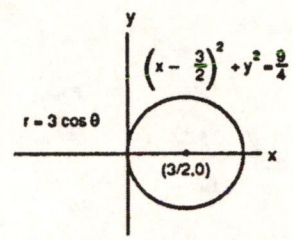

69. $r = \frac{2}{1 + \cos \theta} \Rightarrow e = 1 \Rightarrow$ parabola with vertex at $(1, 0)$

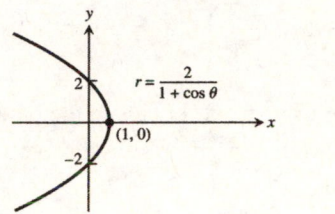

71. $r = \frac{6}{1 - 2 \cos \theta} \Rightarrow e = 2 \Rightarrow$ hyperbola; $ke = 6 \Rightarrow 2k = 6$

$\Rightarrow k = 3 \Rightarrow$ vertices are $(2, \pi)$ and $(6, \pi)$

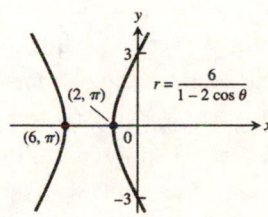

73. $e = 2$ and $r \cos \theta = 2 \Rightarrow x = 2$ is directrix $\Rightarrow k = 2$; the conic is a hyperbola; $r = \frac{ke}{1 + e \cos \theta} \Rightarrow r = \frac{(2)(2)}{1 + 2 \cos \theta}$

$\Rightarrow r = \frac{4}{1 + 2 \cos \theta}$

75. $e = \frac{1}{2}$ and $r \sin \theta = 2 \Rightarrow y = 2$ is directrix $\Rightarrow k = 2$; the conic is an ellipse; $r = \frac{ke}{1 + e \sin \theta} \Rightarrow r = \frac{(2)\left(\frac{1}{2}\right)}{1 + \left(\frac{1}{2}\right) \sin \theta}$

$\Rightarrow r = \frac{2}{2 + \sin \theta}$

77. $A = 2 \int_0^\pi \frac{1}{2} r^2 \, d\theta = \int_0^\pi (2 - \cos \theta)^2 \, d\theta = \int_0^\pi (4 - 4 \cos \theta + \cos^2 \theta) \, d\theta = \int_0^\pi \left(4 - 4 \cos \theta + \frac{1 + \cos 2\theta}{2}\right) d\theta$

$= \int_0^\pi \left(\frac{9}{2} - 4 \cos \theta + \frac{\cos 2\theta}{2}\right) d\theta = \left[\frac{9}{2} \theta - 4 \sin \theta + \frac{\sin 2\theta}{4}\right]_0^\pi = \frac{9}{2} \pi$

79. $r = 1 + \cos 2\theta$ and $r = 1 \Rightarrow 1 = 1 + \cos 2\theta \Rightarrow 0 = \cos 2\theta \Rightarrow 2\theta = \frac{\pi}{2} \Rightarrow \theta = \frac{\pi}{4}$; therefore

$A = 4 \int_0^{\pi/4} \frac{1}{2} \left[(1 + \cos 2\theta)^2 - 1^2\right] d\theta = 2 \int_0^{\pi/4} (1 + 2 \cos 2\theta + \cos^2 2\theta - 1) \, d\theta$

$= 2 \int_0^{\pi/4} \left(2 \cos 2\theta + \frac{1}{2} + \frac{\cos 4\theta}{2}\right) d\theta = 2 \left[\sin 2\theta + \frac{1}{2} \theta + \frac{\sin 4\theta}{8}\right]_0^{\pi/4} = 2 \left(1 + \frac{\pi}{8} + 0\right) = 2 + \frac{\pi}{4}$

81. $r = -1 + \cos \theta \Rightarrow \frac{dr}{d\theta} = -\sin \theta$; Length $= \int_0^{2\pi} \sqrt{(-1 + \cos \theta)^2 + (-\sin \theta)^2} \, d\theta = \int_0^{2\pi} \sqrt{2 - 2 \cos \theta} \, d\theta$

$= \int_0^{2\pi} \sqrt{\frac{4(1 - \cos \theta)}{2}} \, d\theta = \int_0^{2\pi} 2 \sin \frac{\theta}{2} \, d\theta = \left[-4 \cos \frac{\theta}{2}\right]_0^{2\pi} = (-4)(-1) - (-4)(1) = 8$

83. $r = 8 \sin^3 \left(\frac{\theta}{3}\right), 0 \le \theta \le \frac{\pi}{4} \Rightarrow \frac{dr}{d\theta} = 8 \sin^2 \left(\frac{\theta}{3}\right) \cos \left(\frac{\theta}{3}\right); r^2 + \left(\frac{dr}{d\theta}\right)^2 = \left[8 \sin^3 \left(\frac{\theta}{3}\right)\right]^2 + \left[8 \sin^2 \left(\frac{\theta}{3}\right) \cos \left(\frac{\theta}{3}\right)\right]^2$

$= 64 \sin^4 \left(\frac{\theta}{3}\right) \Rightarrow L = \int_0^{\pi/4} \sqrt{64 \sin^4 \left(\frac{\theta}{3}\right)} \, d\theta = \int_0^{\pi/4} 8 \sin^2 \left(\frac{\theta}{3}\right) d\theta = \int_0^{\pi/4} 8 \left[\frac{1 - \cos \left(\frac{2\theta}{3}\right)}{2}\right] d\theta$

$= \int_0^{\pi/4} \left[4 - 4 \cos \left(\frac{2\theta}{3}\right)\right] d\theta = \left[4\theta - 6 \sin \left(\frac{2\theta}{3}\right)\right]_0^{\pi/4} = 4 \left(\frac{\pi}{4}\right) - 6 \sin \left(\frac{\pi}{6}\right) - 0 = \pi - 3$

85. $r = \sqrt{\cos 2\theta} \Rightarrow \frac{dr}{d\theta} = \frac{-\sin 2\theta}{\sqrt{\cos 2\theta}}$; Surface Area $= \int_0^{\pi/4} 2\pi (r \sin \theta) \sqrt{r^2 + \left(\frac{dr}{d\theta}\right)^2} \, d\theta$

$= \int_0^{\pi/4} 2\pi \sqrt{\cos 2\theta} \, (\sin \theta) \sqrt{\cos 2\theta + \frac{\sin^2 2\theta}{\cos 2\theta}} \, d\theta = \int_0^{\pi/4} 2\pi \sqrt{\cos 2\theta} \, (\sin \theta) \sqrt{\frac{1}{\cos 2\theta}} \, d\theta = \int_0^{\pi/4} 2\pi \sin \theta \, d\theta$

$= [2\pi(-\cos \theta)]_0^{\pi/4} = 2\pi \left(1 - \frac{\sqrt{2}}{2}\right) = \left(2 - \sqrt{2}\right) \pi$

87. (a) Around the x-axis: $9x^2 + 4y^2 = 36 \Rightarrow y^2 = 9 - \frac{9}{4} x^2 \Rightarrow y = \pm \sqrt{9 - \frac{9}{4} x^2}$ and we use the positive root:

$$V = 2 \int_0^2 \pi \left(\sqrt{9 - \frac{9}{4} x^2}\right)^2 dx = 2 \int_0^2 \pi \left(9 - \frac{9}{4} x^2\right) dx = 2\pi \left[9x - \frac{3}{4} x^3\right]_0^2 = 24\pi$$

(b) Around the y-axis: $9x^2 + 4y^2 = 36 \Rightarrow x^2 = 4 - \frac{4}{9} y^2 \Rightarrow x = \pm \sqrt{4 - \frac{4}{9} y^2}$ and we use the positive root:

$$V = 2 \int_0^3 \pi \left(\sqrt{4 - \frac{4}{9} y^2}\right)^2 dy = 2 \int_0^3 \pi \left(4 - \frac{4}{9} y^2\right) dy = 2\pi \left[4y - \frac{4}{27} y^3\right]_0^3 = 16\pi$$

89. Each portion of the wave front reflects to the other focus, and since the wave front travels at a constant speed as it expands, the different portions of the wave arrive at the second focus simultaneously, from all directions, causing a spurt at the second focus.

91. The time for the bullet to hit the target remains constant, say $t = t_0$. Let the time it takes for sound to travel from the target to the listener be t_2. Since the listener hears the sounds simultaneously, $t_1 = t_0 + t_2$ where t_1 is the time for the sound to travel from the rifle to the listener. If v is the velocity of sound, then $vt_1 = vt_0 + vt_2$ or $vt_1 - vt_2 = vt_0$. Now vt_1 is the distance from the rifle to the listener and vt_2 is the distance from the target to the listener. Therefore the difference of the distances is constant since vt_0 is constant so the listener is on a branch of a hyperbola with foci at the rifle and the target. The branch is the one with the target as focus.

93. (a) $r = \frac{k}{1 + e \cos \theta} \Rightarrow r + er \cos \theta = k \Rightarrow \sqrt{x^2 + y^2} + ex = k \Rightarrow \sqrt{x^2 + y^2} = k - ex \Rightarrow x^2 + y^2$

$= k^2 - 2kex + e^2 x^2 \Rightarrow x^2 - e^2 x^2 + y^2 + 2kex - k^2 = 0 \Rightarrow (1 - e^2) x^2 + y^2 + 2kex - k^2 = 0$

(b) $e = 0 \Rightarrow x^2 + y^2 - k^2 = 0 \Rightarrow x^2 + y^2 = k^2 \Rightarrow$ circle;

$0 < e < 1 \Rightarrow e^2 < 1 \Rightarrow e^2 - 1 < 0 \Rightarrow B^2 - 4AC = 0^2 - 4 (1 - e^2) (1) = 4 (e^2 - 1) < 0 \Rightarrow$ ellipse;

$e = 1 \Rightarrow B^2 - 4AC = 0^2 - 4(0)(1) = 0 \Rightarrow$ parabola;

$e > 1 \Rightarrow e^2 > 1 \Rightarrow B^2 - 4AC = 0^2 - 4 (1 - e^2) (1) = 4e^2 - 4 > 0 \Rightarrow$ hyperbola

CHAPTER 10 ADDITIONAL AND ADVANCED EXERCISES

1. Directrix $x = 3$ and focus $(4, 0) \Rightarrow$ vertex is $\left(\frac{7}{2}, 0\right)$

$\Rightarrow p = \frac{1}{2} \Rightarrow$ the equation is $x - \frac{7}{2} = \frac{y^2}{2}$

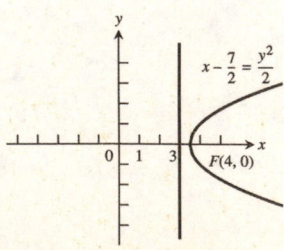

3. $x^2 = 4y \Rightarrow$ vertex is $(0, 0)$ and $p = 1 \Rightarrow$ focus is $(0, 1)$; thus the distance from $P(x, y)$ to the vertex is $\sqrt{x^2 + y^2}$ and the distance from P to the focus is $\sqrt{x^2 + (y - 1)^2} \Rightarrow \sqrt{x^2 + y^2} = 2\sqrt{x^2 + (y - 1)^2}$

$\Rightarrow x^2 + y^2 = 4\left[x^2 + (y-1)^2\right] \Rightarrow x^2 + y^2 = 4x^2 + 4y^2 - 8y + 4 \Rightarrow 3x^2 + 3y^2 - 8y + 4 = 0$, which is a circle

5. Vertices are $(0, \pm 2) \Rightarrow a = 2; e = \frac{c}{a} \Rightarrow 0.5 = \frac{c}{2} \Rightarrow c = 1 \Rightarrow$ foci are $(0, \pm 1)$

7. Let the center of the hyperbola be $(0, y)$.

(a) Directrix $y = -1$, focus $(0, -7)$ and $e = 2 \Rightarrow c - \frac{a}{e} = 6 \Rightarrow \frac{a}{e} = c - 6 \Rightarrow a = 2c - 12$. Also $c = ae = 2a$

$\Rightarrow a = 2(2a) - 12 \Rightarrow a = 4 \Rightarrow c = 8; y - (-1) = \frac{a}{e} = \frac{4}{2} = 2 \Rightarrow y = 1 \Rightarrow$ the center is $(0, 1); c^2 = a^2 + b^2$

$\Rightarrow b^2 = c^2 - a^2 = 64 - 16 = 48$; therefore the equation is $\frac{(y-1)^2}{16} - \frac{x^2}{48} = 1$

(b) $e = 5 \Rightarrow c - \frac{a}{e} = 6 \Rightarrow \frac{a}{e} = c - 6 \Rightarrow a = 5c - 30$. Also, $c = ae = 5a \Rightarrow a = 5(5a) - 30 \Rightarrow 24a = 30 \Rightarrow a = \frac{5}{4}$

$\Rightarrow c = \frac{25}{4}; y - (-1) = \frac{a}{e} = \frac{\left(\frac{5}{4}\right)}{5} = \frac{1}{4} \Rightarrow y = -\frac{3}{4} \Rightarrow$ the center is $\left(0, -\frac{3}{4}\right); c^2 = a^2 + b^2 \Rightarrow b^2 = c^2 - a^2$

$= \frac{625}{16} - \frac{25}{16} = \frac{75}{2}$; therefore the equation is $\frac{\left(y + \frac{3}{4}\right)^2}{\left(\frac{25}{16}\right)} - \frac{x^2}{\left(\frac{75}{2}\right)} = 1$ or $\frac{16\left(y + \frac{3}{4}\right)^2}{25} - \frac{2x^2}{75} = 1$

9. (a) $b^2x^2 + a^2y^2 = a^2b^2 \Rightarrow \frac{dy}{dx} = -\frac{b^2x}{a^2y}$; at (x_1, y_1) the tangent line is $y - y_1 = \left(-\frac{b^2x_1}{a^2y_1}\right)(x - x_1)$

$\Rightarrow a^2yy_1 + b^2xx_1 = b^2x_1^2 + a^2y_1^2 = a^2b^2 \Rightarrow b^2xx_1 + a^2yy_1 - a^2b^2 = 0$

(b) $b^2x^2 - a^2y^2 = a^2b^2 \Rightarrow \frac{dy}{dx} = \frac{b^2x}{a^2y}$; at (x_1, y_1) the tangent line is $y - y_1 = \left(\frac{b^2x_1}{a^2y_1}\right)(x - x_1)$

$\Rightarrow b^2xx_1 - a^2yy_1 = b^2x_1^2 - a^2y_1^2 = a^2b^2 \Rightarrow b^2xx_1 - a^2yy_1 - a^2b^2 = 0$

11.

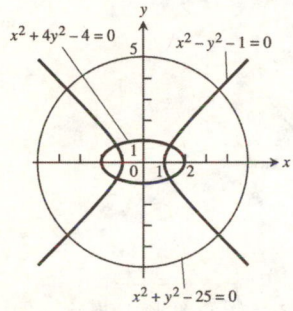

13.

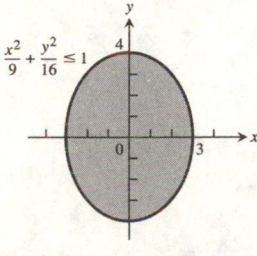

15. $(9x^2 + 4y^2 - 36)(4x^2 + 9y^2 - 16) \le 0$

$\Rightarrow 9x^2 + 4y^2 - 36 \le 0$ and $4x^2 + 9y^2 - 16 \ge 0$

or $9x^2 + 4y^2 - 36 \ge 0$ and $4x^2 + 9y^2 - 16 \le 0$

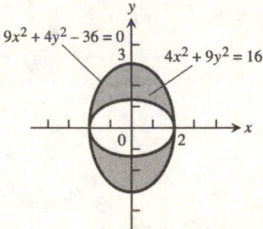

17. $x^4 - (y^2 - 9)^2 = 0 \Rightarrow x^2 - (y^2 - 9) = 0$ or

$x^2 + (y^2 - 9) = 0 \Rightarrow y^2 - x^2 = 9$ or $x^2 + y^2 = 9$

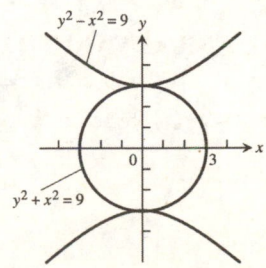

19. Arc PF = Arc AF since each is the distance rolled;

$\angle PCF = \frac{\text{Arc PF}}{b} \Rightarrow$ Arc $PF = b(\angle PCF)$; $\theta = \frac{\text{Arc AF}}{a}$

$\Rightarrow$ Arc $AF = a\theta \Rightarrow a\theta = b(\angle PCF) \Rightarrow \angle PCF = \left(\frac{a}{b}\right)\theta$;

$\angle OCB = \frac{\pi}{2} - \theta$ and $\angle OCB = \angle PCF - \angle PCE$

$= \angle PCF - \left(\frac{\pi}{2} - \alpha\right) = \left(\frac{a}{b}\right)\theta - \left(\frac{\pi}{2} - \alpha\right) \Rightarrow \frac{\pi}{2} - \theta$

$= \left(\frac{a}{b}\right)\theta - \left(\frac{\pi}{2} - \alpha\right) \Rightarrow \frac{\pi}{2} - \theta = \left(\frac{a}{b}\right)\theta - \frac{\pi}{2} + \alpha$

$\Rightarrow \alpha = \pi - \theta - \left(\frac{a}{b}\right)\theta \Rightarrow \alpha = \pi - \left(\frac{a+b}{b}\right)\theta$.

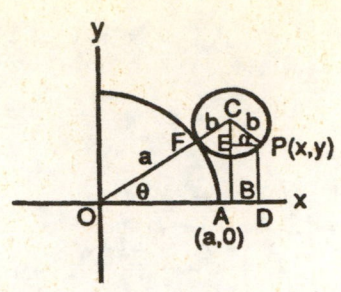

Now $x = OB + BD = OB + EP = (a+b)\cos\theta + b\cos\alpha = (a+b)\cos\theta + b\cos\left(\pi - \left(\frac{a+b}{b}\right)\theta\right)$

$= (a+b)\cos\theta + b\cos\pi\cos\left(\left(\frac{a+b}{b}\right)\theta\right) + b\sin\pi\sin\left(\left(\frac{a+b}{b}\right)\theta\right) = (a+b)\cos\theta - b\cos\left(\left(\frac{a+b}{b}\right)\theta\right)$ and

$y = PD = CB - CE = (a+b)\sin\theta - b\sin\alpha = (a+b)\sin\theta - b\sin\left(\left(\frac{a+b}{b}\right)\theta\right)$

$= (a+b)\sin\theta - b\sin\pi\cos\left(\left(\frac{a+b}{b}\right)\theta\right) + b\cos\pi\sin\left(\left(\frac{a+b}{b}\right)\theta\right) = (a+b)\sin\theta - b\sin\left(\left(\frac{a+b}{b}\right)\theta\right)$;

therefore $x = (a+b)\cos\theta - b\cos\left(\left(\frac{a+b}{b}\right)\theta\right)$ and $y = (a+b)\sin\theta - b\sin\left(\left(\frac{a+b}{b}\right)\theta\right)$

21. (a) $x = e^{2t}\cos t$ and $y = e^{2t}\sin t \Rightarrow x^2 + y^2 = e^{4t}\cos^2 t + e^{4t}\sin^2 t = e^{4t}$. Also $\frac{y}{x} = \frac{e^{2t}\sin t}{e^{2t}\cos t} = \tan t$

$\Rightarrow t = \tan^{-1}\left(\frac{y}{x}\right) \Rightarrow x^2 + y^2 = e^{4\tan^{-1}(y/x)}$ is the Cartesian equation. Since $r^2 = x^2 + y^2$ and

$\theta = \tan^{-1}\left(\frac{y}{x}\right)$, the polar equation is $r^2 = e^{4\theta}$ or $r = e^{2\theta}$ for $r > 0$

(b) $ds^2 = r^2 \, d\theta^2 + dr^2$; $r = e^{2\theta} \Rightarrow dr = 2e^{2\theta} \, d\theta$

$\Rightarrow ds^2 = r^2 \, d\theta^2 + \left(2e^{2\theta} \, d\theta\right)^2 = \left(e^{2\theta}\right)^2 d\theta^2 + 4e^{4\theta} \, d\theta^2$

$= 5e^{4\theta} \, d\theta^2 \Rightarrow ds = \sqrt{5}\, e^{2\theta} \, d\theta \Rightarrow L = \int_0^{2\pi} \sqrt{5}\, e^{2\theta} \, d\theta$

$= \left[\frac{\sqrt{5}\, e^{2\theta}}{2}\right]_0^{2\pi} = \frac{\sqrt{5}}{2}\left(e^{4\pi} - 1\right)$

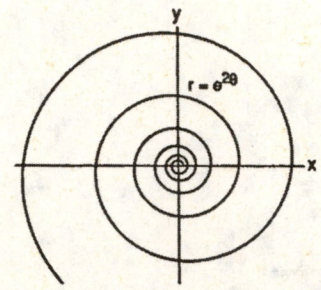

23. $r = 1 + \cos\theta$ and $S = \int 2\pi\rho \, ds$, where $\rho = y = r\sin\theta$; $ds = \sqrt{r^2 \, d\theta^2 + dr^2}$

$= \sqrt{(1 + \cos\theta)^2 \, d\theta^2 + \sin^2\theta \, d\theta^2} \sqrt{1 + 2\cos\theta + \cos^2\theta + \sin^2\theta} \, d\theta = \sqrt{2 + 2\cos\theta} \, d\theta = \sqrt{4\cos^2\left(\frac{\theta}{2}\right)} \, d\theta$

$= 2\cos\left(\frac{\theta}{2}\right) d\theta$ since $0 \le \theta \le \frac{\pi}{2}$. Then $S = \int_0^{\pi/2} 2\pi(r\sin\theta) \cdot 2\cos\left(\frac{\theta}{2}\right) d\theta = \int_0^{\pi/2} 4\pi(1 + \cos\theta) \cdot \sin\theta\cos\left(\frac{\theta}{2}\right) d\theta$

$= \int_0^{\pi/2} 4\pi\left[2\cos^2\left(\frac{\theta}{2}\right)\right]\left[2\sin\left(\frac{\theta}{2}\right)\cos\left(\frac{\theta}{2}\right)\cos\left(\frac{\theta}{2}\right)\right] d\theta = \int_0^{\pi/2} 16\pi\cos^4\left(\frac{\theta}{2}\right)\sin\left(\frac{\theta}{2}\right) d\theta = \left[\frac{-32\pi\cos^5\left(\frac{\theta}{2}\right)}{5}\right]_0^{\pi/2}$

$= \frac{(-32\pi)\left(\frac{\sqrt{2}}{2}\right)^5}{5} - \left(-\frac{32\pi}{5}\right) = \frac{32\pi - 4\pi\sqrt{2}}{5}$

25. $e = 2$ and $r\cos\theta = 2 \Rightarrow x = 2$ is the directrix $\Rightarrow k = 2$; the conic is a hyperbola with $r = \frac{ke}{1 + e\cos\theta}$

$\Rightarrow r = \frac{(2)(2)}{1 + 2\cos\theta} = \frac{4}{1 + 2\cos\theta}$

27. $e = \frac{1}{2}$ and $r\sin\theta = 2 \Rightarrow y = 2$ is the directrix $\Rightarrow k = 2$; the conic is an ellipse with $r = \frac{ke}{1 + e\sin\theta}$

$\Rightarrow r = \frac{2\left(\frac{1}{2}\right)}{1 + \left(\frac{1}{2}\right)\sin\theta} = \frac{2}{2 + \sin\theta}$

29. The length of the rope is $L = 2x + 2c + y \geq 8c$.

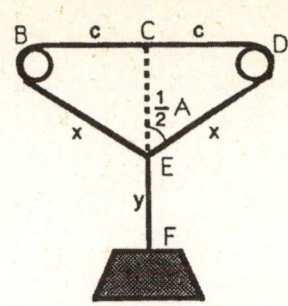

(a) The angle A ($\angle BED$) occurs when the distance

$CF = \ell$ is maximized. Now $\ell = \sqrt{x^2 - c^2} + y$

$\Rightarrow \ell = \sqrt{x^2 - c^2} + L - 2x - 2c$

$\Rightarrow \frac{d\ell}{dx} = \frac{1}{2}(x^2 - c^2)^{-1/2}(2x) - 2 = \frac{x}{\sqrt{x^2 - c^2}} - 2$.

Thus $\frac{d\ell}{dx} = 0 \Rightarrow \frac{x}{\sqrt{x^2 - c^2}} - 2 = 0 \Rightarrow x = 2\sqrt{x^2 - c^2}$

$\Rightarrow x^2 = 4x^2 - 4c^2 \Rightarrow 3x^2 = 4c^2 \Rightarrow \frac{c^2}{x^2} = \frac{3}{4}$

$\Rightarrow \frac{c}{x} = \frac{\sqrt{3}}{2}$. Since $\frac{c}{x} = \sin\frac{A}{2}$ we have $\sin\frac{A}{2} = \frac{\sqrt{3}}{2}$

$\Rightarrow \frac{A}{2} = 60° \Rightarrow A = 120°$

(b) If the ring is fixed at E (i.e., y is held constant) and E is moved to the right, for example, the rope will slip around the pegs so that BE lengthens and DE becomes shorter $\Rightarrow$ BE + ED is always $2x = L - y - 2c$, which is constant $\Rightarrow$ the point E lies on an ellipse with the pegs as foci.

(c) Minimal potential energy occurs when the weight is at its lowest point $\Rightarrow$ E is at the intersection of the ellipse and its minor axis.

31. If the vertex is $(0, 0)$, then the focus is $(p, 0)$. Let $P(x, y)$ be the present position of the comet. Then
$\sqrt{(x - p)^2 + y^2} = 4 \times 10^7$. Since $y^2 = 4px$ we have $\sqrt{(x - p)^2 + 4px} = 4 \times 10^7 \Rightarrow (x - p)^2 + 4px = 16 \times 10^{14}$.
Also, $x - p = 4 \times 10^7 \cos 60° = 2 \times 10^7 \Rightarrow x = p + 2 \times 10^7$. Therefore $(2 \times 10^7)^2 + 4p(p + 2 \times 10^7) = 16 \times 10^{14}$
$\Rightarrow 4 \times 10^{14} + 4p^2 + 8p \times 10^7 = 16 \times 10^{14} \Rightarrow 4p^2 + 8p \times 10^7 - 12 \times 10^{14} = 0 \Rightarrow p^2 + 2p \times 10^7 - 3 \times 10^{14} = 0$
$\Rightarrow (p + 3 \times 10^7)(p - 10^7) = 0 \Rightarrow p = -3 \times 10^7$ or $p = 10^7$. Since p is positive we obtain $p = 10^7$ miles.

33. $\cot 2\alpha = \frac{A - C}{B} = 0 \Rightarrow \alpha = 45°$ is the angle of rotation $\Rightarrow A' = \cos^2 45° + \cos 45° \sin 45° + \sin^2 45° = \frac{3}{2}$, $B' = 0$,

and $C' = \sin^2 45° - \sin 45° \cos 45° + \cos^2 45° = \frac{1}{2} \Rightarrow \frac{3}{2}x'^2 + \frac{1}{2}y'^2 = 1 \Rightarrow b = \sqrt{\frac{2}{3}}$ and $a = \sqrt{2} \Rightarrow c^2 = a^2 - b^2$

$= 2 - \frac{2}{3} = \frac{4}{3} \Rightarrow c = \frac{2}{\sqrt{3}}$. Therefore the eccentricity is $e = \frac{c}{a} = \frac{\left(\frac{2}{\sqrt{3}}\right)}{\sqrt{2}} = \sqrt{\frac{2}{3}} \approx 0.82$.

35. $\sqrt{x} + \sqrt{y} = 1 \Rightarrow x + 2\sqrt{xy} + y = 1 \Rightarrow 2\sqrt{xy} = 1 - (x + y) \Rightarrow 4xy = 1 - 2(x + y) + (x + y)^2$

$\Rightarrow 4xy = x^2 + 2xy + y^2 - 2x - 2y + 1 \Rightarrow x^2 - 2xy + y^2 - 2x - 2y + 1 = 0 \Rightarrow B^2 - 4AC = (-2)^2 - 4(1)(1) = 0$

$\Rightarrow$ the curve is part of a parabola

37. (a) The equation of a parabola with focus $(0, 0)$ and vertex $(a, 0)$ is $r = \frac{2a}{1 + \cos\theta}$ and rotating this parabola

through $\alpha = 45°$ gives $r = \frac{2a}{1 + \cos\left(\theta - \frac{\pi}{4}\right)}$.

(b) Foci at $(0, 0)$ and $(2, 0) \Rightarrow$ the center is $(1, 0) \Rightarrow a = 3$ and $c = 1$ since one vertex is at $(4, 0)$. Then $e = \frac{c}{a}$

$= \frac{1}{3}$. For ellipses with one focus at the origin and major axis along the x-axis we have $r = \frac{a(1 - e^2)}{1 - e\cos\theta}$

$= \frac{3\left(1 - \frac{1}{9}\right)}{1 - \left(\frac{1}{3}\right)\cos\theta} = \frac{8}{3 - \cos\theta}$.

(c) Center at $\left(2, \frac{\pi}{2}\right)$ and focus at $(0, 0) \Rightarrow c = 2$; center at $\left(2, \frac{\pi}{2}\right)$ and vertex at $\left(1, \frac{\pi}{2}\right) \Rightarrow a = 1$. Then $e = \frac{c}{a}$

$= \frac{2}{1} = 2$. Also $k = ae - \frac{a}{e} = (1)(2) - \frac{1}{2} = \frac{3}{2}$. Therefore $r = \frac{ke}{1 + e\sin\theta} = \frac{\left(\frac{3}{2}\right)(2)}{1 + 2\sin\theta} = \frac{3}{1 + 2\sin\theta}$.

39. Arc PT = Arc TO since each is the same distance rolled. Now Arc PT = $a(\angle TAP)$ and Arc TO = $a(\angle TBO)$
$\Rightarrow \angle TAP = \angle TBO$. Since AP = a = BO we have that $\triangle ADP$ is congruent to $\triangle BCO \Rightarrow CO = DP \Rightarrow OP$ is
parallel to AB $\Rightarrow \angle TBO = \angle TAP = \theta$. Then OPDC is a square $\Rightarrow r = CD = AB - AD - CB = AB - 2CB$
$\Rightarrow r = 2a - 2a\cos\theta = 2a(1 - \cos\theta)$, which is the polar equation of a cardioid.

41. $\beta = \psi_2 - \psi_1 \Rightarrow \tan \beta = \tan (\psi_2 - \psi_1) = \frac{\tan \psi_2 - \tan \psi_1}{1 + \tan \psi_2 \tan \psi_1}$;

the curves will be orthogonal when $\tan \beta$ is undefined, or

when $\tan \psi_2 = \frac{-1}{\tan \psi_1} \Rightarrow \frac{r}{g'(\theta)} = \frac{-1}{\left[\frac{r}{f'(\theta)}\right]}$

$\Rightarrow r^2 = -f'(\theta) g'(\theta)$

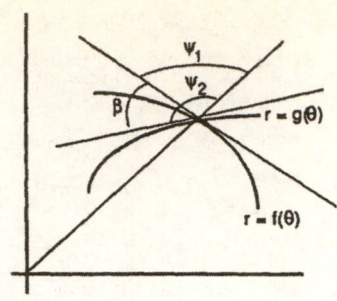

43. $r = 2a \sin 3\theta \Rightarrow \frac{dr}{d\theta} = 6a \cos 3\theta \Rightarrow \tan \psi = \frac{r}{\left(\frac{dr}{d\theta}\right)} = \frac{2a \sin 3\theta}{6a \cos 3\theta} = \frac{1}{3} \tan 3\theta$; when $\theta = \frac{\pi}{6}$, $\tan \psi = \frac{1}{3} \tan \frac{\pi}{2}$

$\Rightarrow \psi = \frac{\pi}{2}$

45. $\tan \psi_1 = \frac{\sqrt{3} \cos \theta}{-\sqrt{3} \sin \theta} = -\cot \theta$ is $-\frac{1}{\sqrt{3}}$ at $\theta = \frac{\pi}{3}$; $\tan \psi_2 = \frac{\sin \theta}{\cos \theta} = \tan \theta$ is $\sqrt{3}$ at $\theta = \frac{\pi}{3}$; since the product of

these slopes is -1, the tangents are perpendicular

47. $r_1 = \frac{1}{1 - \cos \theta} \Rightarrow \frac{dr_1}{d\theta} = -\frac{\sin \theta}{(1 - \cos \theta)^2}$; $r_2 = \frac{3}{1 + \cos \theta} \Rightarrow \frac{dr_2}{d\theta} = \frac{3 \sin \theta}{(1 + \cos \theta)^2}$; $\frac{1}{1 - \cos \theta} = \frac{3}{1 + \cos \theta}$

$\Rightarrow 1 + \cos \theta = 3 - 3 \cos \theta \Rightarrow 4 \cos \theta = 2 \Rightarrow \cos \theta = \frac{1}{2} \Rightarrow \theta = \pm \frac{\pi}{3} \Rightarrow r_1 = r_2 = 2 \Rightarrow$ the curves intersect at the

points $\left(2, \pm \frac{\pi}{3}\right)$; $\tan \psi_1 = \frac{\left(\frac{1}{1 - \cos \theta}\right)}{\left[\frac{-\sin \theta}{(1 - \cos \theta)^2}\right]} = -\frac{1 - \cos \theta}{\sin \theta}$ is $-\frac{1}{\sqrt{3}}$ at $\theta = \frac{\pi}{3}$; $\tan \psi_2 = \frac{\left(\frac{3}{1 + \cos \theta}\right)}{\left[\frac{3 \sin \theta}{(1 + \cos \theta)^2}\right]} = \frac{1 + \cos \theta}{\sin \theta}$ is

$\sqrt{3}$ at $\theta = \frac{\pi}{3}$; therefore $\tan \beta$ is undefined at $\theta = \frac{\pi}{3}$ since $1 + \tan \psi_1 \tan \psi_2 = 1 + \left(-\frac{1}{\sqrt{3}}\right)\left(\sqrt{3}\right) = 0 \Rightarrow \beta = \frac{\pi}{2}$;

$\tan \psi_1|_{\theta = -\pi/3} = -\frac{1 - \cos \left(-\frac{\pi}{3}\right)}{\sin \left(-\frac{\pi}{3}\right)} = \frac{1}{\sqrt{3}}$ and $\tan \psi_2|_{\theta = -\pi/3} = \frac{1 + \cos \left(-\frac{\pi}{3}\right)}{\sin \left(-\frac{\pi}{3}\right)} = -\sqrt{3} \Rightarrow \tan \beta$ is also undefined

at $\theta = -\frac{\pi}{3} \Rightarrow \beta = \frac{\pi}{2}$

49. $r_1 = \frac{a}{1 + \cos \theta} \Rightarrow \frac{dr_1}{d\theta} = \frac{a \sin \theta}{(1 + \cos \theta)^2}$ and $r_2 = \frac{b}{1 - \cos \theta} \Rightarrow \frac{dr_2}{d\theta} = -\frac{b \sin \theta}{(1 - \cos \theta)^2}$; then

$\tan \psi_1 = \frac{\left(\frac{a}{1 + \cos \theta}\right)}{\left[\frac{a \sin \theta}{(1 + \cos \theta)^2}\right]} = \frac{1 + \cos \theta}{\sin \theta}$ and $\tan \psi_2 = \frac{\left(\frac{b}{1 - \cos \theta}\right)}{\left[\frac{-b \sin \theta}{(1 - \cos \theta)^2}\right]} = \frac{1 - \cos \theta}{-\sin \theta} \Rightarrow 1 + \tan \psi_1 \tan \psi_2$

$= 1 + \left(\frac{1 + \cos \theta}{\sin \theta}\right)\left(\frac{1 - \cos \theta}{-\sin \theta}\right) = 1 - \frac{1 - \cos^2 \theta}{\sin^2 \theta} = 0 \Rightarrow \beta$ is undefined $\Rightarrow$ the parabolas are orthogonal at each

point of intersection

51. $r = 3 \sec \theta \Rightarrow r = \frac{3}{\cos \theta}$; $\frac{3}{\cos \theta} = 4 + 4 \cos \theta \Rightarrow 3 = 4 \cos \theta + 4 \cos^2 \theta \Rightarrow (2 \cos \theta + 3)(2 \cos \theta - 1) = 0$

$\Rightarrow \cos \theta = \frac{1}{2}$ or $\cos \theta = -\frac{3}{2} \Rightarrow \theta = \frac{\pi}{3}$ or $\frac{5\pi}{3}$ (the second equation has no solutions); $\tan \psi_2 = \frac{4(1 + \cos \theta)}{-4 \sin \theta}$

$= -\frac{1 + \cos \theta}{\sin \theta}$ is $-\sqrt{3}$ at $\frac{\pi}{3}$ and $\tan \psi_1 = \frac{3 \sec \theta}{3 \sec \theta \tan \theta} = \cot \theta$ is $\frac{1}{\sqrt{3}}$ at $\frac{\pi}{3}$. Then $\tan \beta$ is undefined since

$1 + \tan \psi_1 \tan \psi_2 = 1 + \left(\frac{1}{\sqrt{3}}\right)\left(-\sqrt{3}\right) = 0 \Rightarrow \beta = \frac{\pi}{2}$. Also, $\tan \psi_2|_{5\pi/3} = \sqrt{3}$ and $\tan \psi_1|_{5\pi/3} = -\frac{1}{\sqrt{3}}$

$\Rightarrow 1 + \tan \psi_1 \tan \psi_2 = 1 + \left(-\frac{1}{\sqrt{3}}\right)\left(\sqrt{3}\right) = 0 \Rightarrow \tan \beta$ is also undefined $\Rightarrow \beta = \frac{\pi}{2}$.

53. $\frac{1}{1 - \cos \theta} = \frac{1}{1 - \sin \theta} \Rightarrow 1 - \cos \theta = 1 - \sin \theta \Rightarrow \cos \theta = \sin \theta \Rightarrow \theta = \frac{\pi}{4}$; $\tan \psi_1 = \frac{\left(\frac{1}{1 - \cos \theta}\right)}{\left[\frac{-\sin \theta}{(1 - \cos \theta)^2}\right]} = \frac{1 - \cos \theta}{-\sin \theta}$;

$\tan \psi_2 = \frac{\left(\frac{1}{1 - \sin \theta}\right)}{\left[\frac{\cos \theta}{(1 - \sin \theta)^2}\right]} = \frac{1 - \sin \theta}{\cos \theta}$. Thus at $\theta = \frac{\pi}{4}$, $\tan \psi_1 = \frac{1 - \cos \left(\frac{\pi}{4}\right)}{-\sin \left(\frac{\pi}{4}\right)} = 1 - \sqrt{2}$ and

$\tan \psi_2 = \frac{1 - \sin \left(\frac{\pi}{4}\right)}{\cos \left(\frac{\pi}{4}\right)} = \sqrt{2} - 1$. Then $\tan \beta = \frac{\left(\sqrt{2} - 1\right) - \left(1 - \sqrt{2}\right)}{1 + \left(\sqrt{2} - 1\right)\left(1 - \sqrt{2}\right)} = \frac{2\sqrt{2} - 2}{2\sqrt{2} - 2} = 1 \Rightarrow \beta = \frac{\pi}{4}$

55. (a) $\tan \alpha = \frac{r}{\left(\frac{dr}{d\theta}\right)} \Rightarrow \frac{dr}{r} = \frac{d\theta}{\tan \alpha} \Rightarrow \ln r = \frac{\theta}{\tan \alpha} + C$ (by integration) $\Rightarrow r = Be^{\theta/(\tan \alpha)}$ for some constant B;

$A = \frac{1}{2} \int_{\theta_1}^{\theta_2} B^2 e^{2\theta/(\tan \alpha)} \, d\theta = \left[\frac{B^2 (\tan \alpha) e^{2\theta/(\tan \alpha)}}{4}\right]_{\theta_1}^{\theta_2} = \frac{\tan \alpha}{4} \left[B^2 e^{2\theta_2/(\tan \alpha)} - B^2 e^{2\theta_1/(\tan \alpha)}\right]$

$= \frac{\tan \alpha}{4} (r_2^2 - r_1^2)$ since $r_2^2 = B^2 e^{2\theta_2/(\tan \alpha)}$ and $r_1^2 = B^2 e^{2\theta_1/(\tan \alpha)}$; constant of proportionality $K = \frac{\tan \alpha}{4}$

(b) $\tan \alpha = \frac{r}{\left(\frac{dr}{d\theta}\right)} \Rightarrow \frac{dr}{d\theta} = \frac{r}{\tan \alpha} \Rightarrow \left(\frac{dr}{d\theta}\right)^2 = \frac{r^2}{\tan^2 \alpha} \Rightarrow r^2 + \left(\frac{dr}{d\theta}\right)^2 = r^2 + \frac{r^2}{\tan^2 \alpha} = r^2 \left(\frac{\tan^2 \alpha + 1}{\tan^2 \alpha}\right)$

$= r^2 \left(\frac{\sec^2 \alpha}{\tan^2 \alpha}\right) \Rightarrow$ Length $= \int_{\theta_1}^{\theta_2} r \left(\frac{\sec \alpha}{\tan \alpha}\right) d\theta = \int_{\theta_1}^{\theta_2} Be^{\theta/(\tan \alpha)} \cdot \frac{\sec \alpha}{\tan \alpha} \, d\theta = \left[B (\sec \alpha) e^{\theta/(\tan \alpha)}\right]_{\theta_1}^{\theta_2}$

$= (\sec \alpha) \left[Be^{\theta_2/(\tan \alpha)} - Be^{\theta_1/(\tan \alpha)}\right] = K (r_2 - r_1)$ where $K = \sec \alpha$ is the constant of proportionality

NOTES:

CHAPTER 11 INFINITE SEQUENCES AND SERIES

11.1 SEQUENCES

1. $a_1 = \frac{1-1}{1^2} = 0$, $a_2 = \frac{1-2}{2^2} = -\frac{1}{4}$, $a_3 = \frac{1-3}{3^2} = -\frac{2}{9}$, $a_4 = \frac{1-4}{4^2} = -\frac{3}{16}$

3. $a_1 = \frac{(-1)^2}{2-1} = 1$, $a_2 = \frac{(-1)^3}{4-1} = -\frac{1}{3}$, $a_3 = \frac{(-1)^4}{6-1} = \frac{1}{5}$, $a_4 = \frac{(-1)^5}{8-1} = -\frac{1}{7}$

5. $a_1 = \frac{2}{2^2} = \frac{1}{2}$, $a_2 = \frac{2^2}{2^3} = \frac{1}{2}$, $a_3 = \frac{2^3}{2^4} = \frac{1}{2}$, $a_4 = \frac{2^4}{2^5} = \frac{1}{2}$

7. $a_1 = 1$, $a_2 = 1 + \frac{1}{2} = \frac{3}{2}$, $a_3 = \frac{3}{2} + \frac{1}{2^2} = \frac{7}{4}$, $a_4 = \frac{7}{4} + \frac{1}{2^3} = \frac{15}{8}$, $a_5 = \frac{15}{8} + \frac{1}{2^4} = \frac{31}{16}$, $a_6 = \frac{63}{32}$,

   $a_7 = \frac{127}{64}$, $a_8 = \frac{255}{128}$, $a_9 = \frac{511}{256}$, $a_{10} = \frac{1023}{512}$

9. $a_1 = 2$, $a_2 = \frac{(-1)^2(2)}{2} = 1$, $a_3 = \frac{(-1)^3(1)}{2} = -\frac{1}{2}$, $a_4 = \frac{(-1)^4\left(-\frac{1}{2}\right)}{2} = -\frac{1}{4}$, $a_5 = \frac{(-1)^5\left(-\frac{1}{4}\right)}{2} = \frac{1}{8}$,

   $a_6 = \frac{1}{16}$, $a_7 = -\frac{1}{32}$, $a_8 = -\frac{1}{64}$, $a_9 = \frac{1}{128}$, $a_{10} = \frac{1}{256}$

11. $a_1 = 1$, $a_2 = 1$, $a_3 = 1 + 1 = 2$, $a_4 = 2 + 1 = 3$, $a_5 = 3 + 2 = 5$, $a_6 = 8$, $a_7 = 13$, $a_8 = 21$, $a_9 = 34$, $a_{10} = 55$

13. $a_n = (-1)^{n+1}$, $n = 1, 2, \ldots$

15. $a_n = (-1)^{n+1}n^2$, $n = 1, 2, \ldots$

17. $a_n = n^2 - 1$, $n = 1, 2, \ldots$

19. $a_n = 4n - 3$, $n = 1, 2, \ldots$

21. $a_n = \frac{1 + (-1)^{n+1}}{2}$, $n = 1, 2, \ldots$

23. $\lim\limits_{n \to \infty} 2 + (0.1)^n = 2 \Rightarrow$ converges   (Theorem 5, #4)

25. $\lim\limits_{n \to \infty} \frac{1 - 2n}{1 + 2n} = \lim\limits_{n \to \infty} \frac{\left(\frac{1}{n}\right) - 2}{\left(\frac{1}{n}\right) + 2} = \lim\limits_{n \to \infty} \frac{-2}{2} = -1 \Rightarrow$ converges

27. $\lim\limits_{n \to \infty} \frac{1 - 5n^4}{n^4 + 8n^3} = \lim\limits_{n \to \infty} \frac{\left(\frac{1}{n^4}\right) - 5}{1 + \left(\frac{8}{n}\right)} = -5 \Rightarrow$ converges

29. $\lim\limits_{n \to \infty} \frac{n^2 - 2n + 1}{n - 1} = \lim\limits_{n \to \infty} \frac{(n-1)(n-1)}{n-1} = \lim\limits_{n \to \infty} (n - 1) = \infty \Rightarrow$ diverges

31. $\lim\limits_{n \to \infty} (1 + (-1)^n)$ does not exist $\Rightarrow$ diverges

33. $\lim\limits_{n \to \infty} \left(\frac{n+1}{2n}\right)\left(1 - \frac{1}{n}\right) = \lim\limits_{n \to \infty} \left(\frac{1}{2} + \frac{1}{2n}\right)\left(1 - \frac{1}{n}\right) = \frac{1}{2} \Rightarrow$ converges

35. $\lim\limits_{n \to \infty} \frac{(-1)^{n+1}}{2n - 1} = 0 \Rightarrow$ converges

37. $\lim\limits_{n \to \infty} \sqrt{\frac{2n}{n+1}} = \sqrt{\lim\limits_{n \to \infty} \frac{2n}{n+1}} = \sqrt{\lim\limits_{n \to \infty} \left(\frac{2}{1+\frac{1}{n}}\right)} = \sqrt{2} \Rightarrow$ converges

39. $\lim\limits_{n \to \infty} \sin\left(\frac{\pi}{2} + \frac{1}{n}\right) = \sin\left(\lim\limits_{n \to \infty} \left(\frac{\pi}{2} + \frac{1}{n}\right)\right) = \sin\frac{\pi}{2} = 1 \Rightarrow$ converges

41. $\lim\limits_{n \to \infty} \frac{\sin n}{n} = 0$ because $-\frac{1}{n} \leq \frac{\sin n}{n} \leq \frac{1}{n} \Rightarrow$ converges by the Sandwich Theorem for sequences

43. $\lim\limits_{n \to \infty} \frac{n}{2^n} = \lim\limits_{n \to \infty} \frac{1}{2^n \ln 2} = 0 \Rightarrow$ converges (using l'Hôpital's rule)

45. $\lim\limits_{n \to \infty} \frac{\ln(n+1)}{\sqrt{n}} = \lim\limits_{n \to \infty} \frac{\left(\frac{1}{n+1}\right)}{\left(\frac{1}{2\sqrt{n}}\right)} = \lim\limits_{n \to \infty} \frac{2\sqrt{n}}{n+1} = \lim\limits_{n \to \infty} \frac{\left(\frac{2}{\sqrt{n}}\right)}{1+\left(\frac{1}{n}\right)} = 0 \Rightarrow$ converges

47. $\lim\limits_{n \to \infty} 8^{1/n} = 1 \Rightarrow$ converges   (Theorem 5, #3)

49. $\lim\limits_{n \to \infty} \left(1 + \frac{7}{n}\right)^n = e^7 \Rightarrow$ converges   (Theorem 5, #5)

51. $\lim\limits_{n \to \infty} \sqrt[n]{10n} = \lim\limits_{n \to \infty} 10^{1/n} \cdot n^{1/n} = 1 \cdot 1 = 1 \Rightarrow$ converges   (Theorem 5, #3 and #2)

53. $\lim\limits_{n \to \infty} \left(\frac{3}{n}\right)^{1/n} = \frac{\lim\limits_{n \to \infty} 3^{1/n}}{\lim\limits_{n \to \infty} n^{1/n}} = \frac{1}{1} = 1 \Rightarrow$ converges   (Theorem 5, #3 and #2)

55. $\lim\limits_{n \to \infty} \frac{\ln n}{n^{1/n}} = \frac{\lim\limits_{n \to \infty} \ln n}{\lim\limits_{n \to \infty} n^{1/n}} = \frac{\infty}{1} = \infty \Rightarrow$ diverges   (Theorem 5, #2)

57. $\lim\limits_{n \to \infty} \sqrt[n]{4^n \, n} = \lim\limits_{n \to \infty} 4 \sqrt[n]{n} = 4 \cdot 1 = 4 \Rightarrow$ converges   (Theorem 5, #2)

59. $\lim\limits_{n \to \infty} \frac{n!}{n^n} = \lim\limits_{n \to \infty} \frac{1 \cdot 2 \cdot 3 \cdots (n-1)(n)}{n \cdot n \cdot n \cdots n \cdot n} \leq \lim\limits_{n \to \infty} \left(\frac{1}{n}\right) = 0$ and $\frac{n!}{n^n} \geq 0 \Rightarrow \lim\limits_{n \to \infty} \frac{n!}{n^n} = 0 \Rightarrow$ converges

61. $\lim\limits_{n \to \infty} \frac{n!}{10^{6n}} = \lim\limits_{n \to \infty} \frac{1}{\left(\frac{(10^6)^n}{n!}\right)} = \infty \Rightarrow$ diverges   (Theorem 5, #6)

63. $\lim\limits_{n \to \infty} \left(\frac{1}{n}\right)^{1/(\ln n)} = \lim\limits_{n \to \infty} \exp\left(\frac{1}{\ln n} \ln\left(\frac{1}{n}\right)\right) = \lim\limits_{n \to \infty} \exp\left(\frac{\ln 1 - \ln n}{\ln n}\right) = e^{-1} \Rightarrow$ converges

65. $\lim\limits_{n \to \infty} \left(\frac{3n+1}{3n-1}\right)^n = \lim\limits_{n \to \infty} \exp\left(n \ln\left(\frac{3n+1}{3n-1}\right)\right) = \lim\limits_{n \to \infty} \exp\left(\frac{\ln(3n+1) - \ln(3n-1)}{\frac{1}{n}}\right)$

    $= \lim\limits_{n \to \infty} \exp\left(\frac{\frac{3}{3n+1} - \frac{3}{3n-1}}{\left(-\frac{1}{n^2}\right)}\right) = \lim\limits_{n \to \infty} \exp\left(\frac{6n^2}{(3n+1)(3n-1)}\right) = \exp\left(\frac{6}{9}\right) = e^{2/3} \Rightarrow$ converges

67. $\lim\limits_{n \to \infty} \left(\frac{x^n}{2n+1}\right)^{1/n} = \lim\limits_{n \to \infty} x \left(\frac{1}{2n+1}\right)^{1/n} = x \lim\limits_{n \to \infty} \exp\left(\frac{1}{n} \ln\left(\frac{1}{2n+1}\right)\right) = x \lim\limits_{n \to \infty} \exp\left(\frac{-\ln(2n+1)}{n}\right)$

    $= x \lim\limits_{n \to \infty} \exp\left(\frac{-2}{2n+1}\right) = xe^0 = x, \; x > 0 \Rightarrow$ converges

69. $\lim\limits_{n \to \infty} \frac{3^n \cdot 6^n}{2^{-n} \cdot n!} = \lim\limits_{n \to \infty} \frac{36^n}{n!} = 0 \Rightarrow$ converges   (Theorem 5, #6)

71. $\lim\limits_{n \to \infty} \tanh n = \lim\limits_{n \to \infty} \frac{e^n - e^{-n}}{e^n + e^{-n}} = \lim\limits_{n \to \infty} \frac{e^{2n} - 1}{e^{2n} + 1} = \lim\limits_{n \to \infty} \frac{2e^{2n}}{2e^{2n}} = \lim\limits_{n \to \infty} 1 = 1 \Rightarrow$ converges

73. $\displaystyle\lim_{n \to \infty} \frac{n^2 \sin\left(\frac{1}{n}\right)}{2n-1} = \lim_{n \to \infty} \frac{\sin\left(\frac{1}{n}\right)}{\left(\frac{2}{n} - \frac{1}{n^2}\right)} = \lim_{n \to \infty} \frac{-\left(\cos\left(\frac{1}{n}\right)\right)\left(\frac{1}{n^2}\right)}{\left(-\frac{2}{n^2} + \frac{2}{n^3}\right)} = \lim_{n \to \infty} \frac{-\cos\left(\frac{1}{n}\right)}{-2 + \left(\frac{2}{n}\right)} = \frac{1}{2} \Rightarrow$ converges

75. $\displaystyle\lim_{n \to \infty} \tan^{-1} n = \frac{\pi}{2} \Rightarrow$ converges

77. $\displaystyle\lim_{n \to \infty} \left(\frac{1}{3}\right)^n + \frac{1}{\sqrt{2^n}} = \lim_{n \to \infty} \left(\left(\frac{1}{3}\right)^n + \left(\frac{1}{\sqrt{2}}\right)^n\right) = 0 \Rightarrow$ converges   (Theorem 5, #4)

79. $\displaystyle\lim_{n \to \infty} \frac{(\ln n)^{200}}{n} = \lim_{n \to \infty} \frac{200 (\ln n)^{199}}{n} = \lim_{n \to \infty} \frac{200 \cdot 199 (\ln n)^{198}}{n} = \ldots = \lim_{n \to \infty} \frac{200!}{n} = 0 \Rightarrow$ converges

81. $\displaystyle\lim_{n \to \infty} \left(n - \sqrt{n^2 - n}\right) = \lim_{n \to \infty} \left(n - \sqrt{n^2 - n}\right)\left(\frac{n + \sqrt{n^2-n}}{n + \sqrt{n^2-n}}\right) = \lim_{n \to \infty} \frac{n}{n + \sqrt{n^2-n}} = \lim_{n \to \infty} \frac{1}{1 + \sqrt{1 - \frac{1}{n}}}$

$= \frac{1}{2} \Rightarrow$ converges

83. $\displaystyle\lim_{n \to \infty} \frac{1}{n} \int_1^n \frac{1}{x} \, dx = \lim_{n \to \infty} \frac{\ln n}{n} = \lim_{n \to \infty} \frac{1}{n} = 0 \Rightarrow$ converges   (Theorem 5, #1)

85. $1, 1, 2, 4, 8, 16, 32, \ldots = 1, 2^0, 2^1, 2^2, 2^3, 2^4, 2^5, \ldots \Rightarrow x_1 = 1$ and $x_n = 2^{n-2}$ for $n \geq 2$

87. (a) $f(x) = x^2 - 2$; the sequence converges to $1.414213562 \approx \sqrt{2}$

(b) $f(x) = \tan(x) - 1$; the sequence converges to $0.7853981635 \approx \frac{\pi}{4}$

(c) $f(x) = e^x$; the sequence $1, 0, -1, -2, -3, -4, -5, \ldots$ diverges

89. (a) If $a = 2n + 1$, then $b = \left\lfloor \frac{a^2}{2} \right\rfloor = \left\lfloor \frac{4n^2 + 4n + 1}{2} \right\rfloor = \left\lfloor 2n^2 + 2n + \frac{1}{2} \right\rfloor = 2n^2 + 2n$, $c = \left\lceil \frac{a^2}{2} \right\rceil = \left\lceil 2n^2 + 2n + \frac{1}{2} \right\rceil$

$= 2n^2 + 2n + 1$ and $a^2 + b^2 = (2n+1)^2 + \left(2n^2 + 2n\right)^2 = 4n^2 + 4n + 1 + 4n^4 + 8n^3 + 4n^2$

$= 4n^4 + 8n^3 + 8n^2 + 4n + 1 = \left(2n^2 + 2n + 1\right)^2 = c^2$.

(b) $\displaystyle\lim_{a \to \infty} \frac{\left\lfloor \frac{a^2}{2} \right\rfloor}{\left\lceil \frac{a^2}{2} \right\rceil} = \lim_{a \to \infty} \frac{2n^2 + 2n}{2n^2 + 2n + 1} = 1$ or $\displaystyle\lim_{a \to \infty} \frac{\left\lfloor \frac{a^2}{2} \right\rfloor}{\left\lceil \frac{a^2}{2} \right\rceil} = \lim_{a \to \infty} \sin \theta = \lim_{\theta \to \pi/2} \sin \theta = 1$

91. (a) $\displaystyle\lim_{n \to \infty} \frac{\ln n}{n^c} = \lim_{n \to \infty} \frac{\left(\frac{1}{n}\right)}{c n^{c-1}} = \lim_{n \to \infty} \frac{1}{c n^c} = 0$

(b) For all $\epsilon > 0$, there exists an $N = e^{-(\ln \epsilon)/c}$ such that $n > e^{-(\ln \epsilon)/c} \Rightarrow \ln n > -\frac{\ln \epsilon}{c} \Rightarrow \ln n^c > \ln\left(\frac{1}{\epsilon}\right)$

$\Rightarrow n^c > \frac{1}{\epsilon} \Rightarrow \frac{1}{n^c} < \epsilon \Rightarrow \left|\frac{1}{n^c} - 0\right| < \epsilon \Rightarrow \displaystyle\lim_{n \to \infty} \frac{1}{n^c} = 0$

93. $\displaystyle\lim_{n \to \infty} n^{1/n} = \lim_{n \to \infty} \exp\left(\frac{1}{n} \ln n\right) = \lim_{n \to \infty} \exp\left(\frac{1}{n}\right) = e^0 = 1$

95. Assume the hypotheses of the theorem and let ϵ be a positive number. For all ϵ there exists a N_1 such that when $n > N_1$ then $|a_n - L| < \epsilon \Rightarrow -\epsilon < a_n - L < \epsilon \Rightarrow L - \epsilon < a_n$, and there exists a N_2 such that when $n > N_2$ then $|c_n - L| < \epsilon \Rightarrow -\epsilon < c_n - L < \epsilon \Rightarrow c_n < L + \epsilon$. If $n > \max\{N_1, N_2\}$, then $L - \epsilon < a_n \leq b_n \leq c_n < L + \epsilon \Rightarrow |b_n - L| < \epsilon \Rightarrow \displaystyle\lim_{n \to \infty} b_n = L$.

97. $a_{n+1} \geq a_n \Rightarrow \frac{3(n+1)+1}{(n+1)+1} > \frac{3n+1}{n+1} \Rightarrow \frac{3n+4}{n+2} > \frac{3n+1}{n+1} \Rightarrow 3n^2 + 3n + 4n + 4 > 3n^2 + 6n + n + 2$

$\Rightarrow 4 > 2$; the steps are reversible so the sequence is nondecreasing; $\frac{3n+1}{n+1} < 3 \Rightarrow 3n + 1 < 3n + 3$

$\Rightarrow 1 < 3$; the steps are reversible so the sequence is bounded above by 3

99. $a_{n+1} \leq a_n \Rightarrow \frac{2^{n+1} 3^{n+1}}{(n+1)!} \leq \frac{2^n 3^n}{n!} \Rightarrow \frac{2^{n+1} 3^{n+1}}{2^n 3^n} \leq \frac{(n+1)!}{n!} \Rightarrow 2 \cdot 3 \leq n + 1$ which is true for $n \geq 5$; the steps are reversible so the sequence is decreasing after a_5, but it is not nondecreasing for all its terms; $a_1 = 6, a_2 = 18,$

$a_3 = 36$, $a_4 = 54$, $a_5 = \frac{324}{5} = 64.8 \Rightarrow$ the sequence is bounded from above by 64.8

101. $a_n = 1 - \frac{1}{n}$ converges because $\frac{1}{n} \to 0$ by Example 1; also it is a nondecreasing sequence bounded above by 1

103. $a_n = \frac{2^n - 1}{2^n} = 1 - \frac{1}{2^n}$ and $0 < \frac{1}{2^n} < \frac{1}{n}$; since $\frac{1}{n} \to 0$ (by Example 1) $\Rightarrow \frac{1}{2^n} \to 0$, the sequence converges; also it is a nondecreasing sequence bounded above by 1

105. $a_n = ((-1)^n + 1)\left(\frac{n+1}{n}\right)$ diverges because $a_n = 0$ for n odd, while for n even $a_n = 2\left(1 + \frac{1}{n}\right)$ converges to 2; it diverges by definition of divergence

107. If $\{a_n\}$ is nonincreasing with lower bound M, then $\{-a_n\}$ is a nondecreasing sequence with upper bound $-M$. By Theorem 1, $\{-a_n\}$ converges and hence $\{a_n\}$ converges. If $\{a_n\}$ has no lower bound, then $\{-a_n\}$ has no upper bound and therefore diverges. Hence, $\{a_n\}$ also diverges.

109. $a_n \geq a_{n+1} \Leftrightarrow \frac{1 + \sqrt{2n}}{\sqrt{n}} \geq \frac{1 + \sqrt{2(n+1)}}{\sqrt{n+1}} \Leftrightarrow \sqrt{n+1} + \sqrt{2n^2 + 2n} \geq \sqrt{n} + \sqrt{2n^2 + 2n} \Leftrightarrow \sqrt{n+1} \geq \sqrt{n}$

and $\frac{1 + \sqrt{2n}}{\sqrt{n}} \geq \sqrt{2}$; thus the sequence is nonincreasing and bounded below by $\sqrt{2} \Rightarrow$ it converges

111. $\frac{4^{n+1} + 3^n}{4^n} = 4 + \left(\frac{3}{4}\right)^n$ so $a_n \geq a_{n+1} \Leftrightarrow 4 + \left(\frac{3}{4}\right)^n \geq 4 + \left(\frac{3}{4}\right)^{n+1} \Leftrightarrow \left(\frac{3}{4}\right)^n \geq \left(\frac{3}{4}\right)^{n+1} \Leftrightarrow 1 \geq \frac{3}{4}$ and $4 + \left(\frac{3}{4}\right)^n \geq 4$; thus the sequence is nonincreasing and bounded below by $4 \Rightarrow$ it converges

113. Let $0 < M < 1$ and let N be an integer greater than $\frac{M}{1-M}$. Then $n > N \Rightarrow n > \frac{M}{1-M} \Rightarrow n - nM > M$ $\Rightarrow n > M + nM \Rightarrow n > M(n+1) \Rightarrow \frac{n}{n+1} > M$.

115. The sequence $a_n = 1 + \frac{(-1)^n}{2}$ is the sequence $\frac{1}{2}, \frac{3}{2}, \frac{1}{2}, \frac{3}{2}, \ldots$. This sequence is bounded above by $\frac{3}{2}$, but it clearly does not converge, by definition of convergence.

117. Given an $\epsilon > 0$, by definition of convergence there corresponds an N such that for all $n > N$, $|L_1 - a_n| < \epsilon$ and $|L_2 - a_n| < \epsilon$. Now $|L_2 - L_1| = |L_2 - a_n + a_n - L_1| \leq |L_2 - a_n| + |a_n - L_1| < \epsilon + \epsilon = 2\epsilon$. $|L_2 - L_1| < 2\epsilon$ says that the difference between two fixed values is smaller than any positive number 2ϵ. The only nonnegative number smaller than every positive number is 0, so $|L_1 - L_2| = 0$ or $L_1 = L_2$.

119. $a_{2k} \to L \Leftrightarrow$ given an $\epsilon > 0$ there corresponds an N_1 such that $[2k > N_1 \Rightarrow |a_{2k} - L| < \epsilon]$. Similarly, $a_{2k+1} \to L \Leftrightarrow [2k + 1 > N_2 \Rightarrow |a_{2k+1} - L| < \epsilon]$. Let $N = \max\{N_1, N_2\}$. Then $n > N \Rightarrow |a_n - L| < \epsilon$ whether n is even or odd, and hence $a_n \to L$.

121. $\left|\sqrt[n]{0.5} - 1\right| < 10^{-3} \Rightarrow -\frac{1}{1000} < \left(\frac{1}{2}\right)^{1/n} - 1 < \frac{1}{1000} \Rightarrow \left(\frac{999}{1000}\right)^n < \frac{1}{2} < \left(\frac{1001}{1000}\right)^n \Rightarrow n > \frac{\ln\left(\frac{1}{2}\right)}{\ln\left(\frac{999}{1000}\right)} \Rightarrow n > 692.8$ $\Rightarrow N = 692$; $a_n = \left(\frac{1}{2}\right)^{1/n}$ and $\lim_{n \to \infty} a_n = 1$

123. $(0.9)^n < 10^{-3} \Rightarrow n \ln(0.9) < -3 \ln 10 \Rightarrow n > \frac{-3 \ln 10}{\ln(0.9)} \approx 65.54 \Rightarrow N = 65$; $a_n = \left(\frac{9}{10}\right)^n$ and $\lim_{n \to \infty} a_n = 0$

125. (a) $f(x) = x^2 - a \Rightarrow f'(x) = 2x \Rightarrow x_{n+1} = x_n - \frac{x_n^2 - a}{2x_n} \Rightarrow x_{n+1} = \frac{2x_n^2 - (x_n^2 - a)}{2x_n} = \frac{x_n^2 + a}{2x_n} = \frac{\left(x_n + \frac{a}{x_n}\right)}{2}$

   (b) $x_1 = 2$, $x_2 = 1.75$, $x_3 = 1.732142857$, $x_4 = 1.73205081$, $x_5 = 1.732050808$; we are finding the positive number where $x^2 - 3 = 0$; that is, where $x^2 = 3$, $x > 0$, or where $x = \sqrt{3}$.

127. $x_1 = 1$, $x_2 = 1 + \cos(1) = 1.540302306$, $x_3 = 1.540302306 + \cos(1 + \cos(1)) = 1.570791601$,
$x_4 = 1.570791601 + \cos(1.570791601) = 1.570796327 = \frac{\pi}{2}$ to 9 decimal places. After a few steps, the
arc (x_{n-1}) and line segment $\cos(x_{n-1})$ are nearly the same as the quarter circle.

11.2 INFINITE SERIES

1. $s_n = \frac{a(1 - r^n)}{(1 - r)} = \frac{2\left(1 - \left(\frac{1}{3}\right)^n\right)}{1 - \left(\frac{1}{3}\right)} \Rightarrow \lim_{n \to \infty} s_n = \frac{2}{1 - \left(\frac{1}{3}\right)} = 3$

3. $s_n = \frac{a(1 - r^n)}{(1 - r)} = \frac{1 - \left(-\frac{1}{2}\right)^n}{1 - \left(-\frac{1}{2}\right)} \Rightarrow \lim_{n \to \infty} s_n = \frac{1}{\left(\frac{3}{2}\right)} = \frac{2}{3}$

5. $\frac{1}{(n+1)(n+2)} = \frac{1}{n+1} - \frac{1}{n+2} \Rightarrow s_n = \left(\frac{1}{2} - \frac{1}{3}\right) + \left(\frac{1}{3} - \frac{1}{4}\right) + \ldots + \left(\frac{1}{n+1} - \frac{1}{n+2}\right) = \frac{1}{2} - \frac{1}{n+2} \Rightarrow \lim_{n \to \infty} s_n = \frac{1}{2}$

7. $1 - \frac{1}{4} + \frac{1}{16} - \frac{1}{64} + \ldots$, the sum of this geometric series is $\frac{1}{1 - \left(-\frac{1}{4}\right)} = \frac{1}{1 + \left(\frac{1}{4}\right)} = \frac{4}{5}$

9. $\frac{7}{4} + \frac{7}{16} + \frac{7}{64} + \ldots$, the sum of this geometric series is $\frac{\left(\frac{7}{4}\right)}{1 - \left(\frac{1}{4}\right)} = \frac{7}{3}$

11. $(5 + 1) + \left(\frac{5}{2} + \frac{1}{3}\right) + \left(\frac{5}{4} + \frac{1}{9}\right) + \left(\frac{5}{8} + \frac{1}{27}\right) + \ldots$, is the sum of two geometric series; the sum is
$\frac{5}{1 - \left(\frac{1}{2}\right)} + \frac{1}{1 - \left(\frac{1}{3}\right)} = 10 + \frac{3}{2} = \frac{23}{2}$

13. $(1 + 1) + \left(\frac{1}{2} - \frac{1}{5}\right) + \left(\frac{1}{4} + \frac{1}{25}\right) + \left(\frac{1}{8} - \frac{1}{125}\right) + \ldots$, is the sum of two geometric series; the sum is
$\frac{1}{1 - \left(\frac{1}{2}\right)} + \frac{1}{1 + \left(\frac{1}{5}\right)} = 2 + \frac{5}{6} = \frac{17}{6}$

15. $\frac{4}{(4n-3)(4n+1)} = \frac{1}{4n-3} - \frac{1}{4n+1} \Rightarrow s_n = \left(1 - \frac{1}{5}\right) + \left(\frac{1}{5} - \frac{1}{9}\right) + \left(\frac{1}{9} - \frac{1}{13}\right) + \ldots + \left(\frac{1}{4n-7} - \frac{1}{4n-3}\right)$
$+ \left(\frac{1}{4n-3} - \frac{1}{4n+1}\right) = 1 - \frac{1}{4n+1} \Rightarrow \lim_{n \to \infty} s_n = \lim_{n \to \infty} \left(1 - \frac{1}{4n+1}\right) = 1$

17. $\frac{40n}{(2n-1)^2(2n+1)^2} = \frac{A}{(2n-1)} + \frac{B}{(2n-1)^2} + \frac{C}{(2n+1)} + \frac{D}{(2n+1)^2}$
$= \frac{A(2n-1)(2n+1)^2 + B(2n+1)^2 + C(2n+1)(2n-1)^2 + D(2n-1)^2}{(2n-1)^2(2n+1)^2}$
$\Rightarrow A(2n-1)(2n+1)^2 + B(2n+1)^2 + C(2n+1)(2n-1)^2 + D(2n-1)^2 = 40n$
$\Rightarrow A\left(8n^3 + 4n^2 - 2n - 1\right) + B\left(4n^2 + 4n + 1\right) + C\left(8n^3 - 4n^2 - 2n + 1\right) = D\left(4n^2 - 4n + 1\right) = 40n$
$\Rightarrow (8A + 8C)n^3 + (4A + 4B - 4C + 4D)n^2 + (-2A + 4B - 2C - 4D)n + (-A + B + C + D) = 40n$

$\Rightarrow \begin{cases} 8A + 8C = 0 \\ 4A + 4B - 4C + 4D = 0 \\ -2A + 4B - 2C - 4D = 40 \\ -A + B + C + D = 0 \end{cases} \Rightarrow \begin{cases} 8A + 8C = 0 \\ A + B - C + D = 0 \\ -A + 2B - C - 2D = 20 \\ -A + B + C + D = 0 \end{cases} \Rightarrow \begin{cases} B + D = 0 \\ 2B - 2D = 20 \end{cases} \Rightarrow 4B = 20 \Rightarrow B = 5$

and $D = -5 \Rightarrow \begin{cases} A + C = 0 \\ -A + 5 + C - 5 = 0 \end{cases} \Rightarrow C = 0$ and $A = 0$. Hence, $\sum_{n=1}^{k} \left[\frac{40n}{(2n-1)^2(2n+1)^2}\right]$

$= 5 \sum_{n=1}^{k} \left[\frac{1}{(2n-1)^2} - \frac{1}{(2n+1)^2}\right] = 5\left(\frac{1}{1} - \frac{1}{9} + \frac{1}{9} - \frac{1}{25} + \frac{1}{25} - \ldots - \frac{1}{(2(k-1)+1)^2} + \frac{1}{(2k-1)^2} - \frac{1}{(2k+1)^2}\right)$

$= 5\left(1 - \frac{1}{(2k+1)^2}\right) \Rightarrow$ the sum is $\lim_{n \to \infty} 5\left(1 - \frac{1}{(2k+1)^2}\right) = 5$

19. $s_n = \left(1 - \frac{1}{\sqrt{2}}\right) + \left(\frac{1}{\sqrt{2}} - \frac{1}{\sqrt{3}}\right) + \left(\frac{1}{\sqrt{3}} - \frac{1}{\sqrt{4}}\right) + \ldots + \left(\frac{1}{\sqrt{n-1}} + \frac{1}{\sqrt{n}}\right) + \left(\frac{1}{\sqrt{n}} - \frac{1}{\sqrt{n+1}}\right) = 1 - \frac{1}{\sqrt{n+1}}$
$\Rightarrow \lim_{n \to \infty} s_n = \lim_{n \to \infty} \left(1 - \frac{1}{\sqrt{n+1}}\right) = 1$

21. $s_n = \left(\frac{1}{\ln 3} - \frac{1}{\ln 2}\right) + \left(\frac{1}{\ln 4} - \frac{1}{\ln 3}\right) + \left(\frac{1}{\ln 5} - \frac{1}{\ln 4}\right) + \cdots + \left(\frac{1}{\ln(n+1)} - \frac{1}{\ln n}\right) + \left(\frac{1}{\ln(n+2)} - \frac{1}{\ln(n+1)}\right)$

$= -\frac{1}{\ln 2} + \frac{1}{\ln(n+2)} \Rightarrow \lim_{n \to \infty} s_n = -\frac{1}{\ln 2}$

23. convergent geometric series with sum $\dfrac{1}{1 - \left(\frac{1}{\sqrt{2}}\right)} = \dfrac{\sqrt{2}}{\sqrt{2}-1} = 2 + \sqrt{2}$

25. convergent geometric series with sum $\dfrac{\left(\frac{3}{2}\right)}{1 - \left(-\frac{1}{2}\right)} = 1$

27. $\lim_{n \to \infty} \cos(n\pi) = \lim_{n \to \infty} (-1)^n \neq 0 \Rightarrow$ diverges

29. convergent geometric series with sum $\dfrac{1}{1 - \left(\frac{1}{e^2}\right)} = \dfrac{e^2}{e^2 - 1}$

31. convergent geometric series with sum $\dfrac{2}{1 - \left(\frac{1}{10}\right)} - 2 = \dfrac{20}{9} - \dfrac{18}{9} = \dfrac{2}{9}$

33. difference of two geometric series with sum $\dfrac{1}{1 - \left(\frac{2}{3}\right)} - \dfrac{1}{1 - \left(\frac{1}{3}\right)} = 3 - \dfrac{3}{2} = \dfrac{3}{2}$

35. $\lim_{n \to \infty} \dfrac{n!}{1000^n} = \infty \neq 0 \Rightarrow$ diverges

37. $\sum_{n=1}^{\infty} \ln\left(\frac{n}{n+1}\right) = \sum_{n=1}^{\infty} [\ln(n) - \ln(n+1)] \Rightarrow s_n = [\ln(1) - \ln(2)] + [\ln(2) - \ln(3)] + [\ln(3) - \ln(4)] + \cdots$

$+ [\ln(n-1) - \ln(n)] + [\ln(n) - \ln(n+1)] = \ln(1) - \ln(n+1) = -\ln(n+1) \Rightarrow \lim_{n \to \infty} s_n = -\infty, \Rightarrow$ diverges

39. convergent geometric series with sum $\dfrac{1}{1 - \left(\frac{e}{\pi}\right)} = \dfrac{\pi}{\pi - e}$

41. $\sum_{n=0}^{\infty} (-1)^n x^n = \sum_{n=0}^{\infty} (-x)^n$; $a = 1, r = -x$; converges to $\dfrac{1}{1 - (-x)} = \dfrac{1}{1 + x}$ for $|x| < 1$

43. $a = 3, r = \dfrac{x-1}{2}$; converges to $\dfrac{3}{1 - \left(\frac{x-1}{2}\right)} = \dfrac{6}{3 - x}$ for $-1 < \dfrac{x-1}{2} < 1$ or $-1 < x < 3$

45. $a = 1, r = 2x$; converges to $\dfrac{1}{1 - 2x}$ for $|2x| < 1$ or $|x| < \dfrac{1}{2}$

47. $a = 1, r = -(x+1)^n$; converges to $\dfrac{1}{1 + (x+1)} = \dfrac{1}{2 + x}$ for $|x + 1| < 1$ or $-2 < x < 0$

49. $a = 1, r = \sin x$; converges to $\dfrac{1}{1 - \sin x}$ for $x \neq (2k+1)\dfrac{\pi}{2}$, k an integer

51. $0.\overline{23} = \sum_{n=0}^{\infty} \dfrac{23}{100} \left(\dfrac{1}{10^2}\right)^n = \dfrac{\left(\frac{23}{100}\right)}{1 - \left(\frac{1}{100}\right)} = \dfrac{23}{99}$

53. $0.\overline{7} = \sum_{n=0}^{\infty} \dfrac{7}{10} \left(\dfrac{1}{10}\right)^n = \dfrac{\left(\frac{7}{10}\right)}{1 - \left(\frac{1}{10}\right)} = \dfrac{7}{9}$

55. $0.0\overline{6} = \sum_{n=0}^{\infty} \left(\dfrac{1}{10}\right)\left(\dfrac{6}{10}\right)\left(\dfrac{1}{10}\right)^n = \dfrac{\left(\frac{6}{100}\right)}{1 - \left(\frac{1}{10}\right)} = \dfrac{6}{90} = \dfrac{1}{15}$

57. $1.24\overline{123} = \frac{124}{100} + \sum\limits_{n=0}^{\infty} \frac{123}{10^5} \left(\frac{1}{10^3}\right)^n = \frac{124}{100} + \frac{\left(\frac{123}{10^5}\right)}{1 - \left(\frac{1}{10^3}\right)} = \frac{124}{100} + \frac{123}{10^5 - 10^2} = \frac{124}{100} + \frac{123}{99,900} = \frac{123,999}{99,900} = \frac{41,333}{33,300}$

59. (a) $\sum\limits_{n=-2}^{\infty} \frac{1}{(n+4)(n+5)}$ (b) $\sum\limits_{n=0}^{\infty} \frac{1}{(n+2)(n+3)}$ (c) $\sum\limits_{n=5}^{\infty} \frac{1}{(n-3)(n-2)}$

61. (a) one example is $\frac{1}{2} + \frac{1}{4} + \frac{1}{8} + \frac{1}{16} + \ldots = \frac{\left(\frac{1}{2}\right)}{1 - \left(\frac{1}{2}\right)} = 1$

(b) one example is $-\frac{3}{2} - \frac{3}{4} - \frac{3}{8} - \frac{3}{16} - \ldots = \frac{\left(-\frac{3}{2}\right)}{1 - \left(\frac{1}{2}\right)} = -3$

(c) one example is $1 - \frac{1}{2} - \frac{1}{4} - \frac{1}{8} - \frac{1}{16} - \ldots$; the series $\frac{k}{2} + \frac{k}{4} + \frac{k}{8} + \ldots = \frac{\left(\frac{k}{2}\right)}{1 - \left(\frac{1}{2}\right)} = k$ where k is any positive or negative number.

63. Let $a_n = b_n = \left(\frac{1}{2}\right)^n$. Then $\sum\limits_{n=1}^{\infty} a_n = \sum\limits_{n=1}^{\infty} b_n = \sum\limits_{n=1}^{\infty} \left(\frac{1}{2}\right)^n = 1$, while $\sum\limits_{n=1}^{\infty} \left(\frac{a_n}{b_n}\right) = \sum\limits_{n=1}^{\infty} (1)$ diverges.

65. Let $a_n = \left(\frac{1}{4}\right)^n$ and $b_n = \left(\frac{1}{2}\right)^n$. Then $A = \sum\limits_{n=1}^{\infty} a_n = \frac{1}{3}$, $B = \sum\limits_{n=1}^{\infty} b_n = 1$ and $\sum\limits_{n=1}^{\infty} \left(\frac{a_n}{b_n}\right) = \sum\limits_{n=1}^{\infty} \left(\frac{1}{2}\right)^n = 1 \neq \frac{A}{B}$.

67. Since the sum of a finite number of terms is finite, adding or subtracting a finite number of terms from a series that diverges does not change the divergence of the series.

69. (a) $\frac{2}{1-r} = 5 \Rightarrow \frac{2}{5} = 1 - r \Rightarrow r = \frac{3}{5}$; $2 + 2\left(\frac{3}{5}\right) + 2\left(\frac{3}{5}\right)^2 + \ldots$

(b) $\frac{\left(\frac{13}{2}\right)}{1-r} = 5 \Rightarrow \frac{13}{10} = 1 - r \Rightarrow r = -\frac{3}{10}$; $\frac{13}{2} - \frac{13}{2}\left(\frac{3}{10}\right) + \frac{13}{2}\left(\frac{3}{10}\right)^2 - \frac{13}{2}\left(\frac{3}{10}\right)^3 + \ldots$

71. $s_n = 1 + 2r + r^2 + 2r^3 + r^4 + 2r^5 + \ldots + r^{2n} + 2r^{2n+1}$, $n = 0, 1, \ldots$
$\Rightarrow s_n = (1 + r^2 + r^4 + \ldots + r^{2n}) + (2r + 2r^3 + 2r^5 + \ldots + 2r^{2n+1}) \Rightarrow \lim\limits_{n \to \infty} s_n = \frac{1}{1-r^2} + \frac{2r}{1-r^2}$
$= \frac{1+2r}{1-r^2}$, if $|r^2| < 1$ or $|r| < 1$

73. distance $= 4 + 2\left[(4)\left(\frac{3}{4}\right) + (4)\left(\frac{3}{4}\right)^2 + \ldots\right] = 4 + 2\left(\frac{3}{1 - \left(\frac{3}{4}\right)}\right) = 28$ m

75. area $= 2^2 + \left(\sqrt{2}\right)^2 + (1)^2 + \left(\frac{1}{\sqrt{2}}\right)^2 + \ldots = 4 + 2 + 1 + \frac{1}{2} + \ldots = \frac{4}{1 - \frac{1}{2}} = 8$ m^2

77. (a) $L_1 = 3, L_2 = 3\left(\frac{4}{3}\right), L_3 = 3\left(\frac{4}{3}\right)^2, \ldots, L_n = 3\left(\frac{4}{3}\right)^{n-1} \Rightarrow \lim\limits_{n \to \infty} L_n = \lim\limits_{n \to \infty} 3\left(\frac{4}{3}\right)^{n-1} = \infty$

(b) Using the fact that the area of an equilateral triangle of side length s is $\frac{\sqrt{3}}{4}s^2$, we see that $A_1 = \frac{\sqrt{3}}{4}$,
$A_2 = A_1 + 3\left(\frac{\sqrt{3}}{4}\right)\left(\frac{1}{3}\right)^2 = \frac{\sqrt{3}}{4} + \frac{\sqrt{3}}{12}$, $A_3 = A_2 + 3(4)\left(\frac{\sqrt{3}}{4}\right)\left(\frac{1}{3^2}\right)^2 = \frac{\sqrt{3}}{4} + \frac{\sqrt{3}}{12} + \frac{\sqrt{3}}{27}$,
$A_4 = A_3 + 3(4)^2\left(\frac{\sqrt{3}}{4}\right)\left(\frac{1}{3^3}\right)^2$, $A_5 = A_4 + 3(4)^3\left(\frac{\sqrt{3}}{4}\right)\left(\frac{1}{3^4}\right)^2, \ldots$,
$A_n = \frac{\sqrt{3}}{4} + \sum\limits_{k=2}^{n} 3(4)^{k-2}\left(\frac{\sqrt{3}}{4}\right)\left(\frac{1}{3^2}\right)^{k-1} = \frac{\sqrt{3}}{4} + \sum\limits_{k=2}^{n} 3\sqrt{3}(4)^{k-3}\left(\frac{1}{9}\right)^{k-1} = \frac{\sqrt{3}}{4} + 3\sqrt{3}\left(\sum\limits_{k=2}^{n} \frac{4^{k-3}}{9^{k-1}}\right)$.
$\lim\limits_{n \to \infty} A_n = \lim\limits_{n \to \infty} \left(\frac{\sqrt{3}}{4} + 3\sqrt{3}\left(\sum\limits_{k=2}^{n} \frac{4^{k-3}}{9^{k-1}}\right)\right) = \frac{\sqrt{3}}{4} + 3\sqrt{3}\left(\frac{\frac{1}{36}}{1 - \frac{4}{9}}\right) = \frac{\sqrt{3}}{4} + 3\sqrt{3}\left(\frac{1}{20}\right) = \frac{2\sqrt{3}}{5}$

11.3 THE INTEGRAL TEST

1. converges; a geometric series with $r = \frac{1}{10} < 1$

3. diverges; by the nth-Term Test for Divergence, $\lim\limits_{n \to \infty} \frac{n}{n+1} = 1 \neq 0$

5. diverges; $\sum\limits_{n=1}^{\infty} \frac{3}{\sqrt{n}} = 3 \sum\limits_{n=1}^{\infty} \frac{1}{\sqrt{n}}$, which is a divergent p-series ($p = \frac{1}{2}$)

7. converges; a geometric series with $r = \frac{1}{8} < 1$

9. diverges by the Integral Test: $\int_2^n \frac{\ln x}{x}\, dx = \frac{1}{2}\left(\ln^2 n - \ln 2\right) \Rightarrow \int_2^\infty \frac{\ln x}{x}\, dx \to \infty$

11. converges; a geometric series with $r = \frac{2}{3} < 1$

13. diverges; $\sum\limits_{n=0}^{\infty} \frac{-2}{n+1} = -2 \sum\limits_{n=0}^{\infty} \frac{1}{n+1}$, which diverges by the Integral Test

15. diverges; $\lim\limits_{n \to \infty} a_n = \lim\limits_{n \to \infty} \frac{2^n}{n+1} = \lim\limits_{n \to \infty} \frac{2^n \ln 2}{1} = \infty \neq 0$

17. diverges; $\lim\limits_{n \to \infty} \frac{\sqrt{n}}{\ln n} = \lim\limits_{n \to \infty} \frac{\left(\frac{1}{2\sqrt{n}}\right)}{\left(\frac{1}{n}\right)} = \lim\limits_{n \to \infty} \frac{\sqrt{n}}{2} = \infty \neq 0$

19. diverges; a geometric series with $r = \frac{1}{\ln 2} \approx 1.44 > 1$

21. converges by the Integral Test: $\int_3^\infty \frac{\left(\frac{1}{x}\right)}{(\ln x)\sqrt{(\ln x)^2 - 1}}\, dx$; $\begin{bmatrix} u = \ln x \\ du = \frac{1}{x}\, dx \end{bmatrix} \to \int_{\ln 3}^\infty \frac{1}{u\sqrt{u^2 - 1}}\, du$

$= \lim\limits_{b \to \infty} \left[\sec^{-1}|u|\right]_{\ln 3}^b = \lim\limits_{b \to \infty} \left[\sec^{-1} b - \sec^{-1}(\ln 3)\right] = \lim\limits_{b \to \infty} \left[\cos^{-1}\left(\frac{1}{b}\right) - \sec^{-1}(\ln 3)\right]$

$= \cos^{-1}(0) - \sec^{-1}(\ln 3) = \frac{\pi}{2} - \sec^{-1}(\ln 3) \approx 1.1439$

23. diverges by the nth-Term Test for divergence; $\lim\limits_{n \to \infty} n \sin\left(\frac{1}{n}\right) = \lim\limits_{n \to \infty} \frac{\sin\left(\frac{1}{n}\right)}{\left(\frac{1}{n}\right)} = \lim\limits_{x \to 0} \frac{\sin x}{x} = 1 \neq 0$

25. converges by the Integral Test: $\int_1^\infty \frac{e^x}{1+e^{2x}}\, dx$; $\begin{bmatrix} u = e^x \\ du = e^x\, dx \end{bmatrix} \to \int_e^\infty \frac{1}{1+u^2}\, du = \lim\limits_{n \to \infty} \left[\tan^{-1} u\right]_e^b$

$= \lim\limits_{b \to \infty} \left(\tan^{-1} b - \tan^{-1} e\right) = \frac{\pi}{2} - \tan^{-1} e \approx 0.35$

27. converges by the Integral Test: $\int_1^\infty \frac{8 \tan^{-1} x}{1+x^2}\, dx$; $\begin{bmatrix} u = \tan^{-1} x \\ du = \frac{dx}{1+x^2} \end{bmatrix} \to \int_{\pi/4}^{\pi/2} 8u\, du = \left[4u^2\right]_{\pi/4}^{\pi/2} = 4\left(\frac{\pi^2}{4} - \frac{\pi^2}{16}\right) = \frac{3\pi^2}{4}$

29. converges by the Integral Test: $\int_1^\infty \operatorname{sech} x\, dx = 2 \lim\limits_{b \to \infty} \int_1^b \frac{e^x}{1+(e^x)^2}\, dx = 2 \lim\limits_{b \to \infty} \left[\tan^{-1} e^x\right]_1^b$

$= 2 \lim\limits_{b \to \infty} \left(\tan^{-1} e^b - \tan^{-1} e\right) = \pi - 2\tan^{-1} e \approx 0.71$

31. $\int_1^\infty \left(\frac{a}{x+2} - \frac{1}{x+4}\right) dx = \lim_{b \to \infty} \left[a \ln|x+2| - \ln|x+4|\right]_1^b = \lim_{b \to \infty} \ln \frac{(b+2)^a}{b+4} - \ln\left(\frac{3^a}{5}\right);$

$\lim_{b \to \infty} \frac{(b+2)^a}{b+4} = a \lim_{b \to \infty} (b+2)^{a-1} = \begin{cases} \infty, a > 1 \\ 1, \quad a = 1 \end{cases} \Rightarrow$ the series converges to $\ln\left(\frac{5}{3}\right)$ if $a = 1$ and diverges to ∞ if

$a > 1$. If $a < 1$, the terms of the series eventually become negative and the Integral Test does not apply. From
that point on, however, the series behaves like a negative multiple of the harmonic series, and so it diverges.

33. (a)

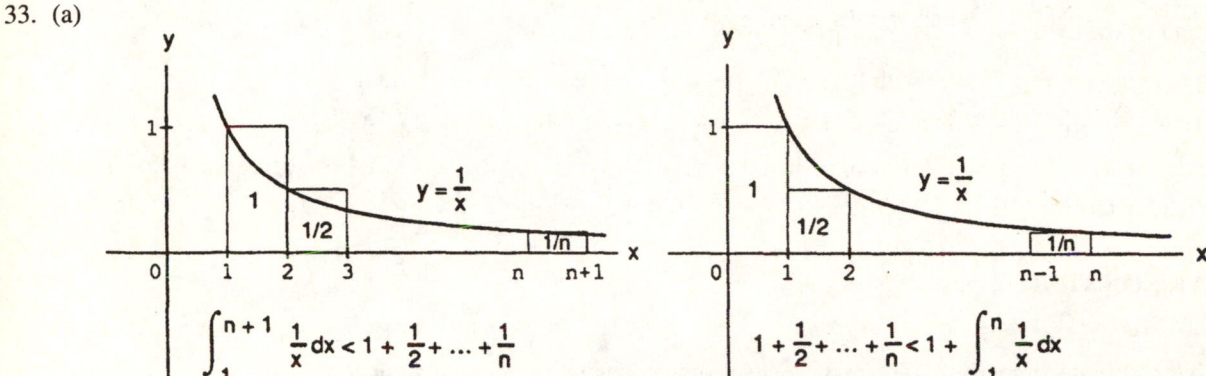

$$\int_1^{n+1} \frac{1}{x} dx < 1 + \frac{1}{2} + \ldots + \frac{1}{n} \qquad\qquad 1 + \frac{1}{2} + \ldots + \frac{1}{n} < 1 + \int_1^n \frac{1}{x} dx$$

(b) There are $(13)(365)(24)(60)(60) \, (10^9)$ seconds in 13 billion years; by part (a) $s_n \le 1 + \ln n$ where
$n = (13)(365)(24)(60)(60) \, (10^9) \Rightarrow s_n \le 1 + \ln\left((13)(365)(24)(60)(60) \, (10^9)\right)$
$= 1 + \ln(13) + \ln(365) + \ln(24) + 2\ln(60) + 9\ln(10) \approx 41.55$

35. Yes. If $\sum_{n=1}^\infty a_n$ is a divergent series of positive numbers, then $\left(\frac{1}{2}\right) \sum_{n=1}^\infty a_n = \sum_{n=1}^\infty \left(\frac{a_n}{2}\right)$ also diverges and $\frac{a_n}{2} < a_n$.

There is no "smallest" divergent series of positive numbers: for any divergent series $\sum_{n=1}^\infty a_n$ of positive

numbers $\sum_{n=1}^\infty \left(\frac{a_n}{2}\right)$ has smaller terms and still diverges.

37. Let $A_n = \sum_{k=1}^n a_k$ and $B_n = \sum_{k=1}^n 2^k a_{(2^k)}$, where $\{a_k\}$ is a nonincreasing sequence of positive terms converging to

0. Note that $\{A_n\}$ and $\{B_n\}$ are nondecreasing sequences of positive terms. Now,
$B_n = 2a_2 + 4a_4 + 8a_8 + \ldots + 2^n a_{(2^n)} = 2a_2 + (2a_4 + 2a_4) + (2a_8 + 2a_8 + 2a_8 + 2a_8) + \ldots$
$+ \underbrace{\left(2a_{(2^n)} + 2a_{(2^n)} + \ldots + 2a_{(2^n)}\right)}_{2^{n-1} \text{ terms}} \le 2a_1 + 2a_2 + (2a_3 + 2a_4) + (2a_5 + 2a_6 + 2a_7 + 2a_8) + \ldots$

$+ \left(2a_{(2^{n-1})} + 2a_{(2^{n-1}+1)} + \ldots + 2a_{(2^n)}\right) = 2A_{(2^n)} \le 2\sum_{k=1}^\infty a_k$. Therefore if $\sum a_k$ converges,

then $\{B_n\}$ is bounded above $\Rightarrow \sum 2^k a_{(2^k)}$ converges. Conversely,

$A_n = a_1 + (a_2 + a_3) + (a_4 + a_5 + a_6 + a_7) + \ldots + a_n < a_1 + 2a_2 + 4a_4 + \ldots + 2^n a_{(2^n)} = a_1 + B_n < a_1 + \sum_{k=1}^\infty 2^k a_{(2^k)}.$

Therefore, if $\sum_{k=1}^\infty 2^k a_{(2^k)}$ converges, then $\{A_n\}$ is bounded above and hence converges.

39. (a) $\int_2^\infty \frac{dx}{x(\ln x)^p}; \begin{bmatrix} u = \ln x \\ du = \frac{dx}{x} \end{bmatrix} \to \int_{\ln 2}^\infty u^{-p} du = \lim_{b \to \infty} \left[\frac{u^{-p+1}}{-p+1}\right]_{\ln 2}^b = \lim_{b \to \infty} \left(\frac{1}{1-p}\right) \left[b^{-p+1} - (\ln 2)^{-p+1}\right]$

$= \begin{cases} \frac{1}{p-1} (\ln 2)^{-p+1}, p > 1 \\ \infty, p < 1 \end{cases} \Rightarrow$ the improper integral converges if $p > 1$ and diverges

if $p < 1$. For $p = 1$: $\int_2^\infty \frac{dx}{x \ln x} = \lim_{b \to \infty} \left[\ln(\ln x)\right]_2^b = \lim_{b \to \infty} \left[\ln(\ln b) - \ln(\ln 2)\right] = \infty$, so the improper
integral diverges if $p = 1$.

(b) Since the series and the integral converge or diverge together, $\sum_{n=2}^{\infty} \frac{1}{n(\ln n)^p}$ converges if and only if $p > 1$.

41. (a) From Fig. 11.8 in the text with $f(x) = \frac{1}{x}$ and $a_k = \frac{1}{k}$, we have $\int_1^{n+1} \frac{1}{x} dx \le 1 + \frac{1}{2} + \frac{1}{3} + \ldots + \frac{1}{n}$

$\le 1 + \int_1^n f(x)\, dx \Rightarrow \ln(n+1) \le 1 + \frac{1}{2} + \frac{1}{3} + \ldots + \frac{1}{n} \le 1 + \ln n \Rightarrow 0 \le \ln(n+1) - \ln n$

$\le \left(1 + \frac{1}{2} + \frac{1}{3} + \ldots + \frac{1}{n}\right) - \ln n \le 1$. Therefore the sequence $\left\{\left(1 + \frac{1}{2} + \frac{1}{3} + \ldots + \frac{1}{n}\right) - \ln n\right\}$ is bounded above by 1 and below by 0.

(b) From the graph in Fig. 11.8(a) with $f(x) = \frac{1}{x}$, $\frac{1}{n+1} < \int_n^{n+1} \frac{1}{x} dx = \ln(n+1) - \ln n$

$\Rightarrow 0 > \frac{1}{n+1} - [\ln(n+1) - \ln n] = \left(1 + \frac{1}{2} + \frac{1}{3} + \ldots + \frac{1}{n+1} - \ln(n+1)\right) - \left(1 + \frac{1}{2} + \frac{1}{3} + \ldots + \frac{1}{n} - \ln n\right)$.

If we define $a_n = 1 + \frac{1}{2} = \frac{1}{3} + \frac{1}{n} - \ln n$, then $0 > a_{n+1} - a_n \Rightarrow a_{n+1} < a_n \Rightarrow \{a_n\}$ is a decreasing sequence of nonnegative terms.

11.4 COMPARISON TESTS

1. diverges by the Limit Comparison Test (part 1) when compared with $\sum_{n=1}^{\infty} \frac{1}{\sqrt{n}}$, a divergent p-series:

$$\lim_{n \to \infty} \frac{\left(\frac{1}{2\sqrt{n} + \sqrt[3]{n}}\right)}{\left(\frac{1}{\sqrt{n}}\right)} = \lim_{n \to \infty} \frac{\sqrt{n}}{2\sqrt{n} + \sqrt[3]{n}} = \lim_{n \to \infty} \left(\frac{1}{2 + n^{-1/6}}\right) = \frac{1}{2}$$

3. converges by the Direct Comparison Test; $\frac{\sin^2 n}{2^n} \le \frac{1}{2^n}$, which is the nth term of a convergent geometric series

5. diverges since $\lim_{n \to \infty} \frac{2n}{3n-1} = \frac{2}{3} \ne 0$

7. converges by the Direct Comparison Test; $\left(\frac{n}{3n+1}\right)^n < \left(\frac{n}{3n}\right)^n = \left(\frac{1}{3}\right)^n$, the nth term of a convergent geometric series

9. diverges by the Direct Comparison Test; $n > \ln n \Rightarrow \ln n > \ln \ln n \Rightarrow \frac{1}{n} < \frac{1}{\ln n} < \frac{1}{\ln(\ln n)}$ and $\sum_{n=3}^{\infty} \frac{1}{n}$

diverges

11. converges by the Limit Comparison Test (part 2) when compared with $\sum_{n=1}^{\infty} \frac{1}{n^2}$, a convergent p-series:

$$\lim_{n \to \infty} \frac{\left[\frac{(\ln n)^2}{n^3}\right]}{\left(\frac{1}{n^2}\right)} = \lim_{n \to \infty} \frac{(\ln n)^2}{n} = \lim_{n \to \infty} \frac{2(\ln n)\left(\frac{1}{n}\right)}{1} = 2 \lim_{n \to \infty} \frac{\ln n}{n} = 0$$

13. diverges by the Limit Comparison Test (part 3) with $\frac{1}{n}$, the nth term of the divergent harmonic series:

$$\lim_{n \to \infty} \frac{\left[\frac{1}{\sqrt{n} \ln n}\right]}{\left(\frac{1}{n}\right)} = \lim_{n \to \infty} \frac{\sqrt{n}}{\ln n} = \lim_{n \to \infty} \frac{\left(\frac{1}{2\sqrt{n}}\right)}{\left(\frac{1}{n}\right)} = \lim_{n \to \infty} \frac{\sqrt{n}}{2} = \infty$$

15. diverges by the Limit Comparison Test (part 3) with $\frac{1}{n}$, the nth term of the divergent harmonic series:

$$\lim_{n \to \infty} \frac{\left(\frac{1}{1 + \ln n}\right)}{\left(\frac{1}{n}\right)} = \lim_{n \to \infty} \frac{n}{1 + \ln n} = \lim_{n \to \infty} \frac{1}{\left(\frac{1}{n}\right)} = \lim_{n \to \infty} n = \infty$$

17. diverges by the Integral Test: $\int_2^{\infty} \frac{\ln(x+1)}{x+1} dx = \int_{\ln 3}^{\infty} u\, du = \lim_{b \to \infty} \left[\frac{1}{2} u^2\right]_{\ln 3}^{b} = \lim_{b \to \infty} \frac{1}{2}(b^2 - \ln^2 3) = \infty$

19. converges by the Direct Comparison Test with $\frac{1}{n^{3/2}}$, the nth term of a convergent p-series: $n^2 - 1 > n$ for

$n \geq 2 \Rightarrow n^2(n^2 - 1) > n^3 \Rightarrow n\sqrt{n^2 - 1} > n^{3/2} \Rightarrow \frac{1}{n^{3/2}} > \frac{1}{n\sqrt{n^2 - 1}}$ or use Limit Comparison Test with $\frac{1}{n^2}$.

21. converges because $\sum_{n=1}^{\infty} \frac{1-n}{n2^n} = \sum_{n=1}^{\infty} \frac{1}{n2^n} + \sum_{n=1}^{\infty} \frac{-1}{2^n}$ which is the sum of two convergent series:

$\sum_{n=1}^{\infty} \frac{1}{n2^n}$ converges by the Direct Comparison Test since $\frac{1}{n2^n} < \frac{1}{2^n}$, and $\sum_{n=1}^{\infty} \frac{-1}{2^n}$ is a convergent geometric

series

23. converges by the Direct Comparison Test: $\frac{1}{3^{n-1}+1} < \frac{1}{3^{n-1}}$, which is the nth term of a convergent geometric

series

25. diverges by the Limit Comparison Test (part 1) with $\frac{1}{n}$, the nth term of the divergent harmonic series:

$\lim_{n \to \infty} \frac{\left(\sin \frac{1}{n}\right)}{\left(\frac{1}{n}\right)} = \lim_{x \to 0} \frac{\sin x}{x} = 1$

27. converges by the Limit Comparison Test (part 1) with $\frac{1}{n^2}$, the nth term of a convergent p-series:

$\lim_{n \to \infty} \frac{\left(\frac{10n+1}{n(n+1)(n+2)}\right)}{\left(\frac{1}{n^2}\right)} = \lim_{n \to \infty} \frac{10n^2+n}{n^2+3n+2} = \lim_{n \to \infty} \frac{20n+1}{2n+3} = \lim_{n \to \infty} \frac{20}{2} = 10$

29. converges by the Direct Comparison Test: $\frac{\tan^{-1}n}{n^{1.1}} < \frac{\frac{\pi}{2}}{n^{1.1}}$ and $\sum_{n=1}^{\infty} \frac{\frac{\pi}{2}}{n^{1.1}} = \frac{\pi}{2} \sum_{n=1}^{\infty} \frac{1}{n^{1.1}}$ is the product of a

convergent p-series and a nonzero constant

31. converges by the Limit Comparison Test (part 1) with $\frac{1}{n^2}$: $\lim_{n \to \infty} \frac{\left(\frac{\coth n}{n^2}\right)}{\left(\frac{1}{n^2}\right)} = \lim_{n \to \infty} \coth n = \lim_{n \to \infty} \frac{e^n + e^{-n}}{e^n - e^{-n}}$

$= \lim_{n \to \infty} \frac{1+e^{-2n}}{1-e^{-2n}} = 1$

33. diverges by the Limit Comparison Test (part 1) with $\frac{1}{n}$: $\lim_{n \to \infty} \frac{\left(\frac{1}{n\sqrt[n]{n}}\right)}{\left(\frac{1}{n}\right)} = \lim_{n \to \infty} \frac{1}{\sqrt[n]{n}} = 1$.

35. $\frac{1}{1+2+3+\ldots+n} = \frac{1}{\left(\frac{n(n+1)}{2}\right)} = \frac{2}{n(n+1)}$. The series converges by the Limit Comparison Test (part 1) with $\frac{1}{n^2}$:

$\lim_{n \to \infty} \frac{\left(\frac{2}{n(n+1)}\right)}{\left(\frac{1}{n^2}\right)} = \lim_{n \to \infty} \frac{2n^2}{n^2+n} = \lim_{n \to \infty} \frac{4n}{2n+1} = \lim_{n \to \infty} \frac{4}{2} = 2$.

37. (a) If $\lim_{n \to \infty} \frac{a_n}{b_n} = 0$, then there exists an integer N such that for all $n > N$, $\left|\frac{a_n}{b_n} - 0\right| < 1 \Rightarrow -1 < \frac{a_n}{b_n} < 1$

$\Rightarrow a_n < b_n$. Thus, if $\sum b_n$ converges, then $\sum a_n$ converges by the Direct Comparison Test.

   (b) If $\lim_{n \to \infty} \frac{a_n}{b_n} = \infty$, then there exists an integer N such that for all $n > N$, $\frac{a_n}{b_n} > 1 \Rightarrow a_n > b_n$. Thus, if

$\sum b_n$ diverges, then $\sum a_n$ diverges by the Direct Comparison Test.

39. $\lim_{n \to \infty} \frac{a_n}{b_n} = \infty \Rightarrow$ there exists an integer N such that for all $n > N$, $\frac{a_n}{b_n} > 1 \Rightarrow a_n > b_n$. If $\sum a_n$ converges,

then $\sum b_n$ converges by the Direct Comparison Test

11.5 THE RATIO AND ROOT TESTS

1. converges by the Ratio Test: $\lim\limits_{n \to \infty} \frac{a_{n+1}}{a_n} = \lim\limits_{n \to \infty} \dfrac{\left[\frac{(n+1)\sqrt{2}}{2^{n+1}}\right]}{\left[\frac{n\sqrt{2}}{2^n}\right]} = \lim\limits_{n \to \infty} \frac{(n+1)\sqrt{2}}{2^{n+1}} \cdot \frac{2^n}{n\sqrt{2}}$

    $= \lim\limits_{n \to \infty} \left(1 + \frac{1}{n}\right)^{\sqrt{2}} \left(\frac{1}{2}\right) = \frac{1}{2} < 1$

3. diverges by the Ratio Test: $\lim\limits_{n \to \infty} \frac{a_{n+1}}{a_n} = \lim\limits_{n \to \infty} \dfrac{\left(\frac{(n+1)!}{e^{n+1}}\right)}{\left(\frac{n!}{e^n}\right)} = \lim\limits_{n \to \infty} \frac{(n+1)!}{e^{n+1}} \cdot \frac{e^n}{n!} = \lim\limits_{n \to \infty} \frac{n+1}{e} = \infty$

5. converges by the Ratio Test: $\lim\limits_{n \to \infty} \frac{a_{n+1}}{a_n} = \lim\limits_{n \to \infty} \dfrac{\left(\frac{(n+1)^{10}}{10^{n+1}}\right)}{\left(\frac{n^{10}}{10^n}\right)} = \lim\limits_{n \to \infty} \frac{(n+1)^{10}}{10^{n+1}} \cdot \frac{10^n}{n^{10}} = \lim\limits_{n \to \infty} \left(1 + \frac{1}{n}\right)^{10} \left(\frac{1}{10}\right)$

    $= \frac{1}{10} < 1$

7. converges by the Direct Comparison Test: $\frac{2 + (-1)^n}{(1.25)^n} = \left(\frac{4}{5}\right)^n [2 + (-1)^n] \le \left(\frac{4}{5}\right)^n (3)$ which is the nth term of a convergent geometric series

9. diverges; $\lim\limits_{n \to \infty} a_n = \lim\limits_{n \to \infty} \left(1 - \frac{3}{n}\right)^n = \lim\limits_{n \to \infty} \left(1 + \frac{-3}{n}\right)^n = e^{-3} \approx 0.05 \neq 0$

11. converges by the Direct Comparison Test: $\frac{\ln n}{n^3} < \frac{n}{n^3} = \frac{1}{n^2}$ for $n \ge 2$, the nth term of a convergent p-series.

13. diverges by the Direct Comparison Test: $\frac{1}{n} - \frac{1}{n^2} = \frac{n-1}{n^2} > \frac{1}{2}\left(\frac{1}{n}\right)$ for $n > 2$ or by the Limit Comparison Test (part 1) with $\frac{1}{n}$.

15. diverges by the Direct Comparison Test: $\frac{\ln n}{n} > \frac{1}{n}$ for $n \ge 3$

17. converges by the Ratio Test: $\lim\limits_{n \to \infty} \frac{a_{n+1}}{a_n} = \lim\limits_{n \to \infty} \frac{(n+2)(n+3)}{(n+1)!} \cdot \frac{n!}{(n+1)(n+2)} = 0 < 1$

19. converges by the Ratio Test: $\lim\limits_{n \to \infty} \frac{a_{n+1}}{a_n} = \lim\limits_{n \to \infty} \frac{(n+4)!}{3!(n+1)!\,3^{n+1}} \cdot \frac{3!\,n!\,3^n}{(n+3)!} = \lim\limits_{n \to \infty} \frac{n+4}{3(n+1)} = \frac{1}{3} < 1$

21. converges by the Ratio Test: $\lim\limits_{n \to \infty} \frac{a_{n+1}}{a_n} = \lim\limits_{n \to \infty} \frac{(n+1)!}{(2n+3)!} \cdot \frac{(2n+1)!}{n!} = \lim\limits_{n \to \infty} \frac{n+1}{(2n+3)(2n+2)} = 0 < 1$

23. converges by the Root Test: $\lim\limits_{n \to \infty} \sqrt[n]{a_n} = \lim\limits_{n \to \infty} \sqrt[n]{\frac{n}{(\ln n)^n}} = \lim\limits_{n \to \infty} \frac{\sqrt[n]{n}}{\ln n} = \lim\limits_{n \to \infty} \frac{1}{\ln n} = 0 < 1$

25. converges by the Direct Comparison Test: $\frac{n!\ln n}{n(n+2)!} = \frac{\ln n}{n(n+1)(n+2)} < \frac{n}{n(n+1)(n+2)} = \frac{1}{(n+1)(n+2)} < \frac{1}{n^2}$

    which is the nth-term of a convergent p-series

27. converges by the Ratio Test: $\lim\limits_{n \to \infty} \frac{a_{n+1}}{a_n} = \lim\limits_{n \to \infty} \frac{\left(\frac{1 + \sin n}{n}\right)a_n}{a_n} = 0 < 1$

29. diverges by the Ratio Test: $\lim\limits_{n \to \infty} \frac{a_{n+1}}{a_n} = \lim\limits_{n \to \infty} \frac{\left(\frac{3n-1}{2n+1}\right)a_n}{a_n} = \lim\limits_{n \to \infty} \frac{3n-1}{2n+1} = \frac{3}{2} > 1$

31. converges by the Ratio Test: $\lim\limits_{n \to \infty} \frac{a_{n+1}}{a_n} = \lim\limits_{n \to \infty} \frac{\left(\frac{2}{n}\right)a_n}{a_n} = \lim\limits_{n \to \infty} \frac{2}{n} = 0 < 1$

33. converges by the Ratio Test: $\lim\limits_{n \to \infty} \frac{a_{n+1}}{a_n} = \lim\limits_{n \to \infty} \frac{\left(\frac{1+\ln n}{n}\right) a_n}{a_n} = \lim\limits_{n \to \infty} \frac{1+\ln n}{n} = \lim\limits_{n \to \infty} \frac{1}{n} = 0 < 1$

35. diverges by the nth-Term Test: $a_1 = \frac{1}{3}$, $a_2 = \sqrt[2]{\frac{1}{3}}$, $a_3 = \sqrt[3]{\sqrt[2]{\frac{1}{3}}} = \sqrt[6]{\frac{1}{3}}$, $a_4 = \sqrt[4]{\sqrt[3]{\sqrt[2]{\frac{1}{3}}}} = \sqrt[4!]{\frac{1}{3}}, \dots,$

$a_n = \sqrt[n!]{\frac{1}{3}} \Rightarrow \lim\limits_{n \to \infty} a_n = 1$ because $\left\{ \sqrt[n!]{\frac{1}{3}} \right\}$ is a subsequence of $\left\{ \sqrt[n]{\frac{1}{3}} \right\}$ whose limit is 1 by Table 8.1

37. converges by the Ratio Test: $\lim\limits_{n \to \infty} \frac{a_{n+1}}{a_n} = \lim\limits_{n \to \infty} \frac{2^{n+1}(n+1)!\,(n+1)!}{(2n+2)!} \cdot \frac{(2n)!}{2^n n!\, n!} = \lim\limits_{n \to \infty} \frac{2(n+1)(n+1)}{(2n+2)(2n+1)}$

$= \lim\limits_{n \to \infty} \frac{n+1}{2n+1} = \frac{1}{2} < 1$

39. diverges by the Root Test: $\lim\limits_{n \to \infty} \sqrt[n]{a_n} \equiv \lim\limits_{n \to \infty} \sqrt[n]{\frac{(n!)^n}{(n^n)^2}} = \lim\limits_{n \to \infty} \frac{n!}{n^2} = \infty > 1$

41. converges by the Root Test: $\lim\limits_{n \to \infty} \sqrt[n]{a_n} = \lim\limits_{n \to \infty} \sqrt[n]{\frac{n^n}{2^{n^2}}} = \lim\limits_{n \to \infty} \frac{n}{2^n} = \lim\limits_{n \to \infty} \frac{1}{2^n \ln 2} = 0 < 1$

43. converges by the Ratio Test: $\lim\limits_{n \to \infty} \frac{a_{n+1}}{a_n} = \lim\limits_{n \to \infty} \frac{1 \cdot 3 \cdots (2n-1)(2n+1)}{4^{n+1} 2^{n+1}(n+1)!} \cdot \frac{4^n 2^n n!}{1 \cdot 3 \cdots (2n-1)}$

$= \lim\limits_{n \to \infty} \frac{2n+1}{(4 \cdot 2)(n+1)} = \frac{1}{4} < 1$

45. Ratio: $\lim\limits_{n \to \infty} \frac{a_{n+1}}{a_n} = \lim\limits_{n \to \infty} \frac{1}{(n+1)^p} \cdot \frac{n^p}{1} = \lim\limits_{n \to \infty} \left(\frac{n}{n+1}\right)^p = 1^p = 1 \Rightarrow$ no conclusion

Root: $\lim\limits_{n \to \infty} \sqrt[n]{a_n} = \lim\limits_{n \to \infty} \sqrt[n]{\frac{1}{n^p}} = \lim\limits_{n \to \infty} \frac{1}{(\sqrt[n]{n})^p} = \frac{1}{(1)^p} = 1 \Rightarrow$ no conclusion

47. $a_n \le \frac{n}{2^n}$ for every n and the series $\sum\limits_{n=1}^{\infty} \frac{n}{2^n}$ converges by the Ratio Test since $\lim\limits_{n \to \infty} \frac{(n+1)}{2^{n+1}} \cdot \frac{2^n}{n} = \frac{1}{2} < 1$

$\Rightarrow \sum\limits_{n=1}^{\infty} a_n$ converges by the Direct Comparison Test

11.6 ALTERNATING SERIES, ABSOLUTE AND CONDITIONAL CONVERGENCE

1. converges absolutely $\Rightarrow$ converges by the Absolute Convergence Test since $\sum\limits_{n=1}^{\infty} |a_n| = \sum\limits_{n=1}^{\infty} \frac{1}{n^2}$ which is a convergent p-series

3. diverges by the nth-Term Test since for $n > 10 \Rightarrow \frac{n}{10} > 1 \Rightarrow \lim\limits_{n \to \infty} \left(\frac{n}{10}\right)^n \ne 0 \Rightarrow \sum\limits_{n=1}^{\infty} (-1)^{n+1} \left(\frac{n}{10}\right)^n$ diverges

5. converges by the Alternating Series Test because $f(x) = \ln x$ is an increasing function of $x \Rightarrow \frac{1}{\ln x}$ is decreasing
   $\Rightarrow u_n \ge u_{n+1}$ for $n \ge 1$; also $u_n \ge 0$ for $n \ge 1$ and $\lim\limits_{n \to \infty} \frac{1}{\ln n} = 0$

7. diverges by the nth-Term Test since $\lim\limits_{n \to \infty} \frac{\ln n}{\ln n^2} = \lim\limits_{n \to \infty} \frac{\ln n}{2 \ln n} = \lim\limits_{n \to \infty} \frac{1}{2} = \frac{1}{2} \ne 0$

9. converges by the Alternating Series Test since $f(x) = \frac{\sqrt{x}+1}{x+1} \Rightarrow f'(x) = \frac{1-x-2\sqrt{x}}{2\sqrt{x}\,(x+1)^2} < 0 \Rightarrow f(x)$ is decreasing
   $\Rightarrow u_n \ge u_{n+1}$; also $u_n \ge 0$ for $n \ge 1$ and $\lim\limits_{n \to \infty} u_n = \lim\limits_{n \to \infty} \frac{\sqrt{n}+1}{n+1} = 0$

11. converges absolutely since $\sum\limits_{n=1}^{\infty} |a_n| = \sum\limits_{n=1}^{\infty} \left(\frac{1}{10}\right)^n$ a convergent geometric series

13. converges conditionally since $\frac{1}{\sqrt{n}} > \frac{1}{\sqrt{n+1}} > 0$ and $\lim_{n \to \infty} \frac{1}{\sqrt{n}} = 0 \Rightarrow$ convergence; but $\sum_{n=1}^{\infty} |a_n| = \sum_{n=1}^{\infty} \frac{1}{n^{1/2}}$ is a divergent p-series

15. converges absolutely since $\sum_{n=1}^{\infty} |a_n| = \sum_{n=1}^{\infty} \frac{n}{n^3+1}$ and $\frac{n}{n^3+1} < \frac{1}{n^2}$ which is the nth-term of a converging p-series

17. converges conditionally since $\frac{1}{n+3} > \frac{1}{(n+1)+3} > 0$ and $\lim_{n \to \infty} \frac{1}{n+3} = 0 \Rightarrow$ convergence; but $\sum_{n=1}^{\infty} |a_n|$

$= \sum_{n=1}^{\infty} \frac{1}{n+3}$ diverges because $\frac{1}{n+3} \geq \frac{1}{4n}$ and $\sum_{n=1}^{\infty} \frac{1}{n}$ is a divergent series

19. diverges by the nth-Term Test since $\lim_{n \to \infty} \frac{3+n}{5+n} = 1 \neq 0$

21. converges conditionally since $f(x) = \frac{1}{x^2} + \frac{1}{x} \Rightarrow f'(x) = -\left(\frac{2}{x^3} + \frac{1}{x^2}\right) < 0 \Rightarrow f(x)$ is decreasing and hence

$u_n > u_{n+1} > 0$ for $n \geq 1$ and $\lim_{n \to \infty} \left(\frac{1}{n^2} + \frac{1}{n}\right) = 0 \Rightarrow$ convergence; but $\sum_{n=1}^{\infty} |a_n| = \sum_{n=1}^{\infty} \frac{1+n}{n^2}$

$= \sum_{n=1}^{\infty} \frac{1}{n^2} + \sum_{n=1}^{\infty} \frac{1}{n}$ is the sum of a convergent and divergent series, and hence diverges

23. converges absolutely by the Ratio Test: $\lim_{n \to \infty} \left(\frac{u_{n+1}}{u_n}\right) = \lim_{n \to \infty} \left[\frac{(n+1)^2 \left(\frac{2}{3}\right)^{n+1}}{n^2 \left(\frac{2}{3}\right)^n}\right] = \frac{2}{3} < 1$

25. converges absolutely by the Integral Test since $\int_1^{\infty} (\tan^{-1} x) \left(\frac{1}{1+x^2}\right) dx = \lim_{b \to \infty} \left[\frac{(\tan^{-1} x)^2}{2}\right]_1^b$

$= \lim_{b \to \infty} \left[(\tan^{-1} b)^2 - (\tan^{-1} 1)^2\right] = \frac{1}{2}\left[\left(\frac{\pi}{2}\right)^2 - \left(\frac{\pi}{4}\right)^2\right] = \frac{3\pi^2}{32}$

27. diverges by the nth-Term Test since $\lim_{n \to \infty} \frac{n}{n+1} = 1 \neq 0$

29. converges absolutely by the Ratio Test: $\lim_{n \to \infty} \left(\frac{u_{n+1}}{u_n}\right) = \lim_{n \to \infty} \frac{(100)^{n+1}}{(n+1)!} \cdot \frac{n!}{(100)^n} = \lim_{n \to \infty} \frac{100}{n+1} = 0 < 1$

31. converges absolutely by the Direct Comparison Test since $\sum_{n=1}^{\infty} |a_n| = \sum_{n=1}^{\infty} \frac{1}{n^2 + 2n + 1}$ and

$\frac{1}{n^2 + 2n + 1} < \frac{1}{n^2}$ which is the nth-term of a convergent p-series

33. converges absolutely since $\sum_{n=1}^{\infty} |a_n| = \sum_{n=1}^{\infty} \left|\frac{(-1)^n}{n\sqrt{n}}\right| = \sum_{n=1}^{\infty} \frac{1}{n^{3/2}}$ is a convergent p-series

35. converges absolutely by the Root Test: $\lim_{n \to \infty} \sqrt[n]{|a_n|} = \lim_{n \to \infty} \left(\frac{(n+1)^n}{(2n)^n}\right)^{1/n} = \lim_{n \to \infty} \frac{n+1}{2n} = \frac{1}{2} < 1$

37. diverges by the nth-Term Test since $\lim_{n \to \infty} |a_n| = \lim_{n \to \infty} \frac{(2n)!}{2^n n! \, n} = \lim_{n \to \infty} \frac{(n+1)(n+2)\cdots(2n)}{2^n}$

$= \lim_{n \to \infty} \frac{(n+1)(n+2)\cdots(n+(n-1))}{2^{n-1}} > \lim_{n \to \infty} \left(\frac{n+1}{2}\right)^{n-1} = \infty \neq 0$

39. converges conditionally since $\frac{\sqrt{n+1} - \sqrt{n}}{1} \cdot \frac{\sqrt{n+1} + \sqrt{n}}{\sqrt{n+1} + \sqrt{n}} = \frac{1}{\sqrt{n+1} + \sqrt{n}}$ and $\left\{\frac{1}{\sqrt{n+1} + \sqrt{n}}\right\}$ is a

decreasing sequence of positive terms which converges to $0 \Rightarrow \sum_{n=1}^{\infty} \frac{(-1)^n}{\sqrt{n+1} + \sqrt{n}}$ converges; but

$\sum_{n=1}^{\infty} |a_n| = \sum_{n=1}^{\infty} \frac{1}{\sqrt{n+1}+\sqrt{n}}$ diverges by the Limit Comparison Test (part 1) with $\frac{1}{\sqrt{n}}$; a divergent p-series:

$$\lim_{n \to \infty} \left(\frac{\frac{1}{\sqrt{n+1}+\sqrt{n}}}{\frac{1}{\sqrt{n}}} \right) = \lim_{n \to \infty} \frac{\sqrt{n}}{\sqrt{n+1}+\sqrt{n}} = \lim_{n \to \infty} \frac{1}{\sqrt{1+\frac{1}{n}}+1} = \frac{1}{2}$$

41. diverges by the nth-Term Test since $\lim_{n \to \infty} \left(\sqrt{n+\sqrt{n}} - \sqrt{n} \right) = \lim_{n \to \infty} \left[\left(\sqrt{n+\sqrt{n}} - \sqrt{n} \right) \left(\frac{\sqrt{n+\sqrt{n}}+\sqrt{n}}{\sqrt{n+\sqrt{n}}+\sqrt{n}} \right) \right]$

$= \lim_{n \to \infty} \frac{\sqrt{n}}{\sqrt{n+\sqrt{n}}+\sqrt{n}} = \lim_{n \to \infty} \frac{1}{\sqrt{1+\frac{1}{\sqrt{n}}}+1} = \frac{1}{2} \neq 0$

43. converges absolutely by the Direct Comparison Test since $\text{sech}(n) = \frac{2}{e^n + e^{-n}} = \frac{2e^n}{e^{2n}+1} < \frac{2e^n}{e^{2n}} = \frac{2}{e^n}$ which is the nth term of a convergent geometric series

45. $|\text{error}| < \left| (-1)^6 \left(\frac{1}{5} \right) \right| = 0.2$

47. $|\text{error}| < \left| (-1)^6 \frac{(0.01)^5}{5} \right| = 2 \times 10^{-11}$

49. $\frac{1}{(2n)!} < \frac{5}{10^6} \Rightarrow (2n)! > \frac{10^6}{5} = 200{,}000 \Rightarrow n \geq 5 \Rightarrow 1 - \frac{1}{2!} + \frac{1}{4!} - \frac{1}{6!} + \frac{1}{8!} \approx 0.54030$

51. (a) $a_n \geq a_{n+1}$ fails since $\frac{1}{3} < \frac{1}{2}$

(b) Since $\sum_{n=1}^{\infty} |a_n| = \sum_{n=1}^{\infty} \left[\left(\frac{1}{3} \right)^n + \left(\frac{1}{2} \right)^n \right] = \sum_{n=1}^{\infty} \left(\frac{1}{3} \right)^n + \sum_{n=1}^{\infty} \left(\frac{1}{2} \right)^n$ is the sum of two absolutely convergent

series, we can rearrange the terms of the original series to find its sum:

$$\left(\frac{1}{3} + \frac{1}{9} + \frac{1}{27} + \dots \right) - \left(\frac{1}{2} + \frac{1}{4} + \frac{1}{8} + \dots \right) = \frac{\left(\frac{1}{3} \right)}{1 - \left(\frac{1}{3} \right)} - \frac{\left(\frac{1}{2} \right)}{1 - \left(\frac{1}{2} \right)} = \frac{1}{2} - 1 = -\frac{1}{2}$$

53. The unused terms are $\sum_{j=n+1}^{\infty} (-1)^{j+1} a_j = (-1)^{n+1} \left(a_{n+1} - a_{n+2} \right) + (-1)^{n+3} \left(a_{n+3} - a_{n+4} \right) + \dots$

$= (-1)^{n+1} \left[\left(a_{n+1} - a_{n+2} \right) + \left(a_{n+3} - a_{n+4} \right) + \dots \right]$. Each grouped term is positive, so the remainder has the same sign as $(-1)^{n+1}$, which is the sign of the first unused term.

55. Theorem 16 states that $\sum_{n=1}^{\infty} |a_n|$ converges $\Rightarrow \sum_{n=1}^{\infty} a_n$ converges. But this is equivalent to $\sum_{n=1}^{\infty} a_n$ diverges $\Rightarrow \sum_{n=1}^{\infty} |a_n|$ diverges.

57. (a) $\sum_{n=1}^{\infty} |a_n + b_n|$ converges by the Direct Comparison Test since $|a_n + b_n| \leq |a_n| + |b_n|$ and hence

$\sum_{n=1}^{\infty} (a_n + b_n)$ converges absolutely

(b) $\sum_{n=1}^{\infty} |b_n|$ converges $\Rightarrow \sum_{n=1}^{\infty} -b_n$ converges absolutely; since $\sum_{n=1}^{\infty} a_n$ converges absolutely and

$\sum_{n=1}^{\infty} -b_n$ converges absolutely, we have $\sum_{n=1}^{\infty} [a_n + (-b_n)] = \sum_{n=1}^{\infty} (a_n - b_n)$ converges absolutely by part (a)

(c) $\sum_{n=1}^{\infty} |a_n|$ converges $\Rightarrow |k| \sum_{n=1}^{\infty} |a_n| = \sum_{n=1}^{\infty} |ka_n|$ converges $\Rightarrow \sum_{n=1}^{\infty} ka_n$ converges absolutely

59. $s_1 = -\frac{1}{2}$, $s_2 = -\frac{1}{2} + 1 = \frac{1}{2}$,

$s_3 = -\frac{1}{2} + 1 - \frac{1}{4} - \frac{1}{6} - \frac{1}{8} - \frac{1}{10} - \frac{1}{12} - \frac{1}{14} - \frac{1}{16} - \frac{1}{18} - \frac{1}{20} - \frac{1}{22} \approx -0.5099,$

$s_4 = s_3 + \frac{1}{3} \approx -0.1766,$

$s_5 = s_4 - \frac{1}{24} - \frac{1}{26} - \frac{1}{28} - \frac{1}{30} - \frac{1}{32} - \frac{1}{34} - \frac{1}{36} - \frac{1}{38} - \frac{1}{40} - \frac{1}{42} - \frac{1}{44} \approx -0.512,$

$s_6 = s_5 + \frac{1}{5} \approx -0.312,$

$s_7 = s_6 - \frac{1}{46} - \frac{1}{48} - \frac{1}{50} - \frac{1}{52} - \frac{1}{54} - \frac{1}{56} - \frac{1}{58} - \frac{1}{60} - \frac{1}{62} - \frac{1}{64} - \frac{1}{66} \approx -0.51106$

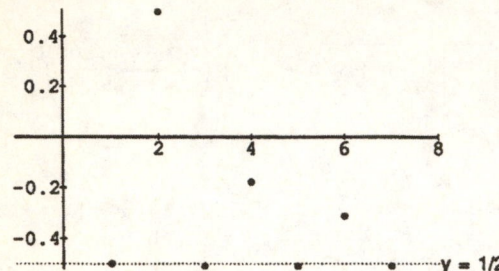

61. (a) If $\sum_{n=1}^{\infty} |a_n|$ converges, then $\sum_{n=1}^{\infty} a_n$ converges and $\frac{1}{2}\sum_{n=1}^{\infty} a_n + \frac{1}{2}\sum_{n=1}^{\infty} |a_n| = \sum_{n=1}^{\infty} \frac{a_n + |a_n|}{2}$

   converges where $b_n = \frac{a_n + |a_n|}{2} = \begin{cases} a_n, & \text{if } a_n \geq 0 \\ 0, & \text{if } a_n < 0 \end{cases}.$

   (b) If $\sum_{n=1}^{\infty} |a_n|$ converges, then $\sum_{n=1}^{\infty} a_n$ converges and $\frac{1}{2}\sum_{n=1}^{\infty} a_n - \frac{1}{2}\sum_{n=1}^{\infty} |a_n| = \sum_{n=1}^{\infty} \frac{a_n - |a_n|}{2}$

   converges where $c_n = \frac{a_n - |a_n|}{2} = \begin{cases} 0, & \text{if } a_n \geq 0 \\ a_n, & \text{if } a_n < 0 \end{cases}.$

63. Here is an example figure when $N = 5$. Notice that $u_3 > u_2 > u_1$ and $u_3 > u_5 > u_4$, but $u_n \geq u_{n+1}$ for $n \geq 5$.

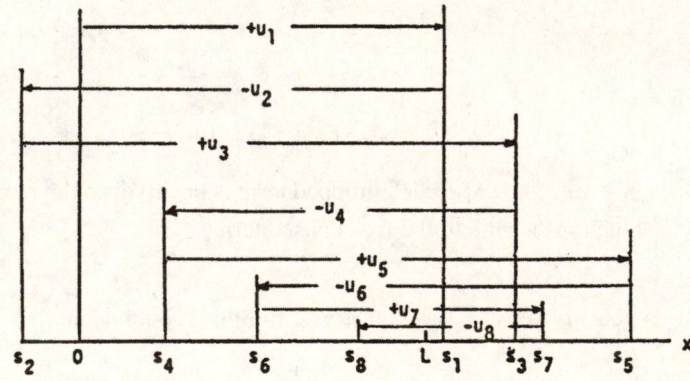

11.7 POWER SERIES

1. $\lim_{n \to \infty} \left|\frac{u_{n+1}}{u_n}\right| < 1 \Rightarrow \lim_{n \to \infty} \left|\frac{x^{n+1}}{x^n}\right| < 1 \Rightarrow |x| < 1 \Rightarrow -1 < x < 1$; when $x = -1$ we have $\sum_{n=1}^{\infty} (-1)^n$, a divergent

   series; when $x = 1$ we have $\sum_{n=1}^{\infty} 1$, a divergent series

   (a) the radius is 1; the interval of convergence is $-1 < x < 1$
   (b) the interval of absolute convergence is $-1 < x < 1$
   (c) there are no values for which the series converges conditionally

3. $\lim\limits_{n \to \infty} \left| \dfrac{u_{n+1}}{u_n} \right| < 1 \Rightarrow \lim\limits_{n \to \infty} \left| \dfrac{(4x+1)^{n+1}}{(4x+1)^n} \right| < 1 \Rightarrow |4x+1| < 1 \Rightarrow -1 < 4x+1 < 1 \Rightarrow -\frac{1}{2} < x < 0$; when $x = -\frac{1}{2}$ we

have $\sum\limits_{n=1}^{\infty} (-1)^n(-1)^n = \sum\limits_{n=1}^{\infty} (-1)^{2n} = \sum\limits_{n=1}^{\infty} 1^n$, a divergent series; when $x = 0$ we have $\sum\limits_{n=1}^{\infty} (-1)^n(1)^n$

$= \sum\limits_{n=1}^{\infty} (-1)^n$, a divergent series

(a) the radius is $\frac{1}{4}$; the interval of convergence is $-\frac{1}{2} < x < 0$

(b) the interval of absolute convergence is $-\frac{1}{2} < x < 0$

(c) there are no values for which the series converges conditionally

5. $\lim\limits_{n \to \infty} \left| \dfrac{u_{n+1}}{u_n} \right| < 1 \Rightarrow \lim\limits_{n \to \infty} \left| \dfrac{(x-2)^{n+1}}{10^{n+1}} \cdot \dfrac{10^n}{(x-2)^n} \right| < 1 \Rightarrow \dfrac{|x-2|}{10} < 1 \Rightarrow |x-2| < 10 \Rightarrow -10 < x-2 < 10$

$\Rightarrow -8 < x < 12$; when $x = -8$ we have $\sum\limits_{n=1}^{\infty} (-1)^n$, a divergent series; when $x = 12$ we have $\sum\limits_{n=1}^{\infty} 1$, a divergent

series

(a) the radius is 10; the interval of convergence is $-8 < x < 12$

(b) the interval of absolute convergence is $-8 < x < 12$

(c) there are no values for which the series converges conditionally

7. $\lim\limits_{n \to \infty} \left| \dfrac{u_{n+1}}{u_n} \right| < 1 \Rightarrow \lim\limits_{n \to \infty} \left| \dfrac{(n+1)x^{n+1}}{(n+3)} \cdot \dfrac{(n+2)}{nx^n} \right| < 1 \Rightarrow |x| \lim\limits_{n \to \infty} \dfrac{(n+1)(n+2)}{(n+3)(n)} < 1 \Rightarrow |x| < 1$

$\Rightarrow -1 < x < 1$; when $x = -1$ we have $\sum\limits_{n=1}^{\infty} (-1)^n \dfrac{n}{n+2}$, a divergent series by the nth-term Test; when $x = 1$ we

have $\sum\limits_{n=1}^{\infty} \dfrac{n}{n+2}$, a divergent series

(a) the radius is 1; the interval of convergence is $-1 < x < 1$

(b) the interval of absolute convergence is $-1 < x < 1$

(c) there are no values for which the series converges conditionally

9. $\lim\limits_{n \to \infty} \left| \dfrac{u_{n+1}}{u_n} \right| < 1 \Rightarrow \lim\limits_{n \to \infty} \left| \dfrac{x^{n+1}}{(n+1)\sqrt{n+1}\,3^{n+1}} \cdot \dfrac{n\sqrt{n}\,3^n}{x^n} \right| < 1 \Rightarrow \dfrac{|x|}{3} \left(\lim\limits_{n \to \infty} \dfrac{n}{n+1} \right) \left(\sqrt{\lim\limits_{n \to \infty} \dfrac{n}{n+1}} \right) < 1$

$\Rightarrow \dfrac{|x|}{3}(1)(1) < 1 \Rightarrow |x| < 3 \Rightarrow -3 < x < 3$; when $x = -3$ we have $\sum\limits_{n=1}^{\infty} \dfrac{(-1)^n}{n^{3/2}}$, an absolutely convergent series;

when $x = 3$ we have $\sum\limits_{n=1}^{\infty} \dfrac{1}{n^{3/2}}$, a convergent p-series

(a) the radius is 3; the interval of convergence is $-3 \le x \le 3$

(b) the interval of absolute convergence is $-3 \le x \le 3$

(c) there are no values for which the series converges conditionally

11. $\lim\limits_{n \to \infty} \left| \dfrac{u_{n+1}}{u_n} \right| < 1 \Rightarrow \lim\limits_{n \to \infty} \left| \dfrac{x^{n+1}}{(n+1)!} \cdot \dfrac{n!}{x^n} \right| < 1 \Rightarrow |x| \lim\limits_{n \to \infty} \left(\dfrac{1}{n+1} \right) < 1$ for all x

(a) the radius is ∞; the series converges for all x

(b) the series converges absolutely for all x

(c) there are no values for which the series converges conditionally

13. $\lim\limits_{n \to \infty} \left| \dfrac{u_{n+1}}{u_n} \right| < 1 \Rightarrow \lim\limits_{n \to \infty} \left| \dfrac{x^{2n+3}}{(n+1)!} \cdot \dfrac{n!}{x^{2n+1}} \right| < 1 \Rightarrow x^2 \lim\limits_{n \to \infty} \left(\dfrac{1}{n+1} \right) < 1$ for all x

(a) the radius is ∞; the series converges for all x

(b) the series converges absolutely for all x

(c) there are no values for which the series converges conditionally

15. $\lim\limits_{n \to \infty} \left| \frac{u_{n+1}}{u_n} \right| < 1 \Rightarrow \lim\limits_{n \to \infty} \left| \frac{x^{n+1}}{\sqrt{(n+1)^2 + 3}} \cdot \frac{\sqrt{n^2 + 3}}{x^n} \right| < 1 \Rightarrow |x| \sqrt{\lim\limits_{n \to \infty} \frac{n^2 + 3}{n^2 + 2n + 4}} < 1 \Rightarrow |x| < 1$

$\Rightarrow -1 < x < 1$; when $x = -1$ we have $\sum\limits_{n=1}^{\infty} \frac{(-1)^n}{\sqrt{n^2 + 3}}$, a conditionally convergent series; when $x = 1$ we have

$\sum\limits_{n=1}^{\infty} \frac{1}{\sqrt{n^2 + 3}}$, a divergent series

(a) the radius is 1; the interval of convergence is $-1 \le x < 1$

(b) the interval of absolute convergence is $-1 < x < 1$

(c) the series converges conditionally at $x = -1$

17. $\lim\limits_{n \to \infty} \left| \frac{u_{n+1}}{u_n} \right| < 1 \Rightarrow \lim\limits_{n \to \infty} \left| \frac{(n+1)(x+3)^{n+1}}{5^{n+1}} \cdot \frac{5^n}{n(x+3)^n} \right| < 1 \Rightarrow \frac{|x+3|}{5} \lim\limits_{n \to \infty} \left(\frac{n+1}{n} \right) < 1 \Rightarrow \frac{|x+3|}{5} < 1$

$\Rightarrow |x+3| < 5 \Rightarrow -5 < x + 3 < 5 \Rightarrow -8 < x < 2$; when $x = -8$ we have $\sum\limits_{n=1}^{\infty} \frac{n(-5)^n}{5^n} = \sum\limits_{n=1}^{\infty} (-1)^n n$, a divergent

series; when $x = 2$ we have $\sum\limits_{n=1}^{\infty} \frac{n5^n}{5^n} = \sum\limits_{n=1}^{\infty} n$, a divergent series

(a) the radius is 5; the interval of convergence is $-8 < x < 2$

(b) the interval of absolute convergence is $-8 < x < 2$

(c) there are no values for which the series converges conditionally

19. $\lim\limits_{n \to \infty} \left| \frac{u_{n+1}}{u_n} \right| < 1 \Rightarrow \lim\limits_{n \to \infty} \left| \frac{\sqrt{n+1}\, x^{n+1}}{3^{n+1}} \cdot \frac{3^n}{\sqrt{n}\, x^n} \right| < 1 \Rightarrow \frac{|x|}{3} \sqrt{\lim\limits_{n \to \infty} \left(\frac{n+1}{n} \right)} < 1 \Rightarrow \frac{|x|}{3} < 1 \Rightarrow |x| < 3$

$\Rightarrow -3 < x < 3$; when $x = -3$ we have $\sum\limits_{n=1}^{\infty} (-1)^n \sqrt{n}$, a divergent series; when $x = 3$ we have

$\sum\limits_{n=1}^{\infty} \sqrt{n}$, a divergent series

(a) the radius is 3; the interval of convergence is $-3 < x < 3$

(b) the interval of absolute convergence is $-3 < x < 3$

(c) there are no values for which the series converges conditionally

21. $\lim\limits_{n \to \infty} \left| \frac{u_{n+1}}{u_n} \right| < 1 \Rightarrow \lim\limits_{n \to \infty} \left| \frac{\left(1 + \frac{1}{n+1} \right)^{n+1} x^{n+1}}{\left(1 + \frac{1}{n} \right)^n x^n} \right| < 1 \Rightarrow |x| \left(\frac{\lim\limits_{t \to \infty} \left(1 + \frac{1}{t} \right)^t}{\lim\limits_{n \to \infty} \left(1 + \frac{1}{n} \right)^n} \right) < 1 \Rightarrow |x| \left(\frac{e}{e} \right) < 1 \Rightarrow |x| < 1$

$\Rightarrow -1 < x < 1$; when $x = -1$ we have $\sum\limits_{n=1}^{\infty} (-1)^n \left(1 + \frac{1}{n} \right)^n$, a divergent series by the nth-Term Test since

$\lim\limits_{n \to \infty} \left(1 + \frac{1}{n} \right)^n = e \ne 0$; when $x = 1$ we have $\sum\limits_{n=1}^{\infty} \left(1 + \frac{1}{n} \right)^n$, a divergent series

(a) the radius is 1; the interval of convergence is $-1 < x < 1$

(b) the interval of absolute convergence is $-1 < x < 1$

(c) there are no values for which the series converges conditionally

23. $\lim\limits_{n \to \infty} \left| \frac{u_{n+1}}{u_n} \right| < 1 \Rightarrow \lim\limits_{n \to \infty} \left| \frac{(n+1)^{n+1} x^{n+1}}{n^n x^n} \right| < 1 \Rightarrow |x| \left(\lim\limits_{n \to \infty} \left(1 + \frac{1}{n} \right)^n \right) \left(\lim\limits_{n \to \infty} (n+1) \right) < 1$

$\Rightarrow e|x| \lim\limits_{n \to \infty} (n+1) < 1 \Rightarrow$ only $x = 0$ satisfies this inequality

(a) the radius is 0; the series converges only for $x = 0$

(b) the series converges absolutely only for $x = 0$

(c) there are no values for which the series converges conditionally

25. $\lim\limits_{n \to \infty} \left| \frac{u_{n+1}}{u_n} \right| < 1 \Rightarrow \lim\limits_{n \to \infty} \left| \frac{(x+2)^{n+1}}{(n+1)\, 2^{n+1}} \cdot \frac{n 2^n}{(x+2)^n} \right| < 1 \Rightarrow \frac{|x+2|}{2} \lim\limits_{n \to \infty} \left(\frac{n}{n+1} \right) < 1 \Rightarrow \frac{|x+2|}{2} < 1 \Rightarrow |x+2| < 2$

$\Rightarrow -2 < x + 2 < 2 \Rightarrow -4 < x < 0$; when $x = -4$ we have $\sum\limits_{n=1}^{\infty} \frac{-1}{n}$, a divergent series; when $x = 0$ we have

$\sum_{n=1}^{\infty} \frac{(-1)^{n+1}}{n}$, the alternating harmonic series which converges conditionally

(a) the radius is 2; the interval of convergence is $-4 < x \leq 0$

(b) the interval of absolute convergence is $-4 < x < 0$

(c) the series converges conditionally at $x = 0$

27. $\lim_{n \to \infty} \left| \frac{u_{n+1}}{u_n} \right| < 1 \Rightarrow \lim_{n \to \infty} \left| \frac{x^{n+1}}{(n+1)(\ln(n+1))^2} \cdot \frac{n(\ln n)^2}{x^n} \right| < 1 \Rightarrow |x| \left(\lim_{n \to \infty} \frac{n}{n+1} \right) \left(\lim_{n \to \infty} \frac{\ln n}{\ln(n+1)} \right)^2 < 1$

$\Rightarrow |x|(1) \left(\lim_{n \to \infty} \frac{\left(\frac{1}{n}\right)}{\left(\frac{1}{n+1}\right)} \right)^2 < 1 \Rightarrow |x| \left(\lim_{n \to \infty} \frac{n+1}{n} \right)^2 < 1 \Rightarrow |x| < 1 \Rightarrow -1 < x < 1$; when $x = -1$ we have

$\sum_{n=1}^{\infty} \frac{(-1)^n}{n(\ln n)^2}$ which converges absolutely; when $x = 1$ we have $\sum_{n=1}^{\infty} \frac{1}{n(\ln n)^2}$ which converges

(a) the radius is 1; the interval of convergence is $-1 \leq x \leq 1$

(b) the interval of absolute convergence is $-1 \leq x \leq 1$

(c) there are no values for which the series converges conditionally

29. $\lim_{n \to \infty} \left| \frac{u_{n+1}}{u_n} \right| < 1 \Rightarrow \lim_{n \to \infty} \left| \frac{(4x-5)^{2n+3}}{(n+1)^{3/2}} \cdot \frac{n^{3/2}}{(4x-5)^{2n+1}} \right| < 1 \Rightarrow (4x-5)^2 \left(\lim_{n \to \infty} \frac{n}{n+1} \right)^{3/2} < 1 \Rightarrow (4x-5)^2 < 1$

$\Rightarrow |4x-5| < 1 \Rightarrow -1 < 4x - 5 < 1 \Rightarrow 1 < x < \frac{3}{2}$; when $x = 1$ we have $\sum_{n=1}^{\infty} \frac{(-1)^{2n+1}}{n^{3/2}} = \sum_{n=1}^{\infty} \frac{-1}{n^{3/2}}$ which is

absolutely convergent; when $x = \frac{3}{2}$ we have $\sum_{n=1}^{\infty} \frac{(1)^{2n+1}}{n^{3/2}}$, a convergent p-series

(a) the radius is $\frac{1}{4}$; the interval of convergence is $1 \leq x \leq \frac{3}{2}$

(b) the interval of absolute convergence is $1 \leq x \leq \frac{3}{2}$

(c) there are no values for which the series converges conditionally

31. $\lim_{n \to \infty} \left| \frac{u_{n+1}}{u_n} \right| < 1 \Rightarrow \lim_{n \to \infty} \left| \frac{(x+\pi)^{n+1}}{\sqrt{n+1}} \cdot \frac{\sqrt{n}}{(x+\pi)^n} \right| < 1 \Rightarrow |x+\pi| \lim_{n \to \infty} \left| \sqrt{\frac{n}{n+1}} \right| < 1$

$\Rightarrow |x+\pi| \sqrt{\lim_{n \to \infty} \left(\frac{n}{n+1} \right)} < 1 \Rightarrow |x+\pi| < 1 \Rightarrow -1 < x + \pi < 1 \Rightarrow -1 - \pi < x < 1 - \pi$;

when $x = -1 - \pi$ we have $\sum_{n=1}^{\infty} \frac{(-1)^n}{\sqrt{n}} = \sum_{n=1}^{\infty} \frac{(-1)^n}{n^{1/2}}$, a conditionally convergent series; when $x = 1 - \pi$ we have

$\sum_{n=1}^{\infty} \frac{1^n}{\sqrt{n}} = \sum_{n=1}^{\infty} \frac{1}{n^{1/2}}$, a divergent p-series

(a) the radius is 1; the interval of convergence is $(-1 - \pi) \leq x < (1 - \pi)$

(b) the interval of absolute convergence is $-1 - \pi < x < 1 - \pi$

(c) the series converges conditionally at $x = -1 - \pi$

33. $\lim_{n \to \infty} \left| \frac{u_{n+1}}{u_n} \right| < 1 \Rightarrow \lim_{n \to \infty} \left| \frac{(x-1)^{2n+2}}{4^{n+1}} \cdot \frac{4^n}{(x-1)^{2n}} \right| < 1 \Rightarrow \frac{(x-1)^2}{4} \lim_{n \to \infty} |1| < 1 \Rightarrow (x-1)^2 < 4 \Rightarrow |x-1| < 2$

$\Rightarrow -2 < x - 1 < 2 \Rightarrow -1 < x < 3$; at $x = -1$ we have $\sum_{n=0}^{\infty} \frac{(-2)^{2n}}{4^n} = \sum_{n=0}^{\infty} \frac{4^n}{4^n} = \sum_{n=0}^{\infty} 1$, which diverges; at $x = 3$

we have $\sum_{n=0}^{\infty} \frac{2^{2n}}{4^n} = \sum_{n=0}^{\infty} \frac{4^n}{4^n} = \sum_{n=0}^{\infty} 1$, a divergent series; the interval of convergence is $-1 < x < 3$; the series

$\sum_{n=0}^{\infty} \frac{(x-1)^{2n}}{4^n} = \sum_{n=0}^{\infty} \left(\left(\frac{x-1}{2} \right)^2 \right)^n$ is a convergent geometric series when $-1 < x < 3$ and the sum is

$\frac{1}{1 - \left(\frac{x-1}{2} \right)^2} = \frac{1}{\left[\frac{4 - (x-1)^2}{4} \right]} = \frac{4}{4 - x^2 + 2x - 1} = \frac{4}{3 + 2x - x^2}$

35. $\lim\limits_{n \to \infty} \left| \frac{u_{n+1}}{u_n} \right| < 1 \Rightarrow \lim\limits_{n \to \infty} \left| \frac{(\sqrt{x}-2)^{n+1}}{2^{n+1}} \cdot \frac{2^n}{(\sqrt{x}-2)^n} \right| < 1 \Rightarrow \left| \sqrt{x} - 2 \right| < 2 \Rightarrow -2 < \sqrt{x} - 2 < 2 \Rightarrow 0 < \sqrt{x} < 4$

$\Rightarrow 0 < x < 16$; when $x = 0$ we have $\sum\limits_{n=0}^{\infty} (-1)^n$, a divergent series; when $x = 16$ we have $\sum\limits_{n=0}^{\infty} (1)^n$, a divergent

series; the interval of convergence is $0 < x < 16$; the series $\sum\limits_{n=0}^{\infty} \left(\frac{\sqrt{x}-2}{2} \right)^n$ is a convergent geometric series when

$0 < x < 16$ and its sum is $\frac{1}{1 - \left(\frac{\sqrt{x}-2}{2} \right)} = \frac{1}{\left(\frac{2-\sqrt{x}+2}{2} \right)} = \frac{2}{4 - \sqrt{x}}$

37. $\lim\limits_{n \to \infty} \left| \frac{u_{n+1}}{u_n} \right| < 1 \Rightarrow \lim\limits_{n \to \infty} \left| \left(\frac{x^2+1}{3} \right)^{n+1} \cdot \left(\frac{3}{x^2+1} \right)^n \right| < 1 \Rightarrow \frac{(x^2+1)}{3} \lim\limits_{n \to \infty} |1| < 1 \Rightarrow \frac{x^2+1}{3} < 1 \Rightarrow x^2 < 2$

$\Rightarrow |x| < \sqrt{2} \Rightarrow -\sqrt{2} < x < \sqrt{2}$; at $x = \pm\sqrt{2}$ we have $\sum\limits_{n=0}^{\infty} (1)^n$ which diverges; the interval of convergence is

$-\sqrt{2} < x < \sqrt{2}$; the series $\sum\limits_{n=0}^{\infty} \left(\frac{x^2+1}{3} \right)^n$ is a convergent geometric series when $-\sqrt{2} < x < \sqrt{2}$ and its sum is

$\frac{1}{1 - \left(\frac{x^2+1}{3} \right)} = \frac{1}{\left(\frac{3-x^2-1}{3} \right)} = \frac{3}{2-x^2}$

39. $\lim\limits_{n \to \infty} \left| \frac{(x-3)^{n+1}}{2^{n+1}} \cdot \frac{2^n}{(x-3)^n} \right| < 1 \Rightarrow |x - 3| < 2 \Rightarrow 1 < x < 5$; when $x = 1$ we have $\sum\limits_{n=1}^{\infty} (1)^n$ which diverges;

when $x = 5$ we have $\sum\limits_{n=1}^{\infty}(-1)^n$ which also diverges; the interval of convergence is $1 < x < 5$; the sum of this

convergent geometric series is $\frac{1}{1 + \left(\frac{x-3}{2} \right)} = \frac{2}{x-1}$. If $f(x) = 1 - \frac{1}{2}(x-3) + \frac{1}{4}(x-3)^2 + \ldots + \left(-\frac{1}{2} \right)^n (x-3)^n + \ldots$

$= \frac{2}{x-1}$ then $f'(x) = -\frac{1}{2} + \frac{1}{2}(x-3) + \ldots + \left(-\frac{1}{2} \right)^n n(x-3)^{n-1} + \ldots$ is convergent when $1 < x < 5$, and diverges

when $x = 1$ or 5. The sum for $f'(x)$ is $\frac{-2}{(x-1)^2}$, the derivative of $\frac{2}{x-1}$.

41. (a) Differentiate the series for $\sin x$ to get $\cos x = 1 - \frac{3x^2}{3!} + \frac{5x^4}{5!} - \frac{7x^6}{7!} + \frac{9x^8}{9!} - \frac{11x^{10}}{11!} + \ldots$

$= 1 - \frac{x^2}{2!} + \frac{x^4}{4!} - \frac{x^6}{6!} + \frac{x^8}{8!} - \frac{x^{10}}{10!} + \ldots$. The series converges for all values of x since

$\lim\limits_{n \to \infty} \left| \frac{x^{2n+2}}{(2n+2)!} \cdot \frac{(2n)!}{x^{2n}} \right| = x^2 \lim\limits_{n \to \infty} \left(\frac{1}{(2n+1)(2n+2)} \right) = 0 < 1$ for all x.

(b) $\sin 2x = 2x - \frac{2^3 x^3}{3!} + \frac{2^5 x^5}{5!} - \frac{2^7 x^7}{7!} + \frac{2^9 x^9}{9!} - \frac{2^{11} x^{11}}{11!} + \ldots = 2x - \frac{8x^3}{3!} + \frac{32x^5}{5!} - \frac{128x^7}{7!} + \frac{512x^9}{9!} - \frac{2048x^{11}}{11!} + \ldots$

(c) $2 \sin x \cos x = 2 \left[(0 \cdot 1) + (0 \cdot 0 + 1 \cdot 1)x + \left(0 \cdot \frac{-1}{2} + 1 \cdot 0 + 0 \cdot 1 \right) x^2 + \left(0 \cdot 0 - 1 \cdot \frac{1}{2} + 0 \cdot 0 - 1 \cdot \frac{1}{3!} \right) x^3 \right.$

$+ \left(0 \cdot \frac{1}{4!} + 1 \cdot 0 - 0 \cdot \frac{1}{2} - 0 \cdot \frac{1}{3!} + 0 \cdot 1 \right) x^4 + \left(0 \cdot 0 + 1 \cdot \frac{1}{4!} + 0 \cdot 0 + \frac{1}{2} \cdot \frac{1}{3!} + 0 \cdot 0 + 1 \cdot \frac{1}{5!} \right) x^5$

$+ \left. \left(0 \cdot \frac{1}{6!} + 1 \cdot 0 + 0 \cdot \frac{1}{4!} + 0 \cdot \frac{1}{3!} + 0 \cdot \frac{1}{2} + 0 \cdot \frac{1}{5!} + 0 \cdot 1 \right) x^6 + \ldots \right] = 2 \left[x - \frac{4x^3}{3!} + \frac{16x^5}{5!} - \ldots \right]$

$= 2x - \frac{2^3 x^3}{3!} + \frac{2^5 x^5}{5!} - \frac{2^7 x^7}{7!} + \frac{2^9 x^9}{9!} - \frac{2^{11} x^{11}}{11!} + \ldots$

43. (a) $\ln|\sec x| + C = \int \tan x \, dx = \int \left(x + \frac{x^3}{3} + \frac{2x^5}{15} + \frac{17x^7}{315} + \frac{62x^9}{2835} + \ldots \right) dx$

$= \frac{x^2}{2} + \frac{x^4}{12} + \frac{x^6}{45} + \frac{17x^8}{2520} + \frac{31x^{10}}{14,175} + \ldots + C; x = 0 \Rightarrow C = 0 \Rightarrow \ln|\sec x| = \frac{x^2}{2} + \frac{x^4}{12} + \frac{x^6}{45} + \frac{17x^8}{2520} + \frac{31x^{10}}{14,175} + \ldots,$

converges when $-\frac{\pi}{2} < x < \frac{\pi}{2}$

(b) $\sec^2 x = \frac{d(\tan x)}{dx} = \frac{d}{dx} \left(x + \frac{x^3}{3} + \frac{2x^5}{15} + \frac{17x^7}{315} + \frac{62x^9}{2835} + \ldots \right) = 1 + x^2 + \frac{2x^4}{3} + \frac{17x^6}{45} + \frac{62x^8}{315} + \ldots,$ converges

when $-\frac{\pi}{2} < x < \frac{\pi}{2}$

(c) $\sec^2 x = (\sec x)(\sec x) = \left(1 + \frac{x^2}{2} + \frac{5x^4}{24} + \frac{61x^6}{720} + \ldots \right) \left(1 + \frac{x^2}{2} + \frac{5x^4}{24} + \frac{61x^6}{720} + \ldots \right)$

$= 1 + \left(\frac{1}{2} + \frac{1}{2} \right) x^2 + \left(\frac{5}{24} + \frac{1}{4} + \frac{5}{24} \right) x^4 + \left(\frac{61}{720} + \frac{5}{48} + \frac{5}{48} + \frac{61}{720} \right) x^6 + \ldots$

$= 1 + x^2 + \frac{2x^4}{3} + \frac{17x^6}{45} + \frac{62x^8}{315} + \ldots, -\frac{\pi}{2} < x < \frac{\pi}{2}$

45. (a) If $f(x) = \sum_{n=0}^{\infty} a_n x^n$, then $f^{(k)}(x) = \sum_{n=k}^{\infty} n(n-1)(n-2)\cdots(n-(k-1)) a_n x^{n-k}$ and $f^{(k)}(0) = k! a_k$

$\Rightarrow a_k = \frac{f^{(k)}(0)}{k!}$; likewise if $f(x) = \sum_{n=0}^{\infty} b_n x^n$, then $b_k = \frac{f^{(k)}(0)}{k!} \Rightarrow a_k = b_k$ for every nonnegative integer k

(b) If $f(x) = \sum_{n=0}^{\infty} a_n x^n = 0$ for all x, then $f^{(k)}(x) = 0$ for all x $\Rightarrow$ from part (a) that $a_k = 0$ for every

nonnegative integer k

47. The series $\sum_{n=1}^{\infty} \frac{x^n}{n}$ converges conditionally at the left-hand endpoint of its interval of convergence $[-1, 1)$; the

series $\sum_{n=1}^{\infty} \frac{x^n}{(n^2)}$ converges absolutely at the left-hand endpoint of its interval of convergence $[-1, 1]$

11.8 TAYLOR AND MACLAURIN SERIES

1. $f(x) = \ln x, f'(x) = \frac{1}{x}, f''(x) = -\frac{1}{x^2}, f'''(x) = \frac{2}{x^3}; f(1) = \ln 1 = 0, f'(1) = 1, f''(1) = -1, f'''(1) = 2 \Rightarrow P_0(x) = 0,$
$P_1(x) = (x-1), P_2(x) = (x-1) - \frac{1}{2}(x-1)^2, P_3(x) = (x-1) - \frac{1}{2}(x-1)^2 + \frac{1}{3}(x-1)^3$

3. $f(x) = \frac{1}{x} = x^{-1}, f'(x) = -x^{-2}, f''(x) = 2x^{-3}, f'''(x) = -6x^{-4}; f(2) = \frac{1}{2}, f'(2) = -\frac{1}{4}, f''(2) = \frac{1}{4}, f'''(x) = -\frac{3}{8}$
$\Rightarrow P_0(x) = \frac{1}{2}, P_1(x) = \frac{1}{2} - \frac{1}{4}(x-2), P_2(x) = \frac{1}{2} - \frac{1}{4}(x-2) + \frac{1}{8}(x-2)^2,$
$P_3(x) = \frac{1}{2} - \frac{1}{4}(x-2) + \frac{1}{8}(x-2)^2 - \frac{1}{16}(x-2)^3$

5. $f(x) = \sin x, f'(x) = \cos x, f''(x) = -\sin x, f'''(x) = -\cos x; f\left(\frac{\pi}{4}\right) = \sin\frac{\pi}{4} = \frac{\sqrt{2}}{2}, f'\left(\frac{\pi}{4}\right) = \cos\frac{\pi}{4} = \frac{\sqrt{2}}{2},$
$f''\left(\frac{\pi}{4}\right) = -\sin\frac{\pi}{4} = -\frac{\sqrt{2}}{2}, f'''\left(\frac{\pi}{4}\right) = -\cos\frac{\pi}{4} = -\frac{\sqrt{2}}{2} \Rightarrow P_0 = \frac{\sqrt{2}}{2}, P_1(x) = \frac{\sqrt{2}}{2} + \frac{\sqrt{2}}{2}\left(x - \frac{\pi}{4}\right),$
$P_2(x) = \frac{\sqrt{2}}{2} + \frac{\sqrt{2}}{2}\left(x - \frac{\pi}{4}\right) - \frac{\sqrt{2}}{4}\left(x - \frac{\pi}{4}\right)^2, P_3(x) = \frac{\sqrt{2}}{2} + \frac{\sqrt{2}}{2}\left(x - \frac{\pi}{4}\right) - \frac{\sqrt{2}}{4}\left(x - \frac{\pi}{4}\right)^2 - \frac{\sqrt{2}}{12}\left(x - \frac{\pi}{4}\right)^3$

7. $f(x) = \sqrt{x} = x^{1/2}, f'(x) = \left(\frac{1}{2}\right)x^{-1/2}, f''(x) = \left(-\frac{1}{4}\right)x^{-3/2}, f'''(x) = \left(\frac{3}{8}\right)x^{-5/2}; f(4) = \sqrt{4} = 2,$
$f'(4) = \left(\frac{1}{2}\right)4^{-1/2} = \frac{1}{4}, f''(4) = \left(-\frac{1}{4}\right)4^{-3/2} = -\frac{1}{32}, f'''(4) = \left(\frac{3}{8}\right)4^{-5/2} = \frac{3}{256} \Rightarrow P_0(x) = 2, P_1(x) = 2 + \frac{1}{4}(x-4),$
$P_2(x) = 2 + \frac{1}{4}(x-4) - \frac{1}{64}(x-4)^2, P_3(x) = 2 + \frac{1}{4}(x-4) - \frac{1}{64}(x-4)^2 + \frac{1}{512}(x-4)^3$

9. $e^x = \sum_{n=0}^{\infty} \frac{x^n}{n!} \Rightarrow e^{-x} = \sum_{n=0}^{\infty} \frac{(-x)^n}{n!} = 1 - x + \frac{x^2}{2!} - \frac{x^3}{3!} + \frac{x^4}{4!} - \dots$

11. $f(x) = (1+x)^{-1} \Rightarrow f'(x) = -(1+x)^{-2}, f''(x) = 2(1+x)^{-3}, f'''(x) = -3!(1+x)^{-4} \Rightarrow \dots f^{(k)}(x)$
$= (-1)^k k!(1+x)^{-k-1}; f(0) = 1, f'(0) = -1, f''(0) = 2, f'''(0) = -3!, \dots, f^{(k)}(0) = (-1)^k k!$

$\Rightarrow \frac{1}{1+x} = 1 - x + x^2 - x^3 + \dots = \sum_{n=0}^{\infty}(-x)^n = \sum_{n=0}^{\infty}(-1)^n x^n$

13. $\sin x = \sum_{n=0}^{\infty} \frac{(-1)^n x^{2n+1}}{(2n+1)!} \Rightarrow \sin 3x = \sum_{n=0}^{\infty} \frac{(-1)^n (3x)^{2n+1}}{(2n+1)!} = \sum_{n=0}^{\infty} \frac{(-1)^n 3^{2n+1} x^{2n+1}}{(2n+1)!} = 3x - \frac{3^3 x^3}{3!} + \frac{3^5 x^5}{5!} - \dots$

15. $7\cos(-x) = 7\cos x = 7\sum_{n=0}^{\infty} \frac{(-1)^n x^{2n}}{(2n)!} = 7 - \frac{7x^2}{2!} + \frac{7x^4}{4!} - \frac{7x^6}{6!} + \dots$, since the cosine is an even function

17. $\cosh x = \frac{e^x + e^{-x}}{2} = \frac{1}{2}\left[\left(1 + x^2 + \frac{x^2}{2!} + \frac{x^3}{3!} + \frac{x^4}{4!} + \dots\right) + \left(1 - x + \frac{x^2}{2!} - \frac{x^3}{3!} + \frac{x^4}{4!} - \dots\right)\right] = 1 + \frac{x^2}{2!} + \frac{x^4}{4!} + \frac{x^6}{6!} + \dots$

$= \sum_{n=0}^{\infty} \frac{x^{2n}}{(2n)!}$

19. $f(x) = x^4 - 2x^3 - 5x + 4 \Rightarrow f'(x) = 4x^3 - 6x^2 - 5, f''(x) = 12x^2 - 12x, f'''(x) = 24x - 12, f^{(4)}(x) = 24$

$\Rightarrow f^{(n)}(x) = 0$ if $n \geq 5$; $f(0) = 4, f'(0) = -5, f''(0) = 0, f'''(0) = -12, f^{(4)}(0) = 24, f^{(n)}(0) = 0$ if $n \geq 5$

$\Rightarrow x^4 - 2x^3 - 5x + 4 = 4 - 5x - \frac{12}{3!}x^3 + \frac{24}{4!}x^4 = x^4 - 2x^3 - 5x + 4$ itself

21. $f(x) = x^3 - 2x + 4 \Rightarrow f'(x) = 3x^2 - 2, f''(x) = 6x, f'''(x) = 6 \Rightarrow f^{(n)}(x) = 0$ if $n \geq 4$; $f(2) = 8, f'(2) = 10,$

$f''(2) = 12, f'''(2) = 6, f^{(n)}(2) = 0$ if $n \geq 4 \Rightarrow x^3 - 2x + 4 = 8 + 10(x - 2) + \frac{12}{2!}(x - 2)^2 + \frac{6}{3!}(x - 2)^3$

$= 8 + 10(x - 2) + 6(x - 2)^2 + (x - 2)^3$

23. $f(x) = x^4 + x^2 + 1 \Rightarrow f'(x) = 4x^3 + 2x, f''(x) = 12x^2 + 2, f'''(x) = 24x, f^{(4)}(x) = 24, f^{(n)}(x) = 0$ if $n \geq 5$;

$f(-2) = 21, f'(-2) = -36, f''(-2) = 50, f'''(-2) = -48, f^{(4)}(-2) = 24, f^{(n)}(-2) = 0$ if $n \geq 5 \Rightarrow x^4 + x^2 + 1$

$= 21 - 36(x + 2) + \frac{50}{2!}(x + 2)^2 - \frac{48}{3!}(x + 2)^3 + \frac{24}{4!}(x + 2)^4 = 21 - 36(x + 2) + 25(x + 2)^2 - 8(x + 2)^3 + (x + 2)^4$

25. $f(x) = x^{-2} \Rightarrow f'(x) = -2x^{-3}, f''(x) = 3!\,x^{-4}, f'''(x) = -4!\,x^{-5} \Rightarrow f^{(n)}(x) = (-1)^n(n + 1)!\,x^{-n-2}$;

$f(1) = 1, f'(1) = -2, f''(1) = 3!, f'''(1) = -4!, f^{(n)}(1) = (-1)^n(n + 1)! \Rightarrow \frac{1}{x^2}$

$= 1 - 2(x - 1) + 3(x - 1)^2 - 4(x - 1)^3 + \ldots = \sum_{n=0}^{\infty} (-1)^n(n + 1)(x - 1)^n$

27. $f(x) = e^x \Rightarrow f'(x) = e^x, f''(x) = e^x \Rightarrow f^{(n)}(x) = e^x$; $f(2) = e^2, f'(2) = e^2, \ldots f^{(n)}(2) = e^2$

$\Rightarrow e^x = e^2 + e^2(x - 2) + \frac{e^2}{2}(x - 2)^2 + \frac{e^3}{3!}(x - 2)^3 + \ldots = \sum_{n=0}^{\infty} \frac{e^2}{n!}(x - 2)^n$

29. If $e^x = \sum_{n=0}^{\infty} \frac{f^{(n)}(a)}{n!}(x - a)^n$ and $f(x) = e^x$, we have $f^{(n)}(a) = e^a$ for all $n = 0, 1, 2, 3, \ldots$

$\Rightarrow e^x = e^a \left[\frac{(x - a)^0}{0!} + \frac{(x - a)^1}{1!} + \frac{(x - a)^2}{2!} + \ldots \right] = e^a \left[1 + (x - a) + \frac{(x - a)^2}{2!} + \ldots \right]$ at $x = a$

31. $f(x) = f(a) + f'(a)(x - a) + \frac{f''(a)}{2}(x - a)^2 + \frac{f'''(a)}{3!}(x - a)^3 + \ldots \Rightarrow f'(x)$

$= f'(a) + f''(a)(x - a) + \frac{f'''(a)}{3!}3(x - a)^2 + \ldots \Rightarrow f''(x) = f''(a) + f'''(a)(x - a) + \frac{f^{(4)}(a)}{4!}4 \cdot 3(x - a)^2 + \ldots$

$\Rightarrow f^{(n)}(x) = f^{(n)}(a) + f^{(n+1)}(a)(x - a) + \frac{f^{(n+2)}(a)}{2}(x - a)^2 + \ldots$

$\Rightarrow f(a) = f(a) + 0, f'(a) = f'(a) + 0, \ldots, f^{(n)}(a) = f^{(n)}(a) + 0$

33. $f(x) = \ln(\cos x) \Rightarrow f'(x) = -\tan x$ and $f''(x) = -\sec^2 x$; $f(0) = 0, f'(0) = 0, f''(0) = -1$

$\Rightarrow L(x) = 0$ and $Q(x) = -\frac{x^2}{2}$

35. $f(x) = (1 - x^2)^{-1/2} \Rightarrow f'(x) = x(1 - x^2)^{-3/2}$ and $f''(x) = (1 - x^2)^{-3/2} + 3x^2(1 - x^2)^{-5/2}$; $f(0) = 1,$

$f'(0) = 0, f''(0) = 1 \Rightarrow L(x) = 1$ and $Q(x) = 1 + \frac{x^2}{2}$

37. $f(x) = \sin x \Rightarrow f'(x) = \cos x$ and $f''(x) = -\sin x$; $f(0) = 0, f'(0) = 1, f''(0) = 0 \Rightarrow L(x) = x$ and $Q(x) = x$

11.9 CONVERGENCE OF TAYLOR SERIES; ERROR ESTIMATES

1. $e^x = 1 + x + \frac{x^2}{2!} + \ldots = \sum_{n=0}^{\infty} \frac{x^n}{n!} \Rightarrow e^{-5x} = 1 + (-5x) + \frac{(-5x)^2}{2!} + \ldots = 1 - 5x + \frac{5^2x^2}{2!} - \frac{5^3x^3}{3!} + \ldots = \sum_{n=0}^{\infty} \frac{(-1)^n 5^n x^n}{n!}$

3. $\sin x = x - \frac{x^3}{3!} + \frac{x^5}{5!} - \ldots = \sum_{n=0}^{\infty} \frac{(-1)^n x^{2n+1}}{(2n+1)!} \Rightarrow 5\sin(-x) = 5 \left[(-x) - \frac{(-x)^3}{3!} + \frac{(-x)^5}{5!} - \ldots \right]$

$= \sum_{n=0}^{\infty} \frac{5(-1)^{n+1} x^{2n+1}}{(2n+1)!}$

5. $\cos x = \sum\limits_{n=0}^{\infty} \frac{(-1)^n x^{2n}}{(2n)!} \Rightarrow \cos\sqrt{x+1} = \sum\limits_{n=0}^{\infty} \frac{(-1)^n \left[(x+1)^{1/2}\right]^{2n}}{(2n)!} = \sum\limits_{n=0}^{\infty} \frac{(-1)^n (x+1)^n}{(2n)!} = 1 - \frac{x+1}{2!} + \frac{(x+1)^2}{4!} - \frac{(x+1)^3}{6!} + \cdots$

7. $e^x = \sum\limits_{n=0}^{\infty} \frac{x^n}{n!} \Rightarrow xe^x = x\left(\sum\limits_{n=0}^{\infty} \frac{x^n}{n!}\right) = \sum\limits_{n=0}^{\infty} \frac{x^{n+1}}{n!} = x + x^2 + \frac{x^3}{2!} + \frac{x^4}{3!} + \frac{x^5}{4!} + \cdots$

9. $\cos x = \sum\limits_{n=0}^{\infty} \frac{(-1)^n x^{2n}}{(2n)!} \Rightarrow \frac{x^2}{2} - 1 + \cos x = \frac{x^2}{2} - 1 + \sum\limits_{n=0}^{\infty} \frac{(-1)^n x^{2n}}{(2n)!} = \frac{x^2}{2} - 1 + 1 - \frac{x^2}{2} + \frac{x^4}{4!} - \frac{x^6}{6!} + \frac{x^8}{8!} - \frac{x^{10}}{10!} + \cdots$

$= \frac{x^4}{4!} - \frac{x^6}{6!} + \frac{x^8}{8!} - \frac{x^{10}}{10!} + \cdots = \sum\limits_{n=2}^{\infty} \frac{(-1)^n x^{2n}}{(2n)!}$

11. $\cos x = \sum\limits_{n=0}^{\infty} \frac{(-1)^n x^{2n}}{(2n)!} \Rightarrow x\cos\pi x = x\sum\limits_{n=0}^{\infty} \frac{(-1)^n(\pi x)^{2n}}{(2n)!} = \sum\limits_{n=0}^{\infty} \frac{(-1)^n \pi^{2n} x^{2n+1}}{(2n)!} = x - \frac{\pi^2 x^3}{2!} + \frac{\pi^4 x^5}{4!} - \frac{\pi^6 x^7}{6!} + \cdots$

13. $\cos^2 x = \frac{1}{2} + \frac{\cos 2x}{2} = \frac{1}{2} + \frac{1}{2}\sum\limits_{n=0}^{\infty} \frac{(-1)^n(2x)^{2n}}{(2n)!} = \frac{1}{2} + \frac{1}{2}\left[1 - \frac{(2x)^2}{2!} + \frac{(2x)^4}{4!} - \frac{(2x)^6}{6!} + \frac{(2x)^8}{8!} - \cdots\right]$

$= 1 - \frac{(2x)^2}{2\cdot 2!} + \frac{(2x)^4}{2\cdot 4!} - \frac{(2x)^6}{2\cdot 6!} + \frac{(2x)^8}{2\cdot 8!} - \cdots = 1 + \sum\limits_{n=1}^{\infty} \frac{(-1)^n(2x)^{2n}}{2\cdot(2n)!} = 1 + \sum\limits_{n=1}^{\infty} \frac{(-1)^n 2^{2n-1} x^{2n}}{(2n)!}$

15. $\frac{x^2}{1-2x} = x^2\left(\frac{1}{1-2x}\right) = x^2\sum\limits_{n=0}^{\infty}(2x)^n = \sum\limits_{n=0}^{\infty} 2^n x^{n+2} = x^2 + 2x^3 + 2^2 x^4 + 2^3 x^5 + \cdots$

17. $\frac{1}{1-x} = \sum\limits_{n=0}^{\infty} x^n = 1 + x + x^2 + x^3 + \cdots \Rightarrow \frac{d}{dx}\left(\frac{1}{1-x}\right) = \frac{1}{(1-x)^2} = 1 + 2x + 3x^2 + \cdots = \sum\limits_{n=1}^{\infty} nx^{n-1}$

$= \sum\limits_{n=0}^{\infty} (n+1)x^n$

19. By the Alternating Series Estimation Theorem, the error is less than $\frac{|x|^5}{5!} \Rightarrow |x|^5 < (5!)\left(5 \times 10^{-4}\right)$

$\Rightarrow |x|^5 < 600 \times 10^{-4} \Rightarrow |x| < \sqrt[5]{6 \times 10^{-2}} \approx 0.56968$

21. If $\sin x = x$ and $|x| < 10^{-3}$, then the error is less than $\frac{(10^{-3})^3}{3!} \approx 1.67 \times 10^{-10}$, by Alternating Series Estimation Theorem;
The Alternating Series Estimation Theorem says $R_2(x)$ has the same sign as $-\frac{x^3}{3!}$. Moreover, $x < \sin x$
$\Rightarrow 0 < \sin x - x = R_2(x) \Rightarrow x < 0 \Rightarrow -10^{-3} < x < 0$.

23. $|R_2(x)| = \left|\frac{e^c x^3}{3!}\right| < \frac{3^{(0.1)}(0.1)^3}{3!} < 1.87 \times 10^{-4}$, where c is between 0 and x

25. $|R_4(x)| < \left|\frac{\cosh c}{5!}x^5\right| = \left|\frac{e^c + e^{-c}}{2}\frac{x^5}{5!}\right| < \frac{1.65 + \frac{1}{1.65}}{2} \cdot \frac{(0.5)^5}{5!} = (1.13)\frac{(0.5)^5}{5!} \approx 0.000294$

27. $|R_1| = \left|\frac{1}{(1+c)^2}\frac{x^2}{2!}\right| < \frac{x^2}{2} = \left|\frac{x}{2}\right||x| < .01|x| = (1\%)|x| \Rightarrow \left|\frac{x}{2}\right| < .01 \Rightarrow 0 < |x| < .02$

29. (a) $\sin x = x - \frac{x^3}{3!} + \frac{x^5}{5!} - \frac{x^7}{7!} + \cdots \Rightarrow \frac{\sin x}{x} = 1 - \frac{x^2}{3!} + \frac{x^4}{5!} - \frac{x^6}{7!} + \cdots$, $s_1 = 1$ and $s_2 = 1 - \frac{x^2}{6}$; if L is the sum of the
series representing $\frac{\sin x}{x}$, then by the Alternating Series Estimation Theorem, $L - s_1 = \frac{\sin x}{x} - 1 < 0$ and
$L - s_2 = \frac{\sin x}{x} - \left(1 - \frac{x^2}{6}\right) > 0$. Therefore $1 - \frac{x^2}{6} < \frac{\sin x}{x} < 1$

(b) The graph of $y = \frac{\sin x}{x}$, $x \neq 0$, is bounded below by the graph of $y = 1 - \frac{x^2}{6}$ and above by the graph of $y = 1$ as derived in part (a).

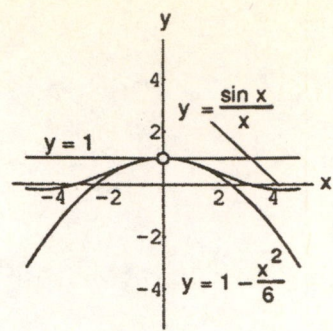

31. $\sin x$ when $x = 0.1$; the sum is $\sin(0.1) \approx 0.099833417$

33. $\tan^{-1} x$ when $x = \frac{\pi}{3}$; the sum is $\tan^{-1}\left(\frac{\pi}{3}\right) \approx 0.808448$

35. $e^x \sin x = 0 + x + x^2 + x^3\left(-\frac{1}{3!} + \frac{1}{2!}\right) + x^4\left(-\frac{1}{3!} + \frac{1}{3!}\right) + x^5\left(\frac{1}{5!} - \frac{1}{2!}\frac{1}{3!} + \frac{1}{4!}\right) + x^6\left(\frac{1}{5!} - \frac{1}{3!}\frac{1}{3!} + \frac{1}{5!}\right) + \cdots$

$= x + x^2 + \frac{1}{3}x^3 - \frac{1}{30}x^5 - \frac{1}{90}x^6 + \cdots$

37. $\sin^2 x = \left(\frac{1 - \cos 2x}{2}\right) = \frac{1}{2} - \frac{1}{2}\cos 2x = \frac{1}{2} - \frac{1}{2}\left(1 - \frac{(2x)^2}{2!} + \frac{(2x)^4}{4!} - \frac{(2x)^6}{6!} + \cdots\right) = \frac{2x^2}{2!} - \frac{2^3 x^4}{4!} + \frac{2^5 x^6}{6!} - \cdots$

$\Rightarrow \frac{d}{dx}\left(\sin^2 x\right) = \frac{d}{dx}\left(\frac{2x^2}{2!} - \frac{2^3 x^4}{4!} + \frac{2^5 x^6}{6!} - \cdots\right) = 2x - \frac{(2x)^3}{3!} + \frac{(2x)^5}{5!} - \frac{(2x)^7}{7!} + \cdots \Rightarrow 2\sin x \cos x$

$= 2x - \frac{(2x)^3}{3!} + \frac{(2x)^5}{5!} - \frac{(2x)^7}{7!} + \cdots = \sin 2x$, which checks

39. A special case of Taylor's Theorem is $f(b) = f(a) + f'(c)(b - a)$, where c is between a and b $\Rightarrow$ $f(b) - f(a) = f'(c)(b - a)$, the Mean Value Theorem.

41. (a) $f'' \leq 0$, $f'(a) = 0$ and $x = a$ interior to the interval I $\Rightarrow$ $f(x) - f(a) = \frac{f''(c_2)}{2}(x - a)^2 \leq 0$ throughout I

$\Rightarrow f(x) \leq f(a)$ throughout I $\Rightarrow$ f has a local maximum at $x = a$

(b) similar reasoning gives $f(x) - f(a) = \frac{f''(c_2)}{2}(x - a)^2 \geq 0$ throughout I $\Rightarrow$ $f(x) \geq f(a)$ throughout I $\Rightarrow$ f has a local minimum at $x = a$

43. (a) $f(x) = (1 + x)^k \Rightarrow f'(x) = k(1 + x)^{k-1} \Rightarrow f''(x) = k(k - 1)(1 + x)^{k-2}$; $f(0) = 1$, $f'(0) = k$, and $f''(0) = k(k - 1)$

$\Rightarrow Q(x) = 1 + kx + \frac{k(k-1)}{2}x^2$

(b) $|R_2(x)| = \left|\frac{3 \cdot 2 \cdot 1}{3!}x^3\right| < \frac{1}{100} \Rightarrow |x^3| < \frac{1}{100} \Rightarrow 0 < x < \frac{1}{100^{1/3}}$ or $0 < x < .21544$

45. If $f(x) = \sum_{n=0}^{\infty} a_n x^n$, then $f^{(k)}(x) = \sum_{n=k}^{\infty} n(n - 1)(n - 2) \cdots (n - k + 1)a_n x^{n-k}$ and $f^{(k)}(0) = k! \, a_k$

$\Rightarrow a_k = \frac{f^{(k)}(0)}{k!}$ for k a nonnegative integer. Therefore, the coefficients of $f(x)$ are identical with the corresponding coefficients in the Maclaurin series of $f(x)$ and the statement follows.

47. (a) Suppose $f(x)$ is a continuous periodic function with period p. Let x_0 be an arbitrary real number. Then f assumes a minimum m_1 and a maximum m_2 in the interval $[x_0, x_0 + p]$; i.e., $m_1 \leq f(x) \leq m_2$ for all x in $[x_0, x_0 + p]$. Since f is periodic it has exactly the same values on all other intervals $[x_0 + p, x_0 + 2p]$, $[x_0 + 2p, x_0 + 3p]$, $\ldots$, and $[x_0 - p, x_0]$, $[x_0 - 2p, x_0 - p]$, $\ldots$, and so forth. That is, for all real numbers $-\infty < x < \infty$ we have $m_1 \leq f(x) \leq m_2$. Now choose $M = \max\{|m_1|, |m_2|\}$. Then $-M \leq -|m_1| \leq m_1 \leq f(x) \leq m_2 \leq |m_2| \leq M \Rightarrow |f(x)| \leq M$ for all x.

(b) The dominate term in the nth order Taylor polynomial generated by $\cos x$ about $x = a$ is $\frac{\sin(a)}{n!}(x - a)^n$ or $\frac{\cos(a)}{n!}(x - a)^n$. In both cases, as $|x|$ increases the absolute value of these dominate terms tends to ∞, causing the graph of $P_n(x)$ to move away from cos x.

49. (a) $e^{-i\pi} = \cos(-\pi) + i\sin(-\pi) = -1 + i(0) = -1$

(b) $e^{i\pi/4} = \cos\left(\frac{\pi}{4}\right) + i\sin\left(\frac{\pi}{4}\right) = \frac{1}{\sqrt{2}} + \frac{i}{\sqrt{2}} = \left(\frac{1}{\sqrt{2}}\right)(1+i)$

(c) $e^{-i\pi/2} = \cos\left(-\frac{\pi}{2}\right) + i\sin\left(-\frac{\pi}{2}\right) = 0 + i(-1) = -i$

51. $e^x = 1 + x + \frac{x^2}{2!} + \frac{x^3}{3!} + \frac{x^4}{4!} + \dots \Rightarrow e^{i\theta} = 1 + i\theta + \frac{(i\theta)^2}{2!} + \frac{(i\theta)^3}{3!} + \frac{(i\theta)^4}{4!} + \dots$ and

$e^{-i\theta} = 1 - i\theta + \frac{(-i\theta)^2}{2!} + \frac{(-i\theta)^3}{3!} + \frac{(-i\theta)^4}{4!} + \dots = 1 - i\theta + \frac{(i\theta)^2}{2!} - \frac{(i\theta)^3}{3!} + \frac{(i\theta)^4}{4!} - \dots$

$\Rightarrow \frac{e^{i\theta} + e^{-i\theta}}{2} = \frac{\left(1 + i\theta + \frac{(i\theta)^2}{2!} + \frac{(i\theta)^3}{3!} + \frac{(i\theta)^4}{4!} + \dots\right) + \left(1 - i\theta + \frac{(i\theta)^2}{2!} - \frac{(i\theta)^3}{3!} + \frac{(i\theta)^4}{4!} - \dots\right)}{2}$

$= 1 - \frac{\theta^2}{2!} + \frac{\theta^4}{4!} - \frac{\theta^6}{6!} + \dots = \cos\theta;$

$\frac{e^{i\theta} - e^{-i\theta}}{2i} = \frac{\left(1 + i\theta + \frac{(i\theta)^2}{2!} + \frac{(i\theta)^3}{3!} + \frac{(i\theta)^4}{4!} + \dots\right) - \left(1 - i\theta + \frac{(i\theta)^2}{2!} - \frac{(i\theta)^3}{3!} + \frac{(i\theta)^4}{4!} - \dots\right)}{2i}$

$= \theta - \frac{\theta^3}{3!} + \frac{\theta^5}{5!} - \frac{\theta^7}{7!} + \dots = \sin\theta$

53. $e^x \sin x = \left(1 + x + \frac{x^2}{2!} + \frac{x^3}{3!} + \frac{x^4}{4!} + \dots\right)\left(x - \frac{x^3}{3!} + \frac{x^5}{5!} - \frac{x^7}{7!} + \dots\right)$

$= (1)x + (1)x^2 + \left(-\frac{1}{6} + \frac{1}{2}\right)x^3 + \left(-\frac{1}{6} + \frac{1}{6}\right)x^4 + \left(\frac{1}{120} - \frac{1}{12} + \frac{1}{24}\right)x^5 + \dots = x + x^2 + \frac{1}{3}x^3 - \frac{1}{30}x^5 + \dots;$

$e^x \cdot e^{ix} = e^{(1+i)x} = e^x(\cos x + i\sin x) = e^x \cos x + i(e^x \sin x) \Rightarrow e^x \sin x$ is the series of the imaginary part

of $e^{(1+i)x}$ which we calculate next; $e^{(1+i)x} = \sum_{n=0}^{\infty} \frac{(x+ix)^n}{n!} = 1 + (x + ix) + \frac{(x+ix)^2}{2!} + \frac{(x+ix)^3}{3!} + \frac{(x+ix)^4}{4!} + \dots$

$= 1 + x + ix + \frac{1}{2!}(2ix^2) + \frac{1}{3!}(2ix^3 - 2x^3) + \frac{1}{4!}(-4x^4) + \frac{1}{5!}(-4x^5 - 4ix^5) + \frac{1}{6!}(-8ix^6) + \dots \Rightarrow$ the imaginary part

of $e^{(1+i)x}$ is $x + \frac{2}{2!}x^2 + \frac{2}{3!}x^3 - \frac{4}{5!}x^5 - \frac{8}{6!}x^6 + \dots = x + x^2 + \frac{1}{3}x^3 - \frac{1}{30}x^5 - \frac{1}{90}x^6 + \dots$ in agreement with our

product calculation. The series for $e^x \sin x$ converges for all values of x.

55. (a) $e^{i\theta_1}e^{i\theta_2} = (\cos\theta_1 + i\sin\theta_1)(\cos\theta_2 + i\sin\theta_2) = (\cos\theta_1\cos\theta_2 - \sin\theta_1\sin\theta_2) + i(\sin\theta_1\cos\theta_2 + \sin\theta_2\cos\theta_1)$

$= \cos(\theta_1 + \theta_2) + i\sin(\theta_1 + \theta_2) = e^{i(\theta_1+\theta_2)}$

(b) $e^{-i\theta} = \cos(-\theta) + i\sin(-\theta) = \cos\theta - i\sin\theta = (\cos\theta - i\sin\theta)\left(\frac{\cos\theta + i\sin\theta}{\cos\theta + i\sin\theta}\right) = \frac{1}{\cos\theta + i\sin\theta} = \frac{1}{e^{i\theta}}$

11.10 APPLICATIONS OF POWER SERIES

1. $(1+x)^{1/2} = 1 + \frac{1}{2}x + \frac{\left(\frac{1}{2}\right)\left(-\frac{1}{2}\right)x^2}{2!} + \frac{\left(\frac{1}{2}\right)\left(-\frac{1}{2}\right)\left(-\frac{3}{2}\right)x^3}{3!} + \dots = 1 + \frac{1}{2}x - \frac{1}{8}x^2 + \frac{1}{16}x^3 - \dots$

3. $(1-x)^{-1/2} = 1 - \frac{1}{2}(-x) + \frac{\left(-\frac{1}{2}\right)\left(-\frac{3}{2}\right)(-x)^2}{2!} + \frac{\left(-\frac{1}{2}\right)\left(-\frac{3}{2}\right)\left(-\frac{5}{2}\right)(-x)^3}{3!} + \dots = 1 + \frac{1}{2}x + \frac{3}{8}x^2 + \frac{5}{16}x^3 + \dots$

5. $\left(1+\frac{x}{2}\right)^{-2} = 1 - 2\left(\frac{x}{2}\right) + \frac{(-2)(-3)\left(\frac{x}{2}\right)^2}{2!} + \frac{(-2)(-3)(-4)\left(\frac{x}{2}\right)^3}{3!} + \dots = 1 - x + \frac{3}{4}x^2 - \frac{1}{2}x^3$

7. $(1+x^3)^{-1/2} = 1 - \frac{1}{2}x^3 + \frac{\left(-\frac{1}{2}\right)\left(-\frac{3}{2}\right)(x^3)^2}{2!} + \frac{\left(-\frac{1}{2}\right)\left(-\frac{3}{2}\right)\left(-\frac{5}{2}\right)(x^3)^3}{3!} + \dots = 1 - \frac{1}{2}x^3 + \frac{3}{8}x^6 - \frac{5}{16}x^9 + \dots$

9. $\left(1+\frac{1}{x}\right)^{1/2} = 1 + \frac{1}{2}\left(\frac{1}{x}\right) + \frac{\left(\frac{1}{2}\right)\left(-\frac{1}{2}\right)\left(\frac{1}{x}\right)^2}{2!} + \frac{\left(\frac{1}{2}\right)\left(-\frac{1}{2}\right)\left(-\frac{3}{2}\right)\left(\frac{1}{x}\right)^3}{3!} + \dots = 1 + \frac{1}{2x} - \frac{1}{8x^2} + \frac{1}{16x^3}$

11. $(1+x)^4 = 1 + 4x + \frac{(4)(3)x^2}{2!} + \frac{(4)(3)(2)x^3}{3!} + \frac{(4)(3)(2)x^4}{4!} = 1 + 4x + 6x^2 + 4x^3 + x^4$

13. $(1-2x)^3 = 1 + 3(-2x) + \frac{(3)(2)(-2x)^2}{2!} + \frac{(3)(2)(1)(-2x)^3}{3!} = 1 - 6x + 12x^2 - 8x^3$

15. Assume the solution has the form $y = a_0 + a_1x + a_2x^2 + \ldots + a_{n-1}x^{n-1} + a_nx^n + \ldots$

$\Rightarrow \frac{dy}{dx} = a_1 + 2a_2x + \ldots + na_nx^{n-1} + \ldots$

$\Rightarrow \frac{dy}{dx} + y = (a_1 + a_0) + (2a_2 + a_1)x + (3a_3 + a_2)x^2 + \ldots + (na_n + a_{n-1})x^{n-1} + \ldots = 0$

$\Rightarrow a_1 + a_0 = 0, 2a_2 + a_1 = 0, 3a_3 + a_2 = 0$ and in general $na_n + a_{n-1} = 0$. Since $y = 1$ when $x = 0$ we have

$a_0 = 1$. Therefore $a_1 = -1, a_2 = \frac{-a_1}{2 \cdot 1} = \frac{1}{2}, a_3 = \frac{-a_2}{3} = -\frac{1}{3 \cdot 2}, \ldots, a_n = \frac{-a_{n-1}}{n} = \frac{(-1)^n}{n!}$

$\Rightarrow y = 1 - x + \frac{1}{2}x^2 - \frac{1}{3!}x^3 + \ldots + \frac{(-1)^n}{n!}x^n + \ldots = \sum_{n=0}^{\infty} \frac{(-1)^n x^n}{n!} = e^{-x}$

17. Assume the solution has the form $y = a_0 + a_1x + a_2x^2 + \ldots + a_{n-1}x^{n-1} + a_nx^n + \ldots$

$\Rightarrow \frac{dy}{dx} = a_1 + 2a_2x + \ldots + na_nx^{n-1} + \ldots$

$\Rightarrow \frac{dy}{dx} - y = (a_1 - a_0) + (2a_2 - a_1)x + (3a_3 - a_2)x^2 + \ldots + (na_n - a_{n-1})x^{n-1} + \ldots = 1$

$\Rightarrow a_1 - a_0 = 1, 2a_2 - a_1 = 0, 3a_3 - a_2 = 0$ and in general $na_n - a_{n-1} = 0$. Since $y = 0$ when $x = 0$ we have

$a_0 = 0$. Therefore $a_1 = 1, a_2 = \frac{a_1}{2} = \frac{1}{2}, a_3 = \frac{a_2}{3} = \frac{1}{3 \cdot 2}, a_4 = \frac{a_3}{4} = \frac{1}{4 \cdot 3 \cdot 2}, \ldots, a_n = \frac{a_{n-1}}{n} = \frac{1}{n!}$

$\Rightarrow y = 0 + 1x + \frac{1}{2}x^2 + \frac{1}{3 \cdot 2}x^3 + \frac{1}{4 \cdot 3 \cdot 2}x^4 + \ldots + \frac{1}{n!}x^n + \ldots$

$= \left(1 + 1x + \frac{1}{2}x^2 + \frac{1}{3 \cdot 2}x^3 + \frac{1}{4 \cdot 3 \cdot 2}x^4 + \ldots + \frac{1}{n!}x^n + \ldots\right) - 1 = \sum_{n=0}^{\infty} \frac{x^n}{n!} - 1 = e^x - 1$

19. Assume the solution has the form $y = a_0 + a_1x + a_2x^2 + \ldots + a_{n-1}x^{n-1} + a_nx^n + \ldots$

$\Rightarrow \frac{dy}{dx} = a_1 + 2a_2x + \ldots + na_nx^{n-1} + \ldots$

$\Rightarrow \frac{dy}{dx} - y = (a_1 - a_0) + (2a_2 - a_1)x + (3a_3 - a_2)x^2 + \ldots + (na_n - a_{n-1})x^{n-1} + \ldots = x$

$\Rightarrow a_1 - a_0 = 0, 2a_2 - a_1 = 1, 3a_3 - a_2 = 0$ and in general $na_n - a_{n-1} = 0$. Since $y = 0$ when $x = 0$ we have

$a_0 = 0$. Therefore $a_1 = 0, a_2 = \frac{1 + a_1}{2} = \frac{1}{2}, a_3 = \frac{a_2}{3} = \frac{1}{3 \cdot 2}, a_4 = \frac{a_3}{4} = \frac{1}{4 \cdot 3 \cdot 2}, \ldots, a_n = \frac{a_{n-1}}{n} = \frac{1}{n!}$

$\Rightarrow y = 0 + 0x + \frac{1}{2}x^2 + \frac{1}{3 \cdot 2}x^3 + \frac{1}{4 \cdot 3 \cdot 2}x^4 + + \ldots + \frac{1}{n!}x^n + \ldots$

$= \left(1 + 1x + \frac{1}{2}x^2 + \frac{1}{3 \cdot 2}x^3 + \frac{1}{4 \cdot 3 \cdot 2}x^4 + \ldots + \frac{1}{n!}x^n + \ldots\right) - 1 - x = \sum_{n=0}^{\infty} \frac{x^n}{n!} - 1 - x = e^x - x - 1$

21. $y' - xy = a_1 + (2a_2 - a_0)x + (3a_3 - a_1)x + \ldots + (na_n - a_{n-2})x^{n-1} + \ldots = 0 \Rightarrow a_1 = 0, 2a_2 - a_0 = 0, 3a_3 - a_1 = 0,$

$4a_4 - a_2 = 0$ and in general $na_n - a_{n-2} = 0$. Since $y = 1$ when $x = 0$, we have $a_0 = 1$. Therefore $a_2 = \frac{a_0}{2} = \frac{1}{2}$,

$a_3 = \frac{a_1}{3} = 0, a_4 = \frac{a_2}{4} = \frac{1}{2 \cdot 4}, a_5 = \frac{a_3}{5} = 0, \ldots, a_{2n} = \frac{1}{2 \cdot 4 \cdot 6 \cdots 2n}$ and $a_{2n+1} = 0$

$\Rightarrow y = 1 + \frac{1}{2}x^2 + \frac{1}{2 \cdot 4}x^4 + \frac{1}{2 \cdot 4 \cdot 6}x^6 + \ldots + \frac{1}{2 \cdot 4 \cdot 6 \cdots 2n}x^{2n} + \ldots = \sum_{n=0}^{\infty} \frac{x^{2n}}{2^n n!} = \sum_{n=0}^{\infty} \frac{\left(\frac{x^2}{2}\right)^n}{n!} = e^{x^2/2}$

23. $(1 - x)y' - y = (a_1 - a_0) + (2a_2 - a_1 - a_1)x + (3a_3 - 2a_2 - a_2)x^2 + (4a_4 - 3a_3 - a_3)x^3 + \ldots$

$+ (na_n - (n-1)a_{n-1} - a_{n-1})x^{n-1} + \ldots = 0 \Rightarrow a_1 - a_0 = 0, 2a_2 - 2a_1 = 0, 3a_3 - 3a_2 = 0$ and in

general $(na_n - na_{n-1}) = 0$. Since $y = 2$ when $x = 0$, we have $a_0 = 2$. Therefore

$a_1 = 2, a_2 = 2, \ldots, a_n = 2 \Rightarrow y = 2 + 2x + 2x^2 + \ldots = \sum_{n=0}^{\infty} 2x^n = \frac{2}{1-x}$

25. $y = a_0 + a_1x + a_2x^2 + \ldots + a_nx^n + \ldots \Rightarrow y'' = 2a_2 + 3 \cdot 2a_3x + \ldots + n(n-1)a_nx^{n-2} + \ldots \Rightarrow y'' - y$

$= (2a_2 - a_0) + (3 \cdot 2a_3 - a_1)x + (4 \cdot 3a_4 - a_2)x^2 + \ldots + (n(n-1)a_n - a_{n-2})x^{n-2} + \ldots = 0 \Rightarrow 2a_2 - a_0 = 0,$

$3 \cdot 2a_3 - a_1 = 0, 4 \cdot 3a_4 - a_2 = 0$ and in general $n(n-1)a_n - a_{n-2} = 0$. Since $y' = 1$ and $y = 0$ when $x = 0$,

we have $a_0 = 0$ and $a_1 = 1$. Therefore $a_2 = 0, a_3 = \frac{1}{3 \cdot 2}, a_4 = 0, a_5 = \frac{1}{5 \cdot 4 \cdot 3 \cdot 2}, \ldots, a_{2n+1} = \frac{1}{(2n+1)!}$ and

$a_{2n} = 0 \Rightarrow y = x + \frac{1}{3!}x^3 + \frac{1}{5!}x^5 + \ldots = \sum_{n=0}^{\infty} \frac{x^{2n+1}}{(2n+1)!} = \sinh x$

27. $y = a_0 + a_1x + a_2x^2 + \ldots + a_nx^n + \ldots \Rightarrow y'' = 2a_2 + 3 \cdot 2a_3x + \ldots + n(n-1)a_nx^{n-2} + \ldots \Rightarrow y'' + y$

$= (2a_2 + a_0) + (3 \cdot 2a_3 + a_1)x + (4 \cdot 3a_4 + a_2)x^2 + \ldots + (n(n-1)a_n + a_{n-2})x^{n-2} + \ldots = x \Rightarrow 2a_2 + a_0 = 0,$

$3 \cdot 2a_3 + a_1 = 1, 4 \cdot 3a_4 + a_2 = 0$ and in general $n(n-1)a_n + a_{n-2} = 0$. Since $y' = 1$ and $y = 2$ when $x = 0$, we have $a_0 = 2$ and $a_1 = 1$. Therefore $a_2 = -1$, $a_3 = 0$, $a_4 = \frac{1}{4\cdot3}$, $a_5 = 0$, ..., $a_{2n} = -2 \cdot \frac{(-1)^{n+1}}{(2n)!}$ and

$a_{2n+1} = 0 \Rightarrow y = 2 + x - x^2 + 2 \cdot \frac{x^4}{4!} + \ldots = 2 + x - 2\sum_{n=1}^{\infty} \frac{(-1)^{n+1}x^{2n}}{(2n)!} = x + \cos 2x$

29. $y = a_0 + a_1(x-2) + a_2(x-2)^2 + \ldots + a_n(x-2)^n + \ldots$

$\Rightarrow y'' = 2a_2 + 3\cdot2a_3(x-2) + \ldots + n(n-1)a_n(x-2)^{n-2} + \ldots \Rightarrow y'' - y$

$= (2a_2 - a_0) + (3\cdot2a_3 - a_1)(x-2) + (4\cdot3a_4 - a_2)(x-2)^2 + \ldots + (n(n-1)a_n - a_{n-2})(x-2)^{n-2} + \ldots = -x$

$= -(x-2) - 2 \Rightarrow 2a_2 - a_0 = -2$, $3\cdot2a_3 - a_1 = -1$, and $n(n-1)a_n - a_{n-2} = 0$ for $n > 3$. Since $y = 0$ when $x = 2$,

we have $a_0 = 0$, and since $y' = -2$ when $x = 2$, we have $a_1 = -2$. Therefore $a_2 = -1$, $a_3 = -\frac{1}{2}$, $a_4 = \frac{1}{4\cdot3}(-1) = \frac{-2}{4\cdot3\cdot2\cdot1}$,

$a_5 = \frac{1}{5\cdot4}\left(-\frac{1}{2}\right) = \frac{-3}{5\cdot4\cdot3\cdot2\cdot1}, \ldots, a_{2n} = \frac{-2}{(2n)!}$, and $a_{2n+1} = \frac{-3}{(2n+1)!}$. Since $a_1 = -2$, we have $a_1(x-2) = (-2)(x-2)$ and

$(-2)(x-2) = (-3+1)(x-2) = (-3)(x-2) + (1)(x-2) = x - 2 - 3(x-2)$.

$\Rightarrow y = x - 2 - 3(x-2) - \frac{2}{2!}(x-2)^2 - \frac{3}{3!}(x-2)^3 - \frac{2}{4!}(x-2)^4 - \frac{3}{5!}(x-2)^5 - \ldots$

$\Rightarrow y = x - 2 - \frac{2}{2!}(x-2)^2 - \frac{2}{4!}(x-2)^4 - \ldots - 3(x-2)x - \frac{3}{3!}(x-2)^3 - \frac{3}{5!}(x-2)^5 - \ldots$

$\Rightarrow y = x - 2\sum_{n=0}^{\infty} \frac{(x-2)^{2n}}{(2n)!} - 3\sum_{n=0}^{\infty} \frac{(x-2)^{2n+1}}{(2n+1)!}$

31. $y'' + x^2 y = 2a_2 + 6a_3 x + (4\cdot3a_4 + a_0)x^2 + \ldots + (n(n-1)a_n + a_{n-4})x^{n-2} + \ldots = x \Rightarrow 2a_2 = 0$, $6a_3 = 1$,

$4\cdot3a_4 + a_0 = 0$, $5\cdot4a_5 + a_1 = 0$, and in general $n(n-1)a_n + a_{n-4} = 0$. Since $y' = b$ and $y = a$ when $x = 0$,

we have $a_0 = a$ and $a_1 = b$. Therefore $a_2 = 0$, $a_3 = \frac{1}{2\cdot3}$, $a_4 = -\frac{a}{3\cdot4}$, $a_5 = -\frac{b}{4\cdot5}$, $a_6 = 0$, $a_7 = \frac{-1}{2\cdot3\cdot6\cdot7}$

$\Rightarrow y = a + bx + \frac{1}{2\cdot3}x^3 - \frac{a}{3\cdot4}x^4 - \frac{b}{4\cdot5}x^5 - \frac{1}{2\cdot3\cdot6\cdot7}x^7 + \frac{ax^8}{3\cdot4\cdot7\cdot8} + \frac{bx^9}{4\cdot5\cdot8\cdot9} + \ldots$

33. $\int_0^{0.2} \sin x^2 \, dx = \int_0^{0.2} \left(x^2 - \frac{x^6}{3!} + \frac{x^{10}}{5!} - \ldots\right) dx = \left[\frac{x^3}{3} - \frac{x^7}{7\cdot3!} + \ldots\right]_0^{0.2} \approx \left[\frac{x^3}{3}\right]_0^{0.2} \approx 0.00267$ with error

$|E| \le \frac{(.2)^7}{7\cdot3!} \approx 0.0000003$

35. $\int_0^{0.1} \frac{1}{\sqrt{1+x^4}} \, dx = \int_0^{0.1} \left(1 - \frac{x^4}{2} + \frac{3x^8}{8} - \ldots\right) dx = \left[x - \frac{x^5}{10} + \ldots\right]_0^{0.1} \approx [x]_0^{0.1} \approx 0.1$ with error

$|E| \le \frac{(0.1)^5}{10} = 0.000001$

37. $\int_0^{0.1} \frac{\sin x}{x} \, dx = \int_0^{0.1} \left(1 - \frac{x^2}{3!} + \frac{x^4}{5!} - \frac{x^6}{7!} + \ldots\right) dx = \left[x - \frac{x^3}{3\cdot3!} + \frac{x^5}{5\cdot5!} - \frac{x^7}{7\cdot7!} + \ldots\right]_0^{0.1} \approx \left[x - \frac{x^3}{3\cdot3!} + \frac{x^5}{5\cdot5!}\right]_0^{0.1}$

≈ 0.0999444611, $|E| \le \frac{(0.1)^7}{7\cdot7!} \approx 2.8 \times 10^{-12}$

39. $(1+x^4)^{1/2} = (1)^{1/2} + \frac{\left(\frac{1}{2}\right)}{1}(1)^{-1/2}(x^4) + \frac{\left(\frac{1}{2}\right)\left(-\frac{1}{2}\right)}{2!}(1)^{-3/2}(x^4)^2 + \frac{\left(\frac{1}{2}\right)\left(-\frac{1}{2}\right)\left(-\frac{3}{2}\right)}{3!}(1)^{-5/2}(x^4)^3$

$+ \frac{\left(\frac{1}{2}\right)\left(-\frac{1}{2}\right)\left(-\frac{3}{2}\right)\left(-\frac{5}{2}\right)}{4!}(1)^{-7/2}(x^4)^4 + \ldots = 1 + \frac{x^4}{2} - \frac{x^8}{8} + \frac{x^{12}}{16} - \frac{5x^{16}}{128} + \ldots$

$\Rightarrow \int_0^{0.1} \left(1 + \frac{x^4}{2} - \frac{x^8}{8} + \frac{x^{12}}{16} - \frac{5x^{16}}{128} + \ldots\right) dx \approx \left[x + \frac{x^5}{10}\right]_0^{0.1} \approx 0.100001$, $|E| \le \frac{(0.1)^9}{72} \approx 1.39 \times 10^{-11}$

41. $\int_0^1 \cos t^2 \, dt = \int_0^1 \left(1 - \frac{t^4}{2} + \frac{t^8}{4!} - \frac{t^{12}}{6!} + \ldots\right) dt = \left[t - \frac{t^5}{10} + \frac{t^9}{9\cdot4!} - \frac{t^{13}}{13\cdot6!} + \ldots\right]_0^1 \Rightarrow |\text{error}| < \frac{1}{13\cdot6!} \approx .00011$

43. $F(x) = \int_0^x \left(t^2 - \frac{t^6}{3!} + \frac{t^{10}}{5!} - \frac{t^{14}}{7!} + \ldots\right) dt = \left[\frac{t^3}{3} - \frac{t^7}{7\cdot3!} + \frac{t^{11}}{11\cdot5!} - \frac{t^{15}}{15\cdot7!} + \ldots\right]_0^x \approx \frac{x^3}{3} - \frac{x^7}{7\cdot3!} + \frac{x^{11}}{11\cdot5!}$

$\Rightarrow |\text{error}| < \frac{1}{15\cdot7!} \approx 0.000013$

45. (a) $F(x) = \int_0^x \left(t - \frac{t^3}{3} + \frac{t^5}{5} - \frac{t^7}{7} + \ldots\right) dt = \left[\frac{t^2}{2} - \frac{t^4}{12} + \frac{t^6}{30} - \ldots\right]_0^x \approx \frac{x^2}{2} - \frac{x^4}{12} \Rightarrow |\text{error}| < \frac{(0.5)^6}{30} \approx .00052$

(b) $|\text{error}| < \frac{1}{33\cdot34} \approx .00089$ when $F(x) \approx \frac{x^2}{2} - \frac{x^4}{3\cdot4} + \frac{x^6}{5\cdot6} - \frac{x^8}{7\cdot8} + \ldots + (-1)^{15}\frac{x^{32}}{31\cdot32}$

47. $\frac{1}{x^2}(e^x - (1+x)) = \frac{1}{x^2}\left(\left(1 + x + \frac{x^2}{2} + \frac{x^3}{3!} + \dots\right) - 1 - x\right) = \frac{1}{2} + \frac{x}{3!} + \frac{x^2}{4!} + \dots \Rightarrow \lim_{x \to 0} \frac{e^x - (1+x)}{x^2}$

$= \lim_{x \to 0}\left(\frac{1}{2} + \frac{x}{3!} + \frac{x^2}{4!} + \dots\right) = \frac{1}{2}$

49. $\frac{1}{t^4}\left(1 - \cos t - \frac{t^2}{2}\right) = \frac{1}{t^4}\left[1 - \frac{t^2}{2} - \left(1 - \frac{t^2}{2} + \frac{t^4}{4!} - \frac{t^6}{6!} + \dots\right)\right] = -\frac{1}{4!} + \frac{t^2}{6!} - \frac{t^4}{8!} + \dots \Rightarrow \lim_{t \to 0}\frac{1 - \cos t - \left(\frac{t^2}{2}\right)}{t^4}$

$= \lim_{t \to 0}\left(-\frac{1}{4!} + \frac{t^2}{6!} - \frac{t^4}{8!} + \dots\right) = -\frac{1}{24}$

51. $\frac{1}{y^3}(y - \tan^{-1} y) = \frac{1}{y^3}\left[y - \left(y - \frac{y^3}{3} + \frac{y^5}{5} - \dots\right)\right] = \frac{1}{3} - \frac{y^2}{5} + \frac{y^4}{7} - \dots \Rightarrow \lim_{y \to 0}\frac{y - \tan^{-1} y}{y^3} = \lim_{y \to 0}\left(\frac{1}{3} - \frac{y^2}{5} + \frac{y^4}{7} - \dots\right)$

$= \frac{1}{3}$

53. $x^2\left(-1 + e^{-1/x^2}\right) = x^2\left(-1 + 1 - \frac{1}{x^2} + \frac{1}{2x^4} - \frac{1}{6x^6} + \dots\right) = -1 + \frac{1}{2x^2} - \frac{1}{6x^4} + \dots \Rightarrow \lim_{x \to \infty} x^2\left(e^{-1/x^2} - 1\right)$

$= \lim_{x \to \infty}\left(-1 + \frac{1}{2x^2} - \frac{1}{6x^4} + \dots\right) = -1$

55. $\frac{\ln(1+x^2)}{1 - \cos x} = \frac{\left(x^2 - \frac{x^4}{2} + \frac{x^6}{3} - \dots\right)}{1 - \left(1 - \frac{x^2}{2!} + \frac{x^4}{4!} - \dots\right)} = \frac{\left(1 - \frac{x^2}{2} + \frac{x^4}{3} - \dots\right)}{\left(\frac{1}{2!} - \frac{x^2}{4!} + \dots\right)} \Rightarrow \lim_{x \to 0}\frac{\ln(1+x^2)}{1 - \cos x} = \lim_{x \to 0}\frac{\left(1 - \frac{x^2}{2} + \frac{x^4}{3} - \dots\right)}{\left(\frac{1}{2!} - \frac{x^2}{4!} + \dots\right)} = 2! = 2$

57. $\ln\left(\frac{1+x}{1-x}\right) = \ln(1+x) - \ln(1-x) = \left(x - \frac{x^2}{2} + \frac{x^3}{3} - \frac{x^4}{4} + \dots\right) - \left(-x - \frac{x^2}{2} - \frac{x^3}{3} - \frac{x^4}{4} - \dots\right) = 2\left(x + \frac{x^3}{3} + \frac{x^5}{5} + \dots\right)$

59. $\tan^{-1} x = x - \frac{x^3}{3} + \frac{x^5}{5} - \frac{x^7}{7} + \frac{x^9}{9} - \dots + \frac{(-1)^{n-1}x^{2n-1}}{2n-1} + \dots \Rightarrow |\text{error}| = \left|\frac{(-1)^{n-1}x^{2n-1}}{2n-1}\right| = \frac{1}{2n-1}$ when $x = 1$;

$\frac{1}{2n-1} < \frac{1}{10^3} \Rightarrow n > \frac{1001}{2} = 500.5 \Rightarrow$ the first term not used is the $501^{\text{st}} \Rightarrow$ we must use 500 terms

61. $\tan^{-1} x = x - \frac{x^3}{3} + \frac{x^5}{5} - \frac{x^7}{7} + \frac{x^9}{9} - \dots + \frac{(-1)^{n-1}x^{2n-1}}{2n-1} + \dots$ and when the series representing $48 \tan^{-1}\left(\frac{1}{18}\right)$ has an

error less than $\frac{1}{3} \cdot 10^{-6}$, then the series representing the sum

$48 \tan^{-1}\left(\frac{1}{18}\right) + 32 \tan^{-1}\left(\frac{1}{57}\right) - 20 \tan^{-1}\left(\frac{1}{239}\right)$ also has an error of magnitude less than 10^{-6}; thus

$|\text{error}| = 48\frac{\left(\frac{1}{18}\right)^{2n-1}}{2n-1} < \frac{1}{3 \cdot 10^6} \Rightarrow n \geq 4$ using a calculator $\Rightarrow$ 4 terms

63. (a) $(1 - x^2)^{-1/2} \approx 1 + \frac{x^2}{2} + \frac{3x^4}{8} + \frac{5x^6}{16} \Rightarrow \sin^{-1} x \approx x + \frac{x^3}{6} + \frac{3x^5}{40} + \frac{5x^7}{112}$; Using the Ratio Test:

$\lim_{n \to \infty}\left|\frac{1 \cdot 3 \cdot 5 \cdots (2n-1)(2n+1)x^{2n+3}}{2 \cdot 4 \cdot 6 \cdots (2n)(2n+2)(2n+3)} \cdot \frac{2 \cdot 4 \cdot 6 \cdots (2n)(2n+1)}{1 \cdot 3 \cdot 5 \cdots (2n-1)x^{2n+1}}\right| < 1 \Rightarrow x^2 \lim_{n \to \infty}\left|\frac{(2n+1)(2n+1)}{(2n+2)(2n+3)}\right| < 1$

$\Rightarrow |x| < 1 \Rightarrow$ the radius of convergence is 1. See Exercise 69.

(b) $\frac{d}{dx}(\cos^{-1} x) = -(1 - x^2)^{-1/2} \Rightarrow \cos^{-1} x = \frac{\pi}{2} - \sin^{-1} x \approx \frac{\pi}{2} - \left(x + \frac{x^3}{6} + \frac{3x^5}{40} + \frac{5x^7}{112}\right) \approx \frac{\pi}{2} - x - \frac{x^3}{6} - \frac{3x^5}{40} - \frac{5x^7}{112}$

65. $\frac{-1}{1+x} = -\frac{1}{1-(-x)} = -1 + x - x^2 + x^3 - \dots \Rightarrow \frac{d}{dx}\left(\frac{-1}{1+x}\right) = \frac{1}{1+x^2} = \frac{d}{dx}\left(-1 + x - x^2 + x^3 - \dots\right)$

$= 1 - 2x + 3x^2 - 4x^3 + \dots$

67. Wallis' formula gives the approximation $\pi \approx 4\left[\frac{2 \cdot 4 \cdot 4 \cdot 6 \cdot 6 \cdot 8 \cdots (2n-2) \cdot (2n)}{3 \cdot 3 \cdot 5 \cdot 5 \cdot 7 \cdot 7 \cdots (2n-1) \cdot (2n-1)}\right]$ to produce the table

n	$\sim \pi$
10	3.221088998
20	3.181104886
30	3.167880758
80	3.151425420
90	3.150331383
93	3.150049112
94	3.149959030
95	3.149870848
100	3.149456425

At n = 1929 we obtain the first approximation accurate to 3 decimals: 3.141999845. At n = 30,000 we still do not obtain accuracy to 4 decimals: 3.141617732, so the convergence to π is very slow. Here is a <u>Maple</u> CAS procedure to produce these approximations:

```
pie :=
    proc(n)
    local i,j;
        a(2) := evalf(8/9);
        for i from 3 to n do a(i) := evalf(2*(2*i−2)*i/(2*i−1)^2*a(i−1)) od;
        [[j,4*a(j)] $ (j = n−5 .. n)]
    end
```

69. $(1 - x^2)^{-1/2} = (1 + (-x^2))^{-1/2} = (1)^{-1/2} + \left(-\frac{1}{2}\right)(1)^{-3/2}(-x^2) + \frac{\left(-\frac{1}{2}\right)\left(-\frac{3}{2}\right)(1)^{-5/2}(-x^2)^2}{2!}$

$+ \frac{\left(-\frac{1}{2}\right)\left(-\frac{3}{2}\right)\left(-\frac{5}{2}\right)(1)^{-7/2}(-x^2)^3}{3!} + \ldots = 1 + \frac{x^2}{2} + \frac{1 \cdot 3 x^4}{2^2 \cdot 2!} + \frac{1 \cdot 3 \cdot 5 x^6}{2^3 \cdot 3!} + \ldots = 1 + \sum_{n=1}^{\infty} \frac{1 \cdot 3 \cdot 5 \cdots (2n-1)x^{2n}}{2^n \cdot n!}$

$\Rightarrow \sin^{-1} x = \int_0^x (1 - t^2)^{-1/2} \, dt = \int_0^x \left(1 + \sum_{n=1}^{\infty} \frac{1 \cdot 3 \cdot 5 \cdots (2n-1)x^{2n}}{2^n \cdot n!}\right) dt = x + \sum_{n=1}^{\infty} \frac{1 \cdot 3 \cdot 5 \cdots (2n-1)x^{2n+1}}{2 \cdot 4 \cdots (2n)(2n+1)},$

where $|x| < 1$

71. (a) $\tan\left(\tan^{-1}(n+1) - \tan^{-1}(n-1)\right) = \frac{\tan(\tan^{-1}(n+1)) - \tan(\tan^{-1}(n-1))}{1 + \tan(\tan^{-1}(n+1))\tan(\tan^{-1}(n-1))} = \frac{(n+1) - (n-1)}{1 + (n+1)(n-1)} = \frac{2}{n^2}$

(b) $\sum_{n=1}^{N} \tan^{-1}\left(\frac{2}{n^2}\right) = \sum_{n=1}^{N} \left[\tan^{-1}(n+1) - \tan^{-1}(n-1)\right] = (\tan^{-1} 2 - \tan^{-1} 0) + (\tan^{-1} 3 - \tan^{-1} 1)$

$+ (\tan^{-1} 4 - \tan^{-1} 2) + \ldots + (\tan^{-1}(N+1) - \tan^{-1}(N-1)) = \tan^{-1}(N+1) + \tan^{-1} N - \frac{\pi}{4}$

(c) $\sum_{n=1}^{\infty} \tan^{-1}\left(\frac{2}{n^2}\right) = \lim_{n \to \infty} \left[\tan^{-1}(N+1) + \tan^{-1} N - \frac{\pi}{4}\right] = \frac{\pi}{2} + \frac{\pi}{2} - \frac{\pi}{4} = \frac{3\pi}{4}$

11.11 FOURIER SERIES

1. $a_0 = \frac{1}{2\pi}\int_0^{2\pi} 1 \, dx = 1$, $a_k = \frac{1}{\pi}\int_0^{2\pi} \cos kx \, dx = \frac{1}{\pi}\left[\frac{\sin kx}{k}\right]_0^{2\pi} = 0$, $b_k = \frac{1}{\pi}\int_0^{2\pi} \sin kx \, dx = \frac{1}{\pi}\left[-\frac{\cos kx}{k}\right]_0^{2\pi} = 0.$

Thus, the Fourier series for $f(x)$ is 1.

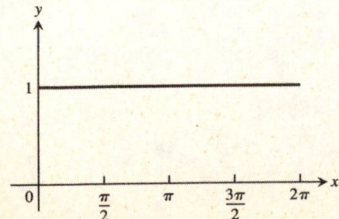

3. $a_0 = \frac{1}{2\pi}\left[\int_0^\pi x\,dx + \int_\pi^{2\pi}(x-2\pi)\,dx\right] = \frac{1}{2\pi}\left[\frac{1}{2}\pi^2 + \frac{1}{2}(4\pi^2-\pi^2) - 2\pi^2\right] = 0.$ Note,

$\int_\pi^{2\pi}(x-2\pi)\cos kx\,dx = -\int_0^\pi u\cos ku\,du$ (Let $u = 2\pi - x$). So $a_k = \frac{1}{\pi}\left[\int_0^\pi x\cos kx\,dx + \int_\pi^{2\pi}(x-2\pi)\cos kx\,dx\right] = 0.$

Note, $\int_\pi^{2\pi}(x-2\pi)\sin kx\,dx = \int_0^\pi u\sin ku\,du$ (Let $u = 2\pi - x$). So $b_k = \frac{1}{\pi}\left[\int_0^\pi x\sin kx\,dx + \int_\pi^{2\pi}(x-2\pi)\sin kx\,dx\right]$

$= \frac{2}{\pi}\int_0^\pi x\sin kx\,dx = \frac{2}{\pi}\left[-\frac{x}{k}\cos kx + \frac{1}{k^2}\sin kx\right]_0^\pi = -\frac{2}{k}\cos k\pi = \frac{2}{k}(-1)^{k+1}.$

Thus, the Fourier series for $f(x)$ is $\sum_{k=1}^\infty (-1)^{k+1}\frac{2\sin kx}{k}.$

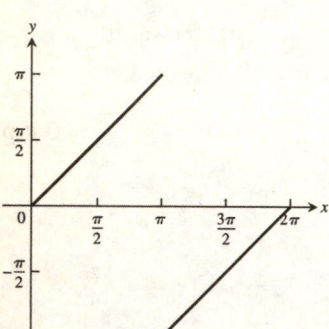

5. $a_0 = \frac{1}{2\pi}\int_0^{2\pi}e^x\,dx = \frac{1}{2\pi}(e^{2\pi}-1),$ $a_k = \frac{1}{\pi}\int_0^{2\pi}e^x\cos kx\,dx = \frac{1}{\pi}\left[\frac{e^x}{1+k^2}(\cos kx + k\sin kx)\right]_0^{2\pi} = \frac{e^{2\pi}-1}{\pi(1+k^2)},$

$b_k = \frac{1}{\pi}\int_0^{2\pi}e^x\sin kx\,dx = \frac{1}{\pi}\left[\frac{e^x}{1+k^2}(\sin kx - k\cos kx)\right]_0^{2\pi} = \frac{k(1-e^{2\pi})}{\pi(1+k^2)}.$

Thus, the Fourier series for $f(x)$ is $\frac{1}{2\pi}(e^{2\pi}-1) + \frac{e^{2\pi}-1}{\pi}\sum_{k=1}^\infty\left(\frac{\cos kx}{1+k^2} - \frac{k\sin kx}{1+k^2}\right).$

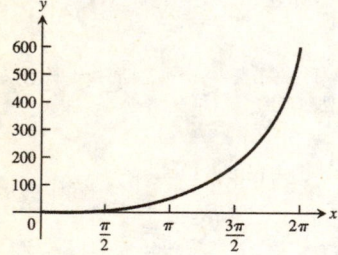

7. $a_0 = \frac{1}{2\pi}\int_0^{2\pi}f(x)\,dx = \frac{1}{2\pi}\int_0^{2\pi}\cos x\,dx = 0,$ $a_k = \frac{1}{\pi}\int_0^{2\pi}\cos x\cos kx\,dx = \begin{cases}\frac{1}{\pi}\left[\frac{\sin(k-1)x}{2(k-1)} + \frac{\sin(k+1)x}{2(k+1)}\right]_0^\pi, & k\neq 1\\ \frac{1}{\pi}\left[\frac{1}{2}x + \frac{1}{4}\sin 2x\right]_0^\pi, & k=1\end{cases}$

$= \begin{cases}0, & k\neq 1\\ \frac{1}{2}, & k=1\end{cases}.$

$b_k = \frac{1}{\pi}\int_0^{2\pi}\cos x\sin kx\,dx = \begin{cases}-\frac{1}{\pi}\left[\frac{\cos(k-1)x}{2(k-1)} + \frac{\cos(k+1)x}{2(k+1)}\right]_0^\pi, & k\neq 1\\ -\frac{1}{4\pi}\cos 2x\Big|_0^\pi, & k=1\end{cases} = \begin{cases}0, & k\text{ odd}\\ \frac{2k}{\pi(k^2-1)}, & k\text{ even}\end{cases}.$

Thus, the Fourier series for $f(x)$ is $\frac{1}{2}\cos x + \sum_{k\text{ even}}\frac{2k}{\pi(k^2-1)}\sin kx.$

9. $\int_0^{2\pi} \cos px\, dx = \frac{1}{p}\sin px \Big|_0^{2\pi} = 0$ if $p \neq 0$.

11. $\int_0^{2\pi} \cos px \cos qx\, dx = \int_0^{2\pi} \frac{1}{2}[\cos(p+q)x + \cos(p-q)x\,]dx = \frac{1}{2}\Big[\frac{1}{p+q}\sin(p+q)x + \frac{1}{p-q}\sin(p-q)x\,\Big]_0^{2\pi} = 0$ if $p \neq q$.

    If $p = q$ then $\int_0^{2\pi} \cos px \cos qx\, dx = \int_0^{2\pi} \cos^2 px\, dx = \int_0^{2\pi} \frac{1}{2}(1 + \cos 2px)\, dx = \frac{1}{2}\Big(x + \frac{1}{2p}\sin 2px\Big)\Big|_0^{2\pi} = \pi$.

13. $\int_0^{2\pi} \sin px \cos qx\, dx = \int_0^{2\pi} \frac{1}{2}[\sin(p+q)x + \sin(p-q)x\,]dx = -\frac{1}{2}\Big[\frac{1}{p+q}\cos(p+q)x + \frac{1}{p-q}\cos(p-q)x\,\Big]_0^{2\pi}$

    $= -\frac{1}{2}\Big[(1-1)\frac{1}{p+q} + (1-1)\frac{1}{p-q}\Big] = 0$. If $p = q$ then $\int_0^{2\pi} \sin px \cos qx\, dx = \int_0^{2\pi} \sin px \cos px\, dx = \int_0^{2\pi} \frac{1}{2}\sin 2px\, dx$

    $= -\frac{1}{4\pi}\cos 2px \Big|_0^{2\pi} = -\frac{1}{4\pi}(1-1) = 0$.

15. (a) $f(x)$ is piecewise continuous on $[0, 2\pi]$ and $f'(x) = 1$ for all $x \neq \pi \Rightarrow f'(x)$ is piecewise continuous on $[0, 2\pi]$. Then
    by Theorem 24, the Fourier series for $f(x)$ converges to $f(x)$ for all $x \neq \pi$ and converges to $\frac{1}{2}(f(\pi^+) + f(\pi^-))$
    $= \frac{1}{2}(-\pi + \pi) = 0$ at $x = \pi$.

    (b) The Fourier series for $f(x)$ is $\displaystyle\sum_{k=1}^{\infty} (-1)^{k+1}\frac{2\sin kx}{k}$. If we differentiate this series term by term we get the series

    $\displaystyle\sum_{k=1}^{\infty} (-1)^{k+1}2\cos kx$, which diverges by the nth term test for divergence for any x since $\displaystyle\lim_{k\to\infty}(-1)^{k+1}2\cos kx \neq 0$.

CHAPTER 11 PRACTICE EXERCISES

1. converges to 1, since $\displaystyle\lim_{n\to\infty} a_n = \lim_{n\to\infty}\Big(1 + \frac{(-1)^n}{n}\Big) = 1$

3. converges to -1, since $\displaystyle\lim_{n\to\infty} a_n = \lim_{n\to\infty}\Big(\frac{1-2^n}{2^n}\Big) = \lim_{n\to\infty}\Big(\frac{1}{2^n} - 1\Big) = -1$

5. diverges, since $\big\{\sin\frac{n\pi}{2}\big\} = \{0, 1, 0, -1, 0, 1, \dots\}$

7. converges to 0, since $\displaystyle\lim_{n\to\infty} a_n = \lim_{n\to\infty}\frac{\ln n^2}{n} = 2\lim_{n\to\infty}\frac{\big(\frac{1}{n}\big)}{1} = 0$

9. converges to 1, since $\displaystyle\lim_{n\to\infty} a_n = \lim_{n\to\infty}\Big(\frac{n + \ln n}{n}\Big) = \lim_{n\to\infty}\frac{1 + \big(\frac{1}{n}\big)}{1} = 1$

11. converges to e^{-5}, since $\displaystyle\lim_{n\to\infty} a_n = \lim_{n\to\infty}\Big(\frac{n-5}{n}\Big)^n = \lim_{n\to\infty}\Big(1 + \frac{(-5)}{n}\Big)^n = e^{-5}$ by Theorem 5

13. converges to 3, since $\displaystyle\lim_{n\to\infty} a_n = \lim_{n\to\infty}\Big(\frac{3^n}{n}\Big)^{1/n} = \lim_{n\to\infty}\frac{3}{n^{1/n}} = \frac{3}{1} = 3$ by Theorem 5

15. converges to ln 2, since $\lim\limits_{n \to \infty} a_n = \lim\limits_{n \to \infty} n\left(2^{1/n} - 1\right) = \lim\limits_{n \to \infty} \frac{2^{1/n} - 1}{\left(\frac{1}{n}\right)} = \lim\limits_{n \to \infty} \frac{\left[\frac{\left(-2^{1/n}\ln 2\right)}{n^2}\right]}{\left(\frac{-1}{n^2}\right)} = \lim\limits_{n \to \infty} 2^{1/n}\ln 2$

$= 2^0 \cdot \ln 2 = \ln 2$

17. diverges, since $\lim\limits_{n \to \infty} a_n = \lim\limits_{n \to \infty} \frac{(n+1)!}{n!} = \lim\limits_{n \to \infty} (n+1) = \infty$

19. $\frac{1}{(2n-3)(2n-1)} = \frac{\left(\frac{1}{2}\right)}{2n-3} - \frac{\left(\frac{1}{2}\right)}{2n-1} \Rightarrow s_n = \left[\frac{\left(\frac{1}{2}\right)}{3} - \frac{\left(\frac{1}{2}\right)}{5}\right] + \left[\frac{\left(\frac{1}{2}\right)}{5} - \frac{\left(\frac{1}{2}\right)}{7}\right] + \ldots + \left[\frac{\left(\frac{1}{2}\right)}{2n-3} - \frac{\left(\frac{1}{2}\right)}{2n-1}\right] = \frac{\left(\frac{1}{2}\right)}{3} - \frac{\left(\frac{1}{2}\right)}{2n-1}$

$\Rightarrow \lim\limits_{n \to \infty} s_n = \lim\limits_{n \to \infty} \left[\frac{1}{6} - \frac{\left(\frac{1}{2}\right)}{2n-1}\right] = \frac{1}{6}$

21. $\frac{9}{(3n-1)(3n+2)} = \frac{3}{3n-1} - \frac{3}{3n+2} \Rightarrow s_n = \left(\frac{3}{2} - \frac{3}{5}\right) + \left(\frac{3}{5} - \frac{3}{8}\right) + \left(\frac{3}{8} - \frac{3}{11}\right) + \ldots + \left(\frac{3}{3n-1} - \frac{3}{3n+2}\right)$

$= \frac{3}{2} - \frac{3}{3n+2} \Rightarrow \lim\limits_{n \to \infty} s_n = \lim\limits_{n \to \infty} \left(\frac{3}{2} - \frac{3}{3n+2}\right) = \frac{3}{2}$

23. $\sum\limits_{n=0}^{\infty} e^{-n} = \sum\limits_{n=0}^{\infty} \frac{1}{e^n}$, a convergent geometric series with $r = \frac{1}{e}$ and $a = 1 \Rightarrow$ the sum is $\frac{1}{1 - \left(\frac{1}{e}\right)} = \frac{e}{e-1}$

25. diverges, a p-series with $p = \frac{1}{2}$

27. Since $f(x) = \frac{1}{x^{1/2}} \Rightarrow f'(x) = -\frac{1}{2x^{3/2}} < 0 \Rightarrow f(x)$ is decreasing $\Rightarrow a_{n+1} < a_n$, and $\lim\limits_{n \to \infty} a_n = \lim\limits_{n \to \infty} \frac{1}{\sqrt{n}} = 0$, the

series $\sum\limits_{n=1}^{\infty} \frac{(-1)^n}{\sqrt{n}}$ converges by the Alternating Series Test. Since $\sum\limits_{n=1}^{\infty} \frac{1}{\sqrt{n}}$ diverges, the given series converges

conditionally.

29. The given series does not converge absolutely by the Direct Comparison Test since $\frac{1}{\ln(n+1)} > \frac{1}{n+1}$, which is

the nth term of a divergent series. Since $f(x) = \frac{1}{\ln(x+1)} \Rightarrow f'(x) = -\frac{1}{(\ln(x+1))^2(x+1)} < 0 \Rightarrow f(x)$ is

decreasing $\Rightarrow a_{n+1} < a_n$, and $\lim\limits_{n \to \infty} a_n = \lim\limits_{n \to \infty} \frac{1}{\ln(n+1)} = 0$, the given series converges conditionally by the

Alternating Series Test.

31. converges absolutely by the Direct Comparison Test since $\frac{\ln n}{n^3} < \frac{n}{n^3} = \frac{1}{n^2}$, the nth term of a convergent p-series

33. $\lim\limits_{n \to \infty} \frac{\left(\frac{1}{n\sqrt{n^2+1}}\right)}{\left(\frac{1}{n^2}\right)} = \sqrt{\lim\limits_{n \to \infty} \frac{n^2}{n^2+1}} = \sqrt{1} = 1 \Rightarrow$ converges absolutely by the Limit Comparison Test

35. converges absolutely by the Ratio Test since $\lim\limits_{n \to \infty} \left[\frac{n+2}{(n+1)!} \cdot \frac{n!}{n+1}\right] = \lim\limits_{n \to \infty} \frac{n+2}{(n+1)^2} = 0 < 1$

37. converges absolutely by the Ratio Test since $\lim\limits_{n \to \infty} \left[\frac{3^{n+1}}{(n+1)!} \cdot \frac{n!}{3^n}\right] = \lim\limits_{n \to \infty} \frac{3}{n+1} = 0 < 1$

39. converges absolutely by the Limit Comparison Test since $\lim\limits_{n \to \infty} \frac{\left(\frac{1}{n^{3/2}}\right)}{\left(\frac{1}{\sqrt{n(n+1)(n+2)}}\right)} = \sqrt{\lim\limits_{n \to \infty} \frac{n(n+1)(n+2)}{n^3}} = 1$

41. $\lim\limits_{n \to \infty} \left|\frac{u_{n+1}}{u_n}\right| < 1 \Rightarrow \lim\limits_{n \to \infty} \left|\frac{(x+4)^{n+1}}{(n+1)3^{n+1}} \cdot \frac{n3^n}{(x+4)^n}\right| < 1 \Rightarrow \frac{|x+4|}{3} \lim\limits_{n \to \infty} \left(\frac{n}{n+1}\right) < 1 \Rightarrow \frac{|x+4|}{3} < 1$

$\Rightarrow |x+4| < 3 \Rightarrow -3 < x+4 < 3 \Rightarrow -7 < x < -1$; at $x = -7$ we have $\sum\limits_{n=1}^{\infty} \frac{(-1)^n 3^n}{n3^n} = \sum\limits_{n=1}^{\infty} \frac{(-1)^n}{n}$, the alternating harmonic series, which converges conditionally; at $x = -1$ we have $\sum\limits_{n=1}^{\infty} \frac{3^n}{n3^n} = \sum\limits_{n=1}^{\infty} \frac{1}{n}$, the divergent

harmonic series

(a) the radius is 3; the interval of convergence is $-7 \le x < -1$

(b) the interval of absolute convergence is $-7 < x < -1$

(c) the series converges conditionally at $x = -7$

43. $\lim\limits_{n \to \infty} \left| \frac{u_{n+1}}{u_n} \right| < 1 \Rightarrow \lim\limits_{n \to \infty} \left| \frac{(3x-1)^{n+1}}{(n+1)^2} \cdot \frac{n^2}{(3x-1)^n} \right| < 1 \Rightarrow |3x-1| \lim\limits_{n \to \infty} \frac{n^2}{(n+1)^2} < 1 \Rightarrow |3x-1| < 1$

$\Rightarrow -1 < 3x - 1 < 1 \Rightarrow 0 < 3x < 2 \Rightarrow 0 < x < \frac{2}{3}$; at $x = 0$ we have $\sum\limits_{n=1}^{\infty} \frac{(-1)^{n-1}(-1)^n}{n^2} = \sum\limits_{n=1}^{\infty} \frac{(-1)^{2n-1}}{n^2}$

$= -\sum\limits_{n=1}^{\infty} \frac{1}{n^2}$, a nonzero constant multiple of a convergent p-series, which is absolutely convergent; at $x = \frac{2}{3}$ we

have $\sum\limits_{n=1}^{\infty} \frac{(-1)^{n-1}(1)^n}{n^2} = \sum\limits_{n=1}^{\infty} \frac{(-1)^{n-1}}{n^2}$, which converges absolutely

(a) the radius is $\frac{1}{3}$; the interval of convergence is $0 \le x \le \frac{2}{3}$

(b) the interval of absolute convergence is $0 \le x \le \frac{2}{3}$

(c) there are no values for which the series converges conditionally

45. $\lim\limits_{n \to \infty} \left| \frac{u_{n+1}}{u_n} \right| < 1 \Rightarrow \lim\limits_{n \to \infty} \left| \frac{x^{n+1}}{(n+1)^{n+1}} \cdot \frac{n^n}{x^n} \right| < 1 \Rightarrow |x| \lim\limits_{n \to \infty} \left| \left(\frac{n}{n+1} \right)^n \left(\frac{1}{n+1} \right) \right| < 1 \Rightarrow \frac{|x|}{e} \lim\limits_{n \to \infty} \left(\frac{1}{n+1} \right) < 1$

$\Rightarrow \frac{|x|}{e} \cdot 0 < 1$, which holds for all x

(a) the radius is ∞; the series converges for all x

(b) the series converges absolutely for all x

(c) there are no values for which the series converges conditionally

47. $\lim\limits_{n \to \infty} \left| \frac{u_{n+1}}{u_n} \right| < 1 \Rightarrow \lim\limits_{n \to \infty} \left| \frac{(n+2)x^{2n+1}}{3^{n+1}} \cdot \frac{3^n}{(n+1)x^{2n-1}} \right| < 1 \Rightarrow \frac{x^2}{3} \lim\limits_{n \to \infty} \left(\frac{n+2}{n+1} \right) < 1 \Rightarrow -\sqrt{3} < x < \sqrt{3}$;

the series $\sum\limits_{n=1}^{\infty} -\frac{n+1}{\sqrt{3}}$ and $\sum\limits_{n=1}^{\infty} \frac{n+1}{\sqrt{3}}$, obtained with $x = \pm\sqrt{3}$, both diverge

(a) the radius is $\sqrt{3}$; the interval of convergence is $-\sqrt{3} < x < \sqrt{3}$

(b) the interval of absolute convergence is $-\sqrt{3} < x < \sqrt{3}$

(c) there are no values for which the series converges conditionally

49. $\lim\limits_{n \to \infty} \left| \frac{u_{n+1}}{u_n} \right| < 1 \Rightarrow \lim\limits_{n \to \infty} \left| \frac{csch(n+1)x^{n+1}}{csch(n)x^n} \right| < 1 \Rightarrow |x| \lim\limits_{n \to \infty} \left| \frac{\left(\frac{2}{e^{n+1}-e^{-n-1}} \right)}{\left(\frac{2}{e^n - e^{-n}} \right)} \right| < 1$

$\Rightarrow |x| \lim\limits_{n \to \infty} \left| \frac{e^{-1}-e^{-2n-1}}{1-e^{-2n-2}} \right| < 1 \Rightarrow \frac{|x|}{e} < 1 \Rightarrow -e < x < e$; the series $\sum\limits_{n=1}^{\infty} (\pm e)^n$ csch n, obtained with $x = \pm e$,

both diverge since $\lim\limits_{n \to \infty} (\pm e)^n$ csch $n \ne 0$

(a) the radius is e; the interval of convergence is $-e < x < e$

(b) the interval of absolute convergence is $-e < x < e$

(c) there are no values for which the series converges conditionally

51. The given series has the form $1 - x + x^2 - x^3 + \dots + (-x)^n + \dots = \frac{1}{1+x}$, where $x = \frac{1}{4}$; the sum is $\frac{1}{1+\left(\frac{1}{4}\right)} = \frac{4}{5}$

53. The given series has the form $x - \frac{x^3}{3!} + \frac{x^5}{5!} - \dots + (-1)^n \frac{x^{2n+1}}{(2n+1)!} + \dots = \sin x$, where $x = \pi$; the sum is

$\sin \pi = 0$

55. The given series has the form $1 + x + \frac{x^2}{2!} + \frac{x^2}{3!} + \dots + \frac{x^n}{n!} + \dots = e^x$, where $x = \ln 2$; the sum is $e^{\ln(2)} = 2$

57. Consider $\frac{1}{1-2x}$ as the sum of a convergent geometric series with $a = 1$ and $r = 2x$ $\Rightarrow$ $\frac{1}{1-2x}$

$$= 1 + (2x) + (2x)^2 + (2x)^3 + \ldots = \sum_{n=0}^{\infty} (2x)^n = \sum_{n=0}^{\infty} 2^n x^n \text{ where } |2x| < 1 \Rightarrow |x| < \tfrac{1}{2}$$

59. $\sin x = \sum_{n=0}^{\infty} \frac{(-1)^n x^{2n+1}}{(2n+1)!}$ $\Rightarrow$ $\sin \pi x = \sum_{n=0}^{\infty} \frac{(-1)^n (\pi x)^{2n+1}}{(2n+1)!} = \sum_{n=0}^{\infty} \frac{(-1)^n \pi^{2n+1} x^{2n+1}}{(2n+1)!}$

61. $\cos x = \sum_{n=0}^{\infty} \frac{(-1)^n x^{2n}}{(2n)!}$ $\Rightarrow$ $\cos \left(x^{5/2}\right) = \sum_{n=0}^{\infty} \frac{(-1)^n \left(x^{5/2}\right)^{2n}}{(2n)!} = \sum_{n=0}^{\infty} \frac{(-1)^n x^{5n}}{(2n)!}$

63. $e^x = \sum_{n=0}^{\infty} \frac{x^n}{n!}$ $\Rightarrow$ $e^{(\pi x/2)} = \sum_{n=0}^{\infty} \frac{\left(\frac{\pi x}{2}\right)^n}{n!} = \sum_{n=0}^{\infty} \frac{\pi^n x^n}{2^n n!}$

65. $f(x) = \sqrt{3+x^2} = (3+x^2)^{1/2}$ $\Rightarrow$ $f'(x) = x(3+x^2)^{-1/2}$ $\Rightarrow$ $f''(x) = -x^2(3+x^2)^{-3/2} + (3+x^2)^{-1/2}$

$\Rightarrow$ $f'''(x) = 3x^3(3+x^2)^{-5/2} - 3x(3+x^2)^{-3/2}$; $f(-1) = 2$, $f'(-1) = -\frac{1}{2}$, $f''(-1) = -\frac{1}{8} + \frac{1}{2} = \frac{3}{8}$,

$f'''(-1) = -\frac{3}{32} + \frac{3}{8} = \frac{9}{32}$ $\Rightarrow$ $\sqrt{3+x^2} = 2 - \frac{(x+1)}{2 \cdot 1!} + \frac{3(x+1)^2}{2^3 \cdot 2!} + \frac{9(x+1)^3}{2^5 \cdot 3!} + \ldots$

67. $f(x) = \frac{1}{x+1} = (x+1)^{-1}$ $\Rightarrow$ $f'(x) = -(x+1)^{-2}$ $\Rightarrow$ $f''(x) = 2(x+1)^{-3}$ $\Rightarrow$ $f'''(x) = -6(x+1)^{-4}$; $f(3) = \frac{1}{4}$,

$f'(3) = -\frac{1}{4^2}$, $f''(3) = \frac{2}{4^3}$, $f'''(2) = \frac{-6}{4^4}$ $\Rightarrow$ $\frac{1}{x+1} = \frac{1}{4} - \frac{1}{4^2}(x-3) + \frac{1}{4^3}(x-3)^2 - \frac{1}{4^4}(x-3)^3 + \ldots$

69. Assume the solution has the form $y = a_0 + a_1 x + a_2 x^2 + \ldots + a_{n-1}x^{n-1} + a_n x^n + \ldots$

$\Rightarrow$ $\frac{dy}{dx} = a_1 + 2a_2 x + \ldots + na_n x^{n-1} + \ldots$ $\Rightarrow$ $\frac{dy}{dx} + y$

$= (a_1 + a_0) + (2a_2 + a_1)x + (3a_3 + a_2)x^2 + \ldots + (na_n + a_{n-1})x^{n-1} + \ldots = 0 \Rightarrow a_1 + a_0 = 0, 2a_2 + a_1 = 0,$

$3a_3 + a_2 = 0$ and in general $na_n + a_{n-1} = 0$. Since $y = -1$ when $x = 0$ we have $a_0 = -1$. Therefore $a_1 = 1$,

$a_2 = \frac{-a_1}{2 \cdot 1} = -\frac{1}{2}$, $a_3 = \frac{-a_2}{3} = \frac{1}{3 \cdot 2}$, $a_4 = \frac{-a_3}{4} = -\frac{1}{4 \cdot 3 \cdot 2}$, $\ldots$, $a_n = \frac{-a_{n-1}}{n} = \frac{-1}{n}\frac{(-1)^n}{(n-1)!} = \frac{(-1)^{n+1}}{n!}$

$\Rightarrow$ $y = -1 + x - \frac{1}{2}x^2 + \frac{1}{3 \cdot 2}x^3 - \ldots + \frac{(-1)^{n+1}}{n!}x^n + \ldots = -\sum_{n=0}^{\infty} \frac{(-1)^n x^n}{n!} = -e^{-x}$

71. Assume the solution has the form $y = a_0 + a_1 x + a_2 x^2 + \ldots + a_{n-1}x^{n-1} + a_n x^n + \ldots$

$\Rightarrow$ $\frac{dy}{dx} = a_1 + 2a_2 x + \ldots + na_n x^{n-1} + \ldots$ $\Rightarrow$ $\frac{dy}{dx} + 2y$

$= (a_1 + 2a_0) + (2a_2 + 2a_1)x + (3a_3 + 2a_2)x^2 + \ldots + (na_n + 2a_{n-1})x^{n-1} + \ldots = 0$. Since $y = 3$ when $x = 0$ we

have $a_0 = 3$. Therefore $a_1 = -2a_0 = -2(3) = -3(2)$, $a_2 = -\frac{2}{2}a_1 = -\frac{2}{2}(-2 \cdot 3) = 3\left(\frac{2^2}{2}\right)$, $a_3 = -\frac{2}{3}a_2$

$= -\frac{2}{3}\left[3\left(\frac{2^2}{2}\right)\right] = -3\left(\frac{2^3}{3 \cdot 2}\right)$, $\ldots$, $a_n = \left(-\frac{2}{n}\right)a_{n-1} = \left(-\frac{2}{n}\right)\left(3\left(\frac{(-1)^{n-1}2^{n-1}}{(n-1)!}\right)\right) = 3\left(\frac{(-1)^n 2^n}{n!}\right)$

$\Rightarrow$ $y = 3 - 3(2x) + 3\frac{(2)^2}{2}x^2 - 3\frac{(2)^3}{3 \cdot 2}x^3 + \ldots + 3\frac{(-1)^n 2^n}{n!}x^n + \ldots$

$= 3\left[1 - (2x) + \frac{(2x)^2}{2!} - \frac{(2x)^3}{3!} + \ldots + \frac{(-1)^n(2x)^n}{n!} + \ldots\right] = 3\sum_{n=0}^{\infty} \frac{(-1)^n(2x)^n}{n!} = 3e^{-2x}$

73. Assume the solution has the form $y = a_0 + a_1 x + a_2 x^2 + \ldots + a_{n-1}x^{n-1} + a_n x^n + \ldots$

$\Rightarrow$ $\frac{dy}{dx} = a_1 + 2a_2 x + \ldots + na_n x^{n-1} + \ldots$ $\Rightarrow$ $\frac{dy}{dx} - y$

$= (a_1 - a_0) + (2a_2 - a_1)x + (3a_3 - a_2)x^2 + \ldots + (na_n - a_{n-1})x^{n-1} + \ldots = 3x \Rightarrow a_1 - a_0 = 0, 2a_2 - a_1 = 3,$

$3a_3 - a_2 = 0$ and in general $na_n - a_{n-1} = 0$ for $n > 2$. Since $y = -1$ when $x = 0$ we have $a_0 = -1$. Therefore

$a_1 = -1$, $a_2 = \frac{3+a_1}{2} = \frac{2}{2}$, $a_3 = \frac{a_2}{3} = \frac{2}{3 \cdot 2}$, $a_4 = \frac{a_3}{4} = \frac{2}{4 \cdot 3 \cdot 2}$, $\ldots$, $a_n = \frac{a_{n-1}}{n} = \frac{2}{n!}$

$\Rightarrow$ $y = -1 - x + \left(\frac{2}{2}\right)x^2 + \frac{3}{3 \cdot 2}x^3 + \frac{2}{4 \cdot 3 \cdot 2}x^4 + \ldots + \frac{2}{n!}x^n + \ldots$

$= 2\left(1 + x + \frac{1}{2}x^2 + \frac{1}{3 \cdot 2}x^3 + \frac{1}{4 \cdot 3 \cdot 2}x^4 + \ldots + \frac{1}{n!}x^n + \ldots\right) - 3 - 3x = 2\sum_{n=0}^{\infty} \frac{x^n}{n!} - 3 - 3x = 2e^x - 3x - 3$

75. Assume the solution has the form $y = a_0 + a_1x + a_2x^2 + \ldots + a_{n-1}x^{n-1} + a_nx^n + \ldots$

$\Rightarrow \frac{dy}{dx} = a_1 + 2a_2x + \ldots + na_nx^{n-1} + \ldots \Rightarrow \frac{dy}{dx} - y$

$= (a_1 - a_0) + (2a_2 - a_1)x + (3a_3 - a_2)x^2 + \ldots + (na_n - a_{n-1})x^{n-1} + \ldots = x \Rightarrow a_1 - a_0 = 0,\ 2a_2 - a_1 = 1,$

$3a_3 - a_2 = 0$ and in general $na_n - a_{n-1} = 0$ for $n > 2$. Since $y = 1$ when $x = 0$ we have $a_0 = 1$. Therefore

$a_1 = 1,\ a_2 = \frac{1+a_1}{2} = \frac{2}{2},\ a_3 = \frac{a_2}{3} = \frac{2}{3\cdot2},\ a_4 = \frac{a_3}{4} = \frac{2}{4\cdot3\cdot2},\ \ldots,\ a_n = \frac{a_{n-1}}{n} = \frac{2}{n!}$

$\Rightarrow y = 1 + x + \left(\frac{2}{2}\right)x^2 + \frac{2}{3\cdot2}x^3 + \frac{2}{4\cdot2\cdot2}x^4 + \ldots + \frac{2}{n!}x^n + \ldots$

$= 2\left(1 + x + \frac{1}{2}x^2 + \frac{1}{3\cdot2}x^3 + \frac{1}{4\cdot3\cdot2}x^4 + \ldots + \frac{1}{n!}x^n + \ldots\right) - 1 - x = 2\sum_{n=0}^{\infty}\frac{x^n}{n!} - 1 - x = 2e^x - x - 1$

77. $\int_0^{1/2} \exp(-x^3)\, dx = \int_0^{1/2}\left(1 - x^3 + \frac{x^6}{2!} - \frac{x^9}{3!} + \frac{x^{12}}{4!} + \ldots\right)dx = \left[x - \frac{x^4}{4} + \frac{x^7}{7\cdot2!} - \frac{x^{10}}{10\cdot3!} + \frac{x^{13}}{13\cdot4!} - \ldots\right]_0^{1/2}$

$\approx \frac{1}{2} - \frac{1}{2^4\cdot4} + \frac{1}{2^7\cdot7\cdot2!} - \frac{1}{2^{10}\cdot10\cdot3!} + \frac{1}{2^{13}\cdot13\cdot4!} - \frac{1}{2^{16}\cdot16\cdot5!} \approx 0.484917143$

79. $\int_1^{1/2} \frac{\tan^{-1}x}{x}\, dx = \int_1^{1/2}\left(1 - \frac{x^2}{3} + \frac{x^4}{5} - \frac{x^6}{7} + \frac{x^8}{9} - \frac{x^{10}}{11} + \ldots\right)dx = \left[x - \frac{x^3}{9} + \frac{x^5}{25} - \frac{x^7}{49} + \frac{x^9}{81} - \frac{x^{11}}{121} + \ldots\right]_0^{1/2}$

$\approx \frac{1}{2} - \frac{1}{9\cdot2^3} + \frac{1}{5^2\cdot2^5} - \frac{1}{7^2\cdot2^7} + \frac{1}{9^2\cdot2^9} - \frac{1}{11^2\cdot2^{11}} + \frac{1}{13^2\cdot2^{13}} - \frac{1}{15^2\cdot2^{15}} + \frac{1}{17^2\cdot2^{17}} - \frac{1}{19^2\cdot2^{19}} + \frac{1}{21^2\cdot2^{21}}$

≈ 0.4872223583

81. $\lim_{x \to 0} \frac{7\sin x}{e^{2x}-1} = \lim_{x \to 0} \frac{7\left(x - \frac{x^3}{3!} + \frac{x^5}{5!} - \ldots\right)}{\left(2x + \frac{2^2x^2}{2!} + \frac{2^3x^3}{3!} + \ldots\right)} = \lim_{x \to 0} \frac{7\left(1 - \frac{x^2}{3!} + \frac{x^4}{5!} - \ldots\right)}{\left(2 + \frac{2^2x}{2!} + \frac{2^3x^2}{3!} + \ldots\right)} = \frac{7}{2}$

83. $\lim_{t \to 0}\left(\frac{1}{2-2\cos t} - \frac{1}{t^2}\right) = \lim_{t \to 0}\frac{t^2 - 2 + 2\cos t}{2t^2(1-\cos t)} = \lim_{t \to 0}\frac{t^2 - 2 + 2\left(1 - \frac{t^2}{2} + \frac{t^4}{4!} - \ldots\right)}{2t^2\left(1 - 1 + \frac{t^2}{2} - \frac{t^4}{4!} + \ldots\right)} = \lim_{t \to 0}\frac{2\left(\frac{t^4}{4!} - \frac{t^6}{6!} + \ldots\right)}{\left(t^4 - \frac{2t^6}{4!} + \ldots\right)}$

$= \lim_{t \to 0}\frac{2\left(\frac{1}{4!} - \frac{t^2}{6!} + \ldots\right)}{\left(1 - \frac{2t^2}{4!} + \ldots\right)} = \frac{1}{12}$

85. $\lim_{z \to 0}\frac{1 - \cos^2 z}{\ln(1-z) + \sin z} = \lim_{z \to 0}\frac{1 - \left(1 - z^2 + \frac{z^4}{3} - \ldots\right)}{\left(-z - \frac{z^2}{2} - \frac{z^3}{3} - \ldots\right) + \left(z - \frac{z^3}{3!} + \frac{z^5}{5!} - \ldots\right)} = \lim_{z \to 0}\frac{\left(z^2 - \frac{z^4}{3} + \ldots\right)}{\left(-\frac{z^2}{2} - \frac{2z^3}{3} - \frac{z^4}{4} - \ldots\right)}$

$= \lim_{z \to 0}\frac{\left(1 - \frac{z^2}{3} + \ldots\right)}{\left(-\frac{1}{2} - \frac{2z}{3} - \frac{z^2}{4} - \ldots\right)} = -2$

87. $\lim_{x \to 0}\left(\frac{\sin 3x}{x^3} + \frac{r}{x^2} + s\right) = \lim_{x \to 0}\left[\frac{\left(3x - \frac{(3x)^3}{6} + \frac{(3x)^5}{120} - \ldots\right)}{x^3} + \frac{r}{x^2} + s\right] = \lim_{x \to 0}\left(\frac{3}{x^2} - \frac{9}{2} + \frac{81x^2}{40} + \ldots + \frac{r}{x^2} + s\right) = 0$

$\Rightarrow \frac{r}{x^2} + \frac{3}{x^2} = 0$ and $s - \frac{9}{2} = 0 \Rightarrow r = -3$ and $s = \frac{9}{2}$

89. (a) $\sum_{n=1}^{\infty}\left(\sin\frac{1}{2n} - \sin\frac{1}{2n+1}\right) = \left(\sin\frac{1}{2} - \sin\frac{1}{3}\right) + \left(\sin\frac{1}{4} - \sin\frac{1}{5}\right) + \left(\sin\frac{1}{6} - \sin\frac{1}{7}\right) + \ldots + \left(\sin\frac{1}{2n} - \sin\frac{1}{2n+1}\right)$

$+ \ldots = \sum_{n=2}^{\infty}(-1)^n\sin\frac{1}{n};\ f(x) = \sin\frac{1}{x} \Rightarrow f'(x) = \frac{-\cos\left(\frac{1}{x}\right)}{x^2} < 0$ if $x \geq 2 \Rightarrow \sin\frac{1}{n+1} < \sin\frac{1}{n}$, and

$\lim_{n \to \infty}\sin\frac{1}{n} = 0 \Rightarrow \sum_{n=2}^{\infty}(-1)^n\sin\frac{1}{n}$ converges by the Alternating Series Test

(b) $|error| < \left|\sin\frac{1}{42}\right| \approx 0.02381$ and the sum is an underestimate because the remainder is positive

91. $\lim_{n \to \infty}\left|\frac{2\cdot5\cdot8\cdots(3n-1)(3n+2)x^{n+1}}{2\cdot4\cdot6\cdots(2n)(2n+2)}\cdot\frac{2\cdot4\cdot6\cdots(2n)}{2\cdot5\cdot8\cdots(3n-1)x^n}\right| < 1 \Rightarrow |x|\lim_{n \to \infty}\left|\frac{3n+2}{2n+2}\right| < 1 \Rightarrow |x| < \frac{2}{3}$

$\Rightarrow$ the radius of convergence is $\frac{2}{3}$

93. $\displaystyle\sum_{k=2}^{n} \ln\left(1 - \tfrac{1}{k^2}\right) = \sum_{k=2}^{n} \left[\ln\left(1 + \tfrac{1}{k}\right) + \ln\left(1 - \tfrac{1}{k}\right)\right] = \sum_{k=2}^{n} \left[\ln(k+1) - \ln k + \ln(k-1) - \ln k\right]$

$= [\ln 3 - \ln 2 + \ln 1 - \ln 2] + [\ln 4 - \ln 3 + \ln 2 - \ln 3] + [\ln 5 - \ln 4 + \ln 3 - \ln 4] + [\ln 6 - \ln 5 + \ln 4 - \ln 5]$

$+ \ldots + [\ln(n+1) - \ln n + \ln(n-1) - \ln n] = [\ln 1 - \ln 2] + [\ln(n+1) - \ln n]$ after cancellation

$\Rightarrow \displaystyle\sum_{k=2}^{n} \ln\left(1 - \tfrac{1}{k^2}\right) = \ln\left(\tfrac{n+1}{2n}\right) \Rightarrow \sum_{k=2}^{\infty} \ln\left(1 - \tfrac{1}{k^2}\right) = \lim_{n \to \infty} \ln\left(\tfrac{n+1}{2n}\right) = \ln\tfrac{1}{2}$ is the sum

95. (a) $\displaystyle\lim_{n \to \infty} \left| \frac{1 \cdot 4 \cdot 7 \cdots (3n-2)(3n+1)x^{3n+3}}{(3n+3)!} \cdot \frac{(3n)!}{1 \cdot 4 \cdot 7 \cdots (3n-2)x^{3n}} \right| < 1 \Rightarrow |x^3| \lim_{n \to \infty} \frac{(3n+1)}{(3n+1)(3n+2)(3n+3)}$

$\quad = |x^3| \cdot 0 < 1 \Rightarrow$ the radius of convergence is ∞

(b) $y = 1 + \displaystyle\sum_{n=1}^{\infty} \frac{1 \cdot 4 \cdot 7 \cdots (3n-2)}{(3n)!} x^{3n} \Rightarrow \frac{dy}{dx} = \sum_{n=1}^{\infty} \frac{1 \cdot 4 \cdot 7 \cdots (3n-2)}{(3n-1)!} x^{3n-1}$

$\Rightarrow \dfrac{d^2y}{dx^2} = \displaystyle\sum_{n=1}^{\infty} \frac{1 \cdot 4 \cdot 7 \cdots (3n-2)}{(3n-2)!} x^{3n-2} = x + \sum_{n=2}^{\infty} \frac{1 \cdot 4 \cdot 7 \cdots (3n-5)}{(3n-3)!} x^{3n-2}$

$\quad = x \left(1 + \displaystyle\sum_{n=1}^{\infty} \frac{1 \cdot 4 \cdot 7 \cdots (3n-2)}{(3n)!} x^{3n}\right) = xy + 0 \Rightarrow a = 1$ and $b = 0$

97. Yes, the series $\displaystyle\sum_{n=1}^{\infty} a_n b_n$ converges as we now show. Since $\displaystyle\sum_{n=1}^{\infty} a_n$ converges it follows that $a_n \to 0 \Rightarrow a_n < 1$

for $n >$ some index $N \Rightarrow a_n b_n < b_n$ for $n > N \Rightarrow \displaystyle\sum_{n=1}^{\infty} a_n b_n$ converges by the Direct Comparison Test with $\displaystyle\sum_{n=1}^{\infty} b_n$

99. $\displaystyle\sum_{n=1}^{\infty} (x_{n+1} - x_n) = \lim_{n \to \infty} \sum_{k=1}^{\infty} (x_{k+1} - x_k) = \lim_{n \to \infty} (x_{n+1} - x_1) = \lim_{n \to \infty} (x_{n+1}) - x_1 \Rightarrow$ both the series and

sequence must either converge or diverge.

101. Newton's method gives $x_{n+1} = x_n - \dfrac{(x_n - 1)^{40}}{40(x_n - 1)^{39}} = \dfrac{39}{40} x_n + \dfrac{1}{40}$, and if the sequence $\{x_n\}$ has the limit L, then

$L = \dfrac{39}{40} L + \dfrac{1}{40} \Rightarrow L = 1$ and $\{x_n\}$ converges since $\left| \dfrac{f(x)f''(x)}{[f'(x)]^2} \right| = \dfrac{39}{40} < 1$

103. $a_n = \dfrac{1}{\ln n}$ for $n \geq 2 \Rightarrow a_2 \geq a_3 \geq a_4 \geq \ldots$, and $\dfrac{1}{\ln 2} + \dfrac{1}{\ln 4} + \dfrac{1}{\ln 8} + \ldots = \dfrac{1}{\ln 2} + \dfrac{1}{2\ln 2} + \dfrac{1}{3\ln 2} + \ldots$

$= \dfrac{1}{\ln 2}\left(1 + \dfrac{1}{2} + \dfrac{1}{3} + \ldots\right)$ which diverges so that $1 + \displaystyle\sum_{n=2}^{\infty} \dfrac{1}{n \ln n}$ diverges by the Integral Test.

105. $a_0 = \dfrac{1}{2\pi} \displaystyle\int_0^{2\pi} f(x)\, dx = \dfrac{1}{2\pi} \int_\pi^{2\pi} 1\, dx = \dfrac{1}{2}$, $a_k = \dfrac{1}{\pi} \int_0^{2\pi} f(x) \cos kx\, dx = \dfrac{1}{\pi} \int_\pi^{2\pi} \cos kx\, dx = 0$.

$b_k = \dfrac{1}{\pi} \displaystyle\int_0^{2\pi} f(x) \sin kx\, dx = \dfrac{1}{\pi} \int_\pi^{2\pi} \sin kx\, dx = -\dfrac{\cos kx}{\pi k}\Big|_\pi^{2\pi} = -\dfrac{1}{\pi k}\left(1 - (-1)^k\right) = \begin{cases} -\dfrac{2}{\pi k}, & k \text{ odd} \\ 0, & k \text{ even} \end{cases}$.

Thus, the Fourier series of $f(x)$ is $\dfrac{1}{2} - \displaystyle\sum_{k \text{ odd}} \dfrac{2}{\pi k} \sin kx$

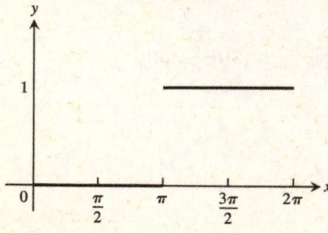

107. $a_0 = \dfrac{1}{2\pi}\left[\displaystyle\int_0^\pi (\pi - x)\, dx + \int_\pi^{2\pi} (x - 2\pi)\, dx\right] = \dfrac{1}{2\pi}\left[\int_0^\pi (\pi - x)\, dx - \int_0^\pi (\pi - u)\, du\right] = 0$ where we used the

substitution $u = x - \pi$ in the second integral. We have $a_k = \dfrac{1}{\pi}\left[\displaystyle\int_0^\pi (\pi - x) \cos kx\, dx + \int_\pi^{2\pi} (x - 2\pi) \cos kx\, dx\right]$. Using the substitution $u = x - \pi$ in the second integral gives $\int_{\pi}^{2\pi} (x - 2\pi) \cos kx \, dx = \int_{0}^{\pi} -(\pi - u) \cos(ku + k\pi) \, du$

$$= \begin{cases} \int_{0}^{\pi} (\pi - u) \cos ku \, du, & k \text{ odd} \\ \int_{0}^{\pi} -(\pi - u) \cos ku \, du, & k \text{ even} \end{cases}.$$

Thus, $a_k = \begin{cases} \frac{2}{\pi} \int_{0}^{\pi} (\pi - x) \cos kx \, dx, & k \text{ odd} \\ 0, & k \text{ even} \end{cases}.$

Now, since k is odd, letting $v = \pi - x \Rightarrow \frac{2}{\pi} \int_{0}^{\pi} (\pi - x) \cos kx \, dx = -\frac{2}{\pi} \int_{0}^{\pi} v \cos kv \, dv = -\frac{2}{\pi} \left(-\frac{2}{k^2} \right) = \frac{4}{\pi k^2}$, k odd. (See

Exercise 106). So, $a_k = \begin{cases} \frac{4}{\pi k^2}, & k \text{ odd} \\ 0, & k \text{ even} \end{cases}.$

Using similar techniques we see that $b_k = \begin{cases} \frac{2}{\pi} \int_{0}^{\pi} (\pi - u) \sin ku \, du, & k \text{ odd} \\ 0, & k \text{ even} \end{cases} = \begin{cases} \frac{2}{k}, & k \text{ odd} \\ 0, & k \text{ even} \end{cases}.$

Thus, the Fourier series of $f(x)$ is $\sum_{k \text{ odd}} \left(\frac{4}{\pi k^2} \cos kx + \frac{2}{k} \sin kx \right)$.

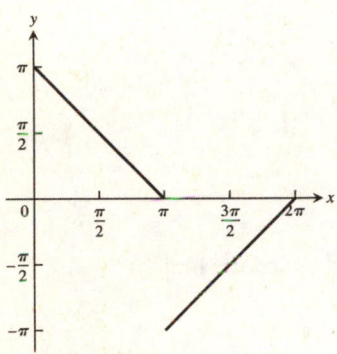

CHAPTER 11 ADDITIONAL AND ADVANCED EXERCISES

1. converges since $\frac{1}{(3n-2)^{(2n+1)/2}} < \frac{1}{(3n-2)^{3/2}}$ and $\sum_{n=1}^{\infty} \frac{1}{(3n-2)^{3/2}}$ converges by the Limit Comparison Test:

$$\lim_{n \to \infty} \frac{\left(\frac{1}{n^{3/2}} \right)}{\left(\frac{1}{(3n-2)^{3/2}} \right)} = \lim_{n \to \infty} \left(\frac{3n-2}{n} \right)^{3/2} = 3^{3/2}$$

3. diverges by the nth-Term Test since $\lim_{n \to \infty} a_n = \lim_{n \to \infty} (-1)^n \tanh n = \lim_{b \to \infty} (-1)^n \left(\frac{1 - e^{-2n}}{1 + e^{-2n}} \right) = \lim_{n \to \infty} (-1)^n$

does not exist

5. converges by the Direct Comparison Test: $a_1 = 1 = \frac{12}{(1)(3)(2)^2}$, $a_2 = \frac{1 \cdot 2}{3 \cdot 4} = \frac{12}{(2)(4)(3)^2}$, $a_3 = \left(\frac{2 \cdot 3}{4 \cdot 5} \right) \left(\frac{1 \cdot 2}{3 \cdot 4} \right)$

$$= \frac{12}{(3)(5)(4)^2}, a_4 = \left(\frac{3 \cdot 4}{5 \cdot 6} \right) \left(\frac{2 \cdot 3}{4 \cdot 5} \right) \left(\frac{1 \cdot 2}{3 \cdot 4} \right) = \frac{12}{(4)(6)(5)^2}, \ldots \Rightarrow 1 + \sum_{n=1}^{\infty} \frac{12}{(n+1)(n+3)(n+2)^2} \text{ represents the}$$

given series and $\frac{12}{(n+1)(n+3)(n+2)^2} < \frac{12}{n^4}$, which is the nth-term of a convergent p-series

7. diverges by the nth-Term Test since if $a_n \to L$ as $n \to \infty$, then $L = \frac{1}{1+L} \Rightarrow L^2 + L - 1 = 0 \Rightarrow L = \frac{-1 \pm \sqrt{5}}{2}$
$\neq 0$

9. $f(x) = \cos x$ with $a = \frac{\pi}{3} \Rightarrow f\left(\frac{\pi}{3} \right) = 0.5, f'\left(\frac{\pi}{3} \right) = -\frac{\sqrt{3}}{2}, f''\left(\frac{\pi}{3} \right) = -0.5, f'''\left(\frac{\pi}{3} \right) = \frac{\sqrt{3}}{2}, f^{(4)}\left(\frac{\pi}{3} \right) = 0.5;$

$\cos x = \frac{1}{2} - \frac{\sqrt{3}}{2} \left(x - \frac{\pi}{3} \right) - \frac{1}{4} \left(x - \frac{\pi}{3} \right)^2 + \frac{\sqrt{3}}{12} \left(x - \frac{\pi}{3} \right)^3 + \ldots$

11. $e^x = 1 + x + \frac{x^2}{2!} + \frac{x^3}{3!} + \ldots$ with $a = 0$

13. $f(x) = \cos x$ with $a = 22\pi \Rightarrow f(22\pi) = 1, f'(22\pi) = 0, \ f''(22\pi) = -1, f'''(22\pi) = 0, f^{(4)}(22\pi) = 1,$
$f^{(5)}(22\pi) = 0, f^{(6)}(22\pi) = -1; \cos x = 1 - \frac{1}{2}(x - 22\pi)^2 + \frac{1}{4!}(x - 22\pi)^4 - \frac{1}{6!}(x - 22\pi)^6 + \dots$

15. Yes, the sequence converges: $c_n = (a^n + b^n)^{1/n} \Rightarrow c_n = b\left(\left(\frac{a}{b}\right)^n + 1\right)^{1/n} \Rightarrow \lim_{n \to \infty} c_n = \ln b + \lim_{n \to \infty} \frac{\ln\left(\left(\frac{a}{b}\right)^n + 1\right)}{n}$

$= \ln b + \lim_{n \to \infty} \frac{\left(\frac{a}{b}\right)^n \ln\left(\frac{a}{b}\right)}{\left(\frac{a}{b}\right)^n + 1} = \ln b + \frac{0 \cdot \ln\left(\frac{a}{b}\right)}{0 + 1} = \ln b$ since $0 < a < b$. Thus, $\lim_{n \to \infty} c_n = e^{\ln b} = b.$

17. $s_n = \sum_{k=0}^{n-1} \int_k^{k+1} \frac{dx}{1+x^2} \Rightarrow s_n = \int_0^1 \frac{dx}{1+x^2} + \int_1^2 \frac{dx}{1+x^2} + \dots + \int_{n-1}^n \frac{dx}{1+x^2} \Rightarrow s_n = \int_0^n \frac{dx}{1+x^2}$
$\Rightarrow \lim_{n \to \infty} s_n = \lim_{n \to \infty} (\tan^{-1} n - \tan^{-1} 0) = \frac{\pi}{2}$

19. (a) Each A_{n+1} fits into the corresponding upper triangular region, whose vertices are:
$(n, f(n) - f(n+1)), (n+1, f(n+1))$ and $(n, f(n))$ along the line whose slope is $f(n+1) - f(n)$.

All the A_n's fit into the first upper triangular region whose area is $\frac{f(1) - f(2)}{2} \Rightarrow \sum_{n=1}^{\infty} A_n < \frac{f(1) - f(2)}{2}$

(b) If $A_k = \frac{f(k+1) + f(k)}{2} - \int_k^{k+1} f(x) \, dx$, then

$\sum_{k=1}^{n-1} A_k = \frac{f(1) + f(2) + f(2) + f(3) + f(3) + \dots + f(n-1) + f(n)}{2} - \int_1^2 f(x) \, dx - \int_2^3 f(x) \, dx - \dots - \int_{n-1}^n f(x) \, dx$

$= \frac{f(1) + f(n)}{2} + \sum_{k=2}^{n-1} f(k) - \int_1^n f(x) \, dx \Rightarrow \sum_{k=1}^{n-1} A_k = \sum_{k=1}^{n} f(k) - \frac{f(1) + f(n)}{2} - \int_1^n f(x) \, dx < \frac{f(1) - f(2)}{2}$, from

part (a). The sequence $\left\{ \sum_{k=1}^{n-1} A_k \right\}$ is bounded above and increasing, so it converges and the limit in
question must exist.

(c) Let $L = \lim_{n \to \infty} \left[\sum_{k=1}^{n} f(k) - \int_1^n f(x) \, dx - \frac{1}{2}(f(1) + f(n)) \right]$, which exists by part (b). Since f is positive and

decreasing $\lim_{n \to \infty} f(n) = M \geq 0$ exists. Thus $\lim_{n \to \infty} \left[\sum_{k=1}^{n} f(k) - \int_1^n f(x) \, dx \right] = L + \frac{1}{2}(f(1) + M).$

21. (a) No, the limit does not appear to depend on the value of the constant a
(b) Yes, the limit depends on the value of b

(c) $s = \left(1 - \frac{\cos\left(\frac{a}{n}\right)}{n}\right)^n \Rightarrow \ln s = \frac{\ln\left(1 - \frac{\cos\left(\frac{a}{n}\right)}{n}\right)}{\left(\frac{1}{n}\right)} \Rightarrow \lim_{n \to \infty} \ln s = \frac{\left(\frac{1}{1 - \frac{\cos\left(\frac{a}{n}\right)}{n}}\right)\left(\frac{-\frac{a}{n}\sin\left(\frac{a}{n}\right) + \cos\left(\frac{a}{n}\right)}{n^2}\right)}{\left(-\frac{1}{n^2}\right)}$

$= \lim_{n \to \infty} \frac{\frac{a}{n}\sin\left(\frac{a}{n}\right) - \cos\left(\frac{a}{n}\right)}{1 - \frac{\cos\left(\frac{a}{n}\right)}{n}} = \frac{0 - 1}{1 - 0} = -1 \Rightarrow \lim_{n \to \infty} s = e^{-1} \approx 0.3678794412$; similarly,

$\lim_{n \to \infty} \left(1 - \frac{\cos\left(\frac{a}{n}\right)}{bn}\right)^n = e^{-1/b}$

23. $\lim_{n \to \infty} \left| \frac{u_{n+1}}{u_n} \right| < 1 \Rightarrow \lim_{n \to \infty} \left| \frac{b^{n+1}x^{n+1}}{\ln(n+1)} \cdot \frac{\ln n}{b^n x^n} \right| < 1 \Rightarrow |bx| < 1 \Rightarrow -\frac{1}{b} < x < \frac{1}{b} = 5 \Rightarrow b = \pm\frac{1}{5}$

25. $\lim_{x \to 0} \frac{\sin(ax) - \sin x - x}{x^3} = \lim_{x \to 0} \frac{\left(ax - \frac{a^3 x^3}{3!} + \dots\right) - \left(x - \frac{x^3}{3!} + \dots\right) - x}{x^3}$

$= \lim_{x \to 0} \left[\frac{a-2}{x^2} - \frac{a^3}{3!} + \frac{1}{3!} - \left(\frac{a^5}{5!} - \frac{1}{5!}\right)x^2 + \dots \right]$ is finite if $a - 2 = 0 \Rightarrow a = 2$;

$\lim_{x \to 0} \frac{\sin 2x - \sin x - x}{x^3} = -\frac{2^3}{3!} + \frac{1}{3!} = -\frac{7}{6}$

27. (a) $\frac{u_n}{u_{n+1}} = \frac{(n+1)^2}{n^2} = 1 + \frac{2}{n} + \frac{1}{n^2} \Rightarrow C = 2 > 1$ and $\sum_{n=1}^{\infty} \frac{1}{n^2}$ converges

(b) $\frac{u_n}{u_{n+1}} = \frac{n+1}{n} = 1 + \frac{1}{n} + \frac{0}{n^2} \Rightarrow C = 1 \le 1$ and $\sum_{n=1}^{\infty} \frac{1}{n}$ diverges

29. (a) $\sum_{n=1}^{\infty} a_n = L \Rightarrow a_n^2 \le a_n \sum_{n=1}^{\infty} a_n = a_n L \Rightarrow \sum_{n=1}^{\infty} a_n^2$ converges by the Direct Comparison Test

(b) converges by the Limit Comparison Test: $\lim_{n \to \infty} \frac{\left(\frac{a_n}{1-a_n}\right)}{a_n} = \lim_{n \to \infty} \frac{1}{1-a_n} = 1$ since $\sum_{n=1}^{\infty} a_n$ converges and

therefore $\lim_{x \to \infty} a_n = 0$

31. $(1-x)^{-1} = 1 + \sum_{n=1}^{\infty} x^n$ where $|x| < 1 \Rightarrow \frac{1}{(1-x)^2} = \frac{d}{dx}(1-x)^{-1} = \sum_{n=1}^{\infty} nx^{n-1}$ and when $x = \frac{1}{2}$ we have

$4 = 1 + 2\left(\frac{1}{2}\right) + 3\left(\frac{1}{2}\right)^2 + 4\left(\frac{1}{2}\right)^3 + \dots + n\left(\frac{1}{2}\right)^{n-1} + \dots$

33. The sequence $\{x_n\}$ converges to $\frac{\pi}{2}$ from below so $\epsilon_n = \frac{\pi}{2} - x_n > 0$ for each n. By the Alternating Series
Estimation Theorem $\epsilon_{n+1} \approx \frac{1}{3!}(\epsilon_n)^3$ with $|\text{error}| < \frac{1}{5!}(\epsilon_n)^5$, and since the remainder is negative this is an
overestimate $\Rightarrow 0 < \epsilon_{n+1} < \frac{1}{6}(\epsilon_n)^3$.

35. (a) $\frac{1}{(1-x)^2} = \frac{d}{dx}\left(\frac{1}{1-x}\right) = \frac{d}{dx}(1 + x + x^2 + x^3 + \dots) = 1 + 2x + 3x^2 + 4x^3 + \dots = \sum_{n=1}^{\infty} nx^{n-1}$

(b) from part (a) we have $\sum_{n=1}^{\infty} n\left(\frac{5}{6}\right)^{n-1}\left(\frac{1}{6}\right) = \left(\frac{1}{6}\right)\left[\frac{1}{1-\left(\frac{5}{6}\right)}\right]^2 = 6$

(c) from part (a) we have $\sum_{n=1}^{\infty} np^{n-1}q = \frac{q}{(1-p)^2} = \frac{q}{q^2} = \frac{1}{q}$

37. (a) $R_n = C_0 e^{-kt_0} + C_0 e^{-2kt_0} + \dots + C_0 e^{-nkt_0} = \frac{C_0 e^{-kt_0}(1-e^{-nkt_0})}{1-e^{-kt_0}} \Rightarrow R = \lim_{n \to \infty} R_n = \frac{C_0 e^{-kt_0}}{1-e^{-kt_0}} = \frac{C_0}{e^{kt_0}-1}$

(b) $R_n = \frac{e^{-1}(1-e^{-n})}{1-e^{-1}} \Rightarrow R_1 = e^{-1} \approx 0.36787944$ and $R_{10} = \frac{e^{-1}(1-e^{-10})}{1-e^{-1}} \approx 0.58195028$;

$R = \frac{1}{e-1} \approx 0.58197671$; $R - R_{10} \approx 0.00002643 \Rightarrow \frac{R-R_{10}}{R} < 0.0001$

(c) $R_n = \frac{e^{-.1}(1-e^{-.1n})}{1-e^{-.1}}$, $\frac{R}{2} = \frac{1}{2}\left(\frac{1}{e^{.1}-1}\right) \approx 4.7541659$; $R_n > \frac{R}{2} \Rightarrow \frac{1-e^{-.1n}}{e^{.1}-1} > \left(\frac{1}{2}\right)\left(\frac{1}{e^{.1}-1}\right)$

$\Rightarrow 1 - e^{-n/10} > \frac{1}{2} \Rightarrow e^{-n/10} < \frac{1}{2} \Rightarrow -\frac{n}{10} < \ln\left(\frac{1}{2}\right) \Rightarrow \frac{n}{10} > -\ln\left(\frac{1}{2}\right) \Rightarrow n > 6.93 \Rightarrow n = 7$

39. The convergence of $\sum_{n=1}^{\infty} |a_n|$ implies that $\lim_{n \to \infty} |a_n| = 0$. Let $N > 0$ be such that $|a_n| < \frac{1}{2} \Rightarrow 1 - |a_n| > \frac{1}{2}$

$\Rightarrow \frac{|a_n|}{1-|a_n|} < 2|a_n|$ for all $n > N$. Now $|\ln(1+a_n)| = \left|a_n - \frac{a_n^2}{2} + \frac{a_n^3}{3} - \frac{a_n^4}{4} + \dots\right| \le |a_n| + \left|\frac{a_n^2}{2}\right| + \left|\frac{a_n^3}{3}\right| + \left|\frac{a_n^4}{4}\right| + \dots$

$< |a_n| + |a_n|^2 + |a_n|^3 + |a_n|^4 + \dots = \frac{|a_n|}{1-|a_n|} < 2|a_n|$. Therefore $\sum_{n=1}^{\infty} \ln(1+a_n)$ converges by the Direct

Comparison Test since $\sum_{n=1}^{\infty} |a_n|$ converges.

41. (a) $s_{2n+1} = \frac{c_1}{1} + \frac{c_2}{2} + \frac{c_3}{3} + \dots + \frac{c_{2n+1}}{2n+1} = \frac{t_1}{1} + \frac{t_2-t_1}{2} + \frac{t_3-t_2}{3} + \dots + \frac{t_{2n+1}-t_{2n}}{2n+1}$

$= t_1\left(1 - \frac{1}{2}\right) + t_2\left(\frac{1}{2} - \frac{1}{3}\right) + \dots + t_{2n}\left(\frac{1}{2n} - \frac{1}{2n+1}\right) + \frac{t_{2n+1}}{2n+1} = \sum_{k=1}^{2n} \frac{t_k}{k(k+1)} + \frac{t_{2n+1}}{2n+1}$.

(b) $\{c_n\} = \{(-1)^n\} \Rightarrow \sum_{n=1}^{\infty} \frac{(-1)^n}{n}$ converges

(c) $\{c_n\} = \{1, -1, -1, 1, 1, -1, -1, 1, 1, \dots\} \Rightarrow$ the series $1 - \frac{1}{2} - \frac{1}{3} + \frac{1}{4} + \frac{1}{5} - \frac{1}{6} - \frac{1}{7} + \dots$ converges

NOTES: